The Mycota

A Comprehensive Treatise on Fungi as Experimental Systems for Basic and Applied Research

Volume 9

The fungi represent a heterogenous assemblage of eukaryotic microorganisms and have become favored organisms for research at the cellular and molecular level. Such research involvement has been stimulated by interest in the biotechnological application of fungi in processes related to industry, agriculture and ecology. Considering both yeasts and mycelial fungi, THE MYCOTA highlights developments in both basic and applied research and presents an overview of fungal systematics and cell structure. Foremost authorities in research on mycology have been assembled to edit and contribute to the volumes.

Yen-Ping Hsueh • Meredith Blackwell
Editors

Fungal Associations

Third Edition

Springer

Editors
Yen-Ping Hsueh
Institute of Molecular Biology,
Academia Sinica
Taipei, Taiwan

Meredith Blackwell
Department of Biological Sciences
Louisiana State University
Baton Rouge, LA, USA

Department of Biological Sciences
University of South Carolina
Columbia, SC, USA

ISSN 2945-8048 ISSN 2945-8056 (electronic)
The Mycota
ISBN 978-3-031-41647-7 ISBN 978-3-031-41648-4 (eBook)
https://doi.org/10.1007/978-3-031-41648-4

This Springer imprint is published by the registered company Springer Nature Switzerland AG
The registered company address is: Gewerbestrasse 11, 6330 Cham, Switzerland

Paper in this product is recyclable.

Series Preface

The Mycota is an encyclopedic book series published by Springer as a comprehensive treatise spanning the Fungal Kingdom through a focus on fungi as experimental systems for basic and applied research. Articles in this series present introductory information in a comprehensive manner to broaden our general knowledge, along with state-of-the-art information on selected topics in fungal biology. Thus, these articles serve as a timeless reference source for historically relevant discoveries in fungal biology to drive further advances in the field.

The Mycota was founded in 1994 by Karl Esser and Paul Lemke. Since then 15 volumes have been published, with several now in their third edition. The steadily growing interest in *The Mycota*, reflected in particular by the high number of e-book downloads, encouraged Springer to continue this publication with a new international Series Editorial Board beginning in 2022.

Historically, the study of fungi originated as a sub-discipline of botany and was a largely descriptive discipline until the early nineteenth century. Subsequent experimental research has propelled the field of mycology, and achievements in the genetics and molecular biology of fungi have benefited studies in the related fields of fungal biochemistry and biophysics, plant pathology, medical mycology, and systematics. Ground-breaking research has been carried out using fungi as recognized by the receipt of nine Nobel Prizes, including the revolutionary 1945 Nobel Prize in Physiology or Medicine for the discovery of penicillin and the foundational 2006 Nobel Prize in Chemistry for establishing the molecular basis of eukaryotic transcription. All such studies expand our knowledge of fungi and of fungal processes and improve our ability to understand, utilize, and control fungi for the benefit of humankind.

Fungi have invaded every conceivable ecological niche. Saprobic forms abound, especially in the decay of organic debris. Pathogenic forms exist with both plant and animal hosts. Fungi even grow on other fungi. They are found in aquatic as well as soil environments, and their spores have been found in the air of extreme environments. Fungi can be variously associated with plants as symbionts in the form of lichens, mycorrhizae, and endophytes and also occur as overt pathogens. Association with animal systems varies; examples include the predaceous fungi that trap nematodes, the micro-fungi that grow in the anaerobic environment of the rumen, and medically important pathogens afflicting humans. Currently, taxonomists have identified about 150,000 species, but it is thought that over 90% of fungal species remain undescribed, with conservative estimates ranging from 2 to 5 million fungal species on earth.

From this perspective, there are new topics in fungal biology to explore, which will expand the current volumes and take them in new directions. Further, there will be new volumes in areas (e.g., Cryptomycota) that were not covered before.

Our understanding of the evolution of fungi is still incomplete and mainly based on species that can be grown in culture. But recent environmental DNA analyses have revealed a highly diverse form of eukaryotic life that branches with the Fungi, and thus the resulting and highly diverse clades were named the cyptomycota. The discovery of novel intermediate forms will redefine the fungal tree of life. Other topics will consider the adaptation of fungi to climate changes, the occurrence of fungal pathogens in the environment, and the dispersal of fungi during global pandemics.

Fungal constituents of the microbiome have received much less attention thus far, yet recent findings from clinical and animal studies clearly establish fungi as a significant component of the oral, gastrointestinal, pulmonary, and skin microbiomes. Finally, the relevance of fungal biology to society is reflected by the increasing number of fungal related human diseases.

In the history of pharmacology, fungi have always been sources of useful molecules for humans, but the diversity of molecules that can be obtained from fungi is still largely under explored. However, efforts in fungal genomics provide resources for data-driven genome mining and large-scale comparisons to explore the molecular repertoires produced by fungi. The result will be new compounds with applications in the pharmaceutical and agricultural science industries.

Fungi can also help the entire planet, and may for example be relevant in specific sectors, such as that linked to pollution from plastic. Fungi produce a wide range of enzymes that have the potential to break down the chemical bonds of plastic polymers, and in this context, the potential role of marine fungi in plastic degradation may be of major relevance. Finally, new biomaterials from fungal species may open the door to alternatives to fossil-based materials, and thus reduce environmental pollution.

For consistency throughout this series of volumes, the names adopted for major groups of fungi should be followed according to the following paper, which gives an overview of all of the orders in the fungal kingdom: https://pubmed.ncbi.nlm.nih.gov/32660385/

We are grateful to Springer for continuing *The Mycota* series and are especially thankful to all the volume editors in selecting topics and assembling experts from diverse fields of fungal biology.

Sydney, Australia — Dee Carter
Delhi, India — Anuradha Chowdhary
Durham, USA — Joseph Heitman
Bochum, Germany — Ulrich Kück
July 15th, 2022

Volume Preface to the Third Edition

Fungi are heterotrophic eukaryotes with chitinous cell walls; some species are single-celled flagellates, yeasts, or yeast-like cells, and many species have complex structures composed of filaments known as hyphae. Except for spore-bearing mushrooms and other large reproductive structures, mycology's charismatic megamycota (Blackwell 2011), most fungi go unseen except for their traces of deterioration and disease. Other characteristics of fungi include osmotropic nutrition whereby enzymes exported to the substrate dissolve organic nutrients that then are absorbed into the organism. Fungi also produce vitamins, sterols, and aromas, often important in their interactions and communication with other organisms. Fungi are an extraordinarily diverse group, in part because of myriad associations (Vega and Biedermann 2019) with almost all kinds of organisms including other fungi.

About a billion years ago populations of ancestral eukaryotic single-celled organisms diverged into fungi and animals, two dominant megadiverse groups we recognize today. Together with plants, fungi and animals constitute the "fauna, flora, and funga," a triumvirate of the most prominent organisms on Earth (Pfister 2020; Berbee and Blackwell 2023). Fungal associations range widely from simple co-occurrence, perhaps the result of only accidental spore dispersal, to complex multipartite symbioses in the broad sense, including mutualistic, commensalistic, and parasitic lifestyles.

This volume looks at the results of a billion years of fungal associations revealed by current methods:

- Many animals and plants are protected from fungi by preadapted and evolved defenses.
- Some fungi have been able to have such intimate contact with other organisms that their DNA has been exchanged.
- The dogma that obligately symbiotic fungi cannot be grown in axenic culture is being investigated with the help of genomes.
- Aromatic volatile compounds lure or repel other organisms in what would seem to be a confusing mélange of bacterial, animal, and plant compounds in the environment.
- Examples of mycoviruses and their influence on fungal phenotypes are remarkable; additional associations and their effects await discovery. [Meredith Blackwell was especially interested to see a reference to virus-

infected chestnut trees in Michigan, perhaps those discovered by an undergraduate at Hope College, Holland, Michigan, where she once taught.]

- Classic textbook examples of fungal associations have been completely updated by the application of traditional and new molecular techniques. These chapters cover lichens, including a model lichen, arbuscular mycorrhizal fungi, ectomycorrizal fungi, and an endophyte rendered much less toxic.
- Also included are fungal predators of nematodes, fungi that control the minds of insects, and even more fungal killers of insects.
- The volume closes with a pharmacopeia of secondary metabolites synthesized by sclerotium-producing Xylarias in termite galleries—metabolites with pharmaceutical potential.

Development of new technologies precedes advancements in our science. Evidence of progress brought about by the polymerase chain reaction (PCR), especially applied to fungi (White et al. 1990), and all the resulting improvements in molecular biology are evident in the research discussed in this volume. As we completed editing these chapters, artificial intelligence (AI) crept up and suddenly burst forth into the scientific literature and mass media, even into the hearing rooms and statutes of governemental bodies. The study of fungal associations and related fields is poised to adopt the use of AI through today's increased access to supercomputers, ever-expanding databases (including the proliferation of scientific literature), and a need to recognize complex patterns; perhaps AI even engenders a willingness to have complicated decisions made for us and eases the demands to undertake repetitive tasks. We anticipate that as successive technologies are developed and improved, future studies of fungi and the organisms with which they associate will continue to reap benefits.

We appreciate the help of our colleagues Donald H. Pfister, Joseph W. Spatafora, and Marc Rapport for editorial assistance. We are especially grateful to our contacts at Springer Nature, Sabine Schwarz, Executive Editor, Life Sciences & Biomedicine Books, and Bibhuti Bhusan Sharma, Project Coordinator (Books), for their unlimited support from beginning to end during the preparation of *Mycota 9: Fungal Associations*. Without them two neophytes would not have managed. Our helpful and patient families are of course gratefully acknowledged.

Columbia, SC Meredith Blackwell
Taipei, Taiwan Yen-Ping Hsueh
11 June 2023

Literature

Blackwell M (2011) The Fungi: 1, 2, 3, … 5.1 million species? Biodiversity special issue. Am J Bot 98:426–438. doi: 10.3732/ajb.1000298

Berbee M, Blackwell M (2023) The origin of fungi: A billion years of functional and morphological diversification. In: Harper D, Seberg O (eds) The origins of all things. Munksgaard, Copenhagen, pp 231–277

Pfister DR (2020) Fantastic Funga: why language matters. Deep Funga Blog. https://fundis.org/resources/blog/127-fantastic-funga-why-language-matters

Vega FE, Biedermann PH (2020) On interactions, associations, mycetangia, mutualists and symbioses in insect-fungus symbioses. Fungal Ecol 44: 100909. doi:10.1016/j.funeco.2019.100909

White TJ, Bruns TD, Lee SB, Taylor JW (1990) Amplification and direct sequencing of fungal ribosomal RNA genes for phylogenetics. In: Innis MA, Gelfand DH, Sninsky JJ, White TJ (eds) PCR protocols: a guide to methods and applications. Academic, New York

Contents

Editors and Contributors

About the Series Editors

Dee Carter is Professor of Microbiology at the University of Sydney. She has a degree in microbiology and biochemistry from Otago University, New Zealand, and a PhD from Imperial College, London, UK. After postdoctoral fellowships in Montpellier, France, and UC Berkeley, California, USA, she moved to Australia in 1995 to take up a lectureship at the University of Sydney. Dee's research has encompassed (1) the population genetics and ecology of medically important fungi, including yeast and mold pathogens; (2) responses of fungi to host and antifungal stress, including the production of variant morphological forms; and (3) the transcriptome and proteome response to antifungal therapy, particularly during synergistic interactions between antifungals and natural products. As well as research, Dee teaches microbiology including mycology, epidemiology, and molecular biology at Sydney University.

Anuradha Chowdhary is Professor of Medical Mycology at the Vallabhbhai Patel Chest Institute, University of Delhi, India. She received her medical degree in 1992 and MD (Microbiology) degree in 1996 from the University of Delhi, India, and PhD from the Faculty of Medical Sciences, Radboud University Medical Center, Netherlands. Dr Chowdhary's research interests include molecular ecology and population genetics of pathogenic fungi, antifungal drug resistance mechanisms, especially *Aspergillus* and *Candida* spp., and the epidemiology of systemic mycoses. She is currently working on molecular epidemiology of *Candida auris*, terbinafine-resistant dermatophytes, and azole-resistant *Aspergillus fumigatus*.

Joseph Heitman is James B. Duke Professor and Chair, Department of Molecular Genetics and Microbiology, Duke University, Durham, NC, USA. His research studies model and pathogenic fungi addressing fundamental questions of scientific and medical importance. Pioneering studies with Baker's yeast revealed how immunosuppressive natural products interdict signaling cascades via FKBP12-drug complexes, and discovered TOR as a globally conserved nutrient sensor targeted by the immunosuppressive, antiproliferative drug rapamycin. His research discovered unisexual reproduction of pathogenic microbes, with implications for pathogen emergence, how sex generates diversity, and how sex evolved. Dr Heitman's lab has further developed genetic and genomic approaches elucidating molecular principles of fungal virulence, identifying therapeutic targets, and illustrating convergent evolution of fungal mating-type loci with mammalian, insect, and plant sex chromosomes, defined the calcium-activated protein phosphatase calcineurin as a globally conserved fungal virulence factor, and elucidated functions of RNAi in microbial pathogen genome integrity, hypervirulent outbreak lineages, and drug resistance via epimutation.

Ulrich Kück is Professor of General and Molecular Botany at the Ruhr-University in Bochum, Germany. He graduated in Biology and Chemistry, and has a long-standing experience in the molecular biology of fungi, algae, and plants. His research with fungi has focused on two general aspects: first, genetic engineering of the secondary metabolism of biotechnically relevant filamentous fungi, including functional genomics and proteomics, and second, molecular genetic analysis of cellular growth in filamentous fungus, with a focus on sexual development. In particular, he is interested in the function of mating-type loci, and the involvement of conserved signaling complexes, such as the striatin-interacting phosphatases and kinases (STRIPAK) complex, in the control of cellular and developmental processes.

About the Volume Editors

Yen-Ping Hsueh was born in Taipei, Taiwan. *Biology* has always been her favorite subject in school so she followed her interest and majored in Plant Pathology and Microbiology at the National Taiwan University as an undergraduate. She received BSc (2001) and MSc (2003) degrees, and moved to Duke University to pursue her PhD under the guidance of Dr Joseph Heitman. Her PhD work mainly focused on the sexual cycle of the human fungal pathogen *Cryptococcus neoformans.* After receiving her PhD in Molecular Genetics and Microbiology from Duke University in 2008, she continued to work with Heitman briefly and then moved to Caltech to start her postdoctoral training with Dr Paul W. Sternberg in 2010. Her postdoctoral work was an investigation of the interactions between nematode-trapping fungi and the model nematode *C. elegans.*

In 2015, Hsueh returned to Taiwan, to begin her independent career at the Institute of Molecular Biology in Academia Sinica where she is currently Associate Research Fellow. Her laboratory studies the predator–prey interactions between carnivorous fungi and the model nematode *Caenorhabditis elegans*. Members of the Hsueh lab explore different aspects of the biology of these predator–prey systems and are striving to understand the mechanisms that underlie their interactions and coevolution. They hope to harness the nematocidal potential of the carnivorous fungi, which will facilitate the discovery of a novel treatment or biocontrol method to combat parasitic nematode infections in plants, animals, and humans in the future. Because of her original research, Hsueh has been recognized as an EMBO Global Investigator, EMBO Young Investigator, and more recently, a Taiwan Outstanding Women in Science Young Talent Award and is an active participant in the international scientific community.

Meredith Blackwell born in Abbeville, Louisiana, spent her early years on the banks of the Vermilion Bayou and attended public schools in Louisiana. As an undergraduate pursuing liberal arts interests, a deep-sea fishing trip led her to become a biology major (BS, University of Louisiana, 1961). She studied ichthyology at the University of Alabama (MS, 1963) and began a PhD at the University of Texas at Austin in 1964. At Texas, she eventually came under the guidance of C. J. Alexopoulos, and her enduring interest in Mycology developed (PhD, 1973). During postdoctoral research with Henry Aldrich and James Kimbrough at the University of Florida, she was introduced to insect-associated fungi. She spent the first 6 years of her career primarily in a teaching position at Hope College, Holland, Michigan.

The next 33 years (1981–2014) were spent in the Department of Biological Sciences, Louisiana State University, where she attained the rank of Boyd Professor. Her research supported by NSF included studies of desert myxomycetes, phylogeny of Laboulbeniomycetes and other insect ectoparasites, and the taxonomic and geographic distribution of insect-associated yeasts. Since 2014, she has lived in Columbia, South Carolina, where she is an affiliate in the Department of Biological Sciences at the University of South Carolina. Blackwell is a former president of the Mycological Society of America (MSA) and the International Mycological Association (IMA). She is a fellow of the British Mycological Association, MSA, IMA, and the American Academy of Arts and Sciences. She has received several mycological awards and is proud to share her name with a number of fungal taxa.

Contributors

Emma Applegate Resilient Agriculture, AgResearch Limited, Grasslands Research Centre, Palmerston North, New Zealand

Daniel Bastias Resilient Agriculture, AgResearch Limited, Grasslands Research Centre, Palmerston North, New Zealand

Joan W. Bennett Department of Plant Biology, Rutgers, The State University of New Jersey, New Brunswick, NJ, USA

Mary Berbee Department of Botany, University of British Columbia, Vancouver, BC, Canada

Paola Bonfante Dipartimento di Scienze della Vita e Biologia dei Sistemi, Università di Torino, Torino, Italy

Stuart Card Resilient Agriculture, AgResearch Limited, Grasslands Research Centre, Palmerston North, New Zealand

Serena Capitanio Dipartimento di Scienze della Vita e Biologia dei Sistemi, Università di Torino, Torino, Italy

Carolyn Elya Department of Organismic and Evolutionary Biology, Harvard University, Cambridge, MA, USA

Andrea Genre Dipartimento di Scienze della Vita e Biologia dei Sistemi, Università di Torino, Torino, Italy

Carla Gonçalves Department of Biological Sciences and Evolutionary Studies Initiative, Vanderbilt University, Nashville, TN, USA

Martin Grube Institute of Biology, University of Graz, Graz, Austria

Pedro Gundel Instituto de Ciencias Biológicas, Talca University, Talca, Chile

Bradley Hillman Plant Biology, Rutgers University, New Brunswick, NJ, USA

Chris Todd Hittinger Laboratory of Genetics, DOE Great Lakes Bioenergy Research Center, Center for Genomic Science Innovation, Wisconsin Energy Institute, J.F. Crow Institute for the Study of Evolution, University of Wisconsin-Madison, WI, USA

Song Hong Laboratory of Insect Developmental and Evolutionary Biology, CAS Center for Excellence in Molecular Plant Sciences, Shanghai Institute of Plant Physiology and Ecology, Chinese Academy of Sciences, Shanghai, China

Huei-Mei Hsieh Institute of Plant and Microbial Biology, Academia Sinica, Nankang, Taipei, Taiwan

Yen-Ping Hsueh Institute of Molecular Biology, Academia Sinica, Nankang, Taipei, Taiwan

Richard Hung Mycorrhizal Applications, St. Louis, MO, USA

Timothy Y. James Department of Ecology and Evolutionary Biology, University of Michigan, Ann Arbor, MI, USA

Linda Johnson Resilient Agriculture, AgResearch Limited, Grasslands Research Centre, Palmerston North, New Zealand

Yu-Ming Ju Institute of Plant and Microbial Biology, Academia Sinica, Nankang, Taipei, Taiwan

Sheng-Chian Juan Institute of Molecular Biology, Academia Sinica, Nankang, Taipei, Taiwan

Annegret Kohler UMR 2UMR INRA/UHP 1136, Interactions Arbres/Micro-organismes, Centre INRA de Nancy, Champenoux, France

Samantha Y. Lee Graduate Women in Science, Mullica Hill, NJ, USA

Tzu-Hsiang Lin Institute of Molecular Biology, Academia Sinica, Nankang, Taipei, Taiwan

Wade Mace Resilient Agriculture, AgResearch Limited, Grasslands Research Centre, Palmerston North, New Zealand

Francis Martin 2UMR INRA/UHP 1136, Interactions Arbres/Micro-organismes, Centre INRA de Nancy, Champenoux, France

Christina Moon Resilient Agriculture, AgResearch Limited, Grasslands Research Centre, Palmerston North, New Zealand

Jonathan M. Plett Hawkesbury Institute for the Environment, Western Sydney University, Richmond, NSW, Australia

Sushma Prakash Resilient Agriculture, AgResearch Limited, Grasslands Research Centre, Palmerston North, New Zealand

Alisha Quandt Department of Ecology and Evolutionary Biology, University of Colorado, Boulder, CO, USA

Saleh Rahimlou Department of Ecology and Evolutionary Biology, University of Michigan, Ann Arbor, MI, USA

Antonis Rokas Department of Biological Sciences and Evolutionary Studies Initiative, Vanderbilt University, Nashville, TN, USA

Junmei Shang Laboratory of Insect Developmental and Evolutionary Biology, CAS Center for Excellence in Molecular Plant Sciences, Shanghai Institute of Plant Physiology and Ecology, Chinese Academy of Sciences, Shanghai, China

Yanlei Sun Laboratory of Insect Developmental and Evolutionary Biology, CAS Center for Excellence in Molecular Plant Sciences, Shanghai Institute of Plant Physiology and Ecology, Chinese Academy of Sciences, Shanghai, China

Emily Trudeau Department of Botany, University of British Columbia, Vancouver, BC, Canada

Massimo Turina IPSP-CNR, Torino, Italy

Guillermo Vidal-Diez de Ulzurrun Institute of Molecular Biology, Academia Sinica, Nankang, Taipei, Taiwan

Chengshu Wang Key Laboratory of Insect Developmental and Evolutionary Biology, CAS Center for Excellence in Molecular Plant Sciences, Shanghai Institute of Plant Physiology and Ecology, Chinese Academy of Sciences, Shanghai, China
School of Life Science and Technology, ShanghaiTech University, Shanghai, China

Yan-Yan Wang State Key Laboratory of Mycology, Institute of Microbiology, Chinese Academy of Sciences, Beijing, China

Jin-Rong Xu Department of Botany and Plant Pathology, Purdue University, West Lafayette, IN, USA

When Plants and Animals First Met Fungi: Insights from the Evolution of Host Immune Systems

1

Emily D. Trudeau and Mary L. Berbee

Abstract

Fungi interact with plants and animals, leading variously to mutually beneficial partnerships or to devastating disease. Reflecting the importance of plant–fungal and animal–fungal interactions, the innate immune systems of terrestrial plants and animals detect fungi using inherited recognition receptors. Upon recognition, the receptor proteins stimulate signaling pathways that foster intimate mutualisms or trigger robust, multi-level innate defenses against parasitism. To provide points of discussion about how the evolutionary history of host recognition may have unfolded, we analyze the phylogenetic distributions of pattern-recognition receptors important in plant and animal immunity, and of two groups of enzymes important in defenses: chitinases in plants and β-glucanases in animals. Our analysis takes advantage of recently sequenced genomes of early-diverging host plants and animals. We draw upon experimental studies using model organisms that reveal how some host plants and animals perceive fungi. To infer functional homology, we explore recent studies using experimental approaches to analyze binding interactions between host proteins and pathogen molecules. We contend that phylogenetic ages of the host receptors provide a window, imperfect but intriguing, into the origins of the host systems that have been detecting fungi over hundreds of millions of years of geological time.

Keywords

Innate immunity · Plant–fungus interactions · Animal–fungus interactions · Microbe-associated molecular patterns · Host pattern-recognition receptor evolution · LysM domain · GH19 chitinase · C-type lectin-like receptor · Geological time · Fungal cell wall

1.1 Introduction

Reconstructing evolutionary histories of plant and animal proteins involved in sensing and responding to fungi provides a genomic basis upon which to speculate about how and when fungi first interacted with plants and animals. When fungi meet plant or animal hosts, myriad complex mechanisms mediate interactions. Host receptors recognize fungal molecules, setting in motion complex signaling cascades to mount a defense or to initiate mutualistic symbiosis.

The mechanisms underlying fungus–plant or fungus–animal interactions have important implications, both positive and negative. Fungal

E. D. Trudeau · M. L. Berbee (✉)
Department of Botany, University of British Columbia, Vancouver, BC, Canada
e-mail: emily.trudeau@botany.ubc.ca; mary.berbee@botany.ubc.ca

Y.-P. Hsueh, M. Blackwell (eds.), *Fungal Associations*, The Mycota 9,
https://doi.org/10.1007/978-3-031-41648-4_1

pathogens devastate over 125 million metric tons of major crop plants including rice, wheat, and maize every year and account for over 25% of crop losses worldwide (Boddy 2016). Arbuscular mycorrhizal symbiosis, thought to have evolved concurrently with the first land plants, likely had a role in helping early plants to become established on land, thereby shaping the first terrestrial ecosystems (Delaux and Schornack 2021).

Fungi in humans range from causative agents of life-threatening, systemic diseases to commensals, such as *Candida* spp., which can be asymptomatic colonizers of mucous membranes. Some mammals, such as ruminants and other herbivores, derive benefits from a mutualistic relationship with gut fungi. These fungi possess the remarkable ability to break down plant cellulose, which is typically indigestible, into digestible sugars (Gordon and Phillips 1998). Mammals and birds are protected from most potentially deadly fungal pathogens by body temperatures above 35 °C, by slightly alkaline body fluids (most fungi prefer slightly acidic substrates), and by skin surfaces that resist fungal attachment (Casadevall 2005). However, selection for additional layers of antifungal protection must have exerted continuing force through animal evolution. This is evident from medical records and laboratory experiments showing that pathogenic fungi rage out of control when our layers of immune protection are compromised (Brown et al. 2018). Especially in parts of the world where availability of sophisticated antiviral drugs is limited, fungal infections kill 500,000 HIV/AIDS patients per year (Limper et al. 2017) because virus-induced faults in adaptive immunity reduce antifungal protection.

Immune systems vary across groups of organisms. Adaptive immunity is widely shared by vertebrates and appears to have originated among early jawed fishes (Buchmann 2014; Rombout et al. 2005). Adaptive immunity results when an individual's exposure to an antigen creates a "molecular memory," which can take the form of antibodies specific to the antigen. Vaccination or exposure to infection triggers adaptive immunity. The molecular memory may last a few months or a lifetime, but it is not passed to the next generation.

Innate immunity, the subject of this chapter, is based on genetically inherited receptors that recognize and bind ligands, which are molecules characteristic of various microbes. Distinguishing among molecules from interacting microbes enables eukaryotic hosts to initiate intracellular signaling, leading to partnerships with mutualists, tolerance of commensals, or the detection and destruction of pathogens. In vertebrates, response to microbes involves coordination between innate and adaptive immune systems. However, the vast majority of eukaryotes including plants and invertebrates have only the more ancient innate immune systems.

We focus on a first line of innate defense that starts when host proteins, pattern-recognition receptors at the cell surface, detect microbe-associated molecular patterns (Sanchez-Vallet et al. 2014; Hatinguais et al. 2020). The hosts' receptors typically target microbial cell coverings. Their ability to distinguish among microbes usually involves their binding preferences. Cell coverings—chitin, mannose, and β-1,3-1,6-glucan in fungi (Gow et al. 2017); lipopolysaccharides and peptidoglycans in Gram-negative bacteria; and peptidoglycans and teichoic acid in Gram-positive bacteria (Fisher and Mobashery 2020)—are ancient molecules, but their organization and components vary and overlap, due in part to continuing selective pressure from pathogens and symbionts. Here, we focus on fungal recognition, but the same or similar host receptors may also detect oomycetes, nematodes, and other organisms that share similar cell surface biochemicals.

Many of the receptors important in recognizing fungi and other microbes are multi-domain proteins with an extracellular lectin domain. Lectin domains are non-catalytic and are defined by function; they bind ligands, usually sugars, without altering them (De Schutter and Van Damme 2015). Lectins are perhaps most familiar as tools used in cell biology to identify polysaccharides through their binding specificities, as when the lectin wheat germ agglutinin, labeled appropriately and added to fungal cells, reveals through its binding the location of chitin in fungal walls (James and Berbee 2012). In animal and plant cells, receptors bearing different lectin domains can distinguish pathogenic or non-pathogenic microbes by their different cell wall components. Enzymes, in contrast to lectins, bind and alter their substrates. Among enzymes, some, such as chitinases in plants and animals and β-glucanases in insects, have roles both in detecting and in destroying pathogens.

After a host protein recognizes a microbe-associated molecular pattern, host responses usually involve cellular signaling, leading variously to phagocytosis or agglutination of the microbe, to the secretion of antimicrobial toxins, or to programmed cell death that limits a pathogen's growth by killing its host cell. This chapter does not explore signaling or how a host ultimately disarms a microbe. For reviews that use elegant diagrams to explain host processes, see Erwig and Gow (2016) (mammals); Hatinguais et al. (2020) (mammals); Hoffmann (2003) (*Drosophila*); Ngou et al. (2022) (plants); and Gauthier et al. (2010) (comparing selected eukaryotes).

Because host receptors and enzymes important in innate immunity are inherited vertically, they can be analyzed phylogenetically. We chose a few of the many host proteins with experimentally demonstrated roles in fungal recognition. Experimental data on the molecular basis for microbial detection come largely from studies of plant models such as *Arabidopsis* and rice, animal models including mice, and human cell cultures. In the following pages, we discuss the appearance of these proteins through host evolution using published phylogenies where applicable. We then use BLAST searches, comparison of protein domain structures, information on ligand binding sites, and some original phylogenetic analyses to apply newly available data to inferences about when proteins with homologous function may have originated. We use published ages of host clades to infer protein ages, with the caveat that estimates of clade ages vary across studies, especially for unicellular, early-diverging taxa without rich fossil records (Morris et al. 2018; Chang et al. 2015, 2019; Nelsen et al. 2020; Lutzoni et al. 2018; Amses et al. 2022). To specify particular proteins, GenBank accession numbers are given in parentheses in the text or are included in figures.

Immunology as a field has burgeoned, and astounding progress brings with it specialized terminology that can be a barrier to understanding for those from other disciplines. We minimized our use of acronyms and abbreviations, apologizing to colleagues who prefer the more concise notation. Our chapter now focusses on phylogenetic interpretation of proteins with roles in recognizing fungal cell walls.

1.2 How Plants Detect and Respond to Fungi

In this section, we speculate on the origins of chitin perception and chitin-degrading enzymes in plants (Fig. 1.1). We review how green algae (chlorophyte and streptophyte algae), bryophytes, and vascular plants perceive fungi, and we explore the diversity in structure and function of plant chitinases—the defense-related enzymes that target the fungal cell wall. We present a phylogenetic analysis of one family (glycoside hydrolase 19) of plant chitinases, reconstructing the evolutionary history of this family spanning bacterial, algal, and plant lineages.

1.2.1 Plants Detect Fungi via Pattern-Recognition Receptors that Contain Extracellular Lysin Motifs (LysM): Lectin Domains that Bind Chitin

The plant immune system allows plants to perceive, initiate defense responses, and mount a resistance to fungal pathogens. Plants evolved to recognize fungi by detecting chitin fragments as microbe-associated molecular patterns. Chitin is an insoluble long-chain polymer of *N*-acetylglucosamines that can be degraded by chitinases to produce soluble fragments of eight or fewer *N*-acetylglucosamine residues (Buendia et al. 2018). Other microbe-associated molecular

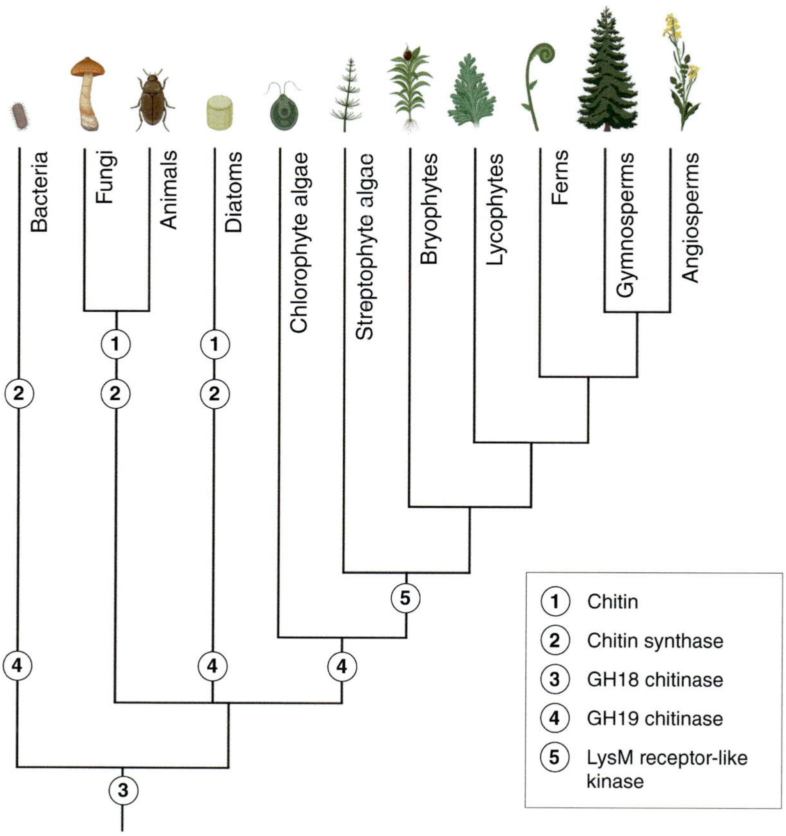

Fig. 1.1 Overview of distributions of chitin, chitin recognition receptors, and defense-related proteins across lineages that are reviewed over the next pages. Icons created with BioRender.com

patterns perceived by plants include peptidoglycan from bacterial cell walls, a polymer made up of alternating residues of *N*-acetylglucosamine and *N*-acetylmuramic acid. In hosts, primary defense responses such as cell wall alterations, production of reactive oxygen species, and expression of defense-related genes like chitinases are initiated by recognition of chitin fragments (Khokhani et al. 2021). A second layer of defense is triggered upon recognition of fungal effectors, which are secreted by some fungi to interfere with plant immune responses, ultimately resulting in hypersensitive responses and programmed cell death, as reviewed by de Wit (2007). The focus of this section will be on primary immune responses to fungi.

LysM-containing receptors can be receptor-like kinases, with a cytoplasmic kinase domain that can signal when the LysM domains have bound to chitin oligomers (Fig. 1.2a), or receptor-like proteins that lack the cytoplasmic kinase domain needed for signal transduction (Shiu and Bleecker 2003). LysM domains can detect and bind to not only chitin fragments but also bacterial peptidoglycan. They were first discovered in peptidoglycan-hydrolyzing enzymes, and the name originates from the bacterial cell lysis that results (Buendia et al. 2018; Willmann et al. 2011).

LysM domains of around 40–50 amino acid residues that belong to the carbohydrate-binding module family 50 are found in most organisms, according to the Carbohydrate-Active Enzymes database (www.cazy.org) (Drula et al. 2022). They are not highly conserved in terms of primary sequence, although cysteine pairs separated by one amino acid are found between the three LysM domains of most plant LysM receptor-like kinases and receptor-like proteins, and are required for disulfide bridging between domains (Lefebvre et al. 2012). Secondary and tertiary structures of LysMs are, however, highly

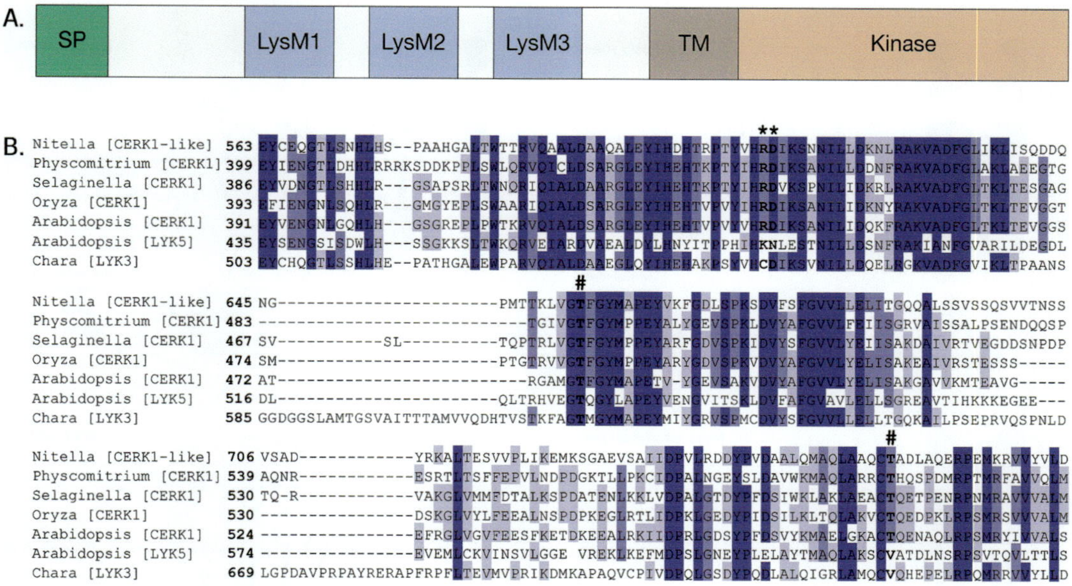

Fig. 1.2 Structure of LysM receptor-like kinases. (**a**) Domain architecture of CERK1 LysM receptor-like kinase (*SP* signal peptide, *LysM1-3* LysM domains, *TM* transmembrane domain). (**b**) Partial alignment of kinase domains in LysM receptor-like kinases from plants and streptophyte algae. Arginine (R) and aspartic acid (D) residues indicative of RD kinase activity are denoted by an asterisk. Threonine (T) residues essential for chitin signaling are denoted by a number sign. LYK5 from *Arabidopsis* and LYK3 from *Chara* lack these residues. Darker shading indicates more highly conserved positions

conserved, consisting of two α-helices stacked onto two antiparallel β-sheets (Buendia et al. 2018).

1.2.1.1 Chitin Perception in Land Plants Typically Involves Complexes of Two or More LysM-Containing Receptors at the Plasma Membrane

These complexes are typically composed of two receptor-like kinases, or one receptor-like kinase and one receptor-like protein. One receptor usually has higher chitin-binding affinity, while the other has an active kinase domain that is necessary for signal transduction (Fig. 1.3). In general, plant LysM-containing receptors have three extracellular LysM domains, with one or more domains binding chitin directly. In *Arabidopsis thaliana*, chitin oligomers are detected by a receptor complex made up of *At*CERK1 and *At*LYK5 (Fig. 1.3a). *At*LYK5 is a LysM receptor-like kinase, but has an inactive kinase domain, and serves as the high-affinity chitin-binding receptor. *At*CERK1 can bind chitin but shows only modest chitin binding compared to *At*LYK5. *At*CERK1 provides kinase activity and is required for chitin-triggered immune responses. Chitin oligomers induce heterodimerization between *At*LYK5 and *At*CERK1, and *At*LYK5 is required for subsequent *At*CERK1 homodimerization (Fig. 1.3a) (Gong et al. 2020). *At*CERK1 phosphorylates the cytoplasmic kinase PBL27, which induces MAPK cascades and immune responses.

The first plant chitin receptor to be identified—*Os*CEBiP—was cloned from rice in 2006 (Kaku et al. 2006). *Os*CEBiP is a LysM receptor-like protein that is anchored to the plasma membrane by a glycophosphatidylinositol. It binds chitin but lacks the cytoplasmic kinase domain needed for signal transduction (Jiang et al. 2019). Two CEBiP monomers homodimerize to bind the same chitin oligomer, showing preference for longer chitin fragments consisting of about seven

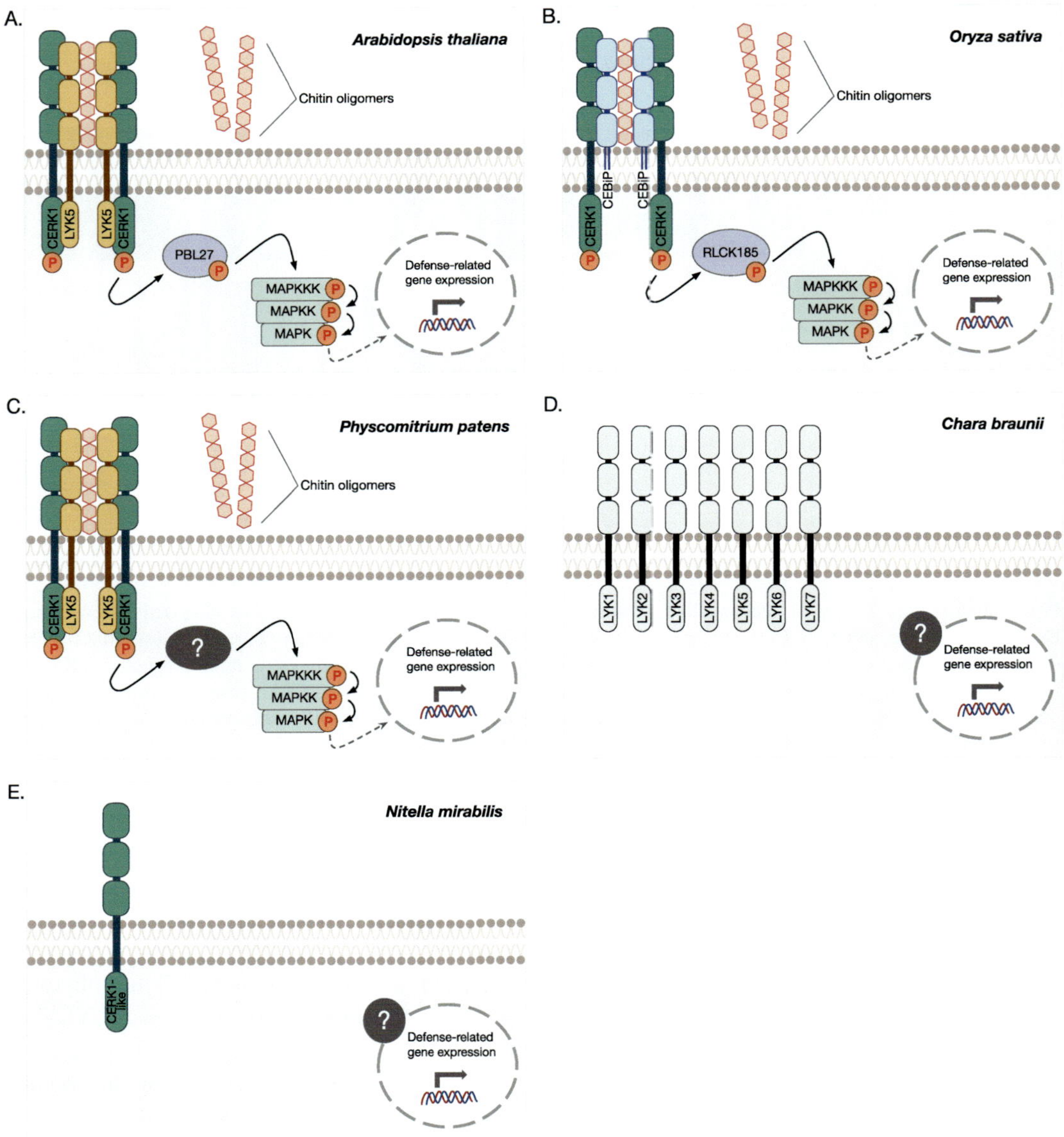

Fig. 1.3 LysM receptor-like kinases and receptor-like proteins detect chitin and initiate immune responses. (**a**) In *A. thaliana* receptor-like kinases CERK1 and LYK5 form a complex to bind chitin, phosphorylate a cytoplasmic kinase (PBL27), activate a mitogen-activated protein kinase (MAPK) cascade, and induce defense-related gene expression. (**b**) In rice, *O. sativa*, CERK1 and receptor-like protein CEBiP form a complex to bind chitin, phosphorylate a cytoplasmic kinase (RLCK185), activate a MAPK cascade, and induce defense responses. (**c**) In the moss *P. patens*, CERK1 activates MAPK signaling cascades. (**d**, **e**) Streptophyte algal genomes have CERK1 or LYK5 homologs with untested roles in chitin perception and defense. (**d**) *Chara braunii*'s genome encodes seven LysM receptor-like kinases. (**e**) *Nitella mirabilis*'s transcriptome encodes a CERK1 homolog

to eight *N*-acetylglucosamine residues (Fig. 1.3b). *Os*CEBiP then heterodimerizes to form a complex with the LysM receptor-like kinase *Os*CERK1, which cannot bind chitin but provides kinase function to the receptor complex (Shimizu et al. 2010). *Os*CERK1 phosphorylates a receptor-like cytoplasmic kinase (*Os*RLCK185), which in turn activates the MAPK cascade resulting in immune responses and defense signaling (Sanchez-Vallet et al. 2014).

In the moss *Physcomitrium patens* (Fig. 1.3c), *Pp*CERK1, an ortholog to *At*CERK1, binds chitin and then initiates MAPK cascade activation (Bressendorff et al. 2016). *Physcomitrium patens*'s three LYK5 genes are homologs to *At*LYK5 (Fig. 1.3a), and silencing of LYK5 genes in *P. patens* renders plants chitin insensitive (Orr et al. 2020). Additionally, one *Pp*LYK5 was shown to be highly expressed upon infection with the fungus *Botrytis cinere*a (Reboledo et al. 2021; Orr et al. 2020).

Representing the streptophyte algae, the closest relatives to land plants, the genome of *Chara braunii* encodes seven LysM receptor-like kinases (LYK1–7) (Fig. 1.3d) and the transcriptome of *Nitella mirabilis* encodes a homolog to CERK1 from land plants (Fig. 1.3e).

1.2.1.2 LysM Receptor-Like Kinases Evolved in the Common Ancestor of Land Plants and Streptophyte Algae

Members of the CERK1 clade of LysM receptor-like kinases are as yet known only from streptophyte algae and land plants (Figs. 1.2 and 1.3). Some chlorophyte, red, and glaucophyte algae have LysM domain-containing proteins of uncharacterized function, but proteins that have both the kinase and the LysM domain are as yet unknown in these lineages (Van Holle and Van Damme 2019; Delaux and Schornack 2021). Distantly related LysM domains are also found across bacteria, fungi, and animals, implying that the LysM domain is even more ancient than the divergence of eukaryotes from bacteria, estimated to have occurred over three billion years ago (Battistuzzi et al. 2004; Knoll 2014).

Receptors like CERK1 with active kinase domains are conserved in land plants, with a possible ortholog in the streptophyte alga *Nitella mirabilis* (Fig. 1.3). *Nitella*, in Charophyceae, diverged from the ancestor of land plants 635 million years ago (Ma) (estimated from Fig. 1.3 in Morris et al. 2018), suggesting that CERK1 may have been present before plant terrestrialization. That it is highly conserved implies importance for mediating interactions between land plants and fungi.

For *Arabidopsis At*CERK1, mutational analysis showed that kinase function, chitin signaling, and defense-related gene expression required phosphorylation sites at threonine (T) residues T479 and T573 (Fig. 1.2b) (Suzuki et al. 2016). CERK1 bears an RD kinase, having a conserved arginine (R) residue immediately preceding a conserved aspartic acid (D) residue in the catalytic loop of the kinase domain (Greeff et al. 2012). These T, R, and D residues are conserved in CERK1 receptors with active kinases in *Arabidopsis* (NP_566689), *Oryza* (XP_015650771), *Selaginella* (XP_002977121), and *Physcomitrium* (XP_024388364) (Fig. 1.2b). In the CERK1 homolog from *N. mirabilis* (JV789971), T, R, and D residues are similarly all conserved.

Arabidopsis LysM receptor-like kinases without kinase activity such as *At*LYK5 (AAB80675) lack the homologous T, R, and D residues. Raising questions about functions of its LysM receptor-like kinases, all seven of the genes in *C. braunii* have a valine (V), not a T that would align at site T573 of *At*CERK1, as shown by *Cb*LYK3 (GBG90350) in Fig. 1.2b. All but one of the LysM receptor-like kinase genes in *C. braunii* lack the R and/or D residues characteristic of RD kinases. Gene family expansion of LysM receptor-like kinases occurred independently in *C. braunii* and land plants, and future experimental testing of the *C. braunii* receptors may give fascinating results (Nishiyama et al. 2018). As Nishiyama et al. (2018) point out, the *C. braunii* receptors' expansion may have allowed adaption to a range of interactions with chitin- and peptidoglycan-containing organisms.

1.2.1.3 Plant Chitin Receptors Can Initiate Both Immune and Symbiotic Responses

The common ancestor to all land plants may have been preadapted for symbiosis with mycorrhizal fungi, based on the many genes important in fungal colonization that are shared between land plants and the streptophyte algae (Delaux et al.

2015). Since microbial symbionts—namely mycorrhizal fungi—produce chitin oligomers, perception of these microbe-associated molecular patterns does not inevitably trigger host defense responses. Longer chitin oligomers of six to eight residues can initiate both immune and symbiotic responses but are stronger activators of host immune responses than are short-chain oligomers (Khokhani et al. 2021). The legume *Medicago truncatula* uses the receptor complex *Mt*LYK9/ *Mt*CERK1 in both its symbiotic and immune responses, and either response can be activated by four- or eight-residue chitin oligomers (Feng et al. 2019). Similarly, in rice, the receptor-like kinase *Os*CERK1 can detect short-chain oligomers in arbuscular mycorrhizal signaling, suggesting a dual role in immunity and symbiosis (Carotenuto et al. 2017). Discriminating between symbiotic hyphae and pathogenic appressoria may involve additional factors such as plant nutrient status and perception of mechanical stimuli (Khokhani et al. 2021). Evidently, plants are balancing activation of immune signaling and symbiotic signaling to respond appropriately to chitin oligomers.

1.2.2 Lipo-Chitooligosaccharides (LCOs), Chitin-Based Signaling Molecules, Are Widely Distributed Among Fungi

The fungal LCOs modulate plant immune responses when produced by mycorrhizal fungi and nitrogen-fixing bacteria. Like chitin oligomers, LCOs are perceived by LysM receptor kinases that activate downstream signaling cascades (Wanke et al. 2020). First characterized in nitrogen-fixing bacteria that form root nodules in legumes (Roche et al. 1996), LCOs were subsequently shown to initiate signaling for mutualistic symbioses in hosts of arbuscular mycorrhizae (Maillet et al. 2011) and ectomycorrhizae (Cope et al. 2019). Land plant mutualisms via perception of LCOs arose convergently. Arbuscular mycorrhizae are the oldest of the mutualisms with LCO recognition; they involve the earliest-diverging lineages of land plants and are represented by 407 Ma fossils. Ectomycorrhizae, restricted to seed plants, are known only from 50 Ma fossils, and the root-associated nitrogen-fixing bacteria, restricted to flowering plants, have a fossil record dating back to ~85 Ma (Bonfante and Requena 2011; van Velzen et al. 2019).

Arbuscular mycorrhizal fungi produce a combination of chitin oligomers and LCOs during colonization of their host plants. Symbiotic outcomes are favored when LCOs suppress the plant immune responses triggered by long-chain chitin oligomers (Maillet et al. 2011). Upon LCO detection, the host initiates signaling pathways that increase the number of lateral roots—the most common sites of arbuscular fungal colonization. Many of the signaling proteins are shared by bryophytes and vascular plants (Delaux and Schornack 2021), implying that they were already in place when these clades of land plants diverged ~515–474 Ma (Morris et al. 2018).

Production of LCOs may represent a fungal pre-adaptation to interactions with plants because it is not restricted to mycorrhizal fungi. Rush et al. (2020) showed that fungi in five phyla, including saprotrophs, parasites, and mutualists, produce LCOs. They demonstrated experimentally that LCOs can function as growth regulating hormones in ascomycetes and basidiomycetes (Rush et al. 2020). A role in regulating fungal development may well be ancestral because LCO production, based on its distribution across chytrids, Blastocladiomycota, and Dikarya, originated some 800 Ma (Chang et al. 2019), hundreds of millions of years before land plants.

1.2.3 Plant Perception of Other Fungal Molecules Is Not as Well Characterized

β-1,3-glucan and β-1,6-glucan are other major components of the fungal cell wall, together making up around half of its dry weight. Wanke et al. (2020) found that response to β-glucans, polymers of glucose, varies among plant species and that the degree of polymerization plays a major role in recognition and response. In

Arabidopsis, perception of short-chain β-glucans requires *At*CERK1. In contrast in rice, perception of long-chain β-glucans requires *Os*CERK1 (Wanke et al. 2020).

Arabidopsis can also recognize other fungal microbe-associated molecular patterns, for example producing ethylene in response to xylanases secreted by fungi and bacteria to degrade hemicellulose. *Arabidopsis* recognizes the fungal endopolygalacturonase enzymes that target pectin in plant cell walls. These molecular patterns are detected at the plasma membrane by receptor-like proteins that contain leucine-rich repeats (Anand et al. 2021; Zeilinger et al. 2016).

Leucine-rich repeats also occur in nucleotide-binding, leucine-rich repeat (NLR) receptors in plants. Well-characterized plant NLR receptors are important in detecting fungal molecules but being cytoplasmic rather than at cell surfaces (Maekawa et al. 2011; Martinon and Tschopp 2005); they are beyond the topics of this chapter. We discuss the roles of leucine-rich repeats in animal toll-like receptors in the upcoming section on detection of fungi by animals.

1.2.4 Chitinases in Plant Defenses Against Fungi

Plants defend themselves against fungi and other microbes by expressing pathogenesis-related proteins including chitinases (Collinge et al. 1993; Pusztahelyi 2018). Plant chitinases compromise fungal cell wall integrity by degrading chitin polymers through the hydrolytic cleavage of glycosidic bonds between β-1,4-linked *N*-acetylglucosamine monomers (Punja and Zhang 1993). Through this process, chitin oligomers are released, stimulating host immune responses (Sanchez-Vallet et al. 2014). Many plant pathogenic fungi make their first contact with the host in the apoplast. The apoplast, the extracellular space beyond the plasma membrane, contains hydrolytic enzymes and toxins that inhibit fungal growth.

The Carbohydrate-Active Enzymes database (www.cazy.org) classifies plant chitinases into two families: glycoside hydrolases 18 and 19 (GH18 and GH19) (Drula et al. 2022). GH18 and GH19 chitinases show no evidence of homology (Hamel et al. 1997) as they exhibit different catalytic mechanisms and differ in their domain architectures and 3D structures (Cheng et al. 2021). While GH18 chitinases are widely distributed across plant, bacterial, animal, and fungal lineages, GH19 chitinases are mainly found in plants and bacteria (Sharma et al. 2020), although GH19 chitinase function has been experimentally characterized in Microsporidia (Han et al. 2016), and using BLAST searches, we found putative GH19 chitinases in some fungi (e.g., GenBank accessions AWI66960 and KAG9575357). Plant chitinases were further divided into seven classes, prior to their classification into GH18 and GH19 families (Fig. 1.4). GH18 includes classes III and V, while GH19 includes classes I, II, IV, VI, and VII (Adrangi and Faramarzi 2013).

Classes were defined based on protein catalytic domains, sequence lengths, and the presence or absence of non-catalytic chitin-binding domains (CBDs) or carbohydrate-binding modules that promote the association of the enzyme with the substrate (Fig. 1.4) (Kitaoku et al. 2022). An alternative framework classifies GH19 chitinases into "loopful" (class I and II) and "loopless" (class IV) chitinases based on the presence or absence of up to six loop regions in the catalytic domain (Hoell et al. 2006; Taira et al. 2011). Recognizing that classes failed to fully reflect phylogeny or capture variation within classes, Orlando et al. (2021) applied a clustering system to divide the catalytic domains of GH19 chitinases into two subfamilies consisting of 51 groups.

There will, however, always be exceptions to the generalized rules involved in classification of multi-domain proteins that evolve with domain shuffling. For example, chitinases in a subclass of class III in the horsetail *Equisetum arvense* and fern *Pteris ryukyuensis* (Inamine et al. 2015; Onaga and Taira 2008) contain LysM domains. This type of LysM–chitinase conjugate is not usually found in plants, but is common in fungal chitinases (Kitaoku et al. 2022). With BLAST searches, we found a putative GH18 class V

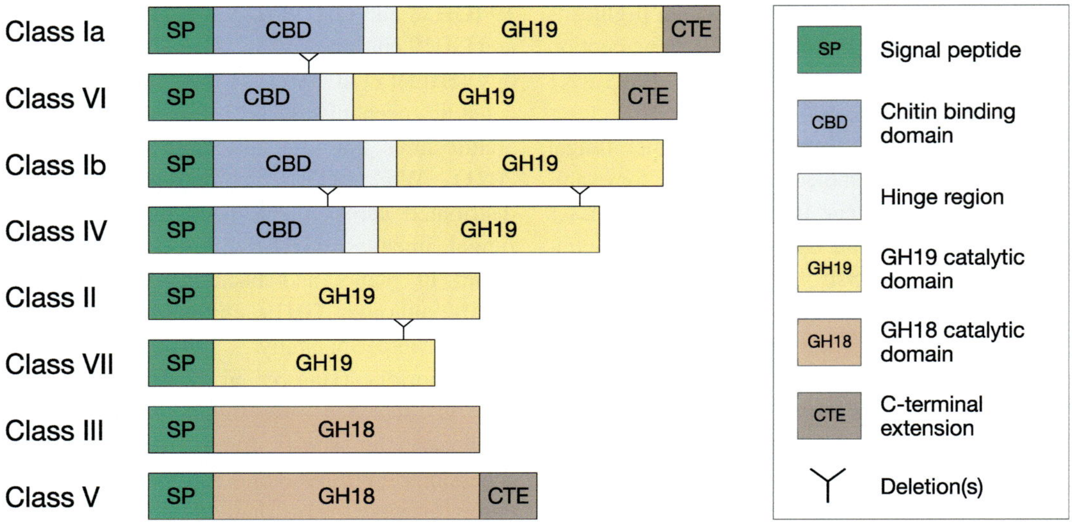

Fig. 1.4 Domain architecture characteristic of classes within GH19 and GH18 chitinases. Classes I, II, IV, VI, and VII are GH19 chitinases that differ in length and presence or absence of a chitin binding domain (CBD), hinge region, and C-terminal extension (Passarinho and de Vries 2002; Orlando et al. 2021). CBDs promote the association of the enzyme with the substrate. Chitinases with a C-terminal extension are transmitted to the vacuole, while those without are transmitted extracellularly to the apoplast (Renner and Specht 2012). Classes III and V are GH18 chitinases that share a catalytic domain with a conserved DxDxE sequence motif but show little sequence similarity, and generally lack CBDs (Ohnuma et al. 2011)

chitinase in the spikemoss *Selaginella moellendorffii* (XP002976579) with a predicted CBD domain more commonly found in class I and IV chitinases of family GH19.

1.2.4.1 Biological Functions and Expression Patterns of Plant Chitinases Vary Among Classes and Species

Since plants do not synthesize chitin, their chitinases can play roles in defense against chitin-containing pathogens such as fungi without compromising their own cellulosic cell walls (Taira et al. 2009). Much research implicates plant chitinases in defense responses against fungi (Kasprzewska 2003; van Loon et al. 2006; Passarinho and de Vries 2002). Some chitinases are synthesized constitutively irrespective of stress conditions, while others can be induced by pathogen infection (Xin et al. 2021). The class I chitinase from rye (*Secale cereale* BAB18519) is an extracellular inhibitor of fungal growth (Yamagami and Funatsu 1993). Improved resistance to the fungal pathogen *Botrytis cinerea* resulted when *Arabidopsis* was transformed with a class IV chitinase from mulberry that is upregulated upon *B. cinerea* infection (Xin et al. 2021). A class VI chitinase in cashew degraded colloidal chitin and inhibited the growth of the fungus *Lasiodiplodia theobromae in vitro* by altering hyphal morphology (Oliveira et al. 2020). Class III chitinases from *Populus trichocarpa* acted on both chitin and chitosan, the deacetylated form of chitin (Zhang et al. 2022).

Many plant chitinases can also contribute to regulating developmental processes such as seed germination and response to abiotic stresses such as freezing and low nutrient conditions (González et al. 2015). A class II chitinase from rye (*Secale cereale* AAG53610) plays a role in extracellular frost resistance (Yeh et al. 2000). Plant chitinases might be involved in developmental regulation by acting on endogenous arabinogalactan proteins that contain *N*-acetylglucosamines (van Hengel et al. 2001), but this has not yet been confirmed experimentally (Hromadova et al. 2021). A role in senescence was suggested when ethylene (a plant hormone associated with senescence) increased expression of a class I chitinase in *Arabidopsis* (Passarinho and de Vries 2002). An ethylene-dependent signaling pathway was required for induction of this chitinase in plants infected with *B. cinerea* (Thomma et al. 1999). This class I chitinase might also serve in host defense as it inhibited growth of the fungus *Trichoderma reesei* in vitro.

1.2.5 Evolutionary History of Plant Chitinases

Since GH18 chitinases can be found across plant, animal, bacterial, and fungal lineages, and since plant GH18 chitinases are more similar to bacterial and fungal GH18 chitinases than other plant chitinases, it follows that GH18 chitinases or "protochitinases" could have been present in the common ancestor of bacteria and eukaryotes. GH19 chitinases, on the other hand, are known mostly from plants and bacteria, although fewer bacterial sequences than plant sequences have been experimentally characterized (Ohno et al. 1996; Sharma et al. 2020; Yano et al. 2021). According to an analysis by Orlando et al. (2021), GH19 chitinases can be divided into a mostly bacterial clade and a mostly plant clade at the basal node of the gene tree. The focus of our analysis is on plant GH19 chitinases, and Orlando et al. (2021) should be consulted for insights into the phylogeny of bacterial GH19 chitinases.

1.2.5.1 Phylogenetic Analysis of GH19 Chitinases

Previous phylogenies of chitinases emphasized seed plants (Taira et al. 2011; Oliveira et al. 2020), but increasingly comprehensive data from chlorophyte algae, streptophyte algae, bryophytes, and seedless vascular plants allow us to explore the deeper origins of GH19 chitinases. To reconstruct the evolutionary history of GH19 chitinases (Fig. 1.5), we analyzed sequences homologous to *Arabidopsis* chitinases from land plants, green algae, bacteria, and diatoms. Glaucophyte algae and red algae appeared to lack homologs.

Figure 1.5 clade 1 shows that angiosperm GH19 sequences previously placed in classes I, II, and VI together form a monophyletic group including the seedless vascular plants, bryophytes, streptophyte algae, chlorophyte algae, and a bacterium, *Chitinophaga pinensis*. Most of these sequences have the chitin-binding sites predicted by Orlando et al. (2021). Some do not, e.g., *Secale cereale* (AAG53610), and comparison of function of proteins with and without predicted binding sites may be useful in further refining bioinformatic predictions. Denser sampling of bacterial chitinases would be needed to test the significance of their relationship to eukaryotic proteins. Together, class I and class II proteins form the well-supported clade 2 (Fig. 1.5), which also appeared in Oliveira et al.'s (2020) analysis of angiosperm sequences. From our broader sampling, clade 2 sequences appear to have originated before the first divergences of vascular plants 451–431 Ma (Morris et al. 2018), as they encompass flowering plants and *Selaginella*. Within clade 2, class I and class II sequences are not separated, indicating that chitin-binding domains (Fig. 1.5) are gained or lost convergently through domain shuffling. A good example of lack of monophyly of class I vs class II comes from *Secale cereale* because one of its class II sequences (BAB18520) is more closely related to its class I sequence (BAB18519), with significant bootstrap support, than it is to another of its class II sequences (AAG53610) (Fig. 1.5).

Our analysis expands the clade of known relatives of class VI chitinases from angiosperms to include other seed plants, bryophytes, and seedless vascular plants, estimated to have diverged starting 515–474 Ma (Morris et al. 2018) (Fig. 1.5, clade 3). Chitinase function has been lost in some angiosperms in class VI, e.g., *Arabidopsis* (AAL37737) (Fig. 1.5). The chitinases from the seedless plants have retained a glutamic acid residue in the enzyme's active site that was lost from the corresponding angiosperm proteins, consistent with possible conservation of

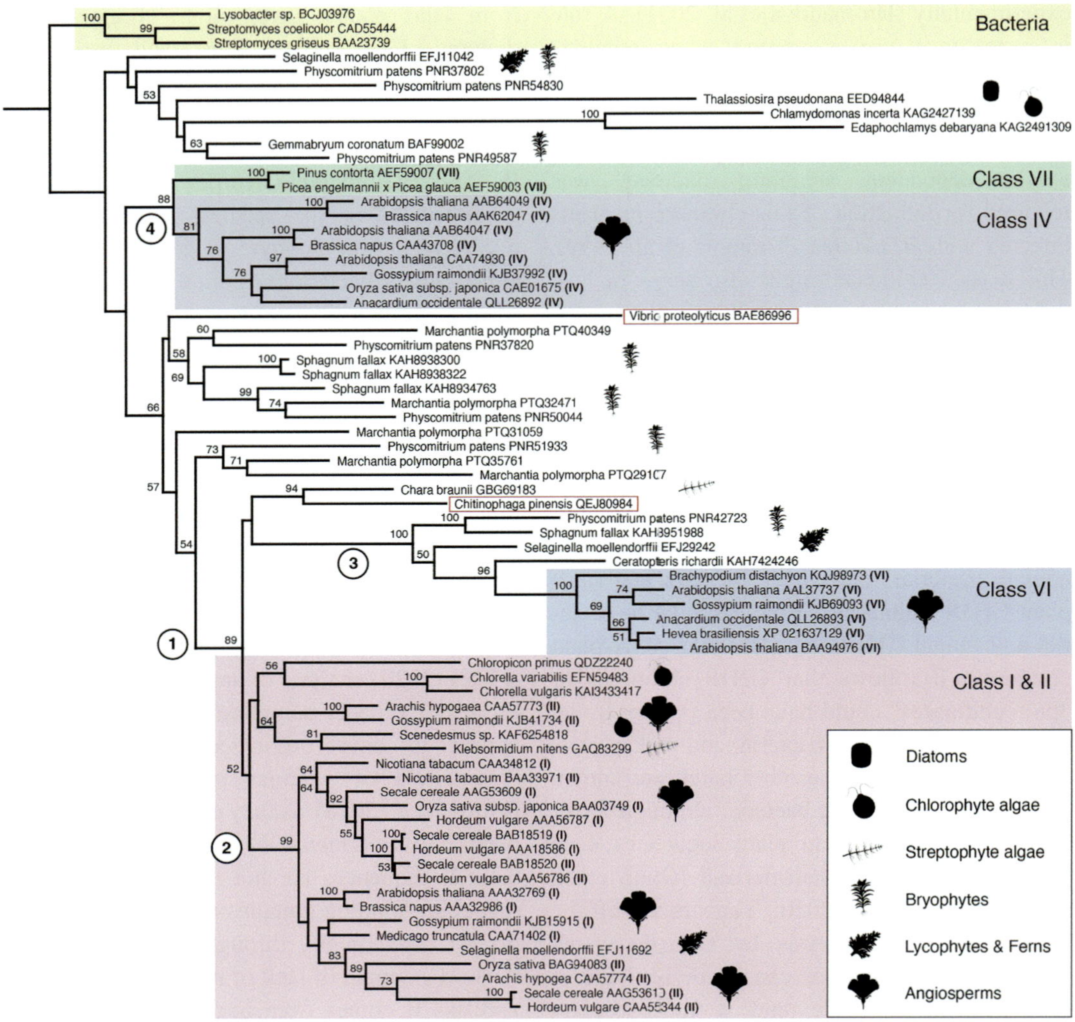

Fig. 1.5 GH19 chitinases in plants have origins in the common ancestor of chlorophyte and streptophyte algae, showing gene duplications in early land plants and retention of gene copies. In this maximum likelihood tree from GH19 catalytic domains, species names are followed by GenBank accession numbers, and Roman numerals in parentheses indicate class as assigned by previous literature. Sequences without an associated class number were obtained through BLAST searches of algal, bryophyte, lycophyte, and fern genomes using different classes of *Arabidopsis* chitinases as queries. The tree was rooted with the bacterial clade as the outgroup. Bacterial sequences nested within plant clades are boxed in red. Bootstrap values greater than 50 are indicated above the nodes. Numbered nodes 1–4 denote well-supported clades that are further discussed in the text. Icons created with BioRender.com

function of class VI chitinases through early land plant evolution. Classes IV and VII form clade 4 (Fig. 1.5), consisting of seed plants that initially diverged ~365–331 Ma (Morris et al. 2018). We found class IV among flowering plant sequences, and class VII only from gymnosperms, as in Oliveira et al. (Table S2, 2020).

Retention of genes implies continued selection for a diversity of chitinases in land plants, and therefore of defense-related enzymes that target chitin-containing cell walls of organisms like fungi. Our analysis supports the presence of at least one GH19 chitinase in chlorophyte algae and in the common ancestor of streptophyte algae and land plants. Gene duplications in early land plants are implied by supported clades that group angiosperms and *Selaginella* (class I and II); angiosperms, seedless plants, and bryophytes (class VI, clade 3); and angiosperms and gymnosperms (class VII and IV) (Fig. 1.5). Relationships of other sequences from bryophytes, seedless vascular plants, diatoms, and algae remain unresolved, leaving the earliest origins of the GH19 family shrouded in the mists of time.

1.2.5.2 Caveats to Reconstructing Ancestral Chitin-Degrading Abilities in Plants

The main target of chitinases in chlorophyte and streptophyte algae in freshwater habitats may have been invertebrates and protists with cell walls of chitin or its deacetylated form chitosan (Mulisch 1993; Steinfeld et al. 2019) rather than fungi. In a terrestrial environment, plant chitinases may defend against insects or nematodes with chitinous mouthparts. Therefore, presence of chitinases in a genome cannot be strictly interpreted as a potential means for plant–fungal interactions. Some plant chitinases also display additional lysozyme-like activity and can degrade peptidoglycan found in bacterial cell walls (Kasprzewska 2003). Furthermore, some plant chitinases are solely involved in endogenous developmental processes and therefore do not play a role in interactions with or response to microbes. Experimental characterization of putative chitinase genes is needed to confirm function in response to fungi.

1.3 How Animals Detect and Respond to Fungi Using Pattern-Recognition Receptors

Pattern-recognition receptors in animals and plants evolved convergently, with shuffling, recruitment, and re-assembly of ancient lectin and signaling domains leading to new, multi-domain receptor proteins. As shown by careful analysis (Urbach and Ausubel 2017; Ausubel 2005), homology between plant and animal receptors may be at the level of individual conserved domains but is not retained across the whole of the multi-domain receptor proteins.

Function constrains plant and animal receptors to similar architecture. Many animal receptors, like plant receptors, bind fungal cell wall polysaccharides including β-glucan or chitin oligosaccharides using extracellular lectin domains. Upon binding, receptors typically form dimers or higher order polymers, and their cytoplasmic domains initiate signaling leading to an immune response.

1.3.1 Evolution of Dectin-1, a C-Type Lectin-Like Receptor with a Central Role in Fungal Detection

By binding β-glucans, the lectin dectin-1, short for "dendritic cell-associated C-type lectin-1," plays a key role in limiting initial invasion by fungi, and in activating human immune systems for further antifungal defense. Figure 1.6 shows dectin-1 spanning the plasma membrane of mobile amoeboid phagocytes including macrophages (Gantner et al. 2003). β-glucans bind at dectin-1's extracellular C-type lectin domain. C-type lectin domains, consisting of about 120 amino acids, are rare in plant recognition receptors but common in animals. Most animal genomes encode dozens of proteins equipped with a domain of this type (Pees et al. 2016).

Host responses after dectin-1 has bound β-glucans can include ingestion and killing of fungal cells by phagocytosis, damaging fungal cells through the production of reactive oxygen species, and secretion of cytokines leading to inflammation, increasing body temperature to disadvantage fungi (Gantner et al. 2003). Phagocytes can engage the adaptive immune system by presenting microbial antigens to T cells (LeibundGut-Landmann et al. 2007).

Disruption of the *CLEC7A* gene encoding dectin-1 in mice reduced immune responses to the fungus *Candida albicans* and led to higher

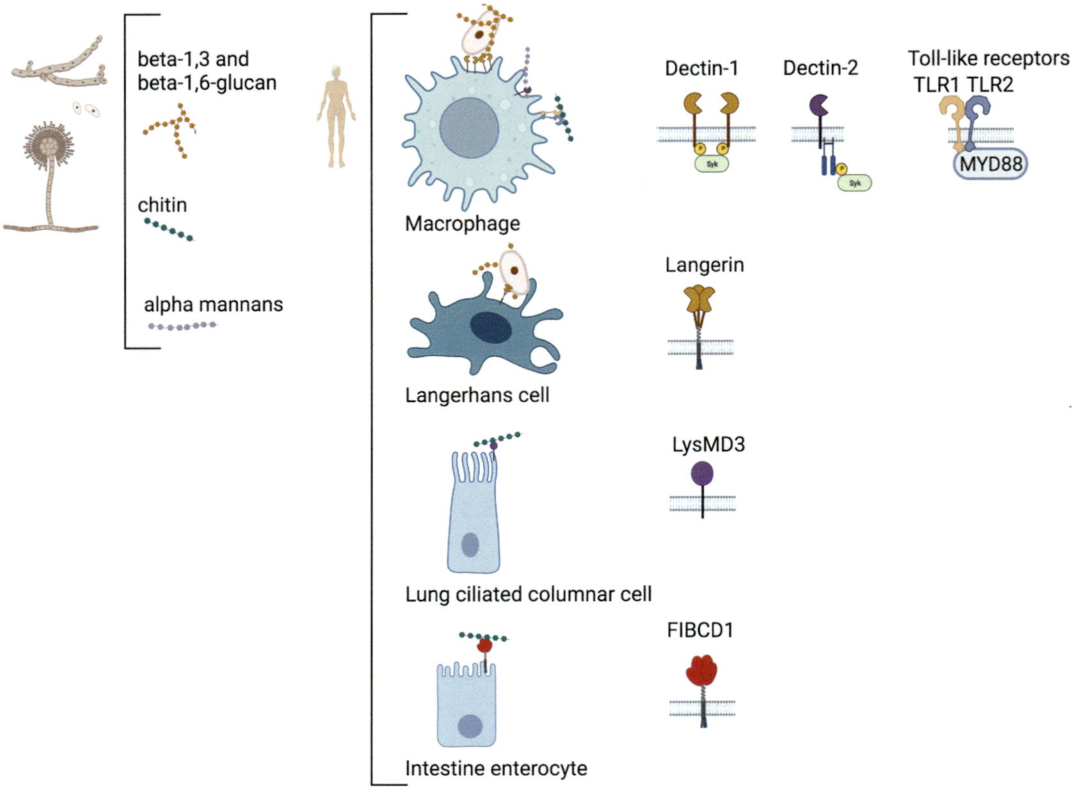

Fig. 1.6 Human pattern-recognition receptors on cells in the immune system, epithelium, or mucous membranes initiate antifungal defenses including the itchy inflammation that is our response to fungal infections such as athlete's foot, ringworm, or candidiasis. As examples of animal recognition of fungi, we focus on the evolution of receptor proteins diagrammed here. Cell wall polysaccharides from fungi or other microbes (left) are recognized by lectin domains of receptors on several human cell types (center). Receptor proteins include dectins, toll-like receptors, langerin, lysMD3, and FIBCD1. The receptors differ in their binding affinities for various polysaccharides, but a receptor that binds fungi can usually also bind some kinds of bacteria. Upon binding wall components, receptors initiate signaling cascades leading to innate immune responses. The receptors also contribute to activating the adaptive immune system for continuing or future defense against the microbes they have detected. Icons created with BioRender.com

fungal burdens and reduced survival (Taylor et al. 2007). In addition to its proven role in fungal detection, dectin-1 has experimentally demonstrated but still unexplained roles in elevating immune response to other microbial pathogens that lack β-glucans, including tuberculosis-causing species of *Mycobacterium* (Yadav and Schorey 2006; Kalia et al. 2021).

1.3.1.1 Comparative Analysis of Dectin-1 Structure Across Vertebrates

Dectin-1 is clearly important in human and mouse recognition of fungi. To infer when dectin-1 may first have appeared, we compared amino acid sequences from whole genomes in GenBank and generated a likelihood phylogeny for its potential orthologs.

Dectin-1's full 250 amino acids were alignable for Amniota, the clade of terrestrial vertebrates including reptiles, birds, and mammals. Within the C-lectin-like domain, Adachi et al. (2004) used mutational analysis to show that a sequence of three amino acids "WIH" is important in β-glucan binding (Fig. 1.7). We found this motif only in mammals, although birds and alligators share similar amino acids in two of the three

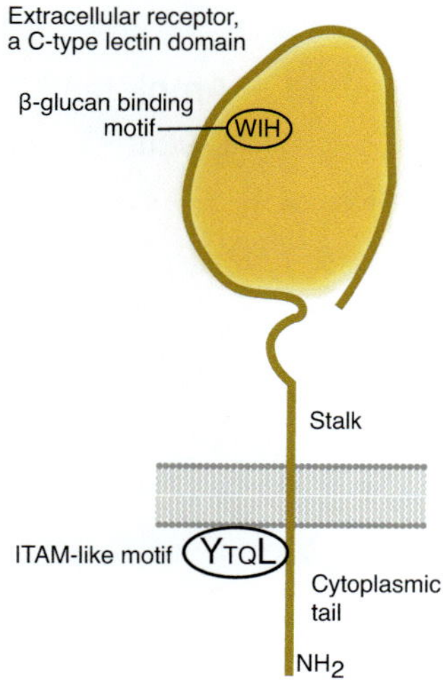

Fig. 1.7 Dectin-1 proteins are pattern-recognition receptors important in detecting fungi (Brown and Gordon 2001). Located on phagocytes, on dendritic cells, or on macrophages, the protein's extracellular C-type lectin domain binds β-glucans of fungal walls. Upon binding, dectin-1 molecules cluster together, initiating phagocytosis of the fungus (Goodridge et al. 2011). The cytoplasmic tail signals detection, leading to a multi-pronged innate and adaptive immune response

homologous positions (WLQ in birds, WIQ in alligators).

In host cell cytoplasm, a short N-terminal motif of the dectin-1 protein signals β-glucan binding (Rogers et al. 2005; Gantner et al. 2003) (Fig. 1.7). The motif has a consensus sequence "YxxL," where x is any amino acid. Categorized as an "immunoreceptor tyrosine-based activation motif (ITAM)-like motif," the motif is phosphorylated and it then recruits a molecule of a spleen tyrosine kinase, initiating a signaling cascade and an immune response (Rogers et al. 2005). Members of the dectin-1 clade from mammals, birds, lizards, and an alligator share the YxxL at a position homologous to the human protein. Stalks of dectin-1 orthologs from birds, reptiles, and fish have predicted coiled-coil domains that are expected to interact, clustering the proteins into dimers or trimers. Here as in other lectins (Feinberg et al. 2010; Shrive et al. 2014), polymer formation may increase the strength or specificity of binding to a polysaccharide ligand.

For our phylogenetic analysis of dectin-1's origins, we restricted our alignment to the conserved ~120 amino acid C-type lectin domain that appears to be homologous across invertebrates and vertebrates. This meant excluding >100 additional but unalignable amino acids from analysis. Figure 1.8 shows that the C-type lectin domain of human dectin-1 formed a strongly supported group with vertebrates such as alligators, birds, and mammals. Homologs from snakes, amphibians, and fishes, and oddly, one purple urchin sequence clustered with dectin-1 with 74% bootstrap support but roles of these homologs are difficult to predict because their sequences lack the functionally important "WIH" binding motif and the YxxL signaling motif (Fig. 1.8). The phylogeny and conserved binding and signaling motifs point to an ancestral dectin-1 that evolved to detect fungal β-glucans before the first Amniota diverged some 332–318 Ma (dos Reis et al. 2015). Fungi are the main β-glucan-containing terrestrial heterotrophs. Dectin-1 likely evolved and persisted over the course of evolution of the predominately terrestrial Amniota due to its importance in protection against fungi.

1.3.1.2 Phylogeny of C-Type Lectin-Like Receptors Related to Dectin-1

We included sequences from C-type lectin domains of dectin-2 (*CLEC6A*) and langerin (*CLECT4k*) as examples of the several paralogs to human dectin-1 that participate in recognition of fungal polysaccharides (Hatinguais et al. 2020) (Fig. 1.6). C-type lectin domains of dectin-2-orthologs from amniotes also form a clade (Fig. 1.8). Through its ability to bind mannan disaccharides, dectin-2 detects bacteria and fungi (Feinberg et al. 2017).

Langerin is expressed in mammalian Langerhans cells, a population of phagocytes that originate in the bone marrow and migrate to the skin and mucous membranes (Fig. 1.6). In mammals, langerin serves as a receptor for mannans, β-glucans, and keratan sulfate (a polymer of alternating galactose and sulfated

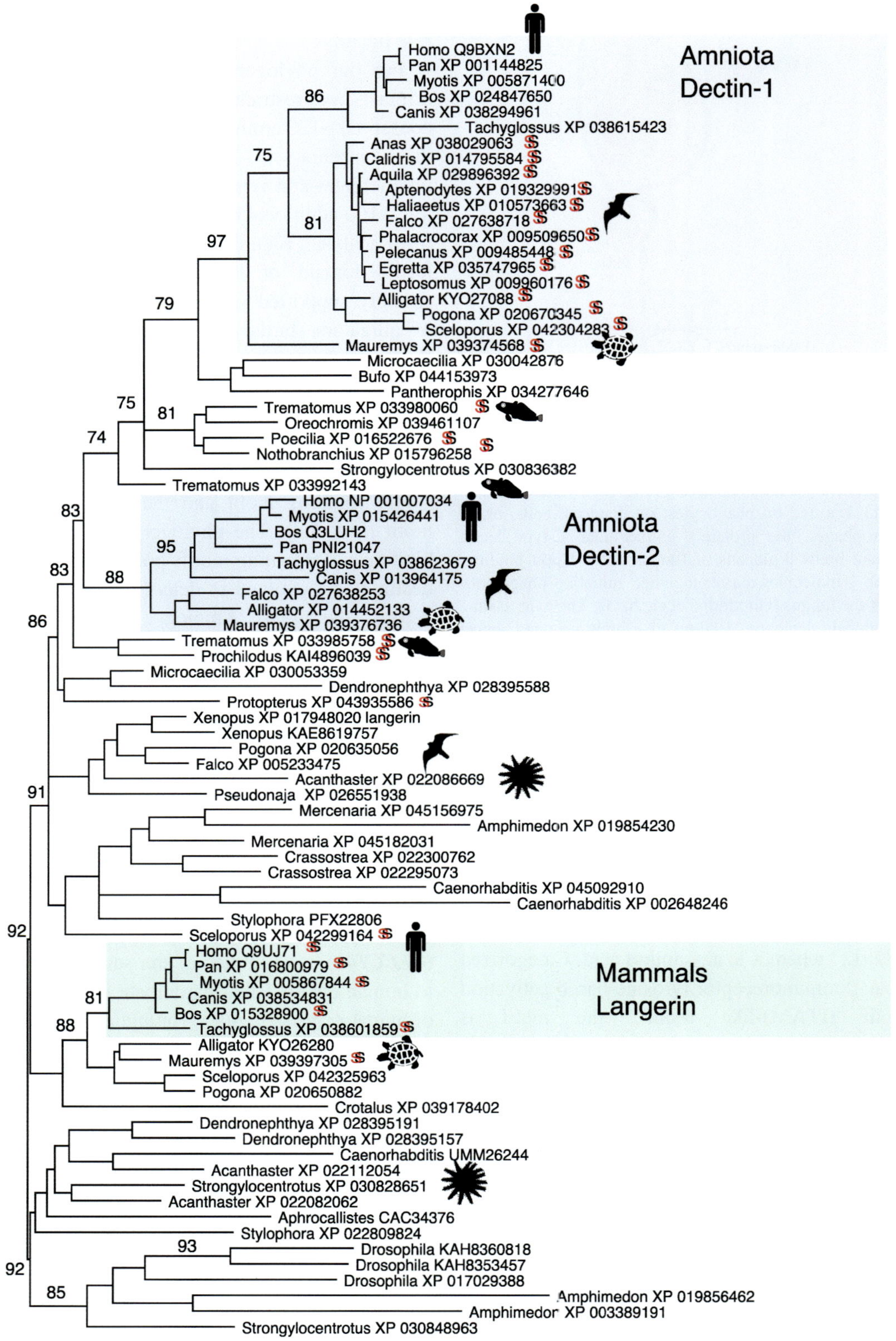

Fig. 1.8 C-type lectin-like receptors dectin-1 and its paralogs dectin-2 and langerin originated early in vertebrate evolution, playing important roles in fungal, bacterial, and viral recognition. C-type lectin-like receptor domains in invertebrates, e.g., *Drosophila* or the spiny starfish *Acanthaster*, are not supported as orthologs to

N-acetylglucosamine units). Langerin therefore detects HIV and other viruses, and bacterial and fungal cells (Ota et al. 2018; de Jong et al. 2010; Tateno et al. 2010). In a likelihood phylogeny, langerin C-type lectin domains from mammals form a clade (Fig. 1.8). Mammalian langerin's conserved "EPN" amino acid motif aligns with the "EPN" of dectin-2, possibly conferring the same preference for binding mannans (Drickamer 1992). Mammal and turtle sequences share an N-terminal coiled-coil domain, consistent with polymerization of langerin into a trimer (Stambach and Taylor 2003).

Outside of mammals, langerin homologs differ in structure and predicted binding specificity, raising questions about functional homology. Turtle, lizard, and alligator orthologs may bind preferentially to galactose rather than mannose, implied by their "QPD" motif instead of the "EPN" in mammals (Drickamer 1992). Lizard and alligator proteins lacked a predicted N-terminal coiled-coil domain.

1.3.2 Toll-Like Receptors of Ancient Origin Duplicated and Specialized for Recognition of Varied Pathogen and Symbiont Cell Wall Polysaccharides

Toll-like receptors (TLRs) have diverse functions including detection of fungi and a wide range of pathogens and other symbionts, and they have been thoroughly reviewed (Romani 2004; Akira et al. 2006; Kawai and Akira 2010; Romani 2011). Toll-like receptors have an extracellular ligand binding region that can be considered a lectin, consisting of leucine-rich repeats. Upon binding a ligand, these proteins typically dimerize. Their conserved intracellular portion is a TIR (Toll–interleukin 1 (IL-1) receptor) domain that is responsible for signal transmission after ligand binding to the extracellular domain (Kawai and Akira 2010). These elements of protein architecture and function are analogous to some of the nucleotide-binding, leucine-rich repeat (NLR) receptors in plants mentioned earlier (Maekawa et al. 2011; Martinon and Tschopp 2005).

Metazoan toll-like receptors are found on mammalian phagocytic cells, neutrophils, macrophages, and dendritic cells (Fig. 1.6) (Romani 2011). Immunological consequences of stimulation of these receptors can include production of antimicrobial molecules such as phenoloxidase in insects and nitric oxide or reactive oxygen species in mammals (Thoma-Uszynski et al. 2001; Gantner et al. 2003). Receptor stimulation also leads to cytokine production and inflammation (Goodridge and Underhill 2008; Romani 2004).

1.3.2.1 How Did Fungal Recognition Evolve

Toll proteins are shared across most Metazoa, but their functions vary (Brennan and Gilmore 2018; Gauthier et al. 2010). Toll was first discovered for its role in establishing dorsoventral polarity in *Drosophila* embryos, but it is also important in *Drosophila* immunity to fungi (Lemaitre et al. 1996). Toll receptors are essential for early development in nematodes and in nerve regeneration in annelids (Brennan and Gilmore 2018). Mammalian toll seem to have lost any ancestral functions in development.

Roach et al. (2005, 2013) analyzed toll-like receptors phylogenetically. Humans have ten of these proteins (e.g., NP_001305716) (Gantner et al. 2003; Roach et al. 2005). Toll proteins duplicated prior to the radiation of vertebrates, giving rise to more than 8 paralogous clades (Roach et al. 2005). Families of orthologs have been retained through vertebrate evolution, and each of 8 clades of toll paralogs (families 1–5 and

Fig. 1.8 (continued) vertebrate receptor clades in a maximum likelihood analysis. Colored boxes enclose predicted orthologs to human receptors. GenBank accession numbers follow genus names. Numbers at branches are bootstrap proportions. "SS" designates predicted coiled-coil regions potentially involved in polymerization of receptors, increasing strength and specificity of ligand binding

7–9) includes an orthologous gene from a human and a fish (Roach et al. 2005).

Among clades, recognition specificities diverged. Toll-like receptor families including 1, 2, 4, 6, and 9 in vertebrates may have evolved to detect bacterial ligands, lipopeptides, peptidoglycan, or viral DNA (Akira et al. 2006), but they also contribute to mammalian recognition of fungal mannans and chitin oligosaccharides (Hatinguais et al. 2020). Toward detection of fungi, Fuchs et al. (2018) showed that purified chitin oligomers, chains of at least six *N*-acetylglucosamine units, triggered an immune response in human and mouse cells via the toll-like receptor 2. However, toll-like receptor 2 is also important in binding bacterial lipopeptides (Roach et al. 2005). Complicating phylogenetic interpretation of ancestral recognition function, the binding specificity of toll depends on interactions with other proteins.

Other proteins in the toll signal transduction pathway are more highly conserved than the toll receptors themselves. Interactions between the TIR domain in toll with a second TIR domain in an adaptor protein, MyD88 (myeloid differentiation primary response protein 88), initiate signaling of ligand reception. TIR domains are ancient, distributed among bacteria, plants, and animals (Nimma et al. 2021), and MyD88 is shared by animals from sponges to mammals (Brennan and Gilmore 2018; Gauthier et al. 2010). The signaling cascade eventually activates nuclear factor κB, which is widely conserved across Metazoa, functioning in activation of transcription of inflammatory cytokine genes (Gauthier et al. 2010).

The example of toll-like receptor 2, which recognizes either bacterial or fungal ligands, together with the great age of signaling pathways, suggests that detection of fungi may originate from the far older perception of varied stimuli in an immune pathway that gains flexibility and adapts to new threats and opportunities through the collaboration of multiple proteins.

1.3.3 Chitinases and Chitin-Binding Proteins in Metazoan Responses to Fungi

While proteins that detect chitin in metazoans may be important in antifungal defense, the supporting evidence is less clear than it is in plants. In animals, roles of chitin recognition proteins include defense against non-fungal parasites, nutrition, development, and modulation of immune responses.

1.3.3.1 Chitinases in Animal Immunity

Under some conditions, chitinases protect animals from chitin-containing pathogens, from nematodes and fungi (Chang et al. 2020; Sutherland 2018). However, in mammals, when produced over too long a time period, or at too high a level, chitinase and chitinase-like proteins result in asthma, harmful inflammation, and damage to hosts. Balance seems to be necessary so that protection from parasites outweighs the damage to the host (Sutherland et al. 2014; Sutherland 2018).

Animals typically have orthologs to chitotriosidase and acidic mammalian chitinase, two enzymatically active GH18 chitinases. In addition, animals have chitinase-like proteins that evolved to become lectins, having lost enzymatic activity while retaining the ancestral chitin-binding ability (Bussink et al. 2007). In mammals, chitotriosidase (e.g., human NP_003456) is secreted by macrophages and other cells important in immune responses. Chitotriosidase is concentrated in lung and gut cells where encounters with pathogenic fungi are frequent. *In vivo*, recombinant human chitotriosidase improved the survival of mice with compromised immune systems upon challenge with pathogenic fungi, and *in vitro*, it reduced growth of human-pathogenic fungi (van Eijk et al. 2005). However, contrary to expectation, **deficiency** in chitotriosidase production improved survival of mice to pathogens including the fungus *Cryptococcus neoformans* (Wiesner et al. 2015).

The acidic mammalian chitinases (e.g., human NP_970615) protected against nematodes in the gastrointestinal tract of mice (Vannella et al. 2016), but they have other roles that may be ancestral. Acidic chitinases seem to enable some insectivorous mammals—bats, mice, and possibly small primates—to digest chitin from insect exoskeletons for food (Ohno et al. 2016; Janiak et al. 2018; Strobel et al. 2013).

A chitinase-like protein, which has lost enzymatic function, contributes to decreasing the parasite load in a mouse that is infected with a nematode (Sutherland et al. 2014). Chitinase-like proteins also seem to moderate immune responses and prolong macrophage life, increasing survival in mice infected with the bacterium *Streptococcus pneumoniae* (Dela Cruz et al. 2012).

1.3.3.2 Fibrinogen C Containing Domain 1 (FIBCD1) and Recognition of Chitin

Fibrinogen-related domains function as lectins, some of them as immune receptors in Metazoa, (for reviews, see Thomsen et al. 2011; Doolittle et al. 2012; Pilecki and Moeller 2020). Somewhat confusingly, the name "fibrinogen" was first applied to vertebrate blood clotting proteins and then applied to a conserved domain shared by many proteins with functions unrelated to blood. Two clades of proteins, membrane-bound FIBCD1, and ficolins from blood serum contain fibrinogen-related domains and have roles in mammalian immune recognition of fungi. Ficolins bind polysaccharides including *N*-acetylglucosamine, *N*-acetylgalactosamine, β-glucans, and peptidoglycans and then activate the complement cascade that results in the production of antimicrobial peptides (Thomsen et al. 2011).

The FIBCD1 protein plays multiple physiological roles, but relevant to fungal recognition, it binds chitin in human intestinal epithelial cells (Schlosser et al. 2009) (Fig. 1.6). It reduces or modulates fungal growth, and Moeller et al. (2019) showed that its overexpression in mice alters the composition of fungi in mouse guts. Its other varied roles include promoting growth of muscle fibers (Graca et al. 2022). Whether and how FIBCD1 initiates cellular signaling is unknown (Pilecki and Moeller 2020).

The conserved structure of FIBCD1 receptors originated in early vertebrates. The human FIBCD1 has an N-terminal cytoplasmic region, a transmembrane motif, and a fibrinogen-like recognition domain (Fig. 1.6) (Shrive et al. 2014). Most vertebrate genomes have only a single orthologous protein that aligns across the length of the human FIBCD1 protein. Amino acid sequences in vertebrates from ray-finned fishes to mammals were readily alignable over 464 sites. The fibrinogen domain of a jawless fish, a lamprey (represented by *Petromyzon* XP_032825085), is perhaps orthologous, but is more distant and its N-terminus is unalignable with the other vertebrates' sequences. Our exploration of related proteins supports Doolittle et al.'s (2012) inference that FIBCD1 may first have appeared in the most recent common ancestor of jawless fishes and mammals, 534–459 Ma (dos Reis et al. 2015). Whatever its ancestral function, FIBCD1's ability to bind chitin suggests that it could be adapted easily and repeatedly for recognition of fungi. From its origin in early fishes, an FIBCD1 chitin recognition receptor may have allowed vertebrates to detect chitin-bearing parasitic arthropods, nematodes, or fungi.

LysMD3 *and Chitin Recognition* LysMD3 is intriguing from an evolutionary perspective because of its central role in the chitin recognition receptors of plants. He et al. (2021) showed that human LysMD3 (NP_938014) in lung epithelial cells binds chitin oligomers and heat-treated *Candida albicans* cells, resulting in the production of cytokines and inflammation. In mud crabs, LysMD3 orthologs bound bacterial peptidoglycans and lipopolysaccharides and contributed to killing bacteria (Wang et al. 2021). No role in immunity was detected for LysMD3 in mice (Yokoyama et al. 2018) or fish (Laroche et al. 2013).

The structure of orthologs to LysMD3 is conserved across animals, and the phylogenies of the proteins are largely congruent with the organismal phylogeny, consistent with vertical transmission of a single conserved gene copy

(Wang et al. 2020). The extracellular LysM domain serves as the receptor (Fig. 1.6). Some animal LysMD3 proteins have predicted coiled-coil domains, suggesting that the monomers interact to form dimers or higher order polymers, as in plant LysM proteins. The animal LysMD3 cytoplasmic C-terminus lacks the kinase domain that is important in LysM signaling in plants, leaving uncertainty about how signaling proceeds in animals.

1.3.4 β-glucanases: Enzymes of Digestion, Recruited by Invertebrates as Immune Pattern-Recognition Receptors, First of Bacteria, and Later of Fungi

β-glucanases in family GH16, subfamily 4 (GH16_4), are carbohydrate-active enzymes important in digestion and in defense (Viborg et al. 2019). Invertebrates lack dectin-1, the β-glucan-binding receptor protein discussed earlier for its role in vertebrate recognition of fungi. Instead, invertebrates detect fungal β-glucans with homologs of GH16_4 β-glucanases (Gottar et al. 2006), an enzyme subfamily that is entirely missing from vertebrates.

The enzymes have endo-1,3-β-glucanase (EC 3.2.1.39) activity and a catalytic site with the amino acid sequence "ExDxxE" (Viborg et al. 2019) (Fig. 1.9). The GH16_4 β-glucanases diversified early. Being found in fungi and widely distributed among invertebrates, they may date back to the divergence of fungi from animals, 1481–1240 Ma (Parfrey et al. 2011). Aquatic invertebrates including filter-feeding bivalves, snails, crustaceans, and sea urchins make use of these enzymes for digestion when feeding on organisms that use β-glucans as storage products or as cell wall materials, including brown or other heterokont algae, fungi, and some bacteria (Fig. 1.9a, b) (Linton 2020; Julius 2018). Gain and loss of carbohydrate-binding module 39, another domain that increases the affinity of the enzyme for its substrate, has occurred repeatedly. One of many possible examples comes from two β-glucanases from the purple sea urchin *Strongylocentrotus purpuratus*. The two proteins arose from recent duplication based on the 92% identity of their glucanase domains, but one protein (NP_999681) has a carbohydrate-binding module while the other, XP_030836386, does not (Fig. 1.9a, b).

A role in immunity for GH16_4 glucanases was suggested when Unestam and Söderhäll (1977) demonstrated an increase in antimicrobial phenoloxidase in serum of crayfish treated with sources of β-glucans including *Saccharomyces cerevisiae* cell wall fragments. One of the several paralogous glucanases in the beetle *Tenebrio molitor* (ACS36221) (Fig. 1.9a) hydrolyzes β-1,3- and β-1,6-glucans and kills yeast cells. It is produced in the beetle midgut, so the fungi may be digested as food while being prevented from

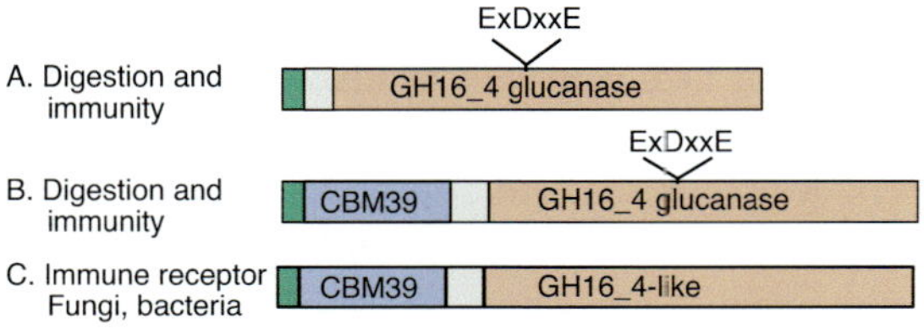

Fig. 1.9 β-glucanases in the glycosyl hydrolase subfamily 16_4 serve as digestive enzymes or immune receptors. (**a**, **b**) Enzymes with catalytic site ExDxxE and with or without a carbohydrate-binding module (CBM39) are shared across invertebrates, functioning in digestion of algae or fungi for nutrition or as receptors to initiate an immune response. (**c**) In early winged insects, duplicated genes lost the catalytic site as they evolved to specialize in immunity to bacteria or fungi. The CBM39 domain is responsible for recognizing the microbial cell walls. In green (left), signal peptide; in gray, region of unknown function

causing disease (Genta et al. 2009). Perović-Ottstadt et al. (2004) found that the sponge *Suberites domuncula* responded to glucan-containing microbes with increasing synthesis of a fibrinogen and an epidermal growth factor, proteins with possible roles in immune responses.

Specialized Pattern-Recognition Receptors Evolved from GH16_4 Glucanases The GH16_4 glucanases from winged insects that retain the catalytic motif ExDxxE (Fig. 1.9a, b) form a clade nested among other invertebrates. (Hughes 2012). Also nested among other invertebrates is a second clade of insect GH16_4 glucanase-like proteins that lost the catalytic motif. Rather than β-glucan digestion, these proteins specialized as receptors that recognize microbial cell walls (Fig. 1.9c) (Hughes 2012; Lee et al. 1996). The clade of receptor proteins spans Orthoptera (*Timema shepardi* CAD7264458) and Diptera (*Drosophila melanogaster* AAF49244), suggesting an age of 413–360 Ma (Misof et al. 2014). Among these specialized receptors, a CBM39 domain has been shown to be essential for glucan binding (Lee et al. 1996).

Additional gene duplications among winged insects gave rise to proteins with even more specialized functions. *Drosophila* Gram-negative binding protein-3 (NP_523986) (Fig. 1.9c) preferentially binds fungal β-glucans (Gottar et al. 2006). Gottar et al.'s (2006) experiments showed that isolated protein-3 binds to β-glucans. Overexpression of "protein-3" induced expression of drosomycin, an antimicrobial toxin. Flies with a null mutation in the binding protein were more sensitive to fungal infection compared with wild-type flies, producing lower levels of antimicrobial proteins (Gottar et al. 2006). Whether related insects share an ability for specific detection of fungi is currently unclear.

1.4 Conclusions

As summarized in Fig. 1.10, domains including LysM, leucine-rich repeats, C-type lectins, and chitinases are ancient building blocks that bind selectively to ligands including fungal or other microbial cell wall polysaccharides. Conserved binding domains of modern plant or animal receptors are variously combined with transmembrane domains, coiled-coil domains, kinase domains, or TIR domains to form multi-domain proteins capable of transducing ligand binding into cellular responses such as antifungal defenses.

We speculate that terrestrialization beginning 450 Ma exposed plants, vertebrates, and insects to increased threats from fungal pathogens while offering increased opportunities for fungal mutualisms. Early plant and animal evolution took place in water, where selective pressure from viruses, bacteria, and other eukaryotes may have been more significant than pressure from fungi. This changed with the invasion of land. The LysM receptor kinases shared among streptophyte algae and land plants may have evolved as part of an early system of defense against aquatic fungi, later bolstered among land plants by expansion of GH19 chitinases paralleling the diversification of terrestrial fungi. Few aquatic fungi target vertebrates, and most fungal pathogens of fishes are derived from terrestrial ancestors (Gozlan et al. 2014). In early terrestrial vertebrates, specificity in fungal recognition increased with the origin of dectin-1 to bind fungal β-glucans. In early winged insects, fungal recognition increased through adaptation of β-glucanases to detect β-glucans (Fig. 1.10).

1.4.1 Limitations to Inferences About Functional Homology

Inferences about ancestral ligand binding rely on knowledge of specificity of modern receptors. However, information on specificity of modern receptors is rarely available for a full range of potential immunogenic compounds from fungi, bacteria, or viruses. Whole genome sequences offer a terrific resource for tracking homologs to well-studied proteins, but experimental evidence for homologous function in non-model species is rarely available. We acknowledge limitations of

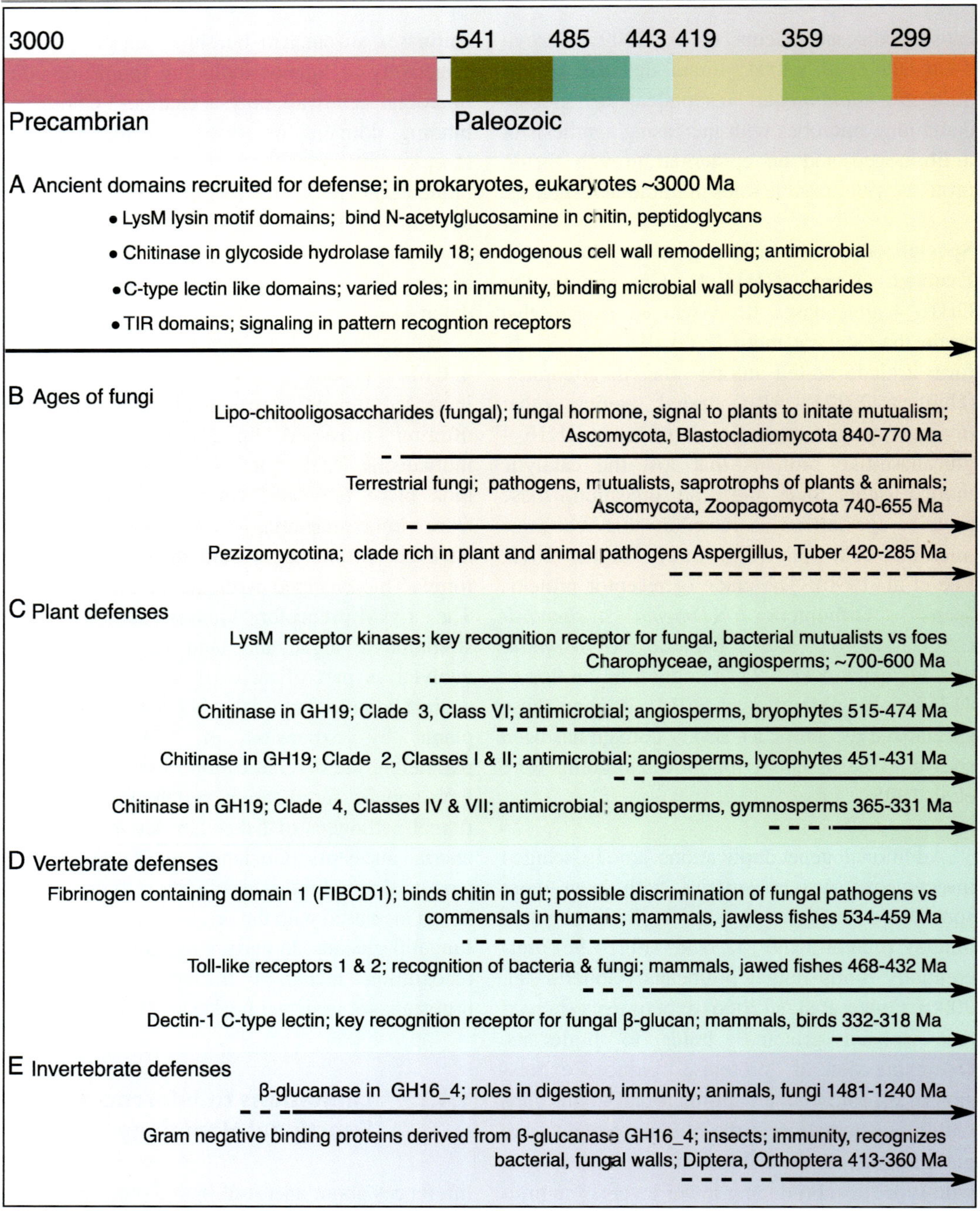

Fig. 1.10 Origins of fungi and their recognition by plant or animal hosts. Arrows indicate age estimates based on shared presence in clades, e.g., LysM motifs are shared by and thus are inferred to have originated in a common ancestor of bacteria and eukaryotes more than three billion years ago (Battistuzzi et al. 2004; Knoll 2014). Dashed lines are confidence intervals, where available, on ages of taxon divergence from the literature. (**a**) Protein domains from these receptors and enzymes may have been present in the common ancestor of cellular life. (**b**) Lipo-chitooligosaccharides, after possible origin as fungal hormones, were later recruited to signal plants to initiate mycorrhizae. The rise of terrestrial fungi stimulated evolution of plant and animal recognition systems to detect new potential fungal pathogens and mutualists. (**c**) In the ancestor of streptophyte algae (in Charophyceae) and land plants, protein domain shuffling combined an ancient LysM domain with a kinase, resulting in LysM receptor kinases that detect bacterial and fungal cell walls. New classes of chitinase paralogs (see Fig. 1.5) important in

this approach but hope for improvement. Mutant studies to test gene function in non-model organisms are becoming easier due to CRISPR-Cas9 technology (Zhang et al. 2014). Increasingly, sophisticated X-ray crystallographic studies of plant or mammal proteins provide information about amino acids in receptor proteins that bind fungal cell wall polysaccharides. Glycan microarrays using libraries of sugar-containing molecules have worked well to test binding profiles of receptors such as human C-type lectins (Vendele et al. 2020; Yamasaki et al. 2009).

1.5 Materials and Methods: Protein Sampling and Phylogenies

1.5.1 Phylogenetic Analysis of Plant GH19 Chitinases

Our chitinase taxon sampling included plants, multiple algal lineages, and bacteria. We started by assembling representatives of seed plant GH19 chitinases from each class from among characterized *Arabidopsis* sequences from the Carbohydrate-Active Enzymes database www.cazy.org (Drula et al. 2022). To search for homologs in glaucophyte, red, chlorophyte, and streptophyte algae, and in bryophyte, lycophyte, and fern embryophytes, we used BLAST searches with the *Arabidopsis* sequences as queries. We also included characterized sequences from bacteria and diatoms. For analysis, we selected sequences from BLAST results using an *e*-value cutoff of 1e-08 that had the appropriate domains as predicted by CDD www.ncbi.nlm.nih.gov/cdd (Lu et al. 2020) and UniProt www.uniprot.org (UniProt Consortium 2021). We aligned the sequences using the L-INS-i model in MAFFT www.mafft.cbrc.jp/alignment/software (Katoh et al. 2019) with default settings and removing sequences that could not be aligned.

The phylogenetic analysis included only the GH19 domains from the aligned sequences. To trim the sequences, we used the *hmmalign* command with *--trim* option in HMMER 3.3.2 www.hmmer.org against the GH19 HMM profile downloaded from the dbCAN2 HMM database v.11 www.bcb.unl.edu/dbCAN2 (Zhang et al. 2018). We carried out a preliminary analysis using IQTree www.iqtree.org (Minh et al. 2020) with 100 search replicates using the program's model testing option. We excluded all but one sequence from among clades of similar sequences from the same species, and carried out a final maximum likelihood analysis using IQTree with 100 search replicates and the WAG+I + G4 model, followed by ultrafast bootstrap analysis with 1000 bootstrap replicates.

1.5.2 Phylogenetic Analysis of Animal C-Type Lectin-Like Domains, Taxon and Gene Sampling

We focused on dectin-1 receptors of β-1,3- and β-1,6-glucans due to their importance in fungal recognition. To evaluate dectin-1's monophyly, we included sequences from two related proteins: dectin-2 (*CLEC6A*) that binds mannose and from langerin (*CLEC4K*) that binds mannose, β-1,3-glucans, and *N*-acetylglucosamine (van

Fig. 1.10 (continued) antifungal defenses arose as land plants evolved. (**d**) Vertebrate receptors for cell wall polysaccharides may have originated for detection of aquatic, non-fungal parasites, but proteins like FIBCD1 and Toll-like receptor 2 also adapted to detect fungal chitin oligomers. Dectin-1 recognizes and signals detection of fungal β-glucan. (**e**) Invertebrates recruited receptor proteins from β-glucanases that had ancestral functions in digestion of food. β-glucanases were lost from early vertebrates. Diptera and other flying insects recognize fungi with paralogs of β-glucanases that lost enzymatic function. Sources of divergence dates for species trees are: Fungi (Chang et al. 2015); plants and algae (Morris et al. 2018); vertebrates (dos Reis et al. 2015); insects (Misof et al. 2014); and opisthokonts (Parfrey et al. 2011). Numbers are millions of years ago (Ma); Precambrian is not to scale with Paleozoic den Berg et al. 2012). This was not an exhaustive study of related proteins; e.g., we did not analyze DC-SIGN or Mannose receptor (van den Berg et al. 2012).

We used BLAST searches with the amino acid sequences from pattern-recognition receptors from well-characterized human or insect hosts as queries. Our ingroup selection consisted of protein sequences of taxa that shared domains and motifs important in recognition function. Graphic representations, scores from the BLAST results, and NCBI distance trees were helpful in recognizing and selecting likely orthologous and paralogous sequences relative to the query. To track early origins of host recognition proteins, we repeated BLAST searches excluding successively broader clades of taxa, starting for example by excluding humans if the query was from a human, then excluding mammals, then excluding amniotes, and so on. We searched for proteins in BLAST searches restricted to species to complete our datasets. As outgroups, we selected paralogous proteins and proteins from distantly related organisms, usually invertebrates.

We aligned amino acid sequences using MAFFT as described above. For domain structure predictions, we used NCBI CDD searches (Lu et al. 2020) and InterPro https://www.ebi.ac.uk/interpro/ (Blum et al. 2021). We used Waggawagga https://waggawagga.motorprotein.de/ to predict coiled-coil protein structures (Simm et al. 2015). Several of the immune receptors that we analyzed were multi-domain proteins. We aligned only the conserved C-type lectin domain of vertebrate dectin-1 with distant homologs from invertebrates, excluding other domains of uncertain homology. We annotated positions of amino acids that are functionally important in substrate binding.

For phylogenetic analysis, we used raxmlGUI 2.0 (Edler et al. 2021), applying the program's model testing, and then using the selected model in a maximum likelihood analysis, followed by 100 bootstrap replicates. Results were generally consistent with monophyly of dectin-1, dectin-2, and langerin, and with the organismal phylogeny.

Acknowledgments Mary Berbee acknowledges funding from a Natural Sciences and Engineering Research Council of Canada Discovery Frontiers Grant RGPIN-2022-03813. We thank Carol Munro (University of Aberdeen) for a review of this manuscript, and Vivian Miao and members of the Berbee lab for constructive comments.

References

Adachi Y, Ishii T, Ikeda Y, Hoshino A, Tamura H, Aketagawa J, Tanaka S, Ohno N (2004) Characterization of β-glucan recognition site on C-type lectin, dectin 1. Infect Immun 72(7):4159–4171. https://doi.org/10.1128/IAI.72.7.4159-4171.2004

Adrangi S, Faramarzi MA (2013) From bacteria to human: a journey into the world of chitinases. Biotechnol Adv 31(8):1786–1795. https://doi.org/10.1016/j.biotechadv.2013.09.012

Akira S, Uematsu S, Takeuchi O (2006) Pathogen recognition and innate immunity. Cell 124(4):783–801. https://doi.org/10.1016/j.cell.2006.02.015

Amses KR, Simmons DR, Longcore JE, Mondo SJ, Seto K, Jerônimo GH, Bonds AE, Quandt CA, Davis WJ, Chang Y (2022) Diploid-dominant life cycles characterize the early evolution of Fungi. Proc Natl Acad Sci U S A 119(36):e2116841119

Anand G, Leibman-Markus M, Elkabetz D, Bar M (2021) Method for the production and purification of plant immune-active xylanase from *Trichoderma*. Int J Mol Sci 22(8). https://doi.org/10.3390/ijms22084214

Ausubel FM (2005) Are innate immune signaling pathways in plants and animals conserved? Nat Immunol 6(10):973–979. https://doi.org/10.1038/ni1253

Battistuzzi FU, Feijao A, Hedges SB (2004) A genomic timescale of prokaryote evolution: insights into the origin of methanogenesis, phototrophy, and the colonization of land. BMC Evol Biol 4(1):44. https://doi.org/10.1186/1471-2148-4-44

Blum M, Chang H-Y, Chuguransky S, Grego T, Kandasamy S, Mitchell A, Nuka G, Paysan-Lafosse T, Qureshi M, Raj S, Richardson L, Salazar GA, Williams L, Bork P, Bridge A, Gough J, Haft DH, Letunic I, Marchler-Bauer A, Mi H, Natale DA, Necci M, Orengo CA, Pandurangan AP, Rivoire C, Sigrist CJA, Sillitoe I, Thanki N, Thomas PD, Tosatto SCE, Wu CH, Bateman A, Finn RD (2021) The InterPro protein families and domains database: 20 years on. Nucleic Acids Res 49(D1):D344–D354. https://doi.org/10.1093/nar/gkaa977

Boddy L (2016) Pathogens of autotrophs. In: Watkinson SC, Boddy L, Money NP (eds) The fungi, 3rd edn. Academic Press, Waltham, MA, pp 245–292. https://doi.org/10.1016/b978-0-12-382034-1.00008-6

Bonfante P, Requena N (2011) Dating in the dark: how roots respond to fungal signals to establish arbuscular mycorrhizal symbiosis. Curr Opin Plant Biol 14(4): 451–457. https://doi.org/10.1016/j.pbi.2011.03.014

Brennan JJ, Gilmore TD (2018) Evolutionary origins of Toll-like receptor signaling. Mol Biol Evol 35(7): 1576–1587. https://doi.org/10.1093/molbev/msy050

Bressendorff S, Azevedo R, Kenchappa CS, Ponce de Leon I, Olsen JV, Rasmussen MW, Erbs G, Newman MA, Petersen M, Mundy J (2016) An innate immunity pathway in the moss *Physcomitrella patens*. Plant Cell 28(6):1328–1342. https://doi.org/10.1105/tpc.15.00774

Brown GD, Gordon S (2001) A new receptor for β-glucans. Nature 413(6851):36–37. https://doi.org/10.1038/35092620

Brown GD, Willment JA, Whitehead L (2018) C-type lectins in immunity and homeostasis. Nat Rev Immunol 18(6):374–389. https://doi.org/10.1038/s41577-018-0004-8

Buchmann K (2014) Evolution of innate immunity: clues from invertebrates via fish to mammals. Front Immunol 5. https://doi.org/10.3389/fimmu.2014.00459

Buendia L, Girardin A, Wang T, Cottret L, Lefebvre B (2018) LysM receptor-like kinase and LysM receptor-like protein families: an update on phylogeny and functional characterization. Front Plant Sci 9:1531. https://doi.org/10.3389/fpls.2018.01531

Bussink AP, Speijer D, Aerts J, Boot RG (2007) Evolution of mammalian chitinase (-like) members of family 18 glycosyl hydrolases. Genetics 177(2):959–970. https://doi.org/10.1534/genetics.107.075846

Carotenuto G, Chabaud M, Miyata K, Capozzi M, Takeda N, Kaku H, Shibuya N, Nakagawa T, Barker DG, Genre A (2017) The rice LysM receptor-like kinase *Os*CERK1 is required for the perception of short-chain chitin oligomers in arbuscular mycorrhizal signaling. New Phytol 214(4):1440–1446. https://doi.org/10.1111/nph.14539

Casadevall A (2005) Fungal virulence, vertebrate endothermy, and dinosaur extinction: Is there a connection? Fungal Genet Biol 42(2):98–106. https://doi.org/10.1016/j.fgb.2004.11.008

Chang Y, Wang S, Sekimoto S, Aerts AL, Choi C, Clum A, LaButti KM, Lindquist EA, Yee Ngan C, Ohm RA (2015) Phylogenomic analyses indicate that early fungi evolved digesting cell walls of algal ancestors of land plants. Genome Biol Evol 7(6): 1590–1601. https://doi.org/10.1093/gbe/evv090

Chang Y, Desiro A, Na H, Sandor L, Lipzen A, Clum A, Barry K, Grigoriev IV, Martin FM, Stajich JE, Smith ME, Bonito G, Spatafora JW (2019) Phylogenomics of Endogonaceae and evolution of mycorrhizas within Mucoromycota. New Phytol 222(1):511–525. https://doi.org/10.1111/nph.15613

Chang D, Sharma L, Dela Cruz CS (2020) Chitotriosidase: a marker and modulator of lung disease. Eur Respir Rev 29(156):190143. https://doi.org/10.1183/16000617.0143-2019

Cheng H, Shao Z, Lu C, Duan D (2021) Genome-wide identification of chitinase genes in *Thalassiosira pseudonana* and analysis of their expression under abiotic stresses. BMC Plant Biol 21(1):87. https://doi.org/10.1186/s12870-021-02849-2

Collinge DB, Kragh KM, Mikkelsen JD, Nielsen KK, Rasmussen U, Vad K (1993) Plant chitinases. Plant J 3(1):31–40. https://doi.org/10.1046/j.1365-313x.1993.t01-1-00999.x

Cope KR, Bascaules A, Irving TB, Venkateshwaran M, Maeda J, Garcia K, Rush TA, Ma C, Labbe J, Jawdy S, Steigerwald E, Setzke J, Fung E, Schnell KG, Wang Y, Schlief N, Bucking H, Strauss SH, Maillet F, Jargeat P, Becard G, Puech-Pages V, Ane JM (2019) The ectomycorrhizal fungus *Laccaria bicolor* produces lipochitooligosaccharides and uses the common symbiosis pathway to colonize *Populus* roots. Plant Cell 31(10):2386–2410. https://doi.org/10.1105/tpc.18.00676

de Jong MA, Vriend LE, Theelen B, Taylor ME, Fluitsma D, Boekhout T, Geijtenbeek TB (2010) C-type lectin Langerin is a β-glucan receptor on human Langerhans cells that recognizes opportunistic and pathogenic fungi. Mol Immunol 47(6):1216–1225. https://doi.org/10.1016/j.molimm.2009.12.016

De Schutter K, Van Damme EJM (2015) Protein-carbohydrate interactions as part of plant defense and animal immunity. Molecules 20(5):9029–9053. https://doi.org/10.3390/molecules20059029

de Wit PJ (2007) How plants recognize pathogens and defend themselves. Cell Mol Life Sci 64(21): 2726–2732. https://doi.org/10.1007/s00018-007-7284-7

Dela Cruz CS, Liu W, He CH, Jacoby A, Gornitzky A, Ma B, Flavell R, Lee CG, Elias JA (2012) Chitinase 3-like-1 promotes *Streptococcus pneumoniae* killing and augments host tolerance to lung antibacterial responses. Cell Host Microbe 12(1):34–46. https://doi.org/10.1016/j.chom.2012.05.017

Delaux PM, Schornack S (2021) Plant evolution driven by interactions with symbiotic and pathogenic microbes. Science 371(6531). https://doi.org/10.1126/science.aba6605

Delaux PM, Radhakrishnan GV, Jayaraman D, Cheema J, Malbreil M, Volkening JD, Sekimoto H, Nishiyama T, Melkonian M, Pokorny L, Rothfels CJ, Sederoff HW, Stevenson DW, Surek B, Zhang Y, Sussman MR, Dunand C, Morris RJ, Roux C, Wong GK, Oldroyd GE, Ane JM (2015) Algal ancestor of land plants was preadapted for symbiosis. Proc Natl Acad Sci U S A 112(43):13390–13395. https://doi.org/10.1073/pnas.1515426112

Doolittle RF, McNamara K, Lin K (2012) Correlating structure and function during the evolution of fibrinogen-related domains. Protein Sci 21(12): 1808–1823. https://doi.org/10.1002/pro.2177

dos Reis M, Thawornwattana Y, Angelis K, Telford MJ, Donoghue PCJ, Yang ZH (2015) Uncertainty in the timing of origin of animals and the limits of precision

in molecular timescales. Curr Biol 25(22):2939–2950. https://doi.org/10.1016/j.cub.2015.09.066
Drickamer K (1992) Engineering galactose-binding activity into a C-type mannose-binding protein. Nature 360(6400):183–186. https://doi.org/10.1038/360183a0
Drula E, Garron ML, Dogan S, Lombard V, Henrissat B, Terrapon N (2022) The carbohydrate-active enzyme database: functions and literature. Nucleic Acids Res 50(D1):D571–D577. https://doi.org/10.1093/nar/gkab1045
Edler D, Klein J, Antonelli A, Silvestro D (2021) raxmlGUI 2.0: a graphical interface and toolkit for phylogenetic analyses using RAxML. Methods Ecol Evol 12(2):373–377. https://doi.org/10.1111/2041-210X.13512
Erwig LP, Gow NAR (2016) Interactions of fungal pathogens with phagocytes. Nat Rev Microbiol 14(3): 163–176. https://doi.org/10.1038/nrmicro.2015.21
Feinberg H, Powlesland AS, Taylor ME, Weis WI (2010) Trimeric structure of langerin. J Biol Chem 285(17): 13285–13293. https://doi.org/10.1074/jbc.M109.086058
Feinberg H, Jégouzo SA, Rex MJ, Drickamer K, Weis WI, Taylor ME (2017) Mechanism of pathogen recognition by human dectin-2. J Biol Chem 292(32): 13402–13414. https://doi.org/10.1074/jbc.M117.799080
Feng F, Sun J, Radhakrishnan GV, Lee T, Bozsoki Z, Fort S, Gavrin A, Gysel K, Thygesen MB, Andersen KR, Radutoiu S, Stougaard J, Oldroyd GED (2019) A combination of chitooligosaccharide and lipochitooligosaccharide recognition promotes arbuscular mycorrhizal associations in *Medicago truncatula*. Nat Commun 10(1):5047. https://doi.org/10.1038/s41467-019-12999-5
Fisher JF, Mobashery S (2020) Constructing and deconstructing the bacterial cell wall. Protein Sci 29(3):629–646. https://doi.org/10.1002/pro.3737
Fuchs K, Cardona Gloria Y, Wolz O-O, Herster F, Sharma L, Dillen CA, Täumer C, Dickhöfer S, Bittner Z, Dang T-M, Singh A, Haischer D, Schlöffel MA, Koymans KJ, Sanmuganantham T, Krach M, Roger T, Le Roy D, Schilling NA, Frauhammer F, Miller LS, Nürnberger T, LeibundGut-Landmann S, Gust AA, Macek B, Frank M, Gouttefangeas C, Dela Cruz CS, Hartl D, Weber AN (2018) The fungal ligand chitin directly binds TLR2 and triggers inflammation dependent on oligomer size. EMBO Rep 19(12): e46065. https://doi.org/10.15252/embr.201846065
Gantner BN, Simmons RM, Canavera SJ, Akira S, Underhill DM (2003) Collaborative induction of inflammatory responses by dectin-1 and toll-like receptor 2. J Exp Med 197(9):1107–1117. https://doi.org/10.1084/jem.20021787
Gauthier MEA, Du Pasquier L, Degnan BM (2010) The genome of the sponge *Amphimedon queenslandica* provides new perspectives into the origin of toll-like and interleukin 1 receptor pathways. Evol Dev 12(5): 519–533. https://doi.org/10.1111/j.1525-142X.2010.00436.x
Genta FA, Bragatto I, Terra WR, Ferreira C (2009) Purification, characterization and sequencing of the major β-1,3-glucanase from the midgut of *Tenebrio molitor* larvae. Insect Biochem Mol Biol 39(12):861–874. https://doi.org/10.1016/j.ibmb.2009.10.003
Gong BQ, Wang FZ, Li JF (2020) Hide-and-seek: Chitin-triggered plant immunity and fungal counterstrategies. Trends Plant Sci 25(8):805–816. https://doi.org/10.1016/j.tplants.2020.03.006
González LMG, El Kayal W, Morris JS, Cooke JEK (2015) Diverse chitinases are invoked during the activity-dormancy transition in spruce. Tree Genet Genom 11:3. https://doi.org/10.1007/s11295-015-0871-0
Goodridge HS, Underhill DM (2008) Fungal recognition by TLR2 and dectin-1. In: Bauer S, Hartmann G (eds) Toll-like receptors (TLRs) and innate immunity. Springer, Berlin, pp 87–109. https://doi.org/10.1007/978-3-540-72167-3_5
Goodridge HS, Reyes CN, Becker CA, Katsumoto TR, Ma J, Wolf AJ, Bose N, Chan AS, Magee AS, Danielson ME, Weiss A, Vasilakos JP, Underhill DM (2011) Activation of the innate immune receptor dectin-1 upon formation of a 'phagocytic synapse'. Nature 472(7344):471–475. https://doi.org/10.1038/nature10071
Gordon GL, Phillips MW (1998) The role of anaerobic gut fungi in ruminants. Nutr Res Rev 11(1):133–168
Gottar M, Gobert V, Matskevich AA, Reichhart J-M, Wang C, Butt TM, Belvin M, Hoffmann JA, Ferrandon D (2006) Dual detection of fungal infections in *Drosophila* via recognition of glucans and sensing of virulence factors. Cell 127(7):1425–1437. https://doi.org/10.1016/j.cell.2006.10.046
Gow NAR, Latge JP, Munro CA (2017) The fungal cell wall: structure, biosynthesis, and function. Microbiology. Spectrum 5:3. https://doi.org/10.1128/microbiolspec.FUNK-0035-2016
Gozlan RE, Marshall WL, Lilje O, Jessop CN, Gleason FH, Andreou D (2014) Current ecological understanding of fungal-like pathogens of fish: What lies beneath? Front Microbiol 5:62. https://doi.org/10.3389/fmicb.2014.00062
Graca FA, Rai M, Hunt LC, Stephan A, Wang Y-D, Gordon B, Wang R, Quarato G, Xu B, Fan Y (2022) The myokine Fibcd1 is an endogenous determinant of myofiber size and mitigates cancer-induced myofiber atrophy. Nat Commun 13(1):1–22. https://doi.org/10.1038/s41467-022-30120-1
Greeff C, Roux M, Mundy J, Petersen M (2012) Receptor-like kinase complexes in plant innate immunity. Front Plant Sci 3:209. https://doi.org/10.3389/fpls.2012.00209
Hamel F, Boivin R, Tremblay C, Bellemare G (1997) Structural and evolutionary relationships among chitinases of flowering plants. J Mol Evol 44(6): 614–624. https://doi.org/10.1007/pl00006184

Han B, Zhou K, Li Z, Sun B, Ni Q, Meng X, Pan G, Li C, Long M, Li T, Zhou C, Li W, Zhou Z (2016) Characterization of the first fungal glycosyl hydrolase family 19 chitinase (*Nb*chiA) from *Nosema bombycis* (*Nb*). J Eukaryot Microbiol 63(1):37–45. https://doi.org/10.1111/jeu.12246

Hatinguais R, Willment JA, Brown GD (2020) PAMPs of the fungal cell wall and mammalian PRRs. Curr Top Microbiol Immunol 425:187–223. https://doi.org/10.1007/82_2020_201

He X, Howard BA, Liu Y, Neumann AK, Li L, Menon N, Roach T, Kale SD, Samuels DC, Li H, Kite T, Kita H, Hu TY, Luo M, Jones CN, Okaa UJ, Squillace DL, Klein BS, Lawrence CB (2021) LYSMD3: a mammalian pattern recognition receptor for chitin. Cell Rep 36(3). https://doi.org/10.1016/j.celrep.2021.109392

Hoell IA, Dalhus B, Heggset EB, Aspmo SI, Eijsink VG (2006) Crystal structure and enzymatic properties of a bacterial family 19 chitinase reveal differences from plant enzymes. FEBS J 273(21):4889–4900. https://doi.org/10.1111/j.1742-4658.2006.05487.x

Hoffmann JA (2003) The immune response of *Drosophila*. Nature 426(6962):33–38. https://doi.org/10.1038/nature02021

Hromadova D, Soukup A, Tylova E (2021) Arabinogalactan proteins in plant roots - an update on possible functions. Front Plant Sci 12:674010. https://doi.org/10.3389/fpls.2021.674010

Hughes AL (2012) Evolution of the βGRP/GNBP/β-1, 3-glucanase family of insects. Immunogenetics 64(7): 549–558. https://doi.org/10.1007/s00251-012-0610-8

Inamine S, Onaga S, Ohnuma T, Fukamizo T, Taira T (2015) Purification, cDNA cloning, and characterization of LysM-containing plant chitinase from horsetail (*Equisetum arvense*). Biosci Biotechnol Biochem 79(8):1296–1304. https://doi.org/10.1080/09168451.2015.1025693

James TY, Berbee ML (2012) No jacket required–new fungal lineage defies dress code: recently described zoosporic fungi lack a cell wall during trophic phase. BioEssays 34(2):94–102. https://doi.org/10.1002/bies.201100110

Janiak MC, Chaney ME, Tosi AJ (2018) Evolution of acidic mammalian chitinase genes (CHIA) is related to body mass and insectivory in primates. Mol Biol Evol 35(3):607–622. https://doi.org/10.1093/molbev/msx312

Jiang X, Bao H, Merzendorfer H, Yang Q (2019) Immune responses of mammals and plants to chitin-containing pathogens. Adv Exp Med Biol 1142:61–81. https://doi.org/10.1007/978-981-13-7318-3_4

Julius ML (2018) Chapter 6 - Carbohydrate diversity in microalgae: a phylogenetically arranged presentation. In: Levine IA, Fleurence J (eds) Microalgae in health and disease prevention. Academic, pp 133–144. https://doi.org/10.1016/B978-0-12-811405-6.00006-2

Kaku H, Nishizawa Y, Ishii-Minami N, Akimoto-Tomiyama C, Dohmae N, Takio K, Minami E, Shibuya N (2006) Plant cells recognize chitin fragments for defense signaling through a plasma membrane receptor. Proc Natl Acad Sci U S A 103(29):11086–11091. https://doi.org/10.1073/pnas.0508882103

Kalia N, Singh J, Kaur M (2021) The role of dectin-1 in health and disease. Immunobiology 226(2):152071. https://doi.org/10.1016/j.imbio.2021.152071

Kasprzewska A (2003) Plant chitinases – regulation and function. Cell Mol Biol Lett 8(3):809–824

Katoh K, Rozewicki J, Yamada KD (2019) MAFFT online service: multiple sequence alignment, interactive sequence choice and visualization. Brief Bioinform 20(4):1160–1166. https://doi.org/10.1093/bib/bbx108

Kawai T, Akira S (2010) The role of pattern-recognition receptors in innate immunity: update on Toll-like receptors. Nat Immunol 11(5):373–384. https://doi.org/10.1038/ni.1863

Khokhani D, Carrera Carriel C, Vayla S, Irving TB, Stonoha-Arther C, Keller NP, Ane JM (2021) Deciphering the chitin code in plant symbiosis, defense, and microbial networks. Annu Rev Microbiol 75:583–607. https://doi.org/10.1146/annurev-micro-051921-114809

Kitaoku Y, Taira T, Numata T, Ohnuma T, Fukamizo T (2022) Structure, mechanism, and phylogeny of LysM-chitinase conjugates specifically found in fern plants. Plant Sci 321:111310. https://doi.org/10.1016/j.plantsci.2022.111310

Knoll AH (2014) Paleobiological perspectives on early eukaryotic evolution. Cold Spring Harb Perspect Biol 6(1):1–14. https://doi.org/10.1101/cshperspect.a016121

Laroche F, Tulotta C, Lamers G, Meijer A, Yang P, Verbeek F, Blaise M, Stougaard J, Spaink H (2013) The embryonic expression patterns of zebrafish genes encoding LysM-domains. Gene Expr Patterns 13(7): 212–224. https://doi.org/10.1016/j.gep.2013.02.007

Lee W-J, Lee J-D, Kravchenko VV, Ulevitch RJ, Brey PT (1996) Purification and molecular cloning of an inducible gram-negative bacteria-binding protein from the silkworm, *Bombyx mori*. Proc Natl Acad Sci U S A 93(15):7888–7893. https://doi.org/10.1073/pnas.93.15.7888

Lefebvre B, Klaus-Heisen D, Pietraszewska-Bogiel A, Herve C, Camut S, Auriac MC, Gasciolli V, Nurisso A, Gadella TW, Cullimore J (2012) Role of N-glycosylation sites and CXC motifs in trafficking of *Medicago truncatula* Nod factor perception protein to plasma membrane. J Biol Chem 287(14): 10812–10823. https://doi.org/10.1074/jbc.M111.281634

LeibundGut-Landmann S, Groß O, Robinson MJ, Osorio F, Slack EC, Tsoni SV, Schweighoffer E, Tybulewicz V, Brown GD, Ruland J, Reis e Sousa C (2007) Syk- and CARD9-dependent coupling of innate immunity to the induction of T helper cells that produce interleukin 17. Nat Immunol 8(6):630–638. https://doi.org/10.1038/ni1460

Lemaitre B, Nicolas E, Michaut L, Reichhart J-M, Hoffmann JA (1996) The dorsoventral regulatory gene cassette spätzle/Toll/cactus controls the potent antifungal response in *Drosophila* adults. Cell 86(6): 973–983. https://doi.org/10.1016/S0092-8674(00)80172-5

Limper AH, Adenis A, Le T, Harrison TS (2017) Fungal infections in HIV/AIDS. Lancet Infect Dis 17(11): e334–e343. https://doi.org/10.1016/S1473-3099(17)30303-1

Linton SM (2020) Review: the structure and function of cellulase (endo-β-1,4-glucanase) and hemicellulase (-β-1,3-glucanase and endo-β-1,4-mannase) enzymes in invertebrates that consume materials ranging from microbes, algae to leaf litter. Comp Biochem Physiol B: Biochem Mol Biol 240:110354. https://doi.org/10.1016/j.cbpb.2019.110354

Lu S, Wang J, Chitsaz F, Derbyshire MK, Geer RC, Gonzales NR, Gwadz M, Hurwitz DI, Marchler GH, Song JS, Thanki N, Yamashita RA, Yang M, Zhang D, Zheng C, Lanczycki CJ, Marchler-Bauer A (2020) CDD/SPARCLE: the conserved domain database in 2020. Nucleic Acids Res 48(D1):D265–d268. https://doi.org/10.1093/nar/gkz991

Lutzoni F, Nowak MD, Alfaro ME, Reeb V, Miadlikowska J, Krug M, Arnold AE, Lewis LA, Swofford DL, Hibbett D (2018) Contemporaneous radiations of fungi and plants linked to symbiosis. Nat Commun 9(1):5451

Maekawa T, Kufer TA, Schulze-Lefert P (2011) NLR functions in plant and animal immune systems: so far and yet so close. Nat Immunol 12(9):817–826. https://doi.org/10.1038/ni.2083

Maillet F, Poinsot V, Andre O, Puech-Pages V, Haouy A, Gueunier M, Cromer L, Giraudet D, Formey D, Niebel A, Martinez EA, Driguez H, Becard G, Denarie J (2011) Fungal lipochitooligosaccharide symbiotic signals in arbuscular mycorrhiza. Nature 469(7328): 58–63. https://doi.org/10.1038/nature09622

Martinon F, Tschopp J (2005) NLRs join TLRs as innate sensors of pathogens. Trends Immunol 26(8):447–454. https://doi.org/10.1016/j.it.2005.06.004

Minh BQ, Schmidt HA, Chernomor O, Schrempf D, Woodhams MD, von Haeseler A, Lanfear R (2020) IQ-TREE 2: new models and efficient methods for phylogenetic inference in the genomic era. Mol Biol Evol 37(5):1530–1534. https://doi.org/10.1093/molbev/msaa015

Misof B, Liu S, Meusemann K, Peters RS, Donath A, Mayer C, Frandsen PB, Ware J, Flouri T, Beutel RG (2014) Phylogenomics resolves the timing and pattern of insect evolution. Science 346(6210):763–767. https://doi.org/10.1126/science.1257570

Moeller JB, Leonardi I, Schlosser A, Flamar A-L, Bessman NJ, Putzel GG, Thomsen T, Hammond M, Jepsen CS, Skjødt K, Füchtbauer E-M, Farber DL, Sorensen GL, Iliev ID, Holmskov U, Artis D (2019) Modulation of the fungal mycobiome is regulated by the chitin-binding receptor FIBCD1. J Exp Med 216(12):2689–2700. https://doi.org/10.1084/jem.20182244

Morris JL, Puttick MN, Clark JW, Edwards D, Kenrick P, Pressel S, Wellman CH, Yang ZH, Schneider H, Donoghue PCJ (2018) The timescale of early land plant evolution. Proc Natl Acad Sci U S A 115(10): E2274–E2283. https://doi.org/10.1073/pnas.1719588115

Mulisch M (1993) Chitin in protistan organisms. Eur J Protistol 29(1):1–18. https://doi.org/10.1016/s0932-4739(11)80291-9

Nelsen MP, Lücking R, Boyce CK, Lumbsch HT, Ree RH (2020) No support for the emergence of lichens prior to the evolution of vascular plants. Geobiology 18(1): 3–13

Ngou BPM, Jones JDG, Ding P (2022) Plant immune networks. Trends Plant Sci 27(3):255–273. https://doi.org/10.1016/j.tplants.2021.08.012

Nimma S, Gu W, Maruta N, Li Y, Pan M, Saikot FK, Lim BYJ, McGuinness HY, Zaoti ZF, Li S, Desa S, Manik MK, Nanson JD, Kobe B (2021) Structural evolution of TIR-domain signalosomes. Front Immunol 12. https://doi.org/10.3389/fimmu.2021.784484

Nishiyama T, Sakayama H, de Vries J, Buschmann H, Saint-Marcoux D, Ullrich KK, Haas FB, Vanderstraeten L, Becker D, Lang D, Vosolsobe S, Rombauts S, Wilhelmsson PKI, Janitza P, Kern R, Heyl A, Rumpler F, Villalobos L, Clay JM, Skokan R, Toyoda A, Suzuki Y, Kagoshima H, Schijlen E, Tajeshwar N, Catarino B, Hetherington AJ, Saltykova A, Bonnot C, Breuninger H, Symeonidi A, Radhakrishnan GV, Van Nieuwerburgh F, Deforce D, Chang C, Karol KG, Hedrich R, Ulvskov P, Glockner G, Delwiche CF, Petrasek J, Van de Peer Y, Friml J, Beilby M, Dolan L, Kohara Y, Sugano S, Fujiyama A, Delaux PM, Quint M, Theissen G, Hagemann M, Harholt J, Dunand C, Zachgo S, Langdale J, Maumus F, Van Der Straeten D, Gould SB, Rensing SA (2018) The *Chara* genome: secondary complexity and implications for plant terrestrialization. Cell 174(2):448–464 e424. https://doi.org/10.1016/j.cell.2018.06.033

Ohno T, Armand S, Hata T, Nikaidou N, Henrissat B, Mitsutomi M, Watanabe T (1996) A modular family 19 chitinase found in the prokaryotic organism *Streptomyces griseus* HUT 6037. J Bacteriol 178(17): 5065–5070. https://doi.org/10.1128/jb.178.17.5065-5070.1996

Ohno M, Kimura M, Miyazaki H, Okawa K, Onuki R, Nemoto C, Tabata E, Wakita S, Kashimura A, Sakaguchi M (2016) Acidic mammalian chitinase is a proteases-resistant glycosidase in mouse digestive system. Sci Rep 6(1):1–9. https://doi.org/10.1038/srep37756

Ohnuma T, Numata T, Osawa T, Mizuhara M, Varum KM, Fukamizo T (2011) Crystal structure and mode of action of a class V chitinase from *Nicotiana tabacum*. Plant Mol Biol 75(3):291–304. https://doi.org/10.1007/s11103-010-9727-z

Oliveira ST, Azevedo MIG, Cunha RMS, Silva CFB, Muniz CR, Monteiro-Junior JE, Carneiro RF, Nagano CS, Girao MS, Freitas CDT, Grangeiro TB (2020) Structural and functional features of a class VI chitinase from cashew (*Anacardium occidentale* L.) with antifungal properties. Phytochemistry 180: 112527. https://doi.org/10.1016/j.phytochem.2020.112527

Onaga S, Taira T (2008) A new type of plant chitinase containing LysM domains from a fern (*Pteris ryukyuensis*): roles of LysM domains in chitin binding and antifungal activity. Glycobiology 18(5):414–423. https://doi.org/10.1093/glycob/cwn018

Orlando M, Buchholz PCF, Lotti M, Pleiss J (2021) The GH19 engineering database: sequence diversity, substrate scope, and evolution in glycoside hydrolase family 19. PLoS One 16(10):e0256817. https://doi.org/10.1371/journal.pone.0256817

Orr RG, Foley SJ, Sherman C, Abreu I, Galotto G, Liu B, Gonzalez-Guerrero M, Vidali L (2020) Robust survival-based RNA interference of gene families using in tandem silencing of adenine phosphoribosyltransferase. Plant Physiol 184(2): 607–619. https://doi.org/10.1104/pp.20.00865

Ota F, Hirayama T, Kizuka Y, Yamaguchi Y, Fujinawa R, Nagata M, Ismanto HS, Lepenies B, Aretz J, Rademacher C (2018) High affinity sugar ligands of C-type lectin receptor langerin. Biochim Biophys Acta, Gen Subj 1862(7):1592–1601. https://doi.org/10.1016/j.bbagen.2018.04.004

Parfrey LW, Lahr DJ, Knoll AH, Katz LA (2011) Estimating the timing of early eukaryotic diversification with multigene molecular clocks. Proc Natl Acad Sci U S A 108(33):13624–13629. https://doi.org/10.1073/pnas.1110633108

Passarinho PA, de Vries SC (2002) *Arabidopsis* chitinases: a genomic survey. Arabidopsis Book 1: e0023. https://doi.org/10.1199/tab.0023

Pees B, Yang W, Zárate-Potes A, Schulenburg H, Dierking K (2016) Innate immune specificity through diversified c-type lectin-like domain proteins in invertebrates. J Innate Immun 8(2):129–142. https://doi.org/10.1159/000441475

Perović-Ottstadt S, Adell T, Proksch P, Wiens M, Korzhev M, Gamulin V, Müller IM, Müller WE (2004) A (1-->3)-beta-D-glucan recognition protein from the sponge *Suberites domuncula*. Mediated activation of fibrinogen-like protein and epidermal growth factor gene expression. Eur J Biochem 271(10): 1924–1937. https://doi.org/10.1111/j.1432-1033.2004.04102.x

Pilecki B, Moeller JB (2020) Fungal recognition by mammalian fibrinogen-related proteins. Scand J Immunol 92(4):e12925. https://doi.org/10.1111/sji.12925

Punja ZK, Zhang Y (1993) Plant chitinases and their roles in resistance to fungal diseases. J Nematol 25(4): 526–540

Pusztahelyi T (2018) Chitin and chitin-related compounds in plant-fungal interactions. Mycology 9(3):189–201. https://doi.org/10.1080/21501203.2018.1473299

Reboledo G, Agorio AD, Vignale L, Batista-Garcia RA, Ponce De Leon I (2021) Transcriptional profiling reveals conserved and species-specific plant defense responses during the interaction of *Physcomitrium patens* with *Botrytis cinerea*. Plant Mol Biol 107(4–5):365–385. https://doi.org/10.1007/s11103-021-01116-0

Renner T, Specht CD (2012) Molecular and functional evolution of class I chitinases for plant carnivory in the Caryophyllales. Mol Biol Evol 29(10):2971–2985. https://doi.org/10.1093/molbev/mss106

Roach JC, Glusman G, Rowen L, Kaur A, Purcell MK, Smith KD, Hood LE, Aderem A (2005) The evolution of vertebrate Toll-like receptors. Proc Natl Acad Sci U S A 102(27):9577. https://doi.org/10.1073/pnas.0502272102

Roach JM, Racioppi L, Jones CD, Masci AM (2013) Phylogeny of toll-like receptor signaling: adapting the innate response. PLoS One 8(1). https://doi.org/10.1371/journal.pone.0054156

Roche P, Maillet F, Plazanet C, Debellé F, Ferro M, Truchet G, Promé JC, Dénarié J (1996) The common *nodABC* genes of *Rhizobium meliloti* are host-range determinants. Proc Natl Acad Sci U S A 93(26): 15305–15310. https://doi.org/10.1073/pnas.93.26.15305

Rogers NC, Slack EC, Edwards AD, Nolte MA, Schulz O, Schweighoffer E, Williams DL, Gordon S, Tybulewicz VL, Brown GD, Reis e Sousa C (2005) Syk-dependent cytokine induction by dectin-1 reveals a novel pattern recognition pathway for C type lectins. Immunity 22(4):507–517. https://doi.org/10.1016/j.immuni.2005.03.004

Romani L (2004) Immunity to fungal infections. Nat Rev Immunol 4(1):11–24. https://doi.org/10.1038/nri1255

Romani L (2011) Immunity to fungal infections. Nat Rev Immunol 11(4):275–288. https://doi.org/10.1038/nri2939

Rombout JHWM, Huttenhuis HBT, Picchietti S, Scapigliati G (2005) Phylogeny and ontogeny of fish leucocytes. Fish Shellfish Immunol 19(5):441–455. https://doi.org/10.1016/j.fsi.2005.03.007

Rush TA, Puech-Pages V, Bascaules A, Jargeat P, Maillet F, Haouy A, Maes AQ, Carriel CC, Khokhani D, Keller-Pearson M, Tannous J, Cope KR, Garcia K, Maeda J, Johnson C, Kleven B, Choudhury QJ, Labbe J, Swift C, O'Malley MA, Bok JW, Cottaz S, Fort S, Poinsot V, Sussman MR, Lefort C, Nett J, Keller NP, Becard G, Ane JM (2020) Lipo-chitooligosaccharides as regulatory signals of fungal growth and development. Nat Commun 11(1):3897. https://doi.org/10.1038/s41467-020-17615-5

Sanchez-Vallet A, Mesters JR, Thomma BP (2014) The battle for chitin recognition in plant-microbe interactions. FEMS Microbiol Rev 39(2):171–183. https://doi.org/10.1093/femsre/fuu003

Schlosser A, Thomsen T, Moeller JB, Nielsen O, Tornøe I, Mollenhauer J, Moestrup SK, Holmskov U (2009) Characterization of FIBCD1 as an acetyl group-binding receptor that binds chitin. J Immunol 183(6):

3800–3809. https://doi.org/10.4049/jimmunol.0901526

Sharma S, Singh R, Kaur R (2020) In silico characterization of a unique plant-like "loopful" GH19 chitinase from newly isolated *Chitinophaga* sp. YS-16. Curr Microbiol 77(9):2248–2257. https://doi.org/10.1007/s00284-020-02022-2

Shimizu T, Nakano T, Takamizawa D, Desaki Y, Ishii-Minami N, Nishizawa Y, Minami E, Okada K, Yamane H, Kaku H, Shibuya N (2010) Two LysM receptor molecules, CEBiP and OsCERK1, cooperatively regulate chitin elicitor signaling in rice. Plant J 64(2):204–214. https://doi.org/10.1111/j.1365-313X.2010.04324.x

Shiu SH, Bleecker AB (2003) Expansion of the receptor-like kinase/Pelle gene family and receptor-like proteins in *Arabidopsis*. Plant Physiol 132(2):530–543. https://doi.org/10.1104/pp.103.021964

Shrive AK, Moeller JB, Burns I, Paterson JM, Shaw AJ, Schlosser A, Sorensen GL, Greenhough TJ, Holmskov U (2014) Crystal structure of the tetrameric fibrinogen-like recognition domain of fibrinogen C domain containing (FIBCD1) protein. J Biol Chem 289(5):2880–2887. https://doi.org/10.1074/jbc.M113.520577

Simm D, Hatje K, Kollmar M (2015) Waggawagga: comparative visualization of coiled-coil predictions and detection of stable single α-helices (SAH domains). Bioinformatics 31(5):767–769. https://doi.org/10.1093/bioinformatics/btu700

Stambach NS, Taylor ME (2003) Characterization of carbohydrate recognition by langerin, a C-type lectin of Langerhans cells. Glycobiology 13(5):401–410. https://doi.org/10.1093/glycob/cwg045

Steinfeld L, Vafaei A, Rosner J, Merzendorfer H (2019) Chitin prevalence and function in bacteria, fungi and protists. Adv Exp Med Biol 1142:19–59. https://doi.org/10.1007/978-981-13-7318-3_3

Strobel S, Roswag A, Becker NI, Trenczek TE, Encarnaçao JA (2013) Insectivorous bats digest chitin in the stomach using acidic mammalian chitinase. PLoS One 8(9):e72770. https://doi.org/10.1371/journal.pone.0072770

Sutherland TE (2018) Chitinase-like proteins as regulators of innate immunity and tissue repair: helpful lessons for asthma? Biochem Soc Trans 46(1):141–151. https://doi.org/10.1042/BST20170108

Sutherland TE, Logan N, Rückerl D, Humbles AA, Allan SM, Papayannopoulos V, Stockinger B, Maizels RM, Allen JE (2014) Chitinase-like proteins promote IL-17-mediated neutrophilia in a tradeoff between nematode killing and host damage. Nat Immunol 15(12):1116–1125. https://doi.org/10.1038/ni.3023

Suzuki M, Shibuya M, Shimada H, Motoyama N, Nakashima M, Takahashi S, Suto K, Yoshida I, Matsui S, Tsujimoto N, Ohnishi M, Ishibashi Y, Fujimoto Z, Desaki Y, Kaku H, Kito K, Shibuya N (2016) Autophosphorylation of specific threonine and tyrosine residues in *Arabidopsis* CERK1 is essential for the activation of chitin-induced immune signaling. Plant Cell Physiol 57(11):2312–2322. https://doi.org/10.1093/pcp/pcw150

Taira T, Hayashi H, Tajiri Y, Onaga S, Uechi G, Iwasaki H, Ohnuma T, Fukamizo T (2009) A plant class V chitinase from a cycad (*Cycas revoluta*): biochemical characterization, cDNA isolation, and posttranslational modification. Glycobiology 19(12):1452–1461. https://doi.org/10.1093/glycob/cwp119

Taira T, Mahoe Y, Kawamoto N, Onaga S, Iwasaki H, Ohnuma T, Fukamizo T (2011) Cloning and characterization of a small family 19 chitinase from moss (*Bryum coronatum*). Glycobiology 21(5):644–654. https://doi.org/10.1093/glycob/cwq212

Tateno H, Ohnishi K, Yabe R, Hayatsu N, Sato T, Takeya M, Narimatsu H, Hirabayashi J (2010) Dual specificity of langerin to sulfated and mannosylated glycans via a single C-type carbohydrate recognition domain. J Biol Chem 285(9):6390–6400. https://doi.org/10.1074/jbc.M109.041863

Taylor PR, Tsoni SV, Willment JA, Dennehy KM, Rosas M, Findon H, Haynes K, Steele C, Botto M, Gordon S, Brown GD (2007) Dectin-1 is required for β-glucan recognition and control of fungal infection. Nat Immunol 8(1):31–38. https://doi.org/10.1038/ni1408

Thoma-Uszynski S, Stenger S, Takeuchi O, Ochoa MT, Engele M, Sieling PA, Barnes PF, Rollinghoff M, Bolcskei PL, Wagner M, Akira S, Norgard MV, Belisle JT, Godowski PJ, Bloom BR, Modlin RL (2001) Induction of direct antimicrobial activity through mammalian toll-like receptors. Science 291(5508):1544–1547. https://doi.org/10.1126/science.291.5508.1544

Thomma BP, Eggermont K, Tierens KF, Broekaert WF (1999) Requirement of functional *ethylene-insensitive 2* gene for efficient resistance of *Arabidopsis* to infection by *Botrytis cinerea*. Plant Physiol 121(4):1093–1101. https://doi.org/10.1104/pp.121.4.1093

Thomsen T, Schlosser A, Holmskov U, Sorensen GL (2011) Ficolins and FIBCD1: soluble and membrane bound pattern recognition molecules with acetyl group selectivity. Mol Immunol 48(4):369–381. https://doi.org/10.1016/j.molimm.2010.09.019

Unestam T, Söderhäll K (1977) Soluble fragments from fungal cell walls elicit defence reactions in crayfish. Nature 267(5606):45–46. https://doi.org/10.1038/267045a0

UniProt Consortium (2021) UniProt: the universal protein knowledgebase in 2021. Nucleic Acids Res 49(D1):D480–D489. https://doi.org/10.1093/nar/gkaa1100

Urbach JM, Ausubel FM (2017) The NBS-LRR architectures of plant R-proteins and metazoan NLRs evolved in independent events. Proc Natl Acad Sci U S A 114(5):1063–1068. https://doi.org/10.1073/pnas.1619730114

van den Berg LM, Gringhuis SI, Geijtenbeek TBH (2012) An evolutionary perspective on C-type lectins in infection and immunity. Ann N Y Acad Sci 1253(1):

149–158. https://doi.org/10.1111/j.1749-6632.2011.06392.x

van Eijk M, van Roomen CPAA, Renkema GH, Bussink AP, Andrews L, Blommaart EFC, Sugar A, Verhoeven AJ, Boot RG, Aerts JMFG (2005) Characterization of human phagocyte-derived chitotriosidase, a component of innate immunity. Int Immunol 17(11): 1505–1512. https://doi.org/10.1093/intimm/dxh328

van Hengel AJ, Tadesse Z, Immerzeel P, Schols H, van Kammen A, de Vries SC (2001) *N*-acetylglucosamine and glucosamine-containing arabinogalactan proteins control somatic embryogenesis. Plant Physiol 125(4): 1880–1890. https://doi.org/10.1104/pp.125.4.1880

Van Holle S, Van Damme EJM (2019) Messages from the past: new insights in plant lectin evolution. Front Plant Sci 10:36. https://doi.org/10.3389/fpls.2019.00036

van Loon LC, Rep M, Pieterse CM (2006) Significance of inducible defense-related proteins in infected plants. Annu Rev Phytopathol 44:135–162. https://doi.org/10.1146/annurev.phyto.44.070505.143425

van Velzen R, Doyle JJ, Geurts R (2019) A resurrected scenario: single gain and massive loss of nitrogen-fixing nodulation. Trends Plant Sci 24(1):49–57. https://doi.org/10.1016/j.tplants.2018.10.005

Vannella KM, Ramalingam TR, Hart KM, de Queiroz PR, Sciurba J, Barron L, Borthwick LA, Smith AD, Mentink-Kane M, White S, Thompson RW, Cheever AW, Bock K, Moore I, Fitz LJ, Urban JF Jr, Wynn TA (2016) Acidic chitinase primes the protective immune response to gastrointestinal nematodes. Nat Immunol 17(5):538–544. https://doi.org/10.1038/ni.3417

Vendele I, Willment JA, Silva LM, Palma AS, Chai W, Liu Y, Feizi T, Spyrou M, Stappers MH, Brown GD (2020) Mannan detecting C-type lectin receptor probes recognise immune epitopes with diverse chemical, spatial and phylogenetic heterogeneity in fungal cell walls. PLoS Pathog 16(1):e1007927. https://doi.org/10.1371/journal.ppat.1007927

Viborg AH, Terrapon N, Lombard V, Michel G, Czjzek M, Henrissat B, Brumer H (2019) A subfamily roadmap of the evolutionarily diverse glycoside hydrolase family 16 (GH16). J Biol Chem 294(44): 15973–15986. https://doi.org/10.1074/jbc.RA119.010619

Wang Y, Wang X-P, Zhang B, Li Z-M, Yang L-G, Li X-C, Ma H (2020) A lysin motif-containing protein (SpLysMD3) functions as a PRR involved in the antibacterial responses of mud crab, *Scylla paramamosain*. Fish Shellfish Immunol 97:257–267. https://doi.org/10.1016/j.fsi.2019.12.036

Wang H, Chen X, Li S, Zhou C, Xu L, Wu Z, Chen X (2021) Identification and expression analysis of Langerhans cells marker Langerin/CD207 in grasscarp, *Ctenopharyngodon idella*. Gene 768: 145315. https://doi.org/10.1016/j.gene.2020.145315

Wanke A, Rovenich H, Schwanke F, Velte S, Becker S, Hehemann JH, Wawra S, Zuccaro A (2020) Plant species-specific recognition of long and short beta-1,3-linked glucans is mediated by different receptor systems. Plant J 102(6):1142–1156. https://doi.org/10.1111/tpj.14688

Wiesner DL, Specht CA, Lee CK, Smith KD, Mukaremera L, Lee ST, Lee CG, Elias JA, Nielsen JN, Boulware DR, Bohjanen PR, Jenkins MK, Levitz SM, Nielsen K (2015) Chitin recognition via chitotriosidase promotes pathologic type-2 helper T cell responses to cryptococcal infection. PLoS Pathog 11(3):e1004701. https://doi.org/10.1371/journal.ppat.1004701

Willmann R, Lajunen HM, Erbs G, Newman MA, Kolb D, Tsuda K, Katagiri F, Fliegmann J, Bono JJ, Cullimore JV, Jehle AK, Gotz F, Kulik A, Molinaro A, Lipka V, Gust AA, Nurnberger T (2011) *Arabidopsis* lysin-motif proteins LYM1 LYM3 CERK1 mediate bacterial peptidoglycan sensing and immunity to bacterial infection. Proc Natl Acad Sci U S A 108(49):19824–19829. https://doi.org/10.1073/pnas.1112862108

Xin Y, Wang D, Han S, Li S, Gong N, Fan Y, Ji X (2021) Characterization of the chitinase gene family in mulberry (*Morus notabilis*) and *Mn*Chi18 involved in resistance to *Botrytis cinerea*. Genes (Basel) 13(1). https://doi.org/10.3390/genes13010098

Yadav M, Schorey JS (2006) The β-glucan receptor dectin-1 functions together with TLR2 to mediate macrophage activation by mycobacteria. Blood 108(9): 3168–3175. https://doi.org/10.1182/blood-2006-05-024406

Yamagami T, Funatsu G (1993) Purification and some properties of three chitinases from the seeds of rye (*Secale cereale*). Biosci Biotechnol Biochem 57(4): 643–647. https://doi.org/10.1271/bbb.57.643

Yamasaki S, Matsumoto M, Takeuchi O, Matsuzawa T, Ishikawa E, Sakuma M, Tateno H, Uno J, Hirabayashi J, Mikami Y (2009) C-type lectin Mincle is an activating receptor for pathogenic fungus, *Malassezia*. Proc Natl Acad Sci U S A 106(6): 1897–1902. https://doi.org/10.1073/pnas.0805177106

Yano S, Kanno H, Tsuhako H, Ogasawara S, Suyotha W, Konno H, Makabe K, Uechi K, Taira T (2021) Cloning, expression, and characterization of a GH 19-type chitinase with antifungal activity from *Lysobacter* sp. MK9-1. J Biosci Bioeng 131(4):348–355. https://doi.org/10.1016/j.jbiosc.2020.11.005

Yeh S, Moffat BA, Griffith M, Xiong F (2000) Chitinase genes responsive to cold encode antifreeze proteins in winter cereals. Plant Physiol 124(3):1251–1264. https://doi.org/10.1104/pp.124.3.1251

Yokoyama CC, Baldridge MT, Leung DW, Zhao G, Desai C, Liu T-C, Diaz-Ochoa VE, Huynh JP, Kimmey JM, Sennott EL (2018) LysMD3 is a type II membrane protein without an in vivo role in the response to a range of pathogens. J Biol Chem 293(16):6022–6038

Zeilinger S, Gupta VK, Dahms TE, Silva RN, Singh HB, Upadhyay RS, Gomes EV, Tsui CK, Nayak SC (2016) Friends or foes? Emerging insights from fungal

interactions with plants. FEMS Microbiol Rev 40(2): 182–207. https://doi.org/10.1093/femsre/fuv045
Zhang F, Wen Y, Guo X (2014) CRISPR/Cas9 for genome editing: progress, implications and challenges. Hum Mol Genet 23(R1):R40–R46. https://doi.org/10.1093/hmg/ddu125
Zhang H, Yohe T, Huang L, Entwistle S, Wu P, Yang Z, Busk PK, Xu Y, Yin Y (2018) dbCAN2: a meta server for automated carbohydrate-active enzyme annotation. Nucleic Acids Res 46(W1):W95–W101. https://doi.org/10.1093/nar/gky418
Zhang YJ, Ren LL, Lin XY, Han XM, Wang W, Yang ZL (2022) Molecular evolution and functional characterization of chitinase gene family in *Populus trichocarpa*. Gene 822:146329. https://doi.org/10.1016/j.gene.2022.146329

2 Metabolic Constraints and Dependencies Between "Uncultivable" Fungi and Their Hosts

Saleh Rahimlou, C. Alisha Quandt, and Timothy Y. James

Abstract

Many host-associated fungi are believed to be uncultivable because they have thus far never been brought into culture. However, the absence of success in bringing a fungus into cultivation in the laboratory should not be interpreted as an indication that the fungus cannot grow axenically without a living host. In this chapter, we review the history of the cultivation of fungi and provide examples where seemingly fastidious fungi have been successfully cultivated, borrowing insights from the literature on bacteria. We define a few of the hypotheses for why fungi become uncultivable and suggest that only organisms with a highly predictable environment will likely lose the ability to produce primary metabolites needed for making cells. Highly host-specific parasites fit this description, while arbuscular mycorrhizae do not. As fungal genomics has blossomed, it provides an unprecedented opportunity to investigate, from a genetic-biochemistry level, particular metabolic requirements that may provide insights into cultivation strategies. Common themes across diverse fungi are the loss of pathways for synthesis of long-chain fatty acids, amino acids, and vitamins. Finally, we discuss novel approaches and experimental systems that may be used to overcome the common barriers to cultivation.

Keywords

Parasite · Biotrophic · Coevolution · Primary metabolites · *Malassezia*

2.1 Introduction

2.1.1 What Are Uncultivable Fungi?

The cultivation of fungi has been essential in their development as model organisms for the study of mating, cell cycle regulation, genetic interactions, and much more (Müller and Grossniklaus 2010). However, most model fungal systems, such as *Saccharomyces*, *Neurospora*, or *Coprinopsis*, are organisms that are considered saprobic while many fungi are symbiotic and are reliant on other species for their nutrition. Because of these nutritional dependencies many of these species of symbiotic fungi have never been brought into culture, making them appear as if they are uncultivable. However, there are multiple reasons to believe that most of these fungi could actually be brought into culture with the correct media, much in the same way that tissue culturing of plants and

S. Rahimlou · T. Y. James (✉)
Department of Ecology and Evolutionary Biology, University of Michigan, Ann Arbor, MI, USA
e-mail: tyjames@umich.edu

C. A. Quandt
Department of Ecology and Evolutionary Biology, University of Colorado Boulder, Boulder, CA, USA

Y.-P. Hsueh, M. Blackwell (eds.), *Fungal Associations*, The Mycota 9,
https://doi.org/10.1007/978-3-031-41648-4_2

animals with complex nutritional requirements is also accomplished (Thorpe 2012; Yao and Asayama 2017).

Saprobes are defined as organisms using dead organic material as food, and thus saprobic growth is independent of a living host. Williams (1992) has provided definitions and terminology for fungal cultivation. In summary, the term "axenic" should be used to refer to the growth of a single species in the absence of whole, live organisms, or living cells of any other species. "Monoxenic" describes a culture containing organisms or cells of two species (e.g., an arbuscular mycorrhizal fungus and a soil bacterium growing together on an agar plate). "Dixenic" should be applied to a culture containing organisms or cells of three species, etc. As stated by Williams (1992), the terms "aseptic," "sterile," "artificial," "pure," "bacteria-free," etc. are ambiguous or imprecise. Thus, uncultivable fungi are defined here as strains that have never been cultured *in vitro*, whether axenic, monoxenic, or polyxenic. In this chapter, we attempt to make a distinction between uncultivable fungi and uncultivated fungi, the former referring to a group of fungi which could not be cultivated under any conceivable conditions, and the latter referring to those fungi which have never been cultivated thus far but may possibly be in the future.

Finally, the speed at which particular fungi grow impacts the assessment of their cultivability. There are examples in the literature of very slow-growing fungi being cultured after months of incubation, with fungi isolated from extreme environments being particularly well represented (Gostinčar et al. 2022). How long researchers should monitor cultures is not standardized—outside of clinical settings (Bosshard 2011)—and a systematic review of the literature on this has not been published.

2.1.2 Major Groups of Uncultivated Fungi

Since fungi are chemoheterotrophic, they must derive their carbon compounds from either non-living organic material, i.e., as saprobes, or living tissues. Uncultivated fungi run the gamut from symbionts to saprobes, and while there is a general tendency for species associated with living organisms, including pathogens and mutualists, to be unable to grow in artificial media, some are readily amenable, for example, plant pathogens such as *Magnaporthe grisea* and *Taphrina deformans*. On the other hand, some saprobes have never been cultured, such as the chytridiomycete *Rhizophydium keratinophilum* and ascomycetes *Rhytisma* spp. In general, the term biotrophic is used to describe organisms that derive their nutrition from living host tissue or cytoplasm (Jeffries and Young 1994), and this applies to species along the entirety of the symbiosis spectrum, such as endophytes and mycorrhizae as well as pathogens. As a result of long-term efforts to cultivate fungi, some biotrophic fungi can be grown with specific nutritional or environmental requirements (Williams et al. 1966; Sugiura et al. 2020). Examples of biotrophic fungi that are cultivated axenically are haustorial mycoparasites such as *Dispira* (Barker and Barnett 1973), nematode parasites such as *Rhopalomyces* (Ellis 1963), plant pathogens such as *Phytophthora infestans*, corn smut *Ustilago maydis*, and leaf curl *Taphrina deformans* (Mix 1949; Day and Anagnostakis 1971; Tumwine et al. 2000), endophytes such as *Epichlöe typhina* (Bacon et al. 1977), ectomycorrhizal fungi (EMF) (Vuorinen et al. 2015), lichens (Stocker-Wörgötter 2001), and recently arbuscular mycorrhizal fungi (Sugiura et al. 2020). Symbiotic fungi that are considered hemibiotrophic are those pathogens that start as biotrophs and then have a period of necrotrophy after the host cells have been killed. Both hemibiotrophs and necrotrophs are considered readily culturable. Heretical as it may seem, it is worth considering whether the ability to grow in pure culture has any meaning. For both *Taphrina deformans* and *Ustilago maydis*, it is the yeast phase that is readily cultured, a phase that is considered saprobic. Therefore, the limitation to cultivation is not hard-wired in the genome, but one of how the genes are being used and at what developmental stage. Another point to consider is whether the ability to culture some biotrophic fungi, for example, the amphibian pathogen *Batrachochytrium dendrobatidis*, belies an innate saprobic ability that is manifested in nature (Piotrowski et al. 2004). In other words, how clear is the relationship between culturability and the ability to grow on non-living nutritive sources? And does the ability to grow readily in culture necessarily suggest some saprobic growth in nature?

It is also worth mentioning that many of the commonly used media for culturing fungi are derived directly from complex and undefined plant-based sources (e.g., potato, corn, barley, and cherry). This necessarily biases culture-based techniques toward growing and identifying fungi which can regularly break down plant-based sugars and may explain why plant pathogens are much more readily "culturable" than fungi which are pathogens and parasites of other non-plant organisms, such as many cordycipitaceous fungi.

Despite recent advances in culturing symbiotic fungi, there are general patterns regarding cultivation that are apparent. Among the major fungal groups, the following mutualists remain largely uncultivated: arbuscular mycorrhizas (Glomeromycotina), mycorrhizal Endogonales, many ectomycorrhizas (including the genus *Cortinarius*), and most lichenized fungi. Of the biotrophic plant pathogens, rusts (Pucciniales), powdery mildews (Erysiphales), and *Physoderma* (Blastocladiomycota) are mostly uncultured. The algal parasites in Aphelidiomycota and some members of Chytridiomycota (e.g., *Zygorhizidium*) are uncultivated, in addition to many mycoparasitic Zoopagomycota, Rozellomycota, and *Caulochytrium* (Chytridiomycota). Uncultured animal symbionts are also common, such as Laboulbeniales, predatory fungi in Zoopagales, some Entomophthorales, the mammal pathogen *Pneumocystis*, and insect pathogenic members of *Cordyceps* s.l. and Blastocladiales, such as *Coelomomyces*.

2.1.3 Where Do Uncultivated Fungi Grow and Should We View Them as Uncultivable?

As seen above, the diversity of hosts of biotrophic fungi has a very broad range that includes other fungi, algae, protists, vascular plants, invertebrates, and vertebrates. All biotrophs have a primary host in their life cycle, but some require a secondary (alternate) host to complete their life cycle when the primary host is unavailable. For instance, some rust fungi may be restricted to a single host (autoecious) or may infect alternate hosts (heteroecious). *Puccinia graminis* Pers. on barberry and cereal grains, *Cronartium ribicola* Dietr. on pines and *Ribes*, and *Gymnosporangium juniperi-virginianae* Schw. on apple and cedar are leading examples of heteroecious rusts, while *Puccinia helianthi* Schw. on sunflower and *Uromyces striatus* Schroet. on alfalfa are autoecious species. The roles of the alternate hosts in disease epidemiology and pathogen variation vary greatly from species to species and from region to region because of different climatic and cropping conditions (Zhao et al. 2016). The distribution of obligate biotrophs is influenced by certain environmental and geographical factors; for example, some graminicolous rusts are negatively affected by higher temperatures and increased concentrations of atmospheric CO_2 (Helfer 2014).

As hosts for both symbionts such as mycorrhizal fungi and pathogens, vascular plants are the most common host for uncultivated fungi, but this likely reflects the rampant co-diversification of fungi with plants as mutualists, pathogens, and saprobes (Lutzoni et al. 2018). Many plant species are susceptible to infection by obligate biotrophs. For example, rusts are pathogenic on vascular plants, including ferns, gymnosperms, and most families of angiosperms. Powdery mildews are among the most common plant diseases, infecting over 650 monocots and over 9000 dicot species (Schulze-Lefert and Vogel 2000). In addition, the majority of land plants form symbiotic associations with mycorrhizal fungi. Ectomycorrhizal fungi, which are mainly associated with forest trees, are the most diverse in temperate biomes, whereas the relative abundance and diversity of agriculturally important arbuscular mycorrhizal fungi are higher in grasslands than in forests with unclear latitudinal patterns (Davison et al. 2015). The majority of mycorrhizal plants are arbuscular mycorrhizal (70%), whereas ectomycorrhizal plants, despite being widely distributed (Read 1991), constitute only a tiny fraction of all vascular plant species (0.7%) (Soudzilovskaia et al. 2020). The orders Pezizales, Agaricales, Helotiales, Boletales, and Cantharellales include the largest number of ectomycorrhizal fungal lineages (Tedersoo et al. 2010).

Environmental metabarcoding has drastically changed our appreciation that a culture-based or specimen-based approach for fungal sampling may be missing the majority of fungal diversity. With environmental metabarcoding, taxonomy is operating metaphorically in the dark, because sequence-based identification cannot be verified with voucher specimens, and culturing is necessary to bring these fungi "to life." Culturing has a long history of use in microbiology and mycology, but it faces various challenges. Since the physiological requirements of dark taxa (unnamed sequences) are not known, culturing requires resource-intensive approaches, a situation that is comparable to prokaryote microbiology. For example, the cultivation of an Asgard archaeon took more than 5 years (Imachi et al. 2020), and initial cultivation of a bacterial strain may cost 10,000 US dollars (Overmann 2015). Even for fungi that can be readily cultured, the quantity of predicted undiscovered species poses a logistical challenge. According to the *World dictionary of culture collections* (http://www.wfcc.info/ccinfo/statistics), there are currently about 892,000 deposited fungal cultures, representing around 30,000 known species. This number represents a maximum of 1.3% of the estimated 2.2–3.8 million fungal species (Tedersoo et al. 2014; Hawksworth and Lücking 2017; Kopytko et al. 2017; Lücking et al. 2021). During the past two decades, about 10,000 new fungi have been described based on cultures (*Index fungorum*; http://www.indexfungorum.org). While this is an impressive number, to formally catalog dark fungal taxa through culturing alone would take millennia or the underlying financial and labor resources would have to be increased by several orders of magnitude (Lücking et al. 2021). To somehow overcome this problem and facilitate the scientific communication of massive eDNA data, the UNITE database has emerged as the most widely used resource for curated fungal ITS sequences, including data from vouchered specimens and reference sequences for so-called "species hypotheses," that is, phylogroups of sequences that are interpreted as species (Kõljalg et al. 2013). Currently, the UNITE database holds around 290,000 species hypotheses with digital object identifiers (DOIs) at a 98.5% similarity threshold (https://unite.ut.ee), representing almost 10% of the estimated unrecognized fungi.

2.2 Why Do Symbionts Become Uncultivable?

We have tried to suggest a link between culturability and symbiosis for fungi. At first glance, this might seem to be a fundamental difference between prokaryotes and eukaryotes; i.e., the relationship between culturability and symbiosis is stronger for fungi than for bacteria, because so-called free-living bacteria seem to be more difficult to cultivate than free-living fungi (Solden et al. 2016). If so, this provides a major insight into the contrasting metabolic evolution and ecosystem ecology of these lineages. What would drive the relationship between culturability and symbiosis for fungi? We present four non-mutually exclusive hypotheses for the relationship between host-associated fungi and lack of cultivability.

The most obvious hypothesis is that uncultivable or fastidious fungi are metabolically in need of nutrients that are provided by the host, either in parasitism or mutualism. This predicts that fungi that are endobiotic will be more likely to be uncultivable than those that are epibiotic, because they are more influenced by the host. Facultative parasites, on the other hand, will be those least subject to these constraints. Heritable bacteria serve as an example of extreme metabolic dependency. These bacteria are known as mutualistic endosymbionts of insects that show transovarial inheritance and provide the host with essential amino acids (McCutcheon and Moran 2012). These associations are highly stable and, from the endosymbiont perspective, foster environmental predictability. An analogous situation may be for the obligate parasite *Pneumocystis*, which is only known from the mammalian lung and has a highly reduced genome (Ma et al. 2016). Rozellids and microsporidia are known as extremely morphologically reduced fungi lacking many of the basic primary metabolism pathways, such as nucleotide and amino acid biosynthesis (James et al. 2013; Keeling and Slamovits 2004; Dean et al. 2016; Quandt et al. 2017). These taxa also appear to have acquired genes (encoding transferases and kinases involved in obtaining these building blocks and energy sources from the host) via horizontal transfer from bacteria (Tsaousis et al. 2008; Alexander et al. 2016). Long-term endosymbiosis has led to the loss of genes to make primary metabolites as they are now obtained from the host.

Using *Escherichia coli* as a model, D'Souza and Kost (2016) provided the first experimental evidence that when metabolites are available in the environment, there is a selection for loss of the ability to produce corresponding metabolites, and these auxotrophic strains occurred much more frequently than by chance in enriched environments than in "normal" ones. Specifically, they grew *E. coli* in amino acid-rich environments and saw loss of genes encoding synthesis and regulation of those amino acids, and, significantly, they showed that the auxotrophs displayed a relative fitness advantage when grown in the presence of amino acids. In support of anecdotes concerning culturing of arbuscular mycorrhizae fungi (AMF) in the presence of other fungi or their filtrates (see Sect. 2.4), these authors also found that auxotrophic *E. coli* could derive amino acids from coexisting bacterial strains. This strongly suggests that organisms growing regularly in close cellular contact (or intracellularly) with one another are more likely to lose functionality involved in the production of metabolites that one host/partner is "providing."

A second hypothesis of why symbiotic fungi become uncultivable is that the cultivation is only possible in a single life stage and the transition to that life stage cannot be triggered. For example, *Ustilago maydis* can only be cultured as a yeast from spores or galls. Isolation of dikaryotic hyphae from leaf tissue has not been reported. Similarly, spores of many ectomycorrhizal fungi will not readily germinate in culture but can be cultured from colonized root tips or sporocarps. The dimorphic facultative pathogen *Paracoccidioides loboi* has never been cultivated, leading Vilela et al. (2021) to hypothesize that the inability to culture the fungus is related to the inability of the fungus to transition from the pathogenic (yeast form) to the saprobic hyphal form. This dimorphic inequality may be hard (also observed in nature) or technical (only related to current methods of culturing). A related phenomenon regarding life cycle transitions is one in which the culturing environment does not allow a fungus to reproduce/sporulate, thus limiting their proliferation in vitro. For example, monocentric chytrids are unicells that grow vegetatively and then convert completely into reproductive spores. Such a failure to sporulate could result from either missing nutrients or missing environmental cues.

A third hypothesis is that host-associated fungi require a chemical signal from a host because growth and specificity are tied together. For example, *Colletotrichum* associated with postharvest disease showed germination and appressorium dependency on host ethylene production, an indication of ripening (Flaishman and Kolattukudy 1994). Chemical signals, mostly undetermined, produced by plant hosts of both AMF and EMF are associated with improved spore germination (Parniske 2008; Nara 2009). The use of specific host-associated signals might be desirable for host-specific plant pathogens such as powdery mildews and rusts; however, such a system could lead to rapid evolution of resistance genes in hosts. It is also hypothesized, with some evidence, that host recognition is important for many insect pathogens to germinate (Gao et al. 2011). Perhaps, without the correct chemical cues from the host, spores may be unable to germinate in the culture.

Finally, physical cues and structure may be missing in culture media. For example, Rozellids lack a cell wall in their endobiotic parasitic phase, allowing them to phagocytose host cytoplasm and to disperse within their coenocytic host, usually a water mold (Powell et al. 2017). After consuming much of the host cytoplasm, *Rozella* forms zoosporangia that are totally dependent on the presence of a host cell wall to generate turgor pressure for discharge. *Rozella* can use its chitin synthase genes to make a resting sporangium wall; however, they are unassociated with their own zoosporangial wall (Held 1980). Haustorial parasites might similarly require physical interactions that are missing in culture to produce the proper assimilative structures. Laboulbeniales, with their complex thallus morphology, may require the host exoskeleton for development, but could also be subject to all of the above constraints on growth in culture.

2.3 The Evolving Approaches for Culturing Fungi

Fungal cultivation dates back to the nineteenth century, contemporaneous with that of bacteria. The development of pure culture techniques, especially those derived from single cells, was critical for the identification of etiological agents of disease, as required for fulfilling Koch's postulates (Hildebrand 1938). The most simple methods of cultivation rely on using extracts from plant products, such as malted barley, potatoes, corn, and oats. Preparations of casein (milk protein) are particularly useful for growing some fungi; for example, the amphibian pathogen *Batrachochytrium dendrobatidis* can grow solely on tryptone, a digestion of casein. Numerous early investigations with fungal cultures improved our understanding of fungal physiology and facilitated development of fully synthetic media. Particular emphasis was placed on forms of nitrogen and carbon available for assimilation.

Nitrate, ammonia, and amino acids are often available to fungi as a source of nitrogen, whereas nitrite is generally not available (Steinberg 1939). Yeast extract is used as a diverse form of nitrogen and trace vitamins suitable for many fungi. Unlike their animal counterparts, fungi are typically able to make most of their growth factors, including all of the amino acids, vitamins, fatty acids, and nucleotides. Therefore, the nutrient requirements for many species are considered minimal, and they may be cultured on a minimal medium with sources of carbon, nitrogen, phosphate, sulfate, and trace elements (Howard 1900; Machlis 1953). Essential vitamins are minimal but most frequently involve biotin, thiamine, and pyridoxine (Steinberg 1950). Thiamine is one of the components of modified Melin-Norkrans (MMN) media which is used to cultivate ECM taxa. Thiamine biosynthesis is lost in many biotrophs (Spanu 2012; Nagae et al. 2016), from the human-pathogenic *Pneumocystis* spp. (Cisse et al. 2014) to the AMF *Rhizophagus irregularis* (Tisserant et al. 2013). Therefore, it is not surprising perhaps that thiamine has long been one of the most commonly added nutrients to non-standard fungal culturing media (Robbins and Kavanagh 1938).

In addition to thiamine, many biotrophic fungi appear to be lacking genes involved in the assimilatory sulfate reductase pathway (ASRP) and some also lack the nitrogen reduction pathway as well. Ahrendt et al. (2018) recorded 7 unrelated biotrophic fungi missing all the genes in ASRP, Duplessis et al. (2011) recorded them missing in the rust *Puccinia graminis*, and Cisse et al. (2014) found ASRP missing in two species of *Pneumocystis*. Whether deficiencies in ASRP impact the ability of fungi to grow in axenic culture remains unknown, but the correlation between the perceived lack of cultivability of many fungi and loss of ASRP genes is now well documented.

Currently, many investigators are focused on applying well-known media that work for described fungal species when cultivating new species. In this sense, exploration of fungal media for cultivating novel species is a less active area of research, and the exploration of fungal diversity using cultivation approaches is biased toward recovering species that have already been cultivated. It is easy to demonstrate such biases by comparison of culture-independent methods, such as DNA metabarcoding or microscopy (Davis et al. 2018; Zhang et al. 2020; Hoyer and Hodkinson 2021), with culturing approaches. A major example of bias in culturing is the example from a very common lineage of soil fungi, the Archaeorhizomycetes. These fungi were not known or cultured until the twenty-first century (Schadt et al. 2003; Porter et al. 2008), and in particular, they were only cultured from a few of many thousands of isolation attempts from individual root tips (Rosling et al. 2011). Culturing fungal communities using rich media is a way to promote biases in growth rates and to favor saprobes with general growth parameters over symbionts which may have more specific nutrient limitations. Interestingly, in the case of *Archaeorhizomyces*, it is unclear whether or not the fungus is symbiotic. *Archaeorhizomyces* can readily grow, albeit at a slow rate, on a number of basic undefined media without complex additions. This example shows that if one can find these fungi and provide the correct cues, the limitation may not be the specifics of the growth medium. On the other hand, as we discuss below, very specific nutrient limitations, e.g., myristate, could be limiting factors that make fungi uncultivated (Sugiura et al. 2020). One lesson from these examples is that if these fungi that were probably considered to be the least amenable to culture can actually be cultivated, could perhaps all fungi? In the following, we will discuss some of the lessons learned from developing methods of culturing symbiotic fungi.

2.4 Major Breakthroughs Cause Reconsideration of Uncultivable Lineages

2.4.1 Big Spores, Big Needs: Arbuscular Mycorrhizal Fungi

Arbuscular mycorrhizal fungi (AMF) are obligate symbionts that cannot grow without their host plants (Schüβler and Walker 2011). It is accepted that spores of AMF possess the suitable genetic information and biosynthetic abilities to germinate once the appropriate, simple, physicochemical conditions of moisture, temperature, and pH (Barea 1986) have been provided. Over 90% germination can be obtained axenically on water agar in the absence of any mineral or organic supply (Azcón-Aguilar et al. 1986). Many hypotheses have been put forward to explain the failure of the continuous axenic growth of these fungi (Hepper 1984; Siqueira 1987; Azcón-Aguilar et al. 1991; Bonfante-Fasolo and Perotto 1992; Williams 1992). In summary, the following five hypotheses related to our previous mechanistic explanations of uncultivability have been postulated: (1) either the requirement for an unknown essential metabolite(s) or the appropriate form/amount has not yet been fulfilled; (2) the continuous synthesis of enzymes or macromolecules necessary during the early stages of the germination process is blocked; (3) nuclear DNA replication during spore germination is absent, suggesting that there is no nuclear division in vitro; (4) the fungi have lost part of their genetic equipment involved in its free-living growth during coevolution with the host plant and, therefore, the corresponding gene function must be supplied by the host in the development of biotrophic symbiosis; and (5) the intraradical feeding structure for the fungal mycelium can be crucial. This latter hypothesis would mean that only an active arbuscule would be able to absorb nutrients, possibly because of its structure and permeability properties, its enzymatic equipment for nutrient transport, or the chemical and physical environment where it is developed and positioned. It has been proposed that success in culturing AM fungi would be achieved by defining the contribution of factors supplied by the host plant to hyphal growth and supplying these factors axenically (Becard and Piche 1989).

The contribution of autotrophic higher plants to the mutualistic relationships of the arbuscular mycorrhiza (AM) appears to begin before the AMF has been established in the root cortex (Barea 1986). Compounds released from plant roots seem to affect the establishment of the association by acting on some pre-infection stages. Obviously, root exudation plays an important role in rhizosphere biology. Roots synthesize many organic compounds, some of which are released into the surrounding soil (Bowen 1980; Curl and Truelove 1986). The compounds released by roots can be classified according to their mobility in the soils as follows: (1) diffusible, water-soluble; (2) diffusible-volatile; and (3) non-diffusible compounds. Water-soluble exudates exert their influence at the root surface and spread only a few millimeters into the soil, whereas volatile exudates act further away from the root surface, often several centimeters into the soil (Koske and Gemma 1992). Becard and Piche (1989) found that the presence of a growing root significantly stimulated the growth of the fungus even when there was no root-fungal contact. They suggested that the root may have exerted its influence by changing the composition of the agar medium or the gaseous phase in the dual culture system. However, since active fungal growth ceased with root removal, it appears that either the relevant compounds in root exudates were unstable or they did not accumulate in sufficient quantities. Gianinazzi-Pearson et al. (1989) demonstrated that two flavanones (hesperetin and naringenin) and a flavone (apigenin), which activate *nod* gene expression in *Rhizobium*, elicited mycelial growth responses from AMF spores. They, therefore, suggested that the presence of such molecules in root exudates may be involved in cell-to-cell contact between the mycorrhizal partners in the early pre-infection phases of the recognition phenomena between symbionts. Since root exudates are known to stimulate microbial activity in the rhizosphere, they may exert an indirect activity on AMF due to the associated microorganisms affecting spore germination (Barea 1986; Linderman 1992). In an early study, Mosse (1959) found that spores of AMF that showed a low level of germination on water agar could be stimulated to germinate in a non-sterile, soil-water agar. She suggested that a water-soluble, heat-labile, dialyzable product of microbial origin was responsible for the enhancement of germination.

Using clues from the biochemistry and morphology of AM symbioses helped uncover how they may be cultured axenically. Extraordinarily large spores of AM fungi hinted that these species have lost independent foraging for sugar acquisition because germinating spores must depend on their own energy sources before a mycorrhizal symbiosis can be established. AM fungi are also known to accumulate lipids that comprise 46–70% of the spore weight (Beilby and Kidby 1980). These high lipid contents are likely stored for germination and initial growth before successful infection. Observations of fatty acid importation from host plants (Jiang et al. 2017; Luginbuehl et al. 2017) and mycorrhizal defects caused by lipid-related plant mutants (Bravo et al. 2017; Keymer et al. 2017) support the hypothesis that AM fungi rely on their hosts for fatty acids. In a breakthrough paper, Sugiura et al. (2020) determined addition of the fatty acid myristate (C14:0) led to an increase in the biomass of *Rhizophagus irregularis*, inducing extensive hyphal growth and formation of infection-competent secondary spores (albeit smaller) in asymbiotic cultures. Furthermore, it has been demonstrated that *R. irregularis* can take up fatty acids in its branched hyphae and use myristate as a carbon and energy source, and myristate also promoted the growth of *Rhizophagus clarus* and *Gigaspora margarita* (Sugiura et al. 2020). It has been reported that the AM fungus *Rhizophagus intraradices* Sy167 can grow and develop fertile spores independently of a host plant in coculture with the bacterium *Paenibacillus Validus* (Hildebrandt et al. 2006). Kameoka et al. (2019) found that (*S*)-12-methyltetradecanoic acid, a methyl branched-chain fatty acid isolated from *Paenibacillus Validus* cultures, stimulates the branching of hyphae germinated from mother spores and the formation of secondary spores in axenic culture of the AM fungus *Rhizophagus irregularis*. Their extensive testing of fatty acids in *R. irregularis* revealed that palmitoleic acid induces more secondary spores than the bacterial fatty acid.

Although AM fungi accumulate lipids, many ECM and saprobic fungi store carbohydrates such as glycogen, trehalose, and mannitol as their carbon sources (Martin et al. 1987; François and Parrou 2001; Nehls 2008). AM fungi lack almost all the genes encoding polysaccharide hydrolases that generate glucose except trehalase and glycogen phosphorylase, whereas most of these polysaccharide hydrolase genes are present in other phyto-biotrophic fungi (Kobayashi et al. 2018). Thus, AM fungi are hypothesized to have completely abolished glucose acquisition except for what is imported from their host plants. This absolute sugar host dependency of AM fungi may be reflective of its long symbiotic history.

The growth of AM fungi is hypothesized to depend on many other compounds in addition to lipids and carbohydrates. The absence of several metabolic pathways associated with the biosynthesis of vitamin B6 derivatives and thiamine synthesis has been confirmed (Kobayashi et al. 2018). The missing thiamine synthetic pathway suggests that AM fungi may have a non-thiamine-dependent infection system. Since these vitamins are essential for events other than infection, the addition of these compounds to media may improve the culture efficiency of AM fungi.

2.4.2 Malassezia: Another Lipid Requiring Fungus

Malassezia species are lipid-dependent and fastidious basidiomycetous yeasts that inhabit the skin and mucosa of humans and other warm-blooded animals and are a major component of the skin microbiome. They occur as skin commensals but are also associated with various skin disorders and bloodstream infections. Under certain circumstances, *Malassezia* species can be involved in relatively common skin diseases, including pityriasis (tinea) versicolor, seborrheic dermatitis, and folliculitis. Their difficulty to culture axenically has long hampered studies of their systematics and diversity (Guého et al. 1996). *Malassezia* species are not able to grow on the mycological media routinely used in clinical laboratories because of their lipid dependency. It is therefore necessary to use specific media such as Dixon agar or Leeming-Notman agar containing lipids. *M. pachydermatis* is a non-lipid-dependent lipophilic *Malassezia* species, which benefits from lipid addition to its medium. This species differs from all other *Malassezia* yeasts by the fact that lipid supplementation is not an absolute requirement and growth does not depend on the presence of long-chain fatty acid precursors in the medium. Human skin is frequently colonized by lipid-dependent *Malassezia* yeasts, but seldom by *M. pachydermatis*. For cultivation of *Malassezia* species most investigators employed media supplemented with olive oil, Tween 20, Tween 80, or glycerol monostearate (Ushijima et al. 1981). Furthermore, in almost all cases of isolation of *Malassezia*, agar plates or slants were overlaid with olive oil (Gordon 1951). It was shown that the pathogenic yeast *M. furfur* (formerly *Pityrosporum orbiculare* and *Pityrosporum ovale*) grows in vitro only if fatty acids from C12 to C24 series are added to the culture medium. Except for elaidinic and nervonic acids, all saturated and unsaturated fatty acids tested support growth (Porro et al. 1976). Until recently, it was assumed that *Malassezia* evolved into a specialized and narrow niche associated with the skin of mammalian hosts. However, culture-independent studies of fungi from environmental samples show that *Malassezia* are exceedingly widespread and ecologically diverse (Amend 2014). Recent studies in little-characterized marine environments point to extensive diversification of *Malassezia*-like organisms, providing exciting opportunities to explore the ecology, evolution, and diversity of this enigmatic group.

2.4.3 Uredinales: Plant Pathogens Requiring Protein

Members of the Uredinales have usually been considered classical obligate parasites among fungal plant pathogens. However, there have been successful attempts to germinate, grow, and maintain rust species. Attempts to culture rusts saprobically have been based on two methods: firstly, the initiation of saprobic growth from germinating uredospores and, secondly, saprobic growth from mycelium formed in infected host callus tissue. The original idea was to form callus tissue of a susceptible host with its rust fungus by aseptic transfer to culture media. Hotson and Cutter (1951) and Cutter (1959) reported saprobic axenic culture from infected callus tissue of three rust species. Because of the failure of others to repeat Cutter's experiments, his work has not received the recognition it deserves.

The greatest amount of work on the reaction of rust spores to nutrient media has been carried out with uredospores. Among species of major interest, they are the most prolific and uniformly produced spores and may be readily obtained by reinfection of plants of the same host species. While relatively little work has been done with the other spore forms, an overview of some of this work is given, since these forms may serve as starting materials for future culture experiments. The first successful axenic culture of rust from uredospores was reported by Williams et al. (1966, 1967) on an Australian isolate of wheat stem rust. It was observed that septate vegetative mycelium was formed from uredospores of *P. graminis* var. *tritici* race 126-ANZ-6,7 incubated on an agar medium containing Czapek's minerals, 0.1% yeast extract, and 3% sucrose. In tracing the development of this mycelium, it was observed that germination and germ tube elongation had occurred after 2 days. Initial branching of germ tubes was observed 4 days after seeding, and this branching continued for 3–4 weeks. In subsequent work (Williams et al. 1967), it was observed that the addition of 0.1% Evan's peptone to the medium enhanced vegetative growth, and some cultures formed stromata bearing uredospores and teliospores 2–4 weeks after inoculation. Bushnell (1968) has subsequently repeated this work, using peptone medium containing glucose and the same Australian isolate of wheat stem rust. He noted that yeast extract medium without the addition of peptone did not support growth. It has been observed that the replacement of sucrose with glucose and the addition of citrate to the yeast extract-peptone medium resulted in more consistent growth (Scott and Maclean 1969). Axenic cultures of other cereal rusts have since been attempted using the procedures described above. It is noted that Bushnell and Stewart (cited in Scott and Maclean 1969) tested 16 North American races of *P. graminis tritici* and obtained vegetative growth from 12 of these races. Most of these cultures survived initial subculturing, but development from subsequent subcultures was erratic. Macroscopic colonies, but no sporulation, of three other Australian races of *P. graminis tritici* and limited post-sporeling growth of *P. graminis secalis* and *P. recondita tritici* were obtained (Scott and Maclean 1969). Singleton and Young (1968) reported vegetative growth of an American isolate of *P. recondita*, using the yeast extract-peptone medium. The dependence on peptone, a digest of protein high in amino acids, suggests that the rusts may be deficient in production of specific amino acids that they may scavenge from their hosts.

Teliospores and basidiospores—Work on the response of teliospores to nutrients, chemicals, moisture, temperature, etc., has been mainly concerned with overcoming dormancy (Arthur et al. 1929; Carleton 1903; Kucharek and Young 1968; Thiel and Weiss 1920). If growth were to be obtained from teliospores presumably the promycelium would form basidiospores which would germinate and grow, or else the promycelium itself would initiate saprobic growth directly. It has been reported (Reed and Crabill 1915) that germinating teliospores of *Gymnosporangium juniperi-virginianae* formed promycelia which occasionally produced hyphae without basidiospore formation. Bailey (1923) mentioned a frequent "indefinite growth" and branching of promycelia of *Puccinia helianthi*. Other workers (Colley 1918; Spaulding 1922) described the formation of a narrow-twisted hypha without basidiospore production when teliospores of *Cronartium ribicola* were germinated underwater. In light of these observations, promycelia may be useful material for studies on the saprobic growth of some rusts.

Aeciospores—Work on these spores has been limited to studies of germination in distilled water or host decoctions (Bailey 1923; Carleton 1903; Klebahn 1904; Reed and Crabill 1915; Spaulding

1922). These spores are often difficult to obtain in abundance.

Spermatia—Although many workers failed in attempts to germinate spermatia (De Bary 1887; Reed and Crabill 1915), Plowright (1889) observed limited yeast-like budding when spermatia of several rusts were placed on nutrient solutions containing sucrose or honey. Other workers (Brefeld 1981; Cornu 1876) observed similar phenomena. Carleton (1903) obtained germ tubes rather than budding, and their formation was stimulated by sucrose or honey. Spaulding (1922) cited the unpublished work of York and Overholts, who investigated the germination of *C. ribicola* spermatia on various nutrient solutions. It was noted that preincubation in the cold for up to 18 days was a necessary prerequisite for germination.

2.5 Use of Genomic Data to Inform Cultivation Strategies

2.5.1 Examples from Bacteria

Microbial diversity is unfathomably deep and getting deeper. Single-cell genomes and metagenomes continue to increase the known diversity of Bacteria and Archaea; however, while 'omics can be used to infer physiological or ecological roles for species in a community, most of these hypothetical roles remain unvalidated. Genome-based metabolic reconstructions might provide insights that underpin the cultivation of uncultured bacteria and archaea, but so far this promise has only been realized for a few species of microbes.

The availability of metagenomics and metabarcoding means we know of the existence of certain taxa that have never been cultivated (Jones et al. 2011; Tedersoo et al. 2017), and one aim is to cultivate those taxa from mixed samples using our ever-improving understanding of translating genome into cellular phenotype. A recently developed approach uses information found in single-cell genomics and metagenomics sequences to design antibodies capable of selectively isolating bacteria or archaea (Cross et al. 2019). The validity of this approach was demonstrated by culturing three diverse species of TM7/*Saccharibacteria* along with their interacting *Actinobacteria* hosts, including one previously uncultured *Actinomyces* species. This method, dubbed "reverse genomics," exploits culture-independent sequence data for physical retrieval of specific microorganisms via antibody affinity and does not depend on any physiological selection criteria. Therefore, it should be broadly applicable to any target organisms, as long as robust cultivation conditions can be identified for isolated cells. Antibody-isolated cells could also be funneled into other platforms that enable high-throughput media selection and optimization protocols, for example, microdroplets or ichips (Nichols et al. 2010; Berdy et al. 2017; Zengler et al. 2002).

Various approaches are used by microbiologists to provide favorable conditions for bacteria to overcome limiting factors. In the following, we present three additional examples of several successful experiments for cultivating fastidious bacteria:

It has been reported that He et al. (2015) have been able to cultivate strain TM7x from the human oral cavity, which is the first taxon of the candidate phylum Saccharibacteria (TM7). This was accomplished by plating saliva samples on a specialized oral culture medium (SHI medium) and targeting the enrichment of TM7 taxa by streptomycin selection. The successful enrichment of strain TM7x in streptomycin-added media might be explained by its association with the streptomycin-resistant *Actinomyces odontolyticus* XH001 species (Bor et al. 2016). Furthermore, a signaling interaction between the two bacterial species has been confirmed by transcriptomic and metabolomic analyses. With an extremely small cell size and reduced genome of 705 kb lacking the capacity to synthesize any amino acids, strain TM7x represents a classic example of auxotrophy.

Coxiella burnetii is the causative agent of Q fever, a worldwide zoonosis that causes polymorphic clinical manifestations such as flu-like syndromes, pneumonia, endocarditis, meningitis, and osteomyelitis (Raoult et al. 1990). This microorganism has previously been isolated from infected animal tissues and a tick in Montana (Cox 1938a). Subsequently, this intracellular bacterium was successfully cultivated in embryonated eggs, mice, and guinea pigs (Cox 1938b). *Coxiella burnetii* could also be propagated in HEL cells using a shell-vial assay (Raoult et al. 1990), which is currently the method used for its isolation from clinical samples (Gouriet et al. 2005). Recently, an axenic medium has been described for *C. Burnetii* (Omsland et al. 2009, 2011). This medium was developed in three steps. First, metabolic pathway deficiencies were identified in *C. burnetii* using *in silico* genomic analysis, and a medium was found that sustained the metabolic activity of this organism, as measured through [35S] Cys-Met incorporation (Omsland et al. 2008). In the second step, the transcript profiles and phenotype microarrays were analyzed, leading to the formulation of ACCM (acidified citrate cysteine medium) (Omsland et al. 2009). This medium has an acidic citrate buffer pH, is incubated in a 2.5% O_2 atmosphere, and is enriched with cysteine, neopeptone, and casamino acids. Optimization through the addition of methyl-β-cyclodextrin has led to the formulation of ACCM 2 (Omsland et al. 2011), a medium that has allowed significant progress in understanding *C. burnetii* pathogenesis.

An effective and straightforward strategy for stimulating the growth of uncultivated bacteria is the supplementation of media with chemical compounds that are likely to be required by the bacteria in question. D'Onofrio et al. (2010) determined the identity of growth factors produced by "helper" strains that were responsible for the growth of dependent uncultivated organisms from marine sediment and were able to stimulate the growth of the latter by supplementing media with the growth factors, namely iron-scavenging siderophores. Several studies have demonstrated the growth-promoting effect of siderophores on uncultivated and "difficult-to-culture" bacteria (Guan and Kamino 2001; Vartoukian et al. 2016). It has been suggested that the ability to autonomously produce siderophores may have been lost in uncultivated bacteria (D'Onofrio et al. 2010; Lewis et al. 2010). Other growth-stimulatory factors include the quorum-sensing molecules acyl-homoserine lactones (AHL), which have been shown to facilitate the growth of an AHL-degrading novel proteobacterium (Chan et al. 2009), and trace elements at levels matching those found in the source environment (Meyer-Dombard et al. 2012).

2.5.2 Examples from Fungi

Taking inspiration from genomics-based work in bacterial systems (Omsland et al. 2013), Ahrendt et al. (2018) used information on the metabolic deficiencies inferred from the first fungal single-cell genome sequencing to try to axenically culture their target mycoparasitic species. The three target species were all obligate parasites which could only be grown in coculture with their fungal hosts (Benny et al. 2016). Functional analysis of the genomes indicated certain commonalities among the uncultivable targets of the single-cell study. Specifically, these fungi lack the gene or genes involved in production of thiamine, spermidine, and biotin. Using media amended with three amino acids (cysteine, methionine, and tryptophan), biotin, spermidine, thiamine, and magnesium sulfate, two of the three species were able to grow axenically for the first time (Ahrendt et al. 2018). One mycoparasite was still unable to grow successfully. While the authors note that even those species which would grow axenically would still not sporulate in culture, this study represents a tale of modest success in using genome-derived metabolic deficiencies as the basis for media creation and new avenues for biotrophic fungal cultivability.

Marine *Malassezia* species represent an interesting case where the fungi have never been cultivated, yet there are apparently distinct taxa that inhabit these ecosystems. One approach to cultivating these fungi would be to obtain a genome sequence first (Fig. 2.1). By comparison of the marine metagenomes to genomes of terrestrial species which have been successfully cultivated, corresponding nutrient deficiencies

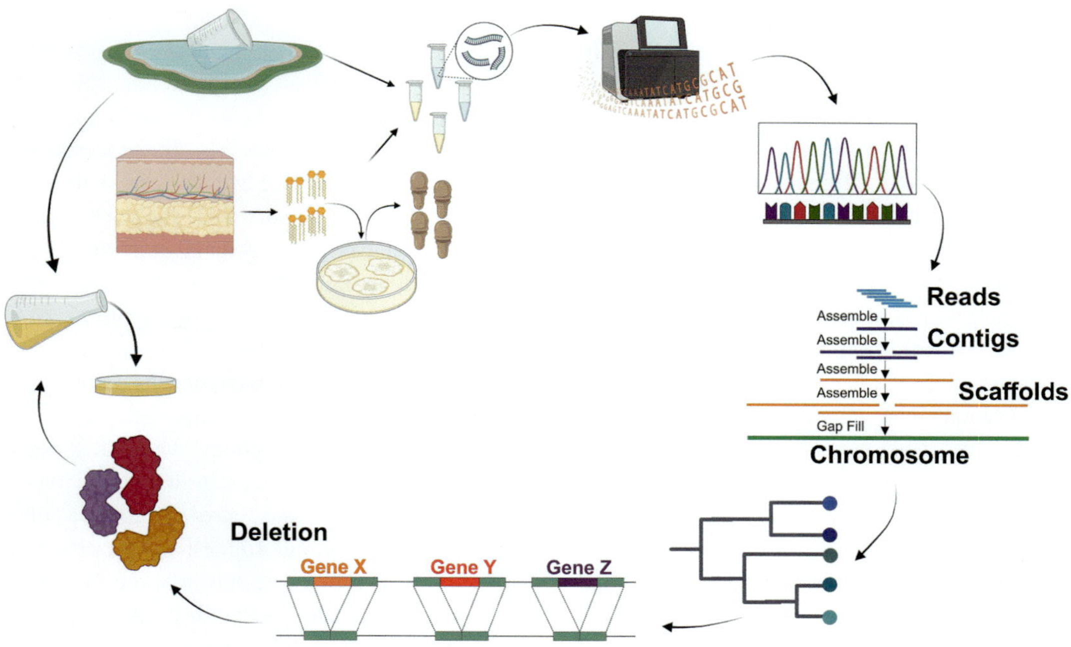

Fig. 2.1 A schematic overview of how genomic data can be used to identify nutritional constraints in marine *Malassezia* species. Sampling aquatic environments and skins from mammals, *Malassezia* species associated with the skin require lipids to grow, extracting DNA from both samples, (meta)genome sequencing, assembling the sequence data, identifying *Malassezia* contigs from aquatic samples, and assessing its phylogenetic affinity to skin-associated *Malassezia*, comparing genomic content of aquatic and skin-associated *Malassezia*, culturing aquatic *Malassezia* by adding protein requirements to media

can be investigated and added to the media as has worked for bacterial species (Tyson et al. 2005).

The Elusive Axenic Culture of *Coelomomyces*, a Biotrophic Pathogen of Microarthropods
Coelomomyces (Blastocladiomycota) has received much attention due to its role as a primary pathogen of dipteran larvae. Because of the importance of mosquitoes as vectors of human disease, there has been considerable interest in using *Coelomomyces* for biocontrol, with a few reports of modestly successful application (Scholte et al. 2004). *Coelomomyces* is generally resistant to culture, which has greatly hindered its development as a biocontrol agent, though it may be readily cultured on its hosts (Bland 1985). A major breakthrough in the persistent cultivation of *Coelomomyces* on its hosts was accomplished with the discovery by Whisler et al. (1975) that *Coelomomyces* has a complex life cycle with alternation of generations between a diploid stage infecting mosquito larvae and a haploid stage infecting crustaceans, especially copepods. Reliable cultures of *Coelomomyces* can be maintained by growing the mosquito larvae in the presence of copepods such as *Cyclops* (Bland 1985). The discovery of the copepod-infecting phase helped to explain why infection of mosquito larvae by field-collected diploid sporangia was unpredictable and reliant on addition of non-sterile substrata.

Limited development in axenic culture has been reported using a number of

(continued)

complex media (Shapiro and Roberts 1976; Castillo and Roberts 1980). For example, Shapiro and Roberts (1976) were able to grow *Coelomomyces* axenically for 2 years in a modified medium for mycoplasma. Importantly, *Coelomomyces* development in axenic culture has not yielded mature sporangia capable of dehiscence which would be necessary for deployment as a biocontrol. Castillo and Roberts (1980) extensively studied the effects of various vertebrate and invertebrate culture media on the development of *Coelomomyces punctatus*. These media were only effective at culturing hyphae from the sporophyte stage and not the gametophyte, though gametes were released in at least one medium condition. The media trials performed by Castillo and Roberts shed some light on metabolic dependencies of *Coelomomyces*. First, conditioning of the media by the appropriate host cell line (*Anopheles stephensi*) could stimulate development of *Coelomomyces*, but conditioning with non-hosts (*A. gambiae* and *A. aegypti*) could not. Second, the addition of the tripeptide glycyl-L-histidyl-L-lysine improved *Coelomomyces* growth in some media, indicative of amino acid dependencies. And third, brain heart infusion (BHI) was one of the better media for cultivation. BHI is known to be rich in lipids, fatty acids, sterols, and vitamins. Importantly, *Coelomomyces* is known to be associated with the fat bodies of the host and is assumed to consume them (Umphlett 1962; Couch 1972).

Given these observations, it is likely that *Coelomomyces* has complex metabolic dependencies on its host, such as fatty acids, vitamins, and amino acids. Very recently, the draft genomes of both gametophyte and sporophyte stages have been produced (Ettinger et al. 2023). The genome is somewhat reduced, with 20–23 Mb and 7416 protein-coding genes. We searched the draft genome for evidence of loss of primary metabolism genes that may suggest dependencies on the hosts for supply. Contrary to our expectations, these searches revealed the presence of fatty acid synthase homologs FAS1 and FAS2, suggesting the ability of *Coelomomyces* to perform fatty acid elongation. However, the genome lacked the genes necessary for synthesis of the vitamins thiamine and biotin and is missing many of the biosynthesis pathways for amino acids, such as histidine, tryptophan, lysine, and arginine. Overall, these data are generally consistent with the previous data on in vitro culturing of *Coelomomyces* and suggest this approach could help with the culturing of the fungus. A major challenge will still be developing methods to stimulate production of resting sporangia or gametes which will be critical for field deployment of a biocontrol.

2.6 The Evolution of Metabolic Constraints in Symbiosis is Shaped by the Host, Lifestyle, and Infection Strategy

Given that symbiotic fungi derive much of their nutrients from the host, it would stand to argue that their primary metabolism should be greatly affected and predicted by the identity of the host as well as the nature of the interaction. Obviously, not all symbionts are the same, and both mutualists and pathogens spend variable amounts of energy building cells separate from their host. Even if a parasite is primarily external to its host, it may yet be completely reliant on the host via absorption through haustoria. Laboulbeniales, for example, are an uncultured group of ectoparasites on arthropods with many haustorial species but with unclear metabolic dependencies on their hosts (Haelewaters et al. 2022). Labouls are notorious for their high host specificity, meaning they may experience a highly predictable environment which could facilitate loss of production of certain metabolites that could be sourced from the host (Benjamin and Shanor 1952). Genome sequencing is needed to address this question.

The parasite-mutualist continuum imposes variable demands on the host for nutrition (Drew et al. 2021). Commensal species like *Malassezia* and Laboulbeniales may maintain high metabolic flexibility as they source diverse nutrients from their hosts. Models of mutualism typically posit movement of sanctioned amounts of specific resources between host and symbiont (Yoder and Tiffin 2017), whereas pathogens, and particularly necrotrophic ones, might be considered less selective in the resources they obtain from their host. Where the host itself is a free-living organism such as a plant, consideration of access to the host cell's nutrients predicts that necrotrophic and hemibiotrophic pathogens would have the greatest access to resources, biotrophic pathogens would have less, and mutualists would have the least. Because of this, mutualists should be the easiest to culture because they are more dependent on resources from their hosts and are required to be able to synthesize a diversity of primary metabolites, and this suggests that mycorrhizae, lichenized fungi, and endophytes will be culturable. Endobiotic mutualists, such as the highly obligate transovarian bacterial endosymbionts, go against this prediction (McCutcheon and Moran 2012). However, there may be a fundamental difference between symbionts that are confined to vertical transmission within a host, versus those that have shown horizontal transmission and require finding new partners (Law and Lewis 1983). This should also be reflected in the transporter protein families that should reflect the need to grow apart from the host, even if only momentarily. For example, the horizontally transferred mycorrhizae should be expected to have a diverse set of transporter genes for supporting growth in the extraradical mycelium. Lichens that reproduce with ascospores should similarly require a greater deal of transporter proteins than those that always co-disperse with their algal partners. These predictions are testable by genome sequencing and annotation as they should be reflected in genome sizes, primary metabolism capabilities, and transporter proteins.

How predictive can the type of host be in patterning fungal symbiont primary metabolism, and can comparative genomics come to the aid of this question? Specifically, when looking at a new genome from, for example, an uncultured fungus known only from metagenomic sequencing, can we perform reverse ecology to determine what type of host it may associate with? Over the last two decades, comparative genomics of fungi has sought the answer to this question by comparing genomes of species that have different ecological niches (Stauber et al. 2020; Wu and Cox 2021; Romero-Olivares et al. 2021; Wang et al. 2022). There are clear indications that animal pathogens and plant pathogens diverge, primarily in their uses of proteinases and carbohydrate-active enzymes that may be at the frontline of extracting energy from their hosts (Wu and Cox 2021). However, the basic building blocks of cells are the same across the eukaryotes, with similar membrane systems, nucleic acids, and proteins. These similarities allow fungi and other pathogens to switch quickly between hosts of different kingdoms (Quandt et al. 2015). On the other hand, pathogens of plants which lack essential vitamins and amino acids might be expected to more often evolve losses of primary metabolism genes more often than animals, whereas animal parasites may need to maintain them as the host must acquire them from the environment. For example, the essential amino acids for mammals, histidine, isoleucine, leucine, lysine, methionine, phenylalanine, threonine, tryptophan, and valine, might be expected to be maintained by their pathogens, while the other amino acids could potentially be taken from the host (Derilus et al. 2021). Other predictions based on the host are that mycoparasites may be most likely to lose primary metabolism relative to animal pathogens, which in turn could happen more than plant pathogens. This logic stems from phylogenetic relatedness such that compatibility of biochemical intermediates and even building blocks is likely increased as parasite and host become more phylogenetically closely related.

What do the genomic data suggest in light of these predictions? We highlight trends by exploring primary metabolism in 8 symbiotic fungi with the results summarized in Fig. 2.2. Firstly, genome size is variable across species and not a great indicator of whether primary metabolism genes will be lost. Most fungi maintain the ability to synthesize most amino acids, with the exception of the water mold endoparasite *Rozella allomycis* and the lung pathogen *Pneumocystis*. Losses were not offset by the presence of additional amino acid permeases, indicating that parasitic fungi are likely using other pathways for scavenging amino acids, perhaps via peptidases. Most fungi, however, retained the ability for uptake of ammonium, which could be used for building nitrogenous molecules, including amino acids. The ability to synthesize the vitamins biotin and thiamine and the ability to make long-chain fatty acids were characteristics readily lost in the obligate symbiotic fungi, with the exception of *Puccinia graminis*. The frequent loss of fatty acid synthase in many symbiotic fungi is curious given the diverse roles of these molecules in the cell (Ciznár and Hostacká 1983); it may be that these molecules are taken from the host as diverse sources of energy through the fatty acid degradation pathway associated with production of acetyl-CoA. Additional analyses of amino acid, fatty acid, and other primary metabolite biosynthesis pathways are warranted to test some of the predictions we have outlined.

2.7 Novel Experimental Systems for Cultivation of Fastidious Microbes

The causes of microbial "uncultivability" are numerous and have been associated with a requirement for factors produced by other microbes, strict interspecies interactions, slow growth, competition/inhibition, and dormancy. Various technical approaches have been developed to alleviate cultivation recalcitrance (many of which were reviewed recently in Rämä and Quandt 2021 and Lewis et al. 2021). Encapsulation and incubation in microdroplets under media flow, for example, retain and allow investigation of interspecies signaling and potential cross-feeding. By integrating multiple media and cultivation conditions, this approach has been successful in expanding the repertoire of cultured species and strains from open environments and human-associated microbiota. Restricting potential competition by stochastic dilutions of complex microbial samples to a few or single cells, isolated in micro-compartments, while exposing them to native environmental conditions through porous membranes (Ichip and similar devices), has also resulted in pure cultures of microbes that were refractory to isolation using standard techniques (Fig. 2.3).

To address each limiting factor, several innovative methods have been developed. Kaeberlein et al. (2002) were among the first to propose the "simulated natural environment" concept. They designed diffusion chambers within which organisms were seeded after extraction from marine sediment. The chambers were incubated under conditions mimicking the natural environment, allowing the passage of growth-stimulatory chemical factors from the external environment across semi-permeable (0.03 μm-pore) membrane walls of the chambers and resulting in the growth, and ultimately pure culture, of previously uncultivated bacteria from the marine environment (Fig. 2.3d). Bacteria of interest may also be exposed to signals or nutrients from the natural habitat, or engineered versions of the latter, after encapsulation of single cells or subsets of the microbial community as microdroplets using microfluidic devices. Zengler et al. (2002) encapsulated single bacterial cells in gel microdroplets, incubated them in amended sterile versions of the natural environmental medium, detected growth as microcolonies in the microdroplets using flow cytometry, and successfully isolated and cultured several novel bacterial strains following transfer to microtiter plates. The process of diluting mixed samples down to single cells prior to attempted cultivation, termed "dilution to extinction," eliminates the potential drawback of competition between bacteria in the community and results ideally in pure cultures (Connon and Giovannoni 2002). Park et al.

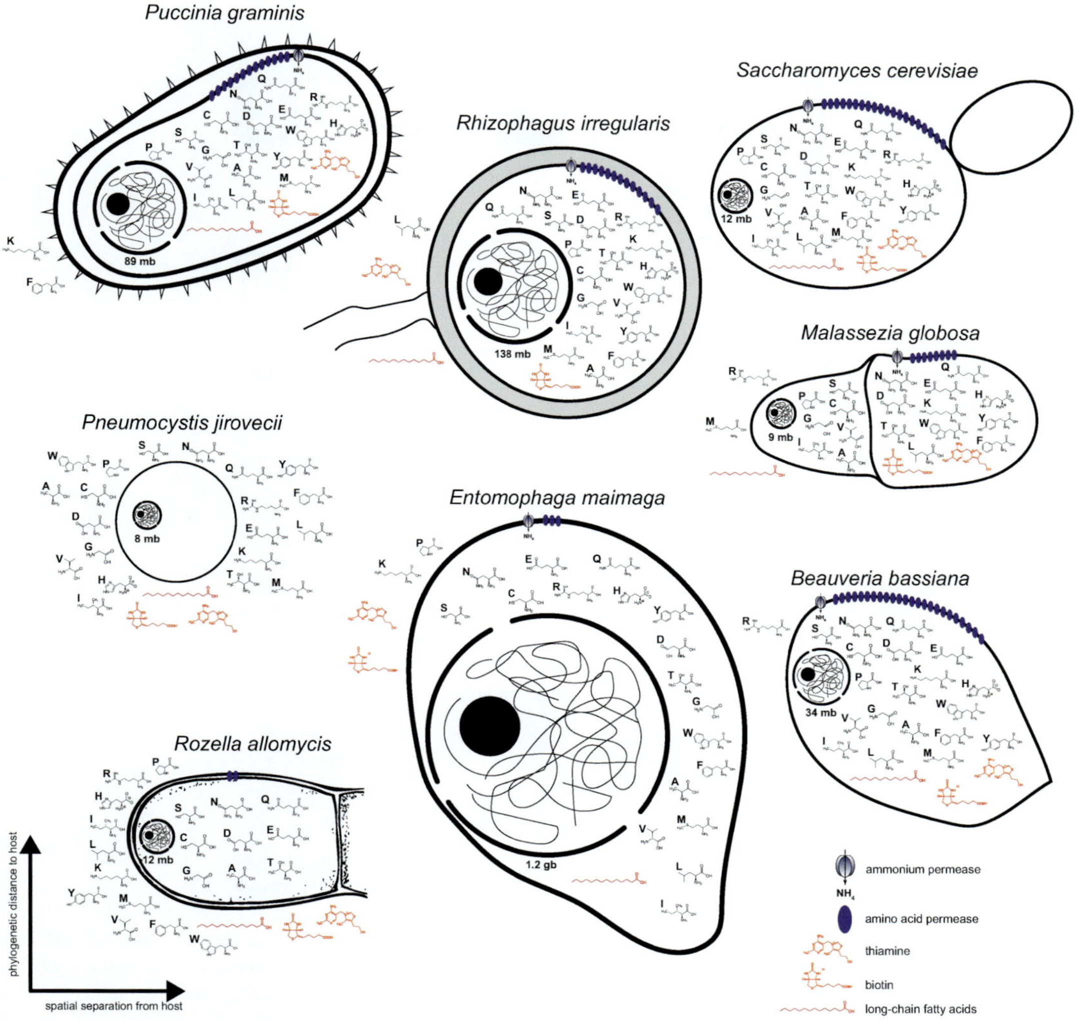

Fig. 2.2 Simplified diagram of several metabolic pathways and transport of nitrogen in fungal species of varying host dependency and phylogenetic distance from host. *Saccharomyces cerevisiae* is shown as a saprobic example. Data were based on genomic analyses using predicted proteins of each species as found on the Joint Genome Institute's MycoCosm web page, accessed on 03/26/2023 (Ahrendt et al. 2022). Production of amino acids was assessed using a BlastKOALA annotation using settings "Fungi" and "family_eukaryotes" (Kanehisa et al. 2016) with pathways reconstructed in KEGG Mapper (Kanehisa and Sato 2020), with the exception of amino acids in *Pneumocystis*, which was obtained from Ma et al. (2016). Synthesis of fatty acids was assessed by BLAST search with *Saccharomyces cerevisiae* FAS1 and FAS2 proteins. Presence of an ammonium permease was assessed with BLAST using *S. cerevisiae* MEP1. Amino acid permease numbers are taken from MycoCosm annotation pages under orthologous gene clusters (KOGs) 1286 and 1287. Thiamine and biotin synthesis capacity was assessed by BLAST searching for THI4 and THI6 and BIO2 and BIO4, respectively, using *S. cerevisiae* as the source

(2011) co-cultured pairs of dependent organisms within oil-coated microdroplets; since the two organisms in each symbiotic pair were auxotrophic for a different amino acid, growth in the minimal medium was achieved only by co-cultivation, indicative of cross-feeding.

The filamentous nature of most fungi complicates the methods described above, which

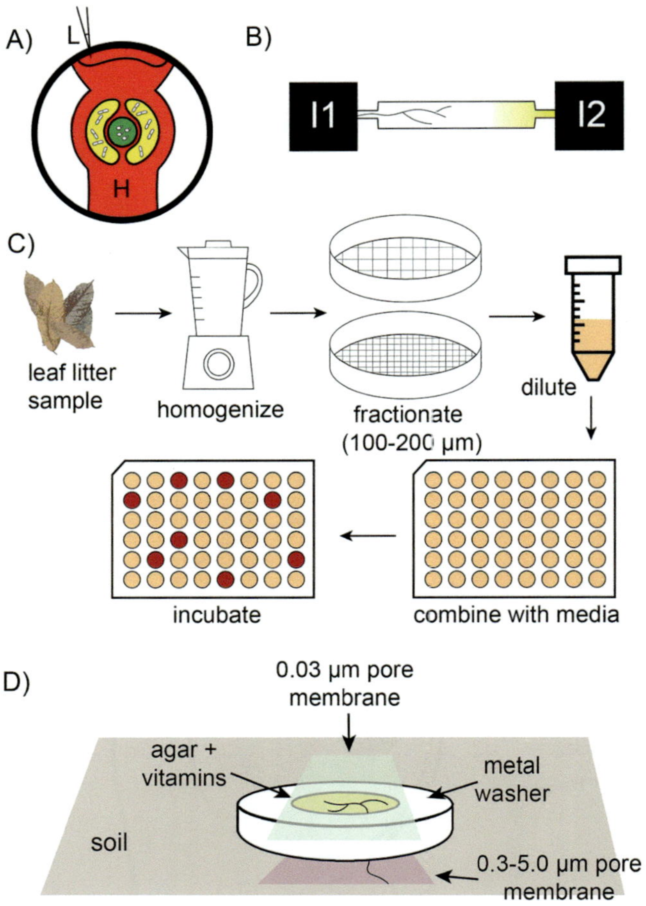

Fig. 2.3 Technological advances in isolating and observing fungi show promise for a better understanding of the nutrient requirements, cultivability, growth kinetics, and interspecies interactions. (**a**) Microscale observation is enabling investigation of the growth of fungi at scale. The Monorail Device (Berry et al. 2017) can be used to establish segregated coculture regions separated by a hydrogel (H) formed in a culture dish, such as a microwell plate. The hydrogel is loaded into a port (L) and forms a segregated set of chambers. The chambers can be loaded with one or more species or nutrient solutions, here depicted as yeast species. Because the Monorail Device can be used to set up many such chambers that fit into a multiwell plate, the whole setup can be replicated allowing numerous nutrient solutions to be examined while monitoring growth via an inverted microscope. (**b**). The growth rate and branching of hyphae can be observed within a microfluidic device that can be observed under a microscope (Uehling et al. 2019). A thin tube can be manufactured in parallel arrays for replication with two inoculation ports (I1 & I2), but one port could be used to load metabolites. Not drawn to scale. (**c**) Dilution to extinction method was successfully used to isolate 73 species of fungi from leaf litter. Homogenized leaf samples are size selected and diluted, and added to media such that less than one colony-forming unit will grow per well. (**d**) *In situ* cultivation has been successful for some microbes using cultivation chambers placed into the environment but maintaining contact. In this example (Gavrish et al. 2008), a metal washer with carbon-free agar is sandwiched between two membranes, one which allows the ingrowth of bacteria and could be adapted for fungi

have been applied to bacteria as highlighted by Rämä and Quandt (2021). Flow cytometry, microdroplet-based, and microfluidic approaches can be used to isolate individual fungal cells and study their physiology, growth, and interactions with other organisms (Delgado-Ramos et al. 2014; Uehling et al. 2019; Samlali et al. 2022; Richter et al. 2022). The major advantages of these methods are the ability to visualize the growth of cells using a stereo- or compound microscope, the ability to sterilize entire systems depending on the material, and the small scale allowing small amounts of experimental chemicals to be used. Microfluidic and microscale approaches show promise for growth of fastidious filamentous fungi and have been used to study the response of hyphae to chemical stimuli and interactions with other species observed using a microscope (Richter et al. 2022) (Fig. 2.3a, b). However, these approaches are not yet amenable to isolating fungal cells from complex environments to use for culturing, and other approaches are necessary. Dilution to extinction uses the idea that the major obstacle for culturing is competition between fastidious species and more rapidly growing ones and has been used successfully to cultivate a greater diversity of leaf litter fungi by isolating them into wells with <1 colony-forming units per well (Fig. 2.3c) (Collado et al. 2007). Such an approach can be scaled up using equipment such as the Prospector (Isolation Bio) that leverages nanoscale chambers to grow microbes in very small volumes (Persyn et al. 2022). The power is in the scale; 6000 wells can be automatically screened using fluorescent reporters for growth. A final strategy that has been successful is to bring instrumentation into the field to conduct in situ baiting or trapping of bacteria from soil or marine habitats (Gavrish et al. 2008; Berdy et al. 2017), where ingrowth of microbes occurs while maintaining exposure to the environment separated by a finely pored <0.03 μm membrane (Fig. 2.3d). Some efforts to implement this were accomplished for soil fungi in which chips are buried and colonized by fungi and other organisms in the soil matrix (Mafla-Endara et al. 2021), and this method shows promise for the development of a strategy to target filamentous fungi.

2.8 Conclusions

The concept of metabolic dependencies of symbionts on their hosts is well couched in examples but little in the way of theory. In this chapter, we have attempted to highlight some of the expectations and predictions that could be extracted from logical extension of how symbioses work. For mutualists, the reciprocal exchange of nutrients can be expected. While this might satisfy energy or mineral requirements, it does not generally lead to the expectation of wholesale support of primary metabolism of one mutualist at the expense of another, except for the case of ancient vertically transmitted endosymbionts. On the other hand, parasites are not constrained by the need for fair exchange. Yet, the evidence points to the fact that they by and large need to retain many of their primary biosynthesis. Evidence that even the rusts can be cultivated axenically shows that most biotrophic, obligate fungal pathogens may retain their primary metabolic capacity. The more telling parameter for determining loss of primary metabolism is environmental constancy, and this may be quite different for heteroecious rusts, for example, in comparison with the mammalian lung parasite *Pneumocystis*. AMF have complex life cycles that require exploration and proliferation in the soil matrix. For this reason, they have likely maintained the ability to produce most of their metabolites. Following breakthroughs in culturing major symbiotic groups, nutritional dependencies do not seem to be the primary obstacle. Nonetheless, many fungi remain uncultured. In part, this mostly reflects the lack of effort, which is considerable, even for a single microbial species.

The rapid expansion of fungal genome sequencing has led to opportunities for reverse ecology or the use of genomes to understand ecological strategies, particularly useful for microbial species (Levy and Borenstein 2012). These same approaches can be used to determine the metabolic dependencies of organisms based on their genomes and to use this information to improve culturing of these microbes (Tyson et al. 2005). One of the weakest links in this approach is that assumptions regarding protein function and the nature of metabolic networks are extracted from model systems, yet we know that protein functions and networks evolve (Wagner 2012).

Beyond metabolic dependencies, there are other aspects of symbionts that may lead to their uncultivated status, which we outline and have been discussed for bacteria (Lewis et al. 2021). They include the absence of physical or chemical cues and the inability to make life cycle transitions in culture, specifically evidenced by dimorphic fungi that transition between saprobic and pathogenic morphotypes, for example in the uncultivated *Paracoccidioides loboi* (Vilela et al. 2021). *P. loboi* yeast cells cannot be transferred from an infected human to a culture plate, yet the cells can readily be used to infect mice. It therefore appears that the major barrier to cultivating this fungus is convincing it to switch its metabolic state into a saprobic one, which is observed in other *Paracoccidioides* species that are opportunistic environmentally acquired infections (Rodrigues et al. 2023). Either an environmental cue, physical stimulation, removal of a blocked pathway, or nutritional requirements could explain the inability of the fungus to be cultivated. For *P. loboi* and other uncultivated pathogens, complete genome sequences can help understand what their growth requirements are. However, a more complete picture may be needed that includes a picture of not only the genes in the genome, but which ones are, and critically, which ones are not expressed, so that such information can enlighten cultivation strategies.

References

Ahrendt SR, Quandt CA, Ciobanu D et al (2018) Leveraging single-cell genomics to expand the fungal tree of life. Nat Microbiol 3(12):1417–1428

Ahrendt SR, Mondo SJ, Haridas S et al (2022) MycoCosm, the JGI's fungal genome portal for comparative genomic and multiomics data analyses. In: Martin F, Uroz S (eds) Microbial environmental genomics (MEG). Springer US, New York, pp 271–291

Alexander WG, Wisecaver JH, Rokas A et al (2016) Horizontally acquired genes in early-diverging pathogenic fungi enable the use of host nucleosides and nucleotides. PANS 113(15):4116–4121

Amend A (2014) From dandruff to deep-sea vents: *Malassezia*-like fungi are ecologically hyper-diverse. PLoS Pathog 10(8):e1004277

Arthur JC, Kern FD, Orton CR et al (1929) The plant rusts (Uredinales). Wiley, New York

Azcón-Aguilar C, Diaz-Rodriguez RM, Barea JM (1986) Effect of soil microorganisms on spore germination and growth of the vesicular-arbuscular mycorrhizal fungus *Glomus mosseae*. Trans Br Mycol Soc 86: 337–340

Azcón-Aguilar C, Garda-Garda F, Barea JM (1991) Germinación y crecimiento axenico de los hongos formadores de micorrizas vesiculo-arbusculares. In: LopezGorge J, Barea JM, Olivares J (eds) Fijación y movilización biológiea de nutrientes, nuevas tendencias. CSIC, Madrid, pp 129–147

Bacon CW, Porter JK, Robbins JD et al (1977) *Epichloe typhina* from toxic tall fescue grasses. Appl Environ Microbiol 34(5):576–581

Bailey DL (1923) Sunflower rust. Minn Univ Agr Exp Sta Tech Bull 16:1–31

Barea JM (1986) Importance of hormones and root exudates in mycorrhizal phenomena. In: Gianinazzi-Pearson V, Gianinazzi S (eds) Physiological and genetical aspects of mycorrhizae. INRA, Paris, pp 77–187

Barker SM, Barnett HL (1973) Nitrogen and vitamin requirements for axenic growth of the haustorial mycoparasite, *Dispira cornuta*. Mycologia 65(1): 21–27

Becard G, Piche Y (1989) New aspects on the acquisition of biotrophic status by a vesicular arbuscular mycorrhizal fungus *Gigaspora margarita*. New Phytol 112: 77–83

Beilby JP, Kidby DK (1980) Biochemistry of ungerminated and germinated spores of the vesicular-arbuscular mycorrhizal fungus, *Glomus caledonius*: changes in neutral and polar lipids. J Lipid Res 21(6): 739–750

Benjamin RK, Shanor L (1952) Sex of host specificity and position specificity of certain species of *Laboulbenia* on *Bembidion picipes*. Am J Bot 39(2):125–131

Benny GL, Smith ME, Kirk PM, Tretter ED, White MM (2016) Challenges and future perspectives in the systematics of Kickxellomycotina, Mortierellomycotina, Mucoromycotina, and Zoopagomycotina. In: Dw L (ed) Biology of Microfungi. Springer, Cham, Heidelberg New York, Dordrecht, London, pp 65–126

Berdy B, Spoering AL, Ling LL et al (2017) In situ cultivation of previously uncultivable microorganisms using the ichip. Nat Protoc 12(10):2232–2242

Berry SB, Zhang T, Day JH et al (2017) Upgrading well plates using open microfluidic patterning. Lab Chip 17(24):4253–4264

Bland CE (1985) Culture. In: Couch JN, Bland CE (eds) The Genus Coelomomyces. Academic, Orlando, FL, pp 349–359

Bonfante-Fasolo P, Perotto S (1992) Plants and endomycorrhizal fungi: the cellular and molecular basis of their interaction. In: Verma DPS (ed) Molecular signals in plant-microbe communications. CRC, Boca Raton, FL, pp 445–470

Bor B, Poweleit N, Bois JS et al (2016) Phenotypic and physiological characterization of the epibiotic interaction between TM7x and its basibiont Actinomyces. Microb Ecol 71(1):243–255

Bosshard PP (2011) Incubation of fungal cultures: how long is long enough? Mycoses 54(5):539–545

Bowen GD (1980) Misconceptions, concepts and approaches in rhizosphere biology. In: Ellwood DC, Latham MJ, Hedger JN, Lynch JM, Slater JM (eds) Contemporary microbial ecology. Academic, London, pp 283–304

Bravo A, Brands M, Wewer V et al (2017) Arbuscular mycorrhiza-specific enzymes FatM and RAM2 fine-tune lipid biosynthesis to promote development of arbuscular mycorrhiza. New Phytol 214(4): 1631–1645

Brefeld O (1981) Untersuchungen aus dem Gesammtgebiete der Mykologie. Heft 9, Die Hemiasci und die Ascomyceten. Munster, p 159

Bushnell WR (1968) In vitro development of an Australian isolate of *puccinia graminis* f. sp. *tritici*. Phytopathology 58(4):526

Carleton MA (1903) Culture methods with Uredineae. J Appl Micros Lab Methods 6(1):2109–2114

Castillo JM, Roberts DW (1980) In vitro studies of *Coelomomyces punctatus* from *Anopheles quadrimaculatus* and *Cyclops vernalis*. J Invertebr Pathol 35(2):144–157

Chan KG, Yin WF, Sam CK et al (2009) A novel medium for the isolation of N-acylhomoserine lactone-degrading bacteria. J Ind Microbiol Biotechnol 36(2): 247

Cisse OH, Pagni M, Hauser PM (2014) Comparative genomics suggests that the human pathogenic fungus *Pneumocystis jirovecii* acquired obligate biotrophy through gene loss. Genome Biol Evol 6(8):1938–1948

Ciznár I, Hostacká A (1983) Toxic properties of *Morganella morganii*. Ceskoslovenska Epidemiol Mikrobiol Imunol 32(3):126–133

Collado J, Platas G, Paulus B et al (2007) High-throughput culturing of fungi from plant litter by a dilution-to-extinction technique. FEMS Microbiol Ecol 60(3): 521–533

Colley RH (1918) Parasitism, morphology, and cytology of *Cronartium ribicola*. J Agric Res 15(6):19–59

Connon SA, Giovannoni SJ (2002) High-throughput methods for culturing microorganisms in very-low-nutrient media yield diverse new marine isolates. Appl Environ Microbiol 68(8):3878–3885

Cornu M (1876) Ou doit-on chercher les organes fécondateurs chez les urédinées et ustilaginées? Bull Soc Bot France 23(3):120–122

Couch JN (1972) Mass production of *Coelomomyces*, a fungus that kills mosquitoes. PNAS 69(8):2043–2047

Cox HR (1938a) A filter-passing infectious agent isolated from ticks. III. Description of organism and cultivation experiments. Public Health Rep 53(52):2270–2276

Cox HR (1938b) Use of yolk sac of developing chick embryo as medium for growing rickettsiae of Rocky Mountain spotted fever and typhus groups. Public Health Rep 1896-1970:2241–2247

Cross KL, Campbell JH, Balachandran M et al (2019) Targeted isolation and cultivation of uncultivated bacteria by reverse genomics. Nat Biotechnol 37(11): 1314–1321

Curl EA, Truelove B (1986) The rhizosphere. Springer, New York

Cutter VM Jr (1959) Studies on the isolation and growth of plant rusts in host tissue cultures and upon synthetic media. I Gymnosporangium. Mycologia 51(2): 248–295

D'Onofrio A, Crawford JM, Stewart EJ et al (2010) Siderophores from neighboring organisms promote the growth of uncultured bacteria. Chem Biol 17(3): 254–264

D'Souza G, Kost C (2016) Experimental evolution of metabolic dependency in bacteria. PLoS Genet 12(11):e1006364

Davis WJ, Picard KT, Antonetti J et al (2018) Inventory of chytrid diversity in two temporary forest ponds using a multiphasic approach. Mycologia 110(5):811–821

Davison J, Moora M, Öpik M et al (2015) Global assessment of arbuscular mycorrhizal fungus diversity reveals very low endemism. Science 349(6251): 970–973

Day PR, Anagnostakis SL (1971) Corn smut dikaryon in culture. Nat New Biol 231(18):19–20

De Bary A (1887) Comparative morphology and biology of the fungi, mycetozoa, and bacteria. Clarendon Press, Oxford

Dean P, Hirt RP, Embley TM (2016) Microsporidia: why make nucleotides if you can steal them? PLoS Pathog 12(11):e1005870

Delgado-Ramos L, Marcos AT, Ramos-Guelfo MS et al (2014) Flow cytometry of microencapsulated colonies for genetics analysis of filamentous fungi. G3: Genes Genom Genet 4(11):2271–2278

Derilus D, Rahman MZ, Serrano AE et al (2021) Proteome size reduction in Apicomplexans is linked with loss of DNA repair and host redundant pathways. Infect Genet Evol 87:104642

Drew GC, Stevens EJ, King KC (2021) Microbial evolution and transitions along the parasite–mutualist continuum. Nat Rev Microbiol 19(10):623–638

Duplessis S, Cuomo CA, Lin YC et al (2011) Obligate biotrophy features unraveled by the genomic analysis of rust fungi. PNAS 108(22):9166–9171

Ellis JJ (1963) A study of *Rhopalomyces elegans* in pure culture. Mycologia 55(2):183–198

Ettinger CL, Ostovar T, Yacoub M et al (2023) Genomic resources enable insight into the developmental transcriptome of the blastoclad fungus, *Coelomomyces lativittatus*, an obligate parasite of mosquitoes and microcrustaceans. bioRxiv 2023-01

Flaishman MA, Kolattukudy PE (1994) Timing of fungal invasion using host's ripening hormone as a signal. PNAS 91(14):6579–6583

François J, Parrou JL (2001) Reserve carbohydrates metabolism in the yeast *Saccharomyces cerevisiae*. FEMS Microbiol Rev 25(1):125–145

Gao Q, Jin K, Ying SH et al (2011) Genome sequencing and comparative transcriptomics of the model entomopathogenic fungi *Metarhizium anisopliae* and *M. acridum*. PLoS Genet 7(1):e1001264

Gavrish E, Bollmann A, Epstein S et al (2008) A trap for in situ cultivation of filamentous actinobacteria. J Microbiol Methods 72(3):257–262

Gianinazzi-Pearson V, Branzanti B, Gianinazzi S (1989) In vitro enhancement of spore germination and early hyphal growth of a vesicular arbuscular mycorrhizal fungus by host root exudates and plant flavonoids. Symbiosis 7:243–255

Gordon MA (1951) Lipophilic yeastlike organisms associated with tinea versicolor. J Invest Dermatol 17(5):267–272

Gostinčar C, Zalar P, Gunde-Cimerman N (2022) No need for speed: slow development of fungi in extreme environments. Fungal Biol Rev 39:1–14

Gouriet F, Fenollar F, Patrice JY et al (2005) Use of shell-vial cell culture assay for isolation of bacteria from clinical specimens: 13 years of experience. J Clin Microbiol 43(10):4993–5002

Guan LL, Kamino K (2001) Bacterial response to siderophore and quorum-sensing chemical signals in the seawater microbial community. BMC Microbiol 1(1):1–11

Guého E, Midgley G, Guillot J (1996) The genus Malassezia with description of four new species. Antonie Van Leeuwenhoek 69:337–355

Haelewaters D, Lubbers M, De Kesel A (2022) The haustorium as a driving force for speciation in thallus-forming Laboulbeniomycetes. IMA Fungus 13(1):1

Hawksworth DL, Lücking R (2017) Fungal diversity revisited: 2.2 to 3.8 million species. Microbiol Spectr 5(4):5–4

He X, McLean JS, Edlund A et al (2015) Cultivation of a human-associated TM7 phylotype reveals a reduced genome and epibiotic parasitic lifestyle. PNAS 112(1):244–249

Held AA (1980) Development of *Rozella* in *Allomyces*: a single zoospore produces numerous zoosporangia and resistant sporangia. Canad J Bot 58(8):959–979

Helfer S (2014) Rust fungi and global change. New Phytol 201(3):770–780

Hepper CM (1984) Isolation and culture of VA mycorrhizal (VAM) fungi. In: Powell CL, Bagyaraj DJ (eds) VA mycorrhiza. CRC Press, Boca Raton, FL, pp 95–112

Hildebrand EM (1938) Techniques for the isolation of single microorganisms. Bot Rev 4:627–664

Hildebrandt U, Ouziad F, Marner FJ et al (2006) The bacterium *Paenibacillus validus* stimulates growth of the arbuscular mycorrhizal fungus *Glomus intraradices* up to the formation of fertile spores. FEMS Microbiol Lett 254(2):258–267

Hotson HH, Cutter VM Jr (1951) The isolation and culture of *Gymnosporangium juniperi-virgininae* schw. upon artificial media. PNAS 37(7):400–403

Howard A (1900) On a disease of tradescantia. Ann Bot 14(53):27–38

Hoyer AK, Hodkinson TR (2021) Hidden fungi: combining culture-dependent and-independent DNA barcoding reveals inter-plant variation in species richness of endophytic root fungi in *Elymus repens*. J Fungi 7(6)

Imachi H, Nobu MK, Nakahara N et al (2020) Isolation of an archaeon at the prokaryote–eukaryote interface. Nature 577(7791):519–525

James TY, Pelin A, Bonen L et al (2013) Shared signatures of parasitism and phylogenomics unite Cryptomycota and Microsporidia. Curr Biol 23(16):1548–1553

Jeffries P, Young TW (1994) Interfungal parasitic relationships. Cab International, Wallingford

Jiang Y, Wang W, Xie Q et al (2017) Plants transfer lipids to sustain colonization by mutualistic mycorrhizal and parasitic fungi. Science 356(6343):1172–1175

Jones MD, Forn I, Gadelha C, Egan MJ et al (2011) Discovery of novel intermediate forms redefines the fungal tree of life. Nature 474(7350):200–203

Kaeberlein T, Lewis K, Epstein SS (2002) Isolating "uncultivable" microorganisms in pure culture in a simulated natural environment. Science 296(5570): 1127–1129

Kameoka H, Tsutsui I, Saito K et al (2019) Stimulation of asymbiotic sporulation in arbuscular mycorrhizal fungi by fatty acids. Nat Microbiol 4(10):1654–1660

Kanehisa M, Sato Y (2020) KEGG Mapper for inferring cellular functions from protein sequences. Protein Sci 29(1):28–35

Kanehisa M, Sato Y, Morishima K (2016) BlastKOALA and GhostKOALA: KEGG tools for functional characterization of genome and metagenome sequences. J Mol Biol 428(4):726–731

Keeling PJ, Slamovits CH (2004) Simplicity and complexity of microsporidian genomes. Eukaryot Cell 3(6): 1363–1369

Keymer A, Pimprikar P, Wewer V et al (2017) Lipid transfer from plants to arbuscular mycorrhiza fungi. elife 6:e29107

Klebahn H (1904) Die Wirtwechselnden Rostpilsen XXXVII. Borntrager, Berlin

Kobayashi Y, Maeda T, Yamaguchi K et al (2018) The genome of *Rhizophagus clarus* HR1 reveals a common genetic basis for auxotrophy among arbuscular mycorrhizal fungi. BMC Genomics 19(1):1–11

Kõljalg U, Nilsson RH, Abarenkov K et al (2013) Towards a unified paradigm for sequence-based identification of fungi. Mol Ecol 21(22):5271–5277

Kopytko M, Correa-Torres SN, Estévez-Gómez MJ (2017) Biodegradación estimulada de los suelos contaminados con pesticidas organoclorados. Revista de investigación agraria y ambiental 8(1):119–130

Koske RW, Gemma JN (1992) Fungal reactions to plants prior to mycorrhizal formation. In: Allen MJ (ed) Mycorrhizal functioning, an integrative plant fungal process. Chapman and Hall, New York, pp 3–36

Kucharek TA, Young HC (1968) Teliospore germination of *Puccinia recondita* f. sp. *tritici*. In: Phytopathology (vol 58(8), 1056). 3340 Pilot Knob Road, St. Paul, MN 55121: American Phytopathological Society

Law R, Lewis DH (1983) Biotic environments and the maintenance of sex–some evidence from mutualistic symbioses. Biol J Linn Soc 20(3):249–276

Levy R, Borenstein E (2012) Reverse ecology: from systems to environments and back. Evol Syst Biol 751:329–345

Lewis K, Epstein S, D'onofrio A et al (2010) Uncultured microorganisms as a source of secondary metabolites. J Antibiot 63(8):468–476

Lewis WH, Tahon G, Geesink P et al (2021) Innovations to culturing the uncultured microbial majority. Nat Rev Microbiol 19(4):225–240

Linderman RG (1992) Vesicular-arbuscular mycorrhizal and soil microbial interactions. In: Bethlenfalvay GJ, Linderman RG (eds) Mycorrhizae in sustainable agriculture. ASA Special Publication, Madison, WI, pp 45–70

Lücking R, Aime MC, Robbertse B et al (2021) Fungal taxonomy and sequence-based nomenclature. Nat Microbiol 6(5):540–548

Luginbuehl LH, Menard GN, Kurup S et al (2017) Fatty acids in arbuscular mycorrhizal fungi are synthesized by the host plant. Science 356(6343):1175–1178

Lutzoni F, Nowak MD, Alfaro ME et al (2018) Contemporaneous radiations of fungi and plants linked to symbiosis. Nat Commun 9(1):5451

Ma L, Chen Z, Huang DW et al (2016) Genome analysis of three *Pneumocystis* species reveals adaptation mechanisms to life exclusively in mammalian hosts. Nat Commun 7(1):10740

Machlis L (1953) Growth and nutrition of water molds in the subgenus Euallomyces. I. Growth factor requirements. Am J Bot 40(3):189–195

Mafla-Endara PM, Arellano-Caicedo C, Aleklett K et al (2021) Microfluidic chips provide visual access to in situ soil ecology. Commun Biol 4(1):889

Martin F, Ramstedt M, Söderhäll K (1987) Carbon and nitrogen metabolism in ectomycorrhizal fungi and ectomycorrhizas. Biochimie 69(6–7):569–581

McCutcheon JP, Moran NA (2012) Extreme genome reduction in symbiotic bacteria. Nat Rev Microbiol 10(1):13–26

Meyer-Dombard DR, Shock EL, Amend JP (2012) Effects of trace element concentrations on culturing thermophiles. Extremophiles 16(2):317–331

Mix AJ (1949) A monograph of the genus *Taphrina*. Univ Kans Sci Bull 33(1):3–167

Mosse B (1959) The regular germination of resting spores and some observations on the growth requirements of an *Endogone* sp. causing vesicular-arbuscular mycorrhiza. Trans Br Mycol Soc 42:273–286

Müller B, Grossniklaus U (2010) Model organisms—a historical perspective. J Proteome 73(11):2054–2063

Nagae M, Parniske M, Kawaguchi M et al (2016) The relationship between thiamine and two symbioses: root nodule symbiosis and arbuscular mycorrhiza. Plant Signal Behav 11(12):e1265723

Nara K (2009) Spores of ectomycorrhizal fungi: ecological strategies for germination and dormancy. New Phytol 181:245–248

Nehls U (2008) Mastering ectomycorrhizal symbiosis: the impact of carbohydrates. J Exp Bot 59(5):1097–1108

Nichols D, Cahoon N, Trakhtenberg EM et al (2010) Use of ichip for high-throughput in situ cultivation of "uncultivable" microbial species. Appl Environ Microbiol 76(8):2445–2450

Omsland A, Cockrell DC, Fischer ER et al (2008) Sustained axenic metabolic activity by the obligate intracellular bacterium *Coxiella burnetii*. J Bacteriol Res 190(9):3203–3212

Omsland A, Cockrell DC, Howe D et al (2009) Host cell-free growth of the Q fever bacterium *Coxiella burnetii*. PNAS 106(11):4430–4434

Omsland A, Beare PA, Hill J et al (2011) Isolation from animal tissue and genetic transformation of *Coxiella burnetii* are facilitated by an improved axenic growth medium. Appl Environ Microbiol 77(11):3720–3725

Omsland A, Hackstadt T, Heinzen RA (2013) Bringing Culture to the Uncultured: Coxiella burnetii and Lessons for Obligate Intracellular Bacterial Pathogens. PLoS Pathog 9(9):e1003540. https://doi.org/10.1371/journal.ppat.1003540

Overmann J (2015) Significance and future role of microbial resource centers. Syst Appl Microbiol 38(4):258–265

Park J, Kerner A, Burns MA et al (2011) Microdroplet-enabled highly parallel co-cultivation of microbial communities. PLoS One 6(2):e17019

Parniske M (2008) Arbuscular mycorrhiza: the mother of plant root endosymbioses. Nat Rev Microbiol 6(10):763–775

Persyn A, Mueller A, Goormachtig S (2022) Drops join to make a stream: high-throughput nanoscale cultivation to grasp the lettuce root microbiome. Environ Microbiol Rep 14(1):60–69

Piotrowski JS, Annis SL, Longcore JE (2004) Physiology of *Batrachochytrium dendrobatidis*, a chytrid pathogen of amphibians. Mycologia 96(1):9–15

Plowright CB (1889) A monograph of the British Uredineae and Ustilagineae. Kegan Paul, Trench & Co., London

Porro MN, Passi S, Caprilli F et al (1976) Growth requirements and lipid metabolism of *Pityrosporum orbiculare*. J Invest Dermatol 66(3):178–182

Porter TM, Schadt CW, Rizvi L et al (2008) Widespread occurrence and phylogenetic placement of a soil clone group adds a prominent new branch to the fungal tree of life. Mol Phylogenet Evol 46(2):635–644

Powell MJ, Letcher PM, James TY (2017) Ultrastructural characterization of the host–parasite interface between *Allomyces anomalus* (Blastocladiomycota) and *Rozella allomycis* (Cryptomycota). Fungal Biol 121(6–7):561–572

Quandt CA, Bushley KE, Spatafora JW (2015) The genome of the truffle-parasite *Tolypocladium*

ophioglossoides and the evolution of antifungal peptaibiotics. BMC Genomics 16(1):1–14
Quandt CA, Beaudet D, Corsaro D et al (2017) The genome of an intranuclear parasite, *Paramicrosporidium saccamoebae*, reveals alternative adaptations to obligate intracellular parasitism. elife 6: e29594
Rämä T, Quandt CA (2021) Improving fungal cultivability for natural products discovery. Front Microbiol 12: 706044
Raoult D, Vestris G, Enea M (1990) Isolation of 16 strains of *Coxiella burnetii* from patients by using a sensitive centrifugation cell culture system and establishment of the strains in HEL cells. J Clin Microbiol 28(11): 2482–2484
Read DJ (1991) Mycorrhizas in ecosystems. Experientia 47:376–391
Reed HS, Crabill CH (1915) Cedar rust disease of apples caused by *Gymnosporangium Juniperi-virginianae* Schw. Virginia Agr Expt Sta Tech Bull 9:1–106
Richter F, Bindschedler S, Calonne-Salmon M et al (2022) Fungi-on-a-Chip: microfluidic platforms for single-cell studies on fungi. FEMS Microbiol Rev 46(6):fuac039
Robbins WJ, Kavanagh F (1938) Vitamin B1 or its intermediates and growth of certain fungi. Am J Bot 25(4):229–236
Rodrigues AM, Hagen F, Puccia R et al (2023) *Paracoccidioides* and Paracoccidioidomycosis in the 21st Century. Mycopathologia 12:1–5
Romero-Olivares AL, Morrison EW, Pringle A et al (2021) Linking genes to traits in fungi. Microb Ecol 82(1):145–155
Rosling A, Cox F, Cruz-Martinez K et al (2011) Archaeorhizomycetes: unearthing an ancient class of ubiquitous soil fungi. Science 333(6044):876–879
Samlali K, Alves CL, Jezernik M et al (2022) Droplet digital microfluidic system for screening filamentous fungi based on enzymatic activity. Microsyst Nanoeng 8(1):123
Schadt CW, Martin AP, Lipson DA et al (2003) Seasonal dynamics of previously unknown fungal lineages in tundra soils. Science 301(5638):1359–1361
Scholte EJ, Knols BG, Samson RA et al (2004) Entomopathogenic fungi for mosquito control: a review. J Insect Sci 4(1):19
Schulze-Lefert P, Vogel J (2000) Closing the ranks to attack by powdery mildew. Trends Plant Sci 5(8): 343–348
Schüβler A, Walker C (2011) Evolution of the 'Plant-Symbiotic' Fungal Phylum, Glomeromycota. In: Pöggeler S, Wöstemeyer J (eds) Evolution of fungi and fungal-like organisms. Springer, pp 164–185
Scott KJ, Maclean DJ (1969) Culturing of rust fungi. Annu Rev Phytopathol 7(1):123–146
Shapiro M, Roberts DW (1976) Growth of *Coelomomyces psorophorae* mycelium in vitro. J Invertebr Pathol 27(3):399–402
Singleton LL, Young HC (1968) The in vitro culture of *Puccinia recondita* f. sp. *tritici*. Phytopathology 58: 1068
Siqueira JO (1987) Cultura axenicae monoxenica dos fungos micorrizicos vesiculoarbusculares. In: II Reuniao Brasileira sobre Micorrizas. Sao Paulo, 2: 44–70
Solden L, Lloyd K, Wrighton K (2016) The bright side of microbial dark matter: lessons learned from the uncultivated majority. Curr Opin Microbiol 31:217–226
Soudzilovskaia NA, Vaessen S, Barcelo M et al (2020) FungalRoot: global online database of plant mycorrhizal associations. New Phytol 227(3):955–966
Spanu PD (2012) The genomics of obligate (and nonobligate) biotrophs. Annu Rev Phytopathol 50: 91–109
Spaulding P (1922) Investigations of the white-pine blister rust. Washington, DC: US Department of Agriculture, Forest Service: Bulletin 957:1–100
Stauber L, Prospero S, Croll D (2020) Comparative genomics analyses of lifestyle transitions at the origin of an invasive fungal pathogen in the genus Cryphonectria. MSphere 5(5):e00737–e00720
Steinberg RA (1939) Growth of fungi in synthetic nutrient solutions. Bot Rev 5:327–350
Steinberg RA (1950) Growth of fungi in synthetic nutrient solutions. II Bot Rev 16(4):208–228
Stocker-Wörgötter E (2001) Experimental lichenology and microbiology of lichens: culture experiments, secondary chemistry of cultured mycobionts, resynthesis, and thallus morphogenesis. Bryologist 104(4): 576–581
Sugiura Y, Akiyama R, Tanaka S et al (2020) Myristate can be used as a carbon and energy source for the asymbiotic growth of arbuscular mycorrhizal fungi. PNAS 117(41):25779–25788
Tedersoo L, May TW, Smith ME (2010) Ectomycorrhizal lifestyle in fungi: global diversity, distribution, and evolution of phylogenetic lineages. Mycorrhiza 20: 217–263
Tedersoo L, Bahram M, Põlme S et al (2014) Global diversity and geography of soil fungi. Science 346(6213):1256688
Tedersoo L, Bahram M, Puusepp R et al (2017) Novel soil-inhabiting clades fill gaps in the fungal tree of life. Microbiome 5(1):1–10
Thiel AF, Weiss F (1920) The effect of citric acid on the germination of the teliospores of *Puccinia graminis tritici*. Phytopathology 10:448–452
Thorpe T (2012) History of plant tissue culture. Plant Cell Cult Protoc 318:9–27
Tisserant E, Malbreil M, Kuo A et al (2013) Genome of an arbuscular mycorrhizal fungus provides insight into the oldest plant symbiosis. PNAS 110(50):20117–20122
Tsaousis AD, Kunji ER, Goldberg AV et al (2008) A novel route for ATP acquisition by the remnant mitochondria of *Encephalitozoon cuniculi*. Nature 453(7194):553–556

Tumwine J, Frinking HD, Jeger MJ (2000) Isolation techniques and cultural media for *Phytophthora infestans* from tomatoes. Mycologist 14(3):137–139

Tyson GW, Lo I, Baker BJ et al (2005) Genome-directed isolation of the key nitrogen fixer *Leptospirillum ferrodiazotrophum* sp. nov. from an acidophilic microbial community. Appl Environ Microbiol 71(10): 6319–6324

Uehling JK, Entler MR, Meredith HR et al (2019) Microfluidics and metabolomics reveal symbiotic bacterial–fungal interactions between *Mortierella elongata* and *Burkholderia* include metabolite exchange. Front Microbiol 10:2163

Umphlett CJ (1962) Morphological and cytological observations on the mycelium of *Coelomomyces*. Mycologia 54(5):540–554

Ushijima T, Takahashi M, Ozaki Y (1981) Selective and differential media for isolation and tentative identification of each species of *Pityrosporum* residing on normal or diseased human skin. Microbiol Immunol 25(11):1109–1118

Vartoukian SR, Adamowska A, Lawlor M et al (2016) In vitro cultivation of 'unculturable' oral bacteria, facilitated by community culture and media supplementation with siderophores. PLoS One 11(1): e0146926

Vilela R, Huebner M, Vilela C et al (2021) The taxonomy of two uncultivated fungal mammalian pathogens is revealed through phylogeny and population genetic analyses. Sci Rep 11(1):18119

Vuorinen I, Hamberg L, Müller M et al (2015) Development of growth media for solid substrate propagation of ectomycorrhizal fungi for inoculation of Norway spruce (*Picea abies*) seedlings. Mycorrhiza 25(4): 311–324

Wagner A (2012) Metabolic networks and their evolution. Evol Sys Biol 751:29–52

Wang Y, Wu J, Yan J et al (2022) Comparative genome analysis of plant ascomycete fungal pathogens with different lifestyles reveals distinctive virulence strategies. BMC Genomics 23(1):34

Whisler HC, Zebold SL, Shemanchuk JA (1975) Life history of Coelomomyces psorophorae. Proc Natl Acad Sci 72(2):693–696. https://doi.org/10.1073/pnas.72.2.693

Williams PG (1992) Axenic culture of arbuscular mycorrhizal fungi. In: Norris JR, Read DJ, Varma AK (eds) Methods in microbiology, vol 24. Academic, London, pp 203–220

Williams PG, Scott KJ, Kuhl JL (1966) Vegetative Growth of *Puccinia graminis* f. sp. *tritici* In Vitro. Phytopathology 56(12):1418

Williams PG, Scott KJ, Kuhl JL et al (1967) Sporulation and pathogenicity of *Puccinia graminis* f. sp. *tritici* grown on an artificial medium. Phytopathology 57(3): 326

Wu B, Cox MP (2021) Comparative genomics reveals a core gene toolbox for lifestyle transitions in Hypocreales fungi. Environ Microbiol 23(6): 3251–3264

Yao T, Asayama Y (2017) Animal-cell culture media: history, characteristics, and current issues. Reprod Med Biol 16(2):99–117

Yoder JB, Tiffin P (2017) Sanctions, partner recognition, and variation in mutualism. Am Nat 190(4):491–505

Zengler K, Toledo G, Rappé M et al (2002) Cultivating the uncultured. PNAS 99(24):15681–15686

Zhang CB, Ren CH, Wang YL et al (2020) Uncovering fungal community composition in natural habitat of *Ophiocordyceps sinensis* using high-throughput sequencing and culture-dependent approaches. BMC Microbiol 20(1):1–16

Zhao J, Wang M, Chen X et al (2016) Role of alternate hosts in epidemiology and pathogen variation of cereal rusts. Annu Rev Phytopathol 54:207–228

3 Horizontal Gene Transfer in Fungi and Its Ecological Importance

Carla Gonçalves, Chris Todd Hittinger, and Antonis Rokas

Abstract

Fungi are one of the most diverse kingdoms in the tree of life, thrive in virtually every habitat on the planet, and are vital for a plethora of ecosystems and human-related activities. The advent of genomic technologies has revolutionized the study of fungal ecology and evolution. Surprisingly, evolutionary genomic investigations have revealed that horizontal gene transfer (HGT)—the movement of genetic material from one organism to another by means other than sexual reproduction—is an important factor influencing the ecology and evolution of fungal genomes. This chapter uses an evolutionary framework to discuss the extent and distribution of HGT across fungal ecologies and lifestyles, focusing on some of the best-supported cases to address fundamental questions such as frequencies, barriers, and mechanisms for eukaryotic HGT. One important conclusion that can be drawn from studies of fungal HGT is that HGT rate varies extensively across lineages and that this variation can generally be associated with distinctive ecologies, lifestyles, or genetic backgrounds. Similarly, not all functional categories of genes are equally likely to result in successful transfers. The factors underlying both frequency and functional variations of HGT across fungi are thoroughly discussed. Advances in genomic and synthetic biology technologies will continue to shed light on these questions and bring many other still unanswered questions into focus. Some of these questions and directions for future research are highlighted at the end of this chapter.

Keywords

Fungal ecology · Fungi · Horizontal gene transfer · Adaptive evolution · Budding yeasts · Fungal pathogens · Lateral gene transfer · Gene cluster · Xenolog

C. Gonçalves (✉) · A. Rokas (✉)
Department of Biological Sciences and Evolutionary Studies Initiative, Vanderbilt University, Nashville, TN, USA
e-mail: carla.goncalves@fct.unl.pt; antonis.rokas@vanderbilt.edu

C. T. Hittinger
Laboratory of Genetics, DOE Great Lakes Bioenergy Research Center, Center for Genomic Science Innovation, Wisconsin Energy Institute, J.F. Crow Institute for the Study of Evolution, University of Wisconsin-Madison, Madison, WI, USA

3.1 Introduction

Fungi populate virtually every habitat on the planet. They play crucial roles in a variety of aquatic and terrestrial ecosystems and exhibit various lifestyles from saprobic to symbiotic or parasitic, engaging in diverse interactions with

Y.-P. Hsueh, M. Blackwell (eds.), *Fungal Associations*, The Mycota 9,
https://doi.org/10.1007/978-3-031-41648-4_3

other microorganisms, algae, animals, or plants (Selbmann et al. 2013; Stajich 2017; Coleine et al. 2022; Hawksworth et al. 2017; Li et al. 2021; Rokas 2022). One key to their success is their highly diverse genetic repertoire that translates into extremely diversified biochemical abilities and morphologies, turning fungi into one of the most adaptable kingdoms in the tree of life. Fungal ecological and evolutionary diversity has been largely attributed to life cycles that entail a combination of sexual and asexual reproduction coupled with major genomic evolutionary processes, such as recombination (Heitman 2010; Magwene et al. 2011; Grandaubert et al. 2019), gene and genome duplication (Wolfe 2015; Sinha et al. 2017; Steenwyk and Rokas 2018; Todd and Selmecki 2020; Stalder et al. 2022), and introgression and hybridization (Libkind et al. 2011; Stukenbrock 2016; Steensels et al. 2021). Yet, analyses of genomic data from certain major ecological adaptations or lifestyle transitions (Rokas 2022; Naranjo-Ortiz and Gabaldón 2019) suggest that other mechanisms might also be important contributors.

One such mechanism is horizontal gene transfer (HGT), which is defined as the movement of genetic material through mechanisms distinct from reproduction and vertical inheritance. The decoding of the genomes of numerous prokaryotes more than two decades ago, coupled with progress in phylogenetic methods, exposed HGT as a major driver of genome evolution in Bacteria and Archaea, playing an indispensable role in functional innovation and speciation and leading to a network of life (Doolittle 1999; Dagan et al. 2008; Ochman et al. 2000; Arnold et al. 2022; Wagner et al. 2017). However, recognition of the importance and magnitude of HGT in the evolution of eukaryotes has only recently gained momentum. Pioneering studies in protists identified several instances of HGT involved in the evolution of organelles or adaptation to anaerobiosis (Richards et al. 2003; Archibald et al. 2003; Andersson et al. 2003); the discovery of other instances of HGT in diverse eukaryotic lineages (Kondrashov et al. 2006; Richardson and Palmer 2007), including fungi (Hall et al. 2005; Slot and Hibbett 2007; Khaldi et al. 2008; Slot and Rokas 2011; Richards et al. 2009), quickly followed.

Despite the skepticism raised by a few surrounding eukaryotic HGT (discussed in Sect. 3.2), in light of the growing number of strongly supported cases across diverse eukaryotic lineages, it is now unquestionable that genes do move horizontally into and even between eukaryotes. HGT events have been convincingly documented in multiple phylogenetically distant eukaryotic lineages ranging from algae (Schönknecht et al. 2013) to plants (Ma et al. 2022) and animals (Dunning Hotopp 2011; Li et al. 2022). Reports of HGT in fungi have dramatically increased in recent years (Wisecaver and Rokas 2015; Gonçalves et al. 2018; Shen et al. 2018; Gonçalves and Gonçalves 2022; Bredeweg and Baker 2020), in part as an outcome of the ever-growing number of available genomes that are empowering high-throughput evolutionary examinations of HGT (Marcet-Houben and Gabaldon 2010; Hittinger et al. 2015; Shen et al. 2018).

Fungi are typically easy to grow under laboratory conditions and have relatively small genomes that are nowadays straightforward to sequence and analyze. In addition, several species in diverse lineages are genetically tractable, which allows for experimental validation of genotype–phenotype associations. These characteristics turn fungi into exceptional models to address some of the major questions related to eukaryotic HGT: how pervasive is it? What are the mechanisms mediating it? Which factors contribute to its success? Is there variation in the frequency of HGT across lineages, ecologies, and gene functions? What is the scale of its functional and ecological importance?

This chapter sheds light on some of these questions with a special focus on the ecological implications of HGT, showing how the advance of high-throughput (gen)omics, supported by an ever-growing body of experimental work, is giving rise to a new view where HGT is a major mechanism of genome evolution in fungi. First, we review the evidence that fungi have been acting as both donors and recipients of genetic material with each other, as well as with other eukaryotic and prokaryotic lineages. Using a state-of-the-art genome-scale phylogeny of the fungal kingdom as a backbone, we draw and discuss a portrait of the distribution and relative quantification of HGT events. The frequency and success of these events seem to be generally governed by the same principles as in bacteria (e.g., ecological overlap between donor and recipient organisms and low interference with the recipient gene network), but HGT rates are generally lower than in bacteria and vary substantially across fungal lineages and among gene functions—the most striking cases and possible reasons underlying those variations are discussed. Using the most recent and well-supported studies as examples, this chapter discusses how several of these HGT events were pivotal for adaptation to new environments and for the transition and modulation of lifestyles, further underscoring the contribution of HGT to the ecological success of fungi.

3.2 Mechanisms Mediating HGT in Eukaryotes

HGT requires both external environmental factors and physical and molecular aspects intrinsic to the host organism. A piece of foreign DNA must enter the host cell, then the nucleus, and finally be stably integrated in the host genome; next, the acquired DNA must spread and become fixed in the population. The probability of each of these events varies across time and lineages (Doolittle 1998; Huang 2013; Doolittle et al. 2003)—for instance, between prokaryotes and eukaryotes, but also among eukaryotes. While the existence of a germ line is a major barrier for HGT in some eukaryotic lineages, access of foreign DNA to the nucleus is facilitated in species carrying endosymbionts (Doolittle 1998; Keeling and Palmer 2008; Husnik and McCutcheon 2017). Environmental factors can also facilitate integration of foreign DNA, as illustrated by bdelloid rotifer species frequently exposed to periods of desiccation (Nowell et al. 2018; Gladyshev et al. 2008; Eyres et al. 2015). Fungal cells can be particularly impermeable to foreign DNA due to their robust cell walls and absence of phagotrophy but are nevertheless quite permissive to HGT (Shen et al. 2018; Richards et al. 2011a).

Despite the growing number of strongly supported cases, HGT in eukaryotes has not been free of controversy. The main arguments that have been put forward against the occurrence of eukaryotic HGT, beyond those occurring within the context of endosymbiosis, are the physical and molecular barriers inherent to the biology and genetics of eukaryotes, the lack of information about the molecular mechanisms mediating HGT, and the observation that eukaryote pangenomes, which would be predicted to result from the cumulative effects of ongoing HGT over time, are smaller than prokaryote pangenomes (Martin 2017). Additionally, the occurrence of phylogenetic conflict between gene trees and the species tree, the gold standard for identifying HGT (Fitzpatrick 2012; Wisecaver and Rokas 2015), is depicted as being either the result of phylogenetic artifacts or a misinterpretation of the alternative hypothesis of differential loss (Martin 2017).

These arguments dissipate in the face of the mounting evidence supporting opportunities and mechanisms mediating HGT in eukaryotes (Lacroix and Citovsky 2016; Sibbald et al. 2020), which do not substantially differ from those already established for bacteria. For instance, natural transformation has been observed in both yeasts and filamentous fungi under certain environmental conditions (Nevoigt et al. 2000; Morogovsky et al. 2022), and fusion of asexual spores, which can result in the exchange of genetic material, is known to occur between different species of filamentous fungi (Mehta and Baghela 2021). Virus-mediated transfers (Bian et al. 2020; Irwin et al. 2022), type IV secretion systems (Zoolkefli et al. 2021), and mobile genetic elements, such as transposons (Ropars et al. 2015; McDonald et al. 2019; Urquhart et al. 2022; Widen et al. 2022), have been shown to facilitate transfer of genetic material into eukaryotic cells. Once inside the cell, integration of foreign DNA into the host genome is commonly mediated by non-homologous end joining but can also be facilitated by transposable elements (Ropars et al. 2015; McDonald et al. 2019; Urquhart et al. 2022), homologous recombination with native homologs (Hao et al. 2010), or mediated by microhomology segments (Ku et al. 2015).

After integration into the host genome, expression of the acquired gene(s) is key for its evolutionary retention. Some of the structural hurdles to successful transcription and translation include the presence or absence of introns, differences in GC content and codon bias, differences in the genetic code or differences in transcriptional promoters, transcription factor-binding sites, and ribosomal-binding sites. The magnitude of these structural differences is expected to be generally a function of evolutionary distance. There are striking differences between bacterial and eukaryotic transcription and translation, suggesting that HGT events between distantly related species are less likely to be successful than those between closely related organisms (Andam and Gogarten 2011). Nevertheless, HGT-derived genes can readily become functional, irrespective of evolutionary distance and even in the absence of selection (Li and Bock 2019). Gene expression can be accomplished by the capturing of native promoters via genome rearrangements (Li and Bock 2019), or by the presence of "leaky" promoters with flexible requirements for activity, such as those found in certain transposable elements in *Drosophila* (Palazzo et al. 2019). Weak expression can also be mitigated by an increase in the gene copy number (Lind et al. 2010). Consistent with these mechanisms, evolution of *de novo* promoters, acquisition of new transcription and polyadenylation sites, and gene fusions have been specifically observed in functional transfers of bacterial pathways into yeasts (Gonçalves and Gonçalves 2019; Kominek et al. 2019). Acquisition of introns (Fan et al. 2020; Husnik and McCutcheon 2017) and GC content (Zhang et al. 2019) or codon bias (Matriano et al. 2021) changes have also been documented. Although accomplishing expression is essential, fine-tuning it in the new genomic context is also needed to achieve proper protein translation (de Reus et al. 2019).

While a newly transferred gene's structural features are key for its successful expression, the protein product of the gene will also interact with the resident proteins inside the cell; those interactions can sometimes be detrimental and lead to its removal by purifying selection. This in part explains why genes that have fewer connections are more likely to be successfully transferred (Cohen et al. 2011). Nonetheless, some types of interacting partners can be co-transferred together. Proteins that interact with each other or take part in the same metabolic pathway are often physically close in fungal genomes and are organized in gene clusters (Rokas et al. 2018). In fact, horizontal transfers of fungal metabolic gene clusters have been well documented and have facilitated the acquisition of new traits encoded by multiple genes (Khaldi et al. 2008; Reynolds et al. 2018; Wisecaver and Rokas 2015; Urquhart et al. 2022). Advances in long-read sequencing technologies are providing evidence for the mechanisms mediating the transfer of some of these clusters, including demonstrating that they are sometimes embedded in massive transposable elements (Gluck-Thaler et al. 2022). These mobile genetic entities can carry a diverse array of genes encoding different functions and are found in many fungi (Gluck-Thaler et al. 2022). Further evidence of recent intra-kingdom HGT events involving these regions, coupled with experimental evidence of the underlying mechanisms (Urquhart et al. 2023), reinforces the view that the actual frequency of HGT between fungal species might be higher than current estimates.

Adding to the evidence deconstructing the arguments denying the occurrence of eukaryotic HGT, pangenomes also occur in eukaryotes, including fungi (Brockhurst et al. 2019; McCarthy and Fitzpatrick 2019; Richard 2020; Fan et al. 2020; Barber et al. 2021). The acquired genes also often show levels of sequence similarity and other genomic signatures (e.g., retention of native genes) between donor and recipient lineages that are much higher than those expected under a differential loss scenario (Slot and Rokas 2011; Slot and Rokas 2010; Wisecaver and Rokas 2015). Moreover, owing to the substantial number of HGT involving donors and recipients that belong to distantly related lineages—such as bacteria and eukaryotes (Husnik and McCutcheon 2017; Li et al. 2022; Gonçalves et al. 2018; Kominek et al. 2019; Shen et al. 2018)—and to the increasingly rich genome sampling available, it is becoming more difficult to postulate differential gene loss as the most likely explanation for the topology inconsistencies observed between species trees and gene trees (Cote-L'Heureux et al. 2022).

3.3 Extent of HGT Involving Fungi

Fungi have been receiving and supplying genes from and to various lineages for millions of years (Fig. 3.1). While fungi have mostly provided genes to other eukaryotic lineages, many different fungal lineages have been recipients of both interdomain and intra-kingdom HGTs with impacts on a variety of biological processes (Wisecaver and Rokas 2015; Shen et al. 2018; Franco et al. 2022; Marsit et al. 2015; Coelho et al. 2013; Gonçalves et al. 2016, 2018; Khaldi et al. 2008; Luo et al. 2022).

Although examples of HGT involving fungi abound, precise quantification of the frequency and main sources of fungal HGT is challenging because the pipelines and criteria used often differ across studies. The focus of these studies also varies. For example, most studies have focused on inference of HGT in particular lineages or on HGT from bacterial donors (Marcet-Houben and Gabaldon 2010; Gonçalves et al. 2018; Shen et al. 2018) because such transfers are easier to detect, especially when employing high-throughput approaches. Therefore, our current appreciation of the origin, distribution, and frequency of HGT across the fungal kingdom (Fig. 3.2) needs to be interpreted in the light of these caveats.

As observed in other eukaryotes (Gladyshev et al. 2008; Pauchet and Heckel 2013; Wybouw et al. 2014; Schönknecht et al. 2013; Eme et al. 2017; Li et al. 2022), a significant fraction of the reported HGT events in fungi involve bacterial sources (Fig. 3.1), which contrasts with the idea that HGT events between distantly related species are less likely to be successful than those between closely related species (Andam and Gogarten 2011). The aforementioned methodological caveats notwithstanding, the preponderance of bacterial genes transferred into fungi and other eukaryotes might also suggest that transfers of bacterial genes are more likely than eukaryotic genes to be fixed in eukaryotic host genomes (Keeling and Palmer 2008). For instance, bacterial genes lack introns, which are generally important for efficient gene expression in eukaryotes. However, variations in intron patterns can also hinder gene expression due to incorrect splicing (Fitzpatrick 2012). Intron patterns vary extensively across fungi (Lim et al. 2021), which could affect the success of some intra-kingdom gene transfers as illustrated by the incorrect splicing of *Fusarium fujikuroi* genes when introduced into *Aspergillus nidulans* (de Reus et al. 2019). Due to their large evolutionary distance from fungi, bacteria might also harbor a wider repertoire of genes that confer ecologically relevant functions or are less likely to disrupt the existing protein network. Perhaps more importantly, bacteria are ubiquitous. Their effective population sizes are considerably larger than fungal or other eukaryotic populations, which means that fungi simply may encounter bacterial donors more frequently than fungal ones.

3.3.1 Some Fungal Lineages Are More Prone to HGT

Rates of HGT vary widely across the fungal tree of life (Fig. 3.2). For instance, members of the Pezizomycotina subphylum have long been thought to be particularly prone to HGT when compared to their sister lineage, the yeasts in the Saccharomycotina subphylum (Wisecaver and Rokas 2015; Marcet-Houben and Gabaldon 2010), or to the Basidiomycota (Marcet-Houben and Gabaldon 2010). Recently, a screen of HGT across ~50 non-Dikarya species (i.e., from phyla other than Ascomycota and Basidiomycota) showed that the frequency of HGT also varies among other fungal phyla, being for instance particularly high in lineages of anaerobic gut

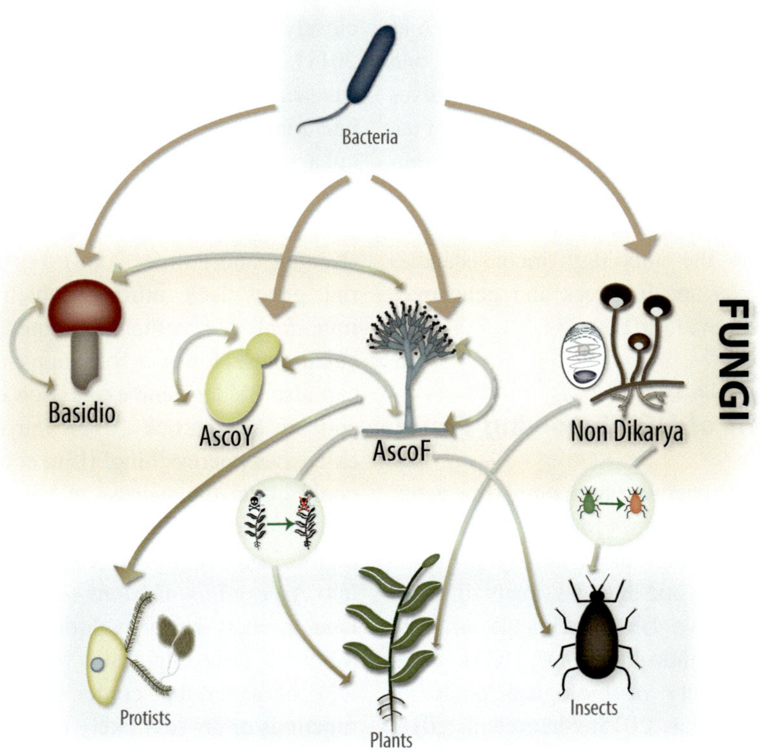

Fig. 3.1 Extent of HGT involving fungi. Fungal lineages were subdivided according to the wealth of information regarding HGT events. As such, fungal lineages belonging to the subkingdom Dikarya, for which more information is publicly available, were subdivided as follows: Basidio (representing the Basidiomycota), AscoY (representing ascomycete yeasts—Saccharomycotina), and AscoF (representing ascomycete filamentous fungi—Pezizomycotina). The Taphrinomycotina were not included in this diagram because of their lower number of species and scarce information regarding HGT (but see Slot and Rokas 2010). "Non Dikarya" represents fungal lineages other than the Ascomycota and the Basidiomycota. Protist lineages mostly refer to the oomycetes, although HGT events from fungi to other protists also have been documented (e.g., Matriano et al. 2021). Directionality of HGT events involving fungi as donors and recipients is represented by the arrows. The thickness of the arrow represents a rough estimate of the relative number of genes inferred to be involved in HGT events between the respective lineages, according to the inspected literature. Examples of fungi-to-insect HGT (involving carotenoid biosynthetic genes) (Nováková and Moran 2011; Moran and Jarvik 2010) and fungi-to-plant HGT (involving glutathione S-transferase Fhb7) (Wang et al. 2020) are illustrated inside the grey circles

fungi (Ciach et al. 2021), such as Neocallimastigomycota and Zoopagomycota (Wang et al. 2019; Murphy et al. 2019; Tabima et al. 2020). This is consistent with the high rates of HGT found in bacteria thriving in the animal gut (Groussin et al. 2021; Shterzer and Mizrahi 2015).

Animal guts are particularly challenging environments dominated by bacteria and characterized by very low levels of oxygen. In the guts of herbivorous animals, these harsh conditions coexist with favorable ones, such as the availability of large amounts of nutrients provided by plant material upon which herbivorous animals feed. Fungi belonging to the Neocallimastigomycota play an important role in the degradation of this plant material, and the acquisition of genes from neighboring rumen bacteria might have impacted this symbiotic relationship (Fig. 3.3) (Murphy et al. 2019). Accounting

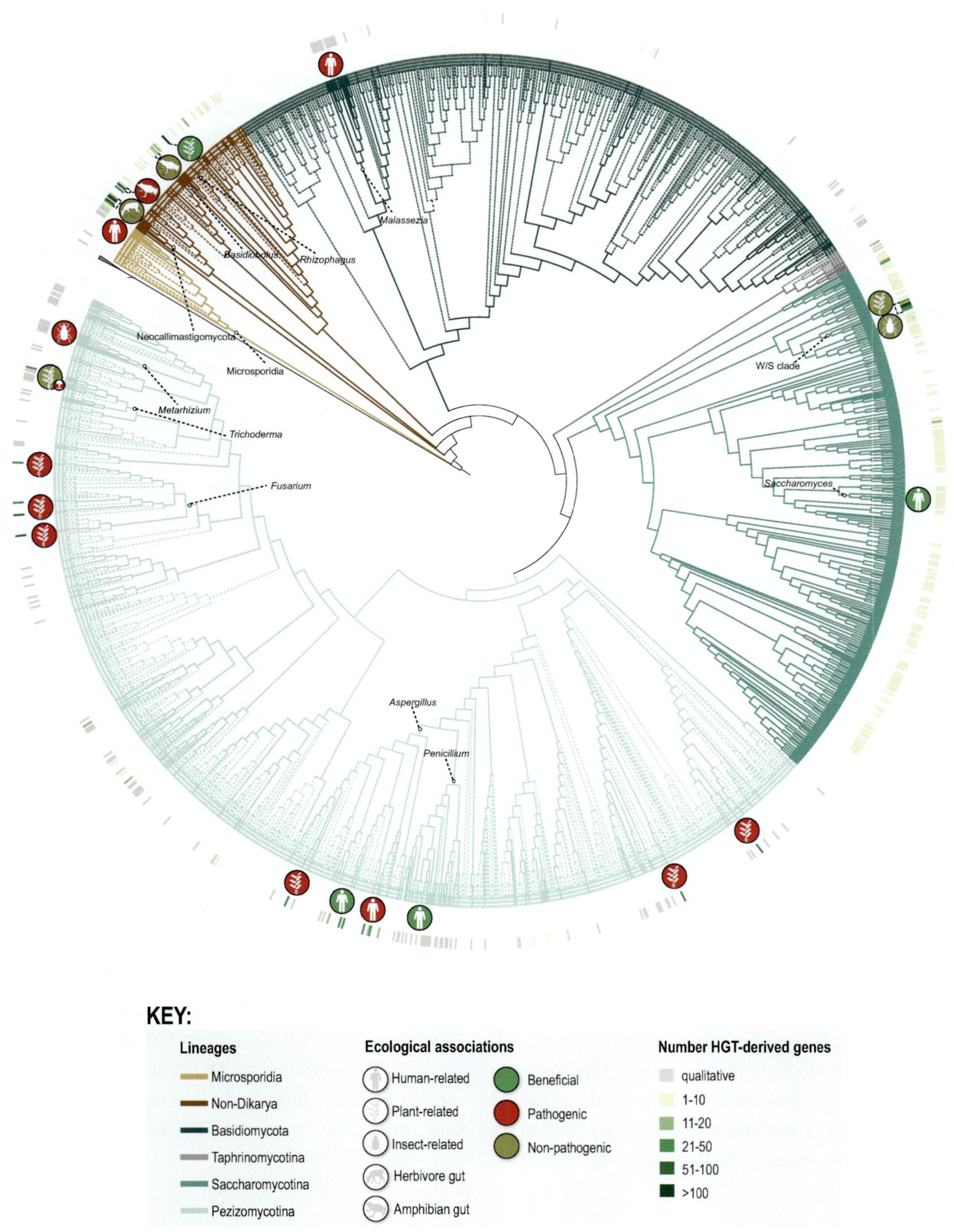

Fig. 3.2 Distribution of HGT across the fungal kingdom. Representation of the phylogenetic relationships among fungi as adapted from the phylogenomic study of the fungal kingdom from Li et al. (2021). Presence of HGT-derived genes is represented by the colored rectangles that form the outermost circle: rectangles with different shades of green denote cases where quantitative assessment was available in the literature from large comparative genomics studies (Marcet-Houben and Gabaldon 2010; Shen et al. 2018; Ciach et al. 2021), and gray rectangles when the reports result from isolated studies. Older HGT events into entire lineages were not considered for simplicity; however, many instances are documented (Richards et al. 2011a). Ecological associations, as well as the nature of such relationships (red being pathogenic, brown being non-pathogenic, and

for multiple duplications of acquired genes post-transfer, approximately 2–3.5% of these species' gene repertoire was inferred to be derived from HGT; this fraction is enriched in ecologically relevant functions (Fig. 3.3), including monosaccharide and polysaccharide degradation, which might have enabled these species to consume new carbon sources (Murphy et al. 2019). Adaptation to the anaerobic lifestyle in the rumen gut was possibly also facilitated by the acquisition of several genes involved in anaerobic nucleotide metabolism, fermentation pathways for energy generation, and cofactor recycling (Murphy et al. 2019; Wiersma et al. 2020; Perli et al. 2021), functions that were previously shown to be also involved in HGT in other anaerobic gut-dwelling microeukaryotes (Eme et al. 2017; Andersson et al. 2003).

The distantly related *Basidiobolus* (Zoopagomycota), an endosymbiont that inhabits the digestive tracts of amphibians, also shows particularly high rates of HGT of genes involved in the production of secondary metabolites (SMs), the biosynthesis of siderophores, and the synthesis of sexual reproduction precursor molecules; its high rates of HGT are hypothesized to have significantly expanded the SM gene repertoire in a lineage that is otherwise a poor SM producer (Tabima et al. 2020).

Parasitic relationships might also favor HGT owing to the close proximity between parasites and their hosts, as demonstrated for animals (Wijayawardena et al. 2013), plants (Yang et al. 2016), and oomycetes (Soanes and Richards 2014). In mycoparasitic *Trichoderma* fungi (Pezizomycotina), the repertoire of plant cell wall-degrading enzymes almost doubled through a series of gene acquisitions from their fungal prey species (Druzhinina et al. 2018). Although *Trichoderma* fungi parasitize species from both Pezizomycotina and Basidiomycota, notably, these transfers only involved genes originating in the Pezizomycotina. Similarly, closely related mycoparasite species belonging to the genus *Escovopsis*, which seemingly feed exclusively on Basidiomycota and exhibit a narrower nutritional spectrum than *Trichoderma*, exhibit negligible evidence for HGT (Druzhinina et al. 2018). This suggests that, although opportunity (parasitism) can enable HGT, other genetic or environmental factors also influence its success. For instance, genomic properties such as the intron-rich nature of basidiomycetous genes might have hampered their successful functionalization, which is consistent with the few reports of HGT involving Basidiomycota as donors to other fungal lineages (Slot and Hibbett 2007; Milner David et al. 2019).

Variation in HGT rates among clades is also observed in yeasts belonging to the Saccharomycotina subphylum (Ascomycota) (Shen et al. 2018) (Fig. 3.2). Species whose nuclear genetic code differs from the universal one have significantly fewer HGT-acquired genes (Shen et al. 2018), which supports the hypothesis that altered genetic codes hamper HGT. On the other end of the spectrum, a budding yeast lineage composed of species thriving in the floral environment (belonging to the genera *Wickerhamiella* and *Starmerella*—also known as the W/S clade) shows unexpectedly high rates of HGT originating in bacteria (Shen et al. 2018; Gonçalves et al. 2018) and impacting multiple metabolic pathways (Fig. 3.3) (Gonçalves et al. 2018; Kominek et al. 2019; Gonçalves and Gonçalves 2019). The reasons behind the particularly high incidence of HGT in the W/S clade remain obscure, but a turbulent evolutionary history marked by numerous ancestral (gene) loss

Fig. 3.2 (continued) green being possibly beneficial), of the species/lineages discussed in this chapter or species for which particularly high rates of HGT were found are represented in the inner circle. In the fungal phylogenetic tree, branches for which no information regarding HGT could be recovered from the literature are represented by dashed lines

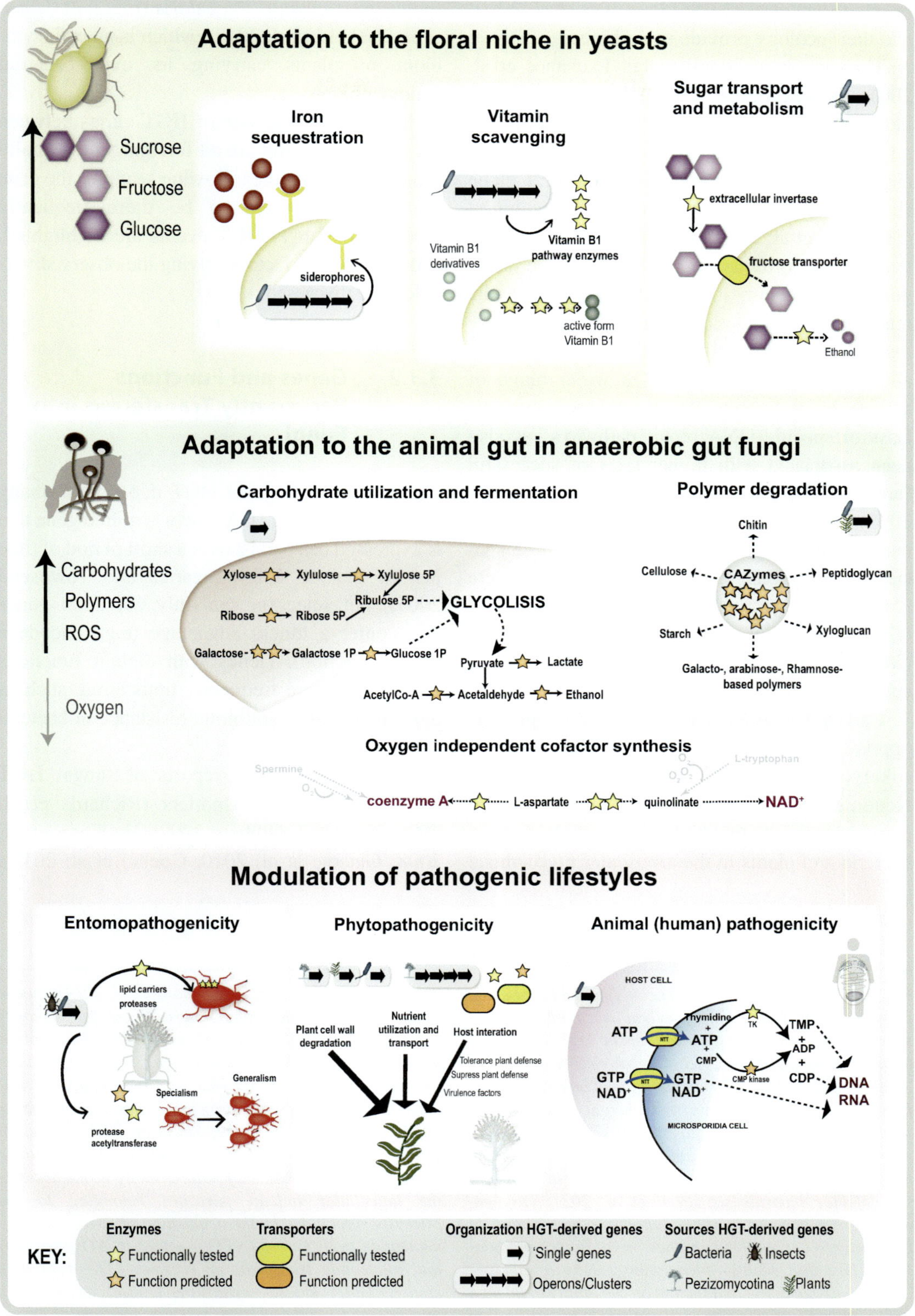

Fig. 3.3 Impact of HGT across fungal ecologies and lifestyles. Pathways and functions impacted by HGT are illustrated for select cases for which a role of HGT-derived genes in adaptation to certain ecological niches or modulation of lifestyles can be envisioned. These examples include the adaptation of floral yeasts (*Wickerhamiella*/

events (Shen et al. 2018; Gonçalves et al. 2020) and their ecology provide some hints.

Thriving in the floral niche (Lachance et al. 2001; de Vega et al. 2017), W/S-clade yeasts likely often travel across this environment in the guts of flower-visiting insects. Both W/S species and other microbes thriving in the floral niche show fingerprints of reductive evolution (Filannino et al. 2019), leading to the hypothesis that both opportunity for HGT (e.g., in the insect gut) and selection might have acted upon the reacquisition and maintenance of ancestrally lost, and ecologically relevant, traits (e.g., allowing the colonization of a wider range of microhabitats within the floral environment). Loss of specific DNA repair pathways has also been associated with higher rates of successful foreign DNA integration (Thomas and Nielsen 2005), but whether these and/or other molecular particularities have facilitated HGT in these species awaits further investigations. Notably, there is little evidence for ongoing HGT in this clade (Gonçalves et al. 2020), suggesting that the circumstances that once permitted extensive foreign DNA acquisition are no longer operating.

Carrying endosymbionts can also provide opportunities for HGT, as showcased across eukaryotes (Qiu et al. 2013; Sloan et al. 2014; Kondo et al. 2002). This model was proposed to partially explain the high rates of HGT from bacteria and plants in the arbuscular mycorrhizal fungus *Rhizophagus irregularis* (Fig. 3.2) (Ciach et al. 2021; Li et al. 2018), which is an endosymbiont of plants carrying its own bacterial endosymbionts.

In summary, the rate of HGT varies between fungal lineages. But are all foreign genes equally prone to HGT? In the following sections, the gene functions that seem to be more frequently involved in fungal HGT events are highlighted, and the possible factors driving the observed variation are discussed.

3.3.2 Genes and Functions Recurrently Transferred into Fungi

The success of a given HGT depends on many factors (see Sect. 3.2). A newly acquired gene that is expressed can persist over a short period of time just by genetic drift (Gogarten and Townsend 2005), but selection can only act upon genes that confer a fitness advantage (e.g., encode a useful function). Genes with certain functions seem to be more frequently transferred, such as those involved in antibiotic resistance in bacteria (Zhou et al. 2021).

Many of the early reports of fungal HGT involved nutrient transporters (Richards et al. 2006; Slot and Hibbett 2007; Hellborg et al. 2008; Galeote et al. 2010; Coelho et al. 2013),

Fig. 3.3 (continued) Starmerella *clade) to the floral niche impacted by the acquisition of several iron uptake, vitamin scavenging, and sugar metabolism related genes (Gonçalves et al. 2016, 2018; Gonçalves and Gonçalves 2019; Kominek et al. 2019); adaptation of anaerobic fungi from the Neocallimastigomycota to the gut of herbivorous animals assisted by the acquisition of genes involved in, but not limited to, carbohydrate assimilation, fermentation (Murphy et al. 2019), and oxygen-independent pathways for cofactor biosynthesis (Perli et al. 2021) (oxygen-dependent pathways in other fungi are represented in light gray to denote their absence in gut fungi); modulation of host shift (plants—phytopathogenicity to insects—entomopathogenicity) and impact on the ability to infect a wider range of insect species (specialism to generalism) in* Metarhizium *entomopathogens through the acquisition of several bacterial and insect-derived genes involved in the degradation of insect cuticles (Zhang et al. 2019); modulation of phytopathogenicity across multiple lineages impacted by the acquisition of genes generally involved in plant infection such as plant cell wall degradation enzymes, transport of nutrients, and production of virulence factors and effector proteins (Soanes and Richards 2014; McDonald et al. 2019; Yin et al. 2016; Qiu et al. 2016; Zhu et al. 2012; Tsavkelova et al. 2012; de Jonge et al. 2012); modulation of pathogenicity in members of the unicellular animal parasites belonging to the Microsporidia aided by the acquisition of genes encoding nucleotide transporters (NTT) involved in ATP transport and nucleotide transport, or enzymes (TK—thymidine kinase, CMP kinase—cytidylate kinase) involved in nucleic acid synthesis and salvage (Alexander et al. 2016; Dean et al. 2018)* which were hypothesized to have been particularly relevant to the osmotrophic mode of fungal feeding (Richards and Talbot 2013; Richards et al. 2011a). Functional data provided further support to the importance of these acquisitions in shaping the nutritional versatility of fungi (Coelho et al. 2013; Gonçalves et al. 2016; Marsit et al. 2015; Milner David et al. 2019). Interestingly, nutrient transporters seem to be so far mostly involved in intra-kingdom transfers, indicating that proper protein targeting to the membrane might be more difficult to accomplish when prokaryotic genes are involved.

Genes participating in diverse metabolic functions are also commonly transferred into fungi (Wisecaver and Rokas 2015; Richards et al. 2011a), as illustrated by early reports of genes involved in pyrimidine (Gojković et al. 2004) or biotin (Hall et al. 2005) biosynthesis in *Saccharomyces cerevisiae*, and as more recently and broadly observed across different fungal lineages (Wisecaver and Rokas 2015; Qiu et al. 2016; Shen et al. 2018; Murphy et al. 2019; Gonçalves et al. 2020).

The higher frequency of HGT involving genes within certain functional categories might be related to their low connectivity and ability to function independently, meaning that a single gene can by itself confer a new trait without disrupting the extant protein network (Cohen et al. 2011). Transporters are an emblematic example of low connectivity proteins that can alone impact the nutritional versatility of the host organism. Similarly, metabolic gene clusters are also examples of standalone metabolic units that can undergo HGT (Slot and Hibbett 2007; Patron et al. 2007; Khaldi et al. 2008; Lind et al. 2017; Reynolds et al. 2018; Slot and Rokas 2011). These examples suggest that function and connectivity are intertwined factors influencing the success of HGT.

The repeated involvement of certain genes and/or functions in fungal HGT suggests that natural selection is involved in their maintenance. Recurrent transfers of genes involved in the same pathway within the same clade (Alexander et al. 2016; Gonçalves et al. 2018; Shen et al. 2018) or among distantly related lineages (Bruto et al. 2014; Naranjo-Ortíz et al. 2016) have been frequently observed. For instance, distinct fungal lineages have independently acquired bacterial nitric oxide-detoxifying flavohemoglobins (Yhb1) (Wisecaver et al. 2016), involved in mediating nitric oxide resistance and nitrosative stress tolerance (Wisecaver et al. 2016; Ianiri et al. 2020). In the basidiomycetous yeast *Malassezia*, two independent acquisitions of *YHB1* were supported by both phylogenetic and synteny analyses, while functional experiments suggest a possible role in the commensal lifestyle of *Malassezia* in the human skin (Ianiri et al. 2020).

While some of these recurrent transfers might encode new functions or enhance existing ones, others are involved in the reacquisition of ancestrally lost traits. For instance, several yeasts belonging to the W/S clade, which had previously lost the entire alcoholic fermentation pathway (Gonçalves et al. 2018), later reacquired three distinct alcohol dehydrogenase encoding genes (*ADH1*) independently from distinct bacterial donors (Gonçalves et al. 2018, 2020). In one species, the acquisition of this single gene, coupled with the co-option of a native enzyme, was sufficient to reinstate alcoholic fermentation (Gonçalves et al. 2018).

Other metabolic functions might depend on the activity of multiple enzymes encoded by different genes. Many of these multigenic metabolic pathways are encoded in gene clusters, which can be transferred across fungal species as a unit (Wisecaver and Rokas 2015; Marcet-Houben and Gabaldón 2019), and have even been found to do so recurrently.

For instance, the catabolic enzymes that are part of the Leloir pathway for the utilization of galactose are encoded in *GAL* gene clusters in ascomycetous yeasts, where they have been lost multiple times in different lineages (Harrison et al. 2022). Interestingly, some of these lineages reacquired *GAL* gene clusters via HGT from distantly related yeast species (Slot and Rokas 2010; Haase et al. 2021), which in most cases resulted in the reinstatement of the ability to utilize galactose (Haase et al. 2021). In the genus *Torulaspora*, both species- and strain-level HGT events involving *GAL* clusters might be related to adaptation to galactose-rich environments (Venkatesh et al. 2021; Silva et al. 2022).

Gene clusters have been commonly transferred across fungi, but their bacterial counterparts (operons) rarely seem to transfer into eukaryotes even though they are recurrently transferred among bacteria (Lawrence and Roth 1996). Building on their high levels of HGT events involving single genes, in the W/S yeast clade two bacterial operons with distinct functions were also horizontally acquired (Kominek et al. 2019; Gonçalves and Gonçalves 2019). The first reported operon transfer introduced an entirely new pathway for siderophore biosynthesis in the common ancestor of this clade and is involved in iron scavenging in many extant species (Fig. 3.3) (Kominek et al. 2019). Most species for which no siderophore production could be detected underwent secondary losses of the operon (Kominek et al. 2019), possibly due to relaxation of selective constraints. The other operon encodes a vitamin B1 salvage and biosynthetic pathway and was acquired on two independent occasions from different bacterial donors by two yeast subclades (Gonçalves and Gonçalves 2019). Independent acquisition of additional genes expanded the spectrum of operon-encoded functions, allowing for the scavenging of a wider range of vitamin B1 derivatives (Fig. 3.3) (Gonçalves and Gonçalves 2019). The absence of native yeast orthologs from all the genomes inspected suggests a loss-prior-to-acquisition scenario, once again underscoring (recurrent) HGT as an important compensatory mechanism for ancestral pathway and trait loss.

Although functional data are relatively scarce for a substantial fraction of HGT events involving fungi, acquired genes tend to generally maintain their original functions. Yet, in some cases HGT-derived genes can functionally diverge (Savory et al. 2018) or be subsequently lost (Campbell et al. 2012; Kominek et al. 2019).

3.3.3 Genes and Functions Transferred from Fungi

Most known examples where fungi have been donors of genes involve other fungi as recipients. However, fungal genes have occasionally jumped into the genomes of other eukaryotes, contributing to important evolutionary innovations (Fig. 3.1). The highest rate of cross-kingdom transfer involving fungi as donors probably occurred in the oomycetes (Richards et al. 2006, 2011b), an independently evolved lineage of osmophilic microbial eukaryotes. Both the period of evolution in which these transfers took place, which coincides with the emergence of phytopathogenicity in the oomycetes (Richards et al. 2011b), and the functions provided by the acquired genes, which include nutrient transport, plant cell wall digestion, and interaction with or adhesion to the host (Richards et al. 2006, 2011b; Soanes and Richards 2014; Savory et al. 2015), hint at an important role of fungal genes in the evolution of oomycete plant pathogens.

Fungi-to-plant HGT might also have significantly contributed to plant evolution. Impacts on the transition from aquatic to terrestrial environments (Ma et al. 2022) and on plant defense mechanisms (Liu et al. 2022) have been hypothesized based on the predicted functions of the acquired genes. A striking example concerns the HGT of a glutathione S-transferase (*Fhb7*) from endophytic *Epichloë* fungi to a close relative of wheat (Wang et al. 2020). The protein shares ~97% of identity with those from *Epichloë* species and displays a patchy distribution across the genus, suggesting that the acquisition of Fhb7 was fairly recent. Most notably, Fhb7 is involved in the detoxification of fungal-produced toxins, namely those produced by the phytopathogen *Fusarium graminearum* (Fig. 3.1) (Wang et al. 2020), indicating that the acquisition of this gene might be important for plant defense against fungal pathogens.

Insects also have horizontally acquired fungal genes, and such events across insect lineages might be substantially more widespread than previously anticipated (Li et al. 2022); however, for most cases the functional role remains elusive. In other cases, such as the ~200-million-year-old transfer of pectin-degrading enzymes from ascomycetous fungi to two herbivorous beetle lineages, the acquired genes might have contributed to insect's ability to degrade plant cell walls (Kirsch et al. 2014). Probably one of the most iconic cases of fungi-to-insect HGT is the transfer of two carotenoid biosynthetic genes to the pea aphid (Fig. 3.1) (Nováková and Moran 2011; Moran and Jarvik 2010). Carotenoids are essential to animals, being involved both in pigmentation and in resistance to oxidative stress, but they are exclusively obtained through their diet. Aphids circumvented this limitation by making use of carotenoid biosynthesis genes originally acquired from fungi, becoming therefore able to produce their own carotenoids. Aphid individuals with this fungal pathway have red body pigmentation instead of the typical green (Fig. 3.1), and this affects aphid susceptibility to predators and parasites (Moran and Jarvik 2010). The acquired genes currently display a typical aphid gene structure (e.g., longer introns), but their original gene orientation where the carotenoid desaturase and the carotenoid synthase–carotenoid cyclase are divergently transcribed has been preserved (Moran and Jarvik 2010), suggesting that this genetic makeup is important for concerted gene expression from a single promoter region. Interestingly, the same genes were later detected in the genome of the spotted spider mite, which also exhibits green and red colorations (Altincicek et al. 2012; Bryon et al. 2017). Whether these genes resulted from an independent HGT event or from a single transfer into the common ancestor of mites and aphids followed by differential loss is unclear.

Finally, so far, HGT between bacteria and fungi seems to chiefly work in one direction (from bacteria to fungi) since fungal or other eukaryotic genes have not been frequently found in bacterial genomes. This might be because opportunity and inherent mechanisms favor intradomain transfers in bacteria or because structural barriers hinder eukaryote-to-prokaryote transfers (Keeling and Palmer 2008).

3.4 Ecological Impact of Fungal HGT

Increasing experimental evidence has been instrumental in examining whether and how HGT contributes to fungal adaptation in distinct environments. As discussed in the previous section, extensive and recurrent HGT can drive major ecological transitions by impacting several aspects of the host biology (Fig. 3.3). Even in cases where the scale of HGT is small, its phenotypic consequences can be substantial. In some cases, the independent emergence of traits in distantly related lineages can be promoted by HGT. For instance, the independent acquisition of a single specific fructose transporter gene was indispensable to the evolution of fructophily, the preference for fructose over glucose, in two distantly related Saccharomycotina yeast clades (Gonçalves et al. 2016). Similarly, the ability to sense gravity (gravitropism), which is important to orient growth of fruiting bodies, is found across the Basidiomycota and also in the distantly related Mucorales (Nguyen et al. 2018). In the latter lineage, the acquisition of a bacterial protein forming octahedral protein crystals might be at the origin of the independent evolution of this trait (Nguyen et al. 2018).

Fungal pathogens affecting humans, other animals, and plants (Rokas 2022; Soanes and Richards 2014; Zhang et al. 2019) also show a mosaic distribution across the fungal tree of life (Fig. 3.2), indicating that pathogenicity evolved from non-pathogenic ancestors on multiple occasions. This repeated evolution of pathogenicity has been driven by diverse genomic changes, such as rearrangements or gene duplication and loss (Rokas 2022; Möller and Stukenbrock 2017), but certain aspects of fungal pathogenicity were shaped by HGT (Soanes and Richards 2014; Alexander et al. 2016; Zhang et al. 2019; Sun et al. 2016; de Vries et al. 2020; Corradi 2015; Gluck-Thaler and Slot 2015). In many instances, the acquired genes are encoded within gene clusters, massive transposable elements, or entire chromosomes encompassing single or multiple functional genes (Slot and Rokas 2011;

Wisecaver and Rokas 2015; Gluck-Thaler and Slot 2015; Ma et al. 2010; McDonald et al. 2019; Dhillon et al. 2015).

Several HGT events, involving both bacterial and insect genes, contributed to the evolution of entomopathogenicity in *Metarhizium* (Pezizomycotina). The acquired genes are generally involved in the degradation of insect cuticles and were shown to be functional and upregulated during cuticle penetration (Fig. 3.3) (Zhang et al. 2019). The HGT-derived repertoire varies among species being larger in species that exhibit a broader host range (generalists). Particularly, an HGT-derived protease was experimentally shown to widen the host range of a specialist *Metarhizium* species (Zhang et al. 2019), suggesting a role for HGT in shifting from a specialist to a generalist lifestyle.

Evolution of pathogenicity in the obligate intracellular parasite phylum of Microsporidia was also fueled by several HGT events, mostly impacting their ability to survive in and to scavenge nucleotides and nucleosides from host cells (Corradi 2015; Alexander et al. 2016; Dean et al. 2018). Some HGT-derived nucleotide transporters have diversified to import additional substrates (Dean et al. 2018), while several HGT-derived enzymes have expanded the range of usable nucleic acid components available in host cells (Alexander et al. 2016), allowing for the scavenging of nucleic acid components in species seemingly unable to synthesize them *de novo*. Some Microsporidia species also acquired genes from the host they parasitize (Pombert et al. 2015; Selman et al. 2011).

Generally, for many of the genes transferred into fungal pathogens, an association between the functions of acquired genes and the organism's pathogenic lifestyle can be envisaged (Fig. 3.3). These functions are encoded in genes involved in specialized metabolism influencing interactions with the host and environment (Wisecaver and Rokas 2015; Sun et al. 2016; Ianiri et al. 2020), host cell degradation (Zhang et al. 2019; Dhillon et al. 2015), scavenging of host resources (Alexander et al. 2016, Dean et al. 2018), virulence (McDonald et al. 2019; Friesen et al. 2006), or stress resistance (Ianiri et al. 2020; Corradi 2015). This pattern suggests that, under similar selective pressures, such as those encountered in the same ecological niche, HGT-derived genes encoding similar functions tend to be retained.

While some fungi can be pathogenic, others are beneficial for our daily lives. The most iconic example is arguably the baker's yeast *Saccharomyces cerevisiae*, which is involved in the production of a plethora of food products and medicines. Several aspects of the metabolism of the baker's yeast have been impacted by HGT, such as the early acquisition of a bacterial *URA1* gene by a *S. cerevisiae* ancestor, which is thought to have facilitated growth under low-oxygen conditions (Hall et al. 2005), commonly faced in the natural environments where this organism thrives. More recent events have influenced other aspects of the metabolism of *S. cerevisiae* strains associated with human-related activities. For example, *S. cerevisiae* strains associated with wine-making acquired three distinct regions from closely related yeast species; these regions encompass several genes that might provide a selective advantage during wine fermentation, such as sugar, vitamins, and amino acid or oligopeptide transporters (Novo et al. 2009; Marsit et al. 2015). HGT-derived regions are particularly prevalent in other strains associated with human-related environments, suggesting that they might be signatures of domestication (Barbosa et al. 2018; Peter et al. 2018).

In cheese-making *Penicillium* species, multiple large genomic regions, some of which embedded in massive transposable elements (Gluck-Thaler et al. 2022), show evidence of interspecies HGT and might be involved in conferring a competitive advantage in the cheese environment (Ropars et al. 2015). For example, genes involved in lactose metabolism were found to be strongly upregulated during cheese rind maturation (Ropars et al. 2015). Lactose is especially abundant in the first steps of cheese maturation when it is readily converted into lactate by the lactic acid bacteria present; therefore, being able to rapidly utilize it would provide a competitive advantage.

In summary, some of the HGTs into fungi are ancient, reflecting ancestral selective pressures that facilitated major evolutionary transitions and affected entire lineages such as demonstrated for some fungal pathogens; others are very recent, reflecting short but intense periods of selection such as those imposed by recent human-related activities. Finally, it is worth noting that ecological environments are complex and dynamic and that reproducing their key aspects in the laboratory is often impractical, hindering our ability to dissect the evolutionary forces that contributed to the acquisition and retention of HGT-derived genes across fungi.

3.5 Concluding Remarks and Outstanding Questions

We set out to discuss several major questions concerning HGT in eukaryotes based on what has been learned from HGTs involving fungi. The mode, tempo, and magnitude of HGT in fungi seem to be, overall, a result of a complex equation involving environmental conditions, cellular states (e.g., permissivity to foreign DNA), genetic backgrounds (e.g., gene content, genetic code), and selective pressures. Although advances in fungal HGT in the last decade have been tremendous, several outstanding questions remain.

Why Do Some Clades/Species Seem to Be More Prone to HGT than Others? Species thriving in certain ecological niches seem to be more prone to HGT, but the extent of the association between HGT and fungal ecology remains unknown. The wealth of data from fungal species with particularly high rates of HGT that are distantly related and thrive in distinct ecological contexts opens avenues for understanding the relationship between fungal HGT and ecology. Examples of such lineages of particular interest for future research include the *Wickerhamiella/Starmerella* clade of yeasts and the gut-dwelling fungi belonging to the Neocallimastigomycota because of the high percentages of foreign genes encoded in their genomes and their well-defined ecologies. Can the repertoires of foreign and native genes in these lineages provide hints? Can the acquisition of certain genes facilitate the acquisition of others? Are there other molecular particularities that make these lineages more permeable to HGT? Are some species naturally more competent at DNA uptake than others? Which factors underlie the propensity to acquire foreign DNA?

How Much HGT and from Whom? Bacterial donors seem to be the most common source of foreign genes in fungal genomes, but nutrient transporters seem to be more frequently involved in transfers within eukaryotes than in transfers from bacteria. Are these observations resulting from sampling biases in genome sequencing or methodological ascertainment biases, or do they reflect biological factors? These questions would benefit from comparative genomic analyses. The wealth of genomic data currently available for fungi (Shen et al. 2020; Li et al. 2021) provides an encouraging launching pad for such analyses. So far, a handful of studies have focused on high-throughput inference of HGT from bacterial donors, yet the more challenging inference of HGT events between closely related species has been less frequently addressed. This bias hinders our ability to evaluate the real extent of HGT across fungi or whether all lineages are equally likely to serve as donors. However, both rich genome sampling and improved methodologies can now be employed. For instance, phylogenomic pipelines starting with fast screening of potential candidates, such as the alien index scores (Gladyshev et al. 2008), followed by robust phylogenetic reconstruction using state-of-the-art statistical methods and meticulous species tree and gene tree reconciliation analyses, can be employed for robust high-throughput HGT inference across fungi (Wisecaver and Rokas 2015; Gonçalves et al. 2018; Shen et al. 2018; Alexander et al. 2016).

What Happens Immediately After an HGT Event? While some HGT events in fungi seem to have occurred rather recently, most of the cases reported entail relatively old events, making it difficult to draw conclusions about, for instance, the processes involved in the accommodation of a newly arrived gene to the new host cell. Which mechanisms are more commonly used by fungi when acquiring foreign DNA? What happens to the genes immediately after being transferred? Does the genome location where a foreign gene lands affect the success of an HGT event? How does it impact the extant gene and protein network? What is the relative role of selection and drift early in an HGT event? Does the donor affect the success of an HGT event? Although ambitious, state-of-the-art approaches such as strain engineering and experimental evolution can help address most of these questions.

What Are the Population Dynamics of HGT? Population-level variability in the presence/absence of HGT-derived genes has only been studied in a few fungi (McDonald et al. 2019; Urquhart et al. 2022; Marsit et al. 2015). For instance, certain HGT-acquired regions are only found in some *S. cerevisiae* populations. In some cases, very recent acquisitions might explain this intraspecific patchy distribution, while in other cases, fluctuating selective pressures might either promote maintenance or gradual decay of previously acquired genes. These dynamics are still blurry, awaiting population-level studies, but they constitute an interesting aspect that might inform us about the evolutionary forces that govern HGT-derived genes, how they respond to environmental changes, and how HGT impacts (and is impacted by) fungal ecology at the population level.

Acknowledgments Research in A.R.'s lab is supported by grants from the National Science Foundation (DEB-2110404), the National Institutes of Health/National Institute of Allergy and Infectious Diseases (R56 AI146096 and R01 AI153356), and the Burroughs Wellcome Fund. Research in C.T.H.'s lab is supported by the National Science Foundation (DEB-1442148 and DEB-2110403), in part by the DOE Great Lakes Bioenergy Research Center (DOE BER Office of Science DE-SC0018409), the USDA National Institute of Food and Agriculture (Hatch Project 1020204), and an H. I. Romnes Faculty Fellowship from the Office of the Vice Chancellor for Research and Graduate Education with funding from the Wisconsin Alumni Research Foundation.

Conflict of Interest Statement A.R. is a scientific consultant for LifeMine Therapeutics, Inc.

References

Alexander WG, Wisecaver JH, Rokas A, Hittinger CT (2016) Horizontally acquired genes in early-diverging pathogenic fungi enable the use of host nucleosides and nucleotides. Proc Natl Acad Sci U S A 113: 4116–4121

Altincicek B, Kovacs JL, Gerardo NM (2012) Horizontally transferred fungal carotenoid genes in the two-spotted spider mite *Tetranychus urticae*. Biol Lett 8:253–257

Andam CP, Gogarten JP (2011) Biased gene transfer in microbial evolution. Nat Rev Microbiol 9:543–555

Andersson JO, Sjögren AM, Davis LA, Embley TM, Roger AJ (2003) Phylogenetic analyses of diplomonad genes reveal frequent lateral gene transfers affecting eukaryotes. Curr Biol 13:94–104

Archibald JM, Rogers MB, Toop M, Ishida K-I, Keeling PJ (2003) Lateral gene transfer and the evolution of plastid-targeted proteins in the secondary plastid-containing alga *Bigelowiella natans*. Proc Natl Acad Sci U S A 100:7678–7683

Arnold BJ, Huang IT, Hanage WP (2022) Horizontal gene transfer and adaptive evolution in bacteria. Nat Rev Microbiol 20:206–218

Barber AE, Sae-Ong T, Kang K, Seelbinder B, Li J, Walther G, Panagiotou G, Kurzai O (2021) *Aspergillus fumigatus* pan-genome analysis identifies genetic variants associated with human infection. Nat Microbiol 6:1526–1536

Barbosa R, Pontes A, Santos RO, Montandon GG, de Ponzzes-Gomes CM, Morais PB, Gonçalves P, Rosa CA, Sampaio JP (2018) Multiple rounds of artificial selection promote microbe secondary domestication—the case of cachaça yeasts. Genome Biol Evol 10: 1939–1955

Bian R, Andika IB, Pang T, Lian Z, Wei S, Niu E, Wu Y, Kondo H, Liu X, Sun L (2020) Facilitative and synergistic interactions between fungal and plant viruses. Proc Natl Acad Sci U S A 117:3779–3788

Bredeweg EL, Baker SE (2020) Horizontal Gene Transfer in Fungi. In: Nevalainen H (ed) Grand challenges in fungal biotechnology. Springer International Publishing, Cham

Brockhurst MA, Harrison E, Hall JPJ, Richards T, McNally A, MacLean C (2019) The ecology and evolution of pangenomes. Curr Biol 29:R1094–r1103

Bruto M, Prigent-Combaret C, Luis P, Moënne-Loccoz Y, Muller D (2014) Frequent, independent transfers of a catabolic gene from bacteria to contrasted filamentous eukaryotes. Proc Biol Sci 281:20140848

Bryon A, Kurlovs AH, Dermauw W, Greenhalgh R, Riga M, Grbić M, Tirry L, Osakabe M, Vontas J, Clark RM, Van Leeuwen T (2017) Disruption of a horizontally transferred phytoene desaturase abolishes carotenoid accumulation and diapause in *Tetranychus urticae*. Proc Natl Acad Sci 114:E5871–E5880

Campbell MA, Rokas A, Slot JC (2012) Horizontal transfer and death of a fungal secondary metabolic gene cluster. Genome Biol Evol 4:289–293

Ciach M, Pawłowska J, Muszewska A (2021) Horizontal gene transfer in 44 early diverging fungi favors short, metabolic, extracellular proteins from associated bacteria. *bioRxiv* 2021.12.02.471044

Coelho MA, Gonçalves C, Sampaio JP, Gonçalves P (2013) Extensive intra-kingdom horizontal gene transfer converging on a fungal fructose transporter gene. PLoS Genet 9:e1003587

Cohen O, Gophna U, Pupko T (2011) The complexity hypothesis revisited: connectivity rather than function constitutes a barrier to horizontal gene transfer. Mol Biol Evol 28:1481–1489

Coleine C, Stajich JE, Selbmann L (2022) Fungi are key players in extreme ecosystems. Trends Ecol Evol 37: 517–528

Corradi N (2015) Microsporidia: eukaryotic intracellular parasites shaped by gene loss and horizontal gene transfers. Annu Rev Microbiol 69:167–183

Cote-L'Heureux A, Maurer-Alcalá XX, Katz LA (2022) Old genes in new places: a taxon-rich analysis of interdomain lateral gene transfer events. PLoS Genet 18:e1010239

Dagan T, Artzy-Randrup Y, Martin W (2008) Modular networks and cumulative impact of lateral transfer in prokaryote genome evolution. Proc Natl Acad Sci U S A 105:10039–10044

de Jonge R, Peter van Esse H, Maruthachalam K, Bolton MD, Santhanam P, Saber MK, Zhang Z, Usami T, Lievens B, Subbarao KV, Thomma BPHJ (2012) Tomato immune receptor Ve1 recognizes effector of multiple fungal pathogens uncovered by genome and RNA sequencing. Proc Natl Acad Sci U S A 109: 5110–5115

de Reus E, Nielsen MR, Frandsen RJN (2019) Metabolic and regulatory insights from the experimental horizontal gene transfer of the aurofusarin and bikaverin gene clusters to *Aspergillus nidulans*. Mol Microbiol 112: 1684–1700

de Vega C, Albaladejo RG, Guzman B, Steenhuisen SL, Johnson SD, Herrera CM, Lachance MA (2017) Flowers as a reservoir of yeast diversity: description of *Wickerhamiella nectarea* f.a. sp. nov., and *Wickerhamiella natalensis* f.a. sp. nov. from South African flowers and pollinators, and transfer of related *Candida* species to the genus *Wickerhamiella* as new combinations. FEMS Yeast Res 17

de Vries S, Stukenbrock EH, Rose LE (2020) Rapid evolution in plant-microbe interactions - an evolutionary genomics perspective. New Phytol 226:1256–1262

Dean P, Sendra KM, Williams TA, Watson AK, Major P, Nakjang S, Kozhevnikova E, Goldberg AV, Kunji ERS, Hirt RP, Embley TM (2018) Transporter gene acquisition and innovation in the evolution of Microsporidia intracellular parasites. Nat Commun 9: 1709

Dhillon B, Feau N, Aerts AL, Beauseigle S, Bernier L, Copeland A, Foster A, Gill N, Henrissat B, Herath P, LaButti KM, Levasseur A, Lindquist EA, Majoor E, Ohm RA, Pangilinan JL, Pribowo A, Saddler JN, Sakalidis ML, de Vries RP, Grigoriev IV, Goodwin SB, Tanguay P, Hamelin RC (2015) Horizontal gene transfer and gene dosage drives adaptation to wood colonization in a tree pathogen. Proc Natl Acad Sci U S A 112:3451–3456

Doolittle WF (1998) You are what you eat: a gene transfer ratchet could account for bacterial genes in eukaryotic nuclear genomes. Trends Genet 14:307–311

Doolittle WF (1999) Lateral genomics. Trends Cell Biol 9: M5–M8

Doolittle WF, Boucher Y, Nesbø CL, Douady CJ, Andersson JO, Roger AJ (2003) How big is the iceberg of which organellar genes in nuclear genomes are but the tip? Philos Trans R Soc Lond Ser B Biol Sci 358: 39–57 discussion 57–8

Druzhinina IS, Chenthamara K, Zhang J, Atanasova L, Yang D, Miao Y, Rahimi MJ, Grujic M, Cai F, Pourmehdi S, Salim KA, Pretzer C, Kopchinskiy AG, Henrissat B, Kuo A, Hundley H, Wang M, Aerts A, Salamov A, Lipzen A, LaButti K, Barry K, Grigoriev IV, Shen Q, Kubicek CP (2018) Massive lateral transfer of genes encoding plant cell wall-degrading enzymes to the mycoparasitic fungus *Trichoderma* from its plant-associated hosts. PLoS Genet 14: e1007322

Dunning Hotopp JC (2011) Horizontal gene transfer between bacteria and animals. Trends Genet 27:157–163

Eme L, Gentekaki E, Curtis B, Archibald JM, Roger AJ (2017) Lateral gene transfer in the adaptation of the anaerobic parasite *Blastocystis* to the gut. Curr Biol 27: 807–820

Eyres I, Boschetti C, Crisp A, Smith TP, Fontaneto D, Tunnacliffe A, Barraclough TG (2015) Horizontal gene transfer in bdelloid rotifers is ancient, ongoing and more frequent in species from desiccating habitats. BMC Biol 13:90

Fan X, Qiu H, Han W, Wang Y, Xu D, Zhang X, Bhattacharya D, Ye N (2020) Phytoplankton pangenome reveals extensive prokaryotic horizontal gene transfer of diverse functions. Science. Advances 6:eaba0111

Filannino P, Di Cagno R, Tlais AZA, Cantatore V, Gobbetti M (2019) Fructose-rich niches traced the evolution of lactic acid bacteria toward fructophilic species. Crit Rev Microbiol 45:65–81

Fitzpatrick DA (2012) Horizontal gene transfer in fungi. FEMS Microbiol Lett 329:1–8

Franco MEE, Wisecaver JH, Arnold AE, Ju Y-M, Slot JC, Ahrendt S, Moore LP, Eastman KE, Scott K, Konkel Z, Mondo SJ, Kuo A, Hayes RD, Haridas S, Andreopoulos B, Riley R, LaButti K, Pangilinan J, Lipzen A, Amirebrahimi M, Yan J, Adam C,

Keymanesh K, Ng V, Louie K, Northen T, Drula E, Henrissat B, Hsieh H-M, Youens-Clark K, Lutzoni F, Miadlikowska J, Eastwood DC, Hamelin RC, Grigoriev IV, U'Ren JM (2022) Ecological generalism drives hyperdiversity of secondary metabolite gene clusters in xylarialean endophytes. New Phytol 233: 1317–1330

Friesen TL, Stukenbrock EH, Liu Z, Meinhardt S, Ling H, Faris JD, Rasmussen JB, Solomon PS, McDonald BA, Oliver RP (2006) Emergence of a new disease as a result of interspecific virulence gene transfer. Nat Genet 38:953–956

Galeote V, Novo M, Salema-Oom M, Brion C, Valério E, Gonçalves P, Dequin S (2010) *FSY1*, a horizontally transferred gene in the *Saccharomyces cerevisiae* EC1118 wine yeast strain, encodes a high-affinity fructose/H+ symporter. Microbiology 156:3754–3761

Gladyshev EA, Meselson M, Arkhipova IR (2008) Massive horizontal gene transfer in bdelloid rotifers. Science 320:1210–1213

Gluck-Thaler E, Slot JC (2015) Dimensions of horizontal gene transfer in eukaryotic microbial pathogens. PLoS Pathog 11:e1005156

Gluck-Thaler E, Ralston T, Konkel Z, Ocampos CG, Ganeshan VD, Dorrance AE, Niblack TL, Wood CW, Slot JC, Lopez-Nicora HD, Vogan AA (2022) Giant starship elements mobilize accessory genes in fungal genomes. Mol Biol Evol 39

Gogarten JP, Townsend JP (2005) Horizontal gene transfer, genome innovation and evolution. Nat Rev Microbiol 3:679–687

Gojković Z, Knecht W, Zameitat E, Warneboldt J, Coutelis JB, Pynyaha Y, Neuveglise C, Møller K, Löffler M, Piskur J (2004) Horizontal gene transfer promoted evolution of the ability to propagate under anaerobic conditions in yeasts. Mol Gen Genomics 271:387–393

Gonçalves C, Gonçalves P (2019) Multilayered horizontal operon transfers from bacteria reconstruct a thiamine salvage pathway in yeasts. Proc Natl Acad Sci U S A 116:22219–22228

Gonçalves P, Gonçalves C (2022) Horizontal gene transfer in yeasts. Curr Opin Genet Dev 76:101950

Gonçalves C, Coelho MA, Salema-Oom M, Gonçalves P (2016) Stepwise functional evolution in a fungal sugar transporter family. Mol Biol Evol 33:352–366

Gonçalves C, Wisecaver JH, Kominek J, Oom MS, Leandro MJ, Shen X-X, Opulente DA, Zhou X, Peris D, Kurtzman CP, Hittinger CT, Rokas A, Gonçalves P (2018) Evidence for loss and reacquisition of alcoholic fermentation in a fructophilic yeast lineage. elife 7:e33034

Gonçalves P, Gonçalves C, Brito PH, Sampaio JP (2020) The *Wickerhamiella/Starmerella* clade—a treasure trove for the study of the evolution of yeast metabolism. Yeast 37(4):313–320

Grandaubert J, Dutheil JY, Stukenbrock EH (2019) The genomic determinants of adaptive evolution in a fungal pathogen. Evol Lett 3:299–312

Groussin M, Poyet M, Sistiaga A, Kearney SM, Moniz K, Noel M, Hooker J, Gibbons SM, Segurel L, Froment A, Mohamed RS, Fezeu A, Juimo VA, Lafosse S, Tabe FE, Girard C, Iqaluk D, Nguyen LTT, Shapiro BJ, Lehtimäki J, Ruokolainen L, Kettunen PP, Vatanen T, Sigwazi S, Mabulla A, Domínguez-Rodrigo M, Nartey YA, Agyei-Nkansah-A, Duah A, Awuku YA, Valles KA, Asibey SO, Afihene MY, Roberts LR, Plymoth A, Onyekwere CA, Summons RE, Xavier RJ, Alm EJ (2021) Elevated rates of horizontal gene transfer in the industrialized human microbiome. Cell 184:2053–2067.e18

Haase MAB, Kominek J, Opulente DA, Shen XX, LaBella AL, Zhou X, DeVirgilio J, Hulfachor AB, Kurtzman CP, Rokas A, Hittinger CT (2021) Repeated horizontal gene transfer of GALactose metabolism genes violates Dollo's law of irreversible loss. Genetics 217

Hall C, Brachat S, Dietrich FS (2005) Contribution of horizontal gene transfer to the evolution of *Saccharomyces cerevisiae.* Eukaryot Cell 4:1102–1115

Hao W, Richardson AO, Zheng Y, Palmer JD (2010) Gorgeous mosaic of mitochondrial genes created by horizontal transfer and gene conversion. Proc Natl Acad Sci U S A 107:21576–21581

Harrison M-C, LaBella AL, Hittinger CT, Rokas A (2022) The evolution of the GALactose utilization pathway in budding yeasts. Trends Genet 38:97–106

Hawksworth DL, Lücking R, Heitman J, James TY (2017) Fungal diversity revisited: 2.2 to 3.8 million species. Microbiology Spectrum, 5, 5.4.10

Heitman J (2010) Evolution of eukaryotic microbial pathogens via covert sexual reproduction. Cell Host Microbe 8:86–99

Hellborg L, Woolfit M, Arthursson-Hellborg M, Piškur J (2008) Complex evolution of the DAL5 transporter family. BMC Genomics 9:164

Hittinger CT, Rokas A, Bai FY, Boekhout T, Gonçalves P, Jeffries TW, Kominek J, Lachance MA, Libkind D, Rosa CA, Sampaio JP, Kurtzman CP (2015) Genomics and the making of yeast biodiversity. Curr Opin Genet Dev 35:100–109

Huang J (2013) Horizontal gene transfer in eukaryotes: the weak-link model. BioEssays 35:868–875

Husnik F, McCutcheon JP (2017) Functional horizontal gene transfer from bacteria to eukaryotes. Nat Rev Microbiol 16:67

Ianiri G, Coelho Marco A, Ruchti F, Sparber F, McMahon Timothy J, Fu C, Bolejack M, Donovan O, Smutney H, Myler P, Dietrich F, Fox D, LeibundGut-Landmann S, Heitman J (2020) HGT in the human and skin commensal *Malassezia*: a bacterially derived flavohemoglobin is required for NO resistance and host interaction. Proc Natl Acad Sci 117:15884–15894

Irwin NAT, Pittis AA, Richards TA, Keeling PJ (2022) Systematic evaluation of horizontal gene transfer between eukaryotes and viruses. Nat Microbiol 7: 327–336

Keeling PJ, Palmer JD (2008) Horizontal gene transfer in eukaryotic evolution. Nat Rev Genet 9:605–618

Khaldi N, Collemare J, Lebrun M-H, Wolfe KH (2008) Evidence for horizontal transfer of a secondary metabolite gene cluster between fungi. Genome Biol 9:R18

Kirsch R, Gramzow L, Theißen G, Siegfried BD, Ffrench-Constant RH, Heckel DG, Pauchet Y (2014) Horizontal gene transfer and functional diversification of plant cell wall degrading polygalacturonases: key events in the evolution of herbivory in beetles. Insect Biochem Mol Biol 52:33–50

Kominek J, Doering DT, Opulente DA, Shen XX, Zhou X, DeVirgilio J, Hulfachor AB, Groenewald M, McGee MA, Karlen SD, Kurtzman CP, Rokas A, Hittinger CT (2019) Eukaryotic Acquisition of a Bacterial Operon. Cell 176:1356–1366.e10

Kondo N, Nikoh N, Ijichi N, Shimada M, Fukatsu T (2002) Genome fragment of *Wolbachia* endosymbiont transferred to X chromosome of host insect. Proc Natl Acad Sci U S A 99:14280–14285

Kondrashov FA, Koonin EV, Morgunov IG, Finogenova TV, Kondrashova MN (2006) Evolution of glyoxylate cycle enzymes in Metazoa: evidence of multiple horizontal transfer events and pseudogene formation. Biol Direct 1:31–31

Ku C, Nelson-Sathi S, Roettger M, Garg S, Hazkani-Covo E, Martin WF (2015) Endosymbiotic gene transfer from prokaryotic pangenomes: inherited chimerism in eukaryotes. Proc Natl Acad Sci 112:10139–10146

Lachance MA, Starmer WT, Rosa CA, Bowles JM, Barker JS, Janzen DH (2001) Biogeography of the yeasts of ephemeral flowers and their insects. FEMS Yeast Res 1:1–8

Lacroix B, Citovsky V (2016) Transfer of DNA from bacteria to eukaryotes. mBio 7

Lawrence JG, Roth JR (1996) Selfish operons: horizontal transfer may drive the evolution of gene clusters. Genetics 143:1843–1860

Li Z, Bock R (2019) Rapid functional activation of a horizontally transferred eukaryotic gene in a bacterial genome in the absence of selection. Nucleic Acids Res 47:6351–6359

Li M, Zhao J, Tang N, Sun H, Huang J (2018) Horizontal gene transfer from bacteria and plants to the arbuscular mycorrhizal fungus *Rhizophagus irregularis*. Front Plant Sci 9

Li Y, Steenwyk JL, Chang Y, Wang Y, James TY, Stajich JE, Spatafora JW, Groenewald M, Dunn CW, Hittinger CT, Shen X-X, Rokas A (2021) A genome-scale phylogeny of the kingdom Fungi. Curr Biol 31:1653–1665.e5

Li Y, Liu Z, Liu C, Shi Z, Pang L, Chen C, Chen Y, Pan R, Zhou W, Chen X-X, Rokas A, Huang J, Shen X-X (2022) HGT is widespread in insects and contributes to male courtship in lepidopterans. Cell 185(16):2975-2987

Libkind D, Hittinger CT, Valério E, Gonçalves C, Dover J, Johnston M, Gonçalves P, Sampaio JP (2011) Microbe domestication and the identification of the wild genetic stock of lager-brewing yeast. Proc Natl Acad Sci U S A 108:14539–14544

Lim CS, Weinstein BN, Roy SW, Brown CM (2021) Analysis of fungal genomes reveals commonalities of intron gain or loss and functions in intron-poor species. Mol Biol Evol 38:4166–4186

Lind PA, Tobin C, Berg OG, Kurland CG, Andersson DI (2010) Compensatory gene amplification restores fitness after inter-species gene replacements. Mol Microbiol 75:1078–1089

Lind AL, Wisecaver JH, Lameiras C, Wiemann P, Palmer JM, Keller NP, Rodrigues F, Goldman GH, Rokas A (2017) Drivers of genetic diversity in secondary metabolic gene clusters within a fungal species. PLoS Biol 15:e2003583

Liu Y, Wang S, Li L, Yang T, Dong S, Wei T, Wu S, Liu Y, Gong Y, Feng X, Ma J, Chang G, Huang J, Yang Y, Wang H, Liu M, Xu Y, Liang H, Yu J, Cai Y, Zhang Z, Fan Y, Mu W, Sahu SK, Liu S, Lang X, Yang L, Li N, Habib S, Yang Y, Lindstrom AJ, Liang P, Goffinet B, Zaman S, Wegrzyn JL, Li D, Liu J, Cui J, Sonnenschein EC, Wang X, Ruan J, Xue J-Y, Shao Z-Q, Song C, Fan G, Li Z, Zhang L, Liu J, Liu Z-J, Jiao Y, Wang X-Q, Wu H, Wang E, Lisby M, Yang H, Wang J, Liu X, Xu X, Li N, Soltis PS, Van de Peer Y, Soltis DE, Gong X, Liu H, Zhang S (2022) The Cycas genome and the early evolution of seed plants. Nat Plants 8:389–401

Luo H, Hallen-Adams HE, Lüli Y, Sgambelluri RM, Li X, Smith M, Yang ZL, Martin FM (2022) Genes and evolutionary fates of the amanitin biosynthesis pathway in poisonous mushrooms. Proc Natl Acad Sci 119: e2201113119

Ma LJ, van der Does HC, Borkovich KA, Coleman JJ, Daboussi MJ, Di Pietro A, Dufresne M, Freitag M, Grabherr M, Henrissat B, Houterman PM, Kang S, Shim WB, Woloshuk C, Xie X, Xu JR, Antoniw J, Baker SE, Bluhm BH, Breakspear A, Brown DW, Butchko RA, Chapman S, Coulson R, Coutinho PM, Danchin EG, Diener A, Gale LR, Gardiner DM, Goff S, Hammond-Kosack KE, Hilburn K, Hua-Van A, Jonkers W, Kazan K, Kodira CD, Koehrsen M, Kumar L, Lee YH, Li L, Manners JM, Miranda-Saavedra D, Mukherjee M, Park G, Park J, Park SY, Proctor RH, Regev A, Ruiz-Roldan MC, Sain D, Sakthikumar S, Sykes S, Schwartz DC, Turgeon BG, Wapinski I, Yoder O, Young S, Zeng Q, Zhou S, Galagan J, Cuomo CA, Kistler HC, Rep M (2010) Comparative genomics reveals mobile pathogenicity chromosomes in *Fusarium*. Nature 464: 367–373

Ma J, Wang S, Zhu X, Sun G, Chang G, Li L, Hu X, Zhang S, Zhou Y, Song C-P, Huang J (2022) Major episodes of horizontal gene transfer drove the evolution of land plants. Mol Plant 15:857–871

Magwene PM, Kayıkçı Ö, Granek JA, Reininga JM, Scholl Z, Murray D (2011) Outcrossing, mitotic recombination, and life-history trade-offs shape genome evolution in *Saccharomyces cerevisiae*. Proc Natl Acad Sci 108:1987–1992

Marcet-Houben M, Gabaldon T (2010) Acquisition of prokaryotic genes by fungal genomes. Trends Genet 26:5–8
Marcet-Houben M, Gabaldón T (2019) Evolutionary and functional patterns of shared gene neighbourhood in fungi. Nat Microbiol 4:2383–2392
Marsit S, Mena A, Bigey F, Sauvage FX, Couloux A, Guy J, Legras JL, Barrio E, Dequin S, Galeote V (2015) Evolutionary advantage conferred by an eukaryote-to-eukaryote gene transfer event in wine yeasts. Mol Biol Evol 32:1695–1707
Martin WF (2017) Too much eukaryote LGT. BioEssays 39:1700115
Matriano DM, Alegado RA, Conaco C (2021) Detection of horizontal gene transfer in the genome of the choanoflagellate *Salpingoeca rosetta*. Sci Rep 11:5993
McCarthy CGP, Fitzpatrick DA (2019) Pan-genome analyses of model fungal species. Microb Genom 5: e000243
McDonald MC, Taranto AP, Hill E, Schwessinger B, Liu Z, Simpfendorfer S, Milgate A, Solomon PS, Pietro AD (2019) Transposon-mediated horizontal transfer of the host-specific virulence protein ToxA between three fungal wheat pathogens. MBio 10: e01515–e01519
Mehta N, Baghela A (2021) Quorum sensing-mediated inter-specific conidial anastomosis tube fusion between *Colletotrichum gloeosporioides* and *C. siamense*. IMA Fungus 12:7
Milner David S, Attah V, Cook E, Maguire F, Savory Fiona R, Morrison M, Müller Carolin A, Foster Peter G, Talbot Nicholas J, Leonard G, Richards Thomas A (2019) Environment-dependent fitness gains can be driven by horizontal gene transfer of transporter-encoding genes. Proc Natl Acad Sci 116: 5613–5622
Möller M, Stukenbrock EH (2017) Evolution and genome architecture in fungal plant pathogens. Nat Rev Microbiol 15:756–771
Moran NA, Jarvik T (2010) Lateral transfer of genes from fungi underlies carotenoid production in aphids. Science 328:624–627
Morogovsky A, Handelman M, Kandil AA, Shadkchan Y, Osherov N, Goldman GH (2022) Horizontal gene transfer of triazole resistance in *Aspergillus fumigatus*. Microbiol Spectr 0:e01112–e01122
Murphy CL, Youssef NH, Hanafy RA, Couger MB, Stajich JE, Wang Y, Baker K, Dagar SS, Griffith GW, Farag IF, Callaghan TM, Elshahed MS (2019) Horizontal gene transfer as an indispensable driver for evolution of neocallimastigomycota into a distinct gut-dwelling fungal lineage. Appl Environ Microbiol 85
Naranjo-Ortiz MA, Gabaldón T (2019) Fungal evolution: major ecological adaptations and evolutionary transitions. Biol Rev 94:1443–1476
Naranjo-Ortíz MA, Brock M, Brunke S, Hube B, Marcet-Houben M, Gabaldón T (2016) Widespread inter- and intra-domain horizontal gene transfer of d-amino acid metabolism enzymes in eukaryotes. Front Microbiol 7
Nevoigt E, Fassbender A, Stahl U (2000) Cells of the yeast *Saccharomyces cerevisiae* are transformable by DNA under non-artificial conditions. Yeast 16:1107–1110
Nguyen TA, Greig J, Khan A, Goh C, Jedd G (2018) Evolutionary novelty in gravity sensing through horizontal gene transfer and high-order protein assembly. PLoS Biol 16:e2004920
Nováková E, Moran NA (2011) Diversification of genes for carotenoid biosynthesis in aphids following an ancient transfer from a fungus. Mol Biol Evol 29: 313–323
Novo M, Bigey F, Beyne E, Galeote V, Gavory F, Mallet S, Cambon B, Legras J-L, Wincker P, Casaregola S, Dequin S (2009) Eukaryote-to-eukaryote gene transfer events revealed by the genome sequence of the wine yeast *Saccharomyces cerevisiae* EC1118. Proc Natl Acad Sci U S A 106:16333–16338
Nowell RW, Almeida P, Wilson CG, Smith TP, Fontaneto D, Crisp A, Micklem G, Tunnacliffe A, Boschetti C, Barraclough TG (2018) Comparative genomics of bdelloid rotifers: insights from desiccating and nondesiccating species. PLoS Biol 16:e2004830
Ochman H, Lawrence JG, Groisman EA (2000) Lateral gene transfer and the nature of bacterial innovation. Nature 405:299–304
Palazzo A, Lorusso P, Miskey C, Walisko O, Gerbino A, Marobbio CMT, Ivics Z, Marsano RM (2019) Transcriptionally promiscuous "blurry" promoters in Tc1/mariner transposons allow transcription in distantly related genomes. Mob DNA 10:13
Patron NJ, Waller RF, Cozijnsen AJ, Straney DC, Gardiner DM, Nierman WC, Howlett BJ (2007) Origin and distribution of epipolythiodioxopiperazine (ETP) gene clusters in filamentous ascomycetes. BMC Evol Biol 7:174
Pauchet Y, Heckel DG (2013) The genome of the mustard leaf beetle encodes two active xylanases originally acquired from bacteria through horizontal gene transfer. Proc R Soc B Biol Sci 280:20131021
Perli T, Vos AM, Bouwknegt J, Dekker WJC, Wiersma SJ, Mooiman C, Ortiz-Merino RA, Daran JM, Pronk JT (2021) Identification of oxygen-independent pathways for pyridine nucleotide and coenzyme A synthesis in anaerobic fungi by expression of candidate genes in yeast. MBio 12:e0096721
Peter J, De Chiara M, Friedrich A, Yue J-X, Pflieger D, Bergström A, Sigwalt A, Barre B, Freel K, Llored A, Cruaud C, Labadie K, Aury J-M, Istace B, Lebrigand K, Barbry P, Engelen S, Lemainque A, Wincker P, Liti G, Schacherer J (2018) Genome evolution across 1,011 *Saccharomyces cerevisiae* isolates. Nature 556:339–344
Pombert JF, Haag KL, Beidas S, Ebert D, Keeling PJ (2015) The *Ordospora colligata* genome: evolution of extreme reduction in microsporidia and host-to-parasite horizontal gene transfer. MBio 6

Qiu H, Yoon HS, Bhattacharya D (2013) Algal endosymbionts as vectors of horizontal gene transfer in photosynthetic eukaryotes. Frontiers. Plant Sci 4

Qiu H, Cai G, Luo J, Bhattacharya D, Zhang N (2016) Extensive horizontal gene transfers between plant pathogenic fungi. BMC Biol 14:41

Reynolds HT, Vijayakumar V, Gluck-Thaler E, Korotkin HB, Matheny PB, Slot JC (2018) Horizontal gene cluster transfer increased hallucinogenic mushroom diversity. Evol Lett 2:88–101

Richard GF (2020) Eukaryotic Pangenomes. In: Tettelin H, Medini D (eds) The pangenome: diversity, dynamics and evolution of genomes. Springer, Cham

Richards TA, Talbot NJ (2013) Horizontal gene transfer in osmotrophs: playing with public goods. Nat Rev Microbiol 11:720–727

Richards TA, Hirt RP, Williams BA, Embley TM (2003) Horizontal gene transfer and the evolution of parasitic protozoa. Protist 154:17–32

Richards TA, Dacks JB, Jenkinson JM, Thornton CR, Talbot NJ (2006) Evolution of filamentous plant pathogens: gene exchange across eukaryotic kingdoms. Curr Biol 16:1857–1864

Richards TA, Soanes DM, Foster PG, Leonard G, Thornton CR, Talbot NJ (2009) Phylogenomic analysis demonstrates a pattern of rare and ancient horizontal gene transfer between plants and fungi. Plant Cell 21:1897–1911

Richards TA, Leonard G, Soanes DM, Talbot NJ (2011a) Gene transfer into the fungi. Fungal Biol Rev 25:98–110

Richards TA, Soanes DM, Jones MDM, Vasieva O, Leonard G, Paszkiewicz K, Foster PG, Hall N, Talbot NJ (2011b) Horizontal gene transfer facilitated the evolution of plant parasitic mechanisms in the oomycetes. Proc Natl Acad Sci U S A 108:15258–15263

Richardson AO, Palmer JD (2007) Horizontal gene transfer in plants. J Exp Bot 58:1–9

Rokas A (2022) Evolution of the human pathogenic lifestyle in fungi. Nat Microbiol 7:607–619

Rokas A, Wisecaver JH, Lind AL (2018) The birth, evolution and death of metabolic gene clusters in fungi. Nat Rev Microbiol 16:731–744

Ropars J, Rodríguez de la Vega RC, López-Villavicencio M, Gouzy J, Sallet E, Dumas É, Lacoste S, Debuchy R, Dupont J, Branca A, Giraud T (2015) Adaptive horizontal gene transfers between multiple cheese-associated fungi. Curr Biol 25:2562–2569

Savory F, Leonard G, Richards TA (2015) The role of horizontal gene transfer in the evolution of the oomycetes. PLoS Pathog 11:e1004805

Savory FR, Milner DS, Miles DC, Richards TA (2018) Ancestral function and diversification of a horizontally acquired oomycete carboxylic acid transporter. Mol Biol Evol 35:1887–1900

Schönknecht G, Chen W-H, Ternes CM, Barbier GG, Shrestha RP, Stanke M, Bräutigam A, Baker BJ, Banfield JF, Garavito RM, Carr K, Wilkerson C, Rensing SA, Gagneul D, Dickenson NE, Oesterhelt C, Lercher MJ, Weber APM (2013) Gene transfer from bacteria and archaea facilitated evolution of an extremophilic eukaryote. Science 339:1207–1210

Selbmann L, Egidi E, Isola D, Onofri S, Zucconi L, de Hoog GS, Chinaglia S, Testa L, Tosi S, Balestrazzi A, Lantieri A, Compagno R, Tigini V, Varese GC (2013) Biodiversity, evolution and adaptation of fungi in extreme environments. Plant Biosyst 147:237–246

Selman M, Pombert JF, Solter L, Farinelli L, Weiss LM, Keeling P, Corradi N (2011) Acquisition of an animal gene by microsporidian intracellular parasites. Curr Biol 21:R576–R577

Shen XX, Opulente DA, Kominek J, Zhou X, Steenwyk JL, Buh KV, Haase MAB, Wisecaver JH, Wang M, Doering DT, Boudouris JT, Schneider RM, Langdon QK, Ohkuma M, Endoh R, Takashima M, Manabe RI, Cadez N, Libkind D, Rosa CA, DeVirgilio J, Hulfachor AB, Groenewald M, Kurtzman CP, Hittinger CT, Rokas A (2018) Tempo and mode of genome evolution in the budding yeast subphylum. Cell 175:1533–1545.e20

Shen X-X, Steenwyk JL, LaBella AL, Opulente DA, Zhou X, Kominek J, Li Y, Groenewald M, Hittinger CT, Rokas A (2020) Genome-scale phylogeny and contrasting modes of genome evolution in the fungal phylum Ascomycota. Sci Adv 6:eabd0079

Shterzer N, Mizrahi I (2015) The animal gut as a melting pot for horizontal gene transfer. Can J Microbiol 61:603–605

Sibbald SJ, Eme L, Archibald JM, Roger AJ (2020) Lateral gene transfer mechanisms and pan-genomes in eukaryotes. Trends Parasitol 36:927–941

Silva M, Pontes A, Franco-Duarte R, Soares P, Sampaio JP, Sousa MJ, Brito PH (2022) A glimpse at an early stage of microbe domestication revealed in the variable genome of *Torulaspora delbrueckii*, an emergent industrial yeast. Mol Ecol

Sinha S, Flibotte S, Neira M, Formby S, Plemenitaš A, Cimerman NG, Lenassi M, Gostinčar C, Stajich JE, Nislow C (2017) Insight into the recent genome duplication of the halophilic yeast *Hortaea werneckii*: combining an improved genome with gene expression and chromatin structure. G3 (Bethesda, MD) 7:2015–2022

Sloan DB, Nakabachi A, Richards S, Qu J, Murali SC, Gibbs RA, Moran NA (2014) Parallel histories of horizontal gene transfer facilitated extreme reduction of endosymbiont genomes in sap-feeding insects. Mol Biol Evol 31:857–871

Slot JC, Hibbett DS (2007) Horizontal transfer of a nitrate assimilation gene cluster and ecological transitions in fungi: a phylogenetic study. PLoS One 2:e1097

Slot JC, Rokas A (2010) Multiple GAL pathway gene clusters evolved independently and by different mechanisms in fungi. Proc Natl Acad Sci U S A 107:10136–10141

Slot JC, Rokas A (2011) Horizontal transfer of a large and highly toxic secondary metabolic gene cluster between fungi. Curr Biol 21:134–139
Soanes D, Richards TA (2014) Horizontal gene transfer in eukaryotic plant pathogens. Annu Rev Phytopathol 52: 583–614
Stajich JE (2017) Fungal genomes and insights into the evolution of the kingdom. Microbiol Spectr 5
Stalder L, Oggenfuss U, Mohd-Assaad N, Croll D (2022) The population genetics of adaptation through copy number variation in a fungal plant pathogen. Mol Ecol
Steensels J, Gallone B, Verstrepen KJ (2021) Interspecific hybridization as a driver of fungal evolution and adaptation. Nat Rev Microbiol 19:485–500
Steenwyk JL, Rokas A (2018) Copy number variation in fungi and its implications for wine yeast genetic diversity and adaptation. Front Microbiol 9
Stukenbrock EH (2016) The role of hybridization in the evolution and emergence of new fungal plant pathogens. Phytopathology 106:104–112
Sun B, Li T, Xiao J, Liu L, Zhang P, Murphy RW, He S, Huang D (2016) Contribution of multiple inter-kingdom horizontal gene transfers to evolution and adaptation of amphibian-killing chytrid, batrachochytrium dendrobatidis. Front Microbiol 7
Tabima JF, Trautman IA, Chang Y, Wang Y, Mondo S, Kuo A, Salamov A, Grigoriev IV, Stajich JE, Spatafora JW (2020) Phylogenomic analyses of non-dikarya fungi supports horizontal gene transfer driving diversification of secondary metabolism in the amphibian gastrointestinal symbiont, basidiobolus. G3 Genes|Genomes|Genetics 10:3417–3433
Thomas CM, Nielsen KM (2005) Mechanisms of, and barriers to, horizontal gene transfer between bacteria. Nat Rev Microbiol 3:711–721
Todd RT, Selmecki A (2020) Expandable and reversible copy number amplification drives rapid adaptation to antifungal drugs. elife 9
Tsavkelova E, Oeser B, Oren-Young L, Israeli M, Sasson Y, Tudzynski B, Sharon A (2012) Identification and functional characterization of indole-3-acetamide-mediated IAA biosynthesis in plant-associated Fusarium species. Fungal Genet Biol 49:48–57
Urquhart AS, Chong NF, Yang Y, Idnurm A (2022) A large transposable element mediates metal resistance in the fungus Paecilomyces variotii. Curr Biol 32:937–950.e5
Urquhart AS, Vogan AA, Gardiner DM, Idnurm A (2023) Starships are active eukaryotic transposable elements mobilized by a new family of tyrosine recombinases. Proc Natl Acad Sci U S A 120 (15):e2214521120
Venkatesh A, Murray AL, Coughlan AY, Wolfe KH (2021) Giant GAL gene clusters for the melibiose-galactose pathway in Torulaspora. Yeast 38:117–126
Wagner A, Whitaker RJ, Krause DJ, Heilers JH, van Wolferen M, van der Does C, Albers SV (2017) Mechanisms of gene flow in archaea. Nat Rev Microbiol 15:492–501
Wang Y, Youssef NH, Couger MB, Hanafy RA, Elshahed MS, Stajich JE (2019) Molecular dating of the emergence of anaerobic rumen fungi and the impact of laterally acquired genes. mSystems 4
Wang H, Sun S, Ge W, Zhao L, Hou B, Wang K, Lyu Z, Chen L, Xu S, Guo J, Li M, Su P, Li X, Wang G, Bo C, Fang X, Zhuang W, Cheng X, Wu J, Dong L, Chen W, Li W, Xiao G, Zhao J, Hao Y, Xu Y, Gao Y, Liu W, Liu Y, Yin H, Li J, Li X, Zhao Y, Wang X, Ni F, Ma X, Li A, Xu SS, Bai G, Nevo E, Gao C, Ohm H, Kong L (2020) Horizontal gene transfer of Fhb7 from fungus underlies *Fusarium* head blight resistance in wheat. Science 368
Widen SA, Campo Bes I, Koreshova A, Krogull D, Burga A (2022) Virus-like transposons cross the species barrier and drive the evolution of genetic incompatibilities. bioRxiv 2022.07.12.499685
Wiersma SJ, Mooiman C, Giera M, Pronk JT (2020) Squalene-tetrahymanol cyclase expression enables sterol-independent growth of *Saccharomyces cerevisiae*. Appl Environ Microbiol 86
Wijayawardena BK, Minchella DJ, DeWoody JA (2013) Hosts, parasites, and horizontal gene transfer. Trends Parasitol 29:329–338
Wisecaver JH, Rokas A (2015) Fungal metabolic gene clusters-caravans traveling across genomes and environments. Front Microbiol 6:161
Wisecaver JH, Alexander WG, King SB, Hittinger CT, Rokas A (2016) Dynamic evolution of nitric oxide detoxifying flavohemoglobins, a family of single-protein metabolic modules in bacteria and eukaryotes. Mol Biol Evol 33:1979–1987
Wolfe KH (2015) Origin of the yeast whole-genome duplication. PLoS Biol 13:e1002221
Wybouw N, Dermauw W, Tirry L, Stevens C, Grbić M, Feyereisen R, Van Leeuwen T (2014) A gene horizontally transferred from bacteria protects arthropods from host plant cyanide poisoning. elife 3:e02365
Yang Z, Zhang Y, Wafula EK, Honaas LA, Ralph PE, Jones S, Clarke CR, Liu S, Su C, Zhang H, Altman NS, Schuster SC, Timko MP, Yoder JI, Westwood JH, dePamphilis CW (2016) Horizontal gene transfer is more frequent with increased heterotrophy and contributes to parasite adaptation. Proc Natl Acad Sci U S A 113:E7010–E7019
Yin Z, Zhu B, Feng H, Huang L (2016) Horizontal gene transfer drives adaptive colonization of apple trees by the fungal pathogen *Valsa mali*. Sci Rep 6:33129
Zhang Q, Chen X, Xu C, Zhao H, Zhang X, Zeng G, Qian Y, Liu R, Guo N, Mi W, Meng Y, Leger RJS, Fang W (2019) Horizontal gene transfer allowed the emergence of broad host range entomopathogens. Proc Natl Acad Sci 116:7982–7989
Zhou H, Beltrán JF, Brito IL (2021) Functions predict horizontal gene transfer and the emergence of antibiotic resistance. Science. Advances 7:eabj5056
Zhu B, Zhou Q, Xie G, Zhang G, Zhang X, Wang Y, Sun G, Li B, Jin G (2012) Interkingdom gene transfer may contribute to the evolution of phytopathogenicity

in *Botrytis cinerea*. Evol Bioinformatics Online 8:105–117

Zoolkefli FIRM, Moriguchi K, Cho Y, Kiyokawa K, Yamamoto S, Suzuki K (2021) Isolation and analysis of donor chromosomal genes whose deficiency is responsible for accelerating bacterial and trans-kingdom conjugations by IncP1 T4SS machinery. Front Microbiol 12:620535–620535

An Overview of Fungal Volatile Organic Compounds (VOCs)

4

Samantha Lee, Richard Hung, and Joan W. Bennett

Abstract

Fungi emit many volatile organic compounds (VOCs) as mixtures of low molecular mass alcohols, aldehydes, esters, terpenoids, thiols, and other small molecules. The study of the gas-phase portion of a metabolome is sometimes called "volatomics." Chemical separation and VOC identification usually rely on gas chromatography-mass spectrometry (GC-MS). Fungal VOC profiles are both complex and dynamic: the specific compounds produced and their abundance vary with the producing species, the age of the colony, water availability, the substrate, the temperature, the presence of other organisms, and other parameters. The single most reported volatile from fungi is 1-octen-3-ol. It functions as a hormone within many fungal species; serves as both an attractant and a deterrent for some species of arthropods; and exhibits toxicity at relatively low concentrations in several model systems. Fungal VOCs have been studied by scientists from a broad range of subdisciplines in both theoretical and applied contexts. In biotechnology, VOCs are exploited for their food and flavor properties as indirect indicators of microbial growth, to stimulate plant growth, and to attract insect pests. Because these compounds can diffuse a long way from their point of origin, they are excellent chemical signaling molecules and facilitate the ability of fungi to engage in "chemical conversations." Volatiles shape ecology at the level of organisms and populations. The physiological effects of fungal VOCs in host–pathogen relationships and mediating interspecific associations in natural ecosystem functioning are emerging frontiers in volatile research.

Keywords

Developmental signals · Infochemical · Molecular communication · Mushroom alcohol · 1-octen-3-ol · Quorum sensing · Volatome/volatilome · Volatomics

4.1 Introduction

In terrestrial ecosystems, fungi occur as decomposers, symbionts, and pathogens living in close association with plants, animals, and other microbes. In these complex, multi-organism ecosystems, interspecific and interkingdom signaling is essential and there is growing scientific awareness that much of this signaling occurs in the gas phase. Nevertheless, the study of fungal volatiles in biological communication is an

S. Lee
Graduate Women in Science, Mullica Hill, NJ, USA

R. Hung
St. Louis, MO, USA

J. W. Bennett (✉)
Department of Plant Biology, Rutgers University, New Brunswick, NJ, USA

Y.-P. Hsueh, M. Blackwell (eds.), *Fungal Associations*, The Mycota 9,
https://doi.org/10.1007/978-3-031-41648-4_4

understudied area of mycology. This chapter provides a general overview of these small, easily volatilized molecules. In terrestrial habitats, especially in soils with low moisture content, volatiles provide a way for microbes to communicate with one another, both within and between species (Massalha et al. 2017). Our goal is to draw attention to the volatile molecules associated with fungal metabolism, to document some of the biological versatility that these molecules display, and to highlight their role in fungal associations. In recent years, there has been an increasing amount of scientific interest in fungal volatiles, and several valuable reviews have been published (Bitas et al. 2013; Hung et al. 2015; Li et al. 2016; Dickschat 2017; Lee et al. 2017; Piechulla et al. 2017; Farh and Jeon 2020; Inamdar et al. 2020; Guo et al. 2021; Vlot and Rosenkranz 2022).

Volatile organic compounds (VOCs) are low molecular weight carbon-containing compounds that evaporate easily at normal temperatures and pressures (Hermann 2010). The best-known volatiles are chemical solvents and other industrial compounds associated with modern life such as paints, cleaning supplies, petroleum fuels, adhesives, photographic solutions, and the like. Exposures to high concentrations of industrial volatiles such as benzene, formaldehyde, methylene chloride, toluene, and xylene are known to have both short- and long-term negative effects on human health (McFee and Zavon 1988; Do et al. 2015). Far less is understood about the physiological impact of biogenic VOCs.

Plants and microbes emit a wide range of volatile acids, alkanes, alkenes, carbonyls, alcohols, esters, terpenoids, and other small molecules into the biosphere (Kesselmeier and Staudt 1999; Baldwin et al. 2006; Vivaldo et al. 2017). Of these, the plant isoprenoids are perhaps the best known. In particular, monoterpenes such as limonene ("citrusy"), menthol ("minty"), and pinene ("resinous") have familiar odors (Yuan et al. 2009). Many non-isoprenoid microbial VOCs also have distinctive aromas: the earthy smell of garden soil, the noxious stench of spoiled foods, the body odors of people who rarely bathe, and the characteristic mustiness of damp basements are all due mainly to gas-phase microbial metabolites.

It is likely that interest in fungal VOCs may have begun with fungal species that humans can smell (Morath et al. 2012). Many mushrooms have distinctive odor profiles, and it is known that the distinct bouquets of edible mushrooms are due to complex blends of VOCs (Breheret et al. 1997; Tirillini et al. 2000; Cho et al. 2008). For example, anise-like odors are characteristic of *Clitocybe odora*, *Lentinellus cochleatus*, and *Agaricus essettei* (Rapior et al. 2002). A "boiled potato" or "farm-feed" odor is characteristic of several species of *Suillus* (de Pinho et al. 2008). Sulfur-containing volatiles are a highly recognizable group of VOCs of which the best known are from plants of the genus *Allium* (chives, garlic, onion, etc.), where they are valued for their flavors in cooking and for their application in traditional medicine. Similar organosulfur "alliaceous" compounds are made by shiitake mushroom (*Lentinula edodes*) and species of *Tricholoma* and *Marasmius* (Bloch and Deorazio 1994; Sneeden et al. 2004).

Several mushrooms have common names that reflect their odors: the mousepee pinkgill, *Entoloma incanum*, smells like mouse urine, the garlic parachute *Marasmius alliaceus* smells like garlic, and the cucumber cap, *Macrocystidia cucumis*, is said to smell like either cucumber sandwiches or cod liver oil. Several *Agaricus* species are said to smell like almonds (Boniface 2020). Stinkhorns in the genus *Phallus* emit dimethyl trisulfide and skatole, yielding a repulsive stench that combines the worst of carrion and feces (Borg-Karlson et al. 1994). Here we must add an important caveat: broad person-to-person variability in odor perception is a hallmark of olfaction research. Not everyone perceives odors the same way (Gilbert 2008). For example, the aromatic pinkgill mushroom, *Entoloma pleopodium*, is said to smell of "pear drops, amyl acetate, bananas, *or* nail polish remover" (Boniface 2020, p. 30).

Microscopic fungi (molds and mildews) similarly emit complex VOC mixtures. Most people know the musty or moldy odors associated with mold-contaminated damp basements and other enclosed indoor spaces. The odor signatures of a given species or strain of filamentous fungus will vary depending on the substrate, the length of incubation, the type of nutrients, temperature, and other environmental parameters. Furthermore, they vary considerably between species (Pasanen et al. 1997; Nilsson et al. 2004; Fiedler et al. 2005). In *Aspergillus* and *Penicillium* species, the concentration of many VOCs increases during sporulation (Börjesson et al. 1993).

Mushroom smells have been reviewed by de Pinho et al. (2008) and Fraatz and Zorn (2010). For a general database of flavors and scents, see Dunkel et al. (2009). Fungal odors are sometimes used as identifying characters in taxonomy, but the subjective nature of this phenotype makes "sniffing taxonomy" a subjective discipline (Larsen 1998). That said, the human nose is an exquisitely sensitive detection system. Humans can distinguish an astonishing range of odors and, for certain odors, are more sensitive than rodents and dogs (McGann 2017).

The structures of a few representative fungal VOCs are given in Fig. 4.1. The single best source resource for microbial volatiles, including fungal volatiles, is the database provided by Lemfack et al. (2018) currently called the "mVOC 3.0 database." This resource aggregates a large amount of information about microbial VOCs from over one thousand species of bacteria and fungi and 2000 compounds. The database provides five different search options such that it can be used to find information on specific VOCs, chemical structures, scaffolds or mass spectra of bacteria, molds, and macroscopic fungi, as well as information on compounds and their volatilization, pathways and species-specific signatures, and relevant literature.

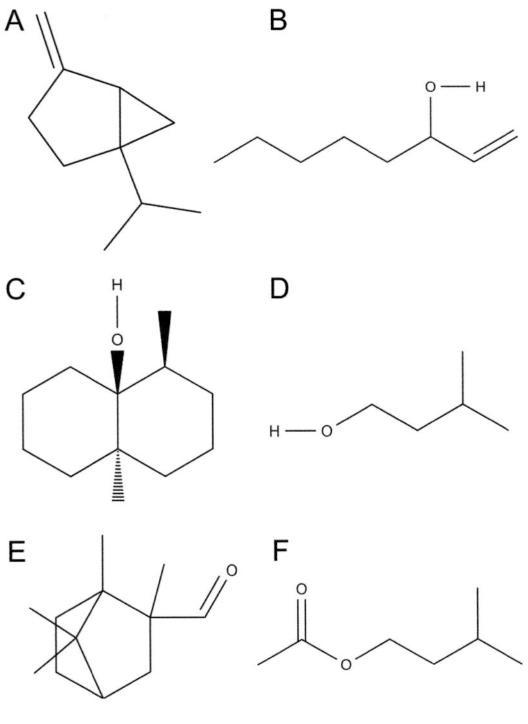

Fig. 4.1 Chemical structures of some representative fungal volatile organic compounds: (**a**) sabinene, (**b**) 1-octen-3-ol, (**c**) geosmin, (**d**) 3-methyl-1-butanol, (**e**) 2-methylisoborneal, (**f**) 3-methyl butyl acetate (isoamyl acetate)

4.2 Analytical Processes for Extraction, Separation, Identification, and Quantification

The first major review on fungal volatiles by Hutchinson (1973) focused almost entirely on carbon dioxide. During the 1970s, fungal VOCs were usually isolated by steam distillation followed by liquid–liquid extraction and concentration of the organic extract (Kaminski et al. 1972; Cronin and Ward 1971). These approaches led to the isolation of several major fungal VOCs, but the technical limitations of the methodology gave an underestimate of their number. Since that time, with more sophisticated trapping, separation, and identification techniques, approximately 300 distinct VOCs have been identified from fungi where they occur as mixtures of simple hydrocarbons, heterocycles, alcohols, phenols, thioalcohols, thioesters, and their derivatives (Chiron and Michelot 2005; Korpi et al. 2009; Lemfack et al. 2018). Not all these VOCs have odors that humans can detect. However, of those that do smell, many of the most familiar ones have extremely low odor thresholds. The human olfactory system is able to identify geosmin, which has an earthy smell, in the range of 150–200 ng/m^3. The mushroom-like odor of

1-octen-3-ol is recognized at 10 μg/m^3, while the musty odor of 2-octen-1-ol is recognized at 16 μg/m^3 (Zogorski et al. 2006). In tests with mice, 1-octen-3-ol and 1-octen-3-one were detected at the part per trillion range. The authors suggested that the high olfactory sensitivity to mold-associated odorants was linked to the behavioral relevance of the odor stimuli, i.e., the avoidance of mold-infested foods (Peixoto et al. 2018).

The large number of different VOCs, their generally low concentrations, and the fact that they tend to occur in mixtures pose challenges for comprehensive sampling and analysis. Different approaches are utilized for sampling, sample preparation, separation, identification, and quantification, depending on the situation and intended application. Traditionally, an initial step of separation into purer components was followed by identification and quantification. In recent decades, purging and trapping of headspace gases have gained favor, especially in studies of odor formation in agricultural settings (Abramson et al. 1980, 1983). Trapping of headspace gases also has been used to obtain VOC profiles of a number of fungal species grown under standardized laboratory conditions (Mattheis and Roberts 1992; Börjesson et al. 1993). The absorbance materials used for trapping (e.g., charcoal, Super Q, and Tenax) can select for compounds with particular binding properties. Similarly, the choice of organic solvents (e.g., dichloromethane and hexane) can affect the VOC profile obtained, sometimes yielding an insufficient resolution of highly volatile, early eluting compounds (Kai et al. 2009). Larsen and Frisvad (1995) showed differences in collecting methods led to vastly distinct profiles of fungal VOCs.

Most determination (separation and identification) of VOCs now relies on gas chromatography-mass spectrometry (GC-MS) because it integrates powerful separation capability and facilitates quantification and identification. Depending on the intended purpose, several methods are currently available. Air sampling onto Tenax desorption tubes followed by thermodesorption allows accurate sampling at one point in time. In order to determine VOCs over an extended time, passive diffusion monitors onto charcoal adsorbents can be used. Matysik et al. (2009) found that this method was particularly suitable in epidemiological studies that attempt to correlate concentrations of specific VOCs and indoor mold exposure. After separation, the individual constituents of VOC mixtures usually are identified by mass spectrometry. Comparison of mass spectra with library spectra or determination of chromatographic retention indices, ideally in conjunction with the parallel determination of authentic standards, is used to confirm identity (Stoppacher et al. 2010).

The means of delivering the sample to the vacuum chamber for analysis have also changed and improved over time. Since its introduction in the 1990s, solid-phase microextraction (SPME) has emerged as an efficient and popular method for assessing VOCs in various contexts (Ackerman 1990; Jeleń 2003). A fiber coated with different types of adsorbents is exposed to the headspace atmosphere, and subsequently, the trapped contents are analyzed by GC-MS. SPME has been applied widely in the flavor and fragrance industry for monitoring freshness, detecting fungal contamination in stored products, and the like (Soria et al. 2015). For good reviews of methodologies for use for separating, identifying, and quantifying biogenic VOCs, see Zhang and Li (2010), Materić et al. (2015), and Tholl et al. (2021). In recent years, there has been a trend toward developing more portable and miniaturized detection methods that can be used in "real-world" situations, such as the early detection of disease outbreaks in the field or mold contamination in storage (Materić et al. 2015). For example, Anishchenko et al. (2018) developed a miniature GC-differential mobility spectrometry (GC-DMS) platform small enough to fit into a shoe box. Smartphone-based VOC sensors also are emerging (Li et al. 2019), and using a small SPME device for environmental VOC sampling through aerial drones has been developed (Ruiz-Jimenez et al. 2019).

Analysis of human exhaled breath is a noninvasive method that shows promise for early-stage diagnostic support in many fields of medicine. Although the volatile analysis in clinical medicine has yet to reach its promise, continuing research interest almost guarantees that some of the best future volatile research will come from the investigation of human diseases (Davis et al. 2020). For example, detection of VOCs emitted by *A. fumigatus* has been suggested as a way for early diagnosis of aspergillosis (Heddergott et al. 2014), and unexpectedly, it was shown that dimethyl sulfide emitted by *Pseudomonas aeruginosa*, a pathogenic bacterium that often co-infects lungs of cystic fibrosis patients, could stimulate growth of *A. fumigatus* (Briard et al. 2016; Scott et al. 2019).

"Artificial olfaction" is an area of sensor technology that has proven useful in certain applied situations. Electronic noses (sometimes called artificial noses) have been used to mimic mammalian olfactory systems and can be applied in healthcare and other biomedical applications. Interestingly, the diseases caused by specific microbial pathogens are associated with characteristic odiferous compounds, e.g., *Pseudomonas* causes a "grape" odor of skin, while the sweat of patients with rubella is said to smell like "freshly plucked feathers" (Pavlou and Turner 2000). In laboratory experiments, electronic noses were able to distinguish between uncontaminated samples and those contaminated with dry-rot wood decay fungi (Kuske et al. 2005). In a clinical setting, E-nose technology was used to detect aspergillosis in patients with cystic fibrosis (De Heer et al. 2016). A comprehensive historical review of electronic nose technology is provided by Wilson and Baietto (2009, 2011).

Finally, it should be noted that dogs can be trained to detect mold growth with 75–94% accuracy (Griffith et al. 2007).

4.3 Classification of VOCs

The anabolic metabolic pathways of bacteria, fungi, and plants can be divided into the dichotomous categories of primary and secondary metabolism. Primary metabolites are essential to the life of the organism and represent the unity of biochemistry. Examples include Krebs cycle intermediates, amino acids, lipids, and nucleic acids (Berg et al. 2007). In contrast, secondary metabolites are the enormous group of diverse natural products not essential to growth, often of highly unusual chemical structure, which almost always are restricted in taxonomic distribution. Antibiotics such as cephalosporins, hallucinogens such as the ergot alkaloids, and mycotoxins such as the trichothecenes are well-known examples of fungal secondary metabolites (Bennett 1983; Bennett and Bentley 1989; Cole and Schweikert 2003). Secondary metabolites are biosynthesized by special pathways (e.g., polyketides, non-ribosomal peptides, and isoprenoids) and constitute the bulk of the field known as natural products chemistry. Advances in genomics facilitate the detection of their signature gene clusters using bioinformatics approaches (Keller et al. 2005; Rokas et al. 2020).

Due to their low molecular weights and phase-dependent appearance, fungal VOCs as a group sometimes are labeled as secondary metabolites. However, there are problems with this simplistic categorization. Secondary metabolites usually are made by only a taxonomically limited number of producing species, while most VOCs are found across a broad range of producing organisms. For example, the production of the toxic secondary metabolite aflatoxin is restricted to a few species within the genus *Aspergillus*, while the production of the volatile 1-octen-3-ol is widespread across many fungi as well as plants and animals (Cole and Schweikert 2003; Chiron and Michelot 2005; Combet et al. 2006; Pennerman et al. 2022). Secondary metabolites are generally produced by complex metabolic pathways encoded by clusters of linked genes (Zhang et al. 2005; Keller et al. 2005). Although less is known about the pathways that produce VOCs, many are either metabolic transformation products of lipids, proteins, heterocyclic metabolites, and other components of living tissues, or are degradation end products ("waste products") of fungal catabolic pathways.

The common fungal VOC, 1-octen-3-ol, is a good case in point. Although 1-octen-3-ol is sometimes called a secondary metabolite, it is derived from the enzymatic or non-enzymatic oxidative breakdown of linoleic acid (Pennerman et al. 2022). In *Psalliota bispora*, both 1-octen-3-

ol and a less volatile 10-oxo-trans-8-decenoic acid are produced through the enzymatic oxidation and cleavage of linoleic (Wurzenberger and Grosch 1984). In *Pleurotus pulmonarius*, two separate lipoxygenases may be involved in the production of 1-octen-3-ol and 10-oxo-trans-8-decenoic acid (Assaf et al. 1997). In *Podospora anserina*, it requires both cyclooxygenase and lipoxygenase genes to produce C_8 volatiles (Ferrari et al. 2018). Enzymatic pathways preferably yield the R enantiomer of 1-octen-3-ol, but auto-oxidation yields racemic mixtures (Tsitsigiannis and Keller 2007; Lee et al. 2020). Another example is 6-pentyl pyrone (6-PP), a lactone responsible for a distinctive coconut aroma produced by certain *Trichoderma* species. It is a degradation product of linoleic acid (Zeilinger et al. 2016).

In summary, not all the "small molecules" outside of the central pathways of intermediary metabolism are secondary metabolites. Therefore, in this chapter, we do not classify fungal VOCs as either primary or secondary metabolites. Instead, we define VOCs according to their number of carbons, their ring structures, and their substituent group and as acids, ketones, aldehydes, terpenes, and the like.

4.4 Overview of Fungal Volatiles

Fungal VOCs are of both theoretical and practical significance within disparate scientific disciplines. They have been studied for their flavor properties, as indicators to detect the presence of fungal growth, as possible contributors to "sick building syndrome," and as signals for fungal development. Moreover, in recent years, the VOCs from endophytes have emerged as of particular interest because some have shown antibiotic activity while others have the potential for possible use as fuel compounds or "biodiesel." Numerous research studies have revealed that fungal VOC profiles are both complex and dynamic: the compounds produced, and their abundance, vary with the producing species, the age of the fungal colony, moisture level, substrate type, temperature, and other environmental parameters. Examples from this enormous and scattered literature are given here.

4.4.1 Exploitation of Fungal VOCs: Flavors and as Indicators of Fungal Growth

Biogenic VOCs can be detected in real time. They provide a way to noninvasively monitor safety applications within the food industry and for potable water supplies. In drinking water, "off" flavors sometimes are caused by very low concentrations of microbial VOCs. Geosmin and 2-methylisoborneol (MIB) are two VOCs that are among the most common compounds involved in malodor problems in surface waters (Watson et al. 2000). They give a musty or earthy odor; both have an exceptionally low odor threshold. MIB is also a factor in cork taints in winemaking. Using isotopic labeling experiments, Bentley and Meganathan (1981) determined that geosmin is a degraded sesquiterpene and MIB is a methylated monoterpene. With improved and inexpensive methods for monitoring VOCs, such as those associated with artificial nose technologies, electronic odor detection will likely become an increasingly useful tool for checking food and water safety.

Mold ripened cheeses are among the best-known food fermentations involving filamentous fungi. Blue cheeses such as Gorgonzola, Roquefort, and Stilton, and white cheeses such as Brie and Camembert, gain their distinctive flavors from methyl ketones and various alcohols produced by fungal metabolism by species of *Penicillium* (Karahadian et al. 1985; Gallois and Langlois 1990). Fungi are also used in various commercial bioconversion products for making flavor products that can be considered "natural aromas" (Berger et al. 1992; Schreier 1992). For example, *Ischnoderma* spp. and other white-rot fungi can produce benzaldehyde, which has a bitter almond flavor, while a variety of yeasts and filamentous basidiomycetes have been used to produce clove (eugenol), jasmine (benzyl acetate), and rose-like (2-phenylethanol) odors (Copetti 2019).

The study of aroma compounds in beers, wines, and other spirits and their relationship with the perceived flavor of alcoholic beverages represents a huge field that addresses the volatile compounds associated with yeast fermentation. For introductions to this vast literature, see (Meilgaard 1975a, b) and Robinson (2006). We will not attempt to cover food and flavor chemistry in this chapter. See Bennett and Feibleman (2001) for an earlier review on mixed bacterial–fungal food fermentations.

VOCs can be used as indirect indicators of mold growth, even in the absence of visible colonies. Methods for remote volatile capture and analysis have allowed an expansion of applications. Considerable research has gone into developing strategies that detect fungal volatiles as an indirect and noninvasive way to indicate the presence of fungal growth in agriculture, water-damaged buildings, and art conservation.

Fungal contamination of stored foods and feeds is a global problem in agriculture. Fungi decrease the nutritive value of stored foods, spoil them by creating off flavors, and can produce mycotoxins that render foods poisonous to human beings and other animals. Compounds such as geosmin, 1-octen-3-ol, 3-octanol, and 3-methyl-1-butanol are regularly found in association with stored grains contaminated with fungi (Börjesson et al. 1989, 1993; Mattheis and Roberts 1992). Volatiles can be used to detect fungal contamination on agricultural products (Wadood et al. 2020; He et al. 2022). Fungal VOCs have been used to monitor favorable and unfavorable food quality (Karlshøj et al. 2007). With advancements in electronic nose technology, "mapping" of volatile compounds may predict the levels of fungi found in agricultural products and identify individual species (Schnürer et al. 1999).

The VOC profiles of common molds grown on building materials have been analyzed in the laboratory (Sunesson et al. 1995, 1996). Mold VOC profiles have also been studied in water-damaged buildings where compounds such as 2-methyl-1-propanol and 3-methyl-1-butanol have been suggested as useful indicators of mold growth (Wilkins et al. 2000; Korpi et al. 2009; Claeson et al. 2002; Matysik et al. 2008). Fungus-like odors can be recognized at concentrations greater than 0.035 μg/m^3. Generally, the background microbial VOC concentrations in mold-free buildings are similar to those found in outdoor air ranging between 2.2 and 8.8 μg/m^3. VOC levels significantly higher than the background ranges may indicate an increase in the active microbial production possibly associated with adverse health effects (Ström et al. 1994). Volatile sampling can be used to look for fungi in small openings in drywall or upholstered furniture and provides a noninvasive method for determining the presence of fungi on valuable objects of cultural heritage such as tapestries, on the wooden frameworks behind paintings, or in libraries (Micheluz et al. 2016). For example, Joblin et al. (2010) developed an index that was used to compare the level of VOCs in the Lascaux caves contaminated with *Fusarium solani* before and after antifungal treatment.

4.4.2 Building-Related Illnesses

Occupants of damp indoor spaces often complain of irritation of eyes and mucous membranes, respiratory discomfort, malaise, headaches, gastrointestinal disturbances, and a variety of other symptoms that are often lumped together and called "nonspecific building-related illness," or by the more controversial name of "sick building syndrome." Many scientists have hypothesized that this spectrum of adverse health is associated with exposure to airborne bacteria and fungi, aerosolized toxins, or bacterial metabolites such as endotoxins (Thorn 2001; Straus et al. 2003; Burge 2004; IOM 2004; Li and Yang 2004; WHO 2009). Mycotoxins have received most of the attention as the hypothetical cause of the symptoms of sick building syndrome (Jarvis and Miller 2005; Straus 2009). However, VOCs may also contribute to the adverse health consequences associated with damp indoor spaces, especially in cases where there is no visible evidence of mold growth (Walinder et al. 2005; Mølhave 2009). High vapor pressures, low to medium water solubility, and low molecular weights allow both fungal and bacterial VOCs to persist and migrate in the environment, and to diffuse through enclosed wall cavities, air conditioning filters, and vapor barriers. The level of VOCs measured in the indoor air of mold-infested building varies with the ventilation rate, moisture levels, composition of mold population, and area of the building/room and is in a state of constant change. One of the highest reported concentrations for a single VOC, 1-octen-3-ol, found in problem buildings is 900 $\mu g/m^3$ or 0.16 ppm (Morey et al. 1997). Moreover, 1-octen-3-ol has been reported to be one of the major fungal VOCs emitted by various species of fungi (*Aspergillus*, *Cladosporium*, *Fusarium*, *Penicillium*, *Stachybotrys*, etc.) regularly found in moldy and water-damaged buildings (Sunesson et al. 1996) or from composting facilities (Fischer et al. 1999). Studies in our laboratory have shown that gas-phase 1-octen-3-ol is more toxic to human embryonic stem cells than toluene (Inamdar et al. 2011) and that it inhibits seed germination and plant growth in *Arabidopsis thaliana* (Lee et al. 2016a). An investigation of fungal VOCs from hospital air showed that 2-heptanone and 2-methyl-1-propanol, not 1-octen-3-ol, were the most frequent compounds isolated (Pantoja et al. 2016).

Our laboratory has pioneered the use of the genetic model *Drosophila melanogaster* for studying the physiological impact of fungal volatiles isolated from indoor environments. When adult *Drosophila* flies were exposed for one week to low concentrations of chemical standards of 2-octanone (0.5%); 2,5 dimethylfuran (0.5%) (DMF), 3-octanol (0.5%), trans-2-octenal (0.5%), and 1-octen-3-ol (0.1%), they exhibited 40%, 35%, 60% 50%, and 100% lethality, respectively (Inamdar et al. 2010). By exposing third instar *D. melanogaster* larvae, we developed an eclosion bioassay and successfully demonstrated the toxicity of VOCs from fungi isolated from flooded homes in the aftermath of Hurricanes Katrina and Sandy (Inamdar and Bennett 2015; Zhao et al. 2017). Other *Drosophila* studies using chemical standards of 1-octen-3-ol showed that exposure of flies to this common fungal VOC decreased dehydrogenase activity as well as levels of nitric oxide but increased levels of reactive oxygen species (ROS). It was hypothesized that the negative consequences of the exposures were related to necrotic mechanisms and that they had implications for human health (Macedo et al. 2022).

4.4.3 VOCs from Endophytes

Endophytes are microorganisms that live intercellularly within plant tissues without causing any evident adverse effects (Bacon and White 2000). They have been found in almost every plant species examined and are likely to play a significant role in plant community structure (Rodriguez et al. 2009). Fungal endophytes have been studied as a source of novel secondary metabolites (Tan and Zou 2001) and also have gained attention as producers of bioactive VOCs. *Muscodor albus* ("stinky white fungus") is an endophyte that produces a blend of inhibitory or lethal VOCs to a wide range of bacteria and pathogenic fungi. Using GC-MS, Strobel et al. (2001) found that *M. albus* produced a mixture of volatile acids, alcohols, esters, ketones, and lipids, which individually had inhibitory but not lethal effects against test species such as *Fusarium solani*, *Pythium ultimum*, and *Rhizoctonia solani*. When applied collectively, these same VOCs acted synergistically to kill a broad range of plant-pathogenic fungi and bacteria (Strobel et al. 2001). Since the original isolation of *Muscodor albus* from the cinnamon tree, several other *Muscodor* strains and species that emit antibiotic mixtures of VOCs have been isolated (Atmosukarto et al. 2005; Zhang et al. 2010). This selective antimicrobial effect can be harnessed against undesirable pathogens and has been termed "mycofumigation." *Muscodor albus* has been used for the biological control of damping-off of broccoli seedlings grown in greenhouse soilless mix (Mercier and Jiménez 2004; Mercier and Manker 2005). *Oxyporus latemarginatus*, an endophyte isolated from pepper plants, has also shown a positive mycofumigation ability against post-harvest decay organisms (Lee et al. 2009). Using genetic screens, Alpha et al. (2015) showed that the toxicity of *M. albus* VOCs was caused by permeabilizing of the cell membrane and an interference of DNA metabolism. Mycofumigation agents have also been isolated from other endophytes including *Fusarium proliferatum*, an unidentified species of *Diaporthe*, and *Geotrichum candidum* (Pangga 2021; Mitra et al. 2023).

VOCs from endophytes may have other biotechnological applications. Growing fungi in the presence of *M. albus* has been used as a selection tool to isolate other fungi that produce bioactive volatiles. Species resistant to the VOCs produced by *M. albus* are then screened for the activity of their own VOC profile. Using this method, an endophytic species of *Gliocladium* was isolated from a Patagonian species of *Eucryphia*. This *Gliocladium* species produced a number of VOCs, of which one of the most interesting was 1,3,5,7 cyclooctatetraene or annulene, an unstable and flammable compound used as a rocket propellant during World War II (Stinson et al. 2003). Strobel et al. (2008) coined the term "mycodiesel" when he found that the endophyte *Gliocladium roseum* produced several VOCs normally associated with diesel fuel.

Endophytic fungi in the genera *Ascocoryne* (Griffin et al. 2010), *Phoma* (Strobel et al. 2011), and *Phomopsis* also synthesize VOCs that have fuel potential. *Phomopsis* species isolated from an orchid in Ecuador produces bicyclic monoterpene sabinene, a monoterpene with a peppery odor first isolated from plants (Singh et al. 2011). Several related monoterpenes are being investigated as possible components of aircraft fuel (Rude and Schirmer 2009). For six species of oleaginous fungi studied, the resultant biodiesel quality reflected the fatty acid profile of the feedstock oil used (Shafiq and Chechan 2019).

In recent years, there has been growing interest in yeasts as endophytes. Species of *Cryptococcus*, *Debaryomyces*, *Sporobolomyces*, and *Rhodotorula* have been reported from many plants (Doty 2013). These endophytic yeasts improve plant vigor by helping host plants resist pests by secreting active metabolites. Moreover, it is important to remember that ethanol is a VOC (Vajzovic et al. 2012) and that ethanol plays a morphogenetic role in the yeast–hyphal transition in *Candida albicans* (Chauhan et al. 2011).

While most mycologists are familiar with the role of fungi as endophytes of plants, we are less familiar with the role of bacteria as residents of fungi. It is becoming increasingly apparent that many fungi carry bacterial lineages that are transmitted vertically (Bonfante and Desirò 2017). The VOCs of these fungal endobacteria are beyond the scope of this chapter, but because fungi and bacteria tend to live in close associations with plants and each other, it is not always clear during volatile analysis which domain of organisms is producing a given metabolite.

4.4.4 Truffles

"Truffle" is the common name for a group of ascomycete fungi that form underground fruiting structures and emit aromatic compounds that attract animals which eat them and then, in turn, disperse their spores through their feces. Mention of these delectable but strange fungi goes back to classical times (Ramsbottom 1953). A truffle is the reproductive structure of an ectomycorrhizal fungus; see Hall et al. (2007) for a list of the plant species that support truffle formation. The most famous truffle species resemble small, misshapen potatoes, hence their genus name *Tuber*, for "lump" (Jeandroz et al. 2008). Despite their undistinguished appearance, *Tuber magnatum*, the "white truffle," and *Tuber melanosporum*, the "black truffle," are deemed the most desirable of all edible fungi. In the world of gourmets, truffles are among the most expensive of food ingredients and therefore are often used as seasonings in other dishes, shaved into paper-thin slices, or minced as a garnish (Hall et al. 1994). In addition to their reputation as the ultimate gourmet fungus, these fungi are purported to serve as aphrodisiacs (Ramsbottom 1953; Bone 2011). In her book simply titled *Truffles*, Elisabeth Luard writes that "when ripe and ready, the truffle reeks of sex" (p. 14) and goes on to say that "the perfume for which it was valued was held to induce the kind of behavior among celibate priests and nuns that could lead to the birth of children" (Luard 2006, p. 36). Other descriptions of the lascivious "intoxicating odor" of truffles and human reactions to this odor are given in *Mycophilia. Revelations from the Weird World of Mushrooms* (Bone 2011).

As with all complex aromas, the prized truffle smell is caused by a mixture of individual volatile molecules that vary with the producing species, the age of the fruiting body, the weather, the general climate, the truffle microbiome, and other environmental factors. Descriptors as varied as "meaty, cheese-like, fruity green apple, malty, and mushroomy" have been used to describe the different truffle odors. However, none of these pedestrian adjectives apparently come close to capturing the alluring scent of a fresh truffle (Allen and Bennett 2021). Not surprisingly, truffle VOCs have been the subject of intense scientific interest and more than one hundred different VOCs have been reported from a range of truffle species using different extraction and identification techniques (Splivallo et al. 2007a; Vita et al. 2015). They encompass many small hydrocarbon molecules, but the most distinctive truffle VOCs are thiols (Mauriello et al. 2004; Splivallo and Ebeler 2015; Mustafa et al. 2020). In other words, the key aroma compounds that make truffles so commercially valuable all contain sulfur (Schmidberger and Schieberle 2017). VOC analysis of truffle fruiting bodies revealed that these thiol derivatives were exclusively synthesized by the resident bacterial microbiome and not by the truffle. When fruiting bodies were treated with antibacterial agents, the sulfur-containing volatiles were fully suppressed; in contrast, fungicide treatment had no inhibitory effect (Splivallo et al. 2015). For *Tuber borchii*, thiophene derivatives are major contributors to the human-sensed aroma, with 3-methyl-4,5-dihydrothiophene as the single most important compound in terms of its contribution to the overall aroma (Splivallo and Ebeler 2015; Splivallo and Culleré 2016). Of the more than 60 volatile compounds detected from *Tuber magnatum*, only 11 contribute to the smell of which 2,4-dithiapentane is considered most important (Niimi et al. 2021). In their meta-analysis of the data from the mVOC database, Elmassry et al. (2020) found that "volatile sulfur metabolism is dominated by bacteria rather than by fungi." Thus, it seems that the most famous odor compounds attributed to fungi may be of bacterial origin!

4.5 VOCs as Ecological Signaling Agents

With each passing year, there is growing recognition of the extent of chemical communication in the biosphere and the role that volatile chemicals play in biological signaling. Chemical signaling occurs within and between cells, between individuals of the same species, and also between different species. Entomologists and chemical ecologists have been at the forefront of studying chemical signaling molecules and have developed an extensive vocabulary that includes terms such as "allelochemicals," "infochemicals," and "semiochemicals" (Table 4.1). Only some of the infochemicals studied to date are found in the gas phase, but the concepts and definitions are useful whatever the physical states of the signaling molecule.

Chemical signaling includes both long- and short-range cellular communication. A major distinction is that pheromones mediate *intra*specific interactions, while allelochemicals mediate *inter*-specific interactions. However, these terms are not mutually exclusive. A single chemical can function in more than one type of interaction (Nordlund and Lewis 1976). For example, the nearly ubiquitous fungal volatile, 1-octen-3-ol, depending on the context and biological system, is described as a hormone, pheromone, semiochemical, or allelochemical.

The revolution in mammalian microbiome research has made all biologists aware of the fact each macroscopic "higher organism" is a complex multi-organism. Similarly, metabolomic research has broadened our awareness that many metabolites used as interspecific signaling agents are in the gas phase (Nicholson et al. 2005;

Table 4.1 Technical terms frequently encountered in discussions of chemical signaling compounds

Term with definition and citations
Allelochemical—Broadly, a compound that mediates chemical interactions between organisms (Hooper and Pickett 2004). Allelochemicals were originally termed "allelochemics" by Whittaker and Feeny (1971) to describe toxic chemicals made by a plant to defend itself against competing plants.
Allomone—An allelochemical that only benefits the organism that emits the chemical signal (Hooper and Pickett 2004) "Originally defined as a chemical substance produced or acquired by an organism, which, when it contacts an individual of another species in the natural context, evokes in the receiver a behavioral or physiological reaction adaptively favorable to the emitter" (Brown 1968; Nordlund and Lewis 1976).
Kairomone—A controversial term for an allelochemical which benefits only the receiving organism, but not the producer; i.e., the emitter does not benefit from the interaction (Nordlund and Lewis 1976; Hooper and Pickett 2004).
Synomone—An allelochemical which benefits both the producing and receiving organisms.
Autoinducer—A compound used by quorum sensing bacteria to monitor cell density.
Auto-inhibitor ("self-inhibition")—a term introduced by Allen (1957) to describe the inhibition of fungal spore germination when spores are overcrowded, a process thought to ensure efficient substrate colonization.
Crowding effect—The delay and inhibition of spore germination resulting from high spore concentrations (Trinci and Whittaker 1968).
Growth regulating substances (GRS)—A term sometimes used as a synonym for "plant hormone." Examples include auxins and gibberellins in plants and possibly 10-oxo-trans-8-decenoic acid in fungi (Takahashi 1986; Mau et al. 1992).
Chemical ecology—The field that studies how plants, animals, and microbes use chemicals to communicate in natural ecosystems.
Hormone—A term originally used by endocrinologists to describe chemicals secreted by a gland or group of cells in vertebrates to regulate specific physiological processes within the organism. The definition has been expanded to describe chemicals in other organisms that control and regulate growth and development within the same organism as the compound is secreted.
Infochemical—Information-conveying chemical. "A chemical that, in the natural context, conveys information in an interaction between two individuals, evoking in the receiver a behavioral or physiological response that is adaptive to either one of the interactants or to both" (Dicke and Sabelis 1988).
Mushroom alcohol—The common name of 1-octen-3-ol, one of the most prevalent fungal VOCs. Another synonym is matsutake alcohol.
PGPB—Plant growth-promoting bacteria.
PGPF—Plant growth-promoting fungi.
Psi factors—"Precocious sexual inducers" first identified from *A. nidulans* by Champe et al. (1987). Champe and el-Zayat (1989) also used the term "sexual sporulation hormone." These oxylipin compounds are involved in several fungal developmental cascades and are classified according to the fatty acids from which they are derived (Tsitsigiannis and Keller 2007; Brodhun and Feussner 2011).
Quorum sensing—Ability of bacterial populations to communicate and coordinate their behavior mediated by the use of chemical signaling molecules (Fuqua et al. 1994; Bassler and Losick 2006). Quorum sensing was originally called "autoinduction" (Nealson and Hastings 1979).
1-oceten-3-ol—An oxylipid C-8 alcohol that is produced by numerous fungi, as well as by many plants, and that can also be emitted by certain arthropods and mammals. It regularly acts as a semiochemical. Depending on the scientific context and literature, it is variously called a pheromone, a kairomone, a synomone, or by other jargon terms.
Oxylipin—A large family of structurally related oxygenated polyenoic fatty acids and the metabolites derived from them. Many oxylipins have physiological activities. They are abundant in mammals but also served as signals of intra- and intercellular communication in other vertebrates, invertebrates, plants, and fungi. In microbes, they regulate growth, differentiation, and apoptosis in addition to the development of the infectious processes caused by some pathogenic microorganisms (Noverr and Erb-Downward 2003; La Camera et al. 2004; Tsitsigiannis and Keller 2007).
Pheromone—An "ecto-hormone," a chemical signal mediating interactions between organisms of the *same* species (Hooper and Pickett 2004). A "substance that is secreted by an organism to the outside and causes a specific reaction in a receiving organism of the same species" (Karlson and Luscher 1959; Nordlund and Lewis 1976). Examples include the trail pheromones and sex pheromones that have been well documented in insects.
Semiochemical—"A chemical involved in the chemical interaction between organisms" (Nordlund and Lewis 1976).
Volatilome (sometimes "**volatome**")—The complete collection of volatile metabolites from a given source under a specific set of environmental parameters, sometimes called the "volatile metabolome" (Kimball 2016).
Volatomics—The volatile part of a metabolome; the study of the kinds and concentrations of volatile organic compounds (VOCs) emitted from a biological system and the study of how changes in metabolism can affect their makeup.

Dietert and Silbergeld 2015). For example, in a review of soil microbiomes, Raza et al. (2021) proposed "that soil microbe-plant communities could be viewed as 'metarhizobiomes,' where VOC-mediated interactions extend the plant rhizobiome further out through interconnected microbial metapopulation networks." Volatile-phase signaling molecules allow organisms to work together at a distance.

4.5.1 Bacterial VOCS

Bacterial species, like fungi, produce complex cocktails of VOCs, often in close collaboration with fungi (e.g., see the section on truffles above). As with fungi, VOC profiles from various bacteria have been analyzed. For example, twenty-six species in the genus *Streptomyces* revealed mixtures of alcohols, alkanes, alkenes, ketones, terpenoids, and thiols in a range of concentrations and combinations (Schöller et al. 2002). Forty-two different volatiles were found from *Myxococcus xanthus* (Dickschat et al. 2005b), and the volatile blend of a marine *Streptomyces* species exhibited antibiotic properties (Dickschat et al. 2005a). See Schulz and Dickschat (2007), Audrain et al. (2015), Schulz-Bohm et al. (2017), and Weisskopf et al. (2021) for comprehensive surveys of bacterial VOCs.

Bacteria communicate with one another using many molecular mechanisms and have interactive effects on the organisms with which they share ecological niches, including plants, animals, fungi, and other bacteria. Bacterial volatiles have been studied for their uses in foods, particularly in the dairy industry; their production of "off" odors in food and water supplies; and other applied areas. Soil bacteria associated with the rhizosphere have been investigated for their growth-promoting activities. There is an extensive literature on plant growth-promoting bacteria (PGPB) (Glick 2020). Similarly, there are plant growth-promoting fungi (PGPF) which include species of *Aspergillus*, *Fusarium*, *Trichoderma*, *Penicillium*, *Piriformospora*, *Phoma*, and *Rhizoctonia*, which have the natural ability to stimulate various growth-related traits of plants (Shoresh et al. 2010; Hossain et al. 2014). VOC-mediated signaling is believed to help plants defend against pathogen attacks (Raza et al. 2021).

When Elmassry et al. (2020) used the mVOC database of microbial volatiles to identify phylogenetic signals, they found that the bacterial domain contained 33 phylogenetically conserved VOCs. The following compounds tended to be from bacteria: acetic acid, dimethyl disulfide, dimethyl trisulfide, indole, isovaleric acid, 2-tridecanone, propanoic acid, and 2-undecanone. When these VOCs were found in a fungal volatome, it was hypothesized that they were due to the presence of "resident" bacteria growing on the fungi. Geosmin was one of 11 VOCs that were conserved from both bacteria and fungi (Elmassry et al. 2020).

4.5.2 Quorum Sensing

Quorum sensing refers to cellular communication systems that generate a population response. Quorum sensing allows microorganisms to communicate and coordinate their behavior via the secretion of signaling molecules that accumulate during their growth and then act in a population-dependent manner. Quorum sensing was originally studied in bacteria where the molecules used to monitor cell density were named "autoinducers." The concentration of autoinducers increased as a function of rising bacterial population density; when a minimum threshold of autoinducers was detected, the bacteria responded by altering gene expression (Bassler and Losick 2006; Camilli and Bassler 2006). Quorum sensing indicated that certain genes were expressed only when bacteria are crowded together (von Bodman et al. 2003). Overall, the study of quorum sensing has transformed our view of "single cell" organisms.

The first quorum sensing system described at the molecular level was in *Vibrio fischeri*, a bioluminescent marine bacterium in which the bacteria grow to high cell density within the light organ of the Hawaiian squid (Nealson et al. 1970; Nealson and Hastings 1979). When *V. fischeri* cells are free-living, they do not luminesce. However, in the nutrient-rich environment of the squid's light organ, they grow to a high concentration and produce an acyl-homoserine lactone which acts as an autoinducer (Eberhard et al. 1981). The population-dependent release of autoinducer leads to transcription of the luciferase gene and bioluminescence (Kaplan and Greenberg 1985). The *V. fischeri* system has become a paradigm for quorum sensing in Gram-negative bacteria; in addition, many other Gram-negative bacteria have cell-to-cell signaling systems that use fatty acid derivatives as autoinducers (Whitehead et al. 2001; Bassler and Losick 2006). Quorum sensing also occurs widely in Gram-positive bacteria, where the best studied autoinduction signals are oligopeptides (Waters and Bassler 2005; Shapiro 1998).

Once thought to be rare, quorum sensing systems are now known to be common. They are involved in a large number of complex environmental responses that include virulence and biofilm formation in *Pseudomonas aeruginosa* (Van Delden and Iglewski 1998; Rumbaugh et al. 2000), fruiting body production in *Myxococcus* (Shimkets 1999), antibiotic and pigment production in *Serratia* (Thomson et al. 2000), pathogenicity in plant-pathogenic bacteria (von Bodman et al. 2003), and many other important phenomena that allow bacterial colonies to behave in a fashion similar to multicellular organisms. These bacterial systems are almost always studied in the context of aqueous environments. In the laboratory, the approach is to use shaken liquid-batch cultures whereby the quorum sensing response is detected at a specific point in the growth curve, coinciding with a threshold concentration of the signal (Horswill et al. 2007). Because of the dominant aqueous experimental model, it is not surprising that only a few bacterial quorum sensing signals that operate through the gas phase have been discovered. Nevertheless, some cases have been reported. *Ralstonia* (*Pseudomonas*) *solanacearum* is a Gram-negative soil-borne plant pathogen that causes bacterial wilt in a wide variety of important crops, including eggplant, potato, and tobacco (Agrios 2008). It uses the volatile signal, 3-hydroxyl palmitic acid methyl ester, for regulating the expression of most of the traits needed for infection and virulence (Flavier et al. 1997; von Bodman et al. 2003). In environments such as soil and plant surfaces, opportunities for signaling in the liquid phase are limited. Therefore, it is highly likely that many other volatile autoinducers will be discovered and that "signaling in the gas phase may be one of the next important frontiers in quorum sensing" (Horswill et al. 2007).

Quorum sensing molecules, both volatile and non-volatile, have also been identified in fungi (Hornby et al. 2001), where they can participate in cross-kingdom communication (Dixon and Hall 2015). The best studied system is *Candida albicans*, a dimorphic fungus, which grows as a commensal on humans in yeast form but functions as an opportunistic pathogen with hyphal morphology. Hornby et al. (2001) showed that farnesol, an acyclic sesquiterpene alcohol, modulates the yeast-to-hyphal transition in *Candida*; this yeast to hyphal transition is essential for causing disseminated disease (Hogan 2006). There is increasing evidence that quorum sensing is widespread in fungi and that in addition to farnesol many of these systems oxylipins such as 1-octen-3-ol play an important role (Kües and Navarro-González 2009; Wongsuk et al. 2016). Aromatic alcohols are involved in quorum sensing in *Saccharomyces cerevisiae* (Chen and Fink 2006). Quorum sensing mechanisms in medically important fungal species have received the most attention and have been reviewed (Tian et al. 2021). In particular, farnesol and other quorum sensing molecules are involved in the yeast-to-hyphal morphological changes that are important in pathogenesis (Chow et al. 2021). *C. albicans* makes aliphatic alcohols (e.g., ethyl alcohol, isoamyl alcohol, 1-dodecanol, 2-dodecanol, and nerolidol) and aromatic alcohols (e.g., 2-phenylethyl alcohol and tryptophol) that inhibit filamentation (Chauhan and Mohan Karuppayil 2021).

Specialized terminology tends to be discipline specific and hinders effective communication between scientists from different fields. For instance, the jargon used to describe the cell density-dependent signaling systems important in fungal spore germination and fruit body development was developed independently from terms used by bacteriologists to describe bacterial quorum sensing systems (Table 4.1). Many fungal processes, such as morphogenesis, mating, biofilm development, and apoptosis, are farnesol mediated (Tian et al. 2021). Hence, more and more mycologists have adopted the quorum sensing terminology to describe the molecules that function as virulence factors, regulate fungal morphological change in fungi, or otherwise influence fungal behaviors (Wongsuk et al. 2016; Tian et al. 2021).

4.5.3 Entomology

Entomologists have been pioneers of chemical ecology, so it is not surprising that the best-known cases of VOC signaling involve arthropods, often studied under "pheromone biology." Many VOCs elicit specific behaviors such as insect mating. A complete survey of this topic would require an encyclopedia of its own; only a few examples are given here as an introduction to this fascinating field. The articles cited give access to the extensive literature describing chemical communication with an emphasis on fungal gas-phase molecules.

Many fungal volatiles, particularly eight-carbon alcohols and ketones, act as semiochemicals and function in insect attraction and deterrence (Pierce et al. 1991a, 1991b; Dowd and Bartelt 1991; Luntz 2003; Nilssen 1998; Ômura et al. 2002; Ramoni et al. 2001; Davis et al. 2013). Many blood-feeding insects have highly developed olfactory systems whereby volatile semiochemicals enter the antennae and later bind to olfactory receptor neurons (Weeks et al. 2011). C-8 alcohols emitted by decay fungi may serve as trophic signals for conifer feeding bark beetle prey and as bypass trophic signals for their predators (Zhang and Schlyter 2004). As assayed using a pitfall olfactometer test, grain beetles showed strong attraction to 1-octen-3-ol (Pierce et al. 1991b).

VOCs are involved in the formation of well-known symbioses between yeasts and insects (Stefanini 2018) or with other putative hosts (Ling et al. 2020). Moreover, many VOCs are involved in arthropod defense. For example, a combination of 1-octen-3-ol and geosmin functioned as defensive allomones in millipedes (Ômura et al. 2002). 1-octen-3-ol also interrupts the response of beetles to their aggregation hormone (Poland et al. 2009). Most true bugs (Heteroptera) have scent glands in what has been described as "overwhelming chemical fortification" (Aldrich 1988). Moreover, many species secrete attractant pheromones. For example, various bugs secrete (E)-2-hexanal, hexanal, hexanoic acid, 2-butyl-2-octenal, β-pinene, limonene, and farnesene as alarm or trail pheromones for defense against their main predators (Aldrich 1988). For those wishing an introduction to the fascinating world of insect defensive secretions, many of which utilize noxious smelling VOCs such as caprylic acid, we recommend Thomas Eisner's beautifully illustrated book *For Love of Insects* (Eisner 2003).

For a long time, it has been known that fungal odors attract insects that live on fungi or on substrates decayed by fungi (Fäldt et al. 1999). In one interesting case in point, female houseflies do not lay their eggs on animal feces that have been colonized by *Fusarium*, *Phoma*, *Rhizopus*, and other fungi. Using chemical standards, six fungal VOCs that consistently inhibit oviposition were determined: dimethyl disulfide, phenylacetaldehyde, 2-phenylethanol, citronellal, and norphytone (Lam et al. 2010).

When the chemical composition of insect attractants is known, in some cases effective artificial pheromones can be prepared to trap insect pests. Moreover, "elucidation of chemical messengers for predatory and weed feeding bugs may lead to pheromonal husbandry of these beneficial insects" (Aldrich 1988). Fungal volatiles may provide innovative, safe pest management strategies for economically problematic insect species (Hayashi 2015; Holighaus and Rohlfs 2016).

4.5.4 Developmental Signals for Plant Growth

Plant growth promotion by microbial species has been studied in several contexts. Microbes influence plant growth by reducing levels of disease (e.g., by antibiosis or competition with pathogens), stimulating growth, and biofertilization. When scientists study these interactions, beneficial effects are usually easiest to demonstrate in the laboratory, with decreasing success in greenhouses, and only a few microbes functioning successfully in field situations (Lugtenberg and Kamilova 2009).

Bacterial volatiles have received the most attention. Many growth-promoting rhizobacteria produce volatile organic compounds that have a positive impact on plant growth (Vespermann et al. 2007). For example, strains of *Bacillus subtilis*, *Bacillus amyloliquefaciens*, and *Enterobacter cloacae* promoted plant growth by releasing several volatiles of which acetoin and 2,3-butanediol gave the highest level of growth promotion (Ryu et al. 2003, 2005). Exposure to *B. subtilis* volatiles induces beneficial effects, including increases in plant size and weight; modulation of root-system architecture; and differential expression of numerous plant genes (Zhang et al. 2007; Xie et al. 2009; Gutiérrez-Luna et al. 2010). Photosynthetic capacity is also increased with higher chlorophyll content and upregulation of genes encoding chloroplast proteins (Zhang et al. 2008; Xie et al. 2009). VOCs stimulate the synthesis of indole-3-acetic acid, cytokinin, and gibberellins. Moreover, genes involved in auxin biosynthesis are upregulated in plants exposed to VOCs from growth-promoting rhizobacteria (Ryu et al. 2003).

When plant growth-promoting bacteria (PGPR) interact with plants, VOCs are important components of the signaling process (Vespermann et al. 2007). Several bacterial volatiles, including alcohols, ammonia, HCN, and phenazine-1-carboxylic acid, have antifungal properties that contribute to the biocontrol properties of PGPR (Choudhary et al. 2008; Kai et al. 2009; Whipps 2001). Kai et al. (2009) showed that volatiles from a given bacterial strain cause different responses in different fungi; i.e., associations of fungi and bacteria interact in different ways. The VOCs of PGPR are associated with induced systemic resistance (Farag et al. 2006). Furthermore, VOCs also lead to an enhancement of aroma compounds in basil (Banchio et al. 2009).

Certain bacterial volatiles promote plant health by inhibiting the growth of fungal pathogens. Negative effects on fungi include the inhibition of both sporulation and spore germination. Positive effects on sclerotial and fruit body development were often observed (Vespermann et al. 2007; Mackie and Wheatley 1999; Fernando et al. 2005). Plants live with complex microbial consortia that include both bacterial and fungal species. The VOCs emitted from these consortia frequently serve as the communication signals used within and between kingdoms (Schulz-Bohm et al. 2017; Heenan-Daly et al. 2021).

Just as bacterial VOCs can enhance or inhibit plant growth, fungal VOCs also have many effects on plant growth and development (Lee et al. 2015; Sánchez-López et al. 2016). For example, *Fusarium oxysporum* and *Verticillium dahliae* both emit volatiles that increase *Arabidopsis thaliana* tolerance to salt stress (Li and Kang 2018; Li et al. 2018). Exposure to volatiles from *Alternaria alternata*, *Penicillium charlesii*, and *Penicillium aurantiogriseum* increases starch accumulation in leaves of *Arabidopsis*, barley, maize, potato, tomato, and tobacco (Ezquer et al. 2010). VOCs from *Alternaria alternata* upregulate sucrose, starch, and glucose transport and modification enzymes, while VOCs from phytopathogenic *Rhizoctonia solani* increase primary root growth in the turnip *Brassica rapa* (Moisan et al. 2021).

Our lab has studied the effect of volatiles from several *Trichoderma* species on the plant genetic model *Arabidopsis thaliana*. Hung et al. (2013) found that plants grown in the presence of volatiles from *T. viride*, without any direct physical contact between the plants and the fungi, were larger, had more chlorophyll, and bloomed earlier than controls, while other volatiles such as 1-octen-3-ol caused negative effects (Hung et al.

2014). In similar studies, *A. thaliana* was grown in the presence of twenty different biocontrol strains of *Trichoderma*. Co-growth with nine of these strains yielded plants with increased biomass and chlorophyll concentration (Lee et al. 2016b). *Arabidopsis* also was grown in the presence of physiological concentrations of chemical standards of 26 VOCs previously identified from *Trichoderma* (Lee et al. 2016a). Increases in fresh weight and chlorophyll contents were observed by exposure to seven of these volatiles: 2-methyl-1-butanol, 3-methyl-1-butanol, 2-methyl-1-propanol, 2-heptylfuran, 1-decene, 1-octen-3-ol, and 2-heptanone. Of compounds tests, plants exposed to vapors of 1-decene had the greatest increase in fresh weight and chlorophyll content. Transcriptome analysis of plants exposed to 1-decene revealed that 123 genes were differentially expressed, including genes involved in cell wall modification, auxin induction, stress, and defense responses (Lee et al. 2019).

Below-ground volatiles produced by a strain of *Trichoderma viride* improved plant growth in okra (*Abelmoschus esculentus*) and induced defense enzyme activities, resulting in reduced lipid peroxidation and cell death. In addition, these VOCs inhibited the growth of *Sclerotium rolfsii*, a rot pathogen. VOCs detected from *T. viride* included 2-butyl-1-octanol, methylbicyclooctan-7-ol, 1-methyl cycloundecanol, tetradecanoic acid, benzestrol, oleic acid, benzenediol, 9-tricosene, hexadecanoic acid, and ascaridole (Singh et al. 2021).

4.5.5 Developmental Signals for Fungal Spore Germination and Growth

Many biological activities, including chemotropic interactions, growth coordination, and inhibition of the growth of other fungi, are mediated by volatile and non-volatile signaling metabolites. See Leeder et al. (2011) for a useful review. Figure 4.2 illustrates some of the interactions associated with 1-octen-3-ol.

The fungal sensing processes that prevent premature germination are called "self-inhibition" or "auto-inhibition." These processes detect spore over-crowding and maximize the chances of survival and colony formation once the spore has germinated. High spore concentrations inhibit germination; with lower spore concentrations, germination proceeds. In general, these effects are reversible and do not affect later mycelial growth. Auto-inhibition has been well studied in *Colletotrichum*, where several germination self-inhibitors have been chemically characterized as non-volatile indole compounds and where these compounds are specific for the producing species (Lax et al. 1985; Tsurushima et al. 1995). The cell density-dependent germination inhibitor systems

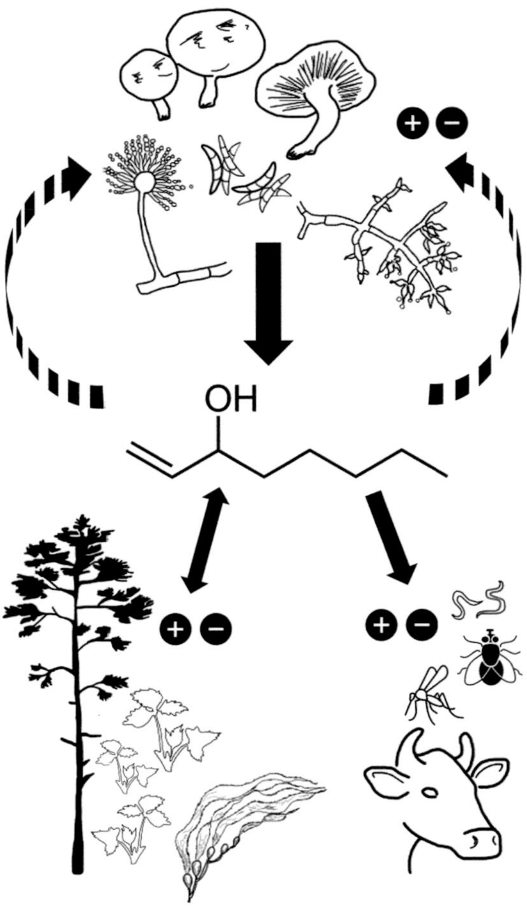

Fig. 4.2 The natural product 1-octen-3-ol is produced by many fungal and plant species. This eight-carbon volatile causes myriad stimulatory and inhibitory responses in fungi, plants, and animals. Specific effects are mediated in a species-specific and concentration-dependent manner

in fungi resemble quorum sensing systems described for bacteria (see above).

The single most commonly reported volatile from both macroscopic and microscopic fungi is 1-octen-3-ol which is also called "mushroom alcohol" or "matsutake alcohol" because it was first identified from *Tricholoma matsutake* (Wood and Fesler 1986). In the straw mushroom, *Volvariella volvacea*, for example, 1-octen-3-ol accounts for the majority (72–83%) of the total volatiles detected (Mau et al. 1997). Mushroom alcohol occurs in two enantiomer forms, (R)-(−)-1-octen-3-ol and (S)-(+)-1-octen-3-ol, of which the most abundant naturally occurring isomer is the R enantiomer, sometimes called "roctenol." Roctenol is produced commercially where it is used as an insect attractant (Bohbot and Dickens 2009; Kline et al. 2007). (R)- (−)-1-octen-3-ol has a characteristic "mushroomy" odor, while (S)-(+)-1-octen-3-ol is more grassy or moldy; (R)-(−)-1-octen-3-ol is the predominant naturally occurring enantiomer of 1-octen-3-ol (Mosandl et al. 1986). It is of interest that the phylogenetic analysis of microbial VOCs by Elmassry et al. (2020) showed that 1-octen-3-ol, 3-octanone, aristolochene, and 2-pentylfuran are primarily synthesized by fungi and not by other microbes.

Mushroom alcohol is produced with 10-oxo-*trans*-8-decenoic acid by the enzymatic breakdown of linoleic acid. Both compounds inhibit the mycelial growth of *Penicillium expansum* at low (1.25 mM) concentrations (Okull et al. 2003). In *Penicillium paneum*, dense suspensions of conidia show poor germination; again the auto-inhibitor is 1-octen-3-ol (Chitarra et al. 2004). Volatile compounds produced by *P. paneum* under high spore density conditions also inhibit the mycelial growth of other fungal species belonging to a variety of genera, suggesting a range of actions beyond auto-inhibition. It has been hypothesized that 1-octen-3-ol interferes, in a reversible manner, with essential metabolic processes involved in the swelling and germination of the conidia (Chitarra et al. 2005).

The effect of 1-octen-3-ol is dependent on its concentration and on the stage of fungal colony formation. In dark-grown cultures of *Trichoderma*, very low concentrations of 1-octen-3-ol, 3-octanol, and 3-octanone induced conidiation. However, at the highest concentrations tested for 1-octen-3-ol (500 μM) and 3-octanol and 3-octanone (both 1000 μM), both conidiation and growth were inhibited (Nemcovic et al. 2008). In another study identifying and profiling the naturally occurring volatile metabolites of *Trichoderma atroviride*, 1-octen-3-ol and 3-octanone reached their maximum concentrations simultaneously to conidiation (Stoppacher et al. 2010).

Similar effects of C-8 compounds have been observed in *Aspergillus nidulans* (Herrero-Garcia et al. 2011). High conidial densities yielded auto-inhibition of germination, which was associated with 1-octen-3-ol produced in association with aerial hyphae. The inhibition effect was reversible; conidiospores germinated after 1-octen-3-ol was removed. The closely related compound 3-octanone enhanced sporulation responses in aerial cultures (Herrero-Garcia et al. 2011). It also is of interest that several C-18 oxylipins derived from linoleic acid (psi factors) can repress conidiation and induce premature sexual sporulation in *A. nidulans* (Champe et al. 1987; Champe and el-Zayat 1989). Oxylipins are broadly involved in the developmental regulation of aspergilli (Tsitsigiannis and Keller 2007), and 10-oxo-trans-8-decenoic acid stimulates mycelia growth, stipe lengthening, and the initiation of fruiting in *Agaricus bisporus* (Mau et al. 1992). Oxylipins have been found from fungal pathogens in both plants and animals. While the oxylipin chemical end products are similar, the fungal enzymes involved in their production are different from those isolated from plants and animals. Several lines of evidence suggest that plant oxylipins are able to mimic fungal oxylipins, suggesting a form of "reciprocal cross-talk" in certain host–pathogen relationships (Brodhagen et al. 2008). See Brodhun and Feussner (2011) for a comprehensive review of fungal oxylipins and the enzymes involved in their production.

Various pheromone communication systems are used to coordinate colony formation, fruit body development, and reproduction in higher fungi (Kües and Navarro-González 2009). The factors that influence sporocarp development in the cultivated mushroom *Agaricus bisporus* have been studied widely. Primordium formation depends on both the presence of a "casing layer" containing appropriate bacteria and on sufficient ventilation. The air exchange is important because it removes inhibitory VOCs, especially 1-octen-3-ol produced by *A. bisporus*, but also 3-ethyl-1-hexanol produced by the rye on which the *A. bisporus* spawn is grown. Pseudomonad populations in the casing metabolize 3-ethyl-1-hexanol, thus removing its inhibitory effect (Noble et al. 2009).

4.5.6 Other Associations

Many organisms produce metabolites that are used in deterring competitors. Humans categorize some of these chemical defenses as antibiotics (if they selectively kill microbes that we wish to see dead) or toxins (if they kill us or our favored plant and animal species). Well-known plant secondary metabolites such as alkaloids are thought to have evolved as defense mechanisms. Plants have exquisitely refined chemical defense mechanisms against insect herbivory, including indirect defenses whereby they emit volatiles that attract natural enemies (Howe and Jander 2008). A survey of 1500 fungal secondary metabolites published between 1993 and 2001 showed that more than half had antibacterial, antifungal, or antitumor activity (Pelaez 2005). The literature on fungal secondary metabolites is extensive and has been reviewed elsewhere (Turner 1971; Turner and Aldridge 1983; Cole and Schweikert 2003). Far less is known about the deterrence activity of plant, fungal, and bacterial VOCs. Nevertheless, certain fungal VOCs are known to have growth inhibitory effects on plants (Splivallo et al. 2007b). When *A. thaliana* is exposed to 1-octen-3-ol, the defense genes that are associated with wounding or ethylene and jasmonic acid signaling are turned on (Kishimoto et al. 2007). Moreover, plants exposed to 1-octen-3-ol can inhibit the expansion of the pathogen, *Botrytis cinerea,* on the infected leaves (Kishimoto et al. 2007). In addition, some soil bacteria emit antifungal VOCs (Liu et al. 2008).

Numerous fungal species engage in symbiotic interactions with plants such as mycorrhizae, endophytes, and lichens, while a large number are plant and insect pathogens. Although fewer fungi are mammalian pathogens, the species that affect humans are becoming increasingly important in modern medicine. In all these interspecific interactions, chemical conversations are taking place, many of them facilitated by VOCs. Because volatile compounds can be sensed at a distance, they are ideal for serving as intra- and interspecific chemical signals or "infochemicals."

Phytopathologists have long recognized the importance of molecular signaling during plant–pathogen interactions. In the rhizosphere, multiple species of bacteria and fungi interact with plant roots and their exudates. Moreover, non-pathogenic microbes such as nitrogen-fixing bacteria and mycorrhizal fungi that establish beneficial relationships with plants rely on complex signaling networks (Oldroyd and Downie 2004; Harrison 2005). Many of the other chapters in this book describe classical and molecular approaches to studying such fungal associations, and Ortiz-Castro et al. (2009) have reviewed the role of microbial signals in plant growth and development.

The soil remains an underexplored habitat that contains a rich diversity of microbial life forms. The relevance of secondary metabolites and volatiles in soil ecosystems has been reviewed in the monograph edited by Karlovsky (2008). It has been estimated that a single gram of soil may contain tens of thousands of different fungal, bacterial, archaeal, and protist species. The soil properties such as nutrient and oxygen availability and the physiological state of the microbial population make dynamic and unique soil-specific communities (McNeal and Herbert 2009). Multispecies microbial populations play a key role in sustaining soil microcosms. Communication within rhizosphere bacterial populations, within mycelial colonies, and between fungi and bacteria is mediated by signaling molecules, but we are only beginning to learn the specifics of this chemical information exchange. Gas-phase molecules are essential components of these chemical conversations. These compounds can diffuse a long way from their point of origin, and they can persist and migrate in soil environments, areas of dense vegetation, and other microhabitats that harbor interacting populations of bacteria and fungi.

Much of the molecular work on biofilm formation has been conducted in laboratory settings in the context of quorum sensing research using single-species biofilms. However, biofilms found in nature are multispecies communities that can harbor several hundred species where various forms of intercellular signaling are being used. By definition, the specialized vocabulary of "autoinducers" is not appropriate in biofilm communities. Nevertheless, many of the same signaling molecules and pathways are in use (Kolter and Greenberg 2006).

The marine alga *Pyropia haitanensis* uses a lipoxygenase enzyme to convert large quantities of C20:4 fatty acids into 1-octen-3-ol (Luo et al. 2014). Production of this important signaling molecule may act as a messenger that induces a "primed" state by upregulating the synthesis of methyl jasmonic acid, indole-3-acetic acid, and gibberellin A3 (Chen et al. 2019).

In an unusual "Trojan horse" form of bacterial pathogenesis against nematodes, *Bacillus nematocida* B16 lures nematodes by emitting several volatile organic compounds, including benzaldehyde and 2-heptanone that are much more attractive to worms than those from ordinary dietary bacteria. Once the bacteria enter the intestine of nematodes, they secrete proteases that preferentially target essential intestinal proteins, leading to nematode death (Niu et al. 2010). An even more striking example of a volatile–nematode interaction concerns the mushroom *Pleurotus ostreatus* (the "oyster mushroom"), which for years has been known to be carnivorous. Lee et al. (2023) showed that 3-octanone accumulates in specialized cells called toxocysts and that the use of chemically pure 3-octanone recapacitates the rapid paralysis, calcium influx, and neuronal death arising from fungal contact.

As these tantalizing cases illustrate, the effect of bacterial and fungal VOCs in both host–pathogen interactions and non-pathogen ecological contexts is an emerging frontier for future research. VOCs produced by a given microbial species can have multiple effects on other microbes and organisms and can be used for defense, environmental monitoring, and nutrient acquisition (Wheatley et al. 1997; Bruce et al. 2004; Minerdi et al. 2009). We agree with Tarkka and Piechulla (2007) that it is likely that many more multifunctional and multi-organismic volatile-based interactions exist in ecosystems that have been envisioned in the past. As the analytical methods for assaying VOCs improve, we will find increasing evidence of the crucial importance of VOC-mediated cross-talk between species, not only for fungal–bacterial associations but also for fungal–plant and fungal–animal associations.

4.6 Summary

This essay has highlighted some of the major ways microbial VOCs have been studied to date and gives examples of how communication in biological systems is mediated chemically through the atmosphere. Numerous plant–fungal, insect–fungal, and bacterial–fungal interactions that involve VOC signaling have been discovered, but many more remain to be elucidated. Technological advances with respect to our ability to detect VOCs and our capacity to monitor their physiological effects in plant, animal, and microbial systems will open the door to new views of the complexities underlying chemical signaling across the biosphere.

Microbial VOCs have been studied by scientists from a broad range of disciplines for both theoretical and applied reasons. As a result, the published information about fungal VOCs is scattered widely among diverse scientific literature. In general, entomologists do not cross-reference the microbiological literature, nor do food and flavor chemists pay attention to the physiological activities of the compounds they study. Nevertheless, in recent years, there has been an increased interdisciplinary interest in fungal volatile research as evidenced by the quantity of review articles on the subject and the breadth of applications and interactions that have emerged. By orchestrating these different kinds of knowledge, it becomes ever more obvious that very small gas-phase molecules from fungi and all biogenic sources can have enormous physiological impacts within and between organisms across the major domains of life. VOCs facilitate the formation and maintenance of mutualistic symbiotic relationships of fungi with plants, animals, and/or bacteria. Nevertheless, their study remains technically challenging. They are produced in complex mixtures, and while the number of volatiles in a biogenic blend can be large, their individual abundance is generally low.

The scientific community has just begun to eavesdrop on the chemical discourse conducted via gas-phase molecules. As we learn more about chemical signaling, it is becoming increasingly apparent that the significance of biogenic VOCs, including fungal VOCs, has been underestimated. It is our hope that this chapter will elicit the interest of fungal biologists to ask questions such as: How do microbes use VOCs as intraspecific developmental signals and as interspecific signaling cues? In what way can we incorporate the insights of the entomological community into the mindset of bacteriologists and mycologists? What kinds of technological tools are needed to expand the study of VOCs? By what means can VOC research provide a platform for productive interdisciplinary collaborations? How many common volatile compounds like 1-octen-3-ol serve unsuspected physiological activities? In what way will metabolomic research revolutionize the study of biogenic gas-phase molecules?

Acknowledgments Research in the Bennett laboratory has been funded by the Northeast Sustainable Agriculture Research and Education Grant (GNE14-084-27806); the U.S. Fish & Wildlife Service Grant (WNS 2016-6); USDA-ARS Cooperative Agreement (Grant No. 2-47012); the Rutgers University Research Fund; and the Sloane and Eppley Foundations. JWB is grateful to the late Karl Esser for his mentorship and to Olivia Froehlich, Arati Inamdar, Shannon Morath, Sally Padhi, Kayla Pennerman, and Guohua Yin for years of fruitful conversations and collaborations.

References

Abramson D, Sinha RN, Mills JT (1980) Mycotoxin and odor formation in moist cereal grain during granary storage. Cereal Chem 57:346–351

Abramson D, Sinha RN, Mills JT (1983) Mycotoxin and odor formation in barley stored at 16 and 20% moisture in Manitoba. Cereal Chem 60:350–355

Ackerman D (1990) A natural history of the senses. Vintage Books, Random House, New York, NY

Agrios GN (2008) Plant pathology, 5th edn. Elsevier, San Diego, CA

Aldrich JR (1988) Chemical ecology of the heteroptera. Annu Rev Entomol 33:211–238

Allen PJ (1957) Properties of a volatile fraction of uredospores of *Puccinia graminis* var. *tritici* affecting their germination and development. I Biological activity. Plant Physiol 32:385–389

Allen K, Bennett JW (2021) Tour of truffles: aromas, aphrodisiacs, adaptogens, and more. Mycobiology 49(3):201–212

Alpha CJ, Campos M, Jacobs-Wagner C, Strobel SA (2015) Mycofumigation by the volatile organic compound-producing fungus *Muscodor albus* induces bacterial cell death through DNA damage. Appl Environ Microbiol 81(3):1147–1156

Anishchenko IM, McCartney MM, Fung AG, Peirano DJ, Schirle MJ, Kenyon NJ, Davis CE (2018) Modular and reconfigurable gas chromatography/differential mobility spectrometry (GC/DMS) package for detection of volatile organic compounds (VOCs). Int J Ion Mobil Spectrom 21:125–136

Assaf S, Hadar Y, Dosoretz CG (1997) 1-Octen-3-ol and 13-hydroperoxylinoleate are products of distinct pathways in the oxidative breakdown of linoleic acid by Pleurotus pulmonarius. Enzym Microb Technol 21: 484–490

Atmosukarto I, Castillo U, Hess W, Sears J, Strobel G (2005) Isolation and characterization of *Muscodor albus* I-41.3s, a volatile antibiotic producing fungus. Plant Sci 169:854–861

Audrain B, Farag MA, Ryu CM, Ghigo JM (2015) Role of bacterial volatile compounds in bacterial biology. FEMS Microbiol Rev 39(2):222–233

Bacon CW, White JW (eds) (2000) Microbial endophytes. Marcel Dekker, New York

Baldwin IT, Halitschke R, Paschold A, von Dahl CC, Preston CA (2006) Volatile signaling in plant-plant interactions: "talking trees" in the genomics era. Science 311:812–815

Banchio E, Xie X, Zhang H, Paré PW (2009) Soil bacteria elevate essential oil accumulation and emissions in sweet basil. J Agric Food Chem 57:653–657

Bassler BL, Losick R (2006) Bacterially speaking. Cell 125:237–246

Bennett JW (1983) Differentiation and secondary metabolism in mycelial fungi. In: Bennett JW, Ciegler A (eds) Secondary metabolism and differentiation in Fungi. Marcel Dekker, New York

Bennett JW, Bentley R (1989) What's in a name? – Microbial secondary metabolism. Adv Appl Microbiol 34:1–28

Bennett JW, Feibleman T (2001) Fungal bacterial interactions. In: Hock B (ed) The Mycota, vol IX. Springer-Verlag, Berlin

Bentley R, Meganathan R (1981) Geosmin and methylisoborneol biosynthesis in streptomycetes: evidence for an isoprenoid pathway and its absence in non-differentiating isolates. FEBS Lett 125(2): 220–222

Berg JM, Tymoch JL, Stryer L (2007) Biochemistry. WH Freeman and Company, New York

Berger RG, Drawert F, Tiefel P (1992) Naturally occuring flavours from fungi, yeasts, and bacteria. In: Patterson RLS, Charlwood BV, MacLeod G, Williams AA (eds) Bioinformation of flavours. The Royal Chemistry Society, Cambridge, pp 1–20

Bitas V, Kim H-S, Bennett J, Kang S (2013) Sniffing on microbes: diverse roles of microbial volatile organic compounds in plant health. Mol Plant-Microbe Interact 26:835–843

Bloch E, Deorazio R (1994) Chemistry in a salad bowl: comparative organosulfur chemistry of garlic, onion, and shitake mushrooms. Pure Appl Chem 66:2205–2206

Bohbot JD, Dickens JC (2009) Characterization of an enantioselective odorant receptor in the yellow fever mosquito *Aedes aegypti*. PLoS One 4:7032

Bone E (2011) Mycophilia: revelations from the weird world of mushrooms. Rodale Books

Bonfante P, Desirò A (2017) Who lives in a fungus? The diversity, origins and functions of fungal endobacteria living in Mucoromycota. ISME J 11(8):1727–1735

Boniface T (2020) The use of odours in the identification of mushrooms and toadstools. Field Mycol 21:28–30

Borg-Karlson A-K, Englund FO, Unelius CR (1994) Dimethyl oligosulphides, major volatiles released from *Sauromatum guttatum* and *Phallus impudicus*. Phytochemistry 35:321–323

Börjesson T, Stöllman UM, Adamek P, Kaspersson A (1989) Analysis of volatile compounds for detection of molds in stored cereals. Cereal Chem 66:300–304

Börjesson T, Stöllman UM, Schnürer J (1993) Off-odorous compounds produced by molds on oatmeal agar: Identification and relation to other growth characteristics. J Agric Food Chem 41:2104–2111

Breheret S, Talou T, Rapior S, Bessiere J-M (1997) Monoterpenes in the aromas of fresh wild mushrooms. J Agric Food Chem 45:831–836

Briard B, Heddergott C, Latge JP (2016) Volatile compounds emitted by *Pseudomonas aeruginosa* stimulate growth of the fungal pathogen *Aspergillus fumigatus*. MBio 7:e00219

Brodhagen M, Tsitsigiannis DI, Hornung E, Goebel C, Feussner I, Keller NP (2008) Reciprocal oxylipin-mediated cross-talk in the *Aspergillus*–seed pathosystem. Mol Microbiol 67:378–391

Brodhun F, Feussner I (2011) Oxylipins in fungi. FEBS J 278:1047–1063

Brown WL (1968) An hypothesis concerning the function of the metapleural glands in ants. Am Nat 102:188–191

Bruce A, Verrall S, Hackett CA, Wheatley RE (2004) Identification of volatile organic compounds (VOCs) from bacteria and yeast causing growth inhibition of sapstain fungi. Holzforschung 58:193–198

Burge PS (2004) Studies on the role of fungi in sick building syndrome. Occup Environ Med 61:185–190

Camilli A, Bassler BL (2006) Bacterial small-molecule signaling pathways. Science 311:1113–1116

Champe SP, el-Zayat AA (1989) Isolation of a sexual sporulation hormone from *Aspergillus nidulans*. J Bacteriol 171:3982–3988

Champe SP, Rao P, Chang A (1987) An endogenous inducer of sexual development in *Aspergillus nidulans*. J Gen Microbiol 133:1383–1387

Chauhan NM, Mohan Karuppayil S (2021) Dual identities for various alcohols in two different yeasts. Mycology 12(1):25–38

Chauhan NM, Raut JS, Karuppayil SM (2011) A morphogenetic regulatory role for ethyl alcohol in *Candida albicans*. Mycoses 54(6):e697–e703

Chen H, Fink GR (2006) Feedback control of morphogenesis in fungi by aromatic alcohols. Genes Dev 20: 1150–1161

Chen H, Yang R, Chen J, Luo Q, Cui X, Yan X, Gerwick WH (2019) 1-Octen-3-ol, a self-stimulating oxylipin messenger, can prime and induce defense of marine alga. BMC Plant Biol 19:37

Chiron N, Michelot D (2005) Odeurs de champignons: chimie et rôle dans les interactions biotiques — une revue. Cryptogam Mycol 26:299–364

Chitarra GS, Abee T, Rombouts FM, Posthumus MA, Dijksterhuis J (2004) Germination of *Penicillium paneum* conidia is regulated by 1-octen-3-ol, a volatile self-inhibitor. Appl Environ Microbiol 70:2823–2829

Chitarra GS, Abee T, Rombouts FM, Dijksterhuis J (2005) 1-octen-3-ol inhibits conidia germination of *Penicillium paneum* despite of mild effects on membrane permeability, respiration, intracellular pH, and changes the protein composition. FEMS Microbiol Ecol 54:67–75

Cho IH, Namgung H-J, Choi H-K, Kim YS (2008) Volatiles and key odorants in the pileus and stipe of pine-mushroom (*Tricholoma matsutake* Sing). Food Chem 106:71–76

Choudhary DK, Johri BN, Prakash A (2008) Volatiles as priming agents that initiate plant growth and defence responses. Curr Sci 94(5):595–604

Chow EWL, Pang LM, Wang Y (2021) From Jekyll to Hyde: the yeast-hyphal transition of *Candida albicans*. Pathogens 10(7):859

Claeson A-S, Levin J-O, Blomquist GR, Sunesson A-L (2002) Volatile metabolites from microorganisms grown on humid building materials and synthetic media. J Environ Monit 4:667–672

Cole R, Schweikert M (2003) Handbook of secondary fungal metabolites, vol 1-3. Academic Press, Amsterdam

Combet E, Henderson J, Eastwood DC, Burton KS (2006) Eight-carbon volatiles in mushrooms and fungi: properties, analysis, and biosynthesis. Mycoscience 47:317–326

Copetti MV (2019) Fungi as industrial producers of food ingredients. Curr Opin Food Sci 25:52–56

Cronin DA, Ward MK (1971) The characterisation of some mushroom volatiles. J Sci Food Agric 22:477–479

Davis TS, Crippen TL, Hofstetter RW, Tomberlin JK (2013) Microbial volatile emissions as insect semiochemicals. J Chem Ecol 39:840–859

Davis C, Pleil J, Beauchamp J (eds) (2020) Breathborne biomarkers and the human volatilome, 2nd edn. Elsevier

De Heer K, Vonk SI, Kok M, Kolader M, Zwinderman AH, Van Oers MHJ, Visser CE (2016) eNose technology can detect and classify human pathogenic molds in vitro: a proof-of-concept study of *Aspergillus fumigatus* and *Rhizopus oryzae*. J Breath Res 10(3): 036008

de Pinho PG, Ribeiro B, Goncalves RF, Baptista P, Valentao P, Seabra RM, Andrade PB (2008) Correlation between the pattern volatiles and the overall aroma of wild edible mushrooms. J Agric Food Chem 56: 1704–1712

Dicke M, Sabelis MW (1988) Infochemical terminology: based on cost-benefit analysis rather than origin of compounds? Funct Ecol 2:131–139

Dickschat JS (2017) Fungal volatiles–a survey from edible mushrooms to moulds. Nat Prod Rep 34(3):310–328

Dickschat JS, Martens R, Brinkhoff T, Simon M, Schulz S (2005a) Volatiles releases by *Streptomyces* species isolated from the North Sea. Chem Biodivers 2:837–865

Dickschat JS, Wenzel SC, Bode HB, Müller R, Schulz S (2005b) Biosynthesis of volatiles by the Myxobacterium *Myxococcus xanthus*. Chembiochem 5:778–787

Dietert RR, Silbergeld EK (2015) Biomarkers for the 21st century: listening to the microbiome. Toxicol Sci 144(2):208–216

Dixon EF, Hall RA (2015) Noisy neighbourhoods: quorum sensing in fungal-polymicrobial infections. Cell Microbiol 17:1431–1441

Do DH, Walgraeve C, Amare AN, Barai KR, Parao AE, Demeestere K, Van Langenhove H (2015) Airborne volatile organic compounds in urban and industrial locations in four developing countries. Atmos Environ 119:330–338

Doty SL (2013) Endophytic yeasts: biology and applications. In: Aroca R (ed) Symbiotic endophytes. Soil biology. Springer, Berlin, pp 335–343

Dowd PF, Bartelt RJ (1991) Host-derived volatiles as attractants and pheromone synergists for dried fruit beetle, *Carpophilus hemipterus*. J Chem Ecol 17: 285–308

Dunkel M, Schmidt U, Struck S, Berger L, Gruening B, Hossbach J, Jaeger IS, Effmert U, Piechulla B, Eriksson R, Knudsen J, Preissner R (2009) SuperScent--a database of flavors and scents. Nucleic Acids Res 37(Database issue):D291–D294

Eberhard A, Burlingame AL, Eberhard C, Kenyon GL, Nealson KH, Oppenheimer NJ (1981) Structural identification of autoinducer of *Photobacterium fischeri* luciferase. Biochemistry 20:2444–2449

Eisner T (2003) For love of insects. Harvard University Press, Cambridge, MA

Elmassry MM, Farag MA, Preissner R, Gohlke BO, Piechulla B, Lemfack MC (2020) Sixty-one volatiles have phylogenetic signals across bacterial domain and fungal kingdom. Front Microbiol 11:557253

Ezquer I, Li J, Ovecka M, Baroja-Fernández E, José Muñoz F, Montero M, de Cerio J, Hidalgo M, Sesma MT, Behaji A, Etxeberria E (2010) Microbial volatile emissions promote accumulation of exceptionally high levels of starch in leaves in mono-and dicotyledonous plants. Plant Cell Physiol 51(10):1674–1693

Fäldt J, Jonsell M, Nordlander G, Borg-Karlson A-K (1999) Volatiles of bracket fungi *Fomitopsis pinicola* and *Fomes fomentarius* and their functions as insect attractants. J Chem Ecol 25:567–590

Farag MA, Ryu CM, Sumner LW, Pare PW (2006) GC-MS SPME profiling of rhizobacterial volatiles reveals prospective inducers of growth promotion and induced systemic resistance in plants. Phytochemistry 67:2262–2268

Farh ME, Jeon J (2020) Roles of fungal volatiles from perspective of distinct lifestyles in filamentous fungi. Plant Pathol J 36(3):193–203

Fernando WGD, Ramarathnam R, Krishnamoorthy AS, Savchuk SC (2005) Identification and use of potential bacterial organic antifungal volatiles in biocontrol. Soil Biol Biochem 37:955–964

Ferrari R, Lacaze I, Le Faouder P, Bertrand-Michel J, Oger C, Galano J-M, Durand T, Moularat S, Tong LCH, Boucher C (2018) Cyclooxygenases and lipoxygenases are used by the fungus *Podospora anserina* to repel nematodes. Biochim Biophys Acta (BBA)-Gen Subj 1862:2174–2182

Fiedler N, Laumbach R, Kelly-McNeil K, Lioy P, Fan Z-H, Zhang J, Ottenweller J, Ohman-Strickland P, Kipen H (2005) Health effects of a mixture of indoor air volatile organics, their ozone oxidation products, and stress. Environ Health Perspect 113:1542–1548

Fischer G, Schwalbe R, Möller M, Ostrowski R, Dott W (1999) Species-specific production of microbial volatile organic compounds (MVOC) by airborne fungi from a compost facility. Chemosphere 39:779–810

Flavier AB, Ganova-Raeva LM, Schell MA, Denny TP (1997) Hierarchial autoinduction in Ralstonia solanacearum: control of actyl- homoserine lactone production by a novel autoregulatory system responsive to 30- hydroxypalmitic acid methyl ester. J Bacteriol 179:7089–7097

Fraatz MA, Zorn H (2010) Fungal flavours. In: Hofrichter M (ed) The Mycota X: industial applications, vol X. industrial applications, 2nd edn. Springer, Berlin, Heidelberg, pp 249–264

Fuqua WC, Winans SC, Greenberg EP (1994) Quorum sensing in bacteria: the LuxR-LuxI family of cell density-responsive transcriptional regulators. J Bact 176:269–275

Gallois A, Langlois D (1990) New results in the volatile odorous compounds of French cheeses. Lait 70:89–106

Gilbert A (2008) What the nose knows. The science of scent in everyday life. Crown Pubs, New York, NY

Glick BR (2020) Beneficial plant-bacterial interactions. Springer

Griffin MA, Spakowicz DJ, Gianoulis TA, Strobel SA (2010) Volatile organic compound production by organisms in the genus *Ascocoryne* and a re-evaluation of myco-diesel production by NRRL 50072. Microbiology 156:3814–3829

Griffith RT, Jayachandran K, Shetty KG, Whitstine W, Furton KG (2007) Differentiation of toxic molds via headspace SPME-GC/MS and canine detection. Sensors 7:1496–1508

Guo Y, Jud W, Weikl F, Ghirardo A, Junker RR, Polle A, Benz JP, Pritsch K, Schnitzler J-P, Rosenkranz M (2021) Volatile organic compound patterns predict fungal trophic mode and lifestyle. Commun Biol 4:673

Gutiérrez-Luna FM, López-Bucio J, Altamirano-Hernández J, Valencia-Cantero E, Cruz HR, Macías-Rodríguez L (2010) Plant growth-promoting rhizobacteria modulate root-system architecture in *Arabidopsis thaliana* through volatile organic compound emission. Symbiosis 51:75–83

Hall IR, Brown G, Byars J (1994) The black truffle: its history, uses and cultivation. New Zealand Institute for Crop & Food Research Limited, Christchurch

Hall IR, Brown GT, Zambonelli A (2007) Taming the truffle. the history, lore, and science of the ultimate mushroom. Timber Press, Portland, OR

Harrison MJ (2005) Signaling in the arbuscular mycorrhizal symbiosis. Annu Rev Microbiol 59:19–42

Hayashi H (2015) Frontier studies on highly selective bio-regulators useful for environmentally benign agricultural production. Biosci Biotechnol Biochem 79: 877–887

He P, Hassan MM, Tang F, Jiang H, Chen M, Liu R, Chen Q (2022) Total fungi counts and metabolic dynamics of volatile organic compounds in paddy contaminated by *Aspergillus niger* during storage employing gas chromatography-ion mobility spectrometry. Food Anal Methods 15(6):1638–1651

Heddergott C, Calvo AM, Latgé JP (2014) The volatome of *Aspergillus fumigatus*. Eukaryot Cell 13:1014–1025

Heenan-Daly D, Coughlan S, Dillane E, Doyle Prestwich B (2021) Volatile compounds from *Bacillus*, *Serratia*, and *Pseudomonas* promote growth and alter the transcriptional landscape of Solanum tuberosum in a passively ventilated growth system. Front Microbiol 1766

Hermann A (2010) The chemistry and biology of volatiles. Wiley, Hoboken, NJ, pp 1–428

Herrero-Garcia E, Garzia A, Cordobés S, Espeso EA, Ugalde U (2011) 8-Carbon oxylipins inhibit germination and growth, and stimulate aerial conidiation in *Aspergillus nidulans*. Fungal Biol 115:393–400

Hogan DA (2006) Talking to themselves: autoregulation and quorum sensing in fungi. Eukaryot Cell 5:613–619

Holighaus G, Rohlfs M (2016) Fungal allelochemicals in insect pest management. Appl Microbiol Biotechnol 100(13):5681–5689

Hooper AM, Pickett JA (2004) Semiochemistry. In: Atwood JL, Steed JW (eds) Encyclopedia of supramolecular chemistry, vol 2. Marcel Dekker, New York, pp 1270–1277

Hornby JM, Jensen EC, Lisec AD, Tasto JJ, Jahnke B, Shoemaker R, Dussault P, Nickerson KW (2001) Quorum sensing in the dimorphic fungus *Candida albicans* is mediated by farnesol. Appl Environ Microbiol 67(7): 2982–2992

Horswill A, Stoodley P, Stewart P, Parsek M (2007) The effect of the chemical, biological, and physical environment on quorum sensing in structured microbial communities. Anal Bioanal Chem 387:371–380

Hossain MM, Sultana F, Miyazawa M, Hyakumachi M (2014) The plant growth promoting fungi *Penicillium* spp. GP15-1 enhances growth and confers protection against damping-off and anthracnose in the cucumber. J Oleo Sci 63(4):391–400

Howe GA, Jander G (2008) Plant immunity to insect herbivores. Annu Rev Plant Biol 59:41–66

Hung R, Lee S, Bennett JW (2013) *Arabidopsis thaliana* as a model system for testing the effect of Trichoderma volatile organic compounds. Fungal Ecol 6:19–26

Hung R, Lee S, Bennett JW (2014) The effects of low concentrations of the enantiomers of mushroom alcohol (1-octen-3-ol) on *Arabidopsis thaliana*. Mycology 5:73–80

Hung R, Lee S, Bennett JW (2015) Fungal volatile organic compounds and their role in ecosystems. Appl Microbiol Biotechnol 99:3395–3405

Hutchinson SA (1973) Biological activities of volatile fungal metabolites. Annu Rev Phytopathol 11:223–246

Inamdar AA, Bennett JW (2015) Volatile organic compounds from fungi isolated after Hurricane Katrina induce developmental defects and apoptosis in a *Drosophila melanogaster* model. Environ Toxicol 30(5): 614–620

Inamdar AA, Masurekar P, Bennett JW (2010) Neurotoxicity of fungal volatile organic compounds in *Drosophila melanogaster*. Toxicol Sci 117:418–426

Inamdar AA, Moore JC, Cohen RI, Bennett JW (2011) A model to evaluate the cytotoxicity of the fungal volatile organic compound 1-octen-3-ol in human embryonic stem cells. Mycopathologia 173:13–20

Inamdar AA, Morath S, Bennett JW (2020) Fungal volatile organic compounds: More than just a funky smell? Annu Rev Microbiol 74:101–116

IOM - Institute of Medicine (2004) Damp indoor spaces and health. National Academies Press, Washington, DC

Jarvis BB, Miller JD (2005) Mycotoxins as harmful indoor air contaminants. Appl Microbiol Biotechnol 66:367–372

Jeandroz S, Murat C, Wang Y, Bonfante P, Tacon FL (2008) Molecular phylogeny and historical biogeography of the genus *Tuber*, the 'true truffles'. J Biogeogr 35(5):815–829

Jeleń HH (2003) Use of solid phase microextraction (SPME) for profiling fungal volatile metabolites. Lett Appl Microbiol 36:263–267

Joblin Y, Moularat S, Anton R, Bousta F, Orial G, Robine E, Picon O, Bourouina T (2010) Detection of moulds by volatile organic compounds; application to heritage conservation. Int Biodeterior Biodegrad 64: 210–217

Kai M, Haustein M, Molina F, Petri A, Scholz B, Piechulla B (2009) Bacterial volatiles and their action potential. Appl Microbiol Biotechnol 81:1001–1012

Kaminski E, Libbey LM, Stawicki S, Wasowicz E (1972) Identification of the predominant volatile compounds produced by *Aspergillus flavus*. Appl Microbiol 24: 721–726

Kaplan HB, Greenberg EP (1985) Diffusion of autoinducer is involved in regulation of the *Vibrio fischeri* luminescence system. J Bacteriol 163:1210–1214

Karahadian C, Josephson DB, Lindsay RC (1985) Contribution of *Penicillium* sp. to the flavour of Brie and Camembert cheese. J Dairy Sci 68:1865–1877

Karlovsky P (ed) (2008) Secondary metabolites in soil ecology. Springer, Berlin

Karlshøj K, Nielsen PV, Larsen TO (2007) Fungal volatiles biomarkers of good and bad food quality. In: Dijksterhuis J, Samson RA (eds) Food mycology. CRC Press, Boca Raton, FL, pp 279–302

Karlson P, Luscher M (1959) 'Pheromones': a new term for a class of biologically active substances. Nature 183:55–56

Keller NP, Turner G, Bennett JW (2005) Fungal secondary metabolism - from biochemistry to genomics. Nat Rev Microbiol 3:937–947

Kesselmeier J, Staudt M (1999) Biogenic volatile organic compounds (VOC): An overview on emission, physiology and ecology. J Atmos Chem 33:23–88

Kimball BA (2016) Volatile metabolome: problems and prospects. Bioanalysis 8(19):1987–1991

Kishimoto K, Matsui K, Ozawa R, Takabayashi J (2007) Volatile 1-octen-3-ol induces a defensive response in *Arabidopsis thaliana*. J Gen Plant Pathol 73:35–37

Kline D, Allan SA, Bernier UR, Welch CH (2007) Evaluation of the enantiomers of 1-octen-3-ol and 1-octyn-3-ol as attractants for mosquitoes associated with a freshwater swamp in Florida, U.S.A. Med Vet Entomol 21:323–331

Kolter R, Greenberg EP (2006) Microbial sciences: the superficial life of microbes. Nature 441:300–302

Korpi A, Järnberg J, Pasanen A-L (2009) Microbial volatile organic compounds. Crit Rev Toxicol 39:139–193

Kües U, Navarro-González M (2009) Communication of fungi on individual, species, kingdom, and above kingdom levels. In: Anke T, Weber D (eds) Physiology and genetics. The Mycota XV. Springer, pp 79–106

Kuske M, Romain A-C, Nicolas J (2005) Microbial volatile organic compounds as indicators of fungi. Can an electronic nose detect fungi in indoor environments? Build Environ 40:824–831

La Camera S, Gouzerh G, Sandrine D, Laurent H, Bernard F, Michel F, Thierry H (2004) Metabolic reprogramming in plant innate immunity: the contributions of phenylpropanoid and oxylipin pathways. Immunol Rev 198:267–284

Lam K, Tsang M, Labrie A, Gries R, Gries G (2010) Semiochemical-mediated oviposition avoidance by female house flies, *Musca domestica*, on animal feces colonized with harmful fungi. J Chem Ecol 36:141–147

Larsen TO (1998) Volatiles in fungal taxonomy. In: Frisvad JC, Bridge PD, Arora DK (eds) Handbook of applied mycology, Chemical fungal taxonomy, vol 6. Marcel Dekker, New York, NY, pp 263–287

Larsen TO, Frisvad JC (1995) Comparison of different methods for collection of volatile chemical markers from fungi. J Microbiol Methods 24:135–144

Lax AR, Templeton GE, Meyer WL (1985) Isolation, purification, and biological activity of a self -inhibotor from conidia of *Colletotrichum gloeosporioides*. Phytopathology 75:386–390

Lee SO, Kim HY, Choi GJ, Lee HB, Jang KS, Choi YH, Kim JC (2009) Mycofumigation with *Oxyporus latemarginatus* EF069 for control of postharvest apple decay and *Rhizoctonia* root rot on moth orchid. J Appl Microbiol 106:1213–1219

Lee S, Hung R, Yap M, Bennett JW (2015) Age matters: the effects of volatile organic compounds emitted by Trichoderma atroviride on plant growth. Arch Microbiol 197:723–727

Lee S, Hung R, Yin G, Klich MA, Grimm C, Bennett JW (2016a) *Arabidopsis thaliana* as bioindicator of fungal VOCs in indoor air. Mycobiology 44:162–170

Lee S, Yap M, Behringer G, Hung R, Bennett JW (2016b) Volatile organic compounds emitted by *Trichoderma* species mediate plant growth. Fungal Biol Biotechnol 3:1–14

Lee S, Yin G, Bennett JW (2017) Airborne signals: volatile-mediated communication between plants, fungi, and microorganisms. In: Dighton J, White JF (eds) The fungal community. Its organization and role in the ecosystem, 4th edn. CRC Press, Taylor & Francis, pp 521–538

Lee S, Behringer G, Hung R, Bennett J (2019) Effects of fungal volatile organic compounds on *Arabidopsis thaliana* growth and gene expression. Fungal Ecol 37: 1–9

Lee NY, Choi DH, Kim MG, Jeong MJ, Kwon HJ, Kim DH, Kim YG, di Luccio E, Arioka M, Yoon HJ, Kim JG (2020) Biosynthesis of (R)-(−)-1-octen-3-ol in recombinant *Saccharomyces cerevisiae* with lipoxygenase-1 and hydroperoxide lyase genes from *Tricholoma matsutake*. J Microbiol Biotechnol 30: 296–305

Lee CH, Lee YY, Chang YC, Pon WL, Lee SP, Wali N, Nakazawa T, Honda Y, Shie JJ, Hsueh YP (2023) A carnivorous mushroom paralyzes and kills nematodes via a volatile ketone. Science. Advances 9(3):eade4809

Leeder AC, Palma-Guerrero J, Glass NL (2011) The social network: deciphering fungal language. Nat Rev Microbiol 9:440–451

Lemfack MC, Gohlke BO, SMT T, Preissner S, Piechulla B, Preissner R (2018) mVOC 2.0: a database of microbial volatiles. Nucleic Acids Res 46(D1): D1261–D1265

Li N, Kang S (2018) Do volatile compounds produced by *Fusarium oxysporum* and *Verticillium dahliae* affect stress tolerance in plants? Mycology 9:166–175

Li DW, Yang CS (2004) Fungal contamination as a major contributor to sick building syndrome. Adv Appl Microbiol 55:31–112

Li N, Alfiky A, Vaughan M, Kang S (2016) Stop and smell the fungi: fungal volatile metabolites are overlooked signals involved in fungal interaction with plants. Fungal Biol Rev 30:134–144

Li N, Wang W, Bitas V, Subbarao K, Liu X, Kang S (2018) Volatile compounds emitted by diverse Verticillium species enhance plant growth by manipulating auxin signaling. Mol Plant-Microbe Interact 31:1021–1031

Li Z, Paul R, Tis TB, Saville AC, Hansel JC, Yu T, Ristaino JB, Qingshan W (2019) Non-invasive plant disease diagnostics enabled by smartphone-based fingerprinting of leaf volatiles. Nat Plants 5:856–866

Ling L, Tu Y, Ma W, Feng S, Yang C, Zhao Y, Wang N, Li Z, Lu L, Zhang J (2020) A potentially important resource: endophytic yeasts. World J Microbiol Biotechnol 36:1–7

Liu W, Mu W, Zhu B, Liu F (2008) Antifungal activities and components of VOCs produced by *Bacillus subtilis* G_8. Curr Res Bacteriol 1:28–34

Luard E (2006) Truffles. Frances Lincoln Limited, Childs Hill, London

Lugtenberg B, Kamilova F (2009) Plant-growth-promoting rhizobacteria. Annu Rev Microbiol 63:541–556

Luntz AJ (2003) Arthropod semiochemicals: mosquitoes, midges and sealice. Biochem Soc Trans 31(Pt 1): 128–133

Luo QJ, Zhu ZG, Zhu ZJ, Yang R, Qian J, Chen HM, Yan XJ (2014) Different responses to heat shock stress revealed heteromorphic adaptation strategy of *Pyropia haitanensis* (Bangiales, Rhodophyta). PLoS One 9(4): e94354

Macedo GE, Vieira PDB, Rodrigues NR, Gomes KK, Rodrigues JF, Franco JL, Posser T (2022) Effect of fungal indoor air pollutant 1-octen-3-ol on levels of reactive oxygen species and nitric oxide as well as dehydrogenases activities in *Drosophila melanogaster* males. J Toxic Environ Health A 85(14):573–585

Mackie AE, Wheatley RE (1999) Effects and incidence of volatile organic compound interactions between soil bacterial and fungal isolates. Soil Biol Biochem 31: 375–385

Massalha H, Korenblum E, Tholl D, Aharoni A (2017) Small molecules below-ground: the role of specialized metabolites in the rhizosphere. Plant J 90:788–807

Materić D, Bruhn D, Turner C, Morgan G, Mason N, Gauci V (2015) Methods in plant foliar volatile organic compounds research. Appl Plant Sci 3(12):1500044

Mattheis JP, Roberts RG (1992) Identification of geosmin as a volatile metabolite of *Penicillium expansum*. Appl Environ Microbiol 58:3170–3172

Matysik S, Herbarth O, Mueller A (2008) Determination of volatile metabolites originating from mould growth on wall paper and synthetic media. J Microbiol Methods 75:182–187

Matysik S, Herbarth O, Mueller A (2009) Determination of microbial volatile organic compounds (MVOCs) by passive sampling onto charcoal sorbents. Chemosphere 76:114–119

Mau J-L, Beelman RB, Ziegler GR (1992) Effect of 10-oxo-trans-8-decenoic acid on growth of *Agaricus bisporus*. Phytochemistry 31:4059–4064

Mau J-L, Chyau C-C, Li J-Y, Tseng Y-H (1997) Flavor compounds in straw mushrooms *Volvariella volvacea* harvested at different stages of maturity. J Agric Food Chem 45:4726–4729

Mauriello G, Marino R, D'Auria M, Cerone G, Rana GL (2004) Determination of volatile organic compounds from truffles via SPME-GC-MS. J Chromatogr Sci 42: 299–305

McFee DR, Zavon P (1988) Solvents. In: Plog BA (ed) Fundamentals of industrial hygiene, 3rd edn. National Safety Council, Chicago, pp 95–121

McGann JP (2017) Poor human olfaction is a 19th-century myth. Science 356(6338):eaam7263

McNeal KS, Herbert BE (2009) Volatile organic metabolites as indicators of soil microbial activity and community composition shifts. Soil Sci Soc Am J 73: 579–588

Meilgaard MC (1975a) Flavor chemistry of beer. I. Flavor interaction between principal volatiles. Technical Quarterly, Master Brewers Association of America 12:107–117

Meilgaard MC (1975b) Flavor chemistry of beer. II. Flavor and threshold of 239 aroma volatiles. Technical Quarterly, Master Brewers Association of America 12:151–168

Mercier J, Jiménez JI (2004) Control of fungal decay of apples and peaches by the biofumigant fungus *Muscodor albus*. Postharvest Biol Technol 31:1–8

Mercier J, Manker D (2005) Biocontrol of soil-borne diseases and plant growth enhancement in greenhouse soilless mix by the volatile-producing fungus *Muscodor albus*. Crop Prot 24:355–362

Micheluz A, Manente S, Rovea M, Slanzi D, Varese GC, Ravagnan G, Formenton G (2016) Detection of volatile metabolites of moulds isolated from a contaminated library. J Microbiol Methods 128:34–41

Minerdi D, Bossi S, Gullino ML, Garibaldi A (2009) Volatile organic compounds: a potential direct long-distance mechanism for antagonistic action of *Fusarium oxysporum* strain MSA 35. Environ Microbiol 11: 844–854

Mitra M, Venkatesh P, Ghissing U, Biswas A, Mitra A, Mandal M, Mishra HN, Maiti MK (2023) Fruity-scented antifungal volatiles from endophytic *Geotrichum candidum* PF005: broad-spectrum bioactivity against stored grain pathogens, mode of action and suitable formulation for mycofumigation. Biol Control 177:105129

Moisan K, Raaijmakers JM, Dicke M, Lucas-Barbosa D, Cordovez V (2021) Volatiles from soil-borne fungi affect directional growth of roots. Plant Cell Environ 44:339–345

Mølhave L (2009) Volatile organic compounds and the sick building syndrome. In: Lippmann M (ed) Environmental toxicants: human exposures and their health effects, 3rd edn. Wiley-Interscience, New York, pp 241–256

Morath SU, Hung R, Bennett JW (2012) Fungal volatile organic compounds: a review with emphasis on their biotechnological potential. Fungal Biol Rev 26(2–3): 73–83

Morey P, Wortham A, Weber A, Horner E, Black M, Muller W (1997) Microbial VOCs in moisture damaged buildings. Healthy Build 1:245–250

Mosandl A, Heusinger G, Gessner M (1986) Analytical and sensory differentiation of 1-octen-3-ol enantiomers. J Agric Food Chem 34:119–122

Mustafa AM, Angeloni S, Nzekoue FK, Abouelenein D, Sagratini G, Caprioli G, Torregiani E (2020) An overview on truffle aroma and main volatile compounds. Molecules 25(24):5948

Nealson KH, Hastings JW (1979) Bacterial bioluminescence: its control and ecological significance. Microbiol Rev 43:496–518

Nealson KH, Platt T, Hastings JW (1970) Cellular control of the synthesis and activity of the bacterial luminescent system. J Bacteriol 104:313–322

Nemcovic M, Jakubikova L, Viden I, Farkas V (2008) Induction of conidiation by endogenous volatile compounds in *Trichoderma* spp. FEMS Microbiol Lett 284:231–236

Nicholson JK, Holmes E, Wilson ID (2005) Gut microorganisms, mammalian metabolism and personalized health care. Nat Rev Microbiol 3:431–438

Niimi J, Deveau A, Splivallo R (2021) Geographical-based variations in white truffle *Tuber magnatum* aroma is explained by quantitative differences in key volatile compounds. New Phytol 230(4):1623–1638

Nilssen AC (1998) Effect of 1-octen-3-ol in field trapping *Aedes spp.* (Dipt., Culicidae) and *Hybomitra spp.* (Dipt., Tabanidae) in subartic Norway. J Appl Entomol 122:465–468

Nilsson A, Kihlstrom E, Lagesson V, Wessen B, Szponar B, Larsson L, Tagesson C (2004) Microorganisms and volatile organic compounds in airborne dust from damp residences. Indoor Air 14: 74–82

Niu Q, Huang X, Zhang L, Xu J, Yang D, Wei K, Niu X, An Z, Bennett JW, Zou C, Yang J, Zhang KQ (2010) A Trojan horse mechanism of bacterial pathogenesis against nematodes. Proc Natl Acad Sci U S A 107: 16631–16636

Noble R, Dobrovin-Pennington A, Hobbs PJ, Pederby J, Rodger A (2009) Volatile C8 compounds and pseudomonads influence primordium formation of *Agaricus bisporus*. Mycologia 101:583–591

Nordlund DA, Lewis WJ (1976) Terminology of chemical releasing stimuli in intraspecific and interspecific interactions. J Chem Ecol 2:211–220

Noverr MC, Erb-Downward JR (2003) Production of eicosanoids and other oxylipins by pathogenic eukaryotic microbes. Clin Microbiol Rev 16:517–533

Okull DO, Beelman RB, Gourama H (2003) Antifungal activity of 10-oxo-trans-8-decenoic acid and 1-octen-3-ol against *Penicillium expansum* in potato dextrose agar medium. J Food Prot 66:1503–1505

Oldroyd GED, Downie JA (2004) Calcium, kinases and nodulation signalling in legumes. Nat Rev Mol Cell Biol 5:566–576

Ômura H, Kuwahara Y, Tanabe T (2002) 1-Octen-3-ol together with geosmin: new secretion compounds from a polydesmid millipede, *Niponia nodulosa*. J Chem Ecol 28:2601–2612

Ortiz-Castro R, Contreras-Cornejo H, Macias-Rodriguez-L, Lopez-Bucio J (2009) The role of microbial signals in plant growth and development. Plant Signal Behav 4:701–712

Pangga IB (2021) Mycofumigation with the endophytic fungi *Fusarium proliferatum* (Matsushima) Nirenberg and Diaporthe sp. for the control of banana and mango anthracnose. Annal Trop Res 43(2):13–24

Pantoja LDM, do Nascimento RF, de Araújo Nunes AB (2016) Investigation of fungal volatile organic compounds in hospital air. Atmos Pollut Res 7(4): 659–663

Pasanen P, Korpi A, Kalliokosi P, Pasanen AL (1997) Growth and volatile metabolite production of *Aspergillus versicolor* in house dust. Environ Int 23:425–432

Pavlou AD, Turner AP (2000) Sniffing out the truth: clinical diagnosis using the electronic nose. Clin Chem Lab Med 38:99–112

Peixoto L, Salazar LTH, Laska M (2018) Olfactory sensitivity for mold-associated odorants in CD-1 mice and spider monkeys. J Comp Physiol A 204:821–833

Pelaez F (2005) Biological activities of fungal metabolites. In: An Z (ed) Handbook of industrial mycology. Marcel Dekker, pp 49–92

Pennerman KK, Yin G, Bennett JW (2022) Eight-carbon volatiles: prominent fungal and plant interaction compounds. J Exp Bot 73(2):487–497

Piechulla B, Lemfack MC, Kai M (2017) Effects of discrete bioactive microbial volatiles on plants and fungi. Plant Cell Environ 40(10):2042–2067

Pierce AM, Pierce HD, Borden JH, Oehlschlager AC (1991a) Fungal volatiles: semiochemicals for stored-product beetles (Coleoptera: Cucujidae). J Chem Ecol 3:567–579

Pierce AM, Pierce HD, Oehlschlager AC, Borden JH (1991b) 1-octen-3-ol, attractive semiochemical for foreign grain beetle, *Ahasverus adevna* (Waltl) (*Coleoptera: Cucujidae*). J Chem Ecol 3:567–579

Poland TM, Pureswaran DS, Ciaramitaro TM, Borden JH (2009) 1-octen-3-ol is repellent to *Ips pini* (Coleoptera: Curculionidae: Scolytinae) in the midwestern United States. Can Entomol 141:158–160

Ramoni R, Vincent F, Grolli S, Conti V, Malosse C, Boyer F, Nagnan-Le Meillour P, Spinelli S, Cambillau C, Tegoni M (2001) The insect attractant 1-octen-3-ol Is the natural ligand of bovine odorant-binding protein. J Biol Chem 276:7150–7155

Ramsbottom J (1953) Mushrooms & toadstools: a study of the activities of fungi. London, Collins, St. James's Place

Rapior S, Breheret S, Talou T, Pelissier Y, Bessiere JM (2002) The anise-like odor of *Clitocybe odora, Lentinellus cochleatus,* and *Agaricus essettie*. Mycologia 94:373–376

Raza W, Wei Z, Jousset A, Shen Q, Friman VP (2021) Extended plant metarhizobiome: understanding volatile organic compound signaling in plant-microbe metapopulation networks. Msystems 6(4):e00849–e00821

Robinson J (ed) (2006) The Oxford companion to wine, 3rd edn. Oxford University Press, Oxford

Rodriguez RJ, White JF, Arnold AE, Redman RS (2009) Fungal endophytes: diversity and functional roles. New Phytol 182:314–330

Rokas A, Mead ME, Steenwyk JL, Raja HA, Oberlies NH (2020) Biosynthetic gene clusters and the evolution of fungal chemodiversity. Nat Prod Rep 37:868–878

Rude MA, Schirmer A (2009) New microbial fuels: a biotech perspective. Curr Opin Microbiol 12:274–281

Ruiz-Jimenez J, Zanca N, Lan H, Jussila M, Hartonen K, Riekkola ML (2019) Aerial drone as a carrier for miniaturized air sampling systems. J Chromatogr A 1597:202–208

Rumbaugh KP, Griswold JA, Hamood AN (2000) The role of quorum sensing in the in vivo virulence of *Pseudomonas aeruginosa*. Microbes Infect 2:1721–1731

Ryu C-M, Farag MA, Hu C-H, Reddy MS, Wei H-X, Paré PW, Kloepper JW (2003) Bacterial volatiles promote growth in *Arabidopsis*. Proc Natl Acad Sci U S A 100: 4927–4932

Ryu C-M, Farag MA, Paré PW, Kloepper JW (2005) Invisible signals from the underground: Bacterial volatiles elicit plant growth promotion and induce systemic resistance. Plant Pathol J 21:7–12

Sánchez-López ÁM, Baslam M, De Diego N, Muñoz FJ, Bahaji A, Almagro G, Ricarte-Bermejo A, García-Gómez P, Li J, Humplík JF, Novák O (2016) Volatile compounds emitted by diverse phytopathogenic microorganisms promote plant growth and flowering through cytokinin action. Plant Cell Environ 39(12): 2592–2608

Schmidberger PC, Schieberle P (2017) Characterization of the key aroma compounds in white Alba truffle (*Tuber magnatum* pico) and Burgundy truffle (*Tuber uncinatum*) by means of the sensomics approach. J Agric Food Chem 65(42):9287–9296

Schnürer J, Olsson J, Börjesson T (1999) Fungal volatiles as indicators of food and feeds spoilage. Fungal Genet Biol 27:209–217

Schöller CEG, Gürtler H, Petersen R, Molin S, Wilkins K (2002) Volatile metabolites from Actinomycetes. J Agric Food Chem 50:2615–2621

Schreier P (1992) Bioflavours: AN overview. In: Patterson RLS, Charlwood BV, MacLeod G, Williams AA (eds) Bioinformation of flavours. The Royal Society of Chemistry, Cambridge, pp 1–20

Schulz S, Dickschat JS (2007) Bacterial volatiles: the smell of small organisms. Nat Prod Rep 24:814–842

Schulz-Bohm K, Lara M-S, Paolina G (2017) Microbial volatiles: Small molecules with an important role in intra- and inter-kingdom interactions. Front Microbiol 8:2484

Scott J, Sueiro-Olivares M, Ahmed W, Heddergott C, Zhao C, Thomas R, Bromley M, Latgé J-P, Krappmann S, Fowler S, Bignell E, Amich J (2019) *Pseudomonas aeruginosa*-derived volatile sulfur compounds promote distal *Aspergillus fumigatus* growth and a synergistic pathogen-pathogen interaction that increases pathogenicity in co-infection. Front Microbiol 10:2311

Shafiq SA, Chechan RA (2019) Influence of different natural media on production of myco-diesel. IOP Conf Ser Earth Environ Sci 388:012054

Shapiro JA (1998) Thinking about bacterial populations as multicellular organisms. Annu Rev Microbiol 52:81–104

Shimkets LJ (1999) Intercellular signaling during fruiting-body development of *Myxococcus xanthus*. Annu Rev Microbiol 53:525–549

Shoresh M, Harman GE, Mastouri F (2010) Induced systemic resistance and plant responses to fungal biocontrol agents. Annu Rev Phytopathol 48:21–43

Singh SK, Strobel GA, Knighton B, Geary B, Sears J, Ezra D (2011) An endophytic *Phomopsis* sp. possessing bioactivity and fuel potential with its volatile organic compounds. Microb Ecol 61:729–739

Singh J, Singh P, Vaishnav A, Ray S, Rajput RS, Singh SM, Singh HB (2021) Belowground fungal volatiles perception in okra (*Abelmoschus esculentus*) facilitates plant growth under biotic stress. Microbiol Res 246: 126721

Sneeden EY, Harris HH, Pickering J, Prince RC, Johnson S, Li X, Block E, George GH (2004) The sulfur chemistry of shitake mushroom. J Am Chem Soc 126:458–459

Soria AC, García-Sarrió MJ, Sanz ML (2015) Volatile sampling by headspace techniques. TrAC Trends Anal Chem 71:85–99

Splivallo R, Culleré L (2016) The smell of truffles: from aroma biosynthesis to product quality. True truffle (tuber spp.) in the world: Soil ecology, systematics and biochemistry, pp 393–407

Splivallo R, Ebeler SE (2015) Sulfur volatiles of microbial origin are key contributors to human-sensed truffle aroma. Appl Microbiol Biotechnol 99:2583–2592

Splivallo R, Bossi S, Maffei M, Bonfante P (2007a) Discrimination of truffle fruiting body versus mycelial aromas by stir bar sorptive extraction. Phytochemistry 68:2584–2598

Splivallo R, Novero M, Bertea CM, Bossi S, Bonfante P (2007b) Truffle volatiles inhibit growth and induce an oxidative burst in *Arabidopsis thaliana*. New Phytol 175:417–424

Splivallo R, Deveau A, Valdez N, Kirchhoff N, Frey-Klett P, Karlovsky P (2015) Bacteria associated with truffle-fruiting bodies contribute to truffle aroma. Environ Microbiol 17(8):2647–2660

Stefanini I (2018) Yeast-insect associations: it takes guts. Yeast 35:315–330

Stinson M, Ezra D, Hess WM, Sears J, Strobel G (2003) An endophytic *Gliocladium* sp. of *Eucryphia cordifolia* producing selective volatile antimicrobial compounds. Plant Sci 165:913–922

Stoppacher N, Kluger B, Zeilinger S, Krska R, Schuhmacher R (2010) Identification and profiling of volatile metabolites of the biocontrol fungus *Trichoderma atroviride* by HS-SPME-GC-MS. J Microbiol Methods 81:187–193

Straus DC (2009) Molds, mycotoxins, and sick building syndrome. Toxicol Ind Health 25:617–635

Straus DC, Cooley JD, Wong WC, Jumper CA (2003) Studies on the role of fungi in sick building syndrome. Arch Environ Health 58:475–478

Strobel GA, Dirkse E, Sears J, Markworth C (2001) Volatile antimicrobials from *Muscodor albus* a novel endophytic fungus. Microbiology 147:2943–2950

Strobel GA, Knighton B, Kluck K, Ren Y, Livinghouse T, Griffin M, Spakowicz D, Sears J (2008) The production of myco-diesel hydrocarbons and their derivatives by the endophytic fungus *Gliocladium roseum* (NRRL 50072). Microbiology 154:3319–3328

Strobel G, Singh SK, Riyaz-Ul-Hassan S, Mitchell AM, Geary B, Sears J (2011) An endophytic/pathogenic *Phoma* sp. from creosote bush producing biologically active volatile compounds having fuel potential. FEMS Microbiol Lett 320:87–94

Ström G, West J, Wessen B, Palmgren U (1994) Health implications of fungi in indoor environments: quantitative analysis of microbial volatiles in damp Swedish houses. Air Qual Monogr 2:291–305

Sunesson AL, Vaes WHJ, Nilsson CA, Blomquist G, Andersson B, Carlson R (1995) Identification of volatile metabolites from five fungal species cultivated on two media. Appl Environ Microbiol 61:2911–2918

Sunesson AL, Nilsson CA, Andersson B, Blomquist G (1996) Volatile metabolites produced by two fungal species cultivated on building materials. Ann Occup Hyg 40(4):397–410

Takahashi N (1986) Chemistry of plant hormones. CRC Press, Boca Raton, FL

Tan RX, Zou WV (2001) Endophytes: a rich source of functional metabolites. Nat Prod Rep 18:448–459

Tarkka MT, Piechulla B (2007) Aromatic weapons: truffles attack plants by the production of volatiles. New Phytol 175:381–383

Tholl D, Hossain O, Weinhold A, Röse USR, Wei Q (2021) Trends and applications in plant volatile sampling and analysis. Plant J 106:314–325

Thomson NR, Crow MA, McGowan SJ, Cox A, Salmond GPC (2000) Biosynthesis of carbapenem antibiotic and prodigiosin pigment in *Serratia* is under quorum sensing control. Mol Microbiol 36:539–556

Thorn J (2001) The inflammatory response in humans after inhalation of bacterial endotoxin: a review. Inflamm Res 50:254–261

Tian X, Ding H, Ke W, Wang L (2021) Quorum sensing in fungal species. Annu Rev Microbiol 75:449–469

Tirillini B, Verdelli G, Paolocci F, Ciccioli P, Frattoni M (2000) The volatile organic compounds from the mycelium of *Tuber borchii* Vitt. Phytochemistry 55: 983–985

Trinci APJ, Whittaker C (1968) Self-inhibition of spore germination in *Aspergillus nidulans*. Trans Brit Mycol Soc 51:594–596

Tsitsigiannis DI, Keller NP (2007) Oxylipins as developmental and host-fungal communication signals. Trends Microbiol 15:109–118

Tsurushima T, Ueno T, Fukami H, Irie H, Inoue M (1995) Germination self-inhibitors from *Colletotrichum gloeosporioides* f.sp *jussiaea*. Mol Plant-Microbe Interact 8:652–657

Turner WB (1971) Fungal metabolites. Academic, London

Turner WB, Aldridge DC (1983) Fungal metabolites II. Academic, London

Vajzovic A, Bura R, Kohlmeier K, Doty SL (2012) Novel endophytic yeast *Rhodotorula mucilaginosa* strain PTD3 II: production of xylitol and ethanol in the presence of inhibitors. J Ind Microbiol Biotechnol 39(10): 1453–1463

Van Delden C, Iglewski BH (1998) Cell-to-cell signaling and *Pseudomonas aeruginosa* infections. Emerg Infect Dis 4(4):551–560

Vespermann A, Kai M, Piechulla B (2007) Rhizobacterial volatiles affect the growth of fungi and *Arabidopsis thaliana*. Appl Environ Microbiol 73(17):5639–5641

Vita F, Taiti C, Pompeiano A, Bazihizina N, Lucarotti V, Mancuso S, Alpi A (2015) Volatile organic compounds in truffle (*Tuber magnatum* Pico): comparison of samples from different regions of Italy and from different seasons. Sci Rep 5:1–15

Vivaldo G, Masi E, Taiti C, Caldarelli G, Mancuso S (2017) The network of plants volatile organic compounds. Sci Rep 7(1):11050

Vlot AC, Rosenkranz M (2022) Volatile compounds-the language of all kingdoms? J Exp Bot 73(2):445–448

von Bodman SB, Bauer WD, Coplin DL (2003) Quorum sensing in plant-pathogenic bacteria. Annu Rev Phytopathol 41:455–482

Wadood SA, Guo B, Zhang X, Raza A, Wei Y (2020) Geographical discrimination of Chinese winter wheat using volatile compound analysis by HS-SPME/GC-MS coupled with multivariate statistical analysis. J Mass Spectrom 55:e4453

Walinder R, Ernstgard L, Johanson G, Venge P, Wieslander G (2005) Acute effects of a fungal volatile compound. Environ Health Perspect 113:1775–1778

Waters CM, Bassler BL (2005) Quorum sensing: cell-to-cell communication in bacteria. Annu Rev Cell Dev Biol 21:319–346

Watson SB, Brownlee B, Satchwill T, Hargesheimer EE (2000) Quantitative analysis of trace levels of geosmin and MIB in source and drinking water using headspace SPME. Water Res 34:2818–2828

Weeks EN, Birkett MA, Cameron MM, Pickett JA, Logan JG (2011) Semiochemicals of the common bed bug, *Cimex lectularius* L. (Hemiptera: Cimicidae), and their potential for use in monitoring and control. Pest Manag Sci 67:10–20

Weisskopf L, Schulz S, Garbeva P (2021) Microbial volatile organic compounds in intra-kingdom and inter-kingdom interactions. Nat Rev Microbiol 19:391–404

Wheatley R, Hackett C, Bruce A, Kundzewicz A (1997) Effect of substrate composition on production of volatile organic compounds from *Trichoderma* spp. inhibitory to wood decay fungi. Int Biodeterior Biodegrad 39(2–3):199–205

Whipps JM (2001) Microbial interactions and biocontrol in the rhizosphere. J Exp Bot 52(suppl 1):487–511

Whitehead NA, Barnard ANL, Slater H, Simpson NJL, Salmond GPC (2001) Quorum-sensing in Gram-negative bacteria. FEMS Microbiol Rev 25:365–404

Whittaker RH, Feeny PP (1971) Allelochemics: chemical interactions between species. Science 171:757–770

WHO (2009) Guidelines for indoor air quality: Dampness and mold. Druckpartner Moser, Rheinbach

Wilkins K, Larsen K, Simkus M (2000) Volatile metabolites from mold growth on building materials and synthetic media. Chemosphere 41:437–446

Wilson AD, Baietto M (2009) Applications and advances in electronic-nose technologies. Sensors 9:5099–5148

Wilson AD, Baietto M (2011) Advances in electronic-nose technologies developed for biomedical applications. Sensors 11:1105–1176

Wongsuk T, Pumeesat P, Luplertlop N (2016) Fungal quorum sensing molecules: role in fungal morphogenesis and pathogenecity. J Basic Microbiol 56:440–447

Wood WF, Fesler M (1986) Mushroom odors: Student synthesis of the odoriferous compounds of the matsutake mushroom. J Chem Educ 63:92

Wurzenberger M, Grosch W (1984) The formation of 1-octen-3-ol from the 10-hydroperoxide isomer of linoleic acid by a hydroperoxide lyase in mushrooms (*Psalliota bispora*). Biochim Biophys Acta - Lipids Lipid Metab 794:25–30

Xie X, Zhang H, Pare P (2009) Sustained growth promotion in *Arabidopsis* with long-term exposure to the beneficial soil bacterium *Bacillus subtilis* (GB03). Plant Signal Behav 4:948–953

Yuan JS, Himanen SJ, Holopainen JK, Chen F, Stewart CN (2009) Smelling global climate change: mitigation of function for plant volatile organic compounds. Trends Ecol Evol 24(6):323–331

Zeilinger S, Gruber S, Bansal R, Mukherjee PK (2016) Secondary metabolism in *Trichoderma*–chemistry meets genomics. Fungal Biol Rev 30(2):74–90

Zhang Z, Li G (2010) A review of advances and new developments in the analysis of biological volatile organic compounds. Microchem J 95:127–139

Zhang QH, Schlyter F (2004) Olfactory recognition and behavioural avoidance of angiosperm nonhost volatiles by conifer-inhabiting bark beetles. Agric For Entomol 6:1–19

Zhang Y-Q, Wilkinson H, Keller NP, Tsitsigiannis D (2005) Secondary metabolite gene clusters. In: An Z (ed) Handbook of industrial microbiology. Marcel Dekker, pp 355–386

Zhang H, Kim M-S, Krishnamachari V, Payton P, Sun Y, Grimson M, Farag M, Ryu C-M, Allen R, Melo I, Paré P (2007) Rhizobacterial volatile emissions regulate auxin homeostasis and cell expansion in *Arabidopsis*. Planta 226:839–851

Zhang H, Xie X, Kim MS, Kornyeyev DA, Holaday S, Pare PW (2008) Soil bacteria augment *Arabidopsis* photosynthesis by decreasing glucose sensing and abscisic acid levels in planta. Plant J 56:264–273

Zhang CL, Wang GP, Mao LJ, Komon-Zelazowska M, Yuan ZL, Lin FC, Druzhinina IS, Kubicek CP (2010) *Muscodor fengyangensis* sp. nov. from southeast China: morphology, physiology and production of volatile compounds. Fungal Biol 114:797–808

Zhao G, Yin G, Inamdar AA, Luo J, Zhang N, Yang I, Buckley B, Bennett JW (2017) Volatile organic compounds emitted by filamentous fungi isolated from flooded homes after Hurricane Sandy show toxicity in a *Drosophila* bioassay. Indoor Air 27:518–528

Zogorski JS, Carter JM, Ivahnenko T, Lapham WW, Moran MJ, Rowe BL, Squillace PJ, Toccalino PL (2006) The quality of our nation's waters – volatile organic compounds in the nation's ground water and drinking-water supply wells. In: U.S. Geological Survey Circular 1292. USGS, Reston, VA

Viruses that Affect Phenotype and Fitness of Fungi

5

Bradley I. Hillman and Massimo Turina

Abstract

Mycoviruses are the viruses that infect fungi and the unrelated stramenopiles such as oomycetes that have morphologies and lifestyles similar to fungi. Our knowledge of mycoviruses and their effects on the biology of their hosts has exploded in recent years. Although the field only started to mature in the 1980s and 1990s, advances in technology coupled with the promise of mycoviruses for biological control of fungi pathogenic to plants, animals, and humans led to increased attention and dedicated research. Mycoviruses are now classified within at least 16 virus families, and that number grows continually as new viruses that do not fit within the current taxonomy are described. This chapter highlights some of the most important fungal virus systems, and some of the advances in our understanding of mycoviruses, especially in the context of their effects on host biology and fitness.

B. I. Hillman (✉)
Plant Biology, Foran Hall, Rutgers University, New Brunswick, NJ, USA
e-mail: bradley.hillman@rutgers.edu

M. Turina
IPSP-CNR, Torino, Italy
e-mail: massimo.turina@ipsp.cnr.it

Keywords

Mycovirus · Fitness · Hypovirulence · dsRNA · Biocontrol · Vegetative incompatibility · RNA silencing

5.1 Introduction

By the time of the discovery and initial characterization of fungal viruses, research with viruses of vertebrates, invertebrates, plants, and bacteria were already well-developed sciences. Why the delay? For one thing, not many capable researchers were looking for them. Most fungal research was done by plant pathologists who were interested in fungi as the most important cause of disease in plants; they were accustomed to seeing sectoring and phenotypic differences in many of their fungal cultures, but few followed up on these aberrant phenotypes, especially given the limited toolset available at the time. Furthermore, their plant virologist colleagues had their hands full characterizing the many different viruses that were themselves plant pathogens. The timeline given below provides an overview of important events in the progress of early mycovirus research:

- **1940s** Viruses that infect filamentous fungi were first identified in cultivated mushrooms as the agent of "La France disease" (Hollings 1962).

Y.-P. Hsueh, M. Blackwell (eds.), *Fungal Associations*, The Mycota 9,
https://doi.org/10.1007/978-3-031-41648-4_5

- **1950s, 1960s** Discovery of *Penicillium stoloniferum* cultures at Lilly Laboratories lead to the characterization of double-stranded RNA as an inducer of interferon in mammals (Kleinschmidt 1988).
- **1970s** As the discovery of fungal viruses became an emerging science in the 1970s, the dogma that most were asymptomatic in their fungal host began to take hold (Ghabrial and Suzuki 2009) and suppress interest in the science of fungal virology.
- **1970s, 1980s** Fungal virology elevated as a science by technology. Virus discovery, particularly viruses with positive-sense (+) RNA or double-stranded (ds) RNA genomes became much easier when rapid methods to purify and analyze dsRNA were developed (Morris and Dodds 1979; Jackson et al. 1971). While this likely led to continued lack of awareness of the importance of negative-sense (−) RNA and perhaps single-stranded (ss) DNA viruses in fungi, these methods have allowed for substantial screening of many fungal cultures and culture collections.
- **1980s, 1990s** Another particularly important milestone in the development of fungal virology as a science was the recognition and characterization of the role of viruses as potential biological control agents of chestnut blight fungus, *Cryphonectria parasitica*. For some time, *C. parasitica* viruses served as the model system for fungal virus research (Hillman and Suzuki 2004). The fungus had many features that rendered it an attractive research model: it caused a disease of historic and ongoing international significance and could be found in temperate forests worldwide; the fungus was easy to isolate and grow stably in culture; it produced conidia abundantly in culture and ascospores could also be produced in culture; the fungus could be transformed easily using molecular methods developed for other filamentous fungi. Most importantly, viruses of *C. parasitica* had a vivid and demonstrable effect on the colony morphology of the fungus and on its virulence to the tree host of the fungus. Thus, when investigators asked questions from a virological perspective, they began to get answers. The authors of this review studied with two of the most important figures in the development of *C. parasitica* as a system for investigation of pathogenic fungal viruses: Drs. Neal Van Alfen and Donald Nuss.

Over the last few years, our knowledge of fungal virology and virus systems has exploded. This expansion has largely been the result of two connected factors: first, technology, particularly sequencing technology, has allowed for rapid and abundant characterization of many new viruses in cultured and uncultured fungi; and second, the number of research groups that now have fungal virus research as at least one major focus area of their research programs has grown immensely. Most fungi that have been studied in the context of their viral partners have plant associations, often but not always pathogenic to the plant host. With the exception of a few examples, relatively little is known of virus impacts on free-living soil fungi, endophytic fungi, and epiphytic fungi, for example, beyond metagenomic searches (Roossinck 2014); (reviewed by Sutela et al. 2019), and these represent fertile ground for further research.

This review will highlight some of the experimentally tractable fungal systems in which viruses are well characterized, and their phenotypic effects on their hosts have been documented and investigated in some depth. Our use of the term "fitness" in the title of the chapter is somewhat loose: only in rare cases have the negative or positive effects of virus infection on a fungal host been unambiguously linked to fitness outcomes in a natural or experimental field setting. Virus taxonomy and nomenclature are undergoing rapid change, and the reader is directed to the excellent website for the International Committee for the Taxonomy of Viruses for most current taxonomy and nomenclature information about individual viruses: https://ictv.global/.

5.2 Diversity of Viruses Infecting Fungi and Oomycetes

Discovery of mycoviruses for many years relied on particle observation at the electron microscopic level or the use of dsRNA extraction (a hallmark of most RNA virus infections) as a tool to find and characterize mycoviruses, thus skewing the representation of the mycovirosphere toward dsRNA viruses (i.e., members of the families *Reoviridae*, *Partitiviridae*, *Totiviridae*, *Chrysoviridae*) and positive (+)strand RNA viruses that greatly accumulate their dsRNA replicative form (members of the families *Hypoviridae*, *Narnaviridae*, and *Endornaviridae*). Up to the end of the twentieth century, there were no DNA or negative (−) strand ssRNA viruses associated with fungi, and only a few plus-strand RNA viruses in the *Flexiviridae* family whose ssRNA genomes were encapsidated in a coat protein-based particle (Ghabrial and Suzuki 2009). Since the first decade of the twentieth century, high throughput sequencing (HTS) approaches began to be applied to the untargeted discovery of mycoviruses while other targeted approaches with characterization methodologies other than dsRNA extraction also achieved some interesting milestones in describing mycovirus biodiversity.

Evidence of minus-strand RNA mycoviruses was first gathered from genomic analysis where endogenous non-retrovirus elements included fragments of RdRP that matched RdRP from negative-stranded RNA viruses (Kondo et al. 2013a). A few years later, the first bona fide minus-strand RNA virus was characterized and remains at the moment the best characterized negative-stranded mycovirus (Liu et al. 2014). Another milestone was the discovery of the first DNA mycovirus: it belongs to the CRESS ssDNA viruses, and at the time of discovery the closest relatives were plant geminiviruses (Yu et al. 2010). Recently different examples of ssDNA mycoviruses with multipartite (multisegmented; i.e., the complete viral genome divided among multiple unique segments of nucleic acid) genomes were also characterized (Li et al. 2020). With the same traditional approach and a well carried out molecular and biological characterization, also members of the hadakavirus (multisegmented, up to 11-naked ss RNA genomes) (Sato et al. 2020) and polymycovirus (infectious as dsRNA) (Kanhayuwa et al. 2015) were described.

After these traditional approaches, in the second half of the second decade of the twenty-first century, both the use of small RNAs (sRNA) (generally enriched in virus-derived sRNA) and ribosomal RNA-depleted total RNA for HTS associated with specific bioinformatic pipelines brought the discovery of a number of new mycovirus clades, greatly widening the mycovirosphere to new uncharted territories. In this respect, after initial steps that still used small RNA (Vainio et al. 2015; Donaire and Ayllon 2017) or total RNA (Deakin et al. 2017) from individual fungal isolates, others looked at metatranscriptomic samples, greatly increasing the diversity of the viruses described (Marzano and Domier 2016; Marzano et al. 2016). In particular, application of ORFan detection pipelines (ORFan-encoding RNA segments without corresponding DNA) led to the discovery of viruses with heretofore unseen genome organizations: the splipalmiviruses (with the palm domain of the RNA-dependent RNA polymerase, RdRP, split between two coding genomic segments); the ambiviruses (ssRNA viruses with bicistronic, ambisense genomes) (Sutela et al. 2020), and the first Mononegavirales with ambisense orientation (now classified in the genus *Penicillimonavirus* in the family *Mymonaviridae*) (Chiapello et al. 2020). Of interest also is the first member of the Bunyavirales with a tricistronic RNA that encodes both the RdRP and the nucleocapsid (Nc) protein on the same RNA segment (now in the family *Tulasviridae*) (Sutela et al. 2020). Finally, another group of RNA viruses, ormycoviruses, encodes putative RdRPs that are so distantly related to currently characterized RNA virus RdRPs that they cannot be reliably positioned phylogenetically, and they display the unusual feature of diverse catalytic triads in the C subdomain of the palm domain (Forgia et al. 2022).

Another group of mycoviruses that was greatly expanded by HTS is that of members of the order *Bunyavirales*, where for the first time a nucleocapsid encoding segment was associated to an RdRP (Nerva et al. 2019) in the genus *Orthodiscovirus* in the family *Discoviridae*. Other recent findings about fungal viruses and replicating RNA elements have demonstrated that they contain catalytic RNA elements (ribozymes) in the same way that plants and animals do: plant viroids replicate and can be maintained in fungi (Wei et al. 2019), unique viroid-like elements (mycoviroids) are found in fungi (Dong et al. 2023), and ambiviruses that replicate through a rolling circle replication cycle that relies on ribozymes, has brought the world of ribozymes to the Mycota. Still at this time, no bona fide dsDNA viruses have been characterized in the fungi.

5.3 Establishing a Link Between Mycovirus Infection and Fungal Phenotype: Clearing (Curing) Infections and Infectivity Systems

One of the most difficult tasks in mycovirus research is the association of a specific phenotype with the presence or absence of a specific mycovirus. In principle, such an association can be established when two isogenic strains differ only in the presence or absence of a specific virus. In common lab practice, currently two approaches can be taken: obtaining a cured strain from an infected isolate or infecting a recipient strain with the mycovirus under study. In both cases, it is important that the initial infected or virus-free isolate contain a clonal population of nuclei. Clonality can be obtained through single conidia in most ascomycetes, since a single conidium generally has a single nucleus. Sterile mycelia (often the case in some fungi in culture) would make this a more difficult task.

One of the most important and often difficult components of fungal virus research is determining specific effects of a particular virus on the host fungus using rigorous methodology. Consider the example of a fungus that is infected in its natural setting. When brought into a laboratory setting, formally disentangling the virus–fungus interaction involves isolating the infected fungus in pure culture, freeing the fungus from its viral parasite, comparing isogenic virus-containing and virus-free colonies, purifying the virus, and putting it back into the virus-free fungal isolate to recapitulate effects observed in the initial comparisons. Any of these steps can be technically challenging, and the real goal of determining the effects of the virus on the fungal host in its natural setting still may not be achieved. With some of the most important fungal systems, such as fungal endophytes, mycorrhizae, or biotrophic pathogens, these steps may constitute monumental challenges, and avenues that provide useful, albeit imperfect information are often necessary. For example, if a fungus cannot be cultured, comparing a number of different biological samples and determining virus presence by total nucleic acid sequencing for comparison may be useful. Some of the methods commonly used to develop virus-free fungal isolates are summarized in Fig. 5.1.

5.3.1 Clearing/Curing Infections

5.3.1.1 Single Asexual Spore Isolation

Colonies derived from single asexual spores, for example single conidia in ascomycetes or single zoospores in oomycetes, have the advantage in a research context because they represent isogenic representatives of their parent cultures. Therefore, if virus (+) and virus (−) single conidium isolates can be generated from an infected culture, direct comparison of virus effects on fungal biology may be possible.

Sometimes, a virus-infected fungus isolated from its natural setting cannot be cleared of its virus infection in order to complete Koch's postulates. The task is easier, however, if a virus has a profound and easily observable effect on the fungal colony morphology. For example, the dramatic white color observed in Cryphonectria

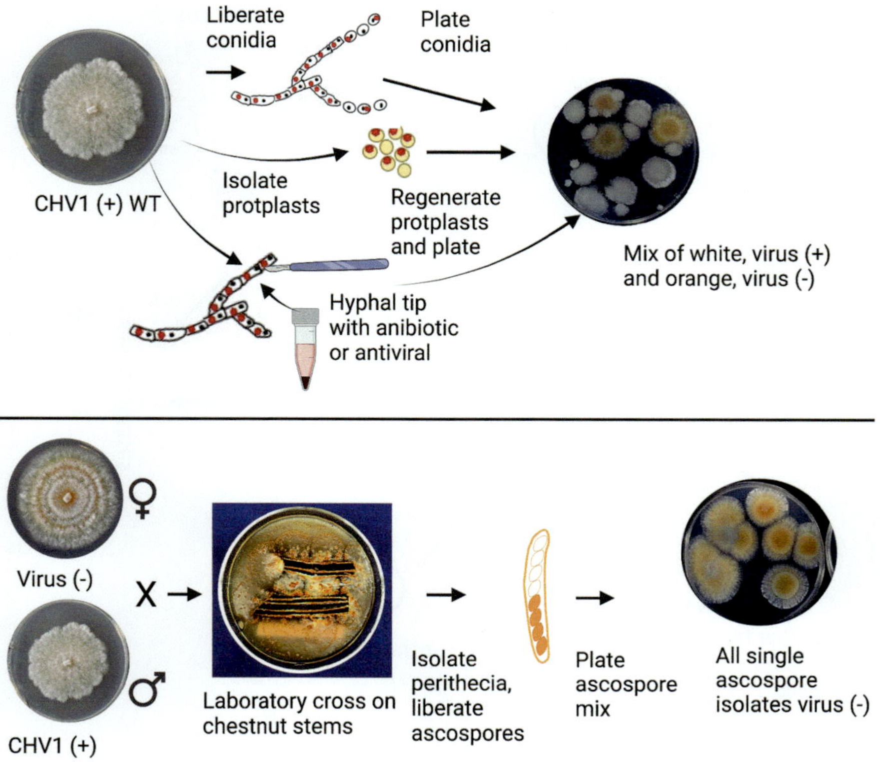

Fig. 5.1 Methods used to clear virus infection from filamentous fungi, with *C. parasitica* as examples. Top panel: CHV1-infected isolate may be cleared of virus by (**a**) liberating and plating conidia on agar; (**b**) isolating, regenerating, and plating protoplasts; or (**c**) excising and plating hyphal tips, possibly with antibiotic (e.g., cycloheximide) or antiviral (e.g., ribavirin) amendment. In each case, virus-free isogenic isolates are identified by colony morphology and confirmed by molecular methods. Bottom panel: CHV1-infected isolate crossed with virus-free isolate. In the resulting cross, single ascospore progeny are all virus free, but are genetically unique, not isogenic

hypovirus 1 (CHV1)-infected *C. parasitica* is immediately discernable from the orange color of its uninfected isogenic counterparts when a conidial suspension derived from CHV1-infected strain EP713 is spread on culture medium and colonies emerge and grow from single conidia (Figs. 5.1 and 5.2). If there is no such profound effect of virus on fungal phenotype, investigators are usually reliant upon nucleic acid detection methods to discern whether single spore-derived colonies contain virus or not. Assuming the viral sequence is known, any of the various polymerase chain reaction (PCR)-based methods for virus identification, including rapid methods, may be deployed (Urayama et al. 2015) or small colony blot methods may be used to screen hundreds of colonies at once (Oh 1995; Oh and Hillman 1995). Often, single conidial isolation fails to yield any virus-free colonies (Oh and Hillman 1995; Polashock and Hillman 1994; Polashock et al. 1997).

5.3.1.2 Single Sexual Spore Isolation

Many fungal viruses are not passed through the sexual cycle, and this therefore represents a route through which to clear a virus from its infected host. Thus, a virus that is recalcitrant to curing by other means may be eliminated by performing crosses and single spore (meiospore) isolations (ascospores, basidiospores, or oospores). Unfortunately, resulting virus-free strains have undergone recombination and are not isogenic with the virus-infected parent. Such methods

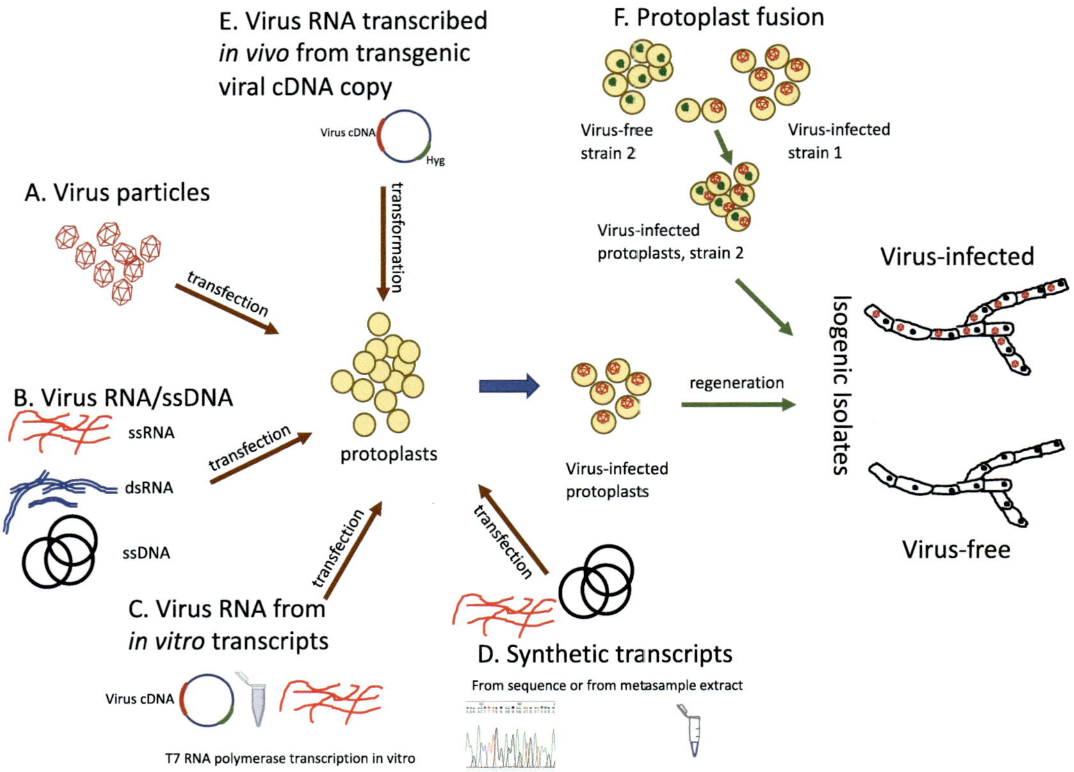

Fig. 5.2 Methods used to initiate virus infection in filamentous fungi. Transfection of protoplasts can be initiated with (**a**) purified virus particles; (**b**) purified viral nucleic acid; (**c**) nucleic acid derived from traditional cDNA clones; or (**d**) chemically synthesized nucleic acid based on sequence data alone. Stable nuclear transformation of protoplasts and regeneration of fungal colonies with a cDNA construct from which infectious viral RNA is transcribed in vivo (**e**) results in an infected fungal isolate that also can transmit virus by anastomosis and can transmit the viral gene itself to progeny in sexual crosses. Finally, fusion of protoplasts from virus-infected and virus-free strains (**f**) circumvents barriers of vegetative incompatibility

serve as a step toward fulfilling Koch's Postulates, but other steps are required to obtain isogenic virus (+) and virus (−) isolates.

When the infected individual serves as male in crosses with an uninfected female partner, virus transmission from the male gametangium to the female often does not occur, and when the infected partner is the female in matings, a virus may suppress female fertility to the point that sexual spores are not produced (Anagnostakis 1982). A limited number of studies aimed at quantifying the transmission rates of specific viruses in controlled crosses have been published, and it appears that some viruses may be transmitted to sexual spore progeny from either the male or the female partner in matings (see Pearson et al. 2009 for discussion). Like many areas of mycovirus research, this is one in need of further work.

Deng and colleagues (2007a) demonstrated that two mycoreoviruses were transmitted at rates of ~60% through ascospores of *C. parasitica* when the infected partner served as female in crosses; this situation is in contrast with the known suppression of female fertility in the same host by different hypoviruses (Anagnostakis 1982; Fulbright et al. 1988). Deng et al. (2007b) further showed that two genes required for mating in *C. parasitica*, the sex pheromone gene *Mf2/1* and the yeast Ste12-like transcription factor *cpst12*, were severely suppressed by hypovirus infection, but not by mycoreovirus infection, revealing a likely mechanism for female fertility in the mycoreovirus-infected isolates. The importance to virus transmission is that a virus-infected female in sexual crosses may transmit virus to ascospore progeny, whereas an infected male more often does not.

Despite the fact that they are not isogenic, virus-free fungal isolates derived from single sexual spores have experimental value and may provide a useful picture of the biological effects of a virus on its fungal host if data are sufficient for such analysis. In the example given above for *C. parasitica* mycoreoviruses, virus-infected single ascospore progeny of *C. parasitica* were easily distinguishable visually from uninfected progeny (Deng et al. 2007a), and the infected and uninfected colony morphologies were consistent. Thus, given sufficient numbers in such systems, a subset of virus-containing and virus-free progeny can serve as a substitute, albeit an imperfect one, for isogenic isolates.

5.3.1.3 Antibiotic Treatment

Antibiotics have been used to eliminate viruses selectively from infected fungal cultures, usually in conjunction with hyphal tip isolation of fungal cultures. Cycloheximide, produced by *Streptomyces* spp., inhibits eukaryotic translation elongation and has been shown to be an effective means of curing some virus-containing isolates. After it had been shown to be effective in yeast, Fulbright used cycloheximide to cure *C. parasitica* of certain viruses (Fulbright 1984). The antiviral ribavirin, used as treatment against a number of human viruses, is a guanosine analog that interferes with nucleic acid synthesis and especially capping. Ribavirin was effective at eliminating Phytophthora infestans RNA virus 2 (PiRV2) (Cai et al. 2019a) and PiRV3 (Cai et al. 2013) from its oomycete host, allowing for isogenic virus (+) and virus (−) isolates to be compared. But ribavirin was not effective at eliminating PiRV1 or PiRV4 from the same host (Cai et al. 2009, 2012). Notably, elimination of PiRV2 and PiRV3 from *P. infestans* required up to 6 months of Ribavirin treatment, hyphal tipping, and subculture (Cai and Hillman 2013), a technically challenging and time-consuming effort. Sun and colleagues were able to cure *Lentinula edodes*, the important edible Shiitake mushroom, of two viruses through a combination of mycelial fragmentation and ribavirin treatment (Sun et al. 2022). When feasible, virus-free isolates should be obtained from more than one curing event, since the use of antivirals can result in unwanted off target mutations on the fungal genomes.

5.3.2 Infectivity Systems

Hypovirulence in *C. parasitica* caused by CHV1 infection is the classic example used to illustrate the CHV1–*C. parasitica* interaction. The genetically tractable fungus can be transformed with high efficiently by using a gene knock out method and infectious cDNA clones that are available for CHV1. Today a number of infectious clones of other mycoviruses are available, and technologies have been developed for many different fungi. Here we summarize some of the approaches for controlled infectivity of a recipient fungal isolate, as depicted in Fig. 5.2:

5.3.2.1 Transfer of the Infectious Agent to a Recipient Strain Through Anastomosis:

Initially, even without knowing the exact nature of the causal agent, a maternal genetic element was shown to be transmissible through anastomosis, the process of hyphal fusion between two fungal individuals of the same species (Fig. 5.3), and to be associated with hypovirulence. This process allows passage of cytoplasmic elements to a recipient strain. Experimentally it is important to distinguish the recipient strain with a marker (e.g., antibiotic resistance, green fluorescent protein (GFP) expression, auxotrophic mutants) in order to ensure that, indeed, only the recipient strain is retrieved after anastomosis. Such an approach was pivotal to demonstrate the role of a transmissible cytoplasmic element in determining hypovirulence for *C. parasitica* (Van Alfen et al. 1975).

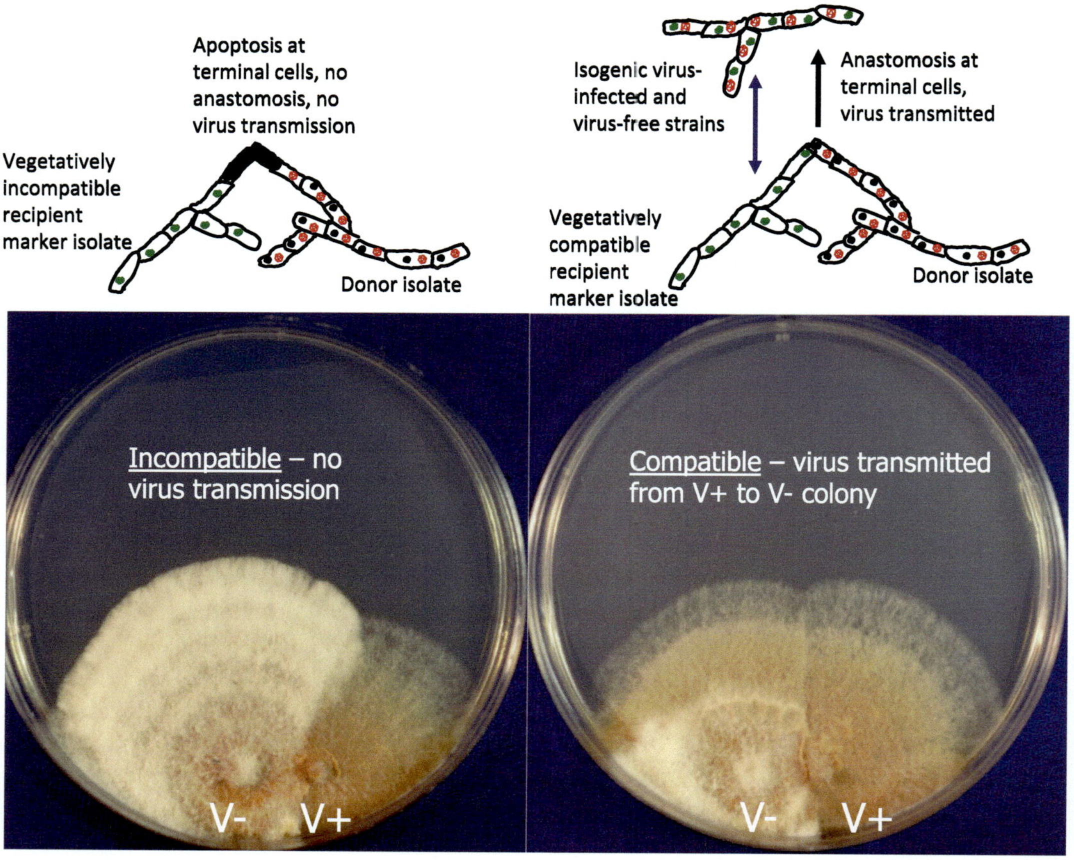

Fig. 5.3 Virus transmission by hyphal anastomosis of vegetatively compatible strains, or restriction of transmission by vegetative incompatibility. Bottom right, vegetatively compatible strains of *C. parasitica* allow for transmission of the virus CHV2. Note that the strains need not be isogenic for anastomosis and transmission to occur. Bottom left, vegetative incompatibility results in failure of transmission and no change in phenotype of virus-free strain. Cartoons showing hyphal interactions are above photos

5.3.2.2 Purified Particles Transfected in Protoplasts:

The first successful attempts at establishing infections of mycovirus from purified particles were considered a major breakthrough. Such was the case for dsRNA viruses of *Penicillium stoloniferum* (Eurotiales) in the early 1970 (Lhoas 1971) and particularly in the killer yeast system, since, previously yeast viruses were thought to be transmitted only through cytoplasmic mixing (budding, mating, cytoduction, or protoplast fusion). In this breakthrough work, yeast protoplasts of a sensitive strain were infected by both the K1 and K2 killer viruses though polyethylene glycol (PEG) inoculation (Elsherbeini and Bostian 1987). In the same years also filamentous fungi were shown to be transfected by particles of dsRNA viruses (Victorivirus) (Ghabrial and Mernaugh 1983) and *Gaeumannomyces graminis* var. tritici (Stanway and Buck 1984). Recently such an approach was used to extend the host range of Rosellinia necatrix victorivirus 1 into *C. parasitica* wild-type (WT) and silencing deficient mutant strains, showing that Victoriviruses are targeted by the antiviral silencing defense (Chiba et al. 2013). The same particle transfection approach was used to test a mix of 6 mycoviruses purified from a single *Penicillium aurantiogriseum* (Eurotiales) strain and to demonstrate that two of them can replicate in plant cell protoplasts (Nerva et al. 2017). When employing this approach, it is important to plate the purified virus preparation to make sure no fungal colonies are born from the original infected isolate. Furthermore, it is important that the transfection mix is well characterized (possibly by next-generation sequencing, NGS) to exclude unforeseen or unwanted contamination from undetected mixed infections. Purified particles from the ssDNA mycovirus infecting *S. sclerotiorum*, were also shown to be infectious, both for fungi and a possible dipteran insect vector *Lycoriella ingenua* (Yu et al. 2010, 2013).

5.3.2.3 Transfection with Nucleic Acid

Use of native nucleic acid from purified or partially purified virus can be used to start mycovirus replication in a new host (Li et al. 2019). A particular application of this approach was used to demonstrate that dsRNA can be infectious in the case Aspergillus fumigatus tetramycovirus 1, one of the first polymycovirus characterized. The researchers treated RNA extracted from a purified virion-like colloidal preparation with proteinase K and the ssRNA-specific S2 nuclease, but not the dsRNA-specific RNase III, demonstrating for the first time that dsRNA from a purified virus can be directly infectious (Kanhayuwa et al. 2015).

5.3.2.4 Transfection with Infectious Clones

A major breakthrough in establishing the link between a host phenotype (mostly disease) and a specific virus, was the ability to start an infection from the cloned DNA or cDNA of the virus. Such clones, generally harbored by plasmids maintained and grown in *E. coli*, are the basis for reverse genetic approaches in virology. The first infectious cDNA clone for an RNA virus was assembled for a bacteriophage, a prokaryotic virus in 1978 (Taniguchi et al. 1978) and for Poliovirus in 1981 (Racaniello and Baltimore 1981). These milestones in the study of RNA viruses were soon replicated for plant viruses with an important modification; rather than transforming cells with the full-length cDNA clone as in the first two examples, Ahlquist and colleagues pioneered the use of infectious RNA transcribed in vitro from full-length cDNA clones to transfect plants (Ahlquist et al. 1984) and then yeast cells (Janda and Ahlquist 1993).

Nuss and colleagues used both DNA transformation and RNA transcript approaches successfully for the CHV1 mycovirus (Choi and Nuss 1992; Chen et al. 1994). Transfection of protoplasts was accomplished with viral RNA transcribed in vitro with the T7 RNA polymerase, with the T7 promoter engineered in front of the 5′ end of the viral sequence, and a restriction site for linearization added after a 20 nt poly A tail (Chen et al. 1994). In the second approach, the viral genome was placed in a fungal transformation plasmid with a selectable hygromycin phosphotransferase resistance gene, between the glyceraldehyde-3-phosphate dehydrogenase (GAPDH) promoter and terminator from *C. parasitica*. In this case, the protoplasts were transformed with the plasmid, but only resistant (transgenic) isolates were retrieved and shown to harbor CHV1 mycovirus infection. This approach allows the virus to be maintained indefinitely in the genome, possibly improving the biocontrol approach because the virus can be spread by mating, thus overcoming vegetative compatibility barriers and will be present in all the conidial progeny without segregation having occurred. For many years, CHV1 was the only mycovirus infecting filamentous fungi that could be studied using an infectious clone. Then in 2003, infectious clones were reported for yeast narnavirus 20S (Esteban and Fujimura 2003) and Diaporthe ambigua RNA virus (Moleleki et al. 2003). More recently, in vitro synthesis and transfection approach were used for an hypovirus (Sclerotinia sclerotiorum Hypovirus 2) from *S. sclerotiorum* (Marzano et al. 2015); the first infectious clone of a member of the *Botourmiaviridae* family was also recently established (Wang et al. 2020).

Even more recently, the first infectious clone for a CRESS-like ssDNA mycovirus was assembled and verified. Tandem repeat of the circular genomic components were cloned in pBluescript II for each of the three genomic components of Fusarium graminearum gemytripvirus 1 (FgGMTV1) and transfected in protoplasts of *F. graminearum* to establish infection; this approach allowed the identification of replication-required virus components and those that are symptom-associated. Finally, another mycovirus infectious transcript allowed the dissection of the very interesting virus symbiotic lifestyle discussed later (see ***Rosellinia* viruses, below**). An infectious clone of the Yado-kari positive-strand virus was verified to be dependent on Yado-nushi co-infection to complete its life cycle (Zhang et al. 2016b).

Still more model cDNA mycovirus infectious clones should be assembled. DsRNA viruses, like mycoreovirus, could in principle be derived, but reoviruses have proven to be the most challenging systems for such engineering. The first such system was not developed until 2007 (see Kanai and Kobayashi 2021 for review). Monopartite or multipartite negative-sense and ambisense RNA viruses are more approachable, and many examples of negative-strand RNA virus infectious clones are described in literature (see Neumann et al. 2002 for review).

5.3.2.5 Synthetic Biology from Metatranscriptomics:

Most of the recently described mycoviruses that greatly increased our view of mycovirus biodiversity and evolution come from metatranscriptomic samples, often complex matrixes, where the biological sample is no longer available. Nevertheless, the principles of synthetic biology can easily be applied to viruses. This approach was used to demonstrate the host of a ssDNA virus (Feng et al. 2021). Infectious clones of a genomovirus were derived from metagenomic samples by amplifying the genome by PCR and used to transfect *S. sclerotiorum*. The virions recovered were also used to support the virus as a cause of hypovirulence, but additionally of infection in *B. cinerea* and *Monilinia fructicola*, and *Spodoptera* insect cells. Given the ease and cost-effectiveness of synthetic DNA, we envision that more viruses will be revived simply by sequence data available in silico, through the synthesis of infectious clones.

5.4 Specific Selected Systems

5.4.1 *Cryphonectria* Viruses

Cryphonectria parasitica is the most thoroughly studied fungus–virus system. While viruses have been identified in other *Cryphonectria* species, they are for the most part of greatest interest in the context of *C. parasitica* comparisons. While we will provide an overview here, we direct readers to many useful reviews on different aspects of the fungus as a pathogen and as virus host, and of the different viruses that have been studied in that fungus (Hillman and Suzuki 2004; Macdonald and Fulbright 1991; Milgroom and Cortesi 2004; Dawe and Nuss 2013; Eusebio-Cope et al. 2015; Nuss 2011; Rigling and Prospero 2018; Van Alfen et al. 1975; Anagnostakis 1982, 1987; Turina and Rostagno 2007).

C. parasitica is the ascomycete that caused the pandemic that wiped out the American chestnut, *Castanea dentata*, as a dominant hardwood species in eastern North America, and also virtually eliminated related chinquapin species in the southeastern U.S. during the early twentieth century (Anagnostakis 1987). Its independent invasion into Europe, while impactful, did not ultimately result in the equivalent dramatic effect on European chestnut forests, owing in part to the slightly greater resistance of the tree host, *C. sativa*, and differences in tree management, as well as the early presence of virulence-suppressing viruses that spread efficiently in the invading populations of the fungus (Rigling and Prospero 2018).

Viruses of *C. parasitica* have been studied extensively and one in particular, CHV1, has been the subject of many hundreds of published studies ranging from epidemiology, natural and introduced biocontrol, transmission, virus molecular biology, and the many effects on host biology. Such exhaustive studies have provided deeper and broader information on that virus than we have on any other fungal virus. Its study, almost to the exclusion of other *C. parasitica* viruses, has had the unintended consequence that many researchers extrapolate results from that singular virus/host system to other virus/fungus systems. It is the equivalent of studying smallpox and its causal variola virus to the exclusion of other human viruses. Nevertheless, its study has provided great insight into fungal viruses, and like other model systems, many of the studies performed with CHV1 would have been much more challenging if attempted with other *C. parasitica* viruses.

The dominance of CHV1 across Europe speaks to its fitness as a virus species, despite the fact that it has marked detrimental effects on the fungus it infects, greatly reducing its virulence and fecundity as measured by asexual sporulation and sexual reproduction (Anagnostakis 1987; Rigling and Prospero 2018). Although few competition experiments have been published, our own anecdotal evidence indicates that co-cultivation of vegetatively compatible *C. parasitica* isolates infected with CHV1 and isolates infected with any of several other viruses, followed by serial passage, usually results in an isolate infected only with CHV1—the other virus does not survive. Furthermore, unlike many other mycoviruses, CHV1 is virtually never lost from cultures in storage. This may be an indication of how well adapted it is to its fungal host.

A few of the many "firsts" for CHV1 are these: (1) it was the first virus associated with and found to be the causal agent of hypovirulence in fungi; indeed, the term hypovirulence was coined based on CHV1 (Grente 1965); (2) it was the first fungal virus for which infectious cDNA clones were made (Choi and Nuss 1992; Chen et al. 1994), leading to immense advances in our understanding of mycovirus host/pathogen interactions [reviewed in (Dawe and Nuss 2013; Nuss 2005, 2011)]; (3) it was the first virus released into nature as a transgenic construct for the purpose of biological control of a plant disease (Anagnostakis et al. 1998); (4) it was the first virus released in the context of a transgenic fungus that had been genetically engineered for universal vegetative compatibility so as to anastomose and transmit virus to a broader range of isolates in nature (Stauder et al. 2019). (5) Furthermore, many mycovirologists worldwide who are currently working on other systems worked with CHV1 at some point in their training.

Although only a few viruses of *C. parasitica* other than CHV1 have been described outside of North America (Forgia et al. 2021), several have been found in North America, where the viral landscape is diverse and interesting [(Turina and Rostagno 2007; Hillman and Suzuki 2004) for reviews]. CHV2, the closest known relative of CHV1 (Hillman et al. 1992, 1994), has been found in North America on resprouted American chestnut only in a small area representing a few hundred square miles in eastern New Jersey (Chung et al. 1994; Peever et al. 1997), close to New York City and close to where the invasion of *C. parasitica* into North America is thought to have occurred. CHV2 was also isolated from infected trees on a single occasion in China (Peever et al. 1998). CHV3 is only distantly related to CHV1 and CHV2, and also has been isolated only from a single localized region of the Great Lakes region of Michigan, USA, and nearby in Canada (Fulbright et al. 1983; Smart et al. 1999). CHV4 dominates the landscape of North America (Linder-Basso et al. 2005). In interesting contrast to CHV1, CHV4 causes little measurable phenotypic effect in the infected fungus (Enebak et al. 1994b), and yet it is present in ~1/4 of *C. parasitica* isolates throughout North America (Peever et al. 1997). CHV4 is relatively closely related to CHV3 (Linder-Basso et al. 2005), but it likely represents an independent entry event into the fungus.

Several other viruses or viral elements have been isolated from *C. parasitica,* but few of them have been examined in detail. A mitochondrial element that had little effect on host phenotype was identified (Polashock and Hillman 1994), leading to the eventual recognition of the large viral family *Mitoviridae*. The two reoviruses identified in *C. parasitica* (Hillman et al. 2004) resulted in the description of a new virus genus, *Mycoreovirus*, and important research on viral RNA recombination in fungi (Sun and Suzuki 2008; Aulia et al. 2019; Eusebio-Cope et al. 2010). Interestingly, the *C. parasitica* mitovirus and the two reoviruses were isolated only a single time and appear not to have spread in chestnut populations. Recently, the first negative-strand RNA virus was identified in *C. parasitica* in isolates from Azerbaijan, along with a novel ambisense RNA virus (Forgia et al. 2021), further expanding the known viral diversity in the fungus.

5.4.2 *Sclerotinia* Viruses

Sclerotinia is the type genus of the ascomycete family *Sclerotiniaceae*. Its most economically important species, *S. sclerotiorum*, an exceptionally broad host range soilborne plant pathogen of great agronomic importance, is often the major pathogen of a particular crop (Xia et al. 2020). Although viruses of *S. sclerotiorum* were not among the subjects of early mycovirus study, they have gained appreciably in importance over the last two decades (for reviews, see (Jiang et al. 2013; Ghabrial et al. 2015; Boland 2004)). Coinfections with more than one virus are common in *S. sclerotiorum* (Ran et al. 2016; Hu et al. 2014; Mu et al. 2021), and several viruses have been shown to cause hypovirulence (Xie et al. 2006; Hu et al. 2005; Yu et al. 2010; Xu et al. 2015; Azhar et al. 2019).

Some years after the initial description of virus-associated hypovirulence of *S. sclerotiorum* by Boland (Boland 1992), research groups at Huazhong University led by Daohong Jiang and then Jiatao Xie began characterizing the virome of *S. sclerotiorum* with the objective of finding viruses that might be useful in its control. To date, members of at least 15 virus families have been identified in *S. sclerotiorum* (Qu et al. 2021). Among the many viruses that have been identified and characterized, several are particularly notable.

The first single-stranded DNA virus identified in a filamentous fungus was found in *S. sclerotiorum* (Yu et al. 2010) and named *S. sclerotiorum* hypovirulence-associated DNA virus 1 (SsHADV1). Although few such viruses have been identified since through classical virology approaches, the ubiquitous presence of similar DNA sequences in environmental samples led to the formation of a new virus family, the *Genomoviridae*, to accommodate such viruses (Krupovic et al. 2016). Although only one other similar virus has been characterized since then, the potential importance of this virus family in the constellation of mycoviruses is evident as you will see below and likely will grow. Indeed, Marzano and Domier (Marzano and Domier 2016) assembled the genome of soybean leaf-associated gemycircularvirus from soybean field-derived metatranscriptomes. The virus was later renamed soybean leaf-associated genomovirus 1 (SlaGemV1), and an infectious cDNA clone was constructed in the laboratory based on the assembled sequence as noted above (Feng et al. 2021). Similarly, construction and deployment of an infectious cDNA clone of the tripartite-genome DNA virus FgGMTV1, the first infectious clone assembled for a DNA mycovirus, allowed for the investigation of viral sequences critical for virus replication and virulence (Li et al. 2020).

SsHADV1 has been a fascinating virus in several respects. Besides being the first fungal ssDNA virus described, the virus was shown to have several other unusual characteristics. The virus was shown to be infectious either as purified particles or as purified viral DNA, first to fungal protoplasts (Yu et al. 2010), and then to whole mycelia either in culture or on plant leaves (Yu et al. 2013). The mechanism of virus uptake into fungal hyphae with their cell walls intact remains unknown, but that property opened the possibility of using SsHADV1 as a biofungicide, albeit with caveats pertaining to efficacy and safety (Yu et al. 2013). Recent studies have shown not only that SsHADV1 is potentially an effective biocontrol agent in field situations, but that it in effect converts the necrotrophic fungus into a beneficial endophyte (Zhang et al. 2020).

Another unexpected finding about SsHADV1 is that it can be vectored from an infected fungus to an uninfected fungus by a mycophagous insect, *Lycoriella ingenua*, and in fact was able to replicate in that insect and was transmitted vertically through insect eggs (Liu et al. 2016). Besides the obvious implications for field spread in agricultural settings, these findings have profound implications for our view of mycovirus ecology and evolution.

5.4.3 *Rosellinia* Viruses

The ascomycete *Rosellinia necatrix* (Xylariales) causes white root rot disease on a wide range of plants, especially perennial woody plants. Although the fungus itself is not so well-known as most of the other exemplar fungi highlighted in this chapter, it has an important place as a mycovirus host [(Kondo et al. 2013b) for review]. This is in large part because of excellent and thorough descriptions of members of several new virus families and clever experimental methods envisioned and planned, especially by Nobuhiro Suzuki at Okayama University and his colleagues. White root rot and similar root diseases incited by *R. necatrix* are particularly difficult to control because the fungus survives well in soil and spreads effectively (ten Hoopen and Krauss 2006). Because of the limited effectiveness of fungicides, biological control of *R. necatrix* with naturally occurring viruses was pursued as a control measure by Matsumoto and colleagues (Ikeda et al. 2004) and later by other plant pathologists, especially in Japan. These efforts led to a large collection of fungal isolates that were then mined for viruses of potential value as biocontrol agents.

Several viruses of importance were discovered in *R. necatrix*, and it continues to be a system of interest to mycovirologists. The first sequence-based evidence for a reovirus in a filamentous fungus came from *R. necatrix* (Osaki et al. 2002). The virus, later named *Mycoreovirus 3-RnW370* (MyRV3-W370) had 12 segments of dsRNA, in contrast to the two reoviruses found in *C. parasitica*, which have 11 segments each. Interestingly, one of the 12 dsRNA segments of MyRV3, segment 8, is often lost during laboratory subculture. Fungal cultures infected with virus strains lacking segment 8 show no discernable phenotype differences compared to their 12-segment counterparts, and they are hypovirulent based on apple fruit assays (Kanematsu et al. 2004; Sasaki et al. 2007). Thus, while the function of this situationally dispensable segment remains unknown, its universal presence in isolates retrieved from naturally infected roots suggests that it is important to virus fitness in natural settings. In light of the examples cited elsewhere for mycovirus replication in plants and insects, it is tempting to speculate that MyRV3 segment 8 may have a role in infection of another host or possible vector (Hillman et al. 2004).

Characterization of a bipartite genome virus in *R. necatrix* with two large RNA segments encapsidated in rigid icosahedral particles (Chiba et al. 2009) led to the description of the new virus family *Megabirnaviridae* (Sato et al. 2019). Only one other such virus has been characterized thoroughly, a closely related virus in *S. sclerotiorum* (Wang et al. 2015), but other possible members of the family have been identified through transcriptome databases (e.g., Nerva et al. 2015). The two characterized viruses cause reduced virulence and phenotypic changes to their respective hosts in culture (Chiba et al. 2009; Wang et al. 2015). Another novel virus discovered in *R. necatrix* that led to description of a new virus family was a four-segment dsRNA virus named *Rosellinia necatrix* quadrivirus 1 (RnQV1) (Lin et al. 2012), resulting in the establishment of the virus family *Quadriviridae* (Chiba et al. 2018). Like the *Megabirnaviridae*, targeted and metagenomic approaches have identified other probable *Quadriviridae* family members, but to date, none has been shown to affect fungal phenotype.

The virus mentioned above (*Transfection with nucleic acid*) characterized by a novel mutualistic lifestyle between two unrelated viruses, was first described in *R. necatri*x by Suzuki and colleagues. In this instance, one virus yado-kari virus 1 (YkV1; yado-kari meaning "borrowing a room"), an ssRNA virus, was hypothesized to use the capsid of an unrelated dsRNA virus, yado-nushi virus 1 (YnV1; yado-nushi meaning "room owner"), and use it as a base for replication and genome encapsidation (Zhang et al. 2016b) (reviewed in Hisano et al. 2018). The yado-kari/yado-nushi hypothesis was supported by later experiments showing that, indeed, YkV1 used its own encoded RdRp enzyme to replicate its RNA, but in YnV1-encoded capsids (Das et al. 2021). Recently, this unique model of virus mutualism has been expanded with the discovery in *S. sclerotiorum* of a virus related to YkV1 partnering in the same manner with an unrelated virus (Jia et al. 2022). Based on the growing constellation of now more than 10 YkV1-related virus sequences in the proposed family Yadokariviridae (Das et al. 2021), it is reasonable to speculate that this virus lifestyle, perhaps unique to fungi and oomycetes, is relatively common in these organisms.

5.4.4 *Fusarium* Viruses

The genus *Fusarium* (Hypocreales) includes a large group of fungal species that are among the most important for plant pathology. The group also includes saprophytes living mostly in the soil. Their economic damage is not only directly to the crop, but also, these fungi affect food and feed indirectly because they produce mycotoxins that cause major adverse effects on human and livestock health. The taxonomy of the genus has evolved considerably over the time when it was based mostly on morphological features until the end of the 1980, then biological (sexual compatibility), to current phylogenetic systematics (Munkvold 2017). Currently more than 300 species can be distinguished through phylogenetic evidence, but our focus is on four main plant pathogenic species complexes based on a RNA polymerase II subunit gene phylogeny: (1) the *Fusarium fujikuroi* species complex that includes important diseases of maize, rice, sugarcane, mango, and pine trees (caused by *F. circinatum*) important for toxic fumonisin contamination; (2) the *Fusarium graminearum* species complex responsible for B trichothecene and zearalenone mycotoxin contamination causing, among other plant diseases, fusarium head blight; (3) *Fusarium oxysporum* species complex, causing vascular wilt and root rot, but not involved in major mycotoxin contamination for humans or livestock; (4) the *Fusarium solani* species complex, mostly causing foot and root rot of many crops but without major mycotoxin contamination. Outside these four complexes, there are other important plant pathogenic Fusariums that produce Type A trichothecenes; *F. langsethiae*, *F. poae,* and *F. sporotrichioides*, often referred to as the *F. sambucinum* species complex that cause Fusarium head blight (Laraba et al. 2021).

Given the importance of this genus to agriculture and human health, it is not surprising that it has been the focus of a number of mycovirus studies (reviewed by Li et al. 2019), so we will summarize some of the best studied virus–fungus interactions in this genus, and update the literature with more recent publications. Looking at the biodiversity of mycoviruses present in *Fusarium* in a recent study, one finds reports of the first ssDNA virus in this genus with the interesting feature of being tripartite; it represents a new genus, distinct from previously described mycovirus genera (Li et al. 2020). Some of the other viruses found in Fusarium are type members of new families, as is the case of the dsRNA genome members of the family *Fusagraviridae*, from Fusarium graminearum virus 3 DK3 (Yu et al. 2009); another mycovirus characterized from *Fusarium*, Fusarium graminearum virus 1 DK21 (Chu et al. 2002), is the type isolate of the newly established family *Fusariviridae* that includes a large number of viruses with a ssRNA genome. DK21 is also one of the five viruses so far described that causes hypovirulence in its host and has potential for biocontrol including a 60-fold decrease in production of trichothecene mycotoxins. Among the other viruses causing hypovirulence, we will mention Fusarium graminearum hypovirus 2 (Li et al. 2015) and Fusarium graminearum virus (FgV) strain ch9 (Darissa et al. 2012), a member of the family *Chrysoviridae*. Fusarium oxysporum dianthi virus 1, also a member of the *Chrysoviridae*, interferes with virulence in its host (Lemus-Minor et al. 2018). The most recent inclusion of a hypovirulence-causing mycovirus in *Fusarium* is a megabirnavirus, which not only causes hypovirulence, but also strongly diminishes deoxynivalenol (DON) mycotoxin production (Li et al. 2022; Xie et al. 2022). Some of the above mycoviruses became the subject of thorough studies of their interactions with the fungal host, particularly regarding the role of RNAi and the mechanism of pathogenesis.

5.4.5 *Botrytis* Viruses

Botrytis cinerea (Helotiales) is a necrotrophic fungus endemic worldwide, affecting both the vegetative stage of a number of crops and fruits (such a grapes) during post-harvest storage. Often known as gray mold, *Botrytis* has the capacity to survive in the environment by production of resistant sclerotia and is considered a model system for the study of plant phytopathogenic fungi (Williamson et al. 2007), particularly the necrotrophic ones. Given the importance of this fungus, it is not surprising that alternatives to fungicides are sought among the mycoviruses that infect it. Over the years, a number of mycoviruses were shown to cause hypovirulence in *B. cinerea* (Hao et al. 2018a, b; Wu et al. 2010; Potgieter et al. 2013; Yu et al. 2015) and reviewed in (Wu et al. 2016). Untargeted approaches have revealed a complex virome associated with this fungus in collections from different regions of the world: the widest virome characterization effort recently reported a number of interesting new viruses for this fungus. Among these are a tripartite member of the negative-strand RNA *Bunyavirales* (Botrytis cinerea bocivirus 1), the report of the type member of a subclade of viruses with palm domains split on two genomic segments (Botrytis cinerea binarnavirus 1, 2, and 3) and a ssDNA virus, named Botrytis cinerea ssDNA virus 1 (BcssDV1) (Ruiz-Padilla et al. 2021).

Other isolates of the same ssDNA virus were also found in China (Hao et al. 2021) and Australia (Khalifa and MacDiarmid 2021). The virus isolate from Australia, tentatively designated Botrytis gemydayirivirus 1 (BGDaV1), was demonstrated to be transfected to new fungal hosts as purified particles, and the resulting infected isolate *B. cinerea* was hypovirulent (Khalifa and MacDiarmid 2021).

Recently an infectious clone of Botrytis virus F was assembled. The clone, infectious through T7-derived RNA transcripts, could potentially be engineered to include fragments of essential host genes to increase the virus pathogenicity through a virus-induced gene silencing method (VIGS) (Cordoba et al. 2022). Also, a potential tool for *Botrytis* manipulation is the infectious clone of the ssDNA virus from *Sclerotinia* assembled synthetically from metatranscriptomic sequence, as detailed below, since it can infect *B. cinerea* as well as *Sclerotinia* (Feng et al. 2021).

A partitivirus was also associated with hypovirulence of *B. cinerea* after the virus was transferred through anastomosis to three recipient virulent strains (Kamaruzzaman et al. 2019). The isolate that originally harbored the partitivirus (QT5–19) was pink and had interesting features related to its biocontrol potential through competitive saprophytic ability (CSA). Volatile organic compounds (VOCs) specifically emitted by this pink *B. cinerea* strain, seemed to be associated with this CSA, although the causal link with the mycovirus presence still needs to be established (Kamaruzzaman et al. 2020). The same VOCs also have an effect on plant growth, stimulated by the interaction of this strain with the plant (Kamaruzzaman et al. 2021).

5.4.6 *Trichoderma* Viruses

Trichoderma (Hypocreales) is a genus of fungi that includes a large number of species that are commonly isolated from soil and rhizosphere where their saprotrophic lifestyle plays a significant role in degradation of plant polysaccharides (Druzhinina et al. 2011). They are particularly interesting for their beneficial interactions with plants, including symbiosis (Harman et al. 2004). *Trichoderma* spp. can also directly attack other fungi (mycoparasitism), and for this reason they are of interest as biocontrol agents (Sood et al. 2020). Mycoparasitism also is a cause of concern for diseases they cause to edible mushrooms.

Given their ecological importance and their biotechnological relevance, it is surprising that the *Trichoderma*-associated mycovirome was only recently investigated. Initially only dsRNA segment polymorphism was associated with *Trichoderma* spp. collections (Jom-in and Akarapisan 2009; Yun et al. 2016). Particularly interesting is the polymorphism described in *Trichoderma* spp. causing green mold disease in 315 cultivated shiitake isolates from contaminated logs and sawdust (Yun et al. 2016). A number of dsRNA genome mycoviruses were characterized, including partitiviruses: Trichoderma atroviride partitivirus 1 (TaPV1) (Chun et al. 2018b), Trichoderma harzianum partitivirus 1 (ThPV1) (Chun et al. 2018a), and Trichoderma harzianum partitivirus 2 (ThPV2) (Wang et al. 2022); Fusagraviridae: Trichoderma atroviride mycovirus 1 (TaMV1) (Lee et al. 2017), Trichoderma asperellum dsRNA Virus 1 (TaRV1) (Zhang et al. 2018); and unclassified viruses belonging to the "bipartite" clade: Trichoderma harzianum bipartite mycovirus 1 (ThBMV1) (Liu et al. 2019b), and Trichoderma harzianum mycovirus 1 (ThMV1) (Liu et al. 2019a) in the Beauveria-like clade as defined in (Nerva et al. 2015). Furthermore, Trichoderma harzianum hypovirus 1 and 2 and two members of the family *Hypoviridae* were also described from *Trichoderma* (You et al. 2019; Chun et al. 2022).

Some of these viruses were cured from the original strains or transferred to new virus-free isolates by anastomosis, and therefore the effect of the virus can be inferred. In particular in the case of ThPV2, the partitivirus has no negative effect on the qualitative biocontrol performance of the *Trichoderma* host, but instead shows a moderate but statistically significant improved biocontrol activity in experiments with cucumber seeds inoculated with *Fusarium oxysporum* f. sp. *cucumerinum* (Wang et al. 2022).

In another study, ThHV1 was discovered in two forms in two different isolates. In one isolate, the virus accumulated with an abundant defective form. Transmission of the two versions of the infectious ThHV1 (with and without defective RNA) showed that the one with defective RNA was highly transmissible and was detrimental to the biocontrol properties of two *Trichoderma* species, raising the concern that successful application of *Trichoderma* might depend over the long term on the presence or absence of mycoviruses (You et al. 2019).

In the case of ThPV1, inhibition of growth in co-cultured *Pleurotus ostreatus* and *Rhizoctonia solani* increased in ThPV1-containing strains compared with ThPV1-cured isogenic strains, associated with a significantly higher β-1,3-glucanase activity but not of chitinases (Chun et al. 2018a); TaPV1 isogenic strains, however, did not show significant changes in colony morphology. In addition, no changes in the enzymatic activities of β-1,3-glucanase and chitinase were observed in virus-cured strains (Chun et al. 2018b). Finally, the bipartite-like ThMV1 on the one hand, increased biomass of the fungal host and its growth-promoting effect on cucumber plants, but at the same time decreased the biocontrol potential (Liu et al. 2019a).

5.4.7 Agaricus and Other Cultivated Mushrooms:

Around 200 species of fruitbodies of fungi (*Agaricales*) are consumed across the world. Interest in edible mushrooms is continuously growing because of the nutraceutical properties associated with both wild harvested and cultivated mushrooms (Kalac 2013). World production from 2010 to 2020 increased steadily from 24.9 to 42.7 million tons (https://www.fao.org/faostat/en/#data/QCL); the highest production is known from China, Japan, USA, Netherland, India, Poland, and Spain. Among those species cultivated, *Agaricus bisporus* (button mushroom, white and brown mushroom, including portobello) is by far the most popular, followed by shiitake (*Lentinula edodes*), hiratake or oyster mushroom (*Pleurotus ostreatus*), and *Flammulina velutipes* (golden needle mushroom, enokitake) (Marcal et al. 2021).

Cultivated mushroom species harbor many mycoviruses, and their number has increased through untargeted HTS of their RNA (Zhang et al. 2022). Nevertheless, their biological role has been difficult to ascertain. Although the first mycoviruses were indeed described as associated with a mushroom disease (Hollings 1962), proper formal establishment of disease etiology through Koch's postulates was not achieved due to the complexity of the infection, and the lack of tractable genetic systems for cultivated mushroom species. Here we will mention examples of the historically important virus-associated diseases of cultivated species, highlighting some of the biologically relevant associations between virus presence and disease.

The primary *Agaricus bisporus* diseases associated with mycovirus etiology are La France disease (originally also known as dieback), described in the mid-1950 in Pennsylvania (Sinden and Hauser 1950) and two distinct syndromes described in the 1990, initially named mushroom virus X disease, now named brown cap mushroom disease (BCMD) and Patch disease (mushroom virus X sensu stricto).

La France disease is characterized by slow mycelial growth, distorted or dramatically delayed fruitbody emergence. A number of published works have associated the particulate isometric virus Agaricus bisporus virus 1 (AbV1) with the disease, consisting of 6 RNA and 3 associated RNA viruses (VanDerLende et al. 1994, 1996; Goodin et al. 1992). Currently AbV1 in the National Center for Biotechnology Information (NCBI) is an unassigned member of the genus *Chrysovirus*, but it is only very distantly related to current members of the *Chrysoviridae* family and its taxonomic position should be revised.

The complexity of the virome associated with Mushroom virus X (MVX) was only recently described, first by elucidation of the dsRNA segments present (Grogan et al. 2003), and later by their molecular description through HTS (Deakin et al. 2017). Recently, this multiple infection was examined in cultures harboring 9 viruses by statistically analyzing their abundance in fruitbodies, distinguishing viruses upregulated in symptomatic fruiting bodies and downregulated in asymptomatic fruiting bodies. AbV16 and Orphan 8 often seems to be associated with the disease (Dobbs et al. 2021).

Lentinula edodes is the second most cultivated edible mushroom. Shiitake mushroom is widely cultivated in East Asia, not only as an edible, but also because of bioactive ingredients shown to be effective in treating tumor and infections (Liu et al. 2015). A number of symptoms (degeneration of mycelia, abnormal browning of the bag cultures, and incapability of fruiting body formation) have been associated with mycovirus presence (Sun et al. 2022), but in most cases, detection of specific mycoviruses in symptomless fruiting bodies complicates the evaluation of the impact of the virus found. In particular, mycelial fragmentation of clonal isolates allowed obtaining mycovirus-free and mycovirus-positive isogenic lines, and is a good alternative to single conidial cultures that are not feasible for edible mushrooms (Kim et al. 2015). Here it is worth pointing out that three similar viruses (LeV-HKB, LeV-HKA, and LeSV) belonging to the recently proposed family *Phlegiviridae* (characterized as monopartite, bicistronic with a genome of 11 kb dsRNA, in one case with associated spherical particles) (Picarelli et al. 2019), are widely present in *L. edodes*. Certain strains have been demonstrated to cause disease in mushroom production (Kim et al. 2015; Sun et al. 2022). A partitivirus (LePV1) was also shown to cause disease in *L. edodes* (Guo et al. 2017), but no biological effect was evident from three negative-stranded viruses (a mymonavirid and two in the *Phenuiviridae* family) (Lin et al. 2019).

Pleurotus ostreatus is also affected by diseases caused by mycoviruses. In particular, a severe dieback epidemic was associated with oyster mushroom spherical virus (OMSV) (Yu et al. 2003). OMSV is an isometric virus with a genome of ~5.7 kb of positive-strand RNA that is related to members of the plant-infecting genus *Tymovirus*. The genome contains 7 open reading frames (ORFs), but only RdRP and coat protein have been assigned a function. Recent work investigated a strain of OMSV from China for which isogenic virus-infected and virus-free lines were obtained, and OMSV-Ch infection did inhibit growth and caused malformation of the host (Hu et al. 2022). Some effects on growth also were established for Pleurotus ostreatus Virus (PoV); (Korean isolate ASI2792). This isolate was studied using virus-free and virus-infected isogenic strains obtained through mycelial fragmentation, and its effect on vegetative growth was demonstrated. PoV is a bipartite virus with two segments of dsRNA of circa 2.3 and 2.2 kB pairs. An association with extracellular enzyme expression and activity was established using this system, providing a molecular mechanism for the diminished capacity of vegetative growth on complex growth substrate (Song et al. 2020).

Flammulina velutipes was linked to a partitivirus, Flamminula velutipes browning virus (FvBV), and was primarily associated with darker fruitbodies (Magae and Sunagawa 2010) even though in this case Koch's postulates were not established.

5.4.8 *Phytophthora* Viruses

Phytophthora spp. (Peronosporales) are oomycetes that are related to some of the algal groups having the accessory chlorophyll *c*, biflagellate zoospores, and cellulose cell walls but usually lacking chitin. Although oomycetes are evolutionarily quite distant from fungi, their structures, morphologies, and lifestyles are similar to fungi; furthermore, there are marked similarities in genome components (Randall et al. 2005) and in host defense mechanisms including vegetative incompatibility (Govers et al. 1997) and RNA silencing systems (Vetukuri et al. 2011). Likely because of these similarities in host properties, viruses that infect oomycetes appear to be generally similar in their properties to the more thoroughly studied viruses of fungi.

Oomycetes including *Phytophthora* spp. are generally more challenging to study in laboratory settings than many fungi. Unlike most filamentous fungi that have been the subjects of virus studies, *Phytophthora* and other oomycetes contain diploid vegetative hyphae that are devoid of cross walls, in contrast to the generally haploid, septate mycelia of true fungi. Oomycetes have generally lagged behind true fungi in terms of molecular manipulations (Judelson et al. 1991). Although dsRNAs, presumably of viral origin, have been identified in many oomycetes (see reviews by Cai and Hillman 2013; Sutela et al. 2019), the absence of an obvious phenotype associated with most of those elements has resulted in little follow-up research on them. Recently, small RNA sequencing has been employed for virus discovery in *Phytophthora* species (Botella and Jung 2021; Raco et al. 2022; Poimala et al. 2021; Sutela et al. 2019). Mining of genomes and follow-up expression analysis showed that giant DNA virus invasion occurred in the past in *Phytophthora* (Hannat et al. 2021; Sharma et al. 2014).

Cai and colleagues undertook investigations of possible RNA viruses infecting *Phytophthora infestans* with the goal of developing a virus-induced gene silencing (VIGS) system for molecular manipulation of the pathogen (Cai et al. 2009; Cai and Hillman 2013).

P. infestans was the organism associated with the plant pandemic that led to the famine in Ireland in the mid-nineteenth century, and it continues to be a pathogen of concern on potato and tomato varieties. Although the goal of developing a VIGS system for *P. infestans* was not ultimately achieved, four unrelated RNA viruses were described in a series of studies (Cai et al. 2009, 2012, 2013, 2019a, b; Cai and Hillman 2013).

Of the four viruses, only one, PiRV2, was shown to have an effect on phenotype of the host (Cai et al. 2019a). Presence of PiRV2 in *P. infestans* was shown to cause an increase in asexual sporulation as measured by sporangium production in culture, and an increase in lesion size induced by infected *P. infestans* in detached potato leaves. Comparative transcriptome analysis suggested downregulation of ammonium and amino acid intake as a possible mechanism through which PiRV2 resulted in increased sporulation. Interestingly, in a sampling of 54 *P. infestans* isolates, PiRV2 was found to be especially abundant in the US-8 lineage of *P. infestans*, one of the dominant lineages of *P. infestans* around the turn of the twenty-first century, consistent with the hypothesis that PiRV2 may have a positive effect on fitness of its host.

5.4.9 RNA Silencing as an Antivirus Defense System

RNA silencing is an ancient host defense system deployed against viruses by fungi, oomycetes, plants, invertebrates, and vertebrates. While RNA silencing appears to be less important to vertebrates with more sophisticated antibody-based adaptive immune systems, it is critically important for organisms without such antibody-based systems. In fungi, RNA silencing was first shown formally to function in antiviral defense of *C. parasitica* against CHV1 by Segers and colleagues (Segers et al. 2007), after the authors had demonstrated that the p29 protein of CHV1 served as a suppressor of RNA silencing (Segers et al. 2006). In the same fungus, roles of the two principal silencing-associated proteins, dicer-like protein 2 (DCL2) and argonaute-like protein 2 (AGL2), in defense against viruses other than CHV1 have been examined (Chiba and Suzuki 2015; Chiba et al. 2016). Antiviral RNA silencing and its suppression by viruses has since been investigated in other fungi and virus types (Mochama et al. 2018). The role of antiviral RNA silencing in the ecology of yeast viruses as situationally dispensable host genes was hypothesized following the observations that agl2 and dcl2 have been lost and gained repeatedly in some fungi, and thus the fitness cost of retaining the genes for antivirus defense is balanced against the fitness cost of losing the genes and allowing for virus infection (Drinnenberg et al. 2011; Nicolás et al. 2013). As noted in the section below, there is growing evidence for the importance of suppression of antiviral RNA silencing as a component of mixed infections of viruses in a single fungal thallus.

5.4.10 Mixed Infections

Examples of a single fungal culture infected with more than one virus are abundant in the literature (Thapa and Roossinck 2019; Hillman et al. 2018). In many instances, multiple RNA species, usually detected as multiple dsRNA bands by gel electrophoresis, must be disentangled to determine whether they represent components of a single virus or multiple viruses. If multiple, how many viruses are in the fungal thallus? Making this determination usually involved methods such as those described above to fulfill Koch's Postulates and obtain singly-infected fungal cultures.

We are really only beginning to learn the importance of mixed viral infections in fungi in the context of virus effects on fungal ecology. Hillman et al. (2018) categorized virus to virus interactions in coinfections as being: (1) synergistic, in which one virus enhances replication of another during coinfection; (2) mutualistic, in which both viruses derive benefit from each other's presence in natural settings; (3) antagonistic, in which viruses impair each other's replication or maintenance; or (4) genome rearrangement-inducing viruses. Examples of each have been discussed (Hillman et al. 2018). A recent example of a synergistic-mutualistic virus infection of *C. parasitica* was initially thought to involve a single mycoreovirus, MyRV2 (Enebak et al. 1994a) but was later found to involve a natural coinfection by MyRV2 and the normally symptomless hypovirus CHV4 (Aulia et al. 2019). In this case, CHV4 was necessary for infection of MyRV2: CHV4 alone was stable in *C. parasitica*, but MyRV2 was lost in a relatively short time without CHV4 presence. Furthermore, presence of CHV4 enhanced the vertical transmission of MyRV2 through conidia by more than two-fold. The mechanism for this trans-stabilization of MyRV2 was suppression of the *C. parasitica* host RNA silencing antiviral defense response by CHV4, which has an identified suppressor of RNA silencing, while MyRV2 apparently does not (Aulia et al. 2019, 2021).

Antagonistic interactions are probably common in natural settings, but likely more difficult to discover, because one of the virus partners leads to elimination of the other during the coinfection. Therefore, these interactions tend to be identified in experimental systems in the laboratory. For example, we have found it difficult to maintain cultures coinfected with CHV1 and CHV2, although the mechanism for possible antagonism has not been investigated (B. Hillman and B. Halpern, unpublished). Chiba and Suzuki found that replication of Rosellinia necatrix victorivirus 1 (RnVV1) was suppressed during coinfection with either the mycoreovirus MyRV1 or a mutant of the hypovirus CHV1 lacking a RNA silencing suppressor (Chiba and Suzuki 2015). In this instance, suppression of RnVV1 likely resulted from induction of the host RNA silencing-associated genes, dcl2 and agl2, by the antagonistic viruses, and the inability of RnVV1 to combat the antiviral response.

Some fungal viruses are able to replicate in plant cells without adaptation to the new host (Nerva et al. 2017), and it is reasonable to conjecture that mixed infection of plant and fungal viruses in single host cells likely occurs in natural settings. The demonstrated replication of fungal viruses in mycophagous insects (Liu et al. 2016) and insect cell lines (Feng et al. 2021) further supports the hypothesis that such heterologous mixed infections may be hotspots of virus recombination and evolution.

5.4.11 Phenotype Changes and Mechanisms

As mentioned briefly above, one of the main dogmas still difficult to debunk is the fact that most mycovirus infections are symptomless, often despite the very high accumulation of the virus, which should somehow come at the expense of basal metabolism, deviated toward viral progeny production. It is now clear that indeed in some cases, the effect of a mycoviral infection is minimal, but in most cases, it is the experimental system that is not developed well enough to measure effects: particularly, we mycovirus researchers should keep in mind that attempts at measuring effects of a mycovirus under "axenic" conditions in laboratory settings, can result in missing the effects of virus infections on "interactions." It is possible that mycovirus effects could be unveiled only in natural ecological niches, under the constant threat of multiple abiotic and biotic stressors. Furthermore, the effect of a mycovirus might not be obvious directly on the fungal host, but even can be dramatic when considering multipartite trophic interactions: this effect on multitrophic interactions has been suggested for the interaction of the mycovirus Curvularia thermal tolerance virus and its endophytic fungal host *Curvularia protuberata* (Pleosporales). The effect of the mycovirus was recognized only after considering the third partner, a plant host, the grass *Dichanthelium lanuginosum*. Only the mycovirus-infected fungal endophyte provided the thermal tolerance phenotype to the grass host necessary to resist high temperature soils close to geothermal activity in Yellowstone Park, (Marquez et al. 2007). Interestingly, the role of the virus was not discovered until 5 years after the original report of high temperature resistant plants. Below we will summarize some of the best studied phenotypical changes caused by mycoviruses, focusing on those where a molecular mechanism of the perturbation has been studied.

5.4.11.1 *Virulence*

As mentioned above, for many years, mycoviruses have been studied as possible biological and biotechnological tools for biocontrol of fungal and oomycetes-caused plant diseases (to a lesser extent for human-caused fungal diseases). The approaches to develop biocontrol tools from mycoviruses are still mostly based on natural transmission of hypovirulence through anastomosis. Nevertheless, newer groundbreaking approaches that rely directly on the mycocidal effect of the mycovirus (used as a proxy of a fungicide molecule) or others that include a change in lifestyle of the host that makes the fungal host protective of further infections have been recently reviewed in detail (Wagemans et al. 2022). Here we will take a deeper dive into the best studied system, CHV1-*C. parasitica*, and the many studies that tried to reveal the precise molecular mechanism of hypovirulence, in order to attempt mimicking it in other interactions.

Initial work on the molecular aspects of the CHV1–*C. parasitica* interaction looked at up or downregulated accumulation of proteins and mRNA, initially with coarse techniques that later developed into more comprehensive omics approaches. Two different pathways were at the center of the investigation: the perturbation of signal transduction, and in particular of G-protein related alterations, and to a lesser of extent of MAP kinases and perturbation of secretion. A detailed review of the main accomplishments of those approaches can be found in previous works (Nuss 2005; Turina and Rostagno 2007). As mentioned earlier, these studies elucidated very important fungal pathways and the effect on virulence when perturbing some of those pathways. Nevertheless, the link to virus-caused hypovirulence remained elusive, and still is after further studies that looked more globally at virus-caused changes. Overall hypovirulence caused by CHV1 infection is probably the result of a multi-faceted complex relationship. Even more recent attempts at describing the perturbation of the whole secretome revealed a massive hypovirus-caused regulation (one-third of the secretome) that, among many others, affect known infection-related proteins, including a long list of plant cell wall-degrading enzymes. At the same time, superoxide dismutase (SOD) is also downregulated; SOD is an enzyme involved in the reduction of reactive oxygen species produced by plant to counteract fungal infection; its downregulation might result in hypovirulent fungal hosts (Wang et al. 2016). More targeted recent approaches based on specific gene knock-outs have revealed the role of both the autophagy pathway with a specific role of ATG-8, likely a susceptibility factor in the virus–fungus interaction (Shi et al. 2019), and the methylation pathway in maintaining hypovirus infection. In fact, knock out of two DNA methyltransferases (DNAMTase) strongly affected the phenotype of CHV1-infected mutants resulting in much smaller colony size and higher mycovirus accumulation, pointing to a role of these two DNAMTases in antiviral defense (Ko et al. 2021).

Regarding regulation of virulence in other mycovirus-host models, it is worth pointing out that while hypovirulence is indeed a preferable outcome from an anthropocentric point of view for plant pathogens or also human pathogens such as *Aspergillus* (Eurotiales), (see detailed reviews, Kotta-Loizou and Coutts 2017a; Kotta-Loizou 2021) in some cases, hypervirulence is of interest, for example enhancing the biocontrol potentials of Trichoderma (see above) or entomopathogenic fungi used for biocontrol, such as mycovirus-infected *Beauveria bassiana* (Kotta-Loizou and Coutts 2017b).

Two other examples also point to the necessity of looking at mycovirus effects on virulence in wider field contexts and not only in pathogenicity assays in laboratory settings. A mycovirus in *Leptosphaeria maculans* (Pleosporales) induces hypervirulence, but at the same times triggers systemic resistance toward *L. maculans*: since only the coinfection causes Phoma stem canker (blackleg), it could be argued that pre-inoculation with the hypervirulent mycovirus-infected isolate of one fungus decreases the overall severity caused by the *L. biglobosa* secondary infection (Shah et al. 2020). In another important biological model, Magnaporthe oryzae chrysovirus 1 strain A (MoCV1-A) caused hypovirulence in its fungal host when rice plants were inoculated by punch-wounding with virus-infected vs. virus-free conidia. In contrast, when a battery of susceptible or resistance rice varieties was inoculated by spraying seedlings with virus-infected or virus-free conidial suspensions, mycovirus infection resulted in variable outcomes depending on the presence of the combination of resistance gene and avirulence determinant, suggesting that mycoviruses could drive physiological diversity, including pathogenic races (Aihara et al. 2018) (Note that this virus infects the fungus now known as *Pyricularia oryzae* (see below, Disruption of known metabolic pathways)).

5.4.11.2 Common Phenotypic Changes Linked to Mycovirus-Caused Development Perturbation:

Given that most fungi are studied after isolation from the complex biome they inhabit into a simplified axenic culture rich in substrate, there are a number of features that are often reported to be impacted by mycovirus infection. These features are the most immediate to evaluate and often are used to screen phenotypic variation in populations, a phenomenon often caused by mycovirus infection. Among these easily scorable phenotypic changes growth rate is likely the easiest to notice, but caution must be paid in drawing more general conclusions from these assays with the tendency to use growth rate as a proxy of hypovirulence. In fact, the two concepts should be well separated, in view of the fact that growth is a component of fitness, and for some models of mycovirus-based biocontrol, the fitness of the fungal host should only be minimally affected. Another simple character that can be judged in vitro is the quantification of conidia and possibly the percentage of mycovirus transmission, both also component of fitness. Mycoviruses often affect conidium production as is the case of CHV1 infection of *C. parasitica*. The trade-offs between hypovirulence, growth, and conidium production are exemplified by studies that compared different isolates of CHV1 that differentially affect these traits and might explain some ineffective attempts at biocontrol intervention (the EURO-7 CHV1 isolate vs EP713 isolate) (Chen and Nuss 1999; Chen et al. 2000).

Finally, another character that is often judged and used for visual screens of mycovirus-infected strains is alteration of pigmentation (although the biological significance of it is not obvious), as exemplified by lack of orange pigment production of CHV1-infected *C. parasitica* (Anagnostakis 1982). In other mycovirus–host interactions, excess pigments are produced.

These phenotypic changes can be reproduced by a number of mutants, but here it is worth pointing to a specific gene that only affects these changes when the fungus is infected by CHV1. In a mutant screen, the *nam-1* gene, encoding for a Mg2+ transporter, was identified as a specific modulator of virus-caused symptoms (Faruk et al. 2008).

5.4.11.3 Sexual Reproduction

The effect of mycovirus infection on sexual reproduction was already mentioned for *C. parasitica* (Anagnostakis 1982; Deng et al. 2007a, b); in particular, it was predicted that virus-caused downregulation of the mating pheromone-encoding genes resulted in limited sexual reproduction of virus-infected isolates (Zhang et al. 1993; Turina et al. 2003). Here we will describe a more recent observation that results from the study of an interesting biological system in which a fungus, *Rhizopus microsporus* (Mucorales) is associated with a virulence factor provided by an endobacterium, *Mycetohabitans* (formerly *Burkholderia*); furthermore, two narnaviruses are part of the system. While narnaviruses by themselves decrease asexual reproduction (sporangiospore production), they are strictly required together with the bacterium for sexual reproduction (zygospore formation); this is an example of conditional mutualism, where the positive effect on sexual reproduction provided by the narnavirus is conditional on the presence of the bacteria (Espino-Vazquez et al. 2020); at the moment the molecular mechanism of such interaction remains unknown.

5.4.12 Disruption of Known Metabolic Pathways

It is noteworthy that, although some mycoviruses accumulate abundantly in their host (as is the case of CHV1-*C. parasitica*), a considerable energetic expense, very limited studies have examined the effect of mycovirus infection on metabolic changes, either from a purely physiological point of view (measuring CO^2 and O^2 exchanges and substrate utilization) or from a profiling of primary metabolites. Again the model system for this approach is CHV1-*C. parasitica*. From a general point of view, mycelium on the periphery of the colony relied heavily on fermentation pathways, while older mycelia shifted to aerobic carbohydrate metabolism and finally lipid metabolism in the interior, older part of the colony. This shift left abundant residual glucose in the substrate, and mycovirus infection affected the metabolism in this older part of the plate where infected mycelium seemed to be unable to utilize lipid-related metabolic pathways, which lead to increased glucose depletion (Dawe et al. 2009). This shift related to lipid or carbohydrate metabolism was confirmed by a more accurate metabolic profiling, where one third of 164 detected metabolites were either up- or down-regulated (Dawe et al. 2009).

Virus effects on secondary metabolite accumulation, including volatile organic compounds (VOCs), represent a major knowledge gap. The fact that a comprehensive metabolomic approach has not yet been applied to mycovirus infection is only minimally alleviated by some specific reports, mostly concentrated on the effect on mycotoxin production. In this regard, the first report was related to the downregulation of aflatoxin production in *A. flavus*: interestingly, curing of resident virus rendered a nontoxigenic strain toxigenic, and transfection with a *Penicillium chrysogenum* (Eurotiales) virus repristinated the repression of aflatoxin production (Schmidt et al. 1986). A number of other studies investigated mycotoxin production in *Aspergillus* with different outcomes (Elias and Cotty 1996; Silva et al. 2001). In another study, upregulation of ochratoxin production by *A. ochraceus* was also linked to the presence of a specific partitivirus (Nerva et al. 2018). Recently tenuazonic acid production was also shown to be induced by a totivirus infection in *Pyricularia oryzae* (*Magnaporthales;* previous name is *Magnaporthe oryzae*) (Zhang et al. 2016b). The upregulation of a specific transcription factor involved in the teA synthetase gene upregulation is the proposed molecular mechanism behind the upregulation (Ninomiya et al. 2020). A different example of a secondary metabolite which is also a virulence factor, AK-toxin, is a host-specific toxin responsible for the necrotic phenotype: in this case, a chrysovirus affects growth of the *Alternaria* (Pleosporales) host, but at the same time it greatly increases AK-toxin production (Okada et al. 2018).

5.4.13 Mechanisms of Fitness Alteration

Considerations of fitness alteration caused by mycovirus infection are particularly relevant when planning natural biocontrol approaches, where the hypovirulent strain is released and expected to establish naturally. In fact, rarely is reproductive fitness (defined as a quantitative representation of reproductive success) measured for fungi infected by viruses in natural environments and compared to their uninfected counterparts. This becomes an essential factor when considering a successful biocontrol agent. In this sense, an interesting recent research paper has applied the concepts of ecology measuring the life-history traits in the CHV1–*C. parasitica* interaction (Brusini et al. 2017). This work measured how hypovirus infection alters different components of fitness (growth, survival, and reproduction) and the necessary trade-offs of resource allocation. A number of parameters were measured for 36 combinations of fungal and virus isolates in cross-inoculation experiments. The authors showed that the classical trade-off between "current reproduction" and "somatic growth" (widespread in nature) is subverted when *C. parasitica* is infected by CHV1: in this case, the faster the growth rate, the greater is sporulation; moreover, the faster the growth rate, the greater vertical transmission of the virus (Brusini et al. 2017).

5.4.14 Concluding Thoughts

The breadth of viruses in the fungal systems that have been investigated in some depth, and the overlap in the types of virus found in those fungi, tells us that the same is true for fungi as with other hosts: when resources available and dedicated virologists set out to look for and characterize the viromes of fungi in detail, we find that each fungus is host to a wealth of viruses, many of which affect host biology and fitness. A few systems have been highlighted here, but there are many other fungal systems that are of long-term or growing interest, including the genera

Saccharomyces (Saccharomycetales), *Aspergillus*, *Colletotrichum* (Glomerellales), *Helminthosporium* (Pleosporales), *Alternaria*, and *Pyricularia*, among many others, that could have sections devoted to them. The difficulties of assessing phenotypes has also excluded an important group of taxonomically diverse fungi that establish mycorrhizal relationships with plant hosts from our review, but we expect great progress in mycovirus–host systems in this field in the near future. We are gratified that the task of summarizing the field today is so much more challenging than it was even a few years ago!

References

Ahlquist PA, French R, Janda MI, Loesch-Fries LS (1984) Multicomponent RNA plant virus infection derived from cloned viral DNA. PNAS 81:7066–7070

Aihara M, Urayama S-i, Minh Tuong L et al (2018) Infection by *Magnaporthe oryzae* chrysovirus 1 strain A triggers reduced virulence and pathogenic race conversion of its host fungus, *Magnaporthe oryzae*. J Gen Plant Pathol 84:92–103. https://doi.org/10.1007/s10327-018-0766-7

Anagnostakis SL (1982) Biological control of chestnut blight. Science 215(4532):466–471. https://doi.org/10.1126/science.215.4532.466

Anagnostakis SL (1987) Chestnut blight – the classical problem of an introduced pathogen. Mycologia 79:23–37

Anagnostakis SL, Chen B, Geletka LM, Nuss DL (1998) Hypovirus transmission to ascospore progeny by field-released transgenic hypovirulent strains of *Cryphonectria parasitica*. Phytopathology 88(7): 598–604. https://doi.org/10.1094/PHYTO.1998.88.7.598

Aulia A, Andika IB, Kondo H et al (2019) A symptomless hypovirus, CHV4, facilitates stable infection of the chestnut blight fungus by a coinfecting reovirus likely through suppression of antiviral RNA silencing. Virology 533:99–107. https://doi.org/10.1016/j.virol.2019.05.004

Aulia A, Hyodo K, Hisano S et al (2021) Identification of an RNA silencing suppressor encoded by a symptomless fungal hypovirus, *Cryphonectria hypovirus* 4. Biology 10(2):100. https://doi.org/10.3390/biology10020100

Azhar A, Mu F, Huang H et al (2019) A novel RNA virus related to sobemoviruses confers hypovirulence on the phytopathogenic fungus *Sclerotinia sclerotiorum*. Viruses-Basel 11(8):759. https://doi.org/10.3390/v11080759

Boland GJ (1992) Hypovirulence and double-stranded-RNA in *Sclerotinia sclerotiorum*. Can J Plant Pathol 14:10–17. https://doi.org/10.1080/07060669209500900

Boland GJ (2004) Fungal viruses, hypovirulence, and biological control of *Sclerotinia* species. Can J Plant Pathol 26:6–18

Botella L, Jung T (2021) Multiple viral infections detected in Phytophthora condilina by total and small RNA sequencing. Viruses 13(4). https://doi.org/10.3390/v13040620

Brusini J, Wayne ML, Franc A, Robin C (2017) The impact of parasitism on resource allocation in a fungal host: the case of *Cryphonectria parasitica* and its mycovirus, Cryphonectria hypovirus 1. Ecol Evol 7: 5967–5976. https://doi.org/10.1002/ece3.3143

Cai G, Hillman BI (2013) *Phytophthora* viruses. Adv Virus Res 86:327–350. https://doi.org/10.1016/B978-0-12-394315-6.00012-X

Cai G, Myers K, Hillman BI, Fry WE (2009) A novel virus of the late blight pathogen, *Phytophthora infestans*, with two RNA segments and a supergroup 1 RNA-dependent RNA polymerase. Virology 392(1):52–61. https://doi.org/10.1016/j.virol.2009.06.040

Cai G, Myers K, Fry WE, Hillman BI (2012) A member of the virus family Narnaviridae from the plant pathogenic oomycete *Phytophthora infestans*. Arch Virol 157:165–169. https://doi.org/10.1007/s00705-011-1126-5

Cai G, Krychiw JF, Myers K et al (2013) A new virus from the plant pathogenic oomycete *Phytophthora infestans* with an 8 kb dsRNA genome: the sixth member of a proposed new virus genus. Virology 435:341–349. https://doi.org/10.1016/j.virol.2012.10.012

Cai G, Fry WE, Hillman BI (2019a) PiRV-2 stimulates sporulation in Phytophthora infestans. Virus Res 271: 197674. https://doi.org/10.1016/j.virusres.2019.197674

Cai G, Myers K, Fry WE, Hillman BI (2019b) *Phytophthora infestans* RNA virus 2, a novel RNA virus from *Phytophthora infestans*, does not belong to any known virus group. Arch Virol 164(2):567–572. https://doi.org/10.1007/s00705-018-4050-0

Chen BS, Nuss DL (1999) Infectious cDNA clone of hypovirus CHV1-Euro7: a comparative virology approach to investigate virus-mediated hypovirulence of the chestnut blight fungus *Cryphonectria parasitica*. J Virol 73:985–992

Chen BS, Choi GH, Nuss DL (1994) Attenuation of fungal virulence by synthetic infectious hypovirus transcripts. Science 264:1762–1764

Chen BS, Geletka LM, Nuss DL (2000) Using chimeric hypoviruses to fine-tune the interaction between a pathogenic fungus and its plant host. J Virol 74: 7562–7567

Chiapello M, Rodriguez-Romero J, Ayllon MA, Turina M (2020) Analysis of the virome associated to grapevine downy mildew lesions reveals new mycovirus lineages. Virus Evol 6. https://doi.org/10.1093/ve/veaa058

Chiba S, Suzuki N (2015) Highly activated RNA silencing via strong induction of dicer by one virus can interfere with the replication of an unrelated virus. PNAS 112(35):E4911–E4918. https://doi.org/10.1073/pnas.1509151112

Chiba S, Salaipeth L, Lin YH et al (2009) A novel bipartite double-stranded RNA mycovirus from the white root rot fungus *Rosellinia necatrix*: molecular and biological characterization, taxonomic considerations, and potential for biological control. J Virol 83:12801–12812. https://doi.org/10.1128/JVI.01830-09

Chiba S, Lin Y-H, Kondo H et al (2013) A novel Victorivirus from a phytopathogenic fungus, *Rosellinia necatrix*, is infectious as particles and targeted by RNA silencing. J Virol 87:6727–6738. https://doi.org/10.1128/jvi.00557-13

Chiba S, Lin YH, Kondo H et al (2016) A novel betapartitivirus RnPV6 from *Rosellinia necatrix* tolerates host RNA silencing but is interfered by its defective RNAs. Virus Res 219:62–72. https://doi.org/10.1016/j.virusres.2015.10.017

Chiba S, Caston JR, Ghabrial SA, Suzuki N, ICTV Report C (2018) ICTV virus taxonomy profile: quadriviridae. J Gen Virol 99 (11):1480–1481. doi:https://doi.org/10.1099/jgv.0.001152

Choi GH, Nuss DL (1992) Hypovirulence of chestnut blight fungus conferred by an infectious viral cDNA. Science 257:800–803

Chu YM, Jeon JJ, Yea SJ et al (2002) Double-stranded RNA mycovirus from *Fusarium graminearum*. Appl Environ Microb 68(5):2529–2534. https://doi.org/10.1128/aem.68.5.2529-2534.2002

Chun J, Yang H-E, Kim D-H (2018a) Identification of a novel partitivirus of *Trichoderma harzianum* NFCF319 and evidence for the related antifungal activity. Front Plant Sci 9. https://doi.org/10.3389/fpls.2018.01699

Chun J, Yang HE, Kim DH (2018b) Identification and molecular characterization of a novel partitivirus from *Trichoderma atroviride* NFCF394. Viruses-Basel 10(11). https://doi.org/10.3390/v10110578

Chun J, So KK, Ko YH, Kim DH (2022) Molecular characteristics of a novel hypovirus from Trichoderma harzianum. Arch Virol 167(1):233–238. https://doi.org/10.1007/s00705-021-05253-1

Chung P-H, Bedker PJ, Hillman BI (1994) Diversity of Cryphonectria parasitica hypovirulence-associated double-stranded RNAs within a chestnut population in New Jersey. Phytopathology 84:984–990

Cordoba L, Ruiz-Padilla A, Rodriguez-Romero J, Ayllon MA (2022) Construction and characterization of a Botrytis Virus F infectious clone. J Fungi 8(5). https://doi.org/10.3390/jof8050459

Darissa O, Adam G, Schafer W (2012) A dsRNA mycovirus causes hypovirulence of *Fusarium graminearum* to wheat and maize. Eur J Plant Pathol 134:181–189. https://doi.org/10.1007/s10658-012-9977-5

Das S, Alam MM, Zhang R et al (2021) Proof of concept of the yadokari nature: a capsidless replicase-encoding but replication-dependent positive-sense single-stranded RNA virus hosted by an unrelated double-stranded RNA virus. J Virol 95:e0046721. https://doi.org/10.1128/JVI.00467-21

Dawe AL, Nuss DL (2013) Hypovirus molecular biology: from Koch's postulates to host self-recognition genes that restrict virus transmission. Adv Virus Res 86:109–147. https://doi.org/10.1016/B978-0-12-394315-6.00005-2

Dawe AL, Van Voorhies WA, Lau TA et al (2009) Major impacts on the primary metabolism of the plant pathogen *Cryphonectria parasitica* by the virulence-attenuating virus CHV1-EP713. Microbiol-Sgm 155:3913–3921. https://doi.org/10.1099/mic.0.029033-0

Deakin G, Dobbs E, Bennett JM et al (2017) Multiple viral infections in *Agaricus bisporus* - Characterisation of 18 unique RNA viruses and 8 ORFans identified by deep sequencing. Sci Rep 7. https://doi.org/10.1038/s41598-017-01592-9

Deng F, Allen TD, Hillman BI, Nuss DL (2007a) Comparative analysis of alterations in host phenotype and transcript accumulation following hypovirus and mycoreovirus infections of the chestnut blight fungus *Cryphonectria parasitica*. Eukaryot Cell 6:1286–1298. https://doi.org/10.1128/EC.00166-07

Deng F, Allen TD, Nuss DL (2007b) Ste12 transcription factor homologue CpST12 is down-regulated by hypovirus infection and required for virulence and female fertility of the chestnut blight fungus *Cryphonectria parasitica*. Eukaryot Cell 6:235–244. https://doi.org/10.1128/EC.00166-07

Dobbs E, Deakin G, Bennett J et al (2021) Viral interactions and pathogenesis during multiple viral infections in *Agaricus bisporus*. MBio 12(1). https://doi.org/10.1128/mBio.03470-20

Donaire L, Ayllon MA (2017) Deep sequencing of mycovirus-derived small RNAs from Botrytis species. Mol Plant Pathol 18:1127–1137. https://doi.org/10.1111/mpp.12466

Dong K, Xu C, Kotta-Loizou I et al (2023) Novel viroid-like RNAs naturally infect a filamentous fungus. Adv Sci 10:202204308

Drinnenberg IA, Fink GR, Bartel DP (2011) Compatibility with killer explains the rise of RNAi-deficient fungi. Science 333:1592–1592. https://doi.org/10.1126/science.1209575

Druzhinina IS, Seidl-Seiboth V, Herrera-Estrella A et al (2011) Trichoderma: the genomics of opportunistic success. Nat Rev Microbiol 9:749–759. https://doi.org/10.1038/nrmicro2637

Elias KS, Cotty PJ (1996) Incidence and stability of infection by double-stranded RNA genetic elements in *Aspergillus* section *flavi* and effects on aflatoxigenicity. Can J Bot 74:716–725. https://doi.org/10.1139/b96-090

Elsherbeini M, Bostian KA (1987) Viruses in fungi - infection of yeast with the K1 and K2 killer viruses. PNAS 84:4293–4297. https://doi.org/10.1073/pnas.84.12.4293

Enebak SA, Hillman BI, MacDonald WL (1994a) A hypovirulent isolate of *Cryphonectria parasitica* with multiple, genetically unique dsRNA segments. Mol Plant Microb Interact 7:590–595

Enebak SA, MacDonald WL, Hillman BI (1994b) Effect of dsRNA associated with isolates of *Cryphonectria parasitica* from the central Appalachians and their relatedness to other dsRNAs from North America and Europe. Phytopathology 84(5):528–534

Espino-Vazquez AN, Bermudez-Barrientos JR, Cabrera-Rangel JF et al (2020) Narnaviruses: novel players in fungal-bacterial symbioses. ISME J. https://doi.org/10.1038/s41396-020-0638-y

Esteban R, Fujimura T (2003) Launching the yeast 23S RNA Narnavirus shows 5' and 3' cis-acting signals for replication. PNAS 100:2568–2573. https://doi.org/10.1073/pnas.0530167100

Eusebio-Cope A, Sun L, Hillman BI, Suzuki N (2010) Mycoreovirus 1 S4-coded protein is dispensable for viral replication but necessary for efficient vertical transmission and normal symptom induction. Virology 397(2):399–408. https://doi.org/10.1016/j.virol.2009.11.035

Eusebio-Cope A, Sun L, Tanaka T et al (2015) The chestnut blight fungus for studies on virus/host and virus/virus interactions: from a natural to a model host. Virology 477:164–175. https://doi.org/10.1016/j.virol.2014.09.024

Faruk MI, Eusebio-Cope A, Suzuki N (2008) A host factor involved in hypovirus symptom expression in the chestnut blight fungus, *Cryphonectria parasitica*. J Virol 82:740–754. https://doi.org/10.1128/jvi.02015-07

Feng CC, Feng JH, Wang ZY et al (2021) Identification of the viral determinant of hypovirulence and host range in Sclerotiniaceae of a Genomovirus reconstructed from the plant metagenome. J Virol 95(17). https://doi.org/10.1128/jvi.00264-21

Forgia M, Isgandarli E, Aghayeva DN et al (2021) Virome characterization of *Cryphonectria parasitica* isolates from Azerbaijan unveiled a new mymonavirus and a putative new RNA virus unrelated to described viral sequences. Virology 553:51–61. https://doi.org/10.1016/j.virol.2020.10.008

Forgia M, Chiapello M, Daghino S et al (2022) Three new clades of putative viral RNA-dependent RNA polymerases with rare or unique catalytic triads discovered in libraries of ORFans from powdery mildews and the yeast of oenological interest Starmerella bacillaris. Virus Evol 8(1). https://doi.org/10.1093/ve/veac038

Fulbright DW (1984) Effect of eliminating dsRNA in hypovirulent *Endothia parasitica*. Phytopathology 74(6):722–724. https://doi.org/10.1094/Phyto-74-722

Fulbright DW, Weidlich WH, Haufler KZ et al (1983) Chestnut blight and recovering American chestnut trees in Michigan. Can J Bot 61(12):3164–3171. https://doi.org/10.1139/b83-354

Fulbright DW, Paul CP, Garrod SW (1988) Hypovirulence: a natural control of chestnut blight. In: Mukerji KG, Garg KL (eds) Biocontrol of plant diseases. CRC Press, Boca Raton, FL, pp 121–138

Ghabrial S, Mernaugh R (1983) Biology and transmission of *Helminthosporium victoriae* mycoviruses, vol 441. Elsevier, Amsterdam, London and New York

Ghabrial SA, Suzuki N (2009) Viruses of plant pathogenic fungi. Annu Rev Phytopathol 47:353–384. https://doi.org/10.1146/annurev-phyto-080508-081932

Ghabrial SA, Caston JR, Jiang D et al (2015) 50-plus years of fungal viruses. Virology 479–480:356–368. https://doi.org/10.1016/j.virol.2015.02.034

Goodin MM, Schlagnhaufer B, Romaine CP (1992) Encapsidation of the La France disease-specific double-stranded RNAs in 36-nm isometric virus-like particles. Phytopathology 82:285–290. https://doi.org/10.1094/Phyto-82-285

Govers F, Drenth A, Pieterse CMJ (1997) The potato late blight pathogen *Phytophthora infestans* and other pathogenic Oomycota. In: Carroll GC, Tudzynski P (eds) The Mycota, vol V Part B. Springer, Berlin, pp 17–36

Grente J (1965) Les formes hypovirulentes *d'Endothia parasitica* et les espoires de lutte contre le chance du chantaingnier. CR Acad Agric France 51:1033–1037

Grogan HM, Adie BAT, Gaze RH et al (2003) Double-stranded RNA elements associated with the MVX disease of *Agaricus bisporus*. Mycol Res 107:147–154. https://doi.org/10.1017/s0953756203007202

Guo MP, Bian YB, Wang JJ et al (2017) Biological and molecular characteristics of a novel partitivirus infecting the edible fungus *Lentinula edodes*. Plant Dis 101(5):726–733. https://doi.org/10.1094/pdis-07-16-0951-re

Hannat S, Pontarotti P, Colson P et al (2021) Diverse trajectories drive the expression of a giant virus in the oomycete plant pathogen *Phytophthora parasitica*. Front Microbiol 12:662762. https://doi.org/10.3389/fmicb.2021.662762

Hao F, Wu M, Li G (2018a) Molecular characterization and geographic distribution of a Mymonavirus in the population of *Botrytis cinerea*. Viruses-Basel 10(8). https://doi.org/10.3390/v10080432

Hao FM, Ding T, Wu MD et al (2018b) Two novel hypovirulence-associated mycoviruses in the phytopathogenic fungus *Botrytis cinerea*: molecular characterization and suppression of infection cushion formation. Viruses-Basel 10(5). https://doi.org/10.3390/v10050254

Hao FM, Wu MD, Li GQ (2021) Characterization of a novel genomovirus in the phytopathogenic fungus *Botrytis cinerea*. Virology 553:111–116. https://doi.org/10.1016/j.virol.2020.11.007

Harman GE, Howell CR, Viterbo A et al (2004) *Trichoderma* species - opportunistic, avirulent plant symbionts. Nat Rev Microbiol 2:43–56. https://doi.org/10.1038/nrmicro797

Hillman BI, Suzuki N (2004) Viruses of the chestnut blight fungus, *Cryphonectria parasitica*. Adv Virus Res 63: 423–472. https://doi.org/10.1016/S0065-3527(04)63007-7

Hillman BI, Tian Y, Bedker PJ, Brown MP (1992) A North American hypovirulent isolate of the chestnut blight fungus with European isolate-related dsRNA. J Gen Virol 73:681–686. https://doi.org/10.1099/0022-1317-73-3-681

Hillman BI, Halpern BT, Brown MP (1994) A viral dsRNA element of the chestnut blight fungus with a distinct genetic organization. Virology 201(2): 241–250. https://doi.org/10.1006/viro.1994.1289

Hillman BI, Supyani S, Kondo H, Suzuki N (2004) A reovirus of the fungus *Cryphonectria parasitica* that is infectious as particles and related to the coltivirus genus of animal pathogens. J Virol 78(2):892–898. https://doi.org/10.1128/jvi.78.2.892-898.2004

Hillman BI, Aulia A, Suzuki N (2018) Viruses of plant-interacting fungi. Adv Virus Res 100:99–116. https://doi.org/10.1016/bs.aivir.2017.10.003

Hisano S, Zhang R, Faruk MI et al (2018) A neo-virus lifestyle exhibited by a (+)ssRNA virus hosted in an unrelated dsRNA virus: taxonomic and evolutionary considerations. Virus Res 244:75–83. https://doi.org/10.1016/j.virusres.2017.11.006

Hollings M (1962) Viruses associated with a die-back disease of cultivated mushroom. Nature 196:962–965. https://doi.org/10.1038/196962a0

Hu X, Roberts DP, Jiang M, Zhang Y (2005) Decreased incidence of disease caused by *Sclerotinia sclerotiorum* and improved plant vigor of oilseed rape with *Bacillus subtilis* Tu-100. Appl Microbiol Biotechnol 68(6):802–807. https://doi.org/10.1007/s00253-005-1938-x

Hu Z, Wu S, Cheng J et al (2014) Molecular characterization of two positive-strand RNA viruses co-infecting a hypovirulent strain of *Sclerotinia sclerotiorum*. Virology 464–465:450–459. https://doi.org/10.1016/j.virol.2014.07.007

Hu HJ, Wang JR, Cheng XH et al (2022) Preliminary studies on the effects of oyster mushroom spherical virus china strain on the mycelial growth and fruiting body yield of the edible mushroom *Pleurotus ostreatus*. Biol-Basel 11(4). https://doi.org/10.3390/biology11040574

Ikeda K, Nakamura H, Arakawa M, Matsumoto N (2004) Diversity and vertical transmission of double-stranded RNA elements in root rot pathogens of trees, *Helicobasidium mompa* and *Rosellinia necatrix*. Mycol Res 108:626–634. https://doi.org/10.1017/s0953756204000061

Jackson AO, Mitchell DM, Siegel A (1971) Replication of tobacco mosaic virus.1. Isolation and characterization of double-stranded forms of ribonucleic acid. Virology 45:182–191. https://doi.org/10.1016/0042-6822(71)90125-5

Janda M, Ahlquist P (1993) RNA-dependent replication, transcription, and persistence of brome mosaic-virus RNA replicons in *Saccharomyces cerevisiae*. Cell 72: 961–970. https://doi.org/10.1016/0092-8674(93)90584-d

Jia JC, Mu F, Fu YP et al (2022) A capsidless virus is trans-encapsidated by a bisegmented botybirnavirus. J Virol 96. https://doi.org/10.1128/jvi.00296-22

Jiang D, Fu Y, Guoqing L, Ghabrial SA (2013) Viruses of the plant pathogenic fungus *Sclerotinia sclerotiorum*. Adv Virus Res 86:215–248. https://doi.org/10.1016/B978-0-12-394315-6.00008-8

Jom-in S, Akarapisan A (2009) Characterization of double-stranded RNA in *Trichoderma* spp. isolates in Chiang Mai province. J Agric Technol 5:261–270

Judelson HS, Tyler BM, Michelmore RW (1991) Transformation of the oomycete pathogen, *Phytophthora infestans*. Mol Plant Microb Interact 4:602–607. https://doi.org/10.1094/mpmi-4-602

Kalac P (2013) A review of chemical composition and nutritional value of wild-growing and cultivated mushrooms. J Sci Food Agr 93:209–218. https://doi.org/10.1002/jsfa.5960

Kamaruzzaman M, He G, Wu M et al (2019) A novel partitivirus in the hypovirulent isolate QT5-19 of the plant pathogenic fungus *Botrytis cinerea*. Viruses-Basel 11(1). https://doi.org/10.3390/v11010024

Kamaruzzaman M, Lyu A, Zhang J et al (2020) Competitive saprophytic ability of the hypovirulent isolate QT5-19 of *Botrytis cinerea* and its importance in biocontrol of necrotrophic fungal pathogens. Biol Control 142. https://doi.org/10.1016/j.biocontrol.2019.104182

Kamaruzzaman M, Wang Z, Wu M et al (2021) Promotion of tomato growth by the volatiles produced by the hypovirulent strain QT5-19 of the plant gray mold fungus *Botrytis cinerea*. Microbiol Res 247. https://doi.org/10.1016/j.micres.2021.126731

Kanai Y, Kobayashi T (2021) FAST Proteins: development and use of reverse genetics systems for Reoviridae viruses. Annu Rev Virol 8:515–536. https://doi.org/10.1146/annurev-virology-091919-070225

Kanematsu S, Arakawa M, Oikawa Y et al (2004) a reovirus causes hypovirulence of *Rosellinia necatrix*. Phytopathology 94:561–568. https://doi.org/10.1094/PHYTO.2004.94.6.561

Kanhayuwa L, Kotta-Loizou I, Oezkan S et al (2015) A novel mycovirus from *Aspergillus fumigatus* contains four unique dsRNAs as its genome and is infectious as dsRNA. PNAS 112:9100–9105. https://doi.org/10.1073/pnas.1419225112

Khalifa ME, MacDiarmid RM (2021) A mechanically transmitted dna mycovirus is targeted by the defence machinery of its host, *Botrytis cinerea*. Viruses-Basel 13(7). https://doi.org/10.3390/v13071315

Kim JM, Song HY, Choi HJ et al (2015) Changes in the mycovirus (LeV) titer and viral effect on the vegetative growth of the edible mushroom *Lentinula edodes*.

Virus Res 197:8–12. https://doi.org/10.1016/j.virusres.2014.11.016

Kleinschmidt WJ (1988) Discovery of a mycovirus: an indirect route. In: Koltin Y, Leibowitz MJ (eds) Viruses of fungi and simple eukaryotes, vol 7. Marcel Dekker, New York, pp 1–14

Ko Y-H, So K-K, Chun J, Kim D-H (2021) Distinct roles of two DNA methyltransferases from *Cryphonectria parasitica* in fungal virulence, responses to hypovirus infection, and viral clearance. MBio 12(1). https://doi.org/10.1128/mBio.02890-20

Kondo H, Chiba S, Toyoda K, Suzuki N (2013a) Evidence for negative-strand RNA virus infection in fungi. Virology 435(2):201–209. https://doi.org/10.1016/j.virol.2012.10.002

Kondo H, Kanematsu S, Suzuki N (2013b) Viruses of the white root rot fungus, *Rosellinia necatrix*. Adv Virus Res 86:177–214. https://doi.org/10.1016/B978-0-12-394315-6.00007-6

Kotta-Loizou I (2021) Mycoviruses and their role in fungal pathogenesis. Curr Opin Microbiol 63:10–18. https://doi.org/10.1016/j.mib.2021.05.007

Kotta-Loizou I, Coutts RH (2017a) Mycoviruses in Aspergilli: a comprehensive review. Front Microbiol 8:1699

Kotta-Loizou I, Coutts RHA (2017b) Studies on the virome of the entomopathogenic fungus *Beauveria bassiana* reveal novel dsRNA elements and mild hypervirulence. PLoS Pathog 13(1). https://doi.org/10.1371/journal.ppat.1006183

Krupovic M, Ghabrial SA, Jiang D, Varsani A (2016) Genomoviridae: a new family of widespread single-stranded DNA viruses. Arch Virol 161:2633–2643. https://doi.org/10.1007/s00705-016-2943-3

Laraba I, McCormick SP, Vaughan MM et al (2021) Phylogenetic diversity, trichothecene potential, and pathogenicity within *Fusarium sambucinum* species complex. PLoS One 16(1):e0245037

Lee SH, Yun S-H, Chun J, Kim D-H (2017) Characterization of a novel dsRNA mycovirus of *Trichoderma atroviride* NFCF028. Arch Virol 162:1073–1077. https://doi.org/10.1007/s00705-016-3214-z

Lemus-Minor CG, Canizares MC, Garcia-Pedrajas MD, Perez-Artes E (2018) *Fusarium oxysporum* f. sp *dianthi* virus 1 accumulation is correlated with changes in virulence and other phenotypic traits of its fungal host. Phytopathology 108:957–963. https://doi.org/10.1094/phyto-06-17-0200-r

Lhoas P (1971) Infection of protoplasts from *Penicillium stoloniferum* with double-stranded RNA viruses. J Gen Virol 13:365–367. https://doi.org/10.1099/0022-1317-13-2-365

Li PF, Zhang HL, Chen XG et al (2015) Molecular characterization of a novel hypovirus from the plant pathogenic fungus *Fusarium graminearum*. Virology 481:151–160. https://doi.org/10.1016/j.virol.2015.02.047

Li H, Bian R, Liu Q et al (2019) Identification of a novel hypovirulence-inducing hypovirus from *Alternaria alternata*. Front Microbiol 10:1076. https://doi.org/10.3389/fmicb.2019.01076

Li P, Wang S, Zhang L et al (2020) A tripartite ssDNA mycovirus from a plant pathogenic fungus is infectious as cloned DNA and purified virions. Sci Adv 6:14. https://doi.org/10.1126/sciadv.aay9634

Li K, Liu DM, Pan X et al (2022) Deoxynivalenol biosynthesis in *Fusarium pseudograminearum* significantly repressed by a megabirnavirus. Toxins 14. https://doi.org/10.3390/toxins14070503

Lin Y-H, Chiba S, Tani A et al (2012) A novel quadripartite dsRNA virus isolated from a phytopathogenic filamentous fungus, *Rosellinia necatrix*. Virology 426:42–50. https://doi.org/10.1016/j.virol.2012.01.013

Lin YH, Fujita M, Chiba S et al (2019) Two novel fungal negative-strand RNA viruses related to mymonaviruses and phenuiviruses in the shiitake mushroom (*Lentinula edodes*). Virology 533:125–136. https://doi.org/10.1016/j.virol.2019.05.008

Linder-Basso D, Dynek JN, Hillman BI (2005) Genome analysis of *Cryphonectria hypovirus* 4, the most common hypovirus species in North America. Virology 337:192–203. https://doi.org/10.1016/j.virol.2005.03.038

Liu L, Xie J, Cheng J et al (2014) Fungal negative-stranded RNA virus that is related to bornaviruses and nyaviruses. PNAS 111:12205–12210. https://doi.org/10.1073/pnas.1401786111

Liu J, Wang ZR, Li C et al (2015) Evaluating genetic diversity and constructing core collections of Chinese *Lentinula edodes* cultivars using ISSR and SRAP markers. J Basic Microbiol 55:749–760. https://doi.org/10.1002/jobm.201400774

Liu S, Xie J, Cheng J et al (2016) Fungal DNA virus infects a mycophagous insect and utilizes it as a transmission vector. PNAS 113:12803–12808. https://doi.org/10.1073/pnas.1608013113

Liu C, Li M, Redda ET et al (2019a) A novel double-stranded RNA mycovirus isolated from *Trichoderma harzianum*. Virol J 16(1). https://doi.org/10.1186/s12985-019-1213-x

Liu CC, Li M, Redda ET et al (2019b) Complete nucleotide sequence of a novel mycovirus from *Trichoderma harzianum* in China. Arch Virol 164:1213–1216. https://doi.org/10.1007/s00705-019-04145-9

Macdonald WL, Fulbright DW (1991) Biological-control of chestnut blight - use and limitations of transmissible hypovirulence. Plant Dis 75:656–661

Magae Y, Sunagawa M (2010) Characterization of a mycovirus associated with the brown discoloration of edible mushroom, *Flammulina velutipes*. Virol J 7. https://doi.org/10.1186/1743-422x-7-342

Marcal S, Sousa AS, Taofiq O et al (2021) Impact of postharvest preservation methods on nutritional value and bioactive properties of mushrooms. Trends Food Sci Technol 110:418–431. https://doi.org/10.1016/j.tifs.2021.02.007

Marquez LM, Redman RS, Rodriguez RJ, Roossinck MJ (2007) A virus in a fungus in a plant: three-way symbiosis required for thermal tolerance. Science 315: 513–515. https://doi.org/10.1126/science.1136237

Marzano S-YL, Domier LL (2016) Novel mycoviruses discovered from metatranscriptomics survey of soybean phyllosphere phytobiomes. Virus Res 213:332–342

Marzano S-YL, Hobbs HA, Nelson BD et al (2015) Transfection of *Sclerotinia sclerotiorum* with in vitro transcripts of a naturally occurring interspecific recombinant of *Sclerotinia sclerotiorum* hypovirus 2 significantly reduces virulence of the fungus. J Virol 89: 5060–5071. https://doi.org/10.1128/jvi.03199-14

Marzano S-YL, Nelson BD, Ajayi-Oyetunde O et al (2016) Identification of diverse mycoviruses through metatranscriptomics characterization of the viromes of five major fungal plant pathogens. J Virol 90:6846–6863. https://doi.org/10.1128/jvi.00357-16

Milgroom MG, Cortesi P (2004) Biological control of chestnut blight with hypovirulence: a critical analysis. Annu Rev Phytopathol 42:311–338. https://doi.org/10.1146/annurev.phyto.42.040803.140325

Mochama P, Jadhav P, Neupane A, Lee Marzano SY (2018) Mycoviruses as triggers and targets of RNA silencing in white mold fungus *Sclerotinia sclerotiorum*. Viruses 10(4). https://doi.org/10.3390/v10040214

Moleleki N, van Heerden SW, Wingfield MJ et al (2003) Transfection of *Diaporthe perjuncta* with *Diaporthe* RNA virus. Appl Environ Microbiol 69:3952–3956

Morris TJ, Dodds JA (1979) Isolation and analysis of double-stranded-RNA from virus-infected plant and fungal tissue. Phytopathology 69:854–858. https://doi.org/10.1094/Phyto-69-854

Mu F, Li B, Cheng SF et al (2021) Nine viruses from eight lineages exhibiting new evolutionary modes that co-infect a hypovirulent phytopathogenic fungus. PLoS Pathog 17(8):e1009823. https://doi.org/10.1371/journal.ppat.1009823

Munkvold GP (2017) *Fusarium* species and their associated mycotoxins. Mycotoxigenic Fungi Method Protocols 1542:51–106. https://doi.org/10.1007/978-1-4939-6707-0_4

Nerva L, Ciuffo M, Vallino M et al (2015) Multiple approaches for the detection and characterization of viral and plasmid symbionts from a collection of marine fungi. Virus Res. https://doi.org/10.1016/j.virusres.2015.10.028

Nerva L, Varese GC, Falk BW, Turina M (2017) Mycoviruses of an endophytic fungus can replicate in plant cells: evolutionary implications. Sci Rep 7(1): 1908. https://doi.org/10.1038/s41598-017-02017-3

Nerva L, Chitarra W, Siciliano I et al (2018) Mycoviruses mediate mycotoxin regulation in *Aspergillus ochraceus*. Environ Microbiol. https://doi.org/10.1111/1462-2920.14436

Nerva L, Forgia M, Ciuffo M et al (2019) The mycovirome of a fungal collection from the sea cucumber *Holothuria polii*. Virus Res 273. https://doi.org/10.1016/j.virusres.2019.197737

Neumann G, Whitt MA, Kawaoka Y (2002) A decade after the generation of a negative-sense RNA virus from cloned cDNA – what have we learned? J Gen Virol 83:2635–2662. https://doi.org/10.3389/fmicb.2020.01641

Nicolás FE, Torres-Martínez S, Ruiz-Vázquez RM (2013) Loss and retention of RNA Interference in fungi and parasites. PLoS Pathog 9(1):e1003089. https://doi.org/10.1371/journal.ppat.1003089

Ninomiya A, Urayama S, Suo R et al (2020) Mycovirus-induced tenuazonic acid production in a rice blast fungus *Magnaporthe oryzae*. Front Microbiol 11. https://doi.org/10.3389/fmicb.2020.01641

Nuss DL (2005) Hypovirulence: mycoviruses at the fungal-plant interface. Nat Rev Microbiol 3(8): 632–642. https://doi.org/10.1038/nrmicro1206

Nuss DL (2011) Mycoviruses, RNA silencing, and viral RNA recombination. Adv Virus Res 80:25–48. https://doi.org/10.1016/B978-0-12-385987-7.00002-6

Oh CS (1995) Characterization of double-stranded RNA viruses and biology of *Atkinsonella hypoxylon*. PhD, Rutgers University, New Brunswick, NJ

Oh CS, Hillman BI (1995) Genome organization of a partitivirus from the filamentous ascomycete *Atkinsonella hypoxylon*. J Gen Virol 76:1461–1470. https://doi.org/10.1099/0022-1317-76-6-1461

Okada R, Ichinose S, Takeshita K et al (2018) Molecular characterization of a novel mycovirus in *Alternaria alternata* manifesting two-sided effects: down-regulation of host growth and up-regulation of host plant pathogenicity. Virology 519:23–32. https://doi.org/10.1016/j.virol.2018.03.027

Osaki H, Wei CZ, Arakawa M et al (2002) Nucleotide sequences of double-stranded RNA segments from a hypovirulent strain of the white root rot fungus *Rosellinia necatrix*: possibility of the first member of the Reoviridae from fungus. Virus Genes 25:101–107. https://doi.org/10.1023/A:1020182427439

Pearson MN, Beever RE, Boine B, Arthur K (2009) Mycoviruses of filamentous fungi and their relevance to plant pathology. Molec Plant Pathol 10:115–128. https://doi.org/10.1111/j.1364-3703.2008.00503.x

Peever TL, Liu YC, Milgroom MG (1997) Diversity of hypoviruses and other double-stranded RNAs in *Cryphonectria parasitica* in North America. Phytopathology 87:1026–1033. https://doi.org/10.1094/PHYTO.1997.87.10.1026

Peever TL, Liu YC, Wang K et al (1998) Incidence and diversity of double-stranded RNAs occurring in the chestnut blight fungus, *Cryphonectria parasitica*, in China and Japan. Phytopathology 88:811–817. https://doi.org/10.1094/PHYTO.1998.88.8.811

Picarelli M, Forgia M, Rivas EB et al (2019) Extreme diversity of mycoviruses present in isolates of *Rhizoctonia solani* AG2-2 LP from *Zoysia japonica* from Brazil. Front Cell Infect Microbiol 9. https://doi.org/10.3389/fcimb.2019.00244

Poimala A, Parikka P, Hantula J, Vainio EJ (2021) Viral diversity in *Phytophthora cactorum* population infecting strawberry. Environ Microbiol 23:5200–5221. https://doi.org/10.1111/1462-2920.15519

Polashock JJ, Hillman BI (1994) A small mitochondrial double-stranded (ds) RNA element associated with a hypovirulent strain of the chestnut blight fungus and ancestrally related to yeast cytoplasmic T and W dsRNAs. PNAS 91(18):8680–8684. https://doi.org/10.1073/pnas.91.18.8680

Polashock JJ, Bedker PJ, Hillman BI (1997) Movement of a small mitochondrial double-stranded RNA element of *Cryphonectria parasitica*: ascospore inheritance and implications for mitochondrial recombination. Mol Gen Genet 256:566–571. https://doi.org/10.1007/s004380050602

Potgieter CA, Castillo A, Castro M et al (2013) A wild-type *Botrytis cinerea* strain co-infected by double-stranded RNA mycoviruses presents hypovirulence-associated traits. Virol J 10. https://doi.org/10.1186/1743-422x-10-220

Qu Z, Fu YP, Lin Y et al (2021) Transcriptional responses of *Sclerotinia sclerotiorum* to the infection by SsHADV-1. J Fungi 7(7):493. https://doi.org/10.3390/jof7070493

Racaniello VR, Baltimore D (1981) Cloned poliovirus complementary-DNA is infectious in mammalian-cells. Science 214(4523):916–919. https://doi.org/10.1126/science.6272391

Raco M, Vainio EJ, Sutela S et al (2022) High diversity of novel viruses in the tree pathogen *Phytophthora castaneae* revealed by high-throughput sequencing of total and small RNA. Front Microbiol 13:911474. https://doi.org/10.3389/fmicb.2022.911474

Ran H, Liu L, Li B et al (2016) Co-infection of a hypovirulent isolate of *Sclerotinia sclerotiorum* with a new botybirnavirus and a strain of a mitovirus. Virol J 13:92. https://doi.org/10.1186/s12985-016-0550-2

Randall TA, Dwyer RA, Huitema E et al (2005) Large-scale gene discovery in the oomycete *Phytophthora infestans* reveals likely components of phytopathogenicity shared with true fungi. Mol Plant-Microbe Interact 18:229–243. https://doi.org/10.1094/MPMI-18-0229

Rigling D, Prospero S (2018) Cryphonectria parasitica, the causal agent of chestnut blight: invasion history, population biology and disease control. Mol Plant Pathol 19(1):7–20. https://doi.org/10.1111/mpp.12542

Roossinck MJ (2014) Metagenomics of plant and fungal viruses reveals an abundance of persistent lifestyles. Front Microbiol 5:767. https://doi.org/10.3389/fmicb.2014.00767

Ruiz-Padilla A, Rodriguez-Romero J, Gomez-Cid I et al (2021) Novel mycoviruses discovered in the mycovirome of a necrotrophic fungus. MBio 12:(3). https://doi.org/10.1128/mBio.03705-20

Sasaki A, Kanematsu S, Onoue M et al (2007) Artificial infection of *Rosellinia necatrix* with purified viral particles of a member of the genus mycoreovirus reveals its uneven distribution in single colonies. Phytopathology 97:278–286. https://doi.org/10.1094/PHYTO-97-3-0278

Sato Y, Miyazaki N, Kanematsu S et al, ICTV Report Committee (2019) ICTV virus taxonomy profile: megabirnaviridae. J Gen Virol 100:1269–1270. doi: https://doi.org/10.1099/jgv.0.001297

Sato Y, Shamsi W, Jamal A et al (2020) Hadaka virus 1: a capsidless eleven-segmented positive-sense single-stranded RNA virus from a phytopathogenic fungus, *Fusarium oxysporum*. mBio 11(3). https://doi.org/10.1128/mBio.00450-20

Schmidt FR, Lemke PA, Esser K (1986) Viral influences on aflatoxin formation by *Aspergillus flavus*. Appl Microbiol Biotech 24:248–252

Segers GC, van Wezel R, Zhang X et al (2006) Hypovirus papain-like protease p29 suppresses RNA silencing in the natural fungal host and in a heterologous plant system. Eukaryot Cell 5(6):896–904. https://doi.org/10.1128/EC.00373-05

Segers GC, Zhang X, Deng F et al (2007) Evidence that RNA silencing functions as an antiviral defense mechanism in fungi. PNAS 104(31):12902–12906. https://doi.org/10.1073/pnas.0702500104

Shah UA, Kotta-Loizou I, Fitt BDL, Coutts RHA (2020) Mycovirus-induced hypervirulence of *Leptosphaeria biglobosa* enhances systemic acquired resistance to leptosphaeria maculans in Brassica napus. Mol Plant Microbe Interact 33:98–107. https://doi.org/10.1094/mpmi-09-19-0254-r

Sharma V, Colson P, Giorgi R et al (2014) DNA-dependent RNA polymerase detects hidden giant viruses in published databanks. Genome Biol Evol 6:1603–1610. https://doi.org/10.1093/gbe/evu128

Shi L, Wang J, Quan R et al (2019) CpATG8, a homolog of yeast autophagy protein ATG8, is required for pathogenesis and hypovirus accumulation in the chestnut blight fungus. Front Cell Infect Microbiol 9. https://doi.org/10.3389/fcimb.2019.00222

Silva VN, Durigon EL, Pires MDC et al (2001) Time course of virus-like particles (VLPS) double-stranded RNA accumulation in toxigenic and non-toxigenic strains of *Aspergillus flavus*. Braz J Microbiol 32:56–60. https://doi.org/10.1590/s1517-83822001000100013

Sinden JW, Hauser E (1950) Report of two new mushroom diseases. Mushroom Sci 1:96–100

Smart CD, Yuan W, Foglia R et al (1999) Cryphonectria hypovirus 3, a virus species in the family hypoviridae with a single open reading frame. Virology 265:66–73. https://doi.org/10.1006/viro.1999.0039

Song HY, Kim N, Kim DH, Kim JM (2020) The PoV mycovirus affects extracellular enzyme expression and fruiting body yield in the oyster mushroom, *Pleurotus ostreatus*. Sci Rep 10(1). https://doi.org/10.1038/s41598-020-58016-4

Sood M, Kapoor D, Kumar V et al (2020) *Trichoderma*: The "secrets" of a multitalented biocontrol agent.

Plants-Basel 9(6). https://doi.org/10.3390/plants9060762

Stanway CA, Buck KW (1984) Infection of protoplasts of the wheat take-all fungus, *Gaeumannomyces graminis* var *tritici*, with double-stranded-RNA viruses. J Gen Virol 65:2061–2065. https://doi.org/10.1099/0022-1317-65-11-2061

Stauder CM, Nuss DL, Zhang D-X et al (2019) Enhanced hypovirus transmission by engineered super donor strains of the chestnut blight fungus, *Cryphonectria parasitica*, into a natural population of strains exhibiting diverse vegetative compatibility genotypes. Virology 528:1–6. https://doi.org/10.1016/j.virol.2018.12.007

Sun L, Suzuki N (2008) Intragenic rearrangements of a mycoreovirus induced by the multifunctional protein p29 encoded by the prototypic hypovirus CHV1-EP713. RNA 14(12):2557–2571. https://doi.org/10.1261/rna.1125408

Sun Y, Guo M, Wang J et al (2022) Curing two predominant viruses occurring in Lentinula edodes by chemotherapy and mycelial fragmentation methods. J Virol Methods 300:114370. https://doi.org/10.1016/j.jviromet.2021.114370

Sutela S, Poimala A, Vainio EJ (2019) Viruses of fungi and oomycetes in the soil environment. FEMS Microbiol Ecol 95(9). https://doi.org/10.1093/femsec/fiz119

Sutela S, Forgia M, Vainio EJ et al (2020) The virome from a collection of endomycorrhizal fungi reveals new viral taxa with unprecedented genome organization. Virus Evol 6(2). https://doi.org/10.1093/ve/veaa076

Taniguchi T, Palmieri M, Weissmann C (1978) Q-beta DNA-containing hybrid plasmids giving rise to Q-beta phage formation in bacterial host. Nature 274:223–228. https://doi.org/10.1038/274223a0

ten Hoopen GM, Krauss U (2006) Biology and control of *Rosellinia bunodes*, *Rosellinia necatrix*, and *Rosellinia pepo*: a review. Crop Prot 25:89–107

Thapa V, Roossinck MJ (2019) Determinants of coinfection in the mycoviruses. Front Cell Infect Microbiol 9. https://doi.org/10.3389/fcimb.2019.00169

Turina M, Rostagno L (2007) Virus-induced hypovirulence in *Cryphonectria parasitica*: still an unresolved conundrum. J Plant Pathol 89(2):165–178

Turina M, Prodi A, Van Alfen NK (2003) Role of the Mf1-1 pheromone precursor gene of the filamentous ascomycete *Cryphonectria parasitica*. Fungal Genet Biol 40:242–251. https://doi.org/10.1016/S1087-1845(03)00084-7

Urayama S, Katoh Y, Fukuhara T et al (2015) Rapid detection of *Magnaporthe oryzae* chrysovirus 1-A from fungal colonies on agar plates and lesions of rice blast. J Gen Plant Pathol 81:97–102. https://doi.org/10.1007/s10327-014-0567-6

Vainio EJ, Jurvansuu J, Streng J et al (2015) Diagnosis and discovery of fungal viruses using deep sequencing of small RNAs. J Gen Virol 96:714–725. https://doi.org/10.1099/jgv.0.000003

Van Alfen NK, Jaynes RA, Anagnostakis SL, Day PR (1975) Chestnut blight: Biological control by transmissible hypovirulence in *Endothia parasitica*. Science 189(4206):890–891. https://doi.org/10.1126/science.189.4206.890

VanDerLende TR, Harmsen MC, Wessels JGH (1994) Double-stranded RNAs and proteins associated with the 34 nm virus-particles of the cultivated mushroom *Agaricus bisporus*. J Gen Virol 75:2533–2536. https://doi.org/10.1099/0022-1317-75-9-2533

VanDerLende TR, Duitman EH, Gunnewijk MGW et al (1996) Functional analysis of dsRNAs (L1, L3, L5, and M2) associated with isometric 34-nm virions of *Agaricus bisporus* (white button mushroom). Virology 217:88–96. https://doi.org/10.1006/viro.1996.0096

Vetukuri RR, Avrova AO, Grenville-Briggs LJ et al (2011) Evidence for involvement of Dicer-like, Argonaute and histone deacetylase proteins in gene silencing in *Phytophthora infestans*. Mol Plant Pathol 12:772–785. https://doi.org/10.1111/j.1364-3703.2011.00710.x

Wagemans J, Holtappels D, Vainio E et al (2022) Going viral: virus-based biological control agents for plant protection. Annu Rev Phytopathol. https://doi.org/10.1146/annurev-phyto-021621-114208

Wang M, Wang Y, Sun X et al (2015) Characterization of a novel Megabirnavirus from *Sclerotinia sclerotiorum* reveals horizontal gene transfer from single-stranded RNA Virus to double-stranded RNA virus. J Virol 89:8567–8579. https://doi.org/10.1128/JVI.00243-15

Wang J, Shi L, He X et al (2016) Comparative secretome analysis reveals perturbation of host secretion pathways by a hypovirus. Sci Rep 6. https://doi.org/10.1038/srep34308

Wang Q, Mu F, Xie J et al (2020) A single ssRNA segment encoding RdRp is sufficient for replication, infection, and transmission of Ourmia-like virus in fungi. Front Microbiol 11:379. https://doi.org/10.3389/fmicb.2020.00379

Wang R, Liu C, Jiang X et al (2022) The newly identified *Trichoderma harzianum* partitivirus (THPV2) does not diminish spore production and biocontrol activity of its host. Viruses-Basel 14(7). https://doi.org/10.3390/v14071532

Wei S, Bian R, Andika IB et al (2019) Symptomatic plant viroid infections in phytopathogenic fungi. PNAS 116(26):13042–13050. https://doi.org/10.1073/pnas.1900762116

Williamson B, Tudzynsk B, Tudzynski P, van Kan JAL (2007) *Botrytis cinerea*: the cause of grey mould disease. Mol Plant Pathol 8(5):561–580. https://doi.org/10.1111/j.1364-3703.2007.00417.x

Wu M, Zhang L, Li G et al (2010) Genome characterization of a debilitation-associated mitovirus infecting the phytopathogenic fungus *Botrytis cinerea*. Virology 406:117–126. https://doi.org/10.1016/j.virol.2010.07.010

Wu M, Zhang J, Yang L, Li G (2016) RNA mycoviruses and their role in *Botrytis* biology. In: *Botrytis*–the fungus, the pathogen and its management in agricultural systems. Springer, pp 71–90

Xia ST, Xu Y, Hoy R et al (2020) The notorious soilborne pathogenic fungus *Sclerotinia sclerotiorum*: An update on genes studied with mutant analysis. Pathogens 9(1): 27. https://doi.org/10.3390/pathogens9010027

Xie J, Wei D, Jiang D et al (2006) Characterization of debilitation-associated mycovirus infecting the plant-pathogenic fungus *Sclerotinia sclerotiorum*. J Gen Virol 87:241–249. https://doi.org/10.1099/vir.0.81522-0

Xie Y, Wang ZF, Li K et al (2022) A Megabirnavirus alleviates the pathogenicity of *Fusarium pseudograminearum* to wheat. Phytopathology 112: 1175–1184. https://doi.org/10.1094/phyto-03-21-0126-r

Xu Z, Wu S, Liu L, Cheng J et al (2015) A mitovirus related to plant mitochondrial gene confers hypovirulence on the phytopathogenic fungus *Sclerotinia sclerotiorum*. Virus Res 197:127–136. https://doi.org/10.1016/j.virusres.2014.12.023

You JQ, Zhou K, Liu XL et al (2019) Defective RNA of a novel mycovirus with high transmissibility detrimental to biocontrol properties of *Trichoderma* spp. Microorganisms 7(11). https://doi.org/10.3390/microorganisms7110507

Yu HJ, Lim D, Lee HS (2003) Characterization of a novel single-stranded RNA mycovirus in *Pleurotus ostreatus*. Virology 314:9–15. https://doi.org/10.1016/s0042-6822(03)00382-9

Yu J, Kwon SJ, Lee KM et al (2009) Complete nucleotide sequence of double-stranded RNA viruses from *Fusarium graminearum* strain DK3. Arch Virol 154:1855–1858. https://doi.org/10.1007/s00705-009-0507-5

Yu X, Li B, Fu Y et al (2010) A geminivirus-related DNA mycovirus that confers hypovirulence to a plant pathogenic fungus. PNAS 107(18):8387–8392. https://doi.org/10.1073/pnas.0913535107

Yu X, Li B, Fu Y et al (2013) Extracellular transmission of a DNA mycovirus and its use as a natural fungicide. PNAS 110(4):1452–1457. https://doi.org/10.1073/pnas.1213755110

Yu L, Sang W, Wu MD et al (2015) Novel hypovirulence-associated RNA mycovirus in the plant-pathogenic fungus *Botrytis cinerea*: molecular and biological characterization. Appl Environ Microb 81(7):2299–2310. https://doi.org/10.1128/aem.03992-14

Yun S-H, Lee SH, So K-K et al (2016) Incidence of diverse dsRNA mycoviruses in *Trichoderma* spp. causing green mold disease of shiitake *Lentinula edodes*. FEMS Microbiol Lett 363(19). https://doi.org/10.1093/femsle/fnw220

Zhang L, Churchill ACL, Kazmierczak P et al (1993) Hypovirulence-associated traits induced by a mycovirus of *Cryphonectria parasitica* are mimicked by targeted inactivation of a host gene. Mol Cellular Biol 13:7782–7792

Zhang R, Hisano S, Tani A et al (2016a) A capsidless ssRNA virus hosted by an unrelated dsRNA virus. Nat Microbiol 1(1):15001. https://doi.org/10.1038/nmicrobiol.2015.1

Zhang N, Luo J, Rossman AY et al (2016b) Generic names in Magnaporthales. IMA Fungus 7:155–159. https://doi.org/10.5598/imafungus.2016.07.01.09

Zhang TT, Zeng XX, Cai XY et al (2018) Molecular characterization of a novel double-stranded RNA mycovirus of *Trichoderma asperellum* strain JLM45-3. Arch Virol 163:3433–3437. https://doi.org/10.1007/s00705-018-3988-2

Zhang H, Xie J, Fu Y et al (2020) A 2-kb mycovirus converts a pathogenic fungus into a beneficial endophyte for *Brassica* protection and yield enhancement. Mol Plant 13:1420–1433. https://doi.org/10.1016/j.molp.2020.08.016

Zhang YJ, Gao J, Li Y (2022) Diversity of mycoviruses in edible fungi. Virus Genes. https://doi.org/10.1007/s11262-022-01908-6

6 Lichens

Martin Grube

Abstract

Lichenized fungi initiate their symbiotic structures from microscopic stages after recognition of compatible algae. The partnerships ultimately emerge as complex macroscopic phenotypes which are unrivaled in the fungal kingdom by their resilience and durability. This chapter presents an overview of lichen symbioses and covers the morphology and systematics of the fungal phenotypes, as well as their associations with diverse photobionts. This is followed by a coarse overview of ecophysiology and the secondary chemistry. A special focus is given to the diversity of and the interactions with additional microorganisms. Finally, a few comments on the effects of pollution and environmental change point to the usefulness of lichens as bioindicators.

Keywords

Symbioses · Morphology · Mycobiont · Photobiont · Microbiome

6.1 Introduction

Lichens are a unique group of organisms in many ways. They are perhaps the only fungal life forms that can be recognized when looking down through the windows of a flying airplane. Lichen symbioses can develop ground-covering carpets in landscapes where plant vegetation is open or absent. In the boreal tundra, along rocky coasts, in alpine habitats, or in coastal deserts they may represent the dominating life form (Fig. 6.1). Although it is impossible to get a precise figure, and only a rough estimate claims 8% coverage of the Earth's land surface (Larson 1987), lichens contribute substantially to biosphere processes and food chains (Elbert et al. 2012; Asplund and Wardle 2017).

In earlier days of systematics, lichens were classified as a separate taxonomic group: "Lichenes," but the etymology of this term is not entirely clear. The Greek word *λειχην* (as it appears in Dioscorides' works), possibly derived from λειχω (Latin lingo), means to lick up or "eating around itself," whereas the word λιχein has the meaning of "poor" or "little." Whether Linnaeus encapsulated this association when calling lichens "rustici pauperrimi" ("peasants") of vegetation is unclear, but he may have appreciated lichens more had he known that each of them represents a pocket-sized ecosystem.

M. Grube (✉)
Institute of Biology, University of Graz, Graz, Austria
e-mail: martin.grube@uni-graz.at

Y.-P. Hsueh, M. Blackwell (eds.), *Fungal Associations*, The Mycota 9,
https://doi.org/10.1007/978-3-031-41648-4_6

Fig. 6.1 (**a**) Alpine landscape tinged by lichens. Switzerland; (**b**) large thallus (c. 1 m diam.) of *Hypogymnia physodes*, Sweden. Photographs: M. Grube

6.2 What Are Lichen Symbioses?

The recognition of lichens as a dual association between a fungus and an alga is attributed to Simon Schwendener, who initially announced his discovery in a talk for the annual general meeting at the Swiss Natural History Society in 1867, and later summarized it in his publication (Schwendener 1869). He concluded that lichen-forming fungi are "parasites, although with the wisdom of statesmen," with their algal partners as "helotes," the class of slaves in ancient Sparta. The term symbiosis (as "Symbiotismus") was later introduced to biology by Albert Bernhard Frank (1876), with lichens as an example. After vigorous rejection of the symbiotic hypothesis by many contemporary lichenologists, it took decades until lichens were classified as fungal life forms and arranged within the taxonomic system of fungi. In textbooks, they are commonly described as a partnership of one fungus with an algal partner (considering also cyanobacteria within the algal lifestyle). While this view has not changed much, evidence shows that lichens evolved as open systems and can be interpreted as miniature ecosystems, which include a variety of organisms operating at different trophic levels (Farrar 1976). A lichen can therefore be regarded as a self-sustaining ecosystem formed by the interaction of a fungus (mycobiont) and one or more microbial photosynthetic partners (photobionts) and an indeterminate number of other microscopic organisms living in association with the fungus and algae (Hawksworth and Grube 2020). In most cases, the fungus acts as the main partner to provide structure. The symbiotic phenotype results from the interaction with the microbial photosynthetic partner, required for the development of fungal sexual structures. Nevertheless, the participating fungus may grow separately under certain conditions—including axenic cultivation (and the same holds true for the algal partners). We may call the fungus a physiologically facultative biotroph, while the resulting symbiotic "lichen" phenotype results from ecologically obligate biotrophy of the lichen-forming fungus (Honegger 2012). As the classification of lichen-forming fungi is fully integrated into the fungal systematics as a whole, we consider the lichen symbiosis as a fungal lifestyle. Therefore, lichens also can be described as fungi that form self-sustained ecosystems containing an extracellular arrangement of one or more microbial photosynthetic partners and an indeterminate number of other microscopic organisms. All organisms within the symbiosis retain their independent taxonomic names, and the collective association itself has no separate name.

Various cases of algal–fungal interactions are not recognized as lichen symbioses, including cases where photoautotrophs are harbored inside fungal cells. For example, *Mortierella elongata* (Mucoromycota), which takes up algae of *Nannochloropsis oceanica* (Du et al. 2019) in its hyphae or *Geosiphon pyriforme* (Glomeromycota) containing *Nostoc* cyanobacteria in bladder-like swollen fungal cells (Schüßler and Kluge 2001). Other examples of fungal–algal associations are harder to distinguish from lichens as defined above. Loose fungal–algal associations have sometimes been dubbed "primitive" or "borderline lichens" (Kohlmeyer et al. 2004). In such cases, the fungal partners do not develop a well-differentiated cortical layer or other fungal structures considered characteristic of lichen thalli, but they develop ascomata and conidiomata following non-pathogenic growth on certain types of algae (as in e.g., *Collemopsidium* spp., *Mastodia tessellata*, *Trizodia acrobia*). Others can live either as saprotrophs on bark or as primitive lichens depending on the details of substrate conditions. This phenomenon has been named "optional lichenization" by Wedin et al. (2004). Also in these cases, characteristic thallus structures are hardly developed. Comparing the genomes of such poorly developed lichen-like life forms with fully developed lichen phenotypes could provide insights into the genomic architecture required for forming lichens.

It has often been claimed that lichens are very old fungal lifestyles. Whether fossils found in c. 600 Ma phosphorites of the Doushan Formation are primitive lichens is unclear (Yuan et al. 2005). It has been suggested that they represent fungal interactions with benthic marine cyanobacterial colonies. Using time-calibrated phylogenies of ascomycete fungi and algae, focusing on lineages with lichen symbionts, Nelsen et al. (2020) estimated ages of several interacting clades and found out that fungal origins of lichenization must have occurred soon after the emergence of land plants (tracheophytes). Clear fossil evidence for this step, which may inform hypotheses of how lichen thalli evolved from unspecific fungal–algal associations, is missing. The younger (micro-) fossils of lichens from the Lower Devonian Welsh Borderland are approximately 415 Ma old (Honegger et al. 2013). They already show leaf-like thalli with characteristic internal stratification. This level of differentiation is also found in younger fossils of lichens, some of which can even be determined to genus level (cited in Honegger 2012).

6.3 The Lichen Phenotype

The formation of more or less sealed, hydroelastic layers of vegetative mycelium appears to be a key evolutionary step for developing a lichen phenotype. Usually, the fungus tightly wraps algal colonies, which enforces functional coherence and lichen thallus integrity. A tight closure of mycelial structures is achieved when fungal hyphae glue together by their more or less gelatinizing outer cell walls, which creates a structure where cells are embedded in a joint matrix. The gelatinizing outer cell wall material has been named variably by different authors "Kittsubstanz" (cementing substance) by Peveling (1970), "gelatinous matrix" by Ahmadjian (1993), "conglutinate zone" by Honegger (1991), or "extracellular interaction matrix" (Spribille et al. 2020). The latter term included the potential roles of additional microorganisms in building this structure. Spribille et al. (2020) provided a comprehensive summary about the current state of knowledge about polysaccharides in lichens. It still needs to be determined whether proteins targeting cell wall polysaccharides could eventually play a role in the functioning of the intercellular matrix (e.g., in cell wall remodeling or as carbohydrate-binding proteins). The tissue-like functional integration of fungal plectenchyma, by conglutination by polysaccharides of the outer cell walls, is otherwise well known from the excipula of fungal sporocarps. Since such sporocarp structures are known from groups which share ancestry with those that contain complex lichen thalli (Díaz-Escandón et al. 2022), it is likely that basic ontogenetic processes evolved before lichenization and progressively adapted to form thalline cortex structures as well.

The proper formation of the extracellular "glue" to establish lichen thalli requires interaction with compatible photobionts. Resynthesis experiments have therefore been conducted to better understand the processes of interplay between algae and fungi. These experiments showed that associations with other than the original algae lead to only loose or temporary attachment of fungi (e.g., Ahmadjian et al. 1980; Schaper and Ott 2003). Resynthesis experiments in agar plates must nevertheless be interpreted with caution, because apart from the identity of the photobiont, medium composition and growth conditions also play a substantial role in the ontogeny of lichens. Mycobionts tend to develop only irregular cell clusters in standard culture media, and even co-culture with the native photobiont does not generally lead to the development of native thallus structures. As an alternative to growth on agar media, Stocker-Wörgötter (2001) used sterilized soil to successfully re-establish the native phenotype of other lichens starting with axenically cultured symbionts. However, combined with appropriate culturing conditions (e.g., under nutrient poor or changing light conditions) the formation of layered thallus structures is possible in certain species. Recently, Kono et al. (2020) managed to grow the native phenotype of *Usnea hakonensis* axenically to study upregulated sets of fungal and algal genes in a symbiotic state. They found evidence for various processes involved in symbiotic establishment, including cell wall remodeling, production of hydrophobins (which seal an apoplastic continuum between interacting cells of fungi and algae) and symbiosis-specific nutrient flow (including polyol transporters).

The required sequential process of lichenization involves five distinguishable steps: (1) the "pre-contact stage," (2) the "contact stage," (3) "envelopment" of algal cells by the fungus, (4) their "incorporation" into a pre-thallus, and (5) "differentiation" into a thallus. It was originally proposed that mycobionts recognize compatible photosynthetic partners by the shape of their cells (Ahmadjian and Jacobs 1981; "thigmotropism hypothesis"). Joneson and Lutzoni (2009) found no support for this hypothesis using glass beads as algal "dummies." Based on these results, Joneson et al. (2011) introduced a "signalling hypothesis" after they found that between 11 and 28% of mycobiont and photobiont genes, including three fungal lipases, are upregulated in the "pre-contact" and the "contact stage" prior to physical contact of the symbionts. Fungal genes upregulated in cocultures of *Cladonia grayi* with its algal partner *Asterochloris glomerata* include membrane transporters, secreted hydrolases, and small proteins, as well as a specific ribitol transporter (Armaleo et al. 2019). The latter is of relevance, as it has been shown that the polyol ribitol enhances both growth of mycobionts and their mucilage production (Meeßen et al. 2013). Interestingly, ribitol is not secreted in aposymbiotic cultures, suggesting more complex upstream signaling processes. Further research of ontogenetic processes ought to consider the role of fluctuating hydric conditions in the natural habitat, since it has been shown that such fluctuations contribute to algal cell wall remodeling (González-Hourcade et al. 2020), and possibly alter the transmission of signaling metabolites as well.

Current literature suggests that certain groups of molecules (including fungal lectins and algal cyclic peptides) are correlated with early contact of symbionts, while others (e.g., phytohormones and molecules involved in carbon exchange) are important throughout all stages of lichen synthesis. In the fully formed thallus, specialized (traditionally called "secondary") lichen metabolites and mineral nutrition (via substrate, microbial, or airborne) stabilize the thallus functionality (Pichler et al. 2023). Poorly water-soluble specialized metabolites can accumulate as substantial amounts of crystalline extracellular deposits on the surface of hyphae, particularly in the peripheric cortex where they act as light filters, among other functions.

6.4 Morphologies and Growth Types

Lichens display diverse morphological shapes and growth styles. Here the focus is given to ascomycetous lichens, with thalli that may be coarsely classified as brush-like, crust-like, or leaf-like forms (Grube and Hawksworth 2007). This is an arbitrary grouping, since there is much greater diversity when studying the details of lichen growth forms. In the following, I would prefer to primarily distinguish lichen growth types which are surface-attached (including those penetrating the substrate surface) and the remainder, which develops surface-detached forms of thalli. Basidiomycetous lichens have thalli which are constructed in simpler ways and lack the typical stratification found in many ascomycetous lichens (Oberwinkler 2012). Most could be assigned to crustose types of thalli and consist of small clusters of algal cells surrounded by fungal hyphae. In other cases (*Dictyonema*), the thallus is apparently derived from the basidiocarps, and thus superficially resembles foliose thalli (Dal-Forno et al. 2013).

6.5 Surface-Attached: Crustose Lichens

Crustose thalli (Fig. 6.2 a–e) include both, more or less mycelial forms that develop within the substrate and those with the main body mass appearing on top of the substrate, where they often develop a closed upper fungal layer (upper cortex). Species developing within the substrate, such as rocks (mostly calcareous rocks) or bark, usually lack such an upper cortex. Crustose thalli attach directly to the substrate with their entire lower surface and usually develop over or inside more or less stable substrates. Microstructure, chemical cues, mechanical parameters, and water holding capacity could be factors determining which species can colonize the substrate. Episubstratic crustose lichens develop typical growth pattern as their symbiotic structures are prone to poikilohydric conditions and recurrent shear forces of swelling and shrinking caused by changing water content. These forces commonly induce the development of fissures in the thalli, which give rise to the formation of areoles. These small thallus patches can become physiologically independent (removal of an areole does not influence the vitality of neighboring areoles). The distinct fissures between the areoles seen in dry stages usually close tightly when thalli of these types are irrigated. The formation of crustose thalli with areole formation is correlated with the development of a specifically composed cortex layer. This type of fungal layer contains the remnants of algal cell walls in the intercellular matrix between living or decaying fungal cells (phenocortex, e.g. in *Rhizocarpon geographicum*, Fig. 6.3a, c, e). Algal cell walls are pushed toward and into the upper cortex from the algal layer, where new algal cells are produced by autospore formation.

The anchoring of crustose and squamulose thalli in the substrate is accomplished either by hyphae produced over the whole lower surface (such as a rhizohyphal felt, which is produced by individual hyphae extending into the soil substrate) or by central rhizinae (when hyphae bound together in a root-like structures).

6.6 Surface-Detached Thalli: Foliose and Fruticose Lichens

While crustose lichens still contain decaying algal cell walls in the cortex, surface-detached lichens exclude algal cells from their upper cortex, and the uniform extracellular matrix apparently represents a mechanical reinforcement allowing them to grow in three dimensions away from the substrate or as flat branches over the substrate surface. Such growth strategies are found in unrelated lineages of lichenized fungi and have also evolved among closely related groups of lichens (Grube and Arup 2001). The upper cortex layer in surface-detached lichens seems more coherent because fragmentation into areoles does not occur (e.g., in *Rusavskia elegans*, Fig. 6.3b, d, f). By exclusion form the cortex, the walls of algal

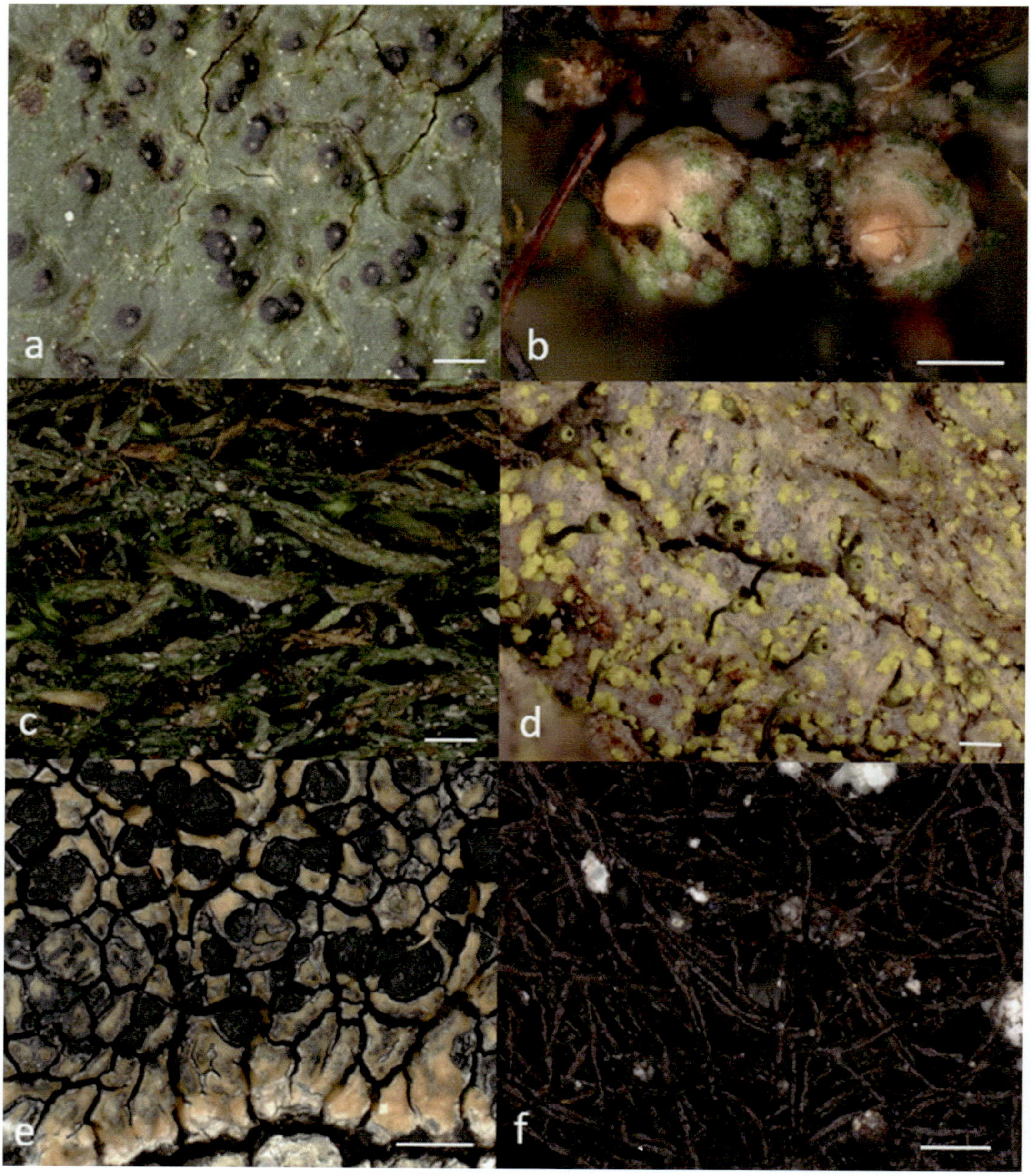

Fig. 6.2 Crustose (**a**–**e**) and microfilamentous (**f**) lichens: (**a**) *Pyrenula nitida* (Austria), bar = 500 μm; (**b**) *Leucocarpia biatorella* (Sweden), bar = 250 μm, (**c**) *Vezdaea aestivalis* (Austria), bar = 1 mm (**d**) *Chaenotheca chrysocephala* (Austria), a pin lichen, bar = 250 μm, (**e**) *Sporastatic testudinea* (Austria), bar = 1 mm, (**f**) *Cystocoleus ebeneus* (Austria), bar = 1 mm. Photographs: M. Grube

cells often keep their shape in the algal layer even after the cell content decayed, reminiscent of a sort of cell "ghosts" cells in that layer. It is still unknown whether the persistence of algal cell walls is due to loss or down-regulation of fungal carbohydrate metabolizing genes.

With increased stability and flexibility, the eucortex facilitates the functional integrity needed

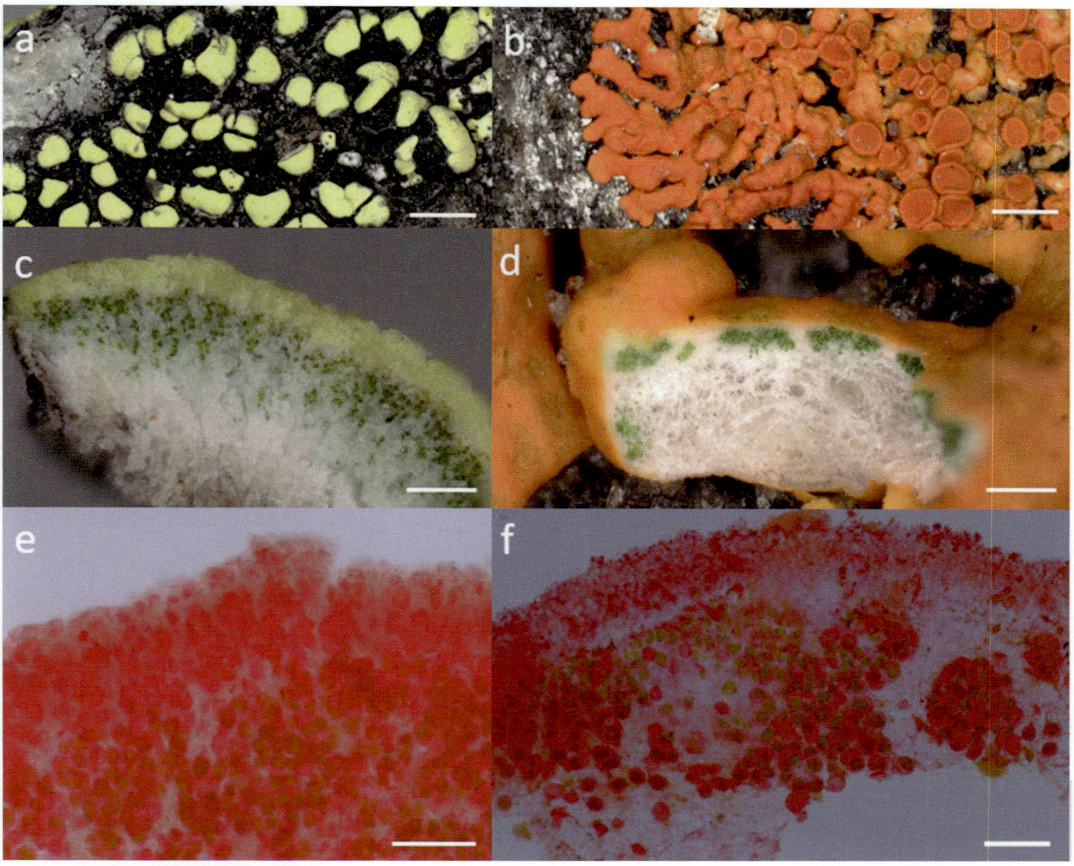

Fig. 6.3 From surface-attached to surface-detached lichens, *Rhizocarpon geographicum* (**a**: habit, **c**: section, **e**: section stained in alkaline Congo red) and *Rusavskia elegans* (**b**: habit, **d**: section, f: section stained in alkaline Congo red) *R. geographicum* is attached to the rock with the entire lower surface and has a phenocortex, whereas *R. elegans* develops a cortex on the lower side, has an eucortex on the upper side, and detaches from the surface with its lobes, **a**, **b**: bar = 1 mm, **c**: bar = 100 µm, **d**: bar = 250 µm, **e**, **f**: bar = 50 µm. Photographs: M. Grube

for development of the larger thalli of morphologically diversified foliose (leaf-like) and fruticose (shrub-like) macrolichens. Many of these lichens also have internal hydrophobic layers to facilitate gas exchange. There are also morphological intermediates between foliose and crustose growth, notably lichens that develop placodioid thalli. Their lower sides partially detach from the substrate at the thallus periphery where they also develop cortex-like plectenchyma on the lower side upper cortex of such transitional morphologies is composed of a mosaic of adjacent portions with either pheno- or eucortex."

The foliose and fruticose lichens, often called "macrolichens," develop the most complex vegetative structures in the kingdom of fungi. Leaf-like (foliose) thalli develop flattened branches of diverse shape (Fig. 6.4a). In most cases, these grow at the tips of the branches, while older parts tend to cease growth and sometimes degenerate at their rear ends (e.-g., *Arctoparmelia centrifuga*, *Brodoa intestiniformis*, and others which can form concentric rings). Foliose lichens represent the typical stratified anatomy of lichens illustrated in textbooks with an upper cortex protecting the algal layer below, a loose medulla delimited by a closed lower cortex layer, which connects to the substrate by rhizines as holdfast structures. Besides rhizines extending from the lower surface

Fig. 6.4 Foliose and terete, surface-detached lichens, (**a**) foliose "lung lichen" *Lobaria pulmonaria* (Austria), bar = 5 cm; (**b**) foliose-terete "California state lichen" *Ramalina menziesii* with typical net-like expansions (California), bar = 5 mm; (**c**) fruticose-terete lichen *Cladia retipora* with net-like growth (New Zealand), bar = 5 mm. Photographs: M. Grube

of the thalli, many other structures of taxonomic value can be formed at the periphery (e.g., cilia which are thick hairs formed by hyphal strands emanating from the thallus margins) or on the surface of thalli, ranging from hair felts to diverse structures for vegetative dispersal, such as isidia (Fig. 6.5a, b) or soredia (Fig. 6.5c, d). Also, special structures for gas exchange, such as pseudocyphellae (patches on upper or lower surfaces where loose hyphae replace the tight cortex for improved gas exchange in many genera), cyphellae (distinct holes on the lower surfaces as in *Sticta* spp.), or holes in the upper surfaces of the thallus (e.g., *Hypogymnia* spp.) are known.

Umbilicate forms (*Umbilicaria* species) represent a special case of foliose growth. Umbilicate thalli differ by being connected to the substrate with a single central, rather rigid holdfast structure, the umbilicus, which is made of strongly conglutinated fungal hyphae. Thalli of *Dermatocarpon* are often also called umbilicate, yet they rather extend laterally from a holdfast and are better called monophyllous. In contrast to most other foliose and terete lichens, the growth of the thalli in Umbilicariaceae can be more or less irregular or "diffuse," (e.g., *Lasallia pustulata*) with its patchy thallus protuberances; growth is, however, inconspicuous at the thallus tips, where thalli may break up or form lacunae and gaps. Sometimes the term intercalary is used synonymously for this type of growth, but this implies discrete non-apical growth and does not apply here.

Terete (shrubby) growth forms vary from upright, brush-like forms to pendant beard-like forms (*Usnea*). They usually lack upper and lower side of the thalli, and their thalli look rather similar from all sides. Typically, developed terete

Fig. 6.5 Vegetative dispersal units, (**a**) *Pseudevernia furfuracea* (Austria), foliose thallus surface with isidia, bar = 1 mm; (**b**) isidia detail, bar =200 μm; (**c**) *Bryoria fuscescens* (Austria), soralia on surface of terete thallus, bar = 5 mm; (**d**) soralium with tiny soredia, bar = 150 μm. Photographs: M. Grube

forms are found, e.g., in the genera *Alectoria*, *Bryoria*, *Usnea*, etc. In certain genera, the thalli may also become more leaf-like, and then the distinction between terete and foliose growth is blurred (e.g., *Ramalina*). Perhaps the most spectacular thalli in that genus are formed by *Ramalina menziesii*, occurring along the North American Pacific coast and perhaps the fastest-growing lichen species (Fig. 6.4b). It has flat branches composed of net-like units that develop by diffuse expansion of perforated tissue produced at the growing branch apex (Sanders and Ascaso 1995). Spectacular net-like growth patterns are also known from the southern hemispheric genus *Cladia* (see Fig. 6.4c). More complex thalli are known from *Cladonia* species, which are dimorphic and composed of a scaly leaf-like primary thallus and terete outgrowths called podetial.

6.7 Jelly Lichens

Lichen thalli are rather brittle when dry and flexible when wet. A special case is the jelly lichens, which become slippery and increase significantly in size when irrigated (Fig. 6.6a, b). Most of them (Collemataceae) are associated with cyanobacteria and most of their gelatinous intercellular matrix originates from the photobionts. Jelly lichens generally do not have hyphae with hydrophobic surfaces and lack internal air-filled cavities in the thalli. In contrast to typical green algal lichens, they often lack a distinct functional layering. *Leptogium* species develop a cortex of a single layer of fungal cells, but the thalli of *Collema* lack such a cortex. Although the fungal hyphae merely extend between the predominant algal polysaccharide matrix, the mycobiont has an

Fig. 6.6 Massive swelling of a jelly lichen, *Enchylium polycarpon* (Austria) with cyanobacterial photobiont (*Nostoc* sp.), when wet, (**a**) dry stage, (**b**) wet stage, bars = 1 mm. Photographs: M. Grube

influence on the overall thallus shape, which differs from free-living cyanobacterial colonies. Apart from these cyanobacterial forms, there are a few more or less jelly-like green algal lichens, but these form thin crusts on soil, mosses, or other lichens (e.g., genera *Epigloea*, *Thrombium*).

6.8 Other Growth Types: Microfilamentous, Microglobose, Leprose, and Byssoid Lichens

Apart from the types presented before, there are other forms which are composites of microscopic thallus structures. The shapes of microfilamentous lichens (Fig. 6.2f) are determined by the photobiont partners which are filamentous forms of cyanobacteria (e.g., *Scytonema* in *Pyrenothrix*, *Scytonema* in *Ephebe*) or trentepohlioid green algae (*Trentepohlia* in *Cystocoleus*, *Racodium,* or many *Coenogonium* species). Microglobose lichens are composed of globose colonies of green algae packed with a coherent, usually single layer of fungal cells (*Micarea* or *Vezdaea*). Similar surface-attached forms are also developed by certain basidiomycetous lichens (e.g., *Lichenomphalia umbellifera* and *Multiclavula mucida*). These mostly hydrophilic thalli contrast with leprose lichens, which have a woolly or powdery appearance, due to loosely interwoven and hydrophobic hyphae developing around small colonies of green algae. These lichens are often sterile and are dispersed by regular fragmentation of the thallus (which usually allows for vegetative dispersal). Thalli that generally do not fragment for dispersal but which have a hydrophobic "fluffy" internal structure without an upper cortex are called byssoid. Such thallus types are frequently found in the tropics (e.g., *Cryptothecia*, *Sagenidiopsis*, *Tania*). Occasionally they break apart into irregular clusters and are then referred to as floccose thalli.

Thallus morphology has always played an important role in lichen classification. Molecular phylogenetic analyses, however, data revealed many cases of convergent evolution of similar growth styles, either by emergence of surface-detached foliose or terete forms from different crustose ancestors within closely related lineages (Grube and Hawksworth 2007), or as similar morphologies that evolved in unrelated lineages (Grube and Kantvilas 2006; Muggia et al. 2011).

6.9 Vegetative Dispersal

Numerous lichens can propagate by specialized vegetative diaspores, which combine fungal and algal symbionts in tiny structures which may easily separate from the parental thallus. This form of propagation keeps the partners together and allows them to start thallus reproduction without a risky search for germinated fungal spores of compatible algal symbionts in the environment. Because vegetative dispersal units are heavier than fungal spores alone, they are predominantly involved in local dispersal (Walser 2004), where successful combinations of symbionts in parental thalli can easily reproduce in the same habitat.

The various types of vegetative propagules are classified by their morphological details, such as isidia (mostly pin-like propagules with terete organization), schizidia (scaly propagules with dorsoventral organization), soredia or other, more special types. Basically, soredia are formed by undifferentiated hyphae which bind a small number of algal cells, whereas many other forms have a covering fungal cortex layer and usually a larger population of algal or cyanobacterial cells. Certain propagules of foliicolous lichens, such as the stalked discs of *Phyllophiale alba*, or the shallow discs of *Chroodiscus mirificus* are unique types of vegetative propagules.

6.10 Conidium Formation

Campylidia, commonly present in Pilocarpaceae and Ectolechiaceae (but also found in Arthoniaceae and Monoblastiaceae), are hood-shaped conidiophores, which may also release algae, but there is no tight connection of the photobionts with the macroconidia (Sanders 2014). Beside the joint dispersal of symbionts there are also purely fungal forms of mitotic propagation, ranging from simple formation of conidia by hyphae at the thallus margins (e.g., thallospores, not uncommon in various crust lichens from arid zones) to more specialized types. Diverse shapes of highly specialized conidium-producing synnemata are characteristic of the Gomphillaceae (Ferraro 2004). While many Gomphillaceae also produce ascospores, there are species in various major lineages, which are only known from sterile stages (e.g., species of *Cheiromycina*, *Dictyocatenulata*, *Reichlingia*, *Sporodochiolichen*).

6.11 Lichen Mycobionts and Their Systematics

The historical development of lichen systematics has been led by methodological advances, which initiated periods of taxonomic rearrangements. In the early days of lichen taxonomy, only bare eyes and simple lenses helped to classify lichens by external characters, while compound microscopes facilitated the observation of anatomical characters—particularly spore characters—in the second half of the nineteenth century. Analyses of secondary chemistry, by thin layer chromatography in particular, were introduced in the second half of nineteenth century, while the analysis of molecular sequence data was introduced in the 1990s to achieve an even more coherent phylogenetic framework. After algal-excluding ("fungal") primers were designed (Gargas and Taylor 1992), it was also possible to use extracts from entire thalli for phylogenetic approaches, avoiding the cumbersome axenic cultivation of mycobionts. At all times, technological advancements were sparked by the hope of better understanding the relationships of lichens and achieve a taxonomy based on their evolutionary history.

The number of independent origins of lichens in the fungal kingdom was one of the prominent questions when molecular methods became popular. In early analyses with limited sampling of lichenized lineages, Gargas et al. (1995) suggested only two lichenization events in the Ascomycota and three such events in Basidiomycota. With a more extended sampling, Lutzoni et al. (2001) found that even a single lichenization event could not be excluded and suggested that important non-lichenized lineages may have had lichenized ancestors (including the molds in Eurotiales).

The total number of lichenized species is given as 19,387 species, which are classified in 39 orders of fungi (Lücking et al. 2016). The number of lichenized species in the fungal classes is unevenly distributed in the fungi. The largest classes Lecanoromycetes (15,131 species), Arthoniomycetes (1541), but also the smaller Lichinomycetes (390), Candelariomycetes (76), and Coniocybomycetes (31) contain primarily lichenized species, whereas other classes such as Eurotiomycetes (1203), Dothideomycetes (812), Agaricomycetes (172), or Sordariomycetes (1) contain a majority of non-lichenized species (Lücking et al. 2016). Recently, the Lichinomycetes have been expanded to include Candelariomycetes and Coniomycetes, as well as some previously unclassified non-lichenized lineages (Díaz-Escandón et al. 2022). The large number of lichenized species in Lecanoromycetes and Arthoniomycetes correlates positively with the phenotypic complexity of their thallus organization, which will be discussed below in greater detail. Although these classes are only distantly related and possibly evolved the lichenized habit independently, there are parallels in the construction of the symbiotic phenotype which suggest that certain general functional constraints play a role in the evolution of the lichen thallus. The similarities involve general anatomical structure and the involvement of the same secondary metabolites in these unrelated lineages.

The remainder of lichens are found in few lineages of the primarily non-lichenized basidiomycetes, in the orders Atheliales, Cantharellales, Corticiales, and Agaricales. Species diversity in this groups still seems to be underestimated. Lücking et al. (2014) provided evidence that *Dictyonema glabratum*, once thought to consist of a single species, is much more diverse than expected. They distinguished at least 126 species and predicted more than 400.

Molecular data have elucidated the large-scale evolutionary relationships of lichens, yet reconciling phylogenetic insights with taxonomy became challenging, in particular at the level of genera, which were previously circumscribed by growth form and other externally visible characters. Many were shown to be non-monophyletic, which required segregation and renaming of lineages. In addition, analyses of traditionally species-rich genera resulted in substantial taxonomic splitting. Both lead to a recent, perhaps too ambitious, proliferation of genus names. Large traditional genera, such as *Lecanora*, *Arthonia*, *Caloplaca*, or *Graphis*, are now split into dozens of smaller entities. For example, starting with a few genus names (Arup et al. 2013), the family Teloschistaceae now contains more than 100 genera.

Technological advances also influenced the way in which species of lichenized fungi are recognized. Early concepts considered variation in the thallus shapes and colors, as well as anatomical characters of the thallus and fruitbody (including spore measurements) for a phenotypic species concept. Simple histochemical differences using Lugol's solution on the thallus have also played a role in some lineages (Lugol staining variation in the ascus walls became very important for recognizing higher categories in Lecanorales by the 1980s), and various chemical spot tests are still quick indicators for recognizing species (e.g., using potassium hydroxide, hypochloride, or p-phenylenediamine). Beside phenotypic characters, which remain important to classify lichen species, chemical compounds were included in species circumscriptions since the 1960s in the framework of chemosystematics. Chemical variation, however, is not trivial to interpret, as the compounds involved have a varying degree of phylogenetic conservation in lichens. Compounds present in the upper cortex are often more conserved than those located below the algal layer. The compounds below the algal layer are often chemically related. Variation in compound presence was often considered as species-specific, or specific for subspecific taxa. However, when chemical patterns were in conflict with inference from phenotypic or molecular sequence data, classification remained controversial.

Soon after first attempts to detect genetic variation using DNA hybridization (Blum and Kashevarov 1986), PCR methodology and DNA sequencing were applied to study the relationships of lichenized fungal species. In the early studies, however, the low number of individuals sampled with only one marker (usually ITS) was not sufficient to clearly resolve species relationships. With the analysis of several independent loci, a gene genealogical approach helped to recognize species in the morphologically variable species complex of *Letharia vulpina* (Kroken and Taylor 2001). Another approach used large numbers of samples from numerous populations in a population genetic approach to draw attention to the slow genetic drift in lichens (Printzen et al. 2003). The resulting ancestral polymorphisms can be problematic for application of sequence-based species recognition (e.g., DNA barcoding). For practical reasons, the systematics of species in lichenized fungi still relies on a combination of phenotypic and genotypic characters. In a few recent cases, the analysis of whole fungal genomes used to address the relationships of closely related species has shown substantial variation across genomes and evidence for hybridization in lichen fungi (Keuler et al. 2020). Hybridization and introgression may be largely underestimated factors in shaping complexes of closely related lineages of lichen species in which the separation of clear-cut species has always been a challenge (Fernandez-Mendoza et al. 2023).

6.12 Lichen Photobionts and Their Systematics

About 90% of the lichenized fungi are associated with eukaryotic algae, and of these, almost all belong in equal numbers in the classes Trebouxiophyceae and Ulvophyceae of the Chlorophyta. Exceptions are the partners of several lichens known or suspected to be members of Chlorophyceae (Sanders and Masumoto 2021). Photobionts of the Trebouxiophyceae are usually unicellular and rarely observed in free-living stages; they are relatives of free-living unicellular algae, except for *Prasiola*, a multicellular, leafy photobiont of the borderline lichen *Mastodia tessellata*. On the other hand, photobionts belonging to Ulvophyceae generally form thread-like or branched thalli in free-living stages. The most prominent genus of Ulvophyceae is *Trentepohlia*, but delimitation of species and genera is not clear (Nelsen et al. 2011). Some of the photobionts are assigned to the closely related genus *Printzina* (Hametner et al. 2014a, b), those of foliicolous lichens are usually assigned to *Cephaleuros* and *Phycopeltis*. The growth form of these algae in lichenized stages on the phylloplane is multicellular and similar to those of nearby free-living forms. The growth type of *Trentepohlia* remains thread-like in members of the lichen genus *Coenogonium* found on bark in the understory of tropical rainforests. In other tropical species, the thread-like morphology of *Trentepohlia* may be modified by the fungus (e.g., variations between thallus and isidia in the foliicolous *Chroodiscus mirificus*, Lücking and Grube 2002). Multicellularity is otherwise more or less absent in the lichenized state in bark- or rock-inhabiting or in surface-detached lichens with *Trentepohlia* symbionts. In these cases, only few concatenated or individual cells remain as remnants of multicellularity. Notably, Resl et al. (2022) found more genes for carbohydrate-active enzymes (CAZymes) in fungal genomes of lichens with *Trentepohlia* than in those associating with unicellular green algae. Could the larger number of CAZymes be linked to an ability to gain control over the algal growth architecture?

Compared with their fungal partners, algal partners are much less diversified morphologically patterns. In their review, Sanders and Masumoto (2021) included about 50 genera of algal and cyanobacterial genera as lichen partners. Since lichen algae are largely microscopic organisms without distinctive features, phenotypic classification has always been difficult. Moreover, the algae lack sexuality and change their ultrastructure in the lichenized state with variations depending on the fungal partner. Phenotypically based classification was therefore impossible without axenic cultures under standardized conditions. Meanwhile, molecular sequence data have massively improved the knowledge about photobiont diversity. Recent work in different algal lineages estimated substantial hidden genetic diversity (e.g., Muggia et al. 2014, 2018, 2020 in *Trebouxia*, Gustavs et al. 2017 in *Coccomyxa*/*Elliptochloris*, or Grube et al. 2017, Kosecka et al. 2020, Borgato et al. 2022 in Trentepohliaceae). Nevertheless, we are still far from describing diversity with species names.

6.13 Cyanobacteria

Approximately 10% of lichens contain cyanobacteria as their primary photobiont (Rikkinen 2017). Cyanobacteria are primary symbionts in about 50 genera and 1000 species of lichens and secondary symbionts in about 20 genera and 500 species (Rai 2002). Most of the cyanobacterial lichens include strains of the genus *Nostoc*, *Rhizonema*, and to lesser extent, lineages of *Calothrix*, *Fischerella* (syn. *Stigonema*), *Scytonema*, or the unicellular *Gloeocapsa*, *Chroococcus*, *Chroococcidiopsis*, and *Anacystis*, all of which are capable of fixing nitrogen. Like eukaryotic photobionts, the cyanobacteria are also modified in the lichenized state and seem to be regulated by the fungal partner both in growth form and in the frequency of heterocysts (Hyvärinen et al. 2002). Whether unculturable strains are also unable to grow free-living in nature still needs to be studied.

6.14 Cyanobacteria as Secondary Symbionts

When occurring as secondary photobionts in addition to green algae, in the so-called tripartite lichens, cyanobacteria are often contained within specialized structures called cephalodia. Different cephalodia of the same thallus may contain the same of different strains of related cyanobacteria. Cephalodiate lichens are an example of division of labor, since the green algal photobionts produce the gross amount of photosynthates whereas the cyanobionts in the cephalodia focus on nitrogen fixation (Nash 2008), which agrees with higher heterocyst frequencies and higher rates of N_2 fixation than found in bipartite cyanolichens. In many species, the cephalodia are external and gall-like, and then more or less scattered on the upper or lower surface of the thallus. Sometimes they are hidden internally in small packets (e.g., *Lobaria* and *Sticta*), while they form a more or less closed layer beneath the green algal layer in *Solorina crocea*.

6.15 Photobiont Flexibility

For a long time, the association of a fungus with an alga was supposed to be stable in a species. Molecular studies with both crustose species (e.g., Blaha et al. 2006) and foliose species (e.g., Fernández-Mendoza et al. 2011; dal Grande et al. 2018; Garrido-Benavent et al. 2020), however, have shown that lichens are able to associate with different strains of closely related algae or cyanobacteria, respectively, depending on the niche, elevation, and geographic location. Similar habitat-adapted patterns of symbiosis have also been observed in basidiomycetous lichens (Gasulla et al. 2020). These results suggest that many widespread lichen species can adapt to local conditions by association with a locally ideal partner. Nevertheless, specificity for the photoautotrophic partner varies among species. There are also species which have high specificity for their algal partners, for example *Lobaria pulmonaria*, which always is found associated with *Dictyochloropsis reticulata*. Whether high specificity coincides with ecological specialization of lichens still needs to be explored. Škvorová et al. (2022) showed that the promiscuity of *Cladonia* mycobionts with their algae is limited by climatic factors and soil chemistry, since most mycobionts analyzed in their study could switch only between algae with similar ecological preferences. The variable extent of photobiont specificity could also influence the patterning of lichen communities (photobiont-mediated guilds; Kaasalainen et al. 2021, Peksa et al. 2022).

Although it has been suspected that an individual thallus is not uniform with respect to the photobiont (e.g., by isolation of several strains from the same lichen, Friedl 1989), the co-occurrence of two algae has been shown in the thallus by Casano et al. (2011) using a combined molecular and microscopic approach. This strategy may increase the ecological flexibility of lichen symbioses. The significance of this phenomenon remains under studied, and it is not known if this is a transient process, a passive phenomenon, or an actively regulated pattern of co-occurrence. Co-occurrence of algal strains has been studied in few species so far but has been found to occur with very different photobionts. Henskens et al. (2012) found that cyanolichens may also incorporate green algae in their thalli occasionally.

Certain lichens have the ability to form thalli with entirely different types of photobionts, the so-called photosymbiodemes. Photosymbiodemes have been long known from certain large leaf-like lichens in the family Peltigeraceae. The photobionts involve cyanobacteria (*Nostoc*) or green algae as primary photobiont (in the latter case *Nostoc* colonies may remain present in cephalodia). The thallus shapes of photosymbiodemes may look fairly similar (*Pseudocyphellaria rufovirescens*—green algal, vs. *P. murrayi*—cyanobacterial) irrespective of the involved photobiont, or completely different. Their representatives were even classified in different genera of lichens (e.g. *Lobaria amplissima*—green algal, vs. *Dendriscocaulon umhausense*—cyanobacterial). The mycobiont switches between different kingdoms of photobiont correlate with ecological preferences of the photobionts involved (usually with cyanobacterial morphs under wetter conditions) and their ability to cope with light stress. Green algal morphs are equipped with a photoprotective xanthophyll cycle that is missing in the cyanobacterial morph (Demmig-Adams et al. 1990b). Also, there are generally differences in the sexual strategy of the mycobiont depending on the photobiont type. In phycosymbiodemes involving cyanobacteria and coccoid green algae, it is usually the green algal morph which develops fungal sporocarps. A fascinating new type of photosymbiodeme was uncovered by Ertz et al. (2018), which is an example of a mycobiont associated with either members of Trebouxiophyceae or Trentepohliaceae, including a switch of its reproduction strategy and phenotypic dimorphism. *Lecanographa amylacea* (Arthoniomycetes) is sexual and lacks soredia with a trentepohlioid photobiont, whereas the same fungus can form a soredial photomorph with *Trebouxia* sp. (formerly known as *Buellia violaceofusca*). These patterns of photobiont flexibility raise fundamental questions of how partnerships and interactions are genetically controlled in different lineages of lichenized fungi and how flexible the variation might be. It also raises the question of how different photobionts influence the morphology of the lichen symbioses and to what extent this could be taxonomically relevant. Steinová et al. (2022) demonstrated that the photobiont species involved in closely related *Cladonia* species was a better marker for species phenotype than the molecular marker of the fungal symbiont, distinguishing a species without soredia from those that do. Again, the specific photobiont seems to influence the reproductive strategy of the fungal partner.

6.16 The Interface of Fungi With Algae

Complex haustorial structures found in certain fungal–plant interactions are missing in lichen fungi, and the contact of fungi with their photobionts is fairly simple; in many cases the algal cell wall is not even penetrated. The main flux of solutes in the apoplast seems to be driven by wetting and drying cycles, whereas active uptake of carbohydrates is presumable via the immediate mycobiont and photobiont contact sites (Honegger 2012).

Simpler, finger-shaped, transparietal (intracellular) haustoria are commonly formed in surface-attached (crustose) lichens, whereas surface-detached lichens commonly develop intraparietal haustoria (Honegger 2012). Three types can be distinguished: type 1 is characterized by appressoria and short infection pegs, which do not penetrate the algal cell wall. Type 2 has short infection pegs sheathed by the algal cell wall, and type 3 develops very short infection pegs with thin fungal cell walls at the contact site within the outer cell walls of still growing algal cells. Honegger (2012) noticed coordinated growth of haustoria with the developing algal cells and suggested this growth pattern shifts algal cells over short distances in the algal layer in Teloschistaceae and Parmeliaceae. Softness and more pronounced hygroscopic swelling of young thallus parts (in contrast with older parts; Grube unpublished) would fit with some flexibility of algal assortment in growing thalli. Appressoria seem more commonly developed in lichens with an eucortex and hydrophobic medullar layer. The interfaces of lichens with trentepohlioid algae have been less well studied, but early reports suggest both intraparietal and transparietal haustoria in an age-dependent pattern, suggesting that early intraparietal stages develop into transparietal haustoria in aging algal cells (Matthews et al. 1989).

Additional interaction structures can be observed in lichen interactions of basidiomycetes. Most of the fungi wrap their algal partner in small globules, but they do not develop distinctly developed algal layers. Basidiomycetes associated with the filamentous cyanobacterial genus *Rhizonema* form haustoria from the so-called mantle hyphae and may penetrate the cyanobacterial cell chains longitudinally with a central hypha (e.g., *Dictyonema*). Basidiomycetes associated with green algae wrap hyphae around clusters of algal cells and may eventually develop appressoria. The rather aggressive mycelia of *Athelia* species, which overgrow and attack a wide range of lichens and algal colonies, develop haustoria within the parasitized algal cells.

6.17 Ecophysiology of Lichens

One the most fascinating aspects of lichen biology is the physiological integration of entirely unrelated organisms, which react in a synchronized manner to environmental conditions. The algae (and cyanobacteria) are kept under fungal control in lichen thalli to sustainably produce more photosynthates than the algae alone require, and which are sufficient to sustain the entire symbiotic system. This is astonishing as the photobiont contributes a minor fraction to the total lichen biomass, sometimes less than 10%. It appears that lichens are—by evolutionary optimization—highly efficient photosynthetic "machines." Growth of the lichen symbiosis, depending on positive carbon gain, depends on a finely balanced share of carbon between the photobiont and mycobiont in excess of respiration (most of which can be attributed to the fungus). Photosynthates taken up by the fungi are called transfer sugars. In lichens with green algae, transfer sugars comprise acyclic polyalcohols such as ribitol, sorbitol, and erythritol, whereas mycobionts in association with cyanobacteria receive glucose from their partner. To finely tune this system, lichenized fungi evolved morphological adaptations to regulate access to light and water, and gas exchange for the algal partner. Widespread species with broad ecological preferences may optimize their water relations *via* anatomical variations within species (Colesie et al. 2017). Such anatomical differences can also lead to different organization of algal layers, as shown by Vondrak and Kubásek (2013). They observed that in morphs from arid or mountain regions, algal cells are organized in thick stacks (algal stacks) separated by vertical channels of likely light-transmitting fungal hyphae (fungal stacks).

Light is the energy source for photosynthesis, while water acts as an electron acceptor molecule and carbon dioxide as carbon source. Light access, as well as protection from excess light, is regulated through positioning of the photobiont beneath a protective fungal cover, and gas exchange in the layer algal beneath is facilitated by water-repelling fungal hyphae. Water transfer is also supported by various hyphal structures, such as the distinctive rhizines and veins on the lower surface of *Peltigera* species. Variations in theses phenotypic structures depend on the ecological circumstances. Unsurprisingly, similar morphologies may emerge convergently in unrelated groups of lichens. For instance, unrelated epiphytic and epiphyllous lichens in the understory of lowland tropical rainforests usually limit investment in fungal biomass under the prevailing conditions of dim light and high temperatures that challenge the balance of carbon fixation of the algae against respiration of the entire lichen. Hence, the thalli of epiphyllous lichens from different lineages (e.g., *Arthonia* and *Porina*) are sometimes confined to a single layer of hyphae running along edges of neighboring algal threads.

Lichens are known for their persistence under environmental conditions that are hostile for most other life forms. Depending on fungal species, lichens are able to survive periodic extremes of heat or cold, and notably, desiccation. The pronounced desiccation tolerance of many lichens is a key feature for surviving such extremes (Kranner et al. 2008). Likely, genomic traces evolved to correlate with desiccation tolerance. For instance, Armaleo and Chiou (2021) found a fascinating link between growth rate, desiccation tolerance, and ribosome biogenesis. Ribosome biogenesis requires splicing of group I introns present in the ribosomal RNA genes. Most introns populating lichen rDNA are unable to self-splice and their insertion into yeast ribosomal RNA genes caused the yeast strains to grow 4.4–6 times slower, and made them 40–1700 times more desiccation tolerant depending on intron position and number. Fast growth rates would indeed poorly reconcile with the effects of recurrent drying on metabolism.

The primary mechanisms of desiccation tolerance must, however, act directly at the cellular level. Desiccation tolerance may generally be considered as the ability to restore metabolism from the air-dried state, or, in a more precise definition, the capability to survive drying at relative humidities below 65% (absolute water content equal to or below 0.1 g H_2O g^{-1} dry mass and a water potential of ≤ -100 MPa; Oliver et al. 2020). The mechanisms conferring desiccation tolerance (Kranner et al. 2008; Pichler et al. 2023), although still not fully understood, are believed to comprise molecules that support "vitrification," i.e., the formation of a "glassy state." For example, non-reducing sugars may substitute for water, maintaining the spacing between and within macromolecules, avoiding molecular crowding and cellular collapse, likely in conjunction with LEA-like proteins ("Late Embryogenesis Abundant" dehydrins). Moreover, a potent antioxidant machinery appears to be essential for desiccation-tolerant organisms (Kranner et al. 2005), serving to protect from oxidative damage by reactive oxygen species produced as inevitable by-products or disrupting electron transport chains upon desiccation. Furthermore, proteins need to be protected in the dry state to retain sufficient molecular order to revive upon rehydration. Osmotic stress is ameliorated by the production of osmoprotectants acting as osmolytes. Various types of small molecules serve this purpose, including various sugars, sugar alcohols (similar or the same as the transfer sugars), and other low-molecular-weight molecules with neutral charge and low or no toxicity at high concentrations, such as certain amino acids and their derivatives, for example mycosporines, and betaines (Gostinčar et al. 2012).

Lichens can resume full metabolic activity within minutes, apparently with proteins already in place and thus not dependent on their complete resynthesis by the ribosomal machinery. Upon rehydration, at least the side groups of carbon backbones must be able to move for enzymatic activity, as is the case in a "rubbery," but not yet a "glassy" state. Transcription, translation, and other more complex processes require transition to the liquid state with full carbon backbone mobility of proteins (Candotto Carniel et al. 2021; Farrant and Hilhorst 2021). Expressed as a percentage of dry weight, lichen water contents can range from 2–15% in a dry state to 100–300% under hydrated conditions, and some cyanobacterial jelly lichens may even take up 2000%. Most of the water is kept in the strongly swollen cell walls (mostly fungal in green algal lichens or cyanobacterial in jelly lichens). Suitable levels of humidity are of profound importance to run photosynthesis. Between 70 and 150% water relative to dry weight appears to be optimal, although lichens may start with net photosynthesis at c. 20% water weight. Productivity decreases with higher amounts of water due to the low solubility of CO_2 leading to gas exchange inhibition. To lower this type of inhibition, the anatomical structure of lichens and particularly the hydrophobicity of certain internal parts of stratified lichen thalli play a significant role to ensure gas exchange for efficient photosynthesis. The hydrophobicity of fungal structures relies on secondary lichen compounds covering the surface of the fungal cell walls, as well as a hydrophobic cell wall surface layer. Being highly significant with regard to translocation of solutes between the partners, this layer covers both mycobiont and photobiont cells as it creates an apoplastic continuum for both symbionts. The sealing is achieved by the self-assembly of hydrophobins, a family of cysteine-rich surface-active proteins forming a continuous rodlet-layer on the symbiont cells. Class 1 hydrophobins have been characterized in *Xanthoria* species (Scherrer et al. 2000; Scherrer and Honegger 2003) and in *Dictyonema glabratum* (dikaryotic hyphae, three different hydrophobins; Trembley et al. 2002).

Many lichens are able to maintain photosynthesis despite wide variation in water content. Photosystem II (Φ, quantum yield) efficiency often used as a proxy of photosynthetic activity, gives a good approximation, but under certain conditions does not directly relate to actual CO_2 uptake considering alternative electron transport routes (e.g., photorespiration, Mehler reaction; Green et al. 1998). High rates of area-related photosynthesis and chlorophyll contents have been reported, in the range of those known from leaves of deciduous trees or evergreen forests (Green et al. 2008). Nevertheless, there seems to be substantial variation of rates among or within thalli of a species and with time. Generally, young thallus parts are much more active than older parts, which agrees well with patterns of microscopically assessed cell vitality and chlorophyll fluorescence (Grube, unpublished observations).

Excess light can cause photoinhibition, i.e., the decrease in photochemical efficiency experienced in response to irradiation exceeding the energy used for photosynthetic ATP and NADPH production and/or photodestruction, i.e. the light-induced damage to the photosynthetic apparatus. Sensitivity of lichens to photoinhibition is species-specific and depends on the present type of algal or cyanobacterial photobiont (Demmig-Adams et al. 1990a) and their capacity for non-photochemical quenching (NPQ), including xanthophyll cycles (Demmig-Adams et al. 1990b; García-Plazaola et al. 2012).

6.18 Nitrogen

Growth of lichens depends on the availability of nitrogen as an essential macronutrient for synthesis of proteins and nucleic acids, among other important roles. For the approximately 90% of green algal lichen species, it is assumed that fixed nitrogen is provided by various forms of deposition from the atmosphere, including aerosols but also in more robust manner by bird manure. It remains to be studied, whether surface-attached bacteria potentially contribute—either dead or alive—to the nitrogen budget of lichens.

Ten percent of lichen species live together with cyanobacteria (mostly *Nostoc*) as the primary symbiont and about 3–5% have specialized organs called cephalodia to grow cyanobacteria in addition to the primary green algal partner (tripartite lichens). Comparison with primarily cyanobacterial lichens suggested a separation of tasks in tripartite lichens. By assessing the number of heterocysts, the cells in which nitrogen fixation takes place, Hyvärinen et al. (2002) found that heterocyst frequency in bipartite lichens ranges from 2 to 8%, and in tripartite lichens between 10 and 55%. Several, mostly soil-inhabiting, green algal lichens with squamulose morphology are known to grow either facultatively or obligately near free-living blue-green algae, mainly *Stigonema*. This little studied phenomenon is known as cyanotrophy (Poelt and Mayrhofer 1988), which represents a form of ecological facilitation. Some of the species occur only on very poor, acidic rocks on *Stigonema*, whereas they may occur independent of *Stigonema* in habitats with high nutrient availability. Nitrogen fixation by cyanobacteria yields NH_4^+, which is metabolized to glutamine, part of which is then transferred to the mycobiont. The mycobiont catalyzes glutamine to glutamate and other amino acids, which can be directed to the green algal photobiont. Supplying the photobiont with higher levels of nitrogen could potentially increase the carbon gain capacity, whereas directing more nitrogen into fungal tissue may increase the rate of maintenance respiration (Lambers 1985; Reich et al. 1998). Thus, a finely tuned balance likely contributed to adjusting growth control of the photobiont and the overinvestment in fungal structures.

Nitrogen fixation requires higher levels of moisture than respiration and photosynthesis, and apparently do not start below water contents of 80% dry weight, while maximal fixation rates were found at 200-400% ODW in species with saturation at 500–800% oven dry weight (ODW) (Kershaw 1985).

Biological fixation occurs via nitrogenase, which uses the relatively rare element molybdenum as a cofactor. Alternatively, nitrogenases also may use vanadium or iron in place of molybdenum in the active site, and genes for those enzymes have been detected in the *Nostoc* symbionts of *Peltigera* species (Hodkinson et al. 2014) in addition to standard molybdenum-dependent nitrogenases. Thus, these alternative possibilities for nitrogen fixation provide flexibility of metabolic processes and play a more prominent role for lichens in habitats where molybdenum is limited. Consistent with this thought, the acquisition of vanadium is strongly regulated by the abundance of molybdenum according to Darnajoux et al. (2017), who also found evidence for the activity of alternative vanadium-based nitrogenase in the *Nostoc* cyanobiont of *Peltigera aphthosa* s.l.

6.19 Phosphorous

The requirements of controlled resource allocation in a symbiotic system extends to phosphorous, which in contrast to nitrogen may be lost by sedimentation. Lichens cope with this limitation by recycling phosphorous, as shown in *Cladonia* podetia by Hyvärinen and Crittenden (2000). Similar recycling was also shown for nitrogen in the reindeer lichen *Cladonia portentosa* by Ellis et al. (2005), who suggested this ability as a reason for the ecological success and landscape-tinging presence of *Cladonia* mats in subarctic regions. Johansson et al. (2011) showed varied effects of simulated phosphorous deposition in dependence of nitrogen deposition, as P supply can both mitigate and intensify the negative effects of excess nitrogen depending on the (epiphytic) lichen species, and apparently allocation to the symbionts.

Transcriptomic analysis of resynthesized *Usnea hakonensis* thalli showed upregulation of photobiont genes involved in photosynthate transport and mycobiont genes involved in nitrogen and phosphorus transport, when compared to isolated and cultured myco- and photobionts (Kono et al. 2020). In this species, genes for acid phosphatase and phosphate transport are upregulated in the mycobiont, but are absent in the photobiont, which instead upregulates genes for ATPase, suggesting dependence of the photobiont from fungal phosphorous supply.

6.20 Secondary Lichen Products

Lichens make the fungal world more colorful with their secondary metabolites. For practical reasons, the term secondary metabolites will be used here, although these compounds fulfill important ecological functions. The compounds, produced by the fungal partner are a metabolic sink for excess carbon but more importantly they serve in the biology of lichen symbioses as photoprotectants against intense radiation, as defense against feeders, or as factors mediating metal homeostasis and pollution tolerance of lichen thalli. For their bioactive properties they are also of interest for biotechnology, while they have long served as taxonomic characters.

Secondary metabolites are deposited as microscopic crystals in the intercellular spaces or are present as insoluble amorphous to melanin-like pigments in hyphal walls or outside of the walls (e.g., as pigment caps or as deposits between fungal cells). The crystallized compounds tend to be water-insoluble and accumulate in the thallus, where they are unevenly distributed. Some are only known as compounds present in the upper cortex. Others may be present in layers below the algal layer, either as medullary compounds or—when a lower cortex is missing—as compounds possibly shielding the lower surface against the environment or chelating ions used during substrate exploitation. The gross distribution of compounds within the thalli could be studied merely by localizing spot reactions, or by fluorescence microscopy (Kauppi and Verseghy-Patay 1990). Progress in mass spectrometry-based imaging methods now resolves metabolic patterns at high spatial and chemical resolution, which makes it possible to visualize the microscopic distribution of each individual compound (Gadea et al. 2020; Garg et al. 2016).

The most widespread lichen substances are depsides and depsidones, followed by representatives of different compound classes, such as yellow-greenish usnic acids, reddish-orange anthraquinones, and bright yellow pulvinic acids (Fig. 6.7a). Of the depsides, atranorin seems to be the most common compound turning the lichens grayish-white (present in perhaps a third of lichen species). The genomes of lichen fungi sequenced so far suggest the presence of dozens of biosynthetic pathway gene clusters, but only a few compounds are produced in a single thallus; these, however, may accumulate in massive amounts in the upper cortex (sometimes up to 30% of dry weight). Only rarely, a higher number of compounds is present in a single thallus. One notable exception is *Lecanora elixii*, which contains more than 10 compounds, most of which are xanthone derivatives. In addition to filtering different wave lengths of light, the upper cortical compounds seem to function as a light switch depending on water content: More light can pass through the cortex to the photosynthetic algal layer beneath when the crystals are more widely spaced in swollen hydrated thalli (Dietz et al. 2000). The enhanced shielding by light reflection from closely spaced crystals in dry stage may prevent the detrimental effects of photoinhibition and radical formation.

Different biochemical pathways lead to a vast diversity of more than 1000 known lichen compounds. The three most important ones are the acetyl-polymalonyl pathway (polyketides), mevalonate pathway (isoprenyls), and shikimic acid pathway (shikimates). The compounds can be classified in major classes of lichen secondary metabolites (summarized in Table 6.1). Some of the compounds can also be dimerized or are able to form heterodimers such as bisanthraquinones, dimeric xanthones, or heterodimers of xanthones with chromones (Nguyen et al. 2020; Tuong et al. 2020). Otherwise, visually detectable compounds present in low amounts (e.g., in ascomata) of many, particularly tropical lichens have poorly been explored (Fig. 6.7b, c; Yamamoto et al. 2002) and still await their structural discovery.

Major compounds can be characteristic of lineages at various taxonomic levels, rarely so at the order level (e.g., most representatives of orange colored Teloschistales produce anthraquinones), but more commonly at the genus level (e.g., the ubiquitous yellowish-green colored species in *Xanthoparmelia* or *Usnea* in Parmeliaceae). Chemistry, however, not always reconciles clearly with phylogeny, as some lineages are chemically more variable than others with respect to their main cortical compounds. Chemical variation of medullary compounds has been most commonly used in the classification of species, either by presence of single compounds or by variation in the patterns of chemically related substances (chemosyndromatic variation).

Because most of the lichen compounds belong to the class of polyketides, their corresponding building enzymes, the polyketide synthases (PKS) raised interest during the early exploration of lichenized fungal genomes. Schmitt and Lumbsch (2009) found that those genes of class 1 polyketide synthases (PKS1) in lichens coding for methylsalicylic acids (MSAS) are more closely related to similar genes from bacteria than to other fungal PKS1 genes. They concluded that horizontal gene transfer is responsible for the capability of lichenized fungi to produce these phenolic compounds. Since then, it became a challenge to find out which of the genes are responsible for the production of the compounds accumulating in the thalli. In an early attempt, Brunauer et al. (2009) characterized the transcript of a polyketide synthase gene (PKS) from the cultured mycobiont of *Xanthoria elegans*

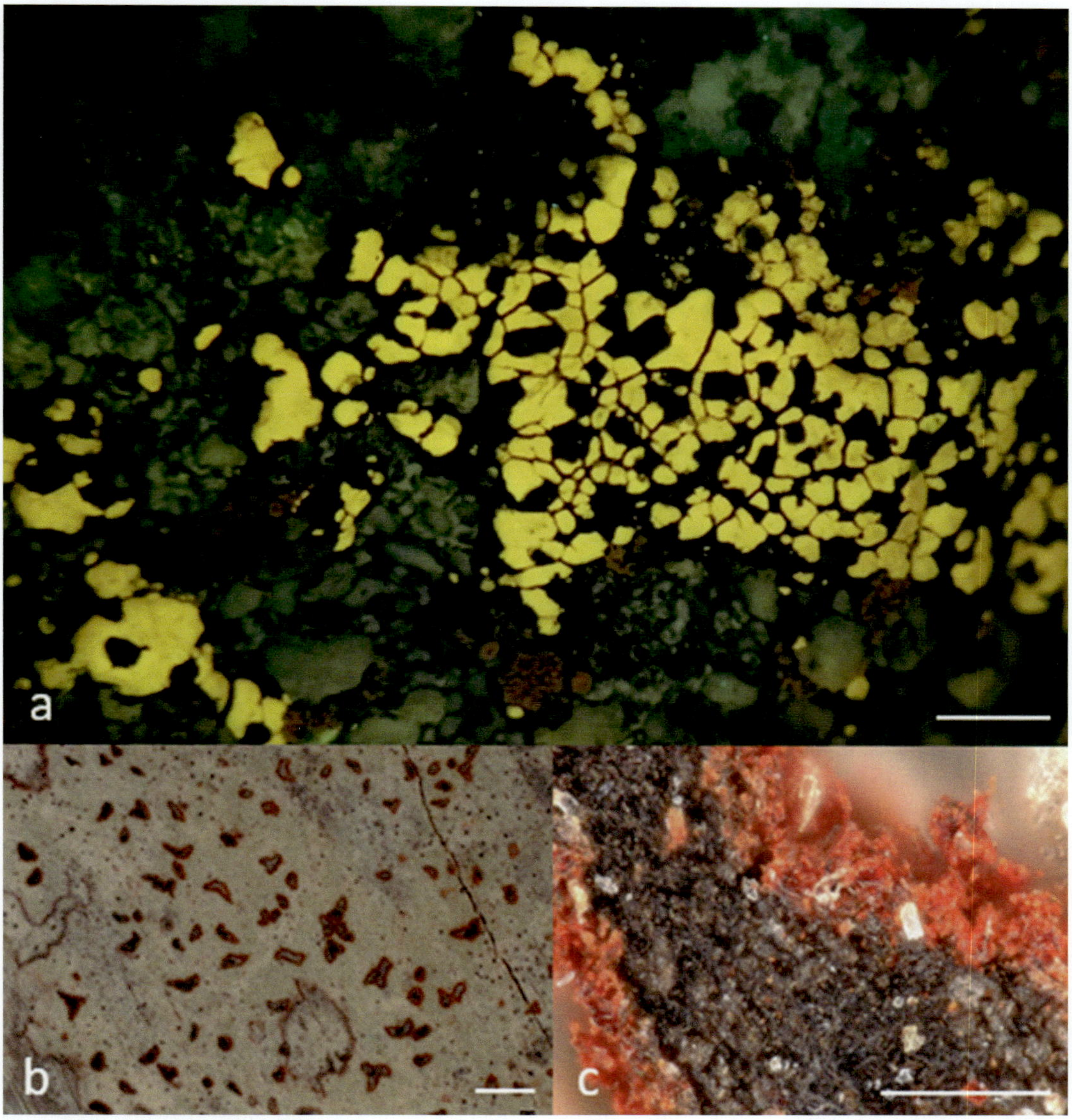

Fig. 6.7 Secondary metabolites in lichens, (**a**) Fluorescence of rhizocarpic acid in thalli of *Rhizocarpon geographicum* under illumination with UV360-light, bar = 2 mm; (**b**) *Coniocarpon rubrocinctum* fruitbodies with crimson margin, bar = 1 mm; (**c**) Crimson crystals (quinoid compounds, likely isofuranonaphthoquinones) on ascomatal margins of *C. rubrocinctum*, bar = 100 μm. Photographs: M. Grube

(XePKS1) using SMART-rapid amplification of cDNA ends (RACE) cDNA synthesis and sequencing of cloned cDNA. They chose conditions under which the cultures exclusively produced anthraquinones to suggest that the detected gene transcript was responsible for the synthesis of anthraquinones. Later, comparative analyses revealed gene clusters producing other major lichen compounds, such as grayanic acid (Armaleo et al. 2011), usnic acid (Abdel-Hameed et al. 2016), atranorin (Kim et al. 2021), or gyrophoric acid (Singh et al. 2022). With the present knowledge about the putative usnic acid biosynthesis cluster, Pizarro et al. (2020) conducted a broader analysis of genomes from Parmeliaceae and found that this gene cluster is

Table 6.1 Survey of the major classes of specialized metabolites in lichens

Acetyl-polymalonyl pathway	Secondary aliphatic acids, esters and related derivatives		
	Polyketide-derived aromatic compounds	Mononuclear phenolic compounds	
		Di- and tri-aryl derivatives of simple phenolic units	Depsides, tridepsides, and benzyl esters
			Depsidones and diphenyl ethers
			Depsones
			Dibenzofurans, usnic acids and derivatives
		Anthraquinones and biogenetically related xanthones	
		Chromones	
		Naphthaquinones	
		Xanthones	
Mevalonate pathway		Di-, sester-, and triterpenes	
		Steroids	
Shikimic acid pathway		Terphenylquinones	
		Pulvinic acid derivatives	

ancestral in Parmeliaceae. All species producing usnic acid contained this gene cluster, but all species without usnic acid apparently lost synthesizing genes quickly during evolution.

Genomic comparison indeed shows considerable gene content dynamics even at low taxonomic levels. Singh et al. (2021) detected biosynthetic gene cluster variation across *Lasallia pustulata* populations from Mediterranean and cold-temperate climate zones. Although all populations produced gyrophoric acid and to minor extent lecanoric and hiascic acid irrespective of the sampling location, the population genomic analyses indicated that *U. pustulata* contains three clusters that are highly differentiated between the Mediterranean and cold-temperate populations, with one entire cluster exclusively present and a putatively dysfunctional second cluster in cold-temperate populations. In a third gene cluster, variation is fixed in all cold-temperate populations. The authors suggested this pattern is shaped by both positive and hitchhiking selection. The pattern of expressed and detected metabolites seems to be unaffected from genomic variation in this case, but the results are of considerable interest for chemosystematics. If pathway genes are present in genomes within a species, the taxonomic significance of chemical differences needs to be critically reassessed when species are distinguished solely by such characters. However, genomic variation of biosynthetic genes not directly reflected by the expressed chemotypes may raise questions about unknown roles of the gene products and unstudied correlations with other characters.

6.21 Additional Interactions: Lichenicolous Lichens, the Lichen Microbiome, and the Lichens as Holobionts

Being widespread and slow-growing organisms with persistent thalli, lichens might be a readily available source of nutrition for associated microorganisms, primarily bacteria and fungi (Fig. 6.9a–d). Past studies about lichen-associated microorganisms have focused on the diversity of lichen-associated fungi, which have been characterized by their phenotypes, while bacterial diversity was studied more thoroughly only in the last two decades by molecular methods.

In a strict sense, lichenicolous fungi live exclusively on lichens, and they may include both, non-lichenized and lichenized fungi, which are also known as lichenicolous lichens (Fig. 6.8a, b). The distinction between these and free-living

Fig. 6.8 Examples of lichenicolous lichens, (**a**) *Rhizocarpon dinothetes* on *Protoparmelia badia* (Switzerland), (**b**) *Rhizocarpon renneri* on *Dimelaena oreina* (Switzerland), bars = 1 mm. Photographs: M. Grube

lichens can sometimes be difficult when they are parasitic on other lichens only in juvenile stages to become independent later (Hafellner 2018) and then develop faster than the initial host thallus. More commonly, lichenicolous lichens grow slowly and remain dependent on the host thallus. Lichenicolous lichens take advantage of their host's algae and may maintain them in the parasitic thallus or, alternatively, switch to another, secondarily acquired photobiont (De Los Ríos et al. 2002b; Wedin et al. 2016; Moya et al. 2020). The take-over of the already lichenized host algae by another fungus requires physiological adjustment for symbiosis with another fungus and was recently called "translichenization" (Pichler et al. 2023).

6.22 Non-lichenized Lichenicolous Fungi and More: The Fungal Microbiome

Fungi have been recognized as lichen inhabitants long before the symbiotic nature of the biology of the host has been revealed. Since the mid-nineteenth century, researchers observed high specificity of many lichenicolous (=living on/in lichens) fungi for their host lichens (Berkeley 1844; Nylander 1857). Zopf (1897) had already referred to them as an additional part of the lichen symbiosis, forming a "parasymbiosis." 2319 species of mostly obligate lichenicolous fungi have been named (Diederich et al. 2018), but their true diversity is still not known. Species may develop local necroses or brain-like outgrowths (galls; Fig. 6.9c) on thalli of their specific hosts but most other species develop more or less without symptoms. The architecture of the overall lichen structure in these cases remains as originally determined by the principal lichen mycobiont.

Since lichens are composite organisms, lichenicolous fungi can have varied affinities to the partners of lichens. Some of the species are clearly associated with the algal partners of the host, whereas others have a mycoparasitic lifestyle, or are preferentially found in the intercellular gels produced by either the fungal partner of the host, or by both fungal and algal partners (Fig. 6.10; De Los Ríos et al. 2002a). Moreover, the cortical layers of lichens are frequently inhabited by dark-septate fungi, with unknown affinity to any partner of the host.

In addition to fungi developing diagnostic spore-producing structures, many asymptomatic fungi can be cultured from crushed lichen thalli (Petrini et al. 1990). According to DNA-sequence analyses, these endolichenic fungi largely belong to families and genera also known as endophytes of plants (Tripathi and Joshi 2019). Using cultivation techniques, up to 48 different fungi were recorded from a single lichen species (U'Ren

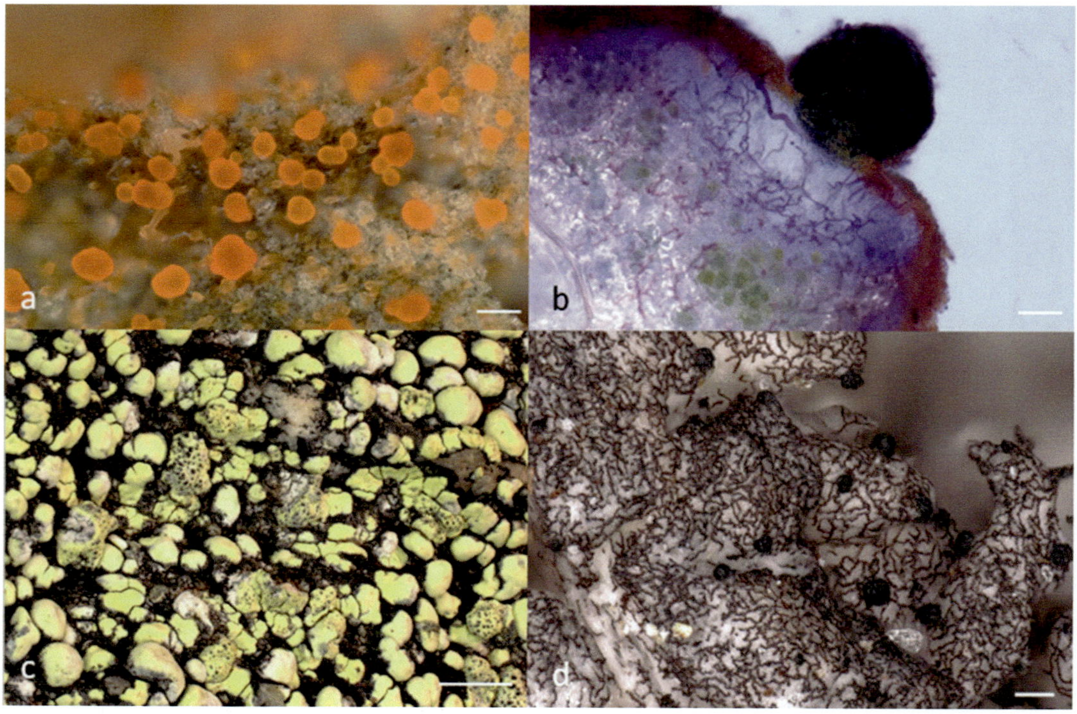

Fig. 6.9 Lichenicolous habit: (**a**) *Melittangium lichenicola* on *Xanthoria parietina*, bar = 50 μm (**b**) *Zwackhiomyces coepulonus* on *Rusavskia elegans*, bar = 50 μm (**c**) *Cercidospora cephalodiorum* on *Rhizocarpon geographicum*, bar = 2 mm (**d**) *Sphaerellothecium contextum* on *Ochrolechia frigida*, bar = 100 μm. Photographs: M. Grube

et al. 2012). Little is known about their role and abundance in their hosts, as they could include trapped resting spores or those with very limited growth, respectively. Microscopic techniques, which differentially visualize endolichenic fungi to distinguish them from the lichen mycobiont, are required to gain more insight into the roles of these associated fungi.

The idea of a unitary role of one main fungal partner in shaping lichen thalli was challenged by the discovery of basidiomycetous yeasts (genus *Cyphobasidium*, Cystobasidiomycetes, Pucciniomycotina). Spribille et al. (2016) visualized these yeasts for the first time in the thallus cortex of *Bryoria fremontii*, a pendent lichen in Parmeliaceae. Those specimens with an abundance of yeast cells had a yellow color, due to vulpinic acid in the branches. These individuals have earlier been classified as a separate species, *Bryoria fremontii*, until molecular work revealed they belong to the same species (Velmala et al. 2009). Further tests need to confirm whether the presence of yeasts or their chemistry could alter the physical properties of the cortex or correlate with long pseudocyphellae developing along infected thalli. Such differences could indicate modified patterns of gas exchange and thus a change in ecological requirements. Because Spribille et al. (2016) detected representatives of this basidiomycetous yeast lineage in a variety of other lichens (particularly Parmeliaceae) using a sensitive PCR assay with highly specific primers, they suggested an integral role for these fungi in the upper cortex of lichens. Their suggestion of a high degree of specificity of these fungi was put in perspective by Mark et al. (2020). It appears that these and other yeasts have a widespread occurrence in lichens (e.g., Fernández-Mendoz et al. 2017; Tuovinen et al.

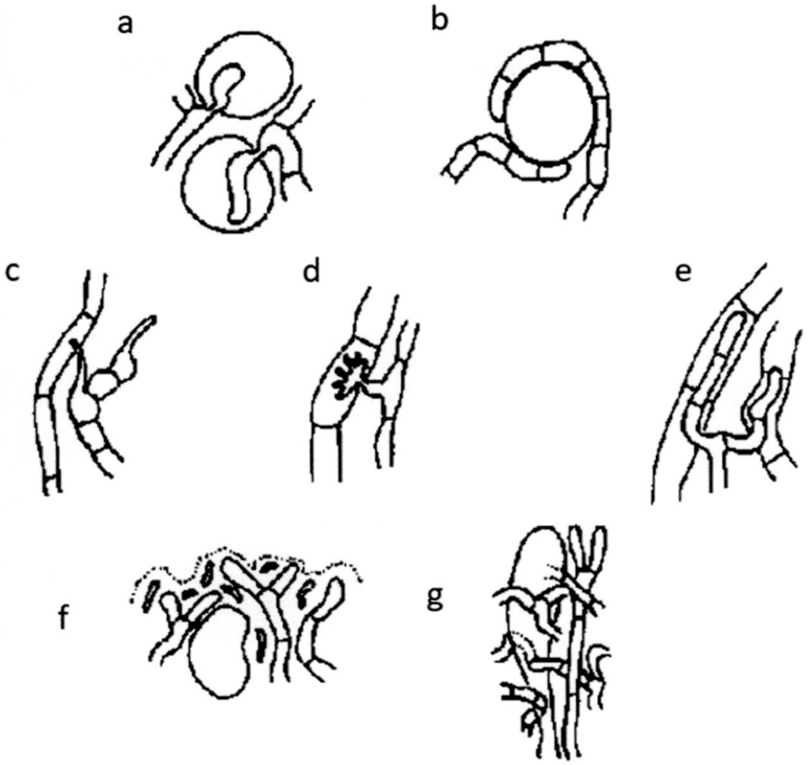

Fig. 6.10 Interaction types of lichenicolous fungi with their hosts, (**a**) haustoria formed in the algal partner of the host, (**b**) appressoria attaching to the algal partner of the host, (**c**) simple haustoria in the host hyphae, (**d**) complex haustoria in the host hyphae, (**e**) intrahyphal hyphae in the host, (**f**) exploitation of the intercellular gels of the host, (**g**) aggressive penetration of host structures. Drawings by M. Grube

2021), while their hyphal and fertile states develop only in certain host species.

Surprisingly few representatives of non-lichenized fungi on lichens are known as devastating pathogens, which suggests that the lichen biology is particularly well prepared to keep the colonizing organisms in balance. The poikilohydric lifestyle of many lichens may limit fast-growing pathogens requiring constant high humidity and instead support osmotolerant and slow-growing microbiota (while lichens may quickly mold when stored under humid conditions). Nonetheless, even in warmer habitats with continued levels of high humidity, which are better suited to promote pathogenic lifestyles, lichens seem to have control over their colonizers, either by producing bioactive compounds or because of extracellular production of reactive oxygen as a defense mechanism (Beckett et al. 2005). Osmotolerance of lichens could help to passively balance commensal associations that either do not express symptoms of disease or that produce localized infections and symptoms by species specialized on their hosts.

6.23 Bacterial Microbiome

Bacteria associated with lichens have been studied more thoroughly during the past two decades, although their presence was recognized long ago by light microscopy form dotted “sprinkles” of cells on the surface of sectioned thalli. The much larger bacterial spore-bearing structures (“fruiting bodies”) of *Melittangium lichenicola* (Fig. 6.9a; Myxococcales, Deltaproteobacteria; commonly known as myxobacteria), however, led to the earliest description of a bacterial species on lichens (Thaxter 1892). Bacteria were later isolated from lichens and cultured (e.g., Cengia Sambo 1926; Henkel and Yuzhakova 1936). Their taxonomic affiliation relied on classic microscopic methods including physiological assays of cultivated strains, while classification

Fig. 6.11 Lichen thallus structure of lung lichen *Lobaria pulmonaria* in cross-section. 3D reconstruction of FISH image stacks. Eubacteria (red) and Alphaproteobacteria (yellow) on both, the upper and the lower cortex, Betaproteobacteria (pink) being less abundant and locally contained. Fungal hyphae (blue) and algae (green) visualized by autofluorescence (from Grube et al. 2015, with permission of Springer *Nature*)

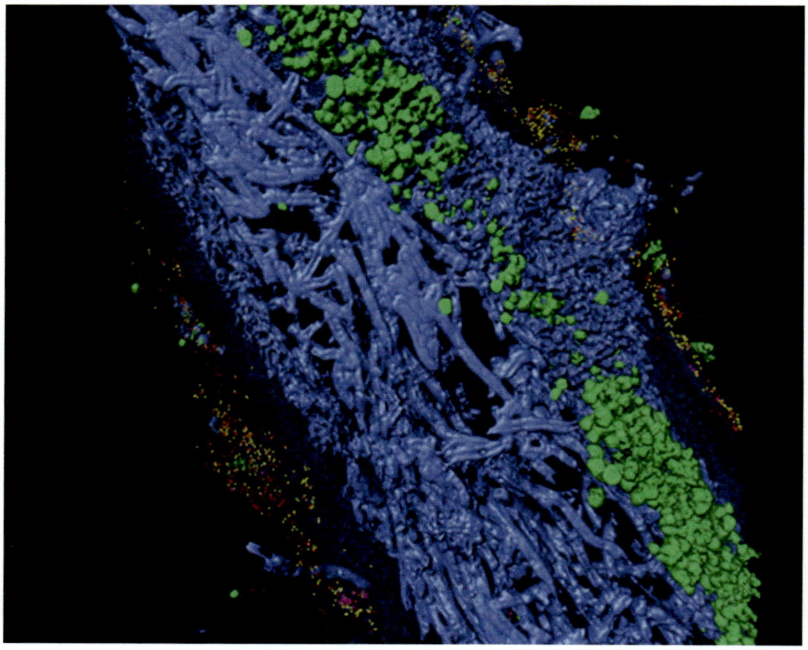

of isolated bacteria by sequence data was accomplished much later (e.g., González et al. 2005; Cardinale et al. 2006; Liba et al. 2006). Counts of bacteria prepared by Cardinale et al. (2008) revealed about 6×10^7 bacteria per gram of lichen and showed that lichens are often densely covered by bacterial colonies in a biofilm-like manner. Using *In situ* hybridization with specific fluorescent probes for ribosomal RNA they also revealed that most bacteria belonged to the Alphaproteobacteria (Fig. 6.11). The diversity and specificity of lichen-associated bacterial communities was subsequently studied in more detail by culture-independent sequencing approaches and more sophisticated-omics technologies. Such studies demonstrated that lichens are furnished with a complex host-specific bacterial microbiome (Grube et al. 2009; Bates et al. 2011; Hodkinson et al. 2012; Aschenbrenner et al. 2014; Leiva et al. 2021). Further support for the specificity of the lichen community was demonstrated in a study of a lichenicolous lichen, when the bacterial community composition shifted dramatically with the onset of the invasion by the parasitic lichen (Wedin et al. 2016). In addition, the effect of thallus age and habitat as parameters driving the composition of bacterial communities was studied by Cardinale et al. (2012), showing that the relative abundance of Alphaproteobacteria "drops" in older thallus parts.

Lichen surfaces vary dramatically in their hydrophobicity/hydrophily that correlate with the numbers of bacteria observed on them: Hydrophobic surfaces generally bear fewer bacteria than hydrophilic surfaces. Microscopic observations also suggest that samples of widespread lichens from hot and dry habitats have fewer bacteria than samples of these lichens from cool and humid habitats (Grube, unpublished).

With complementary-omics approaches, such as metagenomics and metaproteomics, a better understanding of bacterial communities and their potential functions is possible (Grube et al. 2015). Potential vitamin and hormone production, and various other functions suggest that bacteria can be functionally important components of lichen symbioses. It has been suggested that they potentially contribute to the fitness of the lichen holobiont, similar to the situation of microbiomes of plants and animals. In addition, it is suggested that bacterial metabolites could also mediate defensive functions. Direct evidence for production of bacterial defensive metabolites in the lichen thallus came from imaging mass spectrometry analyses of *Peltigera hymenina* by Garg et al. (2016), who were able to localize the production of secondary metabolites *in situ*.

According to metatranscriptomics data, the various bacterial groups also varied in their transcription patterns. Analyses of the metatranscriptome also revealed previously unrecognized groups, such as *Chthoniobacterales,* which contribute biotin and folate synthesis in *Lobaria pulmonaria* (Cernava et al. 2017). Moreover, the hundreds of different bacterial strains on a single thallus of this lichen respond to the poikilohydric conditions. Under dry conditions, upregulation of a specialized ketone metabolism indicated a switch to lipid-based nutrition (Cernava et al. 2019). With respect to the diversity of bacterial species present on lichen thalli, their varied physiological requirements suggest differential reactions under varying hygric conditions of their hosts.

6.24 Other (Eukaryotic) Colonizers of Lichens

Few reports exist on other eukaryotic organisms associated with lichens. Lakatos et al. (2004) reported on diatoms living in association with tropical representatives of the genus *Coenogonium.* Further microbial organisms seem to be regularly present in lichens, including plasmodial myxomycete amoebae, heterotrophic nanoflagellates, and naked and testate amoebae (Anderson 2014; Wilkinson et al. 2015). The influence of these organisms in the lichen symbiosis is unknown, because they seem to be transient.

6.25 Lichen-Associated Viruses

Beside microorganisms, viruses have been found in lichens, including single- (ssRNA) and double-stranded RNA (dsRNA) viruses, apparently algal origin due to their similarities with plant viruses (Petrzik et al. 2014, 2016). DsRNA viruses in the lichens *Chrysothrix chlorina* (Chrysothrix chrysovirus 1; CcCV1) and *Lepraria incana* (Lepraria chrysovirus 1; LiCV1) once classified in the genus *Alphachrysovirus*, rather have a relationship to chrysoviridae known from filamentous ascomycetous fungi (Petrzik et al. 2019). Interestingly, CcCV1 was not found in the lichen mycobiont but in the accompanying endolichenic fungus *Penicillium citreosulfuratum*. The symbiotic partners of lichens may contain several virus species independently and simultaneously: CaMV as well as the capsid protein gene of ApMV was detected in the photobiont of *Xanthoria parietina* and both, the plant cytorhabdovirus and the ApMV were found in *Usnea chaetophora* (Petrzik et al. 2014, 2015). Proteins assigned to rhabdoviruses and betaflexiviruses were found in the metaproteome of *Lobaria pulmonaria* (Eymann et al. 2017), and Grube et al. (2015) found bacteriophage sequences in the metagenome of the same lichen. The presence of bacteriophages in lichens is confirmed by occurrences of bacteriophage proteins assigned to the families Myoviridae and Siphoviridae, which infect Bacteria and Archaea (Eymann et al. 2017). Urayama et al. (2020) characterized the total dsRNA viral community of a lichen species using dsRNA-seq technology and revealed that partitiviruses were dominant and active. Sequences were classified into two genera including both plant- and fungi-infecting partitiviruses. While 17 of 65 OTUs were related to Partitiviridae, viruses related to seven dsRNA virus families (Amalgaviridae, Botybirnaviridae, Chrysoviridae, Endornaviridae, Megabirnaviridae, Picobirnaviridae, and Totiviridae), three ssRNA virus families (Gammaflexviridae, Hypoviridae, and Narnaviridae), and one unclassified RNA virus family (Polymycoviridae) were also identified from a single thallus of lichen (the host lichen species was not indicated, but possibly represents *Cladonia pyxidata* s.l. according to their Fig. 6.1).

These data suggest a high diversity of viruses associated with the different partners in lichens (i.e., mycobiont, photobiont, associated fungi, and bacteria). When Merges et al. (2021) found five novel viruses belonging to *Caulimoviridae*,

Myoviridae, *Podoviridae*, and *Siphoviridae* in metagenomic DNA of *Umbilicaria phaea*, they suggested that the Caulimovirus is associated with green algal photobionts (*Trebouxia*) of the lichen, as its abundance decreased with increasing elevation, reflecting the specific algal lineage hosting this virus. The remaining viruses were attributed to bacterial hosts, with patterns specific to the population of the host lichen. Functional annotation of sequences retrieved from cyanolichens indicate numerous viral-encoded auxiliary metabolic genes (AMGs, involved in amino acid, nucleotide, and carbohydrate metabolism, including AMGs for secondary metabolism and fatty acid biosynthesis). These results raise the question of whether viruses may modulate physiology and interactions within lichens (Ponsero et al. 2021), and thus affect the fitness of their hosts as well.

6.26 Lichens as Bioindicators of Pollution and Environmental Change

Deposition of toxic elements and molecules represents a major constraint for metabolically active lichens because they take up nutrients over their entire thallus surfaces. Hence, lichens seem generally more affected by gaseous emissions than many other organisms. Depending on the sensitivity of species, they respond to air pollution by structural changes and bleaching. Particularly damaging are oxides of sulphur and nitrogen, which affect basic metabolic processes (e.g., decreased electron flow through photosystem II, deactivation of nitrogenase, etc.) and consequently, sexual and asexual reproduction. Lichens have, therefore, been frequently used as bioindicators of air pollution (Conti 2008). A plethora of literature exists about this subject, which cannot be discussed in detail here. Amelioration from acid emissions since the 1980s led to recolonization of lichen-depleted locations, though not leading to a complete restitution of the original composition, but favorizing lichens characteristic for nutrient-rich, basic habitats (Wirth et al. 2013).

Lichens also accumulate pollutants. Particular focus was given to metal accumulation in the past, due to the interest in bioindication of toxic metals ("heavy metals"). Accumulation may occur via particulate entrapment in intercellular spaces of the thallus, by extracellular binding in the cell wall, or by intracellular uptake (Nash 2008). Various elements can be distributed differently in the thalli and accumulation also depends on other factors such as morphology, secondary metabolites, and not least on the species involved (Bačkor and Loppi 2009).

Lichens are affected by environmental change as well. A recent study revealed that climate change proceeds faster than the photobionts of lichens would be able to adapt (Nelsen et al. 2022). Cryophilic species of tundra habitats seem to retract from their southernmost occurrences. However, due to their high dependence on habitat conditions, any alteration in the lichen vegetation could result from a combination of accompanying factors as well. Lichens from exposed rock habitats seem to cope well with warmer temperatures, but more problems are expected for soil-inhabiting lichens, especially as these would be easily outcompeted by higher plant vegetation. Lichens adapted to cool and continuously humid sites are sensitive to climate change and especially to dry conditions. Many of these are also found in forests characterized by years of ecological continuity, such as old-growth forests. Unfortunately, these vulnerable forest habitats are threatened by economic interests as well. Lichen decline is not only a problem of climate change but perhaps more importantly a problem of habitat destruction and other anthropogenic impacts. For instance, lichens suffer from an excess of nitrogen deposition in areas massively influenced by agricultural activities. This impact seems responsible for the recent alterations observed in the lichen flora of both Alps (Nascimbene et al. 2019) and boreal forests (Esseen et al. 2022).

6.27 Concluding Remark

Just as lichens are a prominent example for the "living together" of unrelated organisms, the study of lichens has flourished by cooperation among researchers with different backgrounds. To gain new insights in lichen biology, research requires novel methodological approaches and a continued interest in interdisciplinarity. As such, the study of lichens will also move forward as a symbiotic activity of researchers in the future.

Acknowledgments The author thank Fernando Fernandez Mendoza (Graz) for discussions and Ilse Kranner (Innsbruck) for comments on the text.

References

Abdel-Hameed M, Bertrand RL, Piercey-Normore MD, Sorensen JL (2016) Putative identification of the usnic acid biosynthetic gene cluster by de novo whole-genome sequencing of a lichen-forming fungus. Fungal Biol 120:306–316

Ahmadjian V (1993) The lichen symbiosis. Wiley, New York

Ahmadjian V, Jacobs JB (1981) Relationship between fungus and alga in the lichen *Cladonia cristatella* Tuck. Nature 289:169–172

Ahmadjian V, Russel LA, Hildreth KC (1980) Artificial reestablishment of lichens. I. Morphological interactions between the phycobionts of different lichens and the mycobionts *Cladonia cristatella* and *Lecanora chrysoleuca*. Mycologia 72:73–89

Anderson OR (2014) Microbial communities associated with tree bark foliose lichens: a perspective on their microecology. J Eukar Microbiol 61:364–370

Armaleo D, Chiou L (2021) Modeling in yeast how rDNA introns slow growth and increase desiccation tolerance in lichens. G3 11:jkab279

Armaleo D, Sun X, Culberson C (2011) Insights from the first putative biosynthetic gene cluster for a lichen depside and depsidone. Mycologia 103:741–754

Armaleo D, Müller O, Lutzoni F, Andrésson ÓS, Blanc G, Bode HB, Collart FR, Dal Grande F, Dietrich F, Grigoriev IV, Joneson S, Kuo A, Larsen PE, Logsdon JM Jr, Lopez D, Martin F, May SP, McDonald TR, Merchant SS, Miao V, Morin E, Oono R, Pellegrini M, Rubinstein N, Sanchez-Puerta MV, Savelkoul E, Schmitt I, Slot JC, Soanes D, Szövényi P, Talbot NJ, Veneault-Fourrey C, Xavier BB (2019) The lichen symbiosis re-viewed through the genomes of *Cladonia grayi* and its algal partner *Asterochloris glomerata*. BMC Genomics 20:1–33

Arup U, Søchting U, Frödén P (2013) A new taxonomy of the family Teloschistaceae. Nordic J Bot 31:16–83

Aschenbrenner IA, Cardinale M, Berg G, Grube M (2014) Microbial cargo: do bacteria on symbiotic propagules reinforce the microbiome of lichens? Environ Microbiol 16:3743–3752

Asplund J, Wardle DA (2017) How lichens impact on terrestrial community and ecosystem properties. Biol Rev 92:1720–1738

Bačkor M, Loppi S (2009) Interactions of lichens with heavy metals. Biol Plant 53:214–222

Bates ST, Cropsey GW, Caporaso JG, Knight R, Fierer N (2011) Bacterial communities associated with the lichen symbiosis. Appl Environ Microbiol 77:1309–1314

Beckett RP, Minibayeva FV, Laufer Z (2005) Extracellular reactive oxygen species production by lichens. Lichenologist 37:397–407

Berkeley MJ (1844) Notices of British fungi. Ann Mag Nat Hist 13:340–360

Blaha J, Baloch E, Grube M (2006) High photobiont diversity associated with the euryoecious lichen-forming ascomycete Lecanora rupicola (Lecanoraceae, Ascomycota). Biol J Linn Soc 88:283–293

Blum OB, Kashevarov GP (1986) The DNA homologies as a proof of the legitimacy of the establishment of the lichen genus *Lasallia* Merat (Umbilicariaceae). Doklady Akademii Nauk Ukrainskoi SSR Seriya B 12:61–64

Borgato L, Ertz D, Van Rossum F, Verbeken A (2022) The diversity of lichenized trentepohlioid algal (Ulvophyceae) communities is driven by fungal taxonomy and ecological factors. J Phycol 58(4):582–602

Brunauer G, Muggia L, Stocker-Wörgötter E, Grube M (2009) A transcribed polyketide synthase gene from Xanthoria elegans. Mycol Res 113:82–92

Candotto Carniel F, Fernandez-Marín B, Arc E, Craighero T, Laza JM, Incerti G, Tretiach M, Kranner I (2021) How dry is dry? Molecular mobility in relation to thallus water content in a lichen. J Exp Bot 72:1576–1588

Cardinale M, Puglia AM, Grube M (2006) Molecular analysis of lichen-associated bacterial communities. FEMS Microbiol Ecol 57:484–495

Cardinale M, Vieira de Castro J Jr, Müller H, Berg G, Grube M (2008) *In situ* analysis of the bacterial community associated with the reindeer lichen *Cladonia arbuscula* reveals predominance of Alphaproteobacteria. FEMS Microbiol Ecol 66:63–71

Cardinale M, Steinová J, Rabensteiner J, Berg G, Grube M (2012) Age, sun and substrate: triggers of bacterial communities in lichens. Environ Microbiol Rep 4:23–28

Casano LM, del Campo EM, García-Breijo FJ, Reig-Armiñana J, Gasulla F, Del Hoyo A, Guéro A, Barreno E (2011) Two *Trebouxia* algae with different physiological performances are ever-present in lichen thalli of *Ramalina farinacea*. Coexistence versus competition? Environ Microbiol 13:806–818

Cengia Sambo M (1926) Ancora della polisimbiosi nei licheni ad alghe cianoficee. 1. Batteri simbionti. Atti Soc Ital Sci Nat Mus Civ Storia Nat Milano 64:191–195
Cernava T, Erlacher A, Aschenbrenner IA, Krug L, Lassek C, Riedel K, Grube M, Berg G (2017) Deciphering functional diversification within the lichen microbiota by meta-omics. Microbiome 5:1–13
Cernava T, Aschenbrenner IA, Soh J, Sensen CW, Grube M, Berg G (2019) Plasticity of a holobiont: desiccation induces fasting-like metabolism within the lichen microbiota. ISME J 13:547–556
Colesie C, Williams L, Büdel B (2017) Water relations in the soil crust lichen *Psora decipiens* are optimized via anatomical variability. Lichenologist 49:483–492
Conti ME (2008) Lichens as bioindicators of air pollution. WIT Transact State-of-the-art Sci Engin 30:111–162
Dal Grande F, Rolshausen G, Divakar PK, Crespo A, Otte J, Schleuning M, Schmitt I (2018) Environment and host identity structure communities of green algal symbionts in lichens. New Phytol 217:277–289
Dal-Forno M, Lawrey JD, Sikaroodi M, Bhattarai S, Gillevet PM, Sulzbacher M, Lücking R (2013) Starting from scratch: evolution of the lichen thallus in the basidiolichen *Dictyonema* (Agaricales: Hygrophoraceae). Fungal Biol 117:584–598
Darnajoux R, Zhang X, McRose DL, Miadlikowska J, Lutzoni F, Kraepiel AM, Bellenger JP (2017) Biological nitrogen fixation by alternative nitrogenases in boreal cyanolichens: importance of molybdenum availability and implications for current biological nitrogen fixation estimates. New Phytol 213:680–689
De Los Ríos A, Ascaso C, Grube M (2002a) Infection mechanisms of lichenicolous fungi studied by various microscopic techniques. Biblioth Lichenol 82:153–161
De Los Ríos A, Ascaso C, Grube M (2002b) An ultrastructural, anatomical and molecular study of the lichenicolous lichen *Rimularia insularis*. Mycol Res 106:946–953
Demmig-Adams B, Maguas C, Adams WW, Meyer A, Kilian E, Lange OL (1990a) Effect of high light on the efficiency of photochemical energy conversion in a variety of lichen species with green and blue-green phycobionts. Planta 180:400–409
Demmig-Adams B, Adams WW, Green TGA, Czygan FC, Lange OL (1990b) Differences in the susceptibility to light stress in two lichens forming a phycosymbiodeme, one partner possessing and one lacking the xanthophyll cycle. Oecologia 84:451–456
Díaz-Escandón D, Tagirdzhanova G, Vanderpool D, Allen CCG, Aptroot A, Češka O, Hawksworth DL, Huereca A, Knudsen K, Kocourková J, Lücking R, Resl P, Spribille T (2022) Genome-level analyses resolve an ancient lineage of symbiotic ascomycetes. Curr Biol 32:5209–5218
Diederich P, Lawrey JD, Ertz D (2018) The 2018 classification and checklist of lichenicolous fungi, with 2000 non-lichenized, obligately lichenicolous taxa. Bryologist 121:340–425
Dietz S, Büdel B, Lange OL, Bilger W (2000) Transmittance of light through the cortex of lichens from contrasting habitats. Bibl Lichenol 75:171–182
Du ZY, Zienkiewicz K, Vande Pol N, Ostrom NE, Benning C, Bonito GM (2019) Algal-fungal symbiosis leads to photosynthetic mycelium. elife 8:e47815
Elbert W, Weber B, Burrows S, Steinkamp J, Büdel B, Andreae MO, Pöschl U (2012) Contribution of cryptogamic covers to the global cycles of carbon and nitrogen. Nat Geosci 5:459–462
Ellis CJ, Crittenden PD, Scrimgeour CM, Ashcroft CJ (2005) Translocation of 15^{N} indicates nitrogen recycling in the mat-forming lichen *Cladonia portentosa*. New Phytol 168:423–434
Ertz D, Guzow-Krzemińska B, Thor G, Łubek A, Kukwa M (2018) Photobiont switching causes changes in the reproduction strategy and phenotypic dimorphism in the Arthoniomycetes. Sci Rep 8:1–14
Esseen PA, Ekström M, Grafström JBG, Palmquist K, Westerlund B, Stahl G (2022) Multiple drivers of large-scale decline in boreal forest canopies. Glob Change Biol 28:3293–3309
Eymann C, Lassek C, Wegner U, Bernhardt J, Fritsch OA, Fuch S, Otto A, Albrecht D, Schiefelbein U, Cernava T, Aschenbrenner I, Berg G, Grube M, Riedel K (2017) Symbiotic interplay of fungi, algae, and bacteria within the lung lichen *Lobaria pulmonaria* L. Hoffm. as assessed by state-of-the-art metaproteomics. J Prot Res 16:2160–2173
Farrant JM, Hilhorst HW (2021) What is dry? Exploring metabolism and molecular mobility at extremely low water contents. J Exp Bot 72:1507–1510
Farrar JF (1976) The lichen as an ecosystem: observation and experiment. In: Brown DH, Hawksworth DL, Bailey RH (eds) Lichenology: progress and problems. Academic, London, pp 385–406
Fernández-Mendoz F, Fleischhacker A, Kopun T, Grube M, Muggia L (2017) ITS 1 metabarcoding highlights low specificity of lichen mycobiomes at a local scale. Mol Ecol 26:4811–4830
Fernández-Mendoza F, Domaschke S, García MA, Jordan P, Martín MP, Printzen C (2011) Population structure of mycobionts and photobionts of the widespread lichen *Cetraria aculeata*. Mol Ecol 20:1208–1232
Fernandez-Mendoza F, Strasser E, Frolov I, Vondrak J, Muggia L, Mayrhofer H, Gaya E, Grube M (2023) Introgressive descent and hypersexuality drive the evolution of sexual parasitism and morphological reduction in a fungal species complex. bioRxiv, 2023-01
Ferraro LI (2004) Morphological diversity in the hyphophores of Gomphillaceae (Ostropales, lichenized Ascomycetes). Fungal Div 15:153–169
Frank AB (1876) Über die biologischen Verhältnisse des Thallus einiger Krustenflechten. Cohn, Beitr Biol Pflanzen 2:123–200
Friedl T (1989) Systematik und Biologie von *Trebouxia* (Microthamniales, Chlorophyta) als Phycobiont der

Parmeliaceae (lichenisierte Ascomyceten). 218p. Doctoral thesis, Universität Bayreuth

Gadea A, Fanuel M, Lamer A-CL, Boustie J, Rogniaux H, Charrier M, Devehat FL-L (2020) Mass spectrometry imaging of specialized metabolites for predicting lichen fitness and snail foraging. Plan Theory 9:70

García-Plazaola JI, Esteban R, Fernández-Marín B, Kranner I, Porcar-Castell A (2012) Thermal energy dissipation and xanthophyll cycles beyond the *Arabidopsis* model. Photosynth Res 113:89–103

Garg N, Zeng Y, Edlund A, Melnik AV, Sanchez LM, Mohimani H, Gurevich A, Miao V, Schiffler S, Lim YW, Luzzatto-Knaan T, Cai S, Rohwer F, Pevzner PA, Cichewicz RH, Alexandrov T, Dorrestein PC (2016) Spatial molecular architecture of the microbial community of a *Peltigera* lichen. mSystems 1:e00139–e00116

Gargas A, Taylor JW (1992) Polymerase chain reaction (PCR) primers for amplifying and sequencing nuclear 18s rDNA from lichenized fungi. Mycologia 84:589–592

Gargas A, DePriest PT, Grube M, Tehler A (1995) Multiple origins of lichen symbioses in fungi suggested by SSU rDNA phylogeny. Science 268:1492–1495

Garrido-Benavent I, Pérez-Ortega S, de los Ríos A, Fernández-Mendoza F (2020) Amphitropical variation of the algal partners of *Pseudephebe* (Parmeliaceae, lichenized fungi). Symbiosis 82:35–48

Gasulla F, Barrasa JM, Casano LM, del Campo EM (2020) Symbiont composition of the basidiolichen *Lichenomphalia meridionalis* varies with altitude in the Iberian Peninsula. Lichenologist 52:17–26

González I, Ayuso-Sacido A, Anderson A, Genilloud O (2005) Actinomycetes isolated from lichens: evaluation of their diversity and detection of biosynthetic gene sequences. FEMS Microbiol Ecol 54:401–415

González-Hourcade M, Braga MR, Del Campo EM, Ascaso C, Patiño C, Casano LM (2020) Ultrastructural and biochemical analyses reveal cell wall remodelling in lichen-forming microalgae submitted to cyclic desiccation–rehydration. Ann Bot 125:459–469

Gostinčar C, Muggia L, Grube M (2012) Polyextremotolerant black fungi: oligotrophism, adaptive potential, and a link to lichen symbioses. Front Microbiol 3:390

Green TGA, Schroeter B, Kappen L, Seppelt RD, Maseyk K (1998) An assessment of the relationship between chlorophyll a fluorescence and CO2 gas exchange from field measurements on a moss and lichen. Planta 206: 611–618

Green TGA, Nash TH, Lange OL (2008) Physiological ecology of carbon dioxide exchange. In: Nash TH (ed) Lichen biology. Cambridge University Press, New York, pp 152–181

Grube M, Arup U (2001) Molecular and morphological evolution in the Physciaceae (Lecanorales, lichenized Ascomycotina), with special emphasis on the genus *Rinodina*. Lichenologist 33:63–72

Grube M, Hawksworth DL (2007) Trouble with lichen: the re-evaluation and re-interpretation of thallus form and fruit body types in the molecular era. Mycol Res 111: 1116–1132

Grube M, Kantvilas G (2006) *Siphula* represents a remarkable case of morphological convergence in sterile lichens. Lichenologist 38:241–249

Grube M, Cardinale M, de Castro JV, Müller H, Berg G (2009) Species-specific structural and functional diversity of bacterial communities in lichen symbioses. ISME J 3:1105–1115

Grube M, Cernava T, Soh J, Fuchs S, Aschenbrenner I, Lassek C, Wegner U, Becher D, Riedel K, Sensen CW, Berg G (2015) Exploring functional contexts of symbiotic sustain within lichen-associated bacteria by comparative omics. ISME J 9:412–424

Grube M, Muggia L, Baloch E, Hametner C, Stocker-Wörgötter E (2017) Symbioses of lichen-forming fungi with Trentepohlialean algae. In: Grube M, Seckbach J, Muggia L (eds) Algal and cyanobacteria symbioses. World Scientific, Hackensack, NJ, pp 85–110

Gustavs L, Schiefelbein U, Darienko PT (2017) Symbioses of the green algal genera *Coccomyxa* and *Elliptochloris* (Trebouxiophyceae, Chlorophyta). In: Grube M, Seckbach J, Muggia L (eds) Algal and cyanobacteria symbioses. World Scientific, Hackensack, NJ, pp 169–208

Hafellner J (2018) Focus on lichenicolous fungi: Diversity and taxonomy under the principle "one fungus – one name". In Blanz P (ed) Biodiversity and ecology of fungi, lichens, and mosses, Austrian Academy of Sciences Biosystematics and Ecology Series. 34. Austrian Academy of Sciences, Vienna, pp 227–244

Hametner C, Stocker-Wörgötter E, Grube M (2014a) New insights into diversity and selectivity of trentepohlialean lichen photobionts from the extratropics. Symbiosis 63:31–40

Hametner C, Stocker-Wörgötter E, Rindi F, Grube M (2014b) Phylogenetic position and morphology of lichenized Trentepohliales (Ulvophyceae, Chlorophyta) from selected species of Graphidaceae. Phycol Res 62:170–186

Hawksworth DL, Grube M (2020) Lichens redefined as complex ecosystems. New Phytol 227:1281

Henkel PA, Yuzhakova LA (1936) Azotfiksiruyuschie bakterii v lishaynikah. Izv Biol Inst Permsk Gos Univ 10:9–10

Henskens FL, Green TA, Wilkins A (2012) Cyanolichens can have both cyanobacteria and green algae in a common layer as major contributors to photosynthesis. Ann Bot 110:555–563

Hodkinson BP, Gottel NR, Schadt CW, Lutzoni F (2012) Photoautotrophic symbiont and geography are major factors affecting highly structured and diverse bacterial communities in the lichen microbiome. Environ Microbiol 14:147–161

Hodkinson BP, Allen JL, Forrest LL, Goffinet B, Sérusiaux E, Andrésson ÓS, Miao V, Bellenger JP, Lutzoni F (2014) Lichen-symbiotic cyanobacteria associated with *Peltigera* have an alternative

vanadium-dependent nitrogen fixation system. Eur J Phycol 49:11–19
Honegger R (1991) Functional aspects of the lichen symbiosis. Annu Rev Plant Biol 42:553–578
Honegger R (2012) The symbiotic phenotype of lichen-forming ascomycetes and their endo- and epibionts. In: Hock B (ed) The Mycota IX. Springer, Berlin, pp 287–339
Honegger R, Edwards D, Axe L (2013) The earliest records of internally stratified cyanobacterial and algal lichens from the Lower Devonian of the Welsh Borderland. New Phytol 197:264–275
Hyvärinen M, Crittenden PD (2000) 33^{P} translocation in the thallus of the mat-forming lichen *Cladonia portentosa*. New Phytol 145:281–288
Hyvärinen M, Härdling R, Tuomi J (2002) Cyanobacterial lichen symbiosis: the fungal partner as an optimal harvester. Oikos 98:498–504
Johansson O, Olofsson J, Giesler R, Palmqvist K (2011) Lichen responses to nitrogen and phosphorus additions can be explained by the different symbiont responses. New Phytol 191:795–805
Joneson S, Lutzoni F (2009) Compatibility and thigmotropism in the lichen symbiosis: a reappraisal. Symbiosis 47:109–115
Joneson S, Armaleo D, Lutzoni F (2011) Fungal and algal gene expression in early developmental stages of lichen-symbiosis. Mycologia 103:291–306
Kaasalainen U, Tuovinen V, Mwachala G, Pellikka P, Rikkinen J (2021) Complex interaction networks among cyanolichens of a tropical biodiversity hotspot. Front Microbiol 12:672333
Kauppi M, Verseghy-Patay K (1990) Determination of the distribution of lichen substances in the thallus by fluorescence microscopy. Ann Bot Fenn 27:189–202
Kershaw KA (1985) Physiological ecology of lichens. Cambridge University Press, Cambridge
Keuler R, Garretson A, Saunders T, Erickson RJ, St Andre N, Grewe F, Smith H, Lumbsch HT, Huang J-P, Leavitt SD (2020) Genome-scale data reveal the role of hybridization in lichen-forming fungi. Sci Rep 10:1–14
Kim W, Liu R, Woo S, Kang KB, Park H, Yu YH, Ha HH, Oh SY, Yang JH, Kim H, Yun SH, Hur JS (2021) Linking a gene cluster to atranorin, a major cortical substance of lichens, through genetic dereplication and heterologous expression. MBio 12:e01111–e01121
Kohlmeyer J, Hawksworth DL, Volkmann-Kohlmeyer B (2004) Observations on two marine and maritime "borderline" lichens: *Mastodia tessellata* and *Collemopsidium pelvetiae*. Mycol Prog 3:51–56
Kono M, Kon Y, Ohmura Y, Satta Y, Terai Y (2020) *In vitro* resynthesis of lichenization reveals the genetic background of symbiosis-specific fungal-algal interaction in *Usnea hakonensis*. BMC Genomics 21:1–16
Kosecka M, Jabłońska A, Flakus A, Rodriguez-Flakus P, Kukwa M, Guzow-Krzemińska B (2020) Trentepohlialean algae (Trentepohliales, Ulvophyceae) show preference to selected mycobiont lineages in lichen symbioses. J Phycol 56:979–993
Kranner I, Cram WJ, Zorn M, Wornik S, Yoshimura I, Stabentheiner E, Pfeifhofer H (2005) Antioxidants and photoprotection in a lichen as compared to its isolated symbiotic partners. PNAS 102:3141–3146
Kranner I, Beckett R, Hochman A, Nash TH III (2008) Desiccation-tolerance in lichens: a review. Bryologist 111:576–593
Kroken S, Taylor JW (2001) A gene genealogical approach to recognize phylogenetic species boundaries in the lichenized fungus *Letharia*. Mycologia 93:38–53
Lakatos M, Lange-Bertalot H, Büdel B (2004) Diatoms living inside the thallus of the green algal lichen *Coenogonium linkii* in neotropical lowland rain forests. J Phycol 40:70–73
Lambers H (1985) Respiration in intact plants and tissues: its regulation and dependence on environmental factors, metabolism and invaded organisms. In: Douce R, Day DA (eds) Higher Plant Respiration. Springer, Berlin, pp 418–465
Larson DW (1987) The absorption and release of water by lichens. Bibl Lichenol 25:351–360
Leiva D, Fernández-Mendoza F, Acevedo J, Carú M, Grube M, Orlando J (2021) The bacterial community of the foliose macro-lichen *Peltigera frigida* is more than a mere extension of the microbiota of the subjacent substrate. Microb Ecol 81:965–976
Liba CM, Ferrara FIDS, Manfio GP, Fantinatti-Garboggini F, Albuquerque RC, Pavan C, Ramos PL, Moreira-Filho CA, Barbosa HR (2006) Nitrogen-fixing chemo-organotrophic bacteria isolated from cyanobacteria-deprived lichens and their ability to solubilize phosphate and to release amino acids and phytohormones. J Appl Microbiol 101:1076–1086
Lücking L, Grube M (2002) Facultative parasitism and reproductive strategies in *Chroodiscus* (Ascomycota, Ostropales). Stapfia 80:267–292
Lücking R, Dal-Forno M, Sikaroodi M, Gillevet PM, Bungartz F, Moncada B, Yánez-Ayabaca A, Chaves JL, Coca LF, Lawrey JD (2014) A single macrolichen constitutes hundreds of unrecognized species. PNAS 111:11091–11096
Lücking R, Hodkinson BP, Leavitt SD (2016) The 2016 classification of lichenized fungi in the Ascomycota and Basidiomycota - approaching one thousand genera. Bryologist 119:361–416
Lutzoni F, Pagel M, Reeb V (2001) Major fungal lineages are derived from lichen symbiotic ancestors. Nature 411:937–940
Mark K, Laanisto L, Bueno CG, Niinemets Ü, Keller C, Scheidegger C (2020) Contrasting co-occurrence patterns of photobiont and cystobasidiomycete yeast associated with common epiphytic lichen species. New Phytol 227:1362–1375
Matthews SW, Tucker SC, Chapman RL (1989) Ultrastructural features of mycobionts and

trentepohliaceous phycobionts in selected subtropical crustose lichens. Bot Gaz 150:417–438

Meeßen J, Eppenstein S, Ott S (2013) Recognition mechanisms during the pre-contact state of lichens: II. Influence of algal exudates and ribitol on the response of the mycobiont of *Fulgensia bracteata*. Symbiosis 59:131–143

Merges D, Dal Grande F, Greve C, Otte J, Schmitt I (2021) Virus diversity in metagenomes of a lichen symbiosis (*Umbilicaria phaea*): complete viral genomes, putative hosts and elevational distributions. Environ Microbiol 23:6637–6650

Moya P, Molins A, Chiva S, Bastida J, Barreno E (2020) Symbiotic microalgal diversity within lichenicolous lichens and crustose hosts on Iberian Peninsula gypsum biocrusts. Sci Rep 10:1–14

Muggia L, Nelson P, Wheeler T, Yakovchenko LS, Tønsberg T, Spribille T (2011) Convergent evolution of a symbiotic duet: the case of the lichen genus Polychidium (Peltigerales, Ascomycota). Am J Bot 98:1647–1656

Muggia L, Pérez-Ortega S, Kopun T, Zellnig G, Grube M (2014) Photobiont selectivity leads to ecological tolerance and evolutionary divergence in a polymorphic complex of lichenized fungi. Ann Bot 114:463–475

Muggia L, Leavitt S, Barreno E (2018) The hidden diversity of lichenised Trebouxiophyceae (Chlorophyta). Phycologia 57:503–524

Muggia L, Nelsen MP, Kirika PM, Barreno E, Beck A, Lindgren, H, Lumbsch HT, Leavitt SD, Trebouxia Working Group (2020) Formally described species woefully underrepresent phylogenetic diversity in the common lichen photobiont genus *Trebouxia* (Trebouxiophyceae, Chlorophyta): an impetus for developing an integrated taxonomy. Mol Phylogen Evol 149:106821

Nascimbene J, Benesperi R, Giordani P, Grube M, Marini L, Vallese C, Mayrhofer H (2019) Could hair-lichens of high-elevation forests help detect the impact of global change in the Alps? Diversity 11:45

Nash T (2008) Nutrients, elemental accumulation, and mineral cycling. In: Nash T (ed) Lichen biology, 2nd edn. Cambridge University Press, Cambridge, pp 234–251

Nelsen MP, Plata ER, Andrew CJ, Lücking R, Lumbsch HT (2011) Phylogenetic diversity of trentepohlialean algae associated with lichen-forming fungi. J Phycol 47:282–290

Nelsen MP, Lücking R, Boyce CK, Lumbsch HT, Ree RH (2020) No support for the emergence of lichens prior to the evolution of vascular plants. Geobiology 18:3–13

Nelsen MP, Leavitt SD, Heller K, Muggia L, Lumbsch HT (2022) Contrasting patterns of climatic niche divergence in *Trebouxia*—a clade of lichen-forming algae. Front Microbiol 13:791546

Nguyen VK, Genta-Jouve G, Duong TH, Beniddir MA, Gallard JF, Ferron S, Boustie J, Mouray E, Grellier P, Chavasiri W, Le Pogam P (2020) Eumitrins C-E: structurally diverse xanthone dimers from the Vietnamese lichen *Usnea baileyi*. Fitoterapia 141:104449

Nylander W (1857) De fungillis binis lichenicolis observatio. Bot Not 1857:83–84

Oberwinkler F (2012) Basidiolichens. In: Hock B (ed) Fungal associations. Springer, Berlin, pp 341–362

Oliver MJ, Farrant JM, Hilhorst HWM, Mundree S, Williams B, Bewley JD (2020) Desiccation tolerance: avoiding cellular damage during drying and rehydration. Ann Rev Plant Biol 71:435–460

Peksa O, Gebouská T, Škvorová Z, Vančurová L, Škaloud P (2022) The guilds in green algal lichens—an insight into the life of terrestrial symbiotic communities. FEMS Microbiol Ecol 98(2):fiac008

Petrini O, Hake U, Dreyfuss MM (1990) An analysis of fungal communities isolated from fruticose lichens. Mycologia 82:444–451

Petrzik K, Vondrák J, Barták M, Peksa O, Kubešová O (2014) Lichens, a new source or yet unknown host of herbaceous plant viruses? Eur J Plant Pathol 138:549–559

Petrzik K, Vondrák J, Kvíderová J, Lukavský J (2015) Platinum anniversary: virus and lichen alga together more than 70 years. PLoS One 10:3

Petrzik K, Koloniuk I, Sarkisová T, Èíhal L (2016) Detection of herbaceous-plant pararetrovirus in lichen herbarium samples. Acta Virol 60:196–200

Petrzik K, Koloniuk I, Sehadová H, Sarkisova T (2019) Chrysoviruses inhabited symbiotic fungi of lichens. Viruses 11:12

Peveling E (1970) Die Darstellung der Oberflächenstrukturen von Flechten mit dem Raster-Elektronenmikroskop. Deutsche Bot Ges NF 4:89–101

Pichler G, Muggia L, Candotto Carniel F, Grube M, Kranner I (2023) How to build a lichen: from metabolite release to symbiotic interplay. New Phytol 238: 1362–1378

Pizarro D, Divakar PK, Grewe F, Crespo A, Dal Grande F, Lumbsch HT (2020) Genome-wide analysis of biosynthetic gene cluster reveals correlated gene loss with absence of usnic acid in lichen-forming fungi. Genome Biol Evol 12:1858–1868

Poelt J, Mayrhofer H (1988) Über Cyanotrophie bei Flechten. Plant Syst Evol 158:265–281

Ponsero AJ, Hurwitz BL, Magain N, Miadlikowska J, Lutzoni F, U'Ren JM (2021) Cyanolichen microbiome contains novel viruses that encode genes to promote microbial metabolism. ISME Commun 1:1–4

Printzen C, Ekman S, Tønsberg T (2003) Phylogeography of *Cavernularia hultenii*: evidence of slow genetic drift in a widely disjunct lichen. Mol Ecol 12:1473–1486

Rai AN (2002) Cyanolichens: nitrogen metabolism. In: Rai AM, Bergman B, Rasmusson U (eds) Cyanobacteria in symbiosis. Springer, Dordrecht, pp 97–115

Reich PB, Walters MB, Ellsworth DS, Vose JM, Volin JV, Gresham C, Bowman WD (1998) Relationships of leaf dark respiration to leaf nitrogen, specific leaf area and leaf life-span: a test across biomes and functional groups. Oecologia 114:471–482

Resl P, Bujold AR, Tagirdzhanova G, Meidl P, Freire Rallo S, Kono M, Fernández-Brime S, Guðmundsson H, Andrésson ÓS, Muggia L, Mayrhofer H, McCutcheon JP, Wedin M, Werth S, Willis LM, Spribille T (2022) Large differences in carbohydrate degradation and transport potential among lichen fungal symbionts. Nat Commun 13:2634

Rikkinen J (2017) Cyanobacteria in terrestrial symbiotic systems. In: Hallenbeck PC (ed) Modern topics in the phototrophic prokaryotes. Springer, Cham, pp 243–294

Sanders WB (2014) Complete life cycle of the lichen fungus *Calopadia puiggarii* (Pilocarpaceae, Ascomycetes) documented in situ: Propagule dispersal, establishment of symbiosis, thallus development, and formation of sexual and asexual reproductive structures. Am J Bot 101:1836–1848

Sanders WB, Ascaso C (1995) Reiterative production and deformation of cell walls in expanding thallus nets of the lichen *Ramalina menziesii* (Lecanorales, Ascomycetes). Am J Bot 82:1358–1366

Sanders WB, Masumoto H (2021) Lichen algae: the photosynthetic partners in lichen symbioses. Lichenologist 53:347–393

Schaper GM, Ott S (2003) Photobiont selectivity and interspecific in interactions in lichen communities. Culture experiments with the mycobiont *Fulgensia bracteata*. Plant Biol 5:441–450

Scherrer S, Honegger R (2003) Inter- and intraspecific variation of homologous hydrophobin (H1) gene sequences among *Xanthoria* spp. (lichen-forming ascomycetes). New Phytol 158:375–389

Scherrer S, De Vries OMH, Dudler R, Wessels JGH, Honegger R (2000) Interfacial self-assembly of fungal hydrophobins of the lichen-forming ascomycetes *Xanthoria parietina* and *X. ectaneoides*. Fungal Genet Biol 30:81–93

Schmitt I, Lumbsch HT (2009) Ancient horizontal gene transfer from bacteria enhances biosynthetic capabilities of fungi. PLoS One 4:e4437

Schüßler A, Kluge M (2001) *Geosiphon pyriforme*, an endocytosymbiosis between fungus and cyanobacteria, and its meaning as a model system for arbuscular mycorrhizal research. In: Hock B (ed) The Mycota 9: fungal associations. Springer, Berlin, pp 151–161

Schwendener S (1869) Die Algentypen der Flechtengonidien. C. Schultze

Singh G, Calchera A, Schulz M, Drechsler M, Bode HB, Schmitt I, Dal Grande F (2021) Climate-specific biosynthetic gene clusters in populations of a lichen-forming fungus. Environ Microbiol 23:4260–4275

Singh G, Calchera A, Merges D, Valim H, Otte J, Schmitt I, Dal Grande F (2022) A candidate gene cluster for the bioactive natural product gyrophoric acid in lichen-forming fungi. Microbiol Spectr. https://doi.org/10.1128/spectrum.00109-22

Škvorová Z, Černajová I, Steinová J, Peksa O, Moya P, Škaloud P (2022) Promiscuity in lichens follows clear rules: partner switching in *Cladonia* is regulated by climatic factors and soil chemistry. Front Microbiol 12:781585

Spribille T, Tuovinen V, Resl P, Vanderpool D, Wolinski H, Aime MC, Schneider K, Stabentheiner E, Toome-Heller M, Thor G, Mayrhofer H, Johannesson H, McCutcheon J (2016) Basidiomycete yeasts in the cortex of ascomycete macrolichens. Science 353:488–492

Spribille T, Tagirdzhanova G, Goyette S, Tuovinen V, Case R, Zandberg WF (2020) 3D biofilms: in search of the polysaccharides holding together lichen symbioses. FEMS Microbiol Lett 367:5

Steinová J, Holien H, Košuthová A, Škaloud P (2022) An exception to the rule? Could photobiont identity be a better predictor of lichen phenotype than mycobiont identity? J Fungi 8:275

Stocker-Wörgötter E (2001) Experimental lichenology and microbiology of lichens: culture experiments, secondary chemistry of cultured mycobionts, resynthesis, and thallus morphogenesis. Bryologist 104:576–581

Thaxter R (1892) On the Myxobacteriaceae, a new order of Schizomycetes. Bot Gaz 17:389–406

Trembley ML, Ringli C, Honegger R (2002) Hydrophobins DGH1, DGH2, and DGH3 in the lichen-forming basidiomycete *Dictyonema glabratum*. Fungal Genet Biol 35:247–259

Tripathi M, Joshi Y (2019) Endolichenic fungi: present and future trends. Springer, Singapore

Tuong TL, Do LT, Aree T, Wonganan P, Chavasiri W (2020) Tetrahydroxanthone–chromanone heterodimers from lichen *Usnea aciculifera* and their cytotoxic activity against human cancer cell lines. Fitoterapia 147:104732

Tuovinen V, Millanes AM, Freire-Rallo S, Rosling A, Wedin M (2021) *Tremella macrobasidiata* and *Tremella variae* have abundant and widespread yeast stages in *Lecanora* lichens. Environ Microbiol 23:2484–2498

U'Ren JM, Lutzoni F, Miadlikowska J, Laetsch AD, Arnold AE (2012) Host and geographic structure of endophytic and endolichenic fungi at a continental scale. Am J Bot 99:898–914

Urayama SI, Doi N, Kondo F, Chiba Y, Takaki Y, Hirai M, Minegshi Y, Hagiwara D, Nunoura T (2020) Diverged and active partitiviruses in lichen. Front Microbiol 11:561344

Velmala S, Myllys L, Halonen P, Goward T, Ahti T (2009) Molecular data show that *Bryoria fremontii* and *B. tortuosa* (Parmeliaceae) are conspecific. Lichenologist 41:231–242

Vondrak J, Kubásek J (2013) Algal stacks and fungal stacks as adaptations to high light in lichens. Lichenologist 45:115–124

Walser JC (2004) Molecular evidence for limited dispersal of vegetative propagules in the epiphytic lichen *Lobaria pulmonaria*. Am J Bot 91:1273–1276

Wedin M, Döring H, Gilenstam G (2004) Saprotrophy and lichenization as options for the same fungal species on different substrata: environmental plasticity and fungal lifestyles in the *Stictis–Conotrema* complex. New Phytol 164:459–465

Wedin M, Maier S, Fernandez-Brime S, Cronholm B, Westberg M, Grube M (2016) Microbiome change by symbiotic invasion in lichens. Environ Microbiol 18: 1428–1439

Wilkinson DM, Creevy AL, Kalu CL, Schwartzman DW (2015) Are heterotrophic and silica-rich eukaryotic microbes an important part of the lichen symbiosis? Mycology 6:4–7

Wirth V, Hauck M, Schultz M (2013) Die Flechten Deutschlands, vol Band 1. Ulmer, Stuttgart

Yamamoto Y, Kinoshita Y, Thor G, Hasumi M, Kinoshita K, Koyama K, Takahashi K, Yoshimura I (2002) Isofuranonaphthoquinone derivatives from cultures of the lichen *Arthonia cinnabarina* (DC.) Wallr. Phytochemistry 60:741–745

Yuan X, Xiao S, Taylor TN (2005) Lichen-like symbiosis 600 million years ago. Science 308:1017–1020

Zopf W (1897) Ueber Nebensymbiose (Parasymbiose). Ber Deutsch Bot Ges 15:90–92

Umbilicaria muhlenbergii: A Model for Studying Molecular Mechanisms Regulating Initial Fungal Symbiotic Interactions with Algal Cells

7

Yanyan Wang, Xinli Wei, Yuting Hu, and Jin-Rong Xu

Abstract

Lichens are known for their importance in evolution and ecology but our knowledge of molecular mechanisms regulating symbiont interactions in lichen symbiosis is very limited. Recent studies have shown that *Umbilicaria muhlenbergii* has a relative fast growth rate and is amenable to molecular genetic studies for efficiently generating transformants and targeted gene disruption mutants. It is the only known dimorphic lichen-forming fungus, and pseudohyphal growth is associated with symbiotic interactions with algal cells of *Trebouxia jamesii*, its photobiont. Pharmacological studies and functional characterization of the Gα subunit UmGpa2 of the trimeric G-proteins have shown the involvement of the cAMP-PKA (cyclic AMP-protein kinase A) pathway in regulating yeast-to-pseudohypha transition and fungal–algal interactions. The importance of cAMP-PKA signaling in lichen symbiosis proved for the first time in *U. muhlenbergii* argues for further characterization of this and other well-conserved signal transduction pathways that are known to crosstalk and co-regulate fungal–plant interactions in plant pathogens. The *Umbilicaria muhlenbergii* genome contains key components of mitogen-activated protein (MAP) kinase pathways that may regulate fungal–algal interactions during lichen symbiosis. Although there are still challenges, *U. muhlenbergii* has the potential to be used as a model to study early stages of symbiotic interactions and signal transduction pathways involved in the recognition of compatible symbiont cells and development of highly differentiated masses of fungal–algal cells.

Keywords

Lichen symbiosis · Signal transduction pathways · Fungal–algal interactions · *Trebouxia jamesii* · Early re-lichenization stages

7.1 Introduction

Lichen is mainly composed of a lichen-forming fungus (the mycobiont, commonly an ascomycete) and a photosynthetic organism (the photobiont, often a green alga but can be a cyanobacterium). The fungal–algal association forms unique stratified thalli with morphological and physiological features that are different from the

Y. Wang · X. Wei
State Key Laboratory of Mycology, Institute of Microbiology, Chinese Academy of Sciences, Beijing, China

Y. Hu · J.-R. Xu (✉)
Department of Botany and Plant Pathology, Purdue University, West Lafayette, IN, USA
e-mail: jinrong@purdue.edu

Y.-P. Hsueh, M. Blackwell (eds.), *Fungal Associations*, The Mycota 9,
https://doi.org/10.1007/978-3-031-41648-4_7

mycobiont or photobiont. Besides the primary algal partner, some lichens have a cyanobacterium as a secondary photobiont in cephalodia. Lichens also often harbor endophytic or exophytic bacteria, fungi, and algae. Some of these microorganisms are permanent inhabitants and may play non-negligible roles in lichen symbiosis (Grube et al. 2009; Spribille et al. 2016). The interaction between fungi and photosynthesizing organisms is an ancient adaptation to terrestrial lifestyle. Independent shifts of lichenization and de-lichenization have been observed in different fungal phyla, and many non-lichenized fungal taxa, including some plant pathogenic fungi, are likely to be derived from lichen ancestors (Lutzoni et al. 2018). Approximately 20,000 species of lichen-forming fungi but only fewer than 200 photobiont species have been identified. As a result, the diversity of lichen species refers to diversity of lichen-forming fungi and many of them share the same photobiont, such as the most frequent *Trebouxia*, *Trentepohlia*, and *Nostoc* species (Friedl and Budel 2008).

For many lichens, the mycobionts and photobionts can be isolated and cultured on artificial media separately. For the mycobionts, they normally grow extremely slow in axenic cultures but form typical fungal colonies. Artificial resynthesis of lichen biosynthesis with isolated mycobiont and photobiont cells under controlled laboratory conditions has been used to study the re-lichenization processes, which can be divided into four stages: pre-contact (chemical interactions only), contact or post-contact (chemical and physical interactions), growth of undifferentiated mass of symbiont cells, and differentiation into to a stratified thallus (Nazem-Bokaee et al. 2021). Earlier studies have tested different conditions, including various substrates and nutritional or light conditions, for successful reconstruction of lichens such as *Acarospora fuscata* and *Cladonia cristatella*. However, the co-culture process is time-consuming and it usually takes 2–4 weeks to form a tight fungal–algal contact interface, 3–4 months to develop soredia-like structures, and 8–12 months to form a thallus.

For the initial symbiotic associations, chemical and physical signals must be recognized by both mycobionts and photobionts. Although fungal growth chemotropically toward algal cells may be absent, the pre-contact recognition between compatible symbionts in proximity through diffusible signaling compounds can affect hyphal growth, branching frequency, and mucilage formation in the mycobiont (Joneson and Lutzoni 2009; Meeßen et al. 2013). After the initial contact, compatible photobiont cells are recognized via chemical and physical signals and enveloped by fungal hyphae and mucilage, which then grow into a mass of undifferentiated fungal and algal cells. For the photobionts, some lectins (carbohydrate-binding glycoproteins) secreted by fungal cells may function as algal binding proteins (ABPs) to bind with surface proteins of algal cells, such as lectins of *Xanthoria parietina* and *Evernia prunastri* binding with algal urease proteins, which may affect arginine metabolism, intracellular putrescine level, and glucanase activities in algal cells and symbiotic interactions (Nazem-Bokaee et al. 2021; Singh and Walia 2014). However, none of these assumed signaling reactions in algal cells have been characterized by molecular genetic studies. For the mycobionts, different types of sugar alcohols and other compounds secreted by the photobionts may be recognized but there is no molecular study on photobiont ligands, corresponding fungal receptors, and downstream intracellular signaling pathways in the mycobionts. Although some fungal lectins have been inappropriately described as "receptors," they are secreted and supposed to bind with surface proteins of the photobionts (Nazem-Bokaee et al. 2021; Singh and Walia 2014). More importantly, none of the conserved signaling pathways and upstream receptors identified in the budding yeast and other model fungi (Zhang et al. 2021) have been characterized by their roles in fungal–algal interactions during the establishment of symbiosis.

In the past three decades, many genes and pathways involved in fungal–plant interactions have been characterized in plant pathogenic fungi (Jiang et al. 2018; Zhao et al. 2007). However, due to their slow growth, there are almost no

molecular genetic or functional studies in lichen-forming fungi. The re-lichenization processes are drastically under-investigated in lichen-forming fungi in comparison with plant infection processes in plant pathogenic fungi that are more amenable to molecular and classical genetic studies, such as the rice blast fungus *Magnaporthe oryzae* (Gupta et al. 2021). However, transformation is relatively efficient in *Umbilicaria muhlenbergii* and transformants can be isolated in 1–2 weeks in *Agrobacterium tumefaciens*-mediated transformation (ATMT) or transformation with PEG-treated protoplasts (Park et al. 2013; Wang et al. 2020). Furthermore, homologous recombination frequency is relatively high in this lichen-forming fungus, and targeted gene disruption and knockout mutants have been successfully identified (Wang et al. 2020). In this chapter, we will describe unique features of *U. muhlenbergii* and its suitability as a model for studying the establishment and maintenance of symbiotic relationships between the mycobionts and photobionts. The emphasis will be on signal transduction pathways involved in the initial fungal–algal interactions.

7.2 Unique Features of *U. muhlenbergii*

U. muhlenbergii is a foliose lichen that usually grows on bare stones at high elevations. It is widely distributed in Northern Asia and North America (Wei and Jiang 1993). The genus *Umbilicaria* (in Umbilicariaceae, Umbilicariales, Lecanoromycetes) contains over 70 species that can grow under extreme environmental conditions and often produce heavily melanized thalli. Commonly known as rock tripe, some of them are edible and consumed as an emergency food. Like other *Umbilicaria* species, the preferred propagules for dispersal in *U. muhlenbergii* are ascospores formed in apothecia (sexual fruiting bodies) instead of reproductive mixtures of vegetative hyphae and algal cells (Cao et al. 2015). Besides ascospores, the *U. muhlenbergii* lichen thalli also produce hyaline, rod-like spermatia on spermatiophores in spermogonia (asexual fruiting bodies) on lichen thalli (in the symbiotic state). However, the formation of spermatia and spermogonia has not been observed in cultures, and their biological functions in the life cycle of *U. muhlenbergii* remain to be characterized. Although some *Umbilicaria* species produce multiseptate thalloconidia directly on scleroplectenchymatous tissues or protruding rhizomorphs of the lower cortex of lichen thalli, *U. muhlenbergii* is one of the *Umbilicaria* species that does not form thalloconidia (Davydov et al. 2017; Hestmark 1992).

The photobiont partner for *U. muhlenbergii* is mainly *Trebouxia jamesii* (Fig. 7.1), a green algal species. Species in the genus *Trebouxia* are the most common photobionts and they are found in approximately 80% of all the known lichens (Muggia et al. 2020; Nelsen et al. 2021). Although only nonmotile stages with a reduced chloroplast are observed in lichen thalli, *Trebouxia* species can develop autospores and zoospores in cultures (Friedl and Budel 2008). The features of pyrenoids (micro-compartment of chloroplasts of certain algae and hornworts where the CO_2-fixing enzyme RuBisCO (ribulose-1, 5-bisphosphate carboxylase/oxygenase) is localized and chloroplast morphology are used to identify different species of *Trebouxia*. Cells of *T. jamesii* possess impressa-type pyrenoids (Friedl 1989) and shallowly lobed-type chloroplasts and generally has a diameter of 4- to 20-μm (Bordenave et al. 2022). *T. jamesii* is sensitive to strong light (3000 lux) and is usually found in lichens that inhabit rocks with high metal concentrations. Besides *T. jamesii*, the *U. muhlenbergii* lichen contains an endophytic green algal species. Analysis of the ribosomal DNA (rDNA) sequences showed that it is a novel *Elliptochloris* species closely related to *E. subsphaerica* (Wang et al. 2020). *Elliptochloris* species have been reported as endolichenic algae present in various lichens (Moya et al. 2017) as well as symbionts of sea anemones (Lewis and Muller-Parker 2004) In comparison with *T. jamesii*, the endolichenic *Elliptochloris* isolate has a faster growth rate

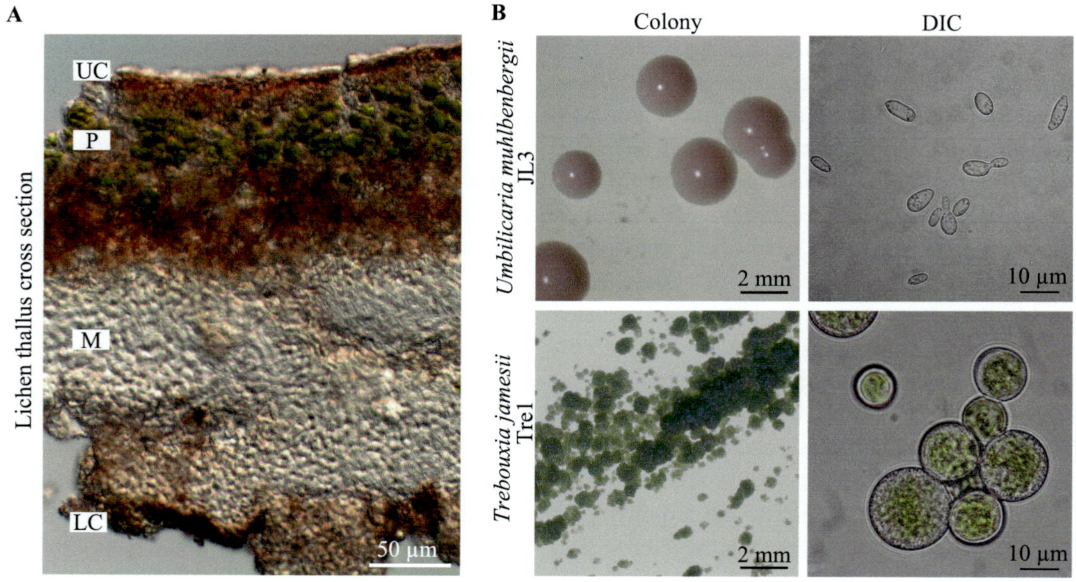

Fig. 7.1 Thallus structures and symbionts of *Umbilicaria muhlenbergii* lichen. (**a**) A typical cross-section of *U. muhlenbergii* lichen thallus. *UC* upper cortex, *P* photobiont zone, *M* medulla, *LC* lower cortex. (**b**) Colonies and cells of *U. muhlenbergii* (JL3) and *Trebouxia jamesii* (Tre1) in axenic cultures

and requires a higher light intensity for optimal growth (Wang et al. 2020).

Unlike all other cultivated lichen-forming ascomycetes that grow only as hyphae, the *U. muhlenbergii* strains isolated from China and the US all grow as yeast cells when cultured on nutritionally rich media (Park et al. 2013; Wang et al. 2020). The unicellular yeast cells of *U. muhlenbergii* mycobiont grow by budding and form typical yeast-like colonies. When cultured in liquid or solid potato dextrose medium at room temperature, typical olive-shaped yeast cells of *U. muhlenbergii* are 3.0–4.5 µm in width and 5.0–6.5 µm in length (Fig. 7.1). On potato dextrose or other common fungal media, it normally takes only 7–10 days for a single yeast cell to form a visible colony, which is faster than other cultivated lichen-forming fungi.

Although *U. muhlenbergii* lives in the yeast form in axenic cultures, it exists in the hyphal or pseudohyphal form in the lichen thallus, indicating that this lichenized fungus must undergo the yeast-to-hypha transition during the establishment of symbiosis. The *U. muhlenbergii* strains isolated from lichens collected at different sites in Asia and North America all have been observed to undergo dimorphic transition in axenic cultures in the presence of algal cells of its photobiont *T. jamesii* but not endolichenic *Elliptochloris* algal cells (Park et al. 2013; Wang et al. 2020). The yeast-to-hypha or pseudohypha dimorphic transition is well known to be associated with pathogenesis in a number of human and plant pathogenic fungi, including *Candida albicans*, *Talaromyces marneffei*, *Blastomyces dermatitidis, Coccidioides immitis, Histoplasma capsulatum*, the corn smut fungus *Ustilago maydis*, and peach leaf curl pathogen *Taphrina deformans* (Boyce and Andrianopoulos 2015; Svetaz et al. 2017). However, to date, *U. muhlenbergii* is the only lichen-forming fungus that has been reported to have dimorphism associated with symbiosis. Nevertheless, it is possible that other *Umbilicaria* species or lichen-forming fungi also undergo dimorphic transition or switching during lichen symbiosis.

Besides its association of dimorphism with symbiosis, *U. muhlenbergii* is amenable for molecular genetic studies because it grows relatively fast as yeast cells. Transformant colonies can be isolated after 10–14 days of incubation. *U. muhlenbergii* is highly sensitive to commonly used antibiotics for selecting fungal transformants. Hygromycin works efficiently at 20–30 μg/ml in *U. muhlenbergii* for selecting transformants carrying the hygromycin phosphotransferase (*hph*) cassette that is commonly used in transformation of *M. oryzae* and other ascomycetous plant pathogens (Park et al. 2013; Wang et al. 2020). Geneticin-resistant transformants carrying the neomycin resistance (*neo*) gene (Zhou et al. 2011) can be selected at 30 μg/ml. More importantly, homologous recombination frequency in *U. muhlenbergii* is also comparable to that of *M. oryzae* and many other plant pathogenic fungi. Targeted gene deletion mutants can be isolated at the rate of approximately 5% of positive transformants by the gene replacement approach (Wang et al. 2020). The *Agrobacterium tumefaciens*-mediated transformation (ATMT) approach also can be used to generate insertional mutants in *U. muhlenbergii* (Park et al. 2013). As the first lichen-forming fungus with gene deletion and disruption mutants being identified and characterized, *Umbilicaria muhlenbergii* brings hope to establish a genetic system for studying molecular mechanisms of lichen symbiosis.

The published draft genome of *U. muhlenbergii* isolate KoLRILF000956 has 297 contigs and contig N50 of 0.21 Mb (Park et al. 2014). The 34.6 Mb assembled genome has a GC content of 47.12%. We have recently sequenced the genome of *U. muhlenbergii* strain JL3 (Wang et al. 2020) by a combination of Illumina PE150 and PacBio sequencing. Strains KoLRILF000956 and JL3 were collected from Mt. Tulaopoding, Ji-Lin Province, China in 2012 and 2017, respectively (Park et al. 2013; Wang et al. 2020). The 35.4 Mb assembled JL3 genome consists of 25 contigs (N50 of 1.58 Mb) and is predicted to encode 8166 proteins that are over 100 amino acids (Wang, unpublished), including 102 glycoside hydrolases, 1251 proteins with transmembrane domains, and 113 cytochrome P450 proteins. Among the predicted protein-coding genes, 1058 of them are related to pathogen–host interactions based on the Pathogen–Host Interactions (PHI) database. Some of these genes, particularly the ones encoding secreted proteins, may be involved in fungal–algal or fungal–microbial interactions in lichens.

7.3 Regulation of *U. muhlenbergii* Dimorphism by the cAMP-PKA Pathway

The yeast-to-hypha or pseudohypha transition has been observed in various mucoromycetes, ascomycetes, and basidiomycetes but best studied in the budding yeast *Saccharomyces cerevisiae*, in which dimorphic switching is important for pseudohyphal growth or filamentation and agar invasiveness (Gimeno et al. 1992). The well-conserved cAMP-protein kinase A (PKA) pathway is involved in the regulation of pseudohyphal growth in yeast and filamentous fungi. In fungal pathogens, this pathway is known for co-regulating dimorphism and pathogenesis. In *C. albicans*, deletion of the key components of the cAMP-PKA pathway results in defects in dimorphic switching and reduced virulence (Sánchez-Martínez and Pérez-Martín 2002; Sonneborn et al. 2000). Likely in *S. cerevisiae* and other dimorphic ascomycetes, elevated intracellular cAMP levels promote hyphal or pseudohyphal growth in *C. albicans*. In *U. maydis*, mutants disrupted in the cAMP-PKA pathway also are defective in plant infection and morphogenesis. Interestingly, *U. maydis* grows by budding as haploid yeast cells in cultures but develops obligate infectious hyphae after mating and penetration of plant cells. However, the *uac1* mutant with reduced intracellular cAMP level displays a constitutively filamentous growth as hyphae, which can be rescued by exogenous cAMP or suppressor mutations in the *ubc1* gene (Choi et al. 2015), indicating that low cAMP levels stimulate hyphal growth in *U. maydis*, a basidiomycete pathogen.

Also interesting, *U. muhlenbergii* underwent dimorphic switching when cultured on nutritionally poor media such as 0.2 × PDA (potato dextrose agar) and BBM (Bold's Basal Medium) for extended times (Wang et al. 2020), suggesting nutritional starvation may affect dimorphism. The cAMP-PKA pathway plays a conserved role in regulating responses to nutritional conditions in fungi. In *U. muhlenbergii*, treatments with cAMP or IBMX (3-isobutyl-1-methylxanthine), an inhibitor of cAMP diphosphoesterase, stimulate the elongation of yeast cells and pseudohyphal growth (Wang et al. 2020). When treated with 10 mM cAMP or 25 nmol of IBMX for 14 d, colonies of *U. muhlenbergii* often become darkly pigmented and have a bumpy (uneven) and hairy appearance associated with pseudohyphal/hyphal growth. Aggregates of pseudohyphae or hyphae and heavily melanized spherical cells (highly differentiated hypha/pseudohyphal compartments) are observed in darkly pigmented bumps. After incubation for 1 month, IBMX induces more significant differentiation in *U. muhlenbergii* (Fig. 7.2a). IBMX-induced bumps have protruding melanized hyphae/pseudohyphae on the surface and become hardened tissue-like structures with a dense epidermal layer in which individual compartments or cells are no longer distinguishable (Wang et al. 2020). This epidermal layer of highly differentiated and melanized hyphae/pseudohyphae on the bumps of IBMX-treated colonies is, to certain degree, similar to the epidermal or cortex layer of lichen thalli. Results from these pharmacological studies suggest that the cAMP-PKA pathway is involved in regulating the yeast-to-pseudohypha transition in *U. muhlenbergii*.

Like in other eukaryotes, adenylate cyclase (AC) is responsible for the synthesis of intracellular cAMP, which is degraded by phosphodiesterase (PDE) in fungi. The only known intracellular target of cAMP is the regulatory subunit of PKA (PKR). Binding of PKR with cAMP results in its disassociation from catalytic subunits of PKA (CPKAs). Free CPKAs are enzymatically active and phosphorylate downstream targets such as transcription factors. In fungi, adenylate cyclase can be activated by upstream trimeric G-proteins (Gα/Gβ/Gγ) or small GTPase (GTP hydrolase) such as RAS in responses to extracellular stimuli recognized by G-protein coupled receptors (GPCRs) or other sensors such as extracellular mucins. All the key components of the cAMP-PKA pathway characterized in *S. cerevisiae* are conserved in the *U. muhlenbergii* genome (Fig. 7.2b) although only functional characterization of the *UmGPA3* Gα subunit gene has been published (Wang et al. 2020). Whereas some lichen-forming fungi such as *Cladonia grayi* appear to have an expanded family of G-alpha subunits, with 5 of them being unique (Armaleo et al. 2019), *U. muhlenbergii* has only three that are well conserved in other ascomycetes. UmGpa3 is homologous to Gpa2 of *C. albicans*, Gpa3 of *U. maydis*, and MagB of *M. oryzae*. The *Umgpa3* deletion mutant is normal in growth in the yeast form but it is defective in switching from yeast cells to pseudohyphal growth when incubated on nutritionally poor media or under hyperosmotic conditions (Wang et al. 2020). Because the cAMP-PKA pathway functions downstream Gα subunits (Fig. 7.2b), these results provide molecular evidence for its involvement in the regulation of dimorphic transition in *U. muhlenbergii*. Recently, we have identified the *Umras2* deletion mutant that appears to be defective in pseudohyphal growth based on our preliminary characterization and responses to IBMX treatments (Wang et al., unpublished), further confirming the role of cAMP signaling in regulating dimorphic transition.

Orthologs of *UmGPA3* are important for pathogenesis in a number of fungal pathogens that lack dimorphic switching during plant infection, such as *M. oryzae* and *Botrytis cinerea*. In *M. oryzae*, the initial surface recognition and early stages of appressorium formation are regulated by cAMP signaling and deletion of the *MAGB* Gα subunit gene affects the formation of appressoria that are necessary for plant penetration (Fang and Dean 2000). In *U. muhlenbergii*, pseudohyphal growth can be induced by its interaction with cells of its compatible photobiont *T. jamesii*, even when they are mixed and co-cultured on nutrient rich medium (Fig. 7.2c). Under the same conditions, co-cultivation with algal cells of endolichenic *Elliptochloris* species fails to induce pseudohyphal growth in

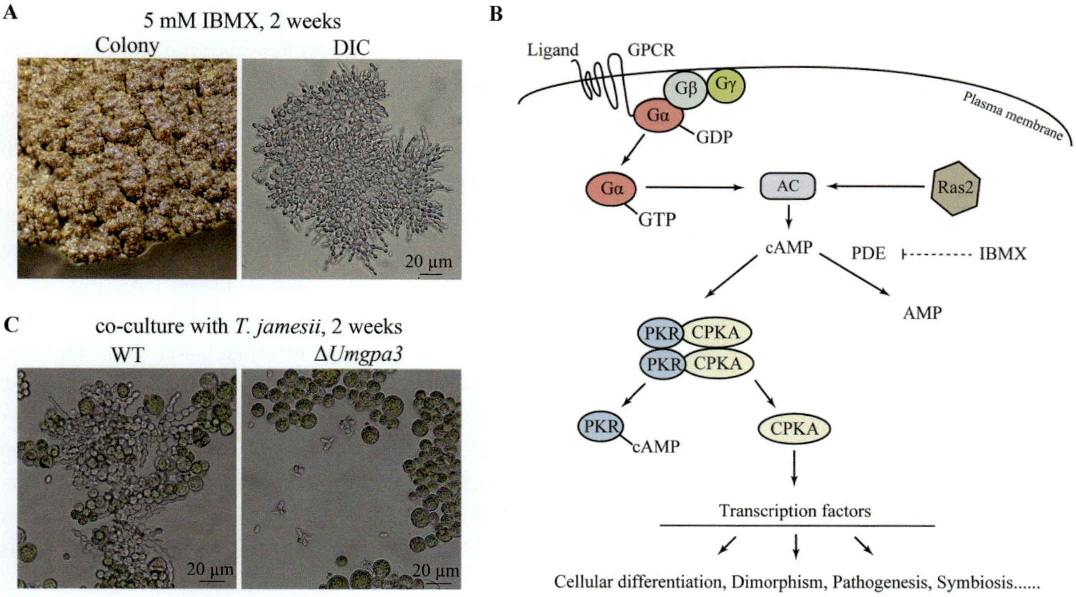

Fig. 7.2 Regulation of yeast-to-hypha dimorphic transition by the cAMP-PKA pathway. (**a**) Colony morphology and pseudohyphae of *U. muhlenbergii* cultures treated with IBMX for 2 weeks. (**b**) A general schematic drawing of the well-conserved cAMP-PKA pathway in fungi. In the budding yeast *Saccharomyces cerevisiae*, both Gpr1 GPCR functioning upstream from the trimeric G-proteins (Gα, Gβ, Gγ) and RAS small GTPase proteins can regulate adenylate cyclase (AC). Intracellular cAMP is synthesized by AC and degraded by cAMP phosphodiesterase (PDE). PKR and CPKA are regulatory and catalytic subunits of PKA. Binding of PKR with cAMP activates the PKA activities by freeing CPKA from its inhibitory binding with PKR. Known downstream transcription factors of PKA include yeast *SFL1*, *FLO11*, *MSN2*, and *MSN4*. (**c**) Algal cells of *T. jamesii* induce pseudohyphal growth in the wild type but not *Umgpa3* mutant of *U. muhlenbergii* in cocultures

U. muhlenbergii, indicating a specific recognition of compatible photobiont (Wang et al. 2020). Longer incubation times lead to the production of darkly pigmented colonies that consist of algal cells and melanized pseudohyphae and have aerial hyphae/pseudohyphae on the surface (Wang et al. 2020). The *UmGpa3* mutant is defective in pseudohyphal growth induced by *T. jamesii* cells. In cocultures of the *UmGpa3* mutant and *T. jamesii* cells, melanized colonies with a mixture of pseudohyphae and algal cells are not observed (Wang et al. 2020), showing the importance of *UmGPA3* and cAMP signaling in *T. jamesii*-induced differentiation in *U. muhlenbergii*. In *M. oryzae*, dominant mutations in *MAGB* lead to appressorium formation on nonconducive conditions but defects in pathogenesis (Fang and Dean 2000). Similarly, expression of the dominant active *UmGPA3*^{G45V} and *UmGPA3*^{Q208L} alleles in *U. muhlenbergii* promotes pseudohyphal growth in nutrient rich media. However, algal cells in close contact with pseudohyphae of the *UmGPA3*^{G45V} and *UmGPA3*^{Q208L} transformants are often damaged, suggesting a destructive effect on symbiotic interactions between fungal and algal cells. Taken together, these pharmacological and molecular studies have shown that the cAMP-PKA pathway plays a critical role in the dimorphic switching and fungal–algal interactions during lichen symbiosis in *U. muhlenbergii*.

7.4 Map Kinase Signaling Pathways in *U. muhlenbergii*

In plant pathogenic fungi, the mitogen-activated protein (MAP) kinase pathways also play important roles in regulating responses to various extracellular cues and different infection processes, such as appressorium formation, plant cell penetration, and invasive growth (Jiang et al. 2018; Zhao et al. 2007). A typical MAP kinase cascade consists of three protein kinases, including a MAP kinase (MAPK), a MAPK kinase (MEK), and a MEK kinase (MEKK) that are activated sequentially. Like most filamentous ascomycetous fungal pathogens, the *U. muhlenbergii* genome has three MAPK pathways consisting of three non-redundant MAPK, MEK, and MEKK genes (Table 7.1). These three MAPK pathways also are conserved in other lichen-forming fungi that have been sequenced, including *Lasallia pustulata* and *C. grayi* (Armaleo et al. 2019; Dal Grande et al. 2017). Although these well-conserved MAPK pathways likely play important roles in regulating fungal–algal interactions and lichen symbiosis, to date, none of them have been functionally characterized in any lichen-forming fungi.

In plant pathogenic fungi, the Pmk1 invasive growth (IG) pathway has a conserved role in regulating the formation of infection structures (such as appressoria and hyphopodia) and infectious growth in infected plant tissues, often in coordination with the cAMP-PKA pathway (Jiang et al. 2018; Zhao et al. 2007). Pmk1 is a MAP kinase orthologous to yeast Fus3 and Kss1 that regulate pheromone response and invasive/pseudohyphal growth in *S. cerevisiae*. This pathway also is known to play a critical role in some human pathogens for virulence and capsule formation. Although some lichen-forming fungi are known to form appressoria or appressorium-like structures such as *Lecidea albocaerulescens, Baeomyces rufus*, and *C. cristatella* (Goodenough et al. 2021; Honegger 1986), it is not clear whether *U. muhlenbergii* form any specialized attachment or penetration structures. Nevertheless, it is likely that the invasive growth (IG) MAPK pathway regulates fungal responses to chemical and physical signals from algal cells, attachment, and appressorium formation if *U. muhlenbergii* forms appressoria. Whereas many lichen-forming fungi do not form haustoria for nutrient uptakes, some species such as *Lecanora conizaeoides* and *Lecidella euphoria* do form intracellular haustoria that are surrounded by thickened cytoplasmic derivatives of algal cells. A few others form intraparietal haustoria that are short, peg-like protrusions compressing but not piercing through algal cell wall (Goodenough et al. 2021; Honegger 1986). Although haustoria are abundant in one Umbilicariaceae species, *L. pustulata*, haustorium formation has not been reported in *U. muhlenbergii* and other Umbilicariaceae species. In haustorium-forming lichenized fungi, the IG MAPK pathway is involved in the regulation of penetration and haustorium formation.

Besides its role in regulating infection structures, the IG MAPK pathway is also important for infectious growth in plant tissues after penetration. For lichen-forming fungi, this MAPK pathway may be also involved in balancing fungal–algal interactions and later stages of the re-lichenization processes. Our preliminary data showed that *U. muhlenbergii* and algal cells form a biofilm-like complex layer upon initial interactions on solid surfaces, which is blocked by deletion of the *PMK1* ortholog (Wang, unpublished). Therefore, this MAPK pathway likely plays an important role in the initial symbiotic associations with algal cells in *U. muhlenbergii*. In human and plant pathogens, this MAPK pathway is known to co-regulate various infection processes with the cAMP signaling pathway. In *U. muhlenbergii,* the IG MAPK and cAMP signaling pathways may co-regulate the initial symbiotic interactions as well as further differentiation of masses of fungal and algal cells in the re-lichenization process. In *C. grayi*, orthologs of the *MGB1* Gβ and *PTH12* (*MoHOX7*) transcription factor genes involved in regulating appressorium formation via the cAMP signaling and *PMK1* MAPK pathways in *M. oryzae* were identified as differentially expressed genes (DEGs) by subtractive hybridization (SSH) analysis during fungal–algal interactions (Joneson et al. 2011).

Studies in a number of plant pathogens have shown that the cell wall integrity (CWI) MAP kinase pathway also has a conserved role in fungal pathogenesis. In the budding yeast, the CWI pathway consisting of the Bck1-Mkk1/Mkk22-

Table 7.1 Key components of conserved MAPK pathways in *U. muhlenbergii*

Functions	*S. cerevisiae*	*M. oryzae*	*U. muhlenbergii*[a]	
			KoLRILF000956	JL3
Kss1/Pmk1 IG pathway				
GPCR	Ste2	Ste2	Contig_79: 15259-16434: +	A2080a
GPCR	Ste3	Ste3	Contig_18: 38439-39570: +	A4507
Mucin	Msb2	MoMsb2	Contig_235: 25655-26374: +	A1409
G-beta	Ste4	Mgb1	Contig_33: 163167-164575: –	A1076
GTPase	Ras2	MoRAS2	Contig_89: 47393-48197: +	A0138
PAK	Ste20	Mst20	Contig_249: 39860-44611: +	A5140
PAK	Cla4	Chm1	Contig_17: 5044-7949: –	A4476
Adaptor	Ste50	Mst50	Contig_143: 62212-63921: –	A1025
MEKK	Ste11	Mst11	Contig_19: 77033-80129:–	A4528
MEK	Ste7	Mst7	Contig_264: 43620-45304: +	A5885
MAPK	Fus3/Kss1	Pmk1	Contig_264: 164503-165751: –	A5439
CWI MAPK pathway				
Sensor	Wsc1	MGG_04325	Contig_266: 184541-185490: +	A5663a
Sensor	Wsc2	MGG_02754	Contig_92: 152719-155985: +	A0313
Sensor	Wsc3	MGG_01466	Contig_240: 136301-141700: –	A2477
Sensor	Wsc4	–	Contig_199: 11029-14777: –	A6922
Sensor	Mid2	MGG_12606	Contig_170: 15231-16130: +	A3494a
Sensor	Mtl1	–		
GTPase	Rho1	MGG_07176	Contig_173: 79123-80189: –	A3616
PKC	Pkc1	MoPkc1	Contig_90: 49578-53475: +	A0159
MEKK	Bck1	Mck1	Contig_268: 104646-108377: –	A5755
MEK	Mkk1/Mkk2	Mkk2	Contig_49: 1494-2850: –	A0560
MAPK	Slt2	Mps1	Contig_264: 279110-280707: +	A5473
HOG pathway				
Osmosensor	Sho1	MoSho1	Contig_163: 400712-401925: –	A7020
Histidine kinase	Sln1	MoSln1	Contig_288: 84078-87819: –	A5981
Histidine kinase	–	MoHik1	Contig_288: 162918-166419: –	A6006
Phosphorelay protein	Ypd1	MoYpd1	Contig_109: 67377-68020: –	A0844
Phosphorelay protein	Ssk1	MoSsk1	Contig_163: 204696-206372: –	A7172
MEKK	Ste11	Mst11	Contig_19: 77033-80129:–	A4528
MEKK	Ste11	Mst11	Contig_19: 77033-80129:–	A4528
MEKK	Ssk1	MoSsk1	Contig_163: 204696-206372: –	A7172
MEKK	Ssk2/Ssk22	MoSsk2	Contig_203: 346910-351089: –	A6737
MEK	Pbs2	MoPbs2	Contig_225: 69255-70877: –	A2301
MAPK	Hog1	Osm1	Contig_70: 173179-174589: +	A3836

[a] Based on unpublished annotations of predicted genes in strains KoLRILF000956 (Genome accession in NCBI: GCA_000611775.1) and JL3 (Genome unpublished) of *U. muhlenbergii* that were isolated from Mt. Tulaopoding, Ji-Lin Province, China in 2012 and 2017, respectively (Park et al. 2013; Wang et al. 2020)

Slt2 MAPK cascade is activated by Rho1 and Pkc1 to regulate gene expression changes via transcription factors Rlm1 and Swi6 (Jiménez-Gutiérrez et al. 2020) for remodeling of the fungal cell wall during growth and development as well as responding to environmental stimuli. In plant pathogens, this MAPK pathway plays important roles in regulating various infection and developmental processes besides responses to cell wall stress. In some fungi, it is also involved in nutrient sensing and adaptations to oxidative or other stresses. The key components of this CWI MAPK pathway are conserved in ascomycetous plant pathogens and lichen-forming fungi, including

U. muhlenbergii (Table 7.1). It is possible that the CWI MAPK pathway regulates nutrient sensing and symbiotic growth with algal cells. In *Cochliobolus heterostrophus*, the CWI MAPK and *ChPMK1* co-regulate melanin biosynthesis. Old cultures of *U. muhlenbergii* and its lichen thalli are heavily melanized. The CWI and IG MAPKs may also co-regulate melanin synthesis in *U. muhlenbergii*. In addition, the Mps1 MAPK regulates the expression of a number of effectors in *M. oryzae* (Zhang et al. 2017). If fungal–algal interactions also involve effectors, the CWI pathway may play a role in regulating the expression of *U. muhlenbergii* effectors to establish and maintain lichen symbiosis. Furthermore, the CWI MAPK is known to regulate hyphal fusion and heterokaryon formation in *Fusarium graminearum* and other ascomycetes (Hou et al. 2002). The differentiation of mature lichen thalli likely involves hyphal fusion and cell wall modifications, which may be regulated by the CWI MAPK pathway in *U. muhlenbergii*. Although lack of functional studies, genomic and expression profiling analyses have implicated the importance of the CWI pathway in other lichen-forming fungi. For example, in 21-day-old cocultures of *C. grayi* and its algal photobiont, the expression level of a phosphatase gene homologous to *MoYVH1* is up-regulated two-fold (Armaleo et al. 2019). In *M. oryzae*, MoYvh1 is functionally related to both the CWI MAPK and cAMP-PKA pathways (Liu et al. 2016). In *Lasallia pustulata*, the orthologs of *PKC1* and *WSC4* that are functionally related to the CWI pathway in *S. cerevisiae* are among the genes identified by gene–environment correlation analysis to be related to adaptations for growth at different elevations (Dal Grande et al. 2017).

Unlike the IG and CWI MAPKs that have the TEY dual phosphorylation site, the third well-conserved MAPK in fungi has the TGY motif, which is similar to p38 stress-activated MAP kinases (SAPKs) in animals. In yeast, this MAPK of the High-Osmolarity Glycerol (HOG) pathway mainly regulates responses to hyperosmotic stress. In plant pathogenic fungi, although its importance for plant infection varies among different fungi, the HOG pathway has a well-conserved role in regulating adaptive responses to hyperosmotic stress. Deletion of the *HOG1* ortholog results in increased sensitivity to hyperosmotic stress in all the plant pathogenic fungi studied. In *M. oryzae*, the *osm1* deletion mutant is hypersensitive to osmotic stress and desiccation (Dixon et al. 1999). The *U. muhlenbergii* genome contains all the key components of the HOG pathway, including orthologs of yeast Ssk1, Ssk2/22, Pbs2, and Hog1 (Table 7.1). The UmHog1 MAPK has the typical TGY dual phosphorylation site and shares 92% identity with Osm1 of *M. oryzae*. Therefore, it is likely that the HOG pathway is important for survival in nature for *U. muhlenbergii* and other lichens that have to endure frequent wetting-drying cycles. In addition, like in other filamentous fungi (Zhang et al. 2021), *U. muhlenbergii* may also use the HOG pathway to regulate adaptations or responses to oxidative, elevated temperature, UV radiation, and other environmental stresses that occur frequently in many lichens. In the desert lichen fungus *Endocarpon pusillum*, orthologs of the Sho1 osmosensor of the HOG pathway (EPUS01068) and *BCK1* MEK1 of the CWI pathway (EPUS4996) are among the differentially expressed genes under PEG-induced drought stress (Wang et al. 2015).

For the upstream receptors or sensors of these signaling pathways, the *U. muhlenbergii* genome has 73 predicted G-protein coupled receptor (GPCR) genes, including several of them with the CFEM (common in several fungal extracellular membrane proteins) domain unique to fungal GPCRs, that are known to function upstream from both the cAMP-PKA and IG MAPK pathways for regulating plant infection processes in fungal pathogens such as *M. oryzae* and *F. graminearum*, including (Jiang et al. 2019; Kulkarni et al. 2005). It is possible that some of these GPCRs in *U. muhlenbergii* are involved in recognizing chemical and physical signals from algal cells of *T. jamesii* in early re-lichenization processes. In *C. grayi*, 13 of the PTH11-like GPCR genes are up-regulated over two-fold in 21-day-old cocultures with its photobiont cells (Armaleo et al. 2019). The genome sequence of *U. muhlenbergii* also has predicted genes orthologous to other receptor genes that are known to function upstream to the MAPK pathway in *S. cerevisiae* and fungal pathogens, such as orthologs of *SLN1*, *MSB2, SHO1, WSC1–3,* and *MID2.* Some of them may be functional during fungal–algal interactions by activating their downstream MAPKs and targets. In cocultures of *C. grayi* with algal cells, the *WSC1* and *MSB2* orthologs are up-regulated, although less than two-fold (Armaleo et al. 2019).

7.5 Conclusions and Perspectives

Our knowledge of regulatory mechanisms controlling symbiont recognition and interactions in lichen symbiosis is very limited, likely due to the slow growth rate of most lichen-forming fungi and lack of a model system for molecular genetic studies. With its relative fast growth rate, efficiency for transformation and gene knockout, and symbiosis-related yeast-pseudohypha transition, *U. muhlenbergii* has the potential to be used as a model to study early stages of symbiotic interactions and signal transduction pathways involved in the recognition of compatible symbiont cells and development of highly differentiated masses of fungal–algal cells. In addition to the cAMP-PKA and MAPK pathways described above, other signaling pathways (such as calcium or nitric oxide signaling) may also play roles in lichen symbiosis. Furthermore, some of these conserved signaling pathways are known to regulate fungal–fungal and fungal–bacterial interactions in ascomycetes (Zhang et al. 2021) and they may affect the interactions of *U. muhlenbergii* with other microbes present in lichen thalli.

Besides genetic regulations, lichen symbiosis may also involve epigenetic mechanisms. Earlier studies with *C. grayi* and other lichen-forming fungi have shown that DNA methylation is increased in the symbiotic condition in comparison with axenic cultures, indicating that high levels of DNA methylation may be associated with lichen formation. Although whether *U. muhlenbergii* is increased in DNA methylation during symbiotic interactions is not clear, its genome has two well-conserved DNA methyltransferase (DNMT) genes orthologous to *RID* and *DIM-5* of *Neurospora crassa*. Besides assaying DNA methylation, functional characterization of these two DNMT genes will clarify the importance of DNA methylation in lichen symbiosis. Along the same line, functional characterization of genes involved in histone modifications, such as histone acetyltransferase or deacetylase genes (Aramayo and Selker 2013; Dubey and Jeon 2017), will determine the importance of epigenetic controls in lichen symbiosis.

Despite recent advances in genetic studies with *U. muhlenbergii*, it is still difficult to pursue molecular genetics studies with its photobiont, which poses a problem for advanced studies of fungal–algal interactions with this lichen system. Nevertheless, genomics, proteomics, and metabolomics analyses are possible with *T. jamesii* to study its interactions with *U. muhlenbergii*. In comparison with fungal pathogens, another lagging area in lichen-forming fungi is advanced cell biology studies with fungal–algal interactions. It will be important to develop transformants of *U. muhlenbergii* expressing fluorescent markers in different cellular components and carefully examine by confocal microscopy for cellular changes and differentiations associated with initial symbiotic interactions.

Acknowledgments The authors are grateful to the National Natural Science Foundation of China for financial support (grant No. 32170082 and 32070096). The authors thank Ms. Qiuxia Yang for her assistance in micromanipulation on the lichen thallus.

References

Aramayo R, Selker EU (2013) *Neurospora crassa*, a model system for epigenetics research. Cold Spring Harb Perspect Biol 5:a017921. https://doi.org/10.1101/cshperspect.a017921

Armaleo D, Müller O, Lutzoni F, Andrésson ÓS, Blanc G, Bode HB, Collart FR, Dal Grande F, Dietrich F, Grigoriev IV et al (2019) The lichen symbiosis re-viewed through the genomes of *Cladonia grayi* and its algal partner *Asterochloris glomerata*. BMC Genomics 20:605–637. https://doi.org/10.1186/s12864-019-5629-x

Bordenave CD, Muggia L, Chiva S, Leavitt SD, Carrasco P, Barreno E (2022) Chloroplast morphology and pyrenoid ultrastructural analyses reappraise the diversity of the lichen phycobiont genus *Trebouxia* (Chlorophyta). Algal Res 61:102561. https://doi.org/10.1016/j.algal.2021.102561

Boyce KJ, Andrianopoulos A (2015) Fungal dimorphism: the switch from hyphae to yeast is a specialized morphogenetic adaptation allowing colonization of a host. FEMS Microbiol Rev 39:797–811. https://doi.org/10.1093/femsre/fuv035

Cao S, Zhang F, Liu C, Hao Z, Tian Y, Zhu L, Zhou Q (2015) Distribution patterns of haplotypes for symbionts from *Umbilicaria esculenta* and *U. muehlenbergii* reflect the importance of reproductive strategy in shaping population genetic structure. BMC Microbiol 15:212. https://doi.org/10.1186/s12866-015-0527-0

Choi J, Jung WH, Kronstad JW (2015) The cAMP/protein kinase A signaling pathway in pathogenic basidiomycete fungi: Connections with iron homeostasis. J Microbiol 53:579–587. https://doi.org/10.1007/s12275-015-5247-5

Dal Grande F, Sharma R, Meiser A, Rolshausen G, Budel B, Mishra B, Thines M, Otte J, Pfenninger M, Schmitt I (2017) Adaptive differentiation coincides with local bioclimatic conditions along an elevational cline in populations of a lichen-forming fungus. BMC Evol Biol 17:93–106. https://doi.org/10.1186/s12862-017-0929-8

Davydov EA, Peršoh D, Rambold G (2017) Umbilicariaceae (lichenized Ascomycota) - Trait evolution and a new generic concept. Taxon 66:1282–1303. https://doi.org/10.12705/666.2

Dixon KP, Xu JR, Smirnoff N, Talbot NJ (1999) Independent signaling pathways regulate cellular turgor during hyperosmotic stress and appressorium-mediated plant infection by *Magnaporthe grisea*. Plant Cell 11:2045–2058. https://doi.org/10.1105/tpc.11.10.2045

Dubey A, Jeon J (2017) Epigenetic regulation of development and pathogenesis in fungal plant pathogens. Mol Plant Pathol 18:887–898. https://doi.org/10.1111/mpp.12499

Fang EGC, Dean RA (2000) Site-directed mutagenesis of the *mag B* gene affects growth and development in *Magnaporthe grisea*. Mol Plant-Microbe Interact 13:1214–1227. https://doi.org/10.1094/Mpmi.2000.13.11.1214

Friedl T (1989) Comparative ultrastructure of pyrenoids in *Trebouxia* (Microthamniales, Chlorophyta). Plant Syst Evol 164:145–159. https://doi.org/10.1007/BF00940435

Friedl T, Budel B (2008) Photobionts. In: Nash TH (ed) Lichen biology. Cambridge University Press, pp 21–38. https://doi.org/10.1017/CBO9780511790478.003

Gimeno CJ, Ljungdahl PO, Styles CA, Fink GR (1992) Unipolar cell divisions in the yeast *S. cerevisiae* lead to filamentous growth: regulation by starvation and RAS. Cell 68:1077–1090. https://doi.org/10.1016/0092-8674(92)90079-r

Goodenough U, Wagner R, Roth R (2021) Lichen 4. The algal layer. Algal Res 58:102355. https://doi.org/10.1016/j.algal.2021.102355

Grube M, Cardinale M, de Castro Jr JV, Muller H, Berg G (2009) Species-specific structural and functional diversity of bacterial communities in lichen symbioses. ISME J 3:1105–1115. https://doi.org/10.1038/ismej.2009.63

Gupta L, Vermani M, Kaur Ahluwalia S, Vijayaraghavan P (2021) Molecular virulence determinants of *Magnaporthe oryzae*: disease pathogenesis and recent interventions for disease management in rice plant. Mycology 12:174–187. https://doi.org/10.1080/21501203.2020.1868594

Hestmark G (1992) Conidiogenesis in five species of *Umbilicaria*. Fungal Biol 96:1033–1043. https://doi.org/10.1016/S0953-7562(09)80113-5

Honegger R (1986) Ultrastructural studies in lichens: I. Haustorial types and their frequencies in a range of lichens with trebouxioid photobionts. New Phytol 103:785–795. https://doi.org/10.1111/j.1469-8137.1986.tb00853.x

Hou ZM, Xue CY, Peng YL, Katan T, Kistler HC, Xu JR (2002) A mitogen-activated protein kinase gene (*MGV1*) in *Fusarium graminearum* is required for female fertility, heterokaryon formation, and plant infection. Mol Plant-Microbe Interact 15:1119–1127. https://doi.org/10.1094/Mpmi.2002.15.11.1119

Jiang C, Zhang X, Liu H, Xu JR (2018) Mitogen-activated protein kinase signaling in plant pathogenic fungi. PLoS Pathog 14:e1006875. https://doi.org/10.1371/journal.ppat.1006875

Jiang C, Cao S, Wang Z, Xu H, Liang J, Liu H, Wang G, Ding M, Wang Q, Gong C et al (2019) An expanded subfamily of G-protein-coupled receptor genes in *Fusarium graminearum* required for wheat infection. Nat Microbiol 4:1582–1591. https://doi.org/10.1038/s41564-019-0468-8

Jiménez-Gutiérrez E, Alegría-Carrasco E, Sellers-Moya Á, Molina M, Martín H (2020) Not just the wall: the other ways to turn the yeast CWI pathway on. Int Microbiol 23:107–119. http://doi.org/0.1007/s10123-019-00092-2

Joneson S, Lutzoni F (2009) Compatibility and thigmotropism in the lichen symbiosis: a reappraisal. Symbiosis 47:109–115. https://doi.org/10.1007/Bf03182294

Joneson S, Armaleo D, Lutzoni F (2011) Fungal and algal gene expression in early developmental stages of lichen-symbiosis. Mycologia 103:291–306. https://doi.org/10.3852/10-064

Kulkarni RD, Thon MR, Pan H, Dean RA (2005) Novel G-protein-coupled receptor-like proteins in the plant pathogenic fungus *Magnaporthe grisea*. Genome Biol 6:R24. https://doi.org/10.1186/gb-2005-6-3-r24

Lewis LA, Muller-Parker G (2004) Phylogenetic placement of "Zoochlorellae" (Chlorophyta), algal symbiont of the temperate sea anemone *Anthopleura elegantissima*. Biol Bull 207:87–92. https://doi.org/10.2307/1543583

Liu X, Qian B, Gao C, Huang S, Cai Y, Zhang H, Zheng X, Wang P, Zhang Z (2016) The putative protein phosphatase MoYvh1 functions upstream of MoPdeH to regulate the development and pathogenicity in *Magnaporthe oryzae*. Mol Plant-Microbe Interact 29:496–507. https://doi.org/10.1094/MPMI-11-15-0259-R

Lutzoni F, Nowak MD, Alfaro ME, Reeb V, Miadlikowska J, Krug M, Arnold AE, Lewis LA, Swofford DL, Hibbett D et al (2018) Contemporaneous radiations of fungi and plants linked to symbiosis. Nat Commun 9:5451–5461. https://doi.org/10.1038/s41467-018-07849-9

Meeßen J, Eppenstein S, Ott S (2013) Recognition mechanisms during the pre-contact state of lichens: II. Influence of algal exudates and ribitol on the response of the mycobiont of *Fulgensia bracteata*. Symbiosis 59:131–143. https://doi.org/10.1007/s13199-012-0219-6

Moya P, Molins A, Martinez-Alberola F, Muggia L, Barreno E (2017) Unexpected associated microalgal diversity in the lichen *Ramalina farinacea* is uncovered by pyrosequencing analyses. PLoS One 12: e0175091. https://doi.org/10.1371/journal.pone.0175091

Muggia L, Nelsen MP, Kirika PM, Barreno E, Beck A, Lindgren H, Lumbsch HT, Leavitt SD, Fernandez-Mendoza F, Perez-Ortega S et al (2020) Formally described species woefully underrepresent phylogenetic diversity in the common lichen photobiont genus *Trebouxia* (Trebouxiophyceae, Chlorophyta): an impetus for developing an integrated taxonomy. Mol Phylogenet Evol 149:106821. https://doi.org/10.1016/j.ympev.2020.106821

Nazem-Bokaee H, Hom EFY, Warden AC, Mathews S, Gueidan C (2021) Towards a systems biology approach to understanding the lichen symbiosis: opportunities and challenges of implementing network modelling. Front Microbiol 12:667864. https://doi.org/10.3389/fmicb.2021.667864

Nelsen MP, Leavitt SD, Heller K, Muggia L, Lumbsch HT (2021) Macroecological diversification and convergence in a clade of keystone symbionts. FEMS Microbiol Ecol 97:fiab072. https://doi.org/10.1093/femsec/fiab072

Park SY, Jeong MH, Wang HY, Kim JA, Yu NH, Kim S, Cheong YH, Kang S, Lee YH, Hur JS (2013) Agrobacterium tumefaciens-mediated transformation of the lichen fungus, *Umbilicaria muehlenbergii*. PLoS One 8:e83896. https://doi.org/10.1371/journal.pone.0083896

Park SY, Choi J, Lee GW, Jeong MH, Kim JA, Oh SO, Lee YH, Hur JS (2014) Draft genome sequence of *Umbilicaria muehlenbergii* KoLRILF000956, a lichen-forming fungus amenable to genetic manipulation. Genome Announc 2:e00357–e00314. https://doi.org/10.1128/genomeA.00357-14

Sánchez-Martínez C, Pérez-Martín J (2002) Gpa2, a G-protein alpha subunit required for hyphal development in *Candida albicans*. Eukaryot Cell 1:865–874. https://doi.org/10.1128/ec.1.6.865-874.2002

Singh RS, Walia AK (2014) Characteristics of lichen lectins and their role in symbiosis. Symbiosis 62: 123–134. https://doi.org/10.1007/s13199-014-0278-y

Sonneborn A, Bockmühl DP, Gerads M, Kurpanek K, Sanglard D, Ernst JF (2000) Protein kinase A encoded by *TPK2* regulates dimorphism of *Candida albicans*. Mol Microbiol 35:386–396. https://doi.org/10.1046/j.1365-2958.2000.01705.x

Spribille T, Tuovinen V, Resl P, Vanderpool D, Wolinski H, Aime MC, Schneider K, Stabentheiner E, Toome-Heller M, Thor G et al (2016) Basidiomycete yeasts in the cortex of ascomycete macrolichens. Science 353:488–492. https://doi.org/10.1126/science.aaf8287

Svetaz LA, Bustamante CA, Goldy C, Rivero N, Muller GL, Valentini GH, Fernie AR, Drincovich MF, Lara MV (2017) Unravelling early events in the Taphrina deformans-Prunus persica interaction: an insight into the differential responses in resistant and susceptible genotypes. Plant Cell Environ 40:1456–1473. https://doi.org/10.1111/pce.12942

Wang Y, Zhang X, Zhou Q, Zhang X, Wei J (2015) Comparative transcriptome analysis of the lichen-forming fungus *Endocarpon pusillum* elucidates its drought adaptation mechanisms. Sci China Life Sci 58:89–100. https://doi.org/10.1007/s11427-014-4760-9

Wang Y, Wei X, Bian Z, Wei J, Xu JR (2020) Coregulation of dimorphism and symbiosis by cyclic AMP signaling in the lichenized fungus *Umbilicaria muhlenbergii*. Proc Natl Acad Sci U S A 117:23847–23858. https://doi.org/10.1073/pnas.2005109117

Wei JC, Jiang YM (1993) The Asian Umbilicariaceae (Ascomycota). International Academic Publishers, Beijing

Zhang X, Liu W, Li Y, Li G, Xu JR (2017) Expression of HopAI interferes with MAP kinase signalling in *Magnaporthe oryzae*. Environ Microbiol 19:4190–4204. https://doi.org/10.1111/1462-2920.13884

Zhang X, Wang Z, Jiang C, Xu JR (2021) Regulation of biotic interactions and responses to abiotic stresses by MAP kinase pathways in plant pathogenic fungi. Stress Biol 1. https://doi.org/10.1007/s44154-021-00004-3

Zhao XH, Mehrabi R, Xu JR (2007) Mitogen-activated protein kinase pathways and fungal pathogenesis. Eukaryot Cell 6:1701–1714. https://doi.org/10.1128/Ec.00216-07

Zhou X, Li G, Xu JR (2011) Efficient approaches for generating GFP fusion and epitope-tagging constructs in filamentous fungi. In: Xu JR, Bluhm BH (eds) Fungal genomics. Springer, pp 199–212. https://doi.org/10.1007/978-1-61779-040-9_15

8 After Air, Light, and Water, the Next Most Important Thing Is Grass: An Introduction to the *Epichloë*–Grass Symbiosis

Daniel A. Bastías, Emma R. Applegate, Pedro E. Gundel, Linda J. Johnson, Wade J. Mace, Christina D. Moon, Sushma Prakash, and Stuart D. Card

Abstract

Epichloë (Clavicipitaceae) is a genus of filamentous fungal endophytes that have co-evolved with cool-season grasses with which they form perpetual, symbiotic associations. In natural ecosystems *Epichloë* endophytes have implications for species diversity, food web structures, and fundamental ecological processes. In many managed pastoral systems, selected *Epichloë* strains are regarded as necessary components of the sward as they confer bioprotective traits to their host grasses. The most agriculturally important associations are those between selected strains of asexual, vertically transmitted *Epichloë* spp. and the widely used pasture species: perennial ryegrass and tall fescue. Selected *Epichloë* strains confer invertebrate, especially insect, pest deterrence to their plant hosts from the production of several alkaloidal secondary metabolites. Additional *Epichloë*-mediated attributes include tolerance against fungal diseases and abiotic stresses, such as those caused by drought and/or nutrient deficiencies. Strains of mutualistic *Epichloë* have been developed into highly efficacious biocontrol products and are widely utilized within the Americas, Australia, and New Zealand for pasture persistence and wildlife deterrence.

Keywords

Alkaloid · Animal toxicosis · Biological control · Endophyte · Fescue · Ryegrass

8.1 Background

This review is focussed primarily on the biology of the *Epichloë*–grass symbioses, and the science driving the selection and development of agriculturally beneficial *Epichloë* (Clavicipitaceae) strains. While several research groups worldwide also study *Epichloë* fungi, this chapter places emphasis on the farming challenges and research performed predominantly in New Zealand. Significant research activity is undertaken by AgResearch Limited, a Crown Research Institute, in conjunction with its wholly owned subsidiary, Grasslanz Technology Limited, which develops and commercializes innovative plant technologies, providing enormous benefits to pastoral agricultural production systems in New Zealand and worldwide.

D. A. Bastías · E. R. Applegate · L. J. Johnson · W. J. Mace · C. D. Moon · S. Prakash · S. D. Card (✉)
Resilient Agriculture, AgResearch Limited, Grasslands Research Centre, Palmerston North, New Zealand
e-mail: Stuart.Card@agresearch.co.nz

P. E. Gundel
Facultad de Agronomía, IFEVA, Universidad de Buenos Aires, CONICET, Buenos Aires, Argentina

Laboratorio de Biología Vegetal, Instituto de Ciencias Biológicas, Universidad de Talca, Talca, Chile

Y.-P. Hsueh, M. Blackwell (eds.), *Fungal Associations*, The Mycota 9,
https://doi.org/10.1007/978-3-031-41648-4_8

Most multicellular life on Earth lives in symbiosis with microorganisms, and plants are no exception (Rodriguez and Redman 2008). The term symbiosis is used to describe a close and often long-term interaction between two different biological species and is derived from the Greek words simply meaning 'living together' and originally popularized by one of the founding fathers of plant pathology, Heinrich Anton de Bary (1879). However, there is still some debate on the exact definition of the term (Martin and Schwab 2012; Schulz et al. 1998). In this chapter, we have accepted the broader definition of the term, which incorporates the spectrum of possible persistent relationships (parasitic, amensalistic, commensalistic, and mutualistic) between all species. These symbioses are complex, ranging from mutualism through commensalism to parasitism in a continuous manner (Partida-Martinez and Heil 2011; Redman et al. 2001), although some authors regard parasitism as an unbalanced symbiosis (Kogel et al. 2006). Plants form symbioses throughout their lifecycle with an aggregation of diverse, interacting microorganisms including archaea, bacteria, fungi, and protists (Mercado-Blanco 2015; Rodriguez et al. 2009) and these intimate associations are important ecological determinants of plant biodiversity (Clay 1988). This microbial community (the plant microbiota) coupled with their surrounding environment (the entire habitat = the microbiome) is crucially important for the health and productivity of the host plant.

For this chapter, we use the definition of the term endophyte in a stricter sense than originally used (also by de Bary). In its original form, endophyte simply meant 'inside the plant', being derived from the Greek words endon (=within) and phyton (=plant). It is suggested that in this original form, the term is vague and meaningless and therefore the term should encompass several other important behavioural traits (Mercado-Blanco 2015). Microorganisms defined as endophytes should therefore also spend all or part of their lifecycle within the plant, and maybe more importantly, should cause no apparent disease symptoms. The term 'mutualistic endophyte' should be used to define a microorganism that has been identified as also displaying a mutualistic behaviour, at least for part of its lifecycle, and in doing so conferring a benefit on its host (for an in-depth discussion, read Wennström 1994; Wilson 1995). As discussed by several authors, the definition of endophyte is subsequently more descriptive and encompasses information regarding the type of association between the microorganism and its plant host (Carroll 1988; Hardoim et al. 2015; Hyde and Soytong 2008; Mercado-Blanco 2015; Rodriguez et al. 2009). The added terms then make this updated definition of the word 'endophyte' more informative and excludes dormant pathogens or saprophytes (read Sinclair and Cerkauskas 1996 for a detailed discussion). This definition would therefore encourage researchers to describe uncharacterized microorganisms isolated from surface-disinfected plant samples more appropriately as leaf, root, or seed 'associated' microorganisms (Bent and Chanway 2002; Mercado-Blanco 2015).

8.2 Historical Summary of *Epichloë* in Agriculture

The historical events that brought *Epichloë* to the attention of the scientific community are both complex and fascinating. Once a relatively obscure group of fungi studied by a few botanists, these Ascomycetes rose to notoriety in the 1970s and early 1980s as being synonymous with livestock toxicity (Bacon et al. 1977; Fletcher and Harvey 1981; Freeman 1904; Lloyd 1959; McLennan 1920; Neil 1940; Sampson 1933). However, several events led to this endophyte-derived toxicity in farmed ruminants, originating with the introduction of forage species into several New World countries. Many European grass species were introduced to the USA and New Zealand from the early 1800s as the new settlers experimented with various foods to feed their farmed ruminants (Caradus et al. 2021c; Star and Brooking 2006; Stewart 2006). In the USA, tall fescue [*Lolium arundinaceum* (Schreb.) S.J. Darbyshire = *Festuca arundinacea* Schreb.] was the forage that was particularly well adapted,

Fig. 8.1 A sheep in summer exhibiting ryegrass staggers, a neuro-muscular disorder caused by the common toxic strain of endophyte in perennial ryegrass (Photo courtesy of Russel Croy, AgResearch Ltd, New Zealand)

although it was not until the 1950s that the species was widely planted (Hoveland 2009; Sleper and West 1996). Perennial ryegrass (*Lolium perenne*) was the most successful forage species in the New Zealand environment and although both these grass species were persistent and productive in their New World habitats, they started to cause serious toxicity problems in livestock. After much investigation of pasture leaves, leaf litter, and soil, *Epichloë* species were identified as the causal agents of mycotoxins present in these grazed pasture species that manifested as serious livestock toxicoses after ingestion (Bacon et al. 1977; Fletcher and Harvey 1981; Hoveland et al. 1980; Latch and Christensen 1982). The most potent of these neurotoxic compounds include the lolitrems (indole diterpenes) that result in ryegrass staggers (Fig. 8.1) and ergot alkaloids responsible for fescue foot, bovine fat necrosis, and fescue toxicosis (summer slump or syndrome) (Bacon et al. 1977; Fletcher et al. 1990; Gallagher et al. 1981; Strickland et al. 2009; White et al. 2003). The most obvious remedy for this agricultural problem was to remove the causal fungal agent of these mammalian syndromes from the fescue and ryegrass hosts (Hoveland 2009; Latch and Christensen 1982). However, this practice was soon found to be largely undesirable as the fungus was unknowingly providing the plant with additional bioprotective traits. For example, in most jurisdictions of New Zealand, the pastures lacking endophytes soon perished due to unforeseen pressure from an introduced insect pest, and an alternative remedy was required.

The situation in New Zealand was unique, whereby a naturalized forage plant and an insect herbivore arrived from different parts of the world and came into direct contact with each other for the first time. The first introduction was deliberate with the importation of perennial ryegrasses from Europe, mainly England (Daly 1973); but the later introduction of the Argentine stem weevil (*Listronotus bonariensis*) in the early part of the twentieth century, was most likely accidental (Stewart 2006; Williams et al. 1994). This event illustrates how a species not considered troublesome in its native ecosystem can manifest as a serious pest when introduced into a foreign ecosystem (Goldson et al. 2005).

A scientific breakthrough came in the mid-1980s when peramine, an alkaloid responsible for insect deterrence, produced by an endophyte now known as *Epichloë festucae* var. *lolii* in its perennial ryegrass host, was discovered (Rowan and Gaynor 1986; Rowan et al. 1986). Further secondary metabolites produced by the grass–endophyte association were linked to bioactivity (Bacon 1995; Gallagher et al. 1981; Lyons et al. 1986; Riedell et al. 1991), and a bioprospecting pipeline was developed to identify *Epichloë* strains that were less toxic to livestock while still producing alkaloids that deterred insect pests (Johnson et al. 2013a).

Since the initial research, scientific programmes were established that centred around the modification of these bioprospecting pipelines to identify, characterize, and select agriculturally beneficial strains of *Epichloë* essentially to exploit this natural phenomenon for agricultural gains (Card et al. 2014a; de Bonth et al. 2015; Johnson et al. 2013a). These selected *Epichloë* strains were then incorporated into elite grass cultivars with the aim to increase pasture persistence and productivity, which is of particular importance to farming systems in New Zealand, Australia, the United States, and parts of South America (Johnson and Caradus 2019). Today, the genus *Epichloë* is synonymous with efficacious biological control of insect pests within applied agriculture (Card et al. 2016) and as a fungus that plays a critical role in host plant defences and grassland ecology (Bastias et al. 2017; Clay 1988; Malinowski and Belesky 2000; Omacini et al. 2001; Saikkonen et al. 2006). For further information, please read di Menna et al. (2012) for the New Zealand historical account and Bacon (1995) and Ball et al. (2019) for the USA perspective.

8.3 Biology of *Epichloë* Endophytes

Many members of the grass family Poaceae, also known as the true grasses, associate with a plethora of endophytic microorganisms (Sánchez Márquez et al. 2012), with the tribe Balansiae (Clavicipitaceae) one of the most researched. Many of the fungi within the Balansiae are endophytic, systematically colonizing their plant hosts while producing an array of secondary metabolites, most notably nitrogen-based compounds known as alkaloids (Diehl 1950; Leuchtmann and Clay 1988; Torres and White 2009). The most intensely studied genus within the Balansiae is *Epichloë*, which forms perpetual symbioses with more than 100 species of cool-season grass (subfamily Pooideae), with the majority originating from Europe and Asia (Caradus et al. 2021b). As *Epichloë* species are ecologically important constituents of many grassland ecosystems, which cover over 30% of the Earth's land area, they are regarded as keystone species (Gibson 2009; Nissinen et al. 2019; Suttie et al. 2005).

Epichloë–grass associations form a range of interactions from mutualistic to antagonistic, with the nature of the symbiosis closely linked to the reproductive lifecycle of the fungus (Brem and Leuchtmann 1999). *Epichloë* is a monophyletic genus that includes anamorphic (asexual) and teleomorphic (sexual) species (Fig. 8.2), with the former previously classified as *Neotyphodium* (read Leuchtmann et al. 2014 for a detailed taxonomic history of the genus). Note that species that produce ascospores usually also produce conidia which may serve as spermatia in sexual reproduction (see below). More than 30 species of *Epichloë* have been described, most exhibiting an asexual lifecycle, being exclusively vertically transmitted via the host's infected seeds (Liu et al. 2017; Zhang et al. 2017). This plant–endophyte association is essentially a hereditary endosymbiosis as the plant's progeny not only inherit their parent's chromosomes and cytoplasm but also their mutualistic symbiont (Wernegreen 2012).

Asexual *Epichloë* species are largely mutualistic, relying on vertical transmission for their reproductive success, while sexual species can transmit horizontally between hosts, and some species can simultaneously transmit both vertically and horizontally in pleiotropic symbioses (Clay 1990). For hosts infected by sexual *Epichloë*, vegetative growth is asymptomatic (as seen also for asexual *Epichloë*); however, the sexual state is characterized by the formation of stromata (Fig. 8.3) on the reproductive tillers of the host, aborting the development of some or all of the inflorescences in a disease descriptively

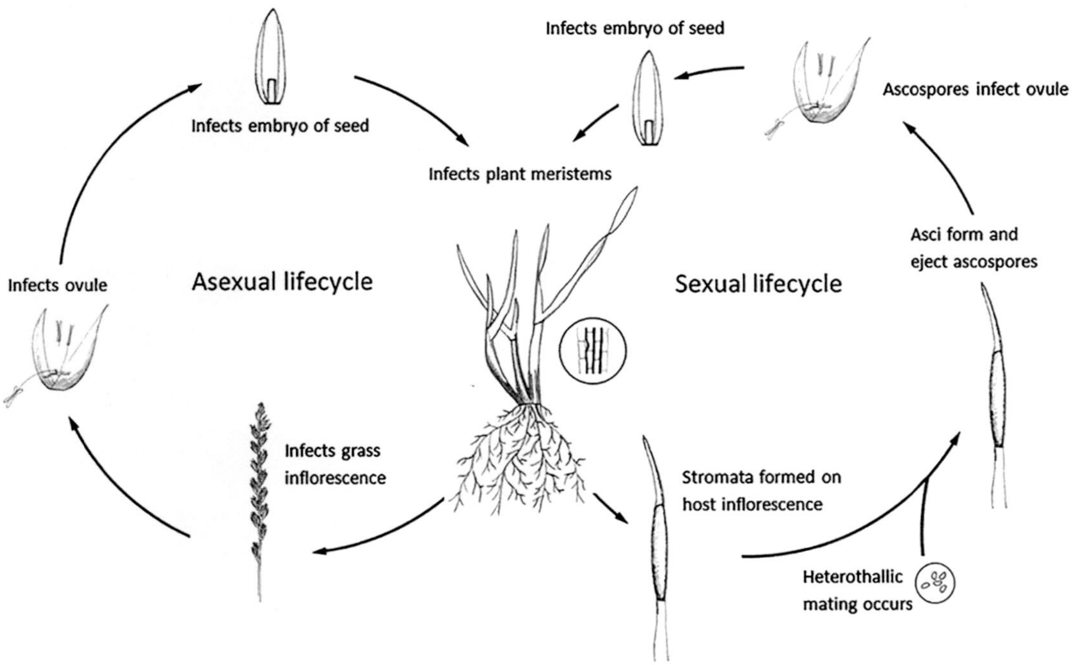

Fig. 8.2 Asexual and sexual lifecycle of *Epichloë* fungal endophytes (Johnson et al. 2013a reprinted by permission of Taylor & Francis Ltd)

known as choke (Kirby 1961; White et al. 1991). The most infamous sexual species is *E. typhina*, which is largely antagonistic to several of its grass hosts, including the economically important orchardgrass (or cocksfoot = *Dactylis glomerata*), as the fungus inhibits plant seed production, impeding the host's ability to reproduce (Chung and Schardl 1997; Leyronas and Raynal 2008; Pfender and Alderman 1999; Western and Cavett 1959).

The mating system of sexual *Epichloë* species is bipolar heterothallic and is further supported by a tripartite relationship, whereby conidia produced on stromata may be vectored by a specialized symbiotic insect between opposite mating types of the stromata in a process of spermatization found in a number of fungi, analogous to plant pollination (Bultman and White 1988; Kohlmeyer and Kohlmeyer 1974; Leuchtmann et al. 2014). Adult flies of the genus *Botanophila* (Anthomyiidae: Diptera) feed on the stromata after being attracted to them by volatile compounds produced by the fungus (Bultman et al. 1995). As feeding ensues, these flies consume spermatia and therefore act as a reservoir of spermatia from the various *Epichloë* strains they have fed upon. When the fly defaecates, a mixture of spermatia from different *Epichloë* strains is laid along with the flies' eggs (Bultman and Leuchtmann 2008), with the stromatal structure forming a cool shelter for *Botanophila* larval development (Lembicz et al. 2021). If these spermatia and stroma are compatible, the stroma will thicken allowing formation of numerous perithecia that contain asci, which in turn produce ascospores (Chung and Schardl 1997). The lifecycle is completed when ascospores are ejected from the perithecia and dispersed by wind to infect new grass hosts. *Botanophila* flies may not be the only vectors of *Epichloë* spermatia. Gastropods, such as slugs, have been observed feeding on *Epichloë* stromata, leading to transmission of viable spermatia and cross-fertilization (Górzyńska et al. 2011; Hoffman and Rao 2013, 2014).

Our understanding of the molecular interactions associated with the symbiosis between grass and endophyte has been

Fig. 8.3 Stromata observed on perennial ryegrass (Photo courtesy of Alan Stewart, PGG Wrightson Seeds Ltd, New Zealand)

significantly enhanced by modern nucleic acid sequencing technologies that allow the genomes and transcriptomes of both partners to be examined under a range of conditions and developmental stages. Studies have revealed that *Epichloë* species are not silent passengers that evade the hosts' defences within the association as earlier thought, but both partners actively interact with each other. *Epichloë* can produce a range of small secreted proteins, termed effectors (Hassing et al. 2019; Wang et al. 2021b), that modulate host defence responses; the effectors show altered expression in endophyte-infected compared to endophyte-free grasses (Dinkins et al. 2017; Dupont et al. 2015; Schmid et al. 2017; Wang et al. 2019). Moreover, the involvement of a myriad of processes contributes to symbiotic compatibility, however as this is a broad area of research, the mechanisms will not be discussed in detail here (for reviews on the subject, please read Johnson et al. 2021; Scott et al. 2018).

While sexual *Epichloë* species are haploid, an intriguing feature of this genus is the prevalence of interspecific hybrids among the asexual species that have ancestries involving two or three distinct *Epichloë* lineages, which contributes to the genetic diversity of these otherwise clonal species (Leuchtmann et al. 2019; Moon et al. 2004; Schardl 2010). The presence of hybrids was first detected in endophytes of tall fescue and perennial ryegrass via sequence analyses of specific marker genes (Schardl et al. 1994; Tsai et al. 1994); however, genome-level analyses have provided far greater insight into the extent that hybridization has had on the genetic diversity of *Epichloë* endophytes (Campbell et al. 2017). Superinfections of established associations followed by anastomosis (hyphal fusion) is likely responsible for the generation of interspecific hybrid endophytes, giving rise to a highly complex evolutionary history (Charlton et al. 2012; Tsai et al. 1994). These unique symbionts can also be defined by several further characters, including their foliar lifestyle, intercalary hyphal growth, transmission via seed and host specificity (see below for more details).

8.3.1 A Foliar Lifestyle

Unlike other well-known plant symbionts such as rhizobia and arbuscular mycorrhizal fungi that are associated with the rhizosphere, asexual *Epichloë* species are only located in the above-ground tissues of their grass hosts, where they spend their entire lifecycle growing systemically within the intercellular spaces between plant cells (Fig. 8.4). Vegetative plants colonized by *Epichloë* remain completely symptomless and cannot be differentiated from plants lacking the symbiosis with the naked eye. *Epichloë* can, however, be easily observed in planta, using basic microscopy techniques that allow the presence of septate, intercellular, infrequently branched hyphae running longitudinally in leaf sheaths to be observed (Bacon and White 1994; Latch et al. 1988). Leaf blade colonization patterns can differ with host and endophyte genotype, and plant age (Christensen et al. 2002). It is speculated that the ligular zone located between the leaf sheath and blade is a physical barrier to the advancement of *Epichloë* hyphae and as grass leaves age, hyphae can increase in diameter, and therefore biomass, but not in frequency (Christensen and Voisey 2006). Critically, *Epichloë* endophytes colonize meristematic tissues and structures such as axillary buds and inflorescence primordia that allow infection of resulting vegetative tillers and reproductive tissues for seed transmission (see below).

8.3.2 Intercalary Hyphal Growth

In their vegetative state, the growth of *Epichloë* is fully synchronized and highly regulated with that of their hosts, colonizing grass leaves by a unique mechanism termed intercalary hyphal extension, and not by the generalized model of hyphal tip growth (Christensen et al. 2002, 2008; Voisey 2010). During intercalary hyphal extension, *Epichloë* hyphae attach to the enlarging and elongating host grass cells through an extracellular matrix of uncertain origin, and cumulative growth along the length of the hyphal filament allows the endophyte to extend at the same rate as its host (Christensen et al. 2008).

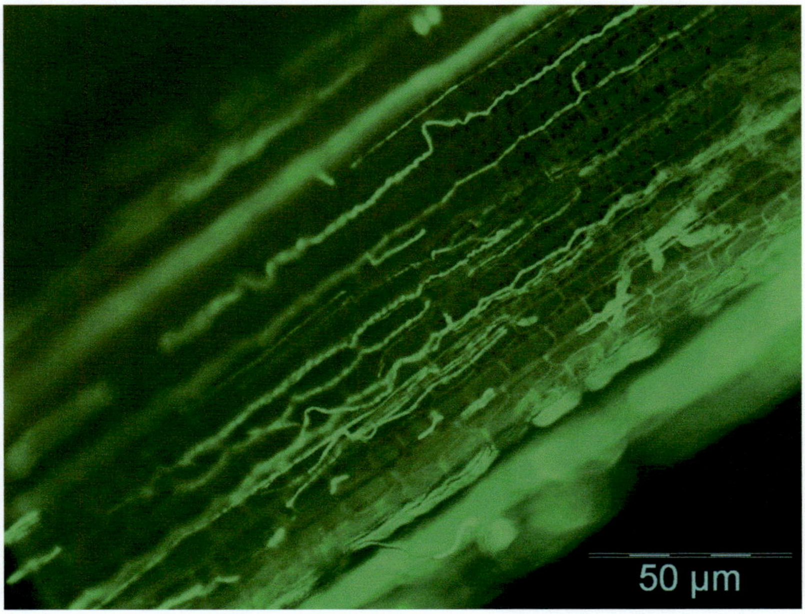

Fig. 8.4 Hyphae of *Epichloë coenophiala* growing within the intercellular spaces of a tall fescue leaf sheath (Card et al. 2013 reprinted by permission of Taylor & Francis Ltd)

8.3.3 Transmission via Seed

As plants and their microbial symbionts have co-evolved, the latter have developed sophisticated mechanisms of transmission. Seed-borne symbionts (e.g. *Alternaria* spp.) may be carried in or on the seed as opportunistic, transient colonizers generally resulting in a low frequency of colonized seed. In contrast, seed transmitted symbionts (e.g. *Epichloë* spp.) have evolved mechanisms that exploit the plants' reproductive processes and use this strategy as an exclusive mechanism to establish themselves within the progeny. These symbionts are therefore totally reliant on the success of the plant's reproductive strategy for their dissemination and survival (Barret et al. 2016). Moreover, the frequency of seed transmission among populations and individual plants can vary widely, and has been shown to be impacted considerably by host genetics (Faeth et al. 2004; Gagic et al. 2018; Leyronas and Raynal 2001; Ravel et al. 1997). For a detailed account of vertical transmission by *Epichloë* spp., read Liu et al. (2017) and Zhang et al. (2017).

8.3.4 Host Specificity

All known species of *Epichloë* are restricted to a single or several host grass species or closely related grass genera within the cool-season grass subfamily Pooideae due to co-speciation over many millennia (Schardl 2010; Schardl et al. 1997; Schardl and Philips 1997). However, the *Epichloë* lineage may have originally emerged from an animal pathogen via an interkingdom host jump (Spatafora et al. 2007).

There is substantial variation between and within different *Epichloë* species, particularly with respect to their in-planta colonization patterns and chemical diversity (discussed later). For example, *Epichloë occultans*, a symbiont of the annual ryegrasses *Lolium multiflorum* and *Lolium rigidum*, is only found in vegetative plants as a dense mycelial mass located at the base of the leaf sheath, while the hyphae of most other *Epichloë* species are found throughout the leaf sheath, aligned to the leaf axis (Christensen et al. 2002). Furthermore, *Epichloë* associations with tall fescue differ in their distribution of fungal hyphae, with continental-type associations exhibiting poor colonization of their leaf blades compared to Mediterranean associations where hyphae can be readily observed in leaf blades and sheaths (Christensen and Voisey 2006; Takach et al. 2012). Many of the advantageous traits conferred by *Epichloë* to their grass hosts are also beneficial for the development and marketing of efficacious biological control agents for the benefit of agriculture (Card et al. 2016).

8.4 Role of Secondary Metabolites in the *Epichloë*–Grass Association

A wide range of agriculturally important traits conferred by *Epichloë* to their grass hosts are due to their production of a diverse range of secondary metabolites. Secondary metabolites are organic compounds produced by life forms that are not directly involved in the organism's growth, development, or reproduction. These chemicals are widely distributed throughout all the biological kingdoms, where they generally mediate ecological interactions. Secondary metabolites produced by fungi generally play an important ecological role, enabling these species to colonize diverse habitats and sustain their existence (Bills and Gloer 2017). Fungal-derived secondary metabolites exhibit a diverse range of activities, and many play a central role in antagonism towards other living organisms (e.g. animal and plant toxins, antibiotics, insecticides, and nematicides), as mineral scavenging agents (e.g. siderophores), in light absorption (e.g. pigments), as agents of symbiosis, and as enzyme inhibitors and autoregulators (Demain and Fang 2000; Keller 2019; Vining 1990). Most fungal-derived secondary metabolites are low molecular weight compounds, categorized into three main classes (non-ribosomal peptides, polyketides, and terpenes), that are encoded by contiguous gene assemblies termed biosynthetic gene clusters (BGCs). Typically, core enzymes within the BGC define the class of secondary metabolite produced. Additional enzymes, transcription factors, and transporters are then involved in the final modification and delivery of the resulting bioactive molecule from the BGC pathway (Keller 2019). The genes within the BGC are usually co-regulated by global or pathway-specific transcription factors which are activated in response to an environmental stimulus. The three common backbone enzymes in BGCs are: polyketide synthases (PKSs), which are involved in the production of polyketides from acetyl CoAs, non-ribosomal peptide synthetases (NRPSs), which can produce non-ribosomal peptides from amino acids, and terpene synthases/cyclases (TSs and TCs) that generate terpenes from activated isoprene units (Cane et al. 1998; Fisch 2013; Keller 2019; Pillay et al. 2022; Staunton and Weissman 2001; Süssmuth and Mainz 2017). However, hybrid pathways such as PKS-TC and PKS-NRPS also exist, and not all fungal secondary metabolites are generated by the three main BGC-type gene clusters. For example, the biosynthesis of the epichloëcyclins (peptides) are ribosomally derived (Johnson et al. 2015).

There is a wide range of chemotypic diversity among and even within species of *Epichloë* underpinned by the resident BGCs, and variation in the presence of middle- and late-pathway genes within alkaloid biosynthesis pathways (Schardl et al. 2012, 2013a). Comprehensive genome sequencing of a range of *Epichloë* endophytes and close relatives revealed the presence of large and pervasive blocks of repeat sequences derived from retroelements, DNA transposons, and miniature inverted-repeat transposable elements (Schardl et al. 2013b). These sequence repeat structures are thought to contribute to genomic instability, including gene duplications and losses, resulting in the production of variant alkaloid compounds (Schardl et al. 2013b). The secondary metabolites derived from *Epichloë* are mostly produced in planta by the grass–endophyte association, although some alkaloids have been detected in fungal cultures at relatively low concentrations (Blankenship et al. 2001; Penn et al. 1993). Many secondary metabolites produced by *Epichloë* spp. exhibit bioactive properties with several acting as feeding deterrents or toxins, protecting the host grass from invertebrate and mammalian herbivores (Bush et al. 1997). These bioactive secondary metabolites confer a selective advantage to the host and are therefore an example of defensive mutualism between plants and fungi (Clay 1988; Saikkonen et al. 2016).

Epichloë-derived secondary metabolites are produced in variable concentrations within the host plant, with host and endophyte genotype, plant age, plant growth regulators, plant nutrition, and environmental conditions (such as air and soil temperature, day length, solar radiation, and precipitation) being critically important factors in their biosynthesis (Agee and Hill 1994; Bastías et al. 2018a, b; Brosi et al. 2011; Bush et al. 1993; Freitas et al. 2020; Fuchs et al. 2017; Hennessy et al. 2016; Krauss et al. 2007; Rasmussen et al. 2007; Reinholz and Paul 2000). *Epichloë*-derived alkaloids also routinely accumulate in specific locations, e.g., in perennial ryegrass, ergovaline accumulates in the pseudostem and basal leaf sheath of intermediate aged tillers while in meadow fescue, lolines are translocated within the plant and accumulate in shoot and root tissues (Patchett et al. 2008; Spiering et al. 2002). Peramine, however, does not progressively accumulate in any plant organ (Koulman et al. 2007). Biosynthesis of some secondary metabolites (e.g., lolines) can also be stimulated after wounding, such as during herbivory by insects (Schardl et al. 2007), indicating that the *Epichloë*–grass symbiosis is able to respond not only to abiotic cues but also biotic ones. Historically, *Epichloë*-derived secondary metabolites have been challenging to isolate, identify, and classify under standard laboratory conditions, although the main pathways are now well defined (Caradus et al. 2022; Schardl et al. 2013a). The most well researched *Epichloë*-derived secondary metabolites are categorized within four main chemical classes; the ergot alkaloids (e.g. ergovaline), the indole diterpenes (e.g. lolitrem B and epoxyjanthitrems), pyrrolizidines (e.g. lolines), and the pyrrolopyrazines (e.g. peramine) (Fig. 8.5).

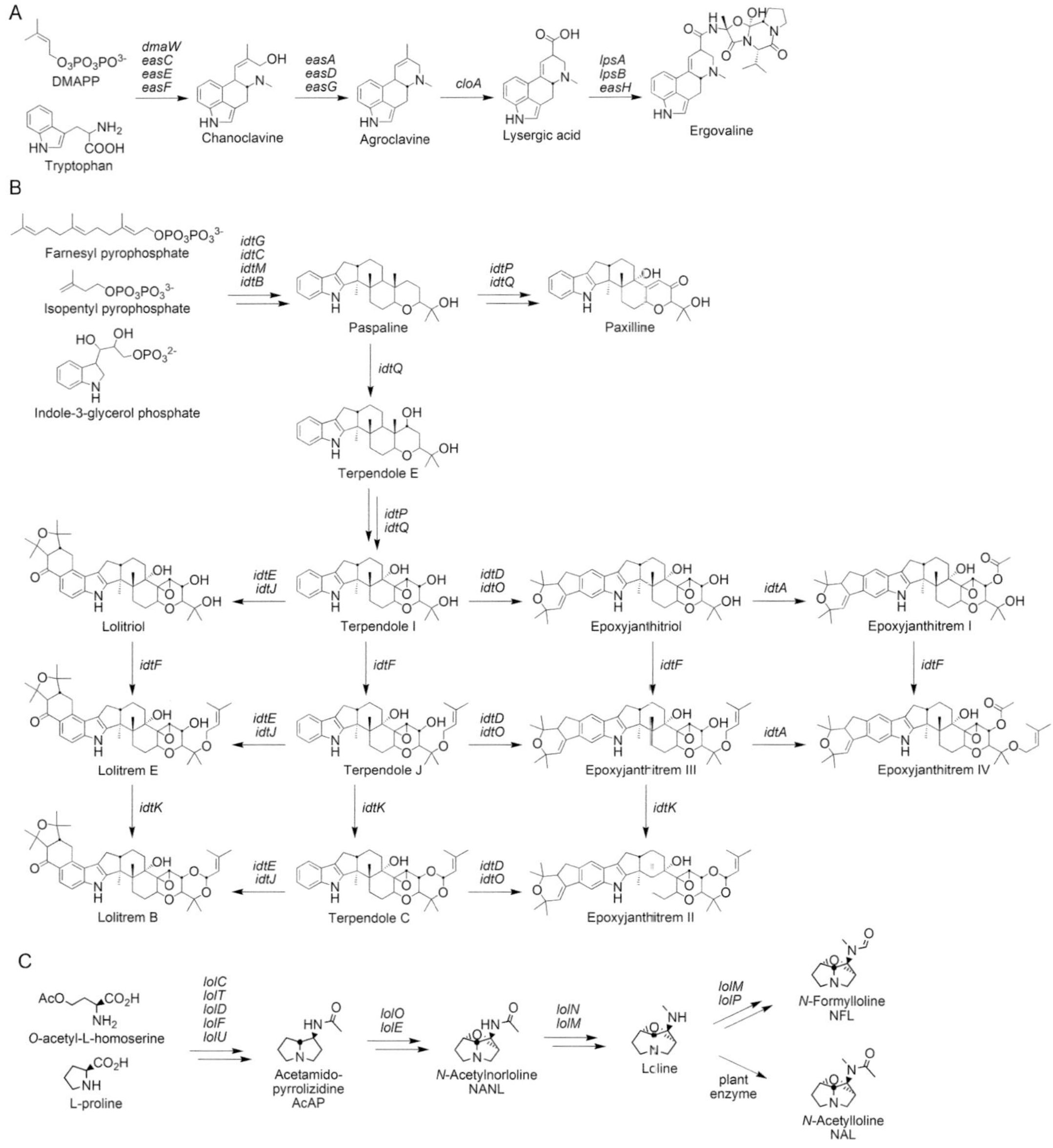

Fig. 8.5 Overview of the key secondary metabolites derived from the three alkaloid biosynthetic gene clusters (BGC) in *Epichloë*: (**a**) Ergot alkaloids from the ergot alkaloid synthesis (EAS) gene cluster, (**b**) indole diterpenes and epoxyjanthitrems from the IDT gene cluster, and (**c**) lolines from the LOL pathway. Proposed genes that encode the enzymes required for the reactions are denoted by arrows

8.4.1 Ergot Alkaloids

Ergot alkaloids are a group of complex, alkaloidal mycotoxins produced by several species of fungi, including phytopathogens (e.g. *Claviceps* spp.) and mutualistic symbionts (e.g. *Epichloë* spp.) (Florea et al. 2017; Gerhards et al. 2014). Although four main groups of ergot alkaloids are recognized: clavines, ergopeptines, lysergic acid and simple lysergic acid amides, different ergot alkaloids are produced by several fungal species within the Ascomycota, with many species producing different end products (Robinson and Panaccione 2015). In *Epichloë*, this diversity is generated by the presence/absence of lysergyl peptide synthetases, encoded by the NRPS EAS gene cluster (Gerhards et al. 2014; Schardl et al. 2006, 2012). Dimethylallyltryptophan synthase catalyzes the first step in ergot alkaloid biosynthesis, and the deletion of the *dmaW* gene resulted in the failure of *Epichloë* isolate Lp1 to produce ergot and clavine alkaloids (Wang et al. 2004). The intermediate secondary metabolites within the ergot alkaloid pathway have insect or mammalian bioactivity although ergovaline, the final product, has the most potent mammalian bioactivity in ruminants (Fig. 8.5a; Bacetty et al. 2009; Bacon et al. 1977; Johnson et al. 2021). Ergovaline within mammals acts as a potent vasoconstrictor and after consumption, grazing animals exhibit several clinical symptoms (Dyer 1993). In general, affected animals experience reduced blood flow to the extremities, and therefore are unable to dissipate heat in hot environments and lose their ability to maintain their peripheral body temperature in cold environments, manifesting as severe animal welfare issues (Klotz and Nicol 2016). Animal performance is subsequently affected, leading to reduced weight gain, milk production, and reproductive performance (Nicol and Klotz 2016). Ergovaline is at the centre of fescue toxicosis syndrome in the USA as it is produced by the symbiosis between *Epichloë coenophiala* and tall fescue, the most common pasture grass in the south-east of the country (Klotz and Nicol 2016). However, evidence exists suggesting that other *Epichloë*-derived secondary metabolites may contribute to the animal toxicosis often attributed to ergovaline (Gadberry et al. 2003). For an in-depth review on ergot alkaloids in pastures, read Caradus et al. (2022).

8.4.2 Indole Diterpenes

Indole diterpenes (IDTs) are a structurally diverse group of secondary metabolites produced by fungi including species of *Aspergillus*, *Claviceps*, *Penicillium*, and *Epichloë* (Ludlow et al. 2019; Saikia et al. 2008). The IDT pathway has intermediate secondary metabolites with invertebrate or mammalian bioactivity, with lolitrem B, the final product, having the most potent mammalian bioactivity (Fletcher et al. 2017; Gallagher et al. 1981; Schardl et al. 2013b). Generally, asexual *Epichloë* spp. are known to produce higher concentrations of IDTs, including lolitrem B, compared to their sexual counterparts (Johnson et al. 2021). In certain *Epichloë* strains, a NRPS BGC located in the sub-telomeric region of the chromosome and assembled into 3 clusters is responsible for the production of various IDTs (Young et al. 2006). A set of four core genes is required for converting the initial diterpene-derived backbone and an indole moiety derived from tryptophan (or a tryptophan derivative) into paspaline (Fig. 8.5b) (Schardl et al. 2012, 2013b). Paspaline is an important and conserved intermediate of this gene cluster, which is then acted on by other enzymes in IDT biosynthetic pathways, including cytochrome P450 monooxygenases, flavin adenine dinucleotide (FAD)-dependent monooxygenases and prenyl transferases that catalyze various modifications to generate a diversity of IDT products (Schardl et al. 2012, 2013b; Young et al. 2009). The common toxic strains of *E. festucae* var. *lolii*, associated with perennial ryegrass, responsible for causing mammalian toxicity in New Zealand, predominantly produce lolitrem B (Johnson et al. 2013a). Ingestion of lolitrem B can manifest as ryegrass staggers, a seasonal neurotoxic disease of grazing livestock, particularly sheep, with symptoms ranging from slight muscular tremors through to staggering, collapse and even death (di Menna et al. 2012). However, compounds other than lolitrem B could make significant contributions to the incidence of ryegrass staggers (di Menna et al. 2012; Finch et al. 2020; Reed et al. 2016).

Further IDTs, known as epoxyjanthitrems, were discovered after a survey of many hundreds of perennial ryegrass accessions for *Epichloë* strains incapable of producing toxic levels of lolitrem B or ergovaline (Tapper and Lane 2004). An *Epichloë* strain, designated AR37, was identified with broad-spectrum insect pest resistance, excellent animal performance but some ryegrass staggers (Fletcher et al. 2017; Fletcher and Sutherland 2009; Hume et al. 2007; Jensen and Popay 2004; Moate et al. 2012; Pennell et al. 2005; Popay and Thom 2009; Popay and Wyatt 1995; Popay and Cox 2016). Evidence from a genomic study shows that the biosynthesis of epoxyjanthitrems branches from the IDT pathway after terpendole I, due to the activity of *idtD* and *idtO* (compared to *idtE* and *idtJ* in the production of lolitrem B) (Fig. 8.3b) (Miller et al. 2022). Furthermore, the structure and bioactivity of five epoxyjanthitrems (epoxyjanthitrems I–IV and epoxyjanthitriol) have been elucidated through their isolation, purification, and characterization using nuclear magnetic resonance (NMR) spectroscopy, liquid chromatography–mass spectrometry, and animal bioassays (Finch et al. 2020). Epoxyjanthitrem I exhibited a sustained tremor response in mice, similar to that experienced with lolitrem B but less potent while also reducing weight gain and feeding in porina (*Wiseana cervinata* Walker), a common pasture pest in New Zealand (Finch et al. 2020). To date, only two species-level taxa of asexual *Epichloë* have been reported to produce epoxyjanthitrems, which are designated LpTG-3 and LpTG-4 (Hettiarachchige et al. 2015; Kaur et al. 2015).

8.4.3 Pyrrolizidines

Hundreds of pyrrolizidine alkaloids have been identified across thousands of plant species and most are produced as a defence mechanism against invertebrate herbivores. The loline alkaloids are a group of saturated pyrrolizidines with an *exo*-1-amine and a strained ether bridge that joins two aliphatic bridgehead carbon atoms (C-2 and C-7) produced by several *Epichloë* spp., including those that form symbioses with meadow fescue (*Festuca pratensis*) and tall fescue grasses (Pan et al. 2014; Schardl et al. 2013a, 2007). The loline BGC (LOL, Fig. 8.5c) consists of 11 genes in 3 clusters interspersed with transposable elements (Schardl et al. 2012; Spiering et al. 2008). A biologically rare feature of the LOL pathway is that the determinant step involves a ϒ-class pyridoxal phosphate enzyme encoded by *lolC* which condenses the precursor molecules L-proline and *O*-acetyl-L-homoserine to form *N*-(3-amino, 3-carboxy) propyl-L-proline. Another unique feature is the potential involvement of host plant acetyltransferases in the conversion of loline to *N*-acetylloline (Pan et al. 2014; Schardl et al. 2013a). Studies using isotope-labelled NMR, infrared spectroscopy, and RNA interference spectroscopy have confirmed the main products of the loline biosynthetic pathway are produced through a sequence of modifications of the 1-amino group (Faulkner et al. 2006; Schardl et al. 2012; Spiering et al. 2008). The most abundant lolines in planta are *N*-formylloline, *N*-acetylloline, and *N*-acetylnorloline, which have a broad spectrum of insecticidal and insect-deterrent activity, and anti-nematode activity, with negligible toxicity to grazing livestock (Bush et al. 1993; Easton et al. 2009; Froehlich et al. 2022; Jensen et al. 2009; Popay et al. 2003, 2009; Wilkinson et al. 2000).

8.4.4 Pyrrolopyrazines

The most researched pyrrolopyrazine is the compound peramine, produced by several *Epichloë* spp., including strain AR1 that is marketed in New Zealand for its insect bioactivity and excellent animal performance (Fletcher 1999; Koulman et al. 2007). Peramine is not toxic to insects but is a strong feeding deterrent against Argentine stem weevil, where it also affects development (Pennell et al. 2005; Rowan and Gaynor 1986; Rowan et al. 1986, 1990). A single gene, *perA*, which codes for a non-ribosomal peptide synthase was identified by a gene deletion study as being responsible for the production of peramine from precursors L-arginine, (S)-pyrroline-5-carboxylate (precursor to L-proline) and a methyl donor (S-adenosylmethionine) (Tanaka et al. 2005). Introduction of *perA* into *Epichloë* strains that do not produce peramine, using genome editing techniques, resulted in peramine production in culture and in planta (Hettiarachchige et al. 2019). Allelic variations to *perA* have been noted in a recent study, although their resulting secondary metabolite bioactivity remains unknown (Berry et al. 2019). Peramine appears to be continuously produced by *Epichloë* spp. in planta, however it does not progressively accumulate as with other *Epichloë*-derived alkaloids (e.g. ergovaline) (Koulman et al. 2007). Instead, peramine is translocated from the endophyte hyphae into the plant's intercellular spaces where it is either metabolized, mobilized, and/or excreted in guttation fluid (Koulman et al. 2007).

Further secondary metabolites have been implicated in the maintenance of the fungal–grass symbiosis, for example the siderophore epichloënin A (produced via an NRPS pathway) is required to maintain iron homeostasis between *E. festucae* and perennial ryegrass (Johnson et al. 2013b). Many novel *Epichloë*-derived secondary metabolites have also been detected using advanced proteomic, metabolomic, and bioinformatic pipelines, however their roles within the *Epichloë*–grass symbiotum have yet to be fully elucidated (Green et al. 2020; Johnson et al. 2015).

8.5 Plant Growth Promotional Traits

Plants infected with *Epichloë* spp. can exhibit an increase in shoot biomass, number of tillers, and seed yield compared to their endophyte-free counterparts (Bastías et al. 2021b; Gundel et al. 2013; Schardl et al. 2004). This growth promotional effect can be linked to an *Epichloë*-mediated increment in photosynthetic processes likely brought about by changes to phytohormone levels and is particularly evident when plants are exposed to stress (Rozpądek et al. 2015; Xu et al. 2021; Zhao et al. 2021). For example, under heavy metal stress (i.e. zinc), *Epichloë sinensis* infection was associated with increased gibberellin content in plants of *Festuca sinensis*, which correlated with enhanced plant growth (Wang et al. 2021a), while *Epichloë gansuensis* infection was associated with increased photosynthesis levels within plants of *Achnatherum inebrians* experiencing drought, which was also correlated with enhanced plant growth (Zhong et al. 2022). This growth promotion is not restricted to abiotic stresses as Ueno et al. (2016) have shown that plants symbiotic with *Epichloë* infested with insect herbivores can exhibit increased biomass compared to plants lacking the symbiosis. *Epichloë* increased the concentration of gibberellins in foliar tissues of *F. sinensis* with elevated levels of this hormone correlated to an increase in plant height and production of tillers (Wang et al. 2021a). However, potentially not all *Epichloë* associations boost gibberellin levels. Tall fescue plants infected with *E. coenophiala* exhibited a general downregulation in the expression of plant genes involved in the early biosynthesis steps of the plant hormone (Dinkins et al. 2019). Auxins are the only growth-promoting plant hormone found within hyphae of *Epichloë* spp.; however, altered levels of these metabolites within *Epichloë*-symbiotic compared to endophyte-free plants have yet to be detected (De Battista et al. 1990; Yue et al. 2000).

8.6 Interactions Between *Epichloë* and Other Microorganisms

Symbiotic microorganisms can alter the phenotype of their plant host in ways that may affect interactions with other species as well as the relative fitness of each partner (Clay et al. 2005). *Epichloë*-derived secondary metabolites can influence herbivorous invertebrates and mammals and this in turn can affect species diversity, food web structures, and fundamental ecological processes like decomposition (Bultman et al. 2012; Fuchs et al. 2013; Omacini et al. 2001; Popay and Bonos 2008). These legacy effects can even persist after the death of the *Epichloë* symbiont and its grass host (Casas et al. 2016; Zhou et al. 2022). Furthermore, *Epichloë* spp. can also modulate the microbiota of their host plants (including the foliar, root, rhizosphere, and seed environments), which subsequently has implications for ecological succession in natural ecosystems (Bastías et al. 2021a; Clay et al. 2005; Ju et al. 2020; Nissinen et al. 2019; Omacini et al. 2001; Roberts and Ferraro 2015; Wakelin et al. 2015). The study of plant microbiomes has received great interest in recent time as molecular approaches have advanced to more readily reveal the composition and dynamics of plant-associated microbial communities, with those associated with *Epichloë*–grass symbioses being no exception (Bastías et al. 2021a; Liu et al. 2022a; Mahmud et al. 2021; Nissinen et al. 2019; Tannenbaum et al. 2020).

A great deal of knowledge has been obtained regarding the bioactivity of *Epichloë* towards mammalian and invertebrate herbivores but far less research has been undertaken on the interactions between *Epichloë* and other microorganisms (Ren et al. 2009). Of the research that has been conducted, the majority has focussed on the biological control of fungal phytopathogens (Card et al. 2021; Xia et al. 2018). Despite this applied focus, and the fact that several *Epichloë*-based grass products are marketed across several continents for their pest deterrent properties (Caradus and Johnson 2020), none has been developed or marketed for their control of plant disease (Card et al. 2021). Furthermore, although a recent meta-analysis has shown that many species of phytopathogen (both biotrophic and necrotrophic) are significantly affected by the presence of *Epichloë* spp. (Pérez et al. 2020), very few published reports describe bioactivity in the field and instead rely on data from crudely designed in vitro dual culture bioassays and greenhouse experiments (Card et al. 2021).

The interactions between *Epichloë*–grass symbioses and other microorganisms are likely to be specific to the endophyte and host genotypes, while also being influenced by environmental factors. For example, a resident strain of *E. festucae* var. *lolii* altered the bacterial communities associated with perennial ryegrass compared to endophyte-free plants but only when infected plants were grown in soil and not sand (Tannenbaum et al. 2020). Furthermore, this *Epichloë* strain increased the abundance of Bacilli and Alphaproteobacteria in shoots but not roots in the second generation of host plants (Tannenbaum et al. 2020). However, in tall fescue, the presence of a strain of *E. coenophiala* strongly shaped the fungal endophyte communities of its hosts leaves but not the bacterial communities (Nissinen et al. 2019).

Epichloë spp. can influence their plant host's microbiota via multiple mechanisms. Although no correlation was detected between the antimicrobial bioactivity expressed by *Epichloë* spp. and their alkaloid profiles, Roberts and Lindow (2014) showed that bacterial communities living on leaf surfaces of *Epichloë*-infected tall fescue were shaped by the presence of loline alkaloids. Lolines promoted bacterial species capable of degrading the alkaloid (Roberts and Lindow 2014). Furthermore, several non-alkaloidal secondary metabolites have been implicated in the antagonism exhibited by *Epichloë* spp. towards phytopathogens. These compounds include aromatic sterols (Koshino et al. 1989), diacetamides (Yue et al. 2000), hydroxyl unsaturated fatty acids (Koshino et al. 1987), phenolic glycerides (Koshino et al. 1988), sesquiterpenes (Yoshihara et al. 1985; Yue et al. 2000), indole derivatives (indole 3 acetic acid (IAA) and indole 3 ethanol), and other volatile compounds such as Chokol K and methyl esters (Bultman and Leuchtmann 2008). Although Chokol K is known for its insect attractant abilities, particularly *Botanophila* flies that are an integral component of the *Epichloë* sexual cycle, the compound is hypothesized to have originated as an antimicrobial agent (Schiestl et al. 2006). Perez et al. (2017) describe an indirect pathway for the protection of *L. multiflorum* from the flower-infecting pathogen *Claviceps purpurea*, the causal agent of ergot disease. The incidence and severity of infection by *C. purpurea* was two-fold lower in *Epichloë*-symbiotic plants than in non-symbiotic ones, but when insects were prohibited from visiting the flowers, this difference disappeared, indicating that endophyte-derived volatile compounds may repel insect vectors of *C. purpurea* and indirectly protect their host plant against this phytopathogen (Perez et al. 2017).

Arbuscular mycorrhizal fungi (AMF) form symbioses with many land plants where they colonize the intracellular spaces of their host's roots, playing a critical role in the acquisition of nutrients while increasing the host's tolerance/

resistance to abiotic and biotic stresses (Averill et al. 2019; Branco et al. 2022; Pozo et al. 2015). *Epichloë* and AMF and can form multi-symbiont interactions with cool-season grasses, simultaneously colonizing the above- and below-ground tissues of the same host plant, respectively. Novel grass–*Epichloë* associations often have a lower diversity and/or population density of AMF within their roots while wild grasses infected with their co-evolved *Epichloë* symbionts usually accommodate a greater abundance of AMF (Arrieta et al. 2015; Chu-Chou et al. 1992; Mack and Rudgers 2008; Müller 2003; Novas et al. 2005, 2009; Omacini et al. 2006; Terlizzi et al. 2022; Vignale et al. 2016, 2018). However, the interactions between these fungal symbionts are complex and poorly understood with the effects of *Epichloë* on AMF not readily generalizable across plant-endophyte symbiota (Kalosa-Kenyon et al. 2018). Grass host genotype, *Epichloë* and AMF genotype, the presence of *Epichloë*-derived bioactive compounds (e.g. exudates), environmental stresses (e.g. drought and saline-alkali stress), and resource supply (e.g. nitrogen, phosphate) are listed as important factors that contribute to the interaction (Guo et al. 2019; Liu et al. 2022b; Novas et al. 2009; Rasmussen et al. 2007; Slaughter et al. 2018; Vignale et al. 2018; Zhang et al. 2022; Zhong et al. 2021).

The relationships between *Epichloë* and the microbiota on and within the host grass tissues are being revealed through detailed culture-independent plant microbiome studies. Members of the phylum Proteobacteria have generally been observed to dominate grass microbiomes (Bastías et al. 2021a), whereby seed transmitted microbiota can colonize and impart beneficial traits to the establishing seedlings. *Epichloë occultans* infection was associated with an increase in the seed bacterial community diversity of *L. multiflorum,* including *Pantoea* and *Pseudomonas*, which contain members that have been shown to impart several benefits to their hosts (Bastías et al. 2021a). The diversity of microbiota associated with mature plants is far greater than that of seed, and the potential of members of the grass and endophyte-associated microbiota to enhance the fitness of the symbiosis and plant persistence is only just being beginning to be explored.

8.7 Factors that Influence the Frequency of Endophyte-Infected Plants in Grass Populations

In pastoral agriculture systems, it is desirable for pastures to contain high infection frequencies of bioprotective and animal safe *Epichloë* endophytes to support production. However, wild (or natural) grass populations have highly variable *Epichloë* infection frequencies that can range from 0 to 100% (Faeth et al. 2004; Leyronas and Raynal 2001; Ravel et al. 1997; Saikkonen et al. 1998). One of the most common approaches to studying the ecology of the grass-endophyte symbiosis is to identify the environmental influences (i.e. abiotic and biotic factors) that correlate (positively or negatively) with the incidence of *Epichloë* symbiosis in grass populations. To establish causality, the performance of endophyte-infected and endophyte-free plants (ideally collected from different locations along the environmental gradient) may be assessed in common garden experiments with contrasting levels of the same abiotic factor as treatments, or directly in the field through full cross-transplant experiments. However, in local populations, the frequency of symbiotic grasses is not only determined by the relative fitness of endophyte-infected and endophyte-free plants but also by the vertical transmission efficiency of endophytes in endophyte-infected plants (Gundel et al. 2008, 2011b; Ravel et al. 1997).

8.7.1 Variation in the Frequency of Endophyte-Symbiotic Plants in Natural Populations

Certain investigations have found negative correlations between symbiosis prevalence and high mean annual precipitation (MAP), or positive ones with areas that experience high summer temperatures, which have led to the hypothesis that symbiosis with *Epichloë* endophytes improves plant tolerance to drought (Lewis et al. 1997; Malinowski and Belesky 2006; Żurek et al. 2013). This has been experimentally supported in species such as *L. perenne*, *Festuca eskia*, and *Bromus laevipes* where endophyte-infected plants collected from sites with different MAP showed higher performance under drought treatments compared to endophyte-free plants (Afkhami et al. 2014; Gibert and Hazard 2013; Gibert et al. 2012). However, a generalization seems to be difficult since the results of studies evaluating the endophyte-mediated plant tolerance to drought are still very variable within and among grass species (Decunta et al. 2021; Gundel et al. 2016). Field surveys carried out in extremely arid environments of Patagonia, which include sites with less than 300 mm MAP in the gradients, have shown opposite relationships between symbiosis prevalence and MAP. For example, the prevalence of *Epichloë* endophytes in populations of *Phleum alpinum*, *Poa spiciformis,* and *Hordeum comosum* was lower in drier sites, i.e. those with a lower MAP (Casas et al. 2022; Victoria et al. 2007). However, no studies have assessed the relative performance of endophyte-infected and endophyte-free plants of these grass species under conditions of extremely low water availability.

The fine fescues that are widely distributed at high latitudes of the Northern Hemisphere have been intensively studied with respect to the incidence of *Epichloë* fungal endophytes. Field surveys, however, have yielded contrasting results when analysed separately or even when trying to identify the factors driving the endophyte incidence (Bazely et al. 1997; Santangelo et al. 2015). For example, while there was a negative correlation between endophyte incidence and elevation in *Festuca rubra* occurring in Scandinavian mountains, there was no pattern for the other two fine fescues (*Festuca ovina* and *Festuca vivipara*) (Bazely et al. 2007; Granath et al. 2007). In west Fennoscandia, the probability of hosting endophytes was higher in *F. rubra* populations from the north than from the south (Saikkonen et al. 2000). In a later survey that covered a broader geographic scale in northern Europe, there was a negative correlation between endophyte incidence in *F. rubra* populations (the higher the latitude the lower the endophyte incidence) (Dirihan et al. 2016). Interestingly, the primary productivity (the normalized difference vegetation index; a dimensionless index that describes the difference between visible and near-infrared reflectance of vegetation cover) was lower with higher latitude; therefore, the endophyte incidence in populations of *F. rubra* was positively associated with primary productivity (Dirihan et al. 2016). Primary productivity can be used to characterize the quality at the site level (where, for example, plant communities in high-quality environments are more productive) and is positively associated with MAP and mean annual temperature (Gaston 2000). Therefore, as an integrative variable of local resources for primary producers to grow and sustain higher trophic consumers, variation in primary productivity may underlie the patterns of variation in grass population symbiosis incidence with geographical features like elevation or latitude. Accordingly, the global review by Semmartin et al. (2015) showed that symbiosis prevalence in grass populations was positively correlated with site primary productivity. The initial prediction in this article was that since herbivory pressure is higher as higher the primary productivity (Oesterheld et al. 1992), the symbiosis would be favourably selected in sites with high productivity (or MAP) (Semmartin et al. 2015). Alternatively, and motivated by previous works (e.g. Lewis et al. 1997; Malinowski and Belesky 2006; Żurek et al. 2013), it was predicted that in the other extreme of the gradient (assuming that low productivity was associated with water deficit), the fungal symbiosis incidence would be high because *Epichloë* can confer tolerance to drought. However, this was not the case, and the probability of hosting these endophytes was lower in sites with low primary productivity (Semmartin et al. 2015). Harsh conditions for growth in very low productivity ecosystems hinder the ability of host plants to maintain the endophyte fungus, thus increasing the probability of symbiont loss (Casas et al. 2022; Gundel et al. 2016). In productive environments, endophyte incidence may respond positively to other factors such as herbivore pressure (Bazely et al. 1997) and/or plant competition (Santangelo et al. 2015) in successional communities.

8.7.2 Variation in the Efficiency of Endophyte Vertical Transmission

Epichloë endophytes evolved the ability to synchronize their growth to that of their host plants' lifecycle (Christensen et al. 2008), and one of the consequences is that most of them are transmitted from mother plants to offspring through the seeds (i.e. vertical transmission). However, the process of vertical transmission is not perfect, and the failures in it can cause endophyte-infected plants to produce endophyte-free seeds (Afkhami et al. 2014; Gundel et al. 2011b). Thus, as highlighted previously, transmission efficiency is important—together with the relative fitness of symbiotic and non-symbiotic individuals, in determining the endophyte incidence in grass populations (Gundel et al. 2008; Ravel et al. 1997). In relation to the lifecycle of the host, two types of failures have been identified (Gundel et al. 2011b): ***(i)*** those that occur during seed production, evaluated as the proportion of endophyte-free seeds of the total seed produced by a plant, and ***(ii)*** those that occur due to the anticipated death of the fungus in the seed, evaluated as the proportion of endophyte-free seedlings produced by a batch of symbiotic seeds.

(i) The fidelity of the vertical transmission process is often imperfect with the fungus being lost from plants at many stages of their life history, including losses among tillers, spikes or panicles, and flowers (Afkhami and Rudgers 2008). In that way, symbiotic plants can produce endophyte-free seeds. Evidence for symbiosis loss has been documented for various North American wild species (*Elymus hystrix*, *Elymus riparius*, *Elymus virginicus*, *Festuca subverticillata*, *Poa alsodes*, *Poa sylvestris,* and *Sphenopholis nitida*) at different stages of their lifecycles (Afkhami and Rudgers 2008). Species like *Festuca arizonica* have a variable rate of tiller colonization by *Epichloë* spp. (Schulthess and Faeth 1998), and thus populations do produce endophyte-free seeds. Endophyte transmission efficiency can respond positively or negatively to environmental factors that affect the host plant. For instance, while strong defoliation late in the growing season of *L. multiflorum*, an annual grass, caused a dramatic depression in endophyte vertical transmission (Garcia Parisi et al. 2012), light defoliations simulating herbivory by small insects in the leaves of *Poa autumnalis* promoted the vertical transmission of the fungus (Gundel et al. 2020). Interestingly, several field surveys found that there was a clear relationship between rate of endophyte loss or failures in vertical transmission and the frequency of endophyte-symbiotic plants in extant populations (Afkhami and Rudgers 2008; Gibert and Hazard 2013; Gundel et al. 2009a). By studying endophyte transmission efficiency of plants collected from different *L. perenne* field populations in a common garden, transmission was determined to be an ecotypic characteristic (Gibert and Hazard 2013). An experiment where *L. multiflorum* plants were crossed with genetically distant pollen donors, concluded that the endophyte transmission efficiency was mainly affected by the host genetic background rather than the level of stress (Gundel et al. 2012). Experiments with genetically diverse breeding populations of *L. perenne* in association with *Epichloë* strain AR37 found that there was a strong influence of host genotype on vertical transmission and smaller host genotype-by-environmental interaction, where the transmission of clonal plants was examined in two locations of New Zealand

(Manawatū and Canterbury) (Gagic et al. 2018). In *L. multiflorum*, there was a clear relationship between the endophyte transmission efficiency and plant seed production (Gundel et al. 2011a). This latter result supports the evolutionary theory that predicts interactions between plants and vertically transmitted symbiotic microorganisms to be mutually beneficial in part because the fitness of both partners is tightly aligned.

(ii) Endophyte-symbiotic seeds can produce endophyte-free seedlings presumably due to the death of the fungus (Gundel et al. 2011b). In fact, the high rates of endophyte viability loss in seeds that are still viable are one of the most documented processes across different host species [e.g. *F. arundinacea* (Welty et al. 1987; Zhang et al. 2019), *L. perenne* (Rolston et al. 1986; Tian et al. 2013), *L. multiflorum* (Gundel et al. 2009b, 2010), *F. pratensis* (Bylin et al. 2016), *Bromus auleticus* (Stefanoni-Rubio et al. 2022)]. Besides, a given *Epichloë* strain can show different dynamics of viability loss in seeds of the same host cultivar, thus showing the existence of endophyte-by-host genotype interaction (Hill and Roach 2009; Rubin et al. 2022; Tian et al. 2013). Despite this overall pattern, the ability of fungal endophytes to maintain viability relative to their host seeds can vary with environmental conditions. For example, no premature deaths of endophytes were observed when symbiotic seeds of *L. rigidum* were buried in the field for 3 months (Canals et al. 2008; Kirkby et al. 2011). Differences in the results of stored seeds in controlled but continuous deteriorative conditions (e.g. accelerated ageing) versus field conditions could be accounted for by the ability of seeds and endophytes to sense and coordinate with the environmental seasonality, and the possibility of repairing damaged cellular structures between dry and wet periods (McDonald 1998). Interestingly, the rate of endophyte loss in seeds is also affected by the environmental conditions that the mother plants have previously encountered (through maternal effects). For example, the exposure of mother plants to ozone increased the rate of endophyte viability loss in *L. multiflorum* (Gundel et al. 2015; Ueno et al. 2020a). Another process that can influence the frequency of symbiotic seedlings produced by a seed batch is the differential fitness of endophyte-infected and endophyte-free seeds. It has been observed that under controlled conditions, the presence of *Epichloë* spp. depresses the longevity of the seeds (i.e. the total seed lifespan). This has been observed in *L. multiflorum* and *F. arundinacea* (Gundel et al. 2009a; Rubin et al. 2022; Ueno et al. 2020b). It has been suggested that seed water content, a parameter that controls seed longevity and depends strongly on environmental temperature and humidity, could be underlying the low endophyte viability in host seed and the differential longevity of endophyte-infected and endophyte-free seeds (Gundel et al. 2010; Rolston et al. 1986; Tian et al. 2013; Welty et al. 1987). However, the relationship between endophyte presence, seed water content, and seed longevity has been elusive (Rubin et al. 2022; Ueno et al. 2020a). Alternatively, Zhang et al. (2019) showed that the presence of *Epichloë* in seed is associated with significant alterations of the biochemical composition of the seed during ageing through changes in the relative abundance of different reserve sugars.

Like the relationship between symbiosis prevalence in populations and ecosystem primary productivity (Semmartin et al. 2015), the relationship between vertical transmission efficiency and host plant fitness was found to be positive (Gundel et al. 2011a). Therefore, *Epichloë* can certainly improve the fitness of host plants, but the relationship is highly regulated by the level of resources available for plant growth. In summary, the evidence suggests that the vertical transmission process is likely to be just as responsive to environmental resources as the relative fitness between endophyte-infected and endophyte-free plants, and that the ecosystem productivity controls the symbiosis prevalence by scaling up the process from the individual to the population level.

8.8 Commercialization of *Epichloë*-Based Products

New Zealand's economy is export-driven and heavily reliant on the productivity of the pastoral sector to support dairy, beef, sheep, and deer farming, supporting the declaration that 'After air, light and water, the next most important thing is grass' (Holford 1933). Perennial ryegrass is regarded as one of the most important forage species in many countries around the world, including New Zealand. Due to the economic importance of forages the New Zealand government, together with the private sector and other technology providers, invested substantially in this area of science from the late 1980s to identify a solution to the animal toxicity problems that plagued the agricultural sector in the previous decades. This significant investment primarily financed the establishment of a scientific programme at AgResearch Limited, a corporatized crown research institute charged with serving the agriculture and biotechnology sectors of New Zealand industry. The programme centred around a bioprospecting pipeline to identify, characterize, and select agriculturally beneficial strains of *Epichloë* that were less toxic to livestock while maintaining their insect-deterrent properties. The pipeline essentially aimed to exploit the natural symbiotic phenomenon between these fungi and their grass hosts for agricultural gains (Card et al. 2014a; de Bonth et al. 2015; Johnson et al. 2013a). Seed accessions, either from germplasm holdings or fresh collections, are first entered into a multiphase screening process to identify ryegrass seed containing viable *Epichloë* mycelium (Card et al. 2014b). Following this, *Epichloë*-infected plants are identified and characterized with respect to their production of secondary metabolites, particularly mammalian toxins (i.e. IDTs and ergot alkaloids). *Epichloë* strains that produce mammalian toxins are not necessarily discarded and may be developed for their wildlife deterrent properties (Pennell et al. 2017a). Once potentially mammalian-friendly strains have been identified, a molecular assay is undertaken with the purpose of uncovering genetic uniqueness. A full screen for known secondary metabolites in planta highlights candidate *Epichloë* strains for inoculation into elite perennial ryegrass grass cultivars. Parallel discovery programmes were developed for other economically important grass species, including tall fescue (Johnson et al. 2013a).

One of the most challenging steps in developing *Epichloë* for commerce is the ability to transfer suitable fungal strains from their original wild host to elite grass cultivars (Easton 2007; Johnson and Caradus 2019). Although *Epichloë* had been previously isolated from grass tissues and grown in culture (Sampson 1937), Latch and Christensen (1985) were the first to devise a repeatable method of infecting elite grasses with selected strains of *Epichloë*. This procedure can however lead to incompatibility manifesting as death of inoculated seedlings, stunted tillers with necrosis in the region of the apical meristem, and chlorotic leaf symptoms (Christensen 1995; Christensen et al. 1997). Symptoms of incompatibility are not limited to the host. Endophyte hyphae can become distorted and collapsed, with degenerated cytoplasm (Koga et al. 1993). *Epichloë* and their grass hosts have co-evolved, and it is now generally accepted that endophyte taxonomic groupings are not host specific, but are confined to forming compatible associations (i.e. symptomless associations) with related grass genera within a tribe (Schardl et al. 1997; Schardl and Philips 1997).

Once compatible associations were developed, and the selected endophytes incorporated into the cultivar, grass populations from subsequent generations were then evaluated through specifically designed bioactivity, agronomic and toxicological screens (Fig. 8.6). Under field conditions, not only were the objectives of insect deterrence and reduced livestock toxicity observed but also additional advantages, including improved grass establishment, higher grass yields, higher tiller densities, and reduced weed invasion, leading to an overall agronomic advantage to these grass plants compared to their endophyte-free counterparts (Hume and Sewell 2014; Popay and Hume 2011; Thom et al. 2012). Additional attributes include tolerance against abiotic stresses such as those caused by drought and/or nutrient deficiencies (Caradus and Johnson 2019; Johnson and Caradus 2019). The high rate of adoption by New Zealand farmers then made this endophyte-technology a scientific, agricultural, and marketing success (Johnson and Caradus 2019; Stewart 2006; Woodfield and Judson 2019). Today, over 70% of the ryegrass seed used in New Zealand pastoral farming contains one of several commercialized, novel *Epichloë* strain–host associations. These associations are regarded as important components of the microbial communities associated with pasture-based production systems in Australia, New Zealand, the USA, South America, and, to a lesser extent, Europe, where they play a critical role in pasture productivity, ruminant nutrition, and food production (Caradus et al. 2021a; Easton and Fletcher 2006; Hill et al. 2005).

Although delivering a quality seed product containing *Epichloë* is not free of challenges (Easton 2007; Rolston and Agee 2006), underpinning the development and application of *Epichloë* technology within the agricultural sector has significantly hinged upon the biological behaviour and characteristics of these endophytic fungi (Card et al. 2016; Johnson et al. 2021), that can be separated into five main themes:

1. **Fungal plasticity**. Most *Epichloë* species can be cultured in the laboratory on simple synthetic media that allows scientists to study them in vitro and develop a reliable method to transfer selected strains of asexual *Epichloë* from wild host grasses into elite breeding

Fig. 8.6 A research trial in New Zealand clearly showing an overall agronomic advantage to perennial ryegrass plants infected with a strain of *Epichloë festucae* var. *lolii* (right-hand side plot) compared with an endophyte-free plot of perennial ryegrass on the left (Photo courtesy of David Hume, AgResearch Ltd, New Zealand)

material and existing cultivars (Latch and Christensen 1985).

2. **An effective mechanism of action**. Although asexual *Epichloë* species are generally restricted to the host's above-ground tissues, their bioactivity (via antibiosis) extends further, protecting most of the plant as many *Epichloë*-derived secondary metabolites (e.g. lolines) are mobilized within the plant's vascular system and translocated to plant organs, such as the roots, where fungal colonization is absent (Bryant et al. 2010; Popay and Bonos 2008).
3. **Chemical and genetic diversity**. A great deal of chemical and genetic diversity exists within *Epichloë* (Hettiarachchige et al. 2015; Schardl et al. 2012; van Zijll de Jong et al. 2008). Therefore, once a culture collection is established and characterized, endophyte strains with suitable chemical profiles (little to no animal toxicity) can be transferred to elite grass cultivars and evaluated in the field (Johnson et al. 2013a). This process also has uncovered new bioactivity from secondary metabolites unknown at the time, i.e. epoxyjanthitrems from strain AR37 (Hume et al. 2009; Johnson and Caradus 2019).
4. **Propagation via seed**. Many other biological control agents (BCA)s are let down by complicated and expensive formulations to captivate, preserve, and disseminate the product. Asexual *Epichloë* spp., however, are exclusively disseminated in seed and this lifecycle adaption has had a profound effect on the marketing, risk assessment, and uptake by farmers to make the technology a success (Card et al. 2016; Johnson et al. 2013a).
5. **Obligate nature**. Asexual *Epichloë* are protected from many environmental variables due to their obligate lifestyle, including damaging ultraviolet light, extreme temperatures, water deficits, and competition for space and nutrients from other faster growing, antagonistic microorganisms (Card et al. 2016).

8.9 Alternative Applications of *Epichloë* Technology

Epichloë endophytes can also impart beneficial attributes to turfgrasses as they enhance the fitness of the grass when under biotic and abiotic stresses (Breen 1994; Funk et al. 1993, 2018; Stewart et al. 2022). Reports include *Epichloë* endophytes providing resistance to several important turfgrass pests [e.g., billbugs (*Sphenophorus* spp.) and sod webworm (Lepidoptera: Pyralidae)] and diseases (e.g. red thread and dollar spot caused by *Laetisaria fuciformis* and *Clarireedia jacksonii*, respectively), while some strains can contribute to enhanced tolerance to heavy metals, shade, heat, cold, and drought (Bonos et al. 2005; Clarke et al. 2006; Meyer and Funk 1989; Murphy et al. 1993; Zaurov et al. 2001).

Researchers in New Zealand have, over many years, accumulated a large collection of genetically diverse *Epichloë* strains for research and commercial development purposes (Johnson et al. 2013a). Based primarily on their secondary metabolite profiles, many of these strains were deemed unsuitable for pastoral use, specifically due to their production of ergovaline, a powerful animal toxin. However, this neurotoxin has been implicated in inducing an avoidance behaviour observed in several mammals and birds (Coley et al. 1995; Conover and Messmer 1996; Panaccione et al. 2006). Coupled with their ability to produce loline alkaloids, potent insecticidal and insect-deterrent compounds, these *Epichloë* strains were deemed suitable from a wildlife deterrent perspective.

Two *Epichloë*-grass products were developed. The first between a turf-type tall fescue cultivar, intentionally infected with a strain of *E. coenophiala*, known to produce high levels of ergovaline and loline alkaloids, which was primarily developed as an avian deterrent at airports. The second, a turf-type ryegrass cultivar, intentionally infected with a strain of *E. festucae* var. *lolii*, known to produce ergovaline, lolitrem B, and peramine, was developed for use as a wildlife deterrent to discourage herbivores, including rabbits, granivorous birds, and rodents at sports fields and recreational areas (Finch et al. 2016b; Pennell et al. 2016, 2017a, b, c).

8.10 Future Opportunities for *Epichloë* Technology

The Poaceae is arguably the most economically important plant family as it not only provides feed for many domesticated animals but also important species suitable for human nutrition [including barley (*Hordeum vulgare*), maize (*Zea mays*), rice (*Oryza* spp.), sugarcane (*Saccharum* spp.), and wheat (*Triticum aestivum*)], building materials [such as bamboo (Bambusoideae)], aesthetics for recreational areas such as parks, lawns, and sports grounds [including bluegrass (*Poa* spp.), bent grass (*Agrostis* spp.), Bermuda grass (*Cynodon dactylon*), and red fescue (*F. rubra*)] and species suitable for the production of cellulosic and non-cellulosic biofuels [including maize, sweet sorghum (*Sorghum bicolor*), and sugarcane)]. While *Epichloë* spp. are not naturally found in modern cereal grasses, they can be found within their wild relatives, including *Elymus* spp. and *Hordeum* spp. (Card et al. 2014b). Selected *Epichloë* strains isolated from these wild Triticeae grasses have been transferred into barley, wheat, and rye (*Secale cereale*) to create artificial symbioses, with the latter association able to inhibit diseases, caused by *Puccinia recondita* and ***Mycosphaerella recutita*** (syn. *Cercosporidium graminis*), in the field (Hume et al. 2018; Li et al. 2021; Simpson et al. 2014).

Further opportunities exist when several agriculturally beneficial microorganisms are combined. For example, Li et al. (2018) describe the association of perennial ryegrass with both *E. festucae* var. *lolii* and the AMF, *Claroideoglomus etunicatum*. Although *C. etunicatum* alone increased the plant's ability to uptake phosphorous and increased its dry weight, this effect was enhanced by the presence of an *Epichloë* strain. Furthermore, the presence of the AMF and *Epichloë* exhibited lower disease incidence of leaf spot caused by *Bipolaris sorokiniana*, compared to plants infected by just one of these mutualistic fungal species (Li et al. 2018). *Epichloë* strains can also closely associate with bacterial species that have potential to synthesize several secondary metabolites, including antibiotics and iron chelating compounds (Bastías et al. 2020a). Bacterial symbionts of fungi have been isolated from many fungal species that form intimate associations with plants, and these bacteria can directly affect plant fitness by modulating plant performance and/or by regulating the performance of their plant-associated fungi (Bastías et al. 2020b, 2022). Symptomless, tripartite associations involving *Epichloë* spp. are not restricted to bacteria and fungi, and *Epichloë* can also be infected with submicroscopic infectious agents (i.e. mycoviruses) (Romo et al. 2007; Zabalgogeazcoa et al. 1998). As these tripartite associations appear to have no detrimental effect on *Epichloë* they may offer selective advantages to the fungus and/or plant hosts, offering a mutualistic tripartite symbiosis that could be exploited for agricultural gains.

Lolines are agriculturally beneficial alkaloids produced by *Epichloë* species that form natural symbiosis with tall fescue (e.g. *E. coenophiala*) and meadow fescue (e.g. *Epichloë uncinatum* and *Epichloë siegelii*) (Schardl et al. 2007). Lolines are not produced from *Epichloë* species that form natural symbioses with perennial ryegrass. Lolines are non-toxic to ruminants and small mammals while exhibiting insect deterring and insecticidal activities (Easton et al. 2009; Finch et al. 2016a, b; Fletcher et al. 2000). An applied motive therefore existed to develop novel symbioses between loline-producing *Epichloë* spp. and elite cultivars of perennial ryegrass to achieve a wider spectrum of invertebrate activity than achieved with *Epichloë* strains naturally present in perrenial ryegrass (Easton 2007; Easton et al. 2009). However, forming compatible associations between these species has been challenging, with their loline production and vertical transmission frequencies being significantly lower than anticipated (Ball and Tapper 1999; Croy et al. 2022; Freitas et al. 2020). To maintain symbiotic compatibilty between elite grass cutivars and asexual *Epichloë* strains that exhibit a greater number of agriculturally beneficial traits, alternative techniques are required. All of the commercial symbiotic associations in cool-season grasses developed to date have utilized naturally occurring *Epichloë* strains, identified from exotic host germplasm. However, while genetic modifications of *Epichloë* strains using traditional cloning-based techniques have demonstrated the roles of key genes in processes such as alkaloid biosynthesis, opportunities now exist to edit secondary metabolite pathways in endophytic fungi using targeted genetic engineering techniques such as CRISPR-Cas9 (Chowdhary et al. 2022). For example, Florea et al. (2021) utilized a CRISPR-based approach to generate non-transgenic deletion mutants of *E. coenophiala*, an endophyte that associates with tall fescue, to remove the gene clusters responsible for the biosynthesis of ergovaline. Additionally, Miller et al. (2022) used CRISPR to demonstrate that new biochemical variation can be created within the epoxyjanthitrem pathway of *Epichloë* strain AR37, a commercialized fungal strain that associates with perennial ryegrass (Caradus et al. 2021a).

The manipulation of secondary metabolite pathways in *Epichloë* using CRISPR-Cas9 gene inactivation therefore has the potential to deliver novel *Epichloë* strains that can be incorporated into elite pasture grasses that eliminate, or significantly reduce, animal health and welfare issues while retaining desirable traits such as pest and disease protection (Miller et al. 2022). Future opportunities to increase the value of *Epichloë* technology will likely be achieved through the inactivation of secondary metabolite pathways and promoting the expression of the most agriculturally beneficial compounds or introducing novel bioactive compounds (Caradus et al. 2021a; Caradus and Johnson 2020; Miller et al. 2022). Genome editing approaches are generally viewed as more socially acceptable than traditional cloning-based genetic modifications, and hold promise for future applications in New Zealand to help tackle future challenges to food production.

The most important challenge currently facing mankind is climate change, and in 2022 many regions around the world experienced severe climatic events. For example, Europe experienced the worst drought in 500 years (Le Page 2022), historic floods submerged more than a third of Pakistan that affected more than 30 million people (Sands 2022), and many countries, from China to the USA, recorded severe heat waves (Lopez 2022). Climate change is predicted to have serious implications on the future of many agricultural systems (Arora 2019). Increasing temperatures, elevated atmospheric CO_2 levels, and changing precipitation patterns are likely to impact agriculturally important insect pests by expanding their geographic range, increasing their overwintering survival rates and affecting their interactions with natural enemies (Skendžić et al. 2021). Plant disease outbreaks are expected to intensify due to invasions of new phytopathogens and/or due to increases in the severity of existing phytopathogens (Kocmánková et al. 2009; Misra et al. 2020). Functional symbiosis is one of the most successful mechanisms by which plants can improve their ability to tolerate stress (Hereme et al. 2020; Morales-Quintana et al. 2022). Climate change, together with current unsustainable agricultural practices that expend valuable resources while degrading the environment (Pareek et al. 2020), highlights the critical importance of integrating knowledge from both agricultural and natural ecosystems, from single plants and multispecies plant communities and their microbiota, and from below-ground and above-ground multitrophic interactions to further improve the resilience and sustainability of worldwide crop production.

Endophytes Everywhere

Endophytic fungi are thought to occur in all species of land plants as well as brown, green, and red marine seaweeds. Ergot alkaloid-producing clavicipitaceous endophytes with narrow host ranges are the subjects of this chapter. These are *Epichloë* and close relatives (Clavicipitaceae) that infect certain grasses and sedges. Another group of alkaloid-

(continued)

producing clavicipitaceous fungi infect some morning glory species (Convolvulaceae), which have been used as entheogens.

Most endophytic fungi, however, are non-clavicipitaceous (NC) species of Ascomycota (Pezizomycotina, especially Sordariomycetes and Dothideomycetes) and a few Basidiomycota (Agaricomycotina, Pucciniomycotina, and Ustilaginomycotina) and Mucoromycota with broad host ranges. The NC-endophytes have been placed in three groups based, not on taxonomy, but largely on life history and ecological traits such as horizontal or vertical transmission, degree, and site of colonization of the hosts (single cells to large areas and not only in leaves and stems, but also roots). They also may have saprobic or parasitic lifestyles under some conditions (Rodriguez et al. 2009).

Interactions of plant hosts and NC-endophytes are not so well studied as clavicipitaceous endophytes, but some are known to convey benefits to the host such as drought tolerance and enhanced growth. A few NC-endophytes benefit their hosts under local conditions of pH and temperature stress. Some root-inhabiting species are prevalent in non-mycorrhizal plants (i.e. Brassicaceae) (Card et al. 2015).

NC-endophytes have been important in bioremediation because the fungi can degrade a variety of recalcitrant chemicals (Dubey et al. 2020). Another applied area of research on these fungi is drug discovery. Endophytic fungi-mediated biosynthesis often results in synthesis of larger quantities of chemical compounds in conjunction with the host plant, a phenomenon shown with the cancer drug taxol (Tiwari and Bae 2022; Yan et al. 2018).

Acknowledgements The authors thank David Hume and Anouck de Bonth for critical revision of the manuscript.

References

Afkhami M, Rudgers J (2008) Symbiosis lost: imperfect vertical transmission of fungal endophytes in grasses. Am Nat 172:405–416. https://doi.org/10.1086/589893

Afkhami ME, McIntyre PJ, Strauss SY (2014) Mutualist-mediated effects on species' range limits across large geographic scales. Ecol Lett 17:1265–1273. https://doi.org/10.1111/ele.12332

Agee C, Hill N (1994) Ergovaline variability in *Acremonium*-infected tall fescue due to environment and plant genotype. Crop Sci 34:221–226

Arora NK (2019) Impact of climate change on agriculture production and its sustainable solutions. Environ Sustain 2:95–96. https://doi.org/10.1007/s42398-019-00078-w

Arrieta AM, Iannone LJ, Scervino JM, Vignale MV, Novas MV (2015) A foliar endophyte increases the diversity of phosphorus-solubilizing rhizospheric fungi and mycorrhizal colonization in the wild grass *Bromus auleticus*. Fungal Ecol 17:146–154

Averill C, Bhatnagar JM, Dietze MC, Pearse WD, Kivlin SN (2019) Global imprint of mycorrhizal fungi on whole-plant nutrient economics. Proc Natl Acad Sci U S A 116:23163–23168. https://doi.org/10.1073/pnas.1906655116

Bacetty AA, Snook ME, Glenn AE, Noe JP, Nagabhyru P, Bacon CW (2009) Chemotaxis disruption in *Pratylenchus Scribneri* by tall fescue root extracts and alkaloids. J Chem Ecol 35:844–850. https://doi.org/10.1007/s10886-009-9657-x

Bacon CW (1995) Toxic endophyte-infected tall fescue and range grasses: historic perspectives. J Anim Sci 73:861–870

Bacon CW, White JF (1994) Stains, media, and procedures for analyzing endophytes. Biotechnology of endophytic fungi of grasses. CRC Press, pp 47–56

Bacon C, Porter J, Robbins J, Luttrell E (1977) Epichloë typhina from toxic tall fescue grasses. Appl Environ Microbiol 34:576–581

Ball O-P, Tapper B (1999) The production of loline alkaloids in artificial and natural grass/endophyte associations. Proceedings of the New Zealand plant protection conference, pp 264–269

Ball DM, Lacefield GD, Hoveland CS (2019) The wonder grass: the story of tall fescue in the United States. Oregon Tall Fescue Commission, Salem, OR

Barret M, Guimbaud JF, Darrasse A, Jacques MA (2016) Plant microbiota affects seed transmission of phytopathogenic microorganisms. Mol Plant Pathol 17:791

Bastias DA, Martínez-Ghersa MA, Ballaré CL, Gundel PE (2017) *Epichloë* fungal endophytes and plant defenses: not just alkaloids. Trends Plant Sci 22:939–948. https://doi.org/10.1016/j.tplants.2017.08.005

Bastías DA, Alejandra Martínez-Ghersa M, Newman JA, Card SD, Mace WJ, Gundel PE (2018a) The plant hormone salicylic acid interacts with the mechanism of anti-herbivory conferred by fungal endophytes in grasses. Plant Cell Environ 41:395–405. https://doi.org/10.1111/pce.13102

Bastías DA, Martínez-Ghersa MA, Newman JA, Card SD, Mace WJ, Gundel PE (2018b) Jasmonic acid regulation of the anti-herbivory mechanism conferred by fungal endophytes in grasses. J Ecol 106:2365–2379. https://doi.org/10.1111/1365-2745.12990

Bastías DA, Jauregui R, Applegate ER, Altermann E, Card SD, Johnson LJ (2020a) Complete genome sequence of *Paenibacillus* sp. strain E222, a bacterial symbiont of an *Epichloë* fungal endophyte of ryegrass. Microbiol Resour Announc 9:e00786–e00720. https://doi.org/10.1128/MRA.00786-20

Bastías DA, Johnson LJ, Card SD (2020b) Symbiotic bacteria of plant-associated fungi: friends or foes? Curr Opin Plant Biol 56:1–8

Bastías DA, Bustos LB, Jáuregui R, Barrera A, Acuña-Rodríguez IS, Molina-Montenegro MA, Gundel PE (2021a) *Epichloë* fungal endophytes influence seed-associated bacterial communities. Front Microbiol 12: 795354–795354

Bastías DA, Gianoli E, Gundel PE (2021b) Fungal endophytes can eliminate the plant growth–defence trade-off. New Phytol 230:2105–2113

Bastías DA, Applegate ER, Johnson LJ, Card SD (2022) Factors controlling the effects of mutualistic bacteria on plants associated with fungi. Ecol Lett 25:1879–1888. https://doi.org/10.1111/ele.14073

Bazely DR, Vicari M, Emmerich S, Filip L, Lin D, Inman A (1997) Interactions between herbivores and endophyte-infected *Festuca rubra* from the Scottish islands of St. Kilda, Benbecula and Rum. J Appl Ecol:847–860

Bazely DR, Ball JP, Vicari M, Tanentzap AJ, Bérenger M, Rakocevic T, Koh S (2007) Broad-scale geographic patterns in the distribution of vertically-transmitted, asexual endophytes in four naturally-occurring grasses in Sweden. Ecography 30:367–374

Bent E, Chanway CP (2002) Potential for misidentification of a spore-forming *Paenibacillus polymyxa* isolate as an endophyte by using culture-based methods. Appl Environ Microbiol 68:4650–4652

Berry D, Mace W, Grage K, Wesche F, Gore S, Schardl CL, Young CA, Dijkwel PP, Leuchtmann A, Bode HB, Scott B (2019) Efficient nonenzymatic cyclization and domain shuffling drive pyrrolopyrazine diversity from truncated variants of a fungal NRPS. Proc Natl Acad Sci U S A 116:25614–25623

Bills GF, Gloer JB (2017) Biologically active secondary metabolites from the fungi. In: The fungal kingdom, pp 1087–1119

Blankenship JD, Spiering MJ, Wilkinson HH, Fannin FF, Bush LP, Schardl CL (2001) Production of loline alkaloids by the grass endophyte, *Neotyphodium uncinatum*, in defined media. Phytochemistry 58: 395–401

Bonos SA, Wilson MM, Meyer WA, Reed FC (2005) Suppression of red thread in fine fescues through endophyte-mediated resistance. Appl Turfgrass Sci 2: 1–7. https://doi.org/10.1094/ATS-2005-0725-01-RS

Branco S, Schauster A, Liao HL, Ruytinx J (2022) Mechanisms of stress tolerance and their effects on the ecology and evolution of mycorrhizal fungi. New Phytol

Breen J (1994) *Acremonium* endophyte interactions with enhanced plant resistance to insects. Annu Rev Entomol 39:401–423

Brem D, Leuchtmann A (1999) High prevalence of horizontal transmission of the fungal endophyte *Epichloë sylvatica*. Bull Geobot Inst ETH 65:12

Brosi GB, McCulley RL, Bush LP, Nelson JA, Classen AT, Norby RJ (2011) Effects of multiple climate change factors on the tall fescue–fungal endophyte symbiosis: infection frequency and tissue chemistry. New Phytol 189:797–805

Bryant R, Cameron N, Edwards G (2010) Response of black beetle and redheaded pasture cockchafer larvae to loline alkaloids in meadow fescue roots. N Z Plant Prot 63:219–223

Bultman TL, Leuchtmann A (2008) Biology of the *Epichloë*-Botanophila interaction: an intriguing association between fungi and insects. Fungal Biol Rev 22: 131–138

Bultman TL, White JF (1988) "Pollination" of a fungus by a fly. Oecologia 75:317–319

Bultman TL, White JF Jr, Bowdish TI, Welch AM, Johnston J (1995) Mutualistic transfer of *Epichloë* spermatia by *Phorbia* flies. Mycologia 87:182–189

Bultman TL, Aguilera A, Sullivan TJ (2012) Influence of fungal isolates infecting tall fescue on multitrophic interactions. Fungal Ecol 5:372–378

Bush LP, Fannin FF, Siegel MR, Dahlman DL, Burton HR (1993) Chemistry, occurrence and biological effects of saturated pyrrolizidine alkaloids associated with endophyte-grass interactions. Agric Ecosyst Environ 44:81–102. https://doi.org/10.1016/0167-8809(93)90040-V

Bush LP, Wilkinson HH, Schardl CL (1997) Bioprotective alkaloids of grass-fungal endophyte symbioses. Plant Physiol 114:1

Bylin A, Card S, Hume D, Lloyd-West C, Huss-Danell K (2016) Endophyte storage and seed germination of *Epichloë*-infected meadow fescue. Seed Sci Technol 44:138–155

Campbell MA, Tapper BA, Simpson WR, Johnson RD, Mace W, Ram A, Lukito Y, Dupont P-Y, Johnson LJ, Scott DB (2017) *Epichloë hybrida*, sp. nov., an emerging model system for investigating fungal allopolyploidy. Mycologia 109:715–729

Canals R, San Emeterio L, Oreja A (2008) Chances of loss of fungal endophytes in agronomic grasses: a case-study for *Lolium rigidum*. Agric Ecosyst Environ 127:146–152

Cane DE, Walsh CT, Khosla C (1998) Harnessing the biosynthetic code: combinations, permutations, and mutations. Science 282:63–68

Caradus JR, Johnson LJ (2019) Improved adaptation of temperate grasses through mutualism with fungal endophytes. In: Schouten A (ed) Endophyte biotechnology: potential for agriculture and pharmacology. CABI, Wageningen, pp 85–108

Caradus JR, Johnson LJ (2020) *Epichloë* fungal endophytes-from a biological curiosity in wild grasses to an essential component of resilient high performing ryegrass and fescue pastures. J Fungi 6:322. https://doi.org/10.3390/jof6040322

Caradus J, Chapman D, Cookson T, Cotching B, Deighton M, Donnelly L, Ferguson J, Finch S, Gard S, Hume D (2021a) *Epichloë* endophytes–new perspectives on a key ingredient for resilient perennial grass pastures. NZGA Res Pract Ser 17:347–360

Caradus JR, Card SD, Hewitt KG, Hume DE, Johnson LJ (2021b) Asexual *Epichloë* fungi—obligate mutualists. Encyclopedia 1:1084–1100

Caradus JR, Goldson SL, Moot DJ, Rowarth JS, Stewart AV (2021c) Pastoral agriculture, a significant driver of New Zealand's economy, based on an introduced grassland ecology and technological advances. J R Soc N Z 1-45. https://doi.org/10.1080/03036758.2021.2008985

Caradus JR, Card SD, Finch SC, Hume DE, Johnson LJ, Mace WJ, Popay AJ (2022) Ergot alkaloids in New Zealand pastures and their impact. N Z J Agric Res 65:1–41

Card SD, Tapper BA, Lloyd-West C, Wright KM (2013) Assessment of fluorescein-based fluorescent dyes for tracing *Neotyphodium* endophytes in planta. Mycologia 105:221–229. https://doi.org/10.3852/12-062

Card S, Johnson L, Bonth Ad, Tapper B, Mace W, Faville M, Pennell C, Caradus J, and Hume D. 2014a. *Epichloë* endophytes from cool season grasses – reaping the rewards from a well-tuned bio-prospecting pipeline. The 10th International Mycological Congress, Bangkok.

Card SD, Faville MJ, Simpson WR, Johnson RD, Voisey CR, de Bonth AC, Hume DE (2014b) Mutualistic fungal endophytes in the Triticeae–survey and description. FEMS Microbiol Ecol 88:94–106

Card SD, Hume DE, Roodi D, McGill CR, Millner JP, Johnson RD (2015) Beneficial endophytic microorganisms of Brassica – a review. Biol Control 90:102–112

Card S, Johnson L, Teasdale S, Caradus J (2016) Deciphering endophyte behaviour: the link between endophyte biology and efficacious biological control agents. FEMS Microbiol Ecol 92

Card SD, Bastías DA, Caradus JR (2021) Antagonism to plant pathogens by *Epichloë* fungal endophytes—a review. Plan Theory 10:1997

Carroll G (1988) Fungal endophytes in stems and leaves: from latent pathogen to mutualistic symbiont. Ecology 69:2–9

Casas C, Torretta JP, Exeler N, Omacini M (2016) What happens next? Legacy effects induced by grazing and grass-endophyte symbiosis on thistle plants and their floral visitors. Plant Soil 405:211–229

Casas C, Gundel PE, Deliens E, Iannone LJ, García Martinez G, Vignale MV, Schnyder H (2022) Loss of fungal symbionts at the arid limit of the distribution range in a native Patagonian grass—resource eco-physiological relations. Funct Ecol 36:583–594

Charlton ND, Craven KD, Mittal S, Hopkins AA, Young CA (2012) *Epichloë canadensis*, a new interspecific epichloid hybrid symbiotic with Canada wildrye (*Elymus canadensis*). Mycologia 104:1187–1199. https://doi.org/10.3852/11-403

Chowdhary K, Arora H, Sharma S (2022) CRISPR/Cas9-based genome editing as a way ahead for inducing production of bioactive metabolites in endophytes. Natl Acad Sci Lett 45:275–280. https://doi.org/10.1007/s40009-022-01107-9

Christensen MJ (1995) Variation in the ability of *Acremonium* endophytes of *Lolium perenne*, *Festuca arundinacea* and *F. pratensis* to form compatible associations in the three grasses. Mycol Res 99:466–470. https://doi.org/10.1016/S0953-7562(09)80647-3

Christensen M, Voisey C (2006) The biology of the endophyte/grass partnership. In: Popay A, Thom ER (eds) 6th international symposium on fungal endophytes of grasses. Christchurch, New Zealand Grassland Association, pp 123–134

Christensen MJ, Ball OJP, Bennett RJ, Schardl CL (1997) Fungal and host genotype effects on compatibility and vascular colonization by *Epichloë festucae*. Mycol Res 101:493–501. https://doi.org/10.1017/S0953756296002833

Christensen MJ, Bennett RJ, Schmid J (2002) Growth of *Epichloë* and *Neotyphodium* and p-endophytes in leaves of *Lolium* and *Festuca* grasses. Mycol Res 106:93–96. https://doi.org/10.1017/S095375620100510X

Christensen MJ, Bennett RJ, Ansari HA, Koga H, Johnson RD, Bryan GT, Simpson WR, Koolaard JP, Nickless EM, Voisey CR (2008) *Epichloë* endophytes grow by intercalary hyphal extension in elongating grass leaves. Fungal Genet Biol 45:84–93. https://doi.org/10.1016/j.fgb.2007.07.013

Chu-Chou M, Guo B, An Z-Q, Hendrix J, Ferriss R, Siegel M, Dougherty C, Burrus P (1992) Suppression of mycorrhizal fungi in fescue by the *Acremonium coenophialum* endophyte. Soil Biol Biochem 24:633–637

Chung KR, Schardl CL (1997) Sexual cycle and horizontal transmission of the grass symbiont, *Epichloë typhina*. Mycol Res 101:295–301. https://doi.org/10.1017/S0953756296002602

Clarke BB, White JF, Hurley RH, Torres MS, Sun S, Huff DR (2006) Endophyte-mediated suppression of dollar spot disease in fine fescues. Plant Dis 90:994–998. https://doi.org/10.1094/pd-90-0994

Clay K (1988) Fungal endophytes of grasses: a defensive mutualism between plants and fungi. Ecology 69:10–16

Clay K (1990) Fungal endophytes of grasses. Annu Rev Ecol Syst 21:275–297

Clay K, Holah J, Rudgers JA (2005) Herbivores cause a rapid increase in hereditary symbiosis and alter plant community composition. Proc Natl Acad Sci 102: 12465–12470

Coley AB, Fribourg HA, Pelton MR, Gwinn KD (1995) Effects of tall fescue endophyte infestation on relative abundance of small mammals. Wiley Online Library

Conover MR, Messmer TA (1996) Feeding preferences and changes in mass of Canada geese grazing endophyte-infected tall fescue, Condor, pp 859–862

Croy RG, Sutherland BL, Hume DE, Mace WJ, van Koten C, Finch SC (2022) Animal safety of a tall fescue endophyte (*Epichloë* sp.) in a perennial ryegrass (*Lolium perenne*) host. N Z Vet J 70:165–176. https://doi.org/10.1080/00480169.2021.2011795

Daly GT (1973) The grasslands of New Zealand. In: Langer RHM (ed) Pastures and pasture plants. Wellington, A H and A W Reed Ltd

de Bary A (1879) Die erscheinung der symbiose. Verlag von Karl J. Trübner, Strassburg

De Battista J, Bacon C, Severson R, Plattner R, Bouton J (1990) Indole acetic acid production by the fungal endophyte of tall fescue. Agron J 82:878–880

de Bonth A, Card S, Briggs L, Faville M, Finch S, Hong W, Johnson L, Liu L, Mace W, Pennell C, Popay A, Schmidt J, Sprosen J, Tapper B, and Hume D. 2015. Fungal foray: the pursuit of beneficial endophyte strains for Australasian pastures. 9th International Symposium on Fungal Endophytes of Grasses (ISFEG 2015), Melbourne, pp 98–99

Decunta FA, Pérez LI, Malinowski DP, Molina-Montenegro MA, Gundel PE (2021) A systematic review on the effects of *Epichloë* fungal endophytes on drought tolerance in cool-season grasses. Front Plant Sci 12:644731

Demain AL, Fang A (2000) The natural functions of secondary metabolites. In: Fiechter A (ed) History of modern biotechnology I. Springer, Berlin, pp 1–39

di Menna ME, Finch SC, Popay AJ, Smith BL (2012) A review of the *Neotyphodium lolii*/*Lolium perenne* symbiosis and its associated effects on animal and plant health, with particular emphasis on ryegrass staggers. N Z Vet J 60:315–328

Diehl WW (1950) Balansia and the Balansiae in America: US Department of Agriculture

Dinkins RD, Nagabhyru P, Graham MA, Boykin D, Schardl CL (2017) Transcriptome response of *Lolium arundinaceum* to its fungal endophyte *Epichloë coenophiala*. New Phytol 213:324–337

Dinkins RD, Nagabhyru P, Young CA, West CP, Schardl CL (2019) Transcriptome analysis and differential expression in tall fescue harboring different endophyte strains in response to water deficit. Plant Genome 12: 180071

Dirihan S, Helander M, Väre H, Gundel PE, Garibaldi LA, Irisarri JGN, Saloniemi I, Saikkonen K (2016) Geographic variation in *Festuca rubra* L. ploidy levels and systemic fungal endophyte frequencies. PLoS One 11: e0166264

Dubey A, Malla MA, Kumar A, Dayanandan S, Khan ML (2020) Plants endophytes: unveiling hidden agenda for bioprospecting toward sustainable agriculture. Crit Rev Biotechnol 40:1210–1231

Dupont P-Y, Eaton CJ, Wargent JJ, Fechtner S, Solomon P, Schmid J, Day RC, Scott B, Cox MP (2015) Fungal endophyte infection of ryegrass reprograms host metabolism and alters development. New Phytol 208:1227–1240. https://doi.org/10.1111/nph.13614

Dyer DC (1993) Evidence that ergovaline acts on serotonin receptors. Life Sci 53:PL223–PL228. https://doi.org/10.1016/0024-3205(93)90555-H

Easton H (2007) Grasses and *Neotyphodium* endophytes: co-adaptation and adaptive breeding. Euphytica 154: 295–306

Easton H, Fletcher L (2006) The importance of endophyte in agricultural systems-changing plant and animal productivity. NZGA Res Pract Ser 13:11–18

Easton H, Lyons T, Cooper B, Mace W (2009) Loline alkaloids for better protection of pastures from insect pests. Proc N Z Grassl Assoc:151–154

Faeth SH, Helander ML, Saikkonen KT (2004) Asexual *Neotyphodium* endophytes in a native grass reduce competitive abilities. Ecol Lett 7:304–313

Faulkner JR, Hussaini SR, Blankenship JD, Pal S, Branan BM, Grossman RB, Schardl CL (2006) On the sequence of bond formation in loline alkaloid biosynthesis. Chembiochem 7:1078–1088

Finch S, Munday J, Munday R, Kerby J (2016a) Short-term toxicity studies of loline alkaloids in mice. Food Chem Toxicol 94:243–249

Finch SC, Pennell CGL, Kerby JWF, Cave VM (2016b) Mice find endophyte-infected seed of tall fescue unpalatable - implications for the aviation industry. Grass Forage Sci 71:659–666. https://doi.org/10.1111/gfs.12203

Finch SC, Prinsep MR, Popay AJ, Wilkins AL, Webb NG, Bhattarai S, Jensen JG, Hawkes AD, Babu JV, Tapper BA, Lane GA (2020) Identification and structure elucidation of epoxyjanthitrems from *Lolium perenne* infected with the endophytic fungus *Epichloë festucae* var. *lolii* and determination of the tremorgenic and anti-Insect activity of epoxyjanthitrem I. Toxins 12:526

Fisch KM (2013) Biosynthesis of natural products by microbial iterative hybrid PKS–NRPS. RSC Adv 3: 18228–18247

Fletcher L (1999) "Non-toxic" endophytes in ryegrass and their effect on livestock health and production. NZGA Res Pract Ser 7:133–139

Fletcher LR, Harvey IC (1981) An association of a *Lolium* endophyte with ryegrass staggers. N Z Vet J 29:185–186

Fletcher L, Sutherland B (2009) Sheep responses to grazing ryegrass with AR37 endophyte. Proc N Z Grassl Assoc:127–132

Fletcher L, Hoglljnd J, Sutherland B (1990) The impact of *Acremonium* endophytes in New Zealand, past, present and future. Proc N Z Grassl Assoc:227–235

Fletcher L, Popay AJ, Stewart AV, Tapper B (2000) Herbage and sheep production from meadow fescue with and without the endophyte *Neotyphodium uncinatum*. Proceedings of the 4th international neotyphodium/grass interactions symposium, 27–29 Sep 2000, Soest, pp 447–453

Fletcher L, Finch S, Sutherland B, deNicolo G, Mace W, Van Koten C, Hume D (2017) The occurrence of ryegrass staggers and heat stress in sheep grazing ryegrass-endophyte associations with diverse alkaloid profiles. N Z Vet J 65:232–241

Florea S, Panaccione DG, Schardl CL (2017) Ergot alkaloids of the family Clavicipitaceae. Phytopathology 107:504–518

Florea S, Jaromczyk J, Schardl CL (2021) Non-transgenic CRISPR-mediated knockout of entire ergot alkaloid gene clusters in slow-growing asexual polyploid fungi. Toxins 13:153

Freeman EM (1904) The seed-fungus of *Lolium temulentum*, L., the Darnel. Philos Trans R Soc Lond Ser B Containing Papers Biol Char 196:1–27

Freitas PP, Hampton JG, Rolston MP, Glare TR, Miller PP, Card SD (2020) A tale of two grass species: temperature affects the symbiosis of a mutualistic *Epichloë* endophyte in both tall fescue and perennial ryegrass. Front Plant Sci 11. https://doi.org/10.3389/fpls.2020.00530

Froehlich KA, McAnulty R, Greer A (2022) Loline alkaloid effects on gastrointestinal nematodes. Animals 12:996

Fuchs B, Krischke M, Mueller MJ, Krauss J (2013) Peramine and lolitrem B from endophyte-grass associations cascade up the food chain. J Chem Ecol 39:1385–1389. https://doi.org/10.1007/s10886-013-0364-2

Fuchs B, Krischke M, Mueller MJ, Krauss J (2017) Plant age and seasonal timing determine endophyte growth and alkaloid biosynthesis. Fungal Ecol 29:52–58

Funk CR, White RH, Breen JP (1993) Importance of *Acremonium* endophytes in turf-grass breeding and management. Agric Ecosyst Environ 44:215–232. https://doi.org/10.1016/0167-8809(93)90048-T

Funk CR, Belanger FC, Murphy JA (2018) Role of endophytes in grasses used for turf and soil conservation. Biotechnology of endophytic fungi of grasses. CRC Press, pp 201–209

Gadberry M, Denard T, Spiers D, Piper E (2003) Effects of feeding ergovaline on lamb performance in a heat stress environment. J Anim Sci 81:1538–1545

Gagic M, Faville MJ, Zhang W, Forester NT, Rolston MP, Johnson RD, Ganesh S, Koolaard JP, Easton HS, Hudson D (2018) Seed transmission of Epichloë endophytes in *Lolium perenne is* heavily influenced by host genetics. Front Plant Sci 9:1580

Gallagher RT, White EP, Mortimer PH (1981) Ryegrass staggers: isolation of potent neurotoxins lolitrem a and lolitrem B From staggers-producing pastures. N Z Vet J 29:189–190. https://doi.org/10.1080/00480169.1981.34843

Garcia Parisi PA, Casas C, Gundel PE, Omacini M (2012) Consequences of grazing on the vertical transmission of a fungal *Neotyphodium* symbiont in an annual grass population. Austral Ecol 37:620–628

Gaston KJ (2000) Global patterns in biodiversity. Nature 405:220–227

Gerhards N, Neubauer L, Tudzynski P, Li S-M (2014) Biosynthetic pathways of ergot alkaloids. *Toxins* 6:3281–3295

Gibert A, Hazard L (2013) Genetically based vertical transmission drives the frequency of the symbiosis between grasses and systemic fungal endophytes. J Ecol 101:743–752

Gibert A, Volaire F, Barre P, Hazard L (2012) A fungal endophyte reinforces population adaptive differentiation in its host grass species. New Phytol 194:561–571. https://doi.org/10.1111/j.1469-8137.2012.04073.x

Gibson DJ (2009) Grasses and grassland ecology. Oxford University Press

Goldson SL, Rowarth JS, Caradus JR (2005) The impact of invasive invertebrate pests in pastoral agriculture: a review. N Z J Agric Res 48:401–415

Górzyńska K, Lembicz M, Olszanowski Z, Leuchtmann A (2011) *Botanophila-epichloë* interaction in a wild grass, *Puccinellia distans*, lacks dependence on the fly vector. Ann Entomol Soc Am 104:841–846

Granath G, Vicari M, Bazely DR, Ball JP, Puentes A, Rakocevic T (2007) Variation in the abundance of fungal endophytes in fescue grasses along altitudinal and grazing gradients. Ecography 30:422–430

Green KA, Berry D, Feussner K, Eaton CJ, Ram A, Mesarich CH, Solomon P, Feussner I, Scott B (2020) *Lolium perenne* apoplast metabolomics for identification of novel metabolites produced by the symbiotic fungus *Epichloë festucae*. New Phytol 227:559–571

Gundel PE, Batista WB, Texeira M, Martínez-Ghersa MA, Omacini M, Ghersa CM (2008) *Neotyphodium* endophyte infection frequency in annual grass populations: relative importance of mutualism and transmission efficiency. Proc R Soc B Biol Sci 275:897–905

Gundel PE, Garibaldi LA, Tognetti PM, Aragón R, Ghersa CM, Omacini M (2009a) Imperfect vertical transmission of the endophyte *Neotyphodium* in exotic grasses in grasslands of the Flooding Pampa. Microb Ecol 57:740–748

Gundel PE, Martínez-Ghersa MA, Garibaldi LA, Ghersa CM (2009b) Viability of *Neotyphodium* endophytic fungus and endophyte-infected and noninfected *Lolium multiflorum* seeds. Botany 87:88–96

Gundel PE, Martínez-Ghersa MA, Batista WB, Ghersa CM (2010) Dynamics of *Neotyphodium* endophyte infection in ageing seed pools: incidence of differential viability loss of endophyte, infected seed and non-infected seed. Ann Appl Biol 156:199–209

Gundel PE, Garibaldi LA, Martínez-Ghersa MA, Ghersa CM (2011a) *Neotyphodium* endophyte transmission to *Lolium multiflorum* seeds depends on the host plant fitness. Environ Exp Bot 71:359–366

Gundel PE, Rudgers J, Ghersa CM (2011b) Incorporating the process of vertical transmission into understanding of host–symbiont dynamics. Oikos 120:1121–1128

Gundel PE, Martínez-Ghersa MA, Omacini M, Cuyeu R, Pagano E, Ríos R, Ghersa CM (2012) Mutualism effectiveness and vertical transmission of symbiotic fungal endophytes in response to host genetic background. Evol Appl 5:838–849

Gundel PE, Pérez LI, Helander M, Saikkonen K (2013) Symbiotically modified organisms: nontoxic fungal endophytes in grasses. Trends Plant Sci 18:420–427

Gundel P, Sorzoli N, Ueno A, Ghersa C, Seal C, Bastías D, Martínez-Ghersa M (2015) Impact of ozone on the viability and antioxidant content of grass seeds is affected by a vertically transmitted symbiotic fungus. Environ Exp Bot 113:40–46

Gundel P, Irisarri J, Fazio L, Casas C, Pérez L (2016) Inferring field performance from drought experiments can be misleading: the case of symbiosis between grasses and *Epichloë* fungal endophytes. J Arid Environ 132:60–62

Gundel PE, Sun P, Charlton ND, Young CA, Miller TE, Rudgers JA (2020) Simulated folivory increases vertical transmission of fungal endophytes that deter herbivores and alter tolerance to herbivory in *Poa autumnalis*. Ann Bot 125:981–991

Guo Y, Gao P, Li F, Duan T (2019) Effects of AM fungi and grass endophytes on perennial ryegrass *Bipolaris sorokiniana* leaf spot disease under limited soil nutrients. Eur J Plant Pathol 154:659–671. https://doi.org/10.1007/s10658-019-01689-z

Hardoim PR, van Overbeek LS, Berg G, Pirttilä AM, Compant S, Campisano A, Döring M, Sessitsch A (2015) The hidden world within plants: ecological and evolutionary considerations for defining functioning of microbial endophytes. Microbiol Mol Biol Rev 79:293–320

Hassing B, Winter D, Becker Y, Mesarich CH, Eaton CJ, Scott B (2019) Analysis of *Epichloë festucae* small secreted proteins in the interaction with *Lolium perenne*. PLoS One 14:e0209463

Hennessy LM, Popay AJ, Finch SC, Clearwater MJ, Cave VM (2016) Temperature and plant genotype alter alkaloid concentrations in ryegrass infected with an *Epichloë* endophyte and this affects an insect herbivore. Front Plant Sci 7:1097

Hereme R, Morales-Navarro S, Ballesteros G, Barrera A, Ramos P, Gundel PE, Molina-Montenegro MA (2020) Fungal endophytes exert positive effects on *Colobanthus quitensis* under water stress but neutral under a projected climate change scenario in Antarctica. Front Microbiol 11. https://doi.org/10.3389/fmicb.2020.00264

Hettiarachchige IK, Ekanayake PN, Mann RC, Guthridge KM, Sawbridge TI, Spangenberg GC, Forster JW (2015) Phylogenomics of asexual *Epichloë* fungal endophytes forming associations with perennial ryegrass. BMC Evol Biol 15:1–14

Hettiarachchige IK, Elkins AC, Reddy P, Mann RC, Guthridge KM, Sawbridge TI, Forster JW, Spangenberg GC (2019) Genetic modification of asexual *Epichloë* endophytes with the perA gene for peramine biosynthesis. Mol Gen Genomics 294:315–328

Hill N, Roach P (2009) Endophyte survival during seed storage: endophyte–host interactions and heritability. Crop Sci 49:1425–1430

Hill N, Bouton J, Hiatt E, Kittle B (2005) Seed maturity, germination, and endophyte relationships in tall fescue. Crop Sci 45:859–863

Hoffman GD, Rao S (2013) Association of slugs with the fungal pathogen *Epichloë typhina* (Ascomycotina: Clavicipitaceae): potential role in stroma fertilisation and disease spread. Ann Appl Biol 162:324–334

Hoffman GD, Rao S (2014) Fertilization of *Epichloë typhina* stromata by mycophagous slugs. Mycologia 106:1–7. https://doi.org/10.3852/13-069

Holford G (1933) Grassland work overseas. Proc N Z Grassl Assoc:1–4

Hoveland CS (2009) Origin and history. Chapter 1. In: Fribourg HA, Hannaway DB and West CP, editors. Tall Fescue for the twenty-first century agronomy monographs 53. Madison, WI

Hoveland CS, Haaland RL, King CC Jr, Anthony WB, Clark EM, McGuire JA, Smith LA, Grimes HW, Holliman JL (1980) Association of *Epichloë typhina* fungus and steer performance on tall fescue pasture. Agron J 72:1064–1065. https://doi.org/10.2134/agronj1980.00021962007200060048x

Hume D, Sewell J (2014) Agronomic advantages conferred by endophyte infection of perennial ryegrass (*Lolium perenne* L.) and tall fescue (*Festuca arundinacea* Schreb.) in Australia. Crop Pasture Sci 65:747–757

Hume D, Ryan D, Cooper B, Popay A (2007) Agronomic performance of AR37-infected ryegrass in northern New Zealand. Proc N Z Grassl Assoc 69:201–205

Hume D, Cooper B, Panckhurst K (2009) The role of endophyte in determining the persistence and productivity of ryegrass, tall fescue and meadow fescue in Northland. Proc N Z Grassl Assoc 71:145–150

Hume DE, Drummond JB, Rolston MP, Simpson WR, Johnson RD (2018) *Epichloë* endophyte improves agronomic performance and grain yield of rye (*Secale cereale*). Proceedings of the 10th international symposium on fungal endophytes of grasses

Hyde KD, Soytong K (2008) The fungal endophyte dilemma. Fungal Divers 33:163–173

Jensen J, Popay A (2004) Perennial ryegrass infected with AR37 endophyte reduces survival of porina larvae. N Z Plant Prot 57:323–328

Jensen J, Popay A, Tapper B (2009) Argentine stem weevil adults are affected by meadow fescue endophyte and its loline alkaloids. N Z Plant Prot 62:12–18

Johnson LJ, Caradus JR (2019) The science required to deliver *Epichloë* endophytes to commerce. Endophytes Grow World:343

Johnson L, de Bonth A, Briggs L, Caradus J, Finch S, Fleetwood D, Fletcher L, Hume D, Johnson R, Popay A, Tapper B, Simpson W, Voisey C, Card S (2013a) The exploitation of *Epichloae* endophytes for

agricultural benefit. Fungal Divers 60:171–188. https://doi.org/10.1007/s13225-013-0239-4
Johnson LJ, Koulman A, Christensen M, Lane GA, Fraser K, Forester N, Johnson RD, Bryan GT, Rasmussen S (2013b) An extracellular siderophore is required to maintain the mutualistic interaction of *Epichloë festucae* with *Lolium perenne*. PLoS Pathog 9:e1003332
Johnson RD, Lane GA, Koulman A, Cao M, Fraser K, Fleetwood DJ, Voisey CR, Dyer JM, Pratt J, Christensen M (2015) A novel family of cyclic oligopeptides derived from ribosomal peptide synthesis of an *in planta*-induced gene, gigA, in *Epichloë* endophytes of grasses. Fungal Genet Biol 85:14–24
Johnson LJ, Bastías DA, Caradus JR, Chettri P, Forester NT, Mace WJ, Miller TA, Moon CD, Voisey CR, Zhang W, Card SD (2021) Chapter 6- The dynamic mechanisms underpinning symbiotic *Epichloë*–grass interactions: implications for sustainable and resilient agriculture. In: White J, Kumar A, Droby S (eds) . Woodhead Publishing, Microbiome stimulants for crops, pp 73–108
Ju Y, Zhong R, Christensen MJ, Zhang X (2020) Effects of *Epichloë gansuensis* endophyte on the root and rhizosphere soil bacteria of *Achnatherum inebrians* under different moisture conditions. Front Microbiol 11:747
Kalosa-Kenyon E, Slaughter LC, Rudgers JA, McCulley RL (2018) Asexual *Epichloë* endophytes do not consistently alter arbuscular mycorrhizal fungi colonization in three grasses. Am Midl Nat 179(2):157–165. https://doi.org/10.1674/0003-0031-179.2.157
Kaur J, Ekanayake PN, Tian P, De Jong EVZ, Dobrowolski MP, Rochfort SJ, Mann RC, Smith KF, Forster JW, Guthridge KM (2015) Discovery and characterisation of novel asexual *Epichloë* endophytes from perennial ryegrass (*Lolium perenne* L.). Crop Pasture Sci 66:1058–1070
Keller NP (2019) Fungal secondary metabolism: regulation, function and drug discovery. Nat Rev Microbiol 17:167–180
Kirby E (1961) Host-parasite relations in the choke disease of grasses. Trans Br Mycol Soc 44:493–503
Kirkby K, Pratley J, Hume D, An M, Wu H (2011) Viability of seed and endophyte (*Neotyphodium occultans*) in annual ryegrass (*Lolium rigidum*) when buried and in long term storage. Seed Sci Technol 39:452–464
Klotz JL, Nicol AM (2016) Ergovaline, an endophytic alkaloid. 1. Animal physiology and metabolism. Anim Prod Sci 56:1761–1774. https://doi.org/10.1071/AN14962
Kocmánková E, Trnka M, Juroch J, Dubrovský M, Semerádová D, Možný M, Žalud Z (2009) Impact of climate change on the occurrence and activity of harmful organisms. Plant Protect Sci 45:S48
Koga H, Christensen MJ, Bennett RJ (1993) Incompatibility of some grass-*Acremonium* endophyte associations. Mycol Res 97:1237–1244. https://doi.org/10.1016/S0953-7562(09)81292-6
Kogel K-H, Franken P, Hückelhoven R (2006) Endophyte or parasite–what decides? Curr Opin Plant Biol 9:358–363
Kohlmeyer J, Kohlmeyer E (1974) Distribution of *Epichloë typhina* (Ascomycetes) and its parasitic fly. Mycologia 66:77–86. https://doi.org/10.1080/00275514.1974.12019575
Koshino H, Togiya S, Yoshihara T, Sakamura S, Shimanuki T, Sato T, Tajimi A (1987) Four fungitoxic C-18 hydroxy unsaturated fatty acids from stromata of *Epichloë typhina*. Tetrahedron Lett 28:73–76. https://doi.org/10.1016/s0040-4039(00)95652-1
Koshino H, Terada S-I, Yoshihara T, Sakamura S, Shimanuki T, Sato T, Tajimi A (1988) Three phenolic acid derivatives from stromata of *Epichloë typhina* on *Phleum pratense*. Phytochemistry 27:1333–1338. https://doi.org/10.1016/0031-9422(88)80188-2
Koshino H, Yoshihara T, Sakamura S, Shimanuki T, Sato T, Tajimi A (1989) A ring B aromatic sterol from stromata of *Epichloë typhina*. Phytochemistry 28:771–772. https://doi.org/10.1016/0031-9422(89)80112-8
Koulman A, Lane GA, Christensen MJ, Fraser K, Tapper BA (2007) Peramine and other fungal alkaloids are exuded in the guttation fluid of endophyte-infected grasses. Phytochemistry 68:355–360. https://doi.org/10.1016/j.phytochem.2006.10.012
Krauss J, Härri SA, Bush L, Husi R, Bigler L, Power SA, Müller CB (2007) Effects of fertilizer, fungal endophytes and plant cultivar on the performance of insect herbivores and their natural enemies. Funct Ecol 21:107–116
Latch GCM, Christensen MJ (1982) Ryegrass endophyte, incidence, and control. N Z J Agric Res 25:443–448
Latch GCM, Christensen MJ (1985) Artificial infection of grasses with endophytes. Ann Appl Biol 107:17–24
Latch GCM, Christensen MJ, Hickson RE (1988) Endophytes of annual and hybrid ryegrasses. N Z J Agric Res 31:57–63. https://doi.org/10.1080/00288233.1988.10421364
Le Page M (2022) Worst drought in 500 years? New Scientist 255:8. https://doi.org/10.1016/S0262-4079(22)01459-2
Lembicz M, Miszalski Z, Kornaś A, Turnau K (2021) Cooling effect of fungal stromata in the *Dactylis-Epichloë-Botanophila* symbiosis. Commun Integr Biol 14:151–157. https://doi.org/10.1080/19420889.2021.1938824
Leuchtmann A, Clay K (1988) *Atkinsonella Hypoxylon* and *Balansia Cyperi*, epiphytic members of the Balansiae. Mycologia 80:192–199. https://doi.org/10.1080/00275514.1988.12025520
Leuchtmann A, Bacon CW, Schardl CL, White JF Jr, Tadych M (2014) Nomenclatural realignment of *Neotyphodium* species with genus *Epichloë*. Mycologia 106:202–215
Leuchtmann A, Young CA, Stewart AV, Simpson WR, Hume DE, Scott B (2019) *Epichloë novae-zelandiae*, a new endophyte from the endemic New Zealand grass *Poa matthewsii*. N Z J Bot 57:271–288
Lewis G, Ravel C, Naffaa W, Astier C, Charmet G (1997) Occurrence of *Acremonium* endophytes in wild populations of *Lolium* spp. in European countries and

a relationship between level of infection and climate in France. Ann Appl Biol 130:227–238

Leyronas C, Raynal G (2001) Presence of *Neotyphodium*-like endophytes in European grasses. Ann Appl Biol 139:119–127

Leyronas C, Raynal G (2008) Role of fungal ascospores in the infection of orchardgrass (*Dactylis glomerata*) by *Epichloë typhina* agent of choke disease. J Plant Pathol 90:15–21

Li F, Guo Y, Christensen MJ, Gao P, Li Y, Duan T (2018) An arbuscular mycorrhizal fungus and *Epichloë festucae* var. *lolii* reduce *Bipolaris sorokiniana* disease incidence and improve perennial ryegrass growth. Mycorrhiza 28:159–169. https://doi.org/10.1007/s00572-017-0813-9

Li C, Wang Z, Chen T, Nan Z (2021) Creation of novel barley germplasm using an *Epichloë* endophyte

Liu J, Nagabhyru P, Schardl CL (2017) *Epichloë festucae* endophytic growth in florets, seeds, and seedlings of perennial ryegrass (*Lolium perenne*). Mycologia 109: 691–700

Liu B, Ju Y, Xia C, Zhong R, Christensen MJ, Zhang X, Nan Z (2022a) The effect of *Epichloë* endophyte on phyllosphere microbes and leaf metabolites in *Achnatherum inebrians*. iScience 25:104144. https://doi.org/10.1016/j.isci.2022.104144

Liu H, Tang H, Ni X, Zhang Y, Wang Y (2022b) Effects of the endophyte *Epichloë coenophiala* on the root microbial community and growth performance of tall fescue in different saline-alkali soils. Fungal Ecol 57-58: 101159. https://doi.org/10.1016/j.funeco.2022.101159

Lloyd AB (1959) The endophytic fungus of perennial ryegrass. N Z J Agric Res 2:1187–1194

Lopez G (2022) A summer of climate disasters. The New York Times

Ludlow EJ, Vassiliadis S, Ekanayake PN, Hettiarachchige IK, Reddy P, Sawbridge TI, Rochfort SJ, Spangenberg GC, Guthridge KM (2019) Analysis of the indole diterpene gene cluster for biosynthesis of the epoxy-janthitrems in *Epichloë* endophytes. Microorganisms 7:560

Lyons PC, Plattner RD, Bacon CW (1986) Occurrence of peptide and clavine ergot alkaloids in tall fescue grass. Science 232:487–489

Mack KM, Rudgers JA (2008) Balancing multiple mutualists: asymmetric interactions among plants, arbuscular mycorrhizal fungi, and fungal endophytes. Oikos 117:310–320

Mahmud K, Lee K, Hill NS, Mergoum A, Missaoui A (2021) Influence of tall fescue *Epichloë* endophytes on rhizosphere soil microbiome. Microorganisms 9:1843

Malinowski DP, Belesky DP (2000) Adaptations of endophyte-infected cool-season grasses to environmental stresses: mechanisms of drought and mineral stress tolerance. Crop Sci 40:923–940. https://doi.org/10.2135/cropsci2000.404923x

Malinowski DP, Belesky DP (2006) Ecological importance of *Neotyphodium* spp. grass endophytes in agroecosystems. Grassl Sci 52:1–14

Martin BD, Schwab E (2012) Symbiosis:"Living together" in chaos. Stud Hist Biol 4:7–25

McDonald MB (1998) Seed quality assessment. Seed Sci Res 8:265–276

McLennan E (1920) The endophytic fungus of *Lolium*. Proc R Soc Vic (XXXII C N S) 32:252–301

Mercado-Blanco J (2015) Life of microbes inside the plant. Principles of plant-microbe interactions. Springer, pp 25–32

Meyer WA, Funk CR (1989) Progress and benefits to humanity from breeding cool-season grasses for turf. Contributions from breeding forage and turf grasses, pp 31–48

Miller TA, Hudson DA, Johnson RD, Singh JS, Mace WJ, Forester NT, Maclean PH, Voisey CR, Johnson LJ (2022) Dissection of the epoxyjanthitrem pathway in *Epichloë* sp. LpTG-3 strain AR37 by CRISPR gene editing. Front Fungal Biol 3. https://doi.org/10.3389/ffunb.2022.944234

Misra A, Yadav S, Mishra S, Tripathi M (2020) Impact of meteorological variables and climate change on plant diseases. Plant pathogens. Apple Academic Press, pp 313–327

Moate PJ, Williams SRO, Grainger C, Hannah MC, Mapleson D, Auldist MJ, Greenwood JS, Popay AJ, Hume DE, Mace WJ, Wales WJ (2012) Effects of wild-type, AR1 and AR37 endophyte-infected perennial ryegrass on dairy production in Victoria, Australia. Anim Prod Sci 52:1117–1130. https://doi.org/10.1071/AN12126

Moon CD, Craven KD, Leuchtmann A, Clement SL, Schardl CL (2004) Prevalence of interspecific hybrids amongst asexual fungal endophytes of grasses. Mol Ecol 13:1455–1467

Morales-Quintana L, Miño R, Mendez-Yañez A, Gundel PE, Ramos P (2022) Do fungal-endosymbionts improve crop nutritional quality and tolerance to stress by boosting flavonoid-mediated responses? Food Res Int 161:111850. https://doi.org/10.1016/j.foodres.2022.111850

Müller J (2003) Artificial infection by endophytes affects growth and mycorrhizal colonisation of *Lolium perenne*. Funct Plant Biol 30:419–424

Murphy JA, Sun S, Betts LL (1993) Endophyte-enhanced resistance to billbug (Coleoptera: Curculionidae), sod webworm (Lepidoptera: Pyralidae), and white Grub (Coleoptera: Scarabeidae) in tall fescue. Environ Entomol 22:699–703. https://doi.org/10.1093/ee/22.3.699

Neil JC (1940) The endophyte of rye-grass (*Lolium perenne*). N Z J Sci Technol A21:280–291

Nicol AM, Klotz JL (2016) Ergovaline, an endophytic alkaloid. 2. Intake and impact on animal production, with reference to New Zealand. Anim Prod Sci 56: 1775–1786. https://doi.org/10.1071/AN14963

Nissinen R, Helander M, Kumar M, Saikkonen K (2019) Heritable *Epichloë* symbiosis shapes fungal but not bacterial communities of plant leaves. Sci Rep 9: 5253. https://doi.org/10.1038/s41598-019-41603-5

Novas MV, Cabral D, Godeas AM (2005) Interaction between grass endophytes and mycorrhizas in *Bromus setifolius* from Patagonia, Argentina. Symbiosis

Novas V, Iannone L, Godeas A, Cabral D (2009) Positive association between mycorrhiza and foliar endophytes in *Poa bonariensis*, a native grass. Mycol Prog 8:75–81

Oesterheld M, Sala O, McNaughton S (1992) Effect of animal husbandry on herbivore-carrying capacity at a regional scale. Nature 356:234–236

Omacini M, Chaneton EJ, Ghersa CM, Müller CB (2001) Symbiotic fungal endophytes control insect host–parasite interaction webs. Nature 409:78–81

Omacini M, Eggers T, Bonkowski M, Gange A, Jones T (2006) Leaf endophytes affect mycorrhizal status and growth of co-infected and neighbouring plants. Funct Ecol 20:226–232

Pan J, Bhardwaj M, Nagabhyru P, Grossman RB, Schardl CL (2014) Enzymes from fungal and plant origin required for chemical diversification of insecticidal loline alkaloids in grass-*Epichloë* symbiota. PLoS One 9:e115590

Panaccione DG, Cipoletti JR, Sedlock AB, Blemings KP, Schardl CL, Machado C, Seidel GE (2006) Effects of ergot alkaloids on food preference and satiety in rabbits, as assessed with gene-knockout endophytes in perennial ryegrass (*Lolium perenne*). J Agric Food Chem 54:4582–4587

Pareek A, Dhankher OP, Foyer CH (2020) Mitigating the impact of climate change on plant productivity and ecosystem sustainability. Oxford University Press, pp 451–456

Partida-Martinez LPP, Heil M (2011) The microbe-free plant: fact or artifact? Front Plant Sci 2:100

Patchett B, Chapman R, Fletcher L, Gooneratne S (2008) Root loline concentration in endophyteinfected meadow fescue (Festuca pratensis) is increased by grass grub (Costelytra zealandica) attack. N Z Plant Prot 61:210–214

Penn J, Garthwaite I, Christensen M, Johnson C, Towers N (1993) The importance of paxilline in screening for potentially tremorgenic *Acremonium* isolates. In: Hume D, Latch G, Easton H (eds) Proceedings of the 2nd international symposium on *acremonium*/grass interactions. AgResearch, Palmerston North, pp 88–93

Pennell C, Popay A, Ball OJP, Hume D, Baird D (2005) Occurrence and impact of pasture mealybug (*Balanococcus poae*) and root aphid (*Aploneura lentisci)* on ryegrass (*Lolium* spp.) with and without infection by *Neotyphodium* fungal endophytes. N Z J Agric Res 48:329–337

Pennell CGL, Popay AJ, Rolston MP, Townsend RJ, Lloyd-West CM, Card SD (2016) Avanex Unique Endophyte Technology – a biological deterrent for the aviation industries. Environ Entomol 45:101–108. https://doi.org/10.1093/ee/nvv145

Pennell C, Rolston M, Baird D, Hume D, Mckenzie C, Card S (2017a) Using novel-grass endophyte associations as an avian deterrent. N Z Plant Prot 70: 255–264

Pennell C, Rolston M, Koten C, Hume D, Card S (2017b) Reducing bird numbers at New Zealand airports using a unique endophyte product. N Z Plant Prot 70:224–234

Pennell C, Rolston M, Latham A, Mace W, Vlaming J, van Koten C, Latham M, Brown S, Card S (2017c) Novel grass-endophyte associations reduce the feeding behaviour of invasive European rabbits (*Oryctolagus cuniculus*). Wildl Res 43:681–690

Perez LI, Gundel PE, Marrero HJ, Arzac AG, Omacini M (2017) Symbiosis with systemic fungal endophytes promotes host escape from vector-borne disease. Oecologia 184:237–245. https://doi.org/10.1007/s00442-017-3850-3

Pérez LI, Gundel PE, Zabalgogeazcoa I, Omacini M (2020) An ecological framework for understanding the roles of *Epichloë* endophytes on plant defenses against fungal diseases. Fungal Biol Rev 34:115–125

Pfender WF, Alderman SC (1999) Geographical distribution and incidence of orchardgrass choke, caused by *Epichloë typhina*, in Oregon. Plant Dis 83:754–758. https://doi.org/10.1094/pdis.1999.83.8.754

Pillay LC, Nekati L, Makhwitine PJ, Ndlovu SI (2022) Epigenetic activation of silent biosynthetic gene clusters in endophytic fungi using small molecular modifiers. Front Microbiol 13. https://doi.org/10.3389/fmicb.2022.815008

Popay AJ, Bonos SA (2008) Biotic responses in endophytic grasses. In: Roberts CA, West CP, Spiers DE (eds) Neotyphodium in cool-season grasses, pp 163–185

Popay AJ, Cox NR (2016) *Aploneura lentisci* (Homoptera: Aphididae) and its interactions with fungal endophytes in perennial ryegrass (*Lolium perenne*). Front Plant Sci 7:1395

Popay A, Hume D (2011) Endophytes improve ryegrass persistence by controlling insects. NZGA Res Pract Ser 15:149–156

Popay A, Thom E (2009) Endophyte effects on major insect pests in Waikato dairy pasture. Proc N Z Grassl Assoc:121–126

Popay A, Wyatt R (1995) Resistance to Argentine stem weevil in perennial ryegrass infected with endophytes producing different alkaloids. Proceedings of the forty eighth New Zealand plant protection conference, Angus Inn, Hastings, August 8–10, 1995: New Zealand Plant Protection Society, pp 229–236

Popay A, Townsend R, Fletcher L (2003) The effect of endophyte (*Neotyphodium uncinatum*) in meadow fescue on grass grub larvae. N Z Plant Prot 56:123–128

Popay A, Tapper B, Podmore C (2009) Endophyte infected meadow fescue and loline alkaloids affect Argentine stem weevil larvae. N Z Plant Prot 62:19–27

Pozo MJ, López-Ráez JA, Azcón-Aguilar C, García-Garrido JM (2015) Phytohormones as integrators of environmental signals in the regulation of mycorrhizal symbioses. New Phytol 205:1431–1436

Rasmussen S, Parsons AJ, Bassett S, Christensen MJ, Hume DE, Johnson LJ, Johnson RD, Simpson WR, Stacke C, Voisey CR (2007) High nitrogen supply and carbohydrate content reduce fungal endophyte and alkaloid concentration in *Lolium perenne*. New Phytol 173:787–797

Ravel C, Michalakis Y, Charmet G (1997) The effect of imperfect transmission on the frequency of mutualistic seed-borne endophytes in natural populations of grasses. Oikos, pp 18–24

Redman RS, Dunigan DD, Rodriguez RJ (2001) Fungal symbiosis from mutualism to parasitism: who controls the outcome, host or invader? New Phytol 151:705–716

Reed KFM, Mace WJ, Walker LV, Fletcher LR (2016) Endophyte metabolites associated with perennial ryegrass toxicosis. Anim Prod Sci 56:895–907. https://doi.org/10.1071/AN14495

Reinholz J, Paul V (2000) Effect of temperature on the lolitrem B content in *Lolium perenne* infected by *Neotyphodium lolii*. The 3rd international conference on harmful and beneficial microorganisms in grassland pastures and turf

Ren A-Z, Wang Y-H, Gao Y-B (2009) Difference in antifungal activity of morphotypes of clavicipitaceous endophytes within and between species. Acta Ecol Sin 29:227–231. https://doi.org/10.1016/j.chnaes.2009.08.005

Riedell W, Kieckhefer R, Petroski R, Powell R (1991) Naturally-occurring and synthetic loline alkaloid derivatives: insect feeding behavior modification and toxicity. J Entomol Sci 26:122–129

Roberts EL, Ferraro A (2015) Rhizosphere microbiome selection by *Epichloë* endophytes of *Festuca arundinacea*. Plant Soil 396:229–239

Roberts E, Lindow S (2014) Loline alkaloid production by fungal endophytes of *Fescue* species select for particular epiphytic bacterial microflora. ISME J 8:359–368

Robinson SL, Panaccione DG (2015) Diversification of ergot alkaloids in natural and modified fungi. Toxins 7: 201–218

Rodriguez R, Redman R (2008) More than 400 million years of evolution and some plants still can't make it on their own: plant stress tolerance via fungal symbiosis. J Exp Bot 59:1109–1114. https://doi.org/10.1093/jxb/erm342

Rodriguez RJ, White JF Jr, Arnold AE, Redman RS (2009) Fungal endophytes: diversity and functional roles. New Phytol 182:314–330. https://doi.org/10.1111/j.1469-8137.2009.02773.x

Rolston M, Hare M, Moore K, Christensen M (1986) Viability of *Lolium* endophyte fungus in seed stored at different moisture contents and temperatures. N Z J Exp Agric 14:297–300

Rolston M, Agee C (2006) Delivering quality seed to specification-the USA and NZ novel endophyte experience. New Zealand Grasslands Association: Research and Practice Series 13:229–231

Romo M, Leuchtmann A, García B, Zabalgogeazcoa I (2007) A totivirus infecting the mutualistic fungal endophyte *Epichloë festucae*. Virus Res 124:38–43

Rowan D, Gaynor D (1986) Isolation of feeding deterrents against Argentine stem weevil from ryegrass infected with the endophyte *Acremonium loliae*. J Chem Ecol 12:647–658

Rowan DD, Hunt MB, Gaynor DL (1986) Peramine, a novel insect feeding deterrent from ryegrass infected with the endophyte *Acremonium loliae*. J Chem Soc Chem Commun:935–936

Rowan DD, Dymock JJ, Brimble MA (1990) Effect of fungal metabolite peramine and analogs on feeding and development of Argentine stem weevil (*Listronotus bonariensis*). J Chem Ecol 16:1683–1695

Rozpądek P, Wężowicz K, Nosek M, Ważny R, Tokarz K, Lembicz M, Miszalski Z, Turnau K (2015) The fungal endophyte *Epichloë typhina* improves photosynthesis efficiency of its host orchard grass (*Dactylis glomerata*). Planta 242:1025–1035

Rubin MAR, Ueno AC, Batlla D, Iannone LJ, Martínez-Ghersa MA, Gundel PE (2022) Seed functional traits in cultivars of tall fescue (*Schedonorus arundinaceus*) are affected by the non-toxic fungal endophyte AR584. Crop Pasture Sci 73(9):1085–1096

Saikia S, Nicholson MJ, Young C, Parker EJ, Scott B (2008) The genetic basis for indole-diterpene chemical diversity in filamentous fungi. Mycol Res 112:184–199. https://doi.org/10.1016/j.mycres.2007.06.015

Saikkonen K, Faeth SH, Helander M, Sullivan T (1998) Fungal endophytes: a continuum of interactions with host plants. Annu Rev Ecol Syst 29:319–343

Saikkonen K, Ahlholm J, Helander M, Lehtimäki S, Niemeläinen O (2000) Endophytic fungi in wild and cultivated grasses in Finland. Ecography 23:360–366

Saikkonen K, Lehtonen P, Helander M, Koricheva J, Faeth SH (2006) Model systems in ecology: dissecting the endophyte–grass literature. Trends Plant Sci 11:428–433

Saikkonen K, Young CA, Helander M, Schardl CL (2016) Endophytic *Epichloë* species and their grass hosts: from evolution to applications. Plant Mol Biol 90: 665–675. https://doi.org/10.1007/s11103-015-0399-6

Sampson K (1933) The systemic infection of grasses by *Epichloë typhina* (Pers.) Tul. Trans Br Mycol Soc 18: 30–47

Sampson K (1937) Further observations on the systemic infection of Lolium. Trans Br Mycol Soc 21:84–97

Sánchez Márquez S, Bills GF, Herrero N, Zabalgogeazcoa Í (2012) Non-systemic fungal endophytes of grasses. Fungal Ecol 5:289–297. https://doi.org/10.1016/j.funeco.2010.12.001

Sands L (2022) Pakistan floods: One third of country is under water – minister. BBC News

Santangelo JS, Turley NE, Johnson MT (2015) Fungal endophytes of *Festuca rubra* increase in frequency following long-term exclusion of rabbits. Botany 93: 233–241

Schardl CL (2010) The epichloae, symbionts of the grass subfamily poöideae. Ann Mo Bot Gard 97:646–665
Schardl CL, Philips TD (1997) Protective grass endophytes: where are they from and where are they going? Plant Dis 81:430–438
Schardl CL, Leuchtmann A, Tsai HF, Collett MA, Watt DM, Scott DB (1994) Origin of a fungal symbiont of perennial ryegrass by interspecific hybridization of a mutualist with the ryegrass choke pathogen, *Epichloë typhina*. Genetics 136:1307–1317
Schardl CL, Leuchtmann A, Chung KR, Penny D, Siegel MR (1997) Coevolution by common descent of fungal symbionts (*Epichloë* spp.) and grass hosts. Mol Biol Evol 14:133–143
Schardl CL, Leuchtmann A, Spiering MJ (2004) Symbioses of grasses with seedborne fungal endophytes. Annu Rev Plant Biol 55:315–340. https://doi.org/10.1146/annurev.arplant.55.031903.141735
Schardl CL, Panaccione DG, Tudzynski P (2006) Ergot alkaloids–biology and molecular biology. Alkaloids Chem Biol 63:45–86
Schardl CL, Grossman RB, Nagabhyru P, Faulkner JR, Mallik UP (2007) Loline alkaloids: currencies of mutualism. Phytochemistry 68:980–996
Schardl CL, Young CA, Faulkner JR, Florea S, Pan J (2012) Chemotypic diversity of epichloae, fungal symbionts of grasses. Fungal Ecol 5:331–344. https://doi.org/10.1016/j.funeco.2011.04.005
Schardl CL, Florea S, Pan J, Nagabhyru P, Bec S, Calie PJ (2013a) The epichloae: alkaloid diversity and roles in symbiosis with grasses. Curr Opin Plant Biol 16:480–488. https://doi.org/10.1016/j.pbi.2013.06.012
Schardl CL, Young CA, Hesse U, Amyotte SG, Andreeva K, Calie PJ, Fleetwood DJ, Haws DC, Moore N, Oeser B (2013b) Plant-symbiotic fungi as chemical engineers: multi-genome analysis of the Clavicipitaceae reveals dynamics of alkaloid loci. PLoS Genet 9
Schiestl FP, Steinbrunner F, Schulz C, von Reub S, Francke W, Weymuth C, Leuchtmann A (2006) Evolution of 'pollinator'-attracting signals in fungi. Biol Lett 2:401–404
Schmid J, Day R, Zhang N, Dupont P-Y, Cox MP, Schardl CL, Minards N, Truglio M, Moore N, Harris DR, Zhou Y (2017) Host tissue environment directs activities of an *Epichloë* endophyte, while it induces systemic hormone and defense responses in its native perennial ryegrass host. Mol Plant-Microbe Interact 30:138–149. https://doi.org/10.1094/MPMI-10-16-0215-R
Schulthess FM, Faeth SH (1998) Distribution, abundances, and associations of the endophytic fungal community of Arizona fescue (*Festuca arizonica*). Mycologia 90:569–578
Schulz B, Guske S, Dammann U, Boyle C (1998) Endophyte-host interactions. II. Defining symbiosis of the endophyte- host interaction. Symbiosis (Rehovot) 25:213–227
Scott B, Green K, Berry D (2018) The fine balance between mutualism and antagonism in the *Epichloë festucae*–grass symbiotic interaction. Curr Opin Plant Biol 44:32–38. https://doi.org/10.1016/j.pbi.2018.01.010
Semmartin M, Omacini M, Gundel PE, Hernández-Agramonte IM (2015) Broad-scale variation of fungal-endophyte incidence in temperate grasses. J Ecol 103:184–190
Simpson WR, Faville MJ, Moraga RA, Williams WM, Mcmanus MT, Johnson RD (2014) *Epichloë* fungal endophytes and the formation of synthetic symbioses in Hordeeae (= Triticeae) grasses. J Syst Evol 52:794–806
Sinclair JB, Cerkauskas RF (1996) Latent infection vs. endophytic colonization by fungi. In: Redlin SC, Carris LM (eds) Endophytic fungi in grasses and woody plants systematics, ecology, and evolution. APS Press, St. Paul, MN
Skendžić S, Zovko M, Živković IP, Lešić V, Lemić D (2021) The impact of climate change on agricultural insect pests. Insects 12:440
Slaughter LC, Nelson JA, Carlisle E, Bourguignon M, Dinkins RD, Phillips TD, McCulley RL (2018) Climate change and *Epichloë coenophiala* association modify belowground fungal symbioses of tall fescue host. Fungal Ecol 31:37–46. https://doi.org/10.1016/j.funeco.2017.10.002
Sleper DA, West CP (1996) Tall fescue. Cool-season forage grasses, pp 471–502
Spatafora J, Sung GH, Sung JM, Hywel-Jones N, White J Jr (2007) Phylogenetic evidence for an animal pathogen origin of ergot and the grass endophytes. Mol Ecol 16:1701–1711
Spiering MJ, Davies E, Tapper BA, Schmid J, Lane GA (2002) Simplified extraction of ergovaline and peramine for analysis of tissue distribution in endophyte-infected grass tillers. J Agric Food Chem 50:5856–5862. https://doi.org/10.1021/jf025602b
Spiering MJ, Faulkner JR, Zhang D-X, Machado C, Grossman RB, Schardl CL (2008) Role of the LolP cytochrome P450 monooxygenase in loline alkaloid biosynthesis. Fungal Genet Biol 45:1307–1314
Star P, Brooking T (2006) Fescue to the rescue: chewings fescue, paspalum, and the application of non-British experience to pastoral practice in New Zealand, 1880-1920. Agric Hist 312–335
Staunton J, Weissman KJ (2001) Polyketide biosynthesis: a millennium review. Nat Prod Rep 18:380–416
Stefanoni-Rubio PJ, Gundel PE, Novas MV, Iannone LJ (2022) Ecotype-specific effects of fungal endophytes on germination responses of seeds of the South American wild forage grass *Bromus auleticus*. Ann Appl Biol 180:247–258
Stewart A (2006) Genetic origins of perennial ryegrass (Lolium perenne) for New Zealand pastures. NZGA Res Pract Ser 12:55–61
Stewart AV, Barcellos G, Brilman L (2022) Use of endophytic fungi in turfgrasses: difficulties in delivery to

the market. Int Turfgrass Soc Res J 14:1070–1073. https://doi.org/10.1002/its2.131

Strickland J, Aiken G, Spiers D, Fletcher L, Oliver J (2009) Physiological basis of fescue toxicosis. Tall Fescue for the Twenty-first Century 53:203–227

Süssmuth RD, Mainz A (2017) Nonribosomal peptide synthesis—principles and prospects. Angew Chem Int Ed 56:3770–3821

Suttie JM, Reynolds SG, Batello C (2005) Grasslands of the World. Food & Agriculture Organization of the United Nations, Rome

Takach JE, Mittal S, Swoboda GA, Bright SK, Trammell MA, Hopkins AA, Young CA (2012) Genotypic and chemotypic diversity of *Neotyphodium* endophytes in tall fescue from Greece. Appl Environ Microbiol 78: 5501–5510

Tanaka A, Tapper BA, Popay A, Parker EJ, Scott B (2005) A symbiosis expressed non-ribosomal peptide synthetase from a mutualistic fungal endophyte of perennial ryegrass confers protection to the symbiotum from insect herbivory. Mol Microbiol 57:1036–1050

Tannenbaum I, Kaur J, Mann R, Sawbridge T, Rodoni B, Spangenberg G (2020) Profiling the *Lolium perenne* microbiome: from seed to seed. Phytobiomes J 4:281–289

Tapper B, Lane G (2004) Janthitrems found in a *Neotyphodium* endophyte of perennial ryegrass. Proceedings of the 5th international symposium on neotyphodium/grass interactions, University of Arkansas Press, Fayetteville, AR

Terlizzi NL, Rodríguez MA, Iannone LJ, Lanari E, Novas MV (2022) *Epichloë* endophyte affects the root colonization pattern of belowground symbionts in a wild grass. Fungal Ecol 57–58:101143. https://doi.org/10.1016/j.funeco.2022.101143

Thom ER, Popay AJ, Hume DE, Fletcher LR (2012) Evaluating the performance of endophytes in farm systems to improve farmer outcomes–a review. Crop Pasture Sci 63:927–943

Tian P, Le T-N, Smith K, Forster J, Guthridge K, Spangenberg G (2013) Stability and viability of novel perennial ryegrass host–*Neotyphodium* endophyte associations. Crop Pasture Sci 64:39–50

Tiwari P, Bae H (2022) Endophytic fungi: key insights, emerging prospects, and challenges in natural product drug discovery. Microorganisms 10:360

Torres M, White J (2009) Clavicipitaceae: free-living and saprotrophs to plant endophytes. Encycl Microbiol 1: 422–430

Tsai HF, Liu JS, Staben C, Christensen MJ, Latch GC, Siegel MR, Schardl CL (1994) Evolutionary diversification of fungal endophytes of tall fescue grass by hybridization with *Epichloë* species. Proc Natl Acad Sci U S A 91:2542–2546. https://doi.org/10.1073/pnas.91.7.2542

Ueno AC, Gundel PE, Omacini M, Ghersa CM, Bush LP, Martínez-Ghersa MA (2016) Mutualism effectiveness of a fungal endophyte in an annual grass is impaired by ozone. Funct Ecol 30:226–234

Ueno AC, Gundel PE, Ghersa CM, Demkura PV, Card SD, Mace WJ, Martínez-Ghersa MA (2020a) Ontogenetic and trans-generational dynamics of a vertically transmitted fungal symbiont in an annual host plant in ozone-polluted settings. Plant Cell Environ 43:2540–2550

Ueno AC, Gundel PE, Seal CE, Ghersa CM, Martínez-Ghersa MA (2020b) The negative effect of a vertically-transmitted fungal endophyte on seed longevity is stronger than that of ozone transgenerational effect. Environ Exp Bot 175:104037

van Zijll de Jong E, Dobrowolski MP, Bannan NR, Stewart AV, Smith KF, Spangenberg GC, Forster JW (2008) Global genetic diversity of the perennial ryegrass fungal endophyte *Neotyphodium lolii*. Crop Sci 48:1487–1501

Victoria N, Collantes M, Cabral D (2007) Environmental effects on grass-endophyte associations in the harsh conditions of south Patagonia. FEMS Microbiol Ecol 61:164–173

Vignale M, Iannone LJ, Pinget AD, De Battista JP, Novas M (2016) Effect of epichloid endophytes and soil fertilization on arbuscular mycorrhizal colonization of a wild grass. Plant Soil 405:279–287

Vignale M, Iannone LJ, Scervino JM, Novas M (2018) *Epichloë* exudates promote *in vitro* and *in vivo* arbuscular mycorrhizal fungi development and plant growth. Plant Soil 422:267–281

Vining LC (1990) Functions of secondary metabolites. Annu Rev Microbiol 44:395–427

Voisey CR (2010) Intercalary growth in hyphae of filamentous fungi. Fungal Biol Rev 24:123–131. https://doi.org/10.1016/j.fbr.2010.12.001

Wakelin S, Harrison S, Mander C, Dignam B, Rasmussen S, Monk S, Fraser K, O'Callaghan M (2015) Impacts of endophyte infection of ryegrass on rhizosphere metabolome and microbial community. Crop Pasture Sci 66:1049–1057

Wang J, Machado C, Panaccione DG, Tsai H-F, Schardl CL (2004) The determinant step in ergot alkaloid biosynthesis by an endophyte of perennial ryegrass. Fungal Genet Biol 41:189–198. https://doi.org/10.1016/j.fgb.2003.10.002

Wang R, Clarke BB, Belanger FC (2019) Transcriptome analysis of choke stroma and asymptomatic inflorescence tissues reveals changes in gene expression in both *Epichloë festucae* and its host plant *Festuca rubra* subsp. *rubra*. Microorganisms 7:567

Wang M, Tian P, Gao M, Li M (2021a) The promotion of *Festuca sinensis* under heavy metal treatment mediated by *Epichloë* endophyte. Agronomy 11:2049

Wang R, Luo S, Clarke BB, Belanger FC (2021b) The *Epichloë festucae* antifungal protein Efe-AfpA is also a possible effector protein required for the interaction of the fungus with its host grass *Festuca rubra* subsp. *rubra*. Microorganisms 9:140

Welty R, Azevedo M, Cooper T (1987) Influence of moisture content, temperature, and length of storage on seed germination and survival of endophytic fungi in seeds

of tall fescue and perennial ryegrass. Phytopathology 77:893–900
Wennström A (1994) Endophyte - the misuse of an old term. Oikos 71:535–536
Wernegreen JJ (2012) Endosymbiosis. Curr Biol 22: R555–R561. https://doi.org/10.1016/j.cub.2012.06.010
Western J, Cavett J (1959) The choke disease of cocksfoot (*Dactylis glomerata*) caused by *Epichloë typhina* (Fr.) Tul. Trans Br Mycol Soc 42:298–307
White JF, Morrow AC, Morgan JG, Chambless DA (1991) Endophyte-host associations in forage grasses. XIV. Primary stromata formation and seed transmission in Epichloë typhina: development and regulatory aspects. Mycologia 83:72–81
White J, James F, Bacon CW, Hywel-Jones NL, Spatafora JW (2003) Historical perspectives: human interactions with Clavicipitalean fungi. CRC Press, Clavicipitalean Fungi, pp 12–25
Wilkinson HH, Siegel MR, Blankenship JD, Mallory AC, Bush LP, Schardl CL (2000) Contribution of fungal loline alkaloids to protection from aphids in a grass-endophyte mutualism. Mol Plant-Microbe Interact 13: 1027–1033. https://doi.org/10.1094/mpmi.2000.13.10.1027
Williams CL, Goldson S, Baird D, Bullock D (1994) Geographical origin of an introduced insect pest, *Listronotus bonariensis* (Kuschel), determined by RAPD analysis. Heredity 72:412–419
Wilson D (1995) Endophyte: the evolution of a term, and clarification of its use and definition. Oikos 73:274–276
Woodfield D, Judson H (2019) Balancing pasture productivity with environmental and animal health requirements. Improving grassland and pasture management in temperate agriculture. Burleigh Dodds Science Publishing, pp 257–272
Xia C, Li N, Zhang Y, Li C, Zhang X, Nan Z (2018) Role of *Epichloë* endophytes in defense responses of cool-season grasses to pathogens: a review. Plant Dis 102: 2061–2073
Xu W, Li M, Lin W, Nan Z, Tian P (2021) Effects of *Epichloë sinensis* endophyte and host ecotype on physiology of *Festuca sinensis* under different soil moisture conditions. Plan Theory 10:1649
Yan L, Zhao H, Zhao X, Xu X, Di Y, Jiang C, Shi J, Shao D, Huang Q, Yang H (2018) Production of bioproducts by endophytic fungi: chemical ecology, biotechnological applications, bottlenecks, and solutions. Appl Microbiol Biotechnol 102:6279–6298
Yoshihara T, Togiya S, Koshino H, Sakamura S, Shimanuki T, Sato T, Tajimi A (1985) Three fungitoxic cyclopentanoid sesquiterpenes from stromata of *Epichloë typhina*. Tetrahedron Lett 26:5551–5554. https://doi.org/10.1016/s0040-4039(01)80885-6
Young CA, Felitti S, Shields K, Spangenberg G, Johnson RD, Bryan GT, Saikia S, Scott B (2006) A complex gene cluster for indole-diterpene biosynthesis in the grass endophyte *Neotyphodium lolii*. Fungal Genet Biol 43:679–693
Young CA, Tapper BA, May K, Moon CD, Schardl CL, Scott B (2009) Indole-diterpene biosynthetic capability of *Epichloë* endophytes as predicted by ltm gene analysis. Appl Environ Microbiol 75:2200–2211
Yue Q, Miller CJ, White JF, Richardson MD (2000) Isolation and characterization of fungal inhibitors from *Epichloë festucae*. J Agric Food Chem 48:4687–4692
Zabalgogeazcoa I, Benito EP, Ciudad AG, Criado BG, Eslava AP (1998) Double-stranded RNA and virus-like particles in the grass endophyte *Epichloë festucae*. Mycol Res 102:914–918
Zaurov DE, Bonos S, Murphy JA, Richardson M, Belanger FC (2001) Endophyte infection can contribute to aluminum tolerance in fine fescues. Crop Sci 41: 1981–1984. https://doi.org/10.2135/cropsci2001.1981
Zhang W, Card SD, Mace WJ, Christensen MJ, McGill CR, Matthew C (2017) Defining the pathways of symbiotic *Epichloë* colonization in grass embryos with confocal microscopy. Mycologia 109:153–161. https://doi.org/10.1080/00275514.2016.1277469
Zhang W, Mace WJ, Matthew C, Card SD (2019) The impact of endophyte infection, seed aging, and imbibition on selected sugar metabolite concentrations in seed. J Agric Food Chem 67:6921–6929
Zhang J, Deng Y, Ge X, Shi X, Fan X, Dong K, Chen L, Zhao N, Gao Y, Ren A (2022) The beneficial effect of *Epichloë* endophytes on the growth of host grasses was affected by arbuscular mycorrhizal fungi, pathogenic fungi and nitrogen addition. Environ Exp Bot 201: 104979. https://doi.org/10.1016/j.envexpbot.2022.104979
Zhao Z, Kou M, Zhong R, Xia C, Christensen MJ, Zhang X (2021) Transcriptome analysis revealed plant hormone biosynthesis and response pathway modification by *Epichloë gansuensis* in *Achnatherum inebrians* under different soil moisture availability. J Fungi 7:640
Zhong R, Xia C, Ju Y, Zhang X, Duan T, Nan Z, Li C (2021) A foliar *Epichloë* endophyte and soil moisture modified belowground arbuscular mycorrhizal fungal biodiversity associated with *Achnatherum inebrians*. Plant Soil 458:105–122. https://doi.org/10.1007/s11104-019-04365-7
Zhong R, Bastías DA, Zhang X, Li C, Nan Z (2022) Vertically transmitted *Epichloë* systemic endophyte enhances drought tolerance of *Achnatherum inebrians* host plants through promoting photosynthesis and biomass accumulation. J Fungi 8:512
Zhou Y, Li X, Gao Y, Ren A (2022) Plant endophytes and arbuscular mycorrhizal fungi alter the decomposition of *Achnatherum sibiricum* litter. Appl Soil Ecol 180. https://doi.org/10.1016/j.apsoil.2022.104616
Żurek G, Wiewióra B, Gozdowski D (2013) Relations between bioclimatic variables and endophyte colonization of grasses in Poland. Fungal Ecol 6:554–556

9 Signals and Host Cell Remodeling in Arbuscular Mycorrhizal Symbiosis

Andrea Genre, Serena Capitanio, and Paola Bonfante

Abstract

Mycorrhizas are mutualistic interactions that the majority of land plants establish with a heterogeneous group of soil fungi; their distribution and diversity have supported the success of plants on the planet. Among all different types of mycorrhizas, arbuscular mycorrhiza (AM) is the most ancient and the most common in host plants of all major crops. The functional core of AM is a finely branched fungal structure called the arbuscule. Arbuscules are hosted inside living root cells, within a specialized cell compartment that is generated through a precise sequence of molecular and cellular events. Over the last 10 years, the application of novel technologies, such as genome sequencing, high-throughput transcriptomics, and live cell imaging, has generated substantial advances in our knowledge of such events. Here, we present a synopsis of the recent literature on the interactions between AM fungi and their hosts, with an evolutionary-developmental focus on the intimate contact that develops between plant cells and fungal hyphae, in terms of molecular signaling, nutrient exchange, and cell organization.

Keywords

Symbiosis · Mycorrhiza · Plant nutrition · Fungal genomics · Signaling · Strigolactone · Chitin · Calcium · Cell cycle · Evo-devo

9.1 Introduction

Our planet hosts around 390,000 plant species. Within their hypogeous and epigeous organs, plants store more than 450 Gigatons (Gt) of carbon, roughly corresponding to 80% of the planet biomass (Bar-On et al. 2018). Interestingly, the belowground biomass is mainly composed of plant roots (130 Gt C) and soil-dwelling microbes. Among them, fungi (with their 12 Gt C) are the second most conspicuous soil microbes, only outnumbered by bacteria (70 Gt. C). Such figures provide solid experimental support to the acknowledged claim that mycorrhizas—symbiotic associations between around 340,000 plant species and 50,000 species of soil fungi—represent one of the most powerful and pervasive ecological processes in terrestrial environments. In these mutually beneficial interactions, fungi receive photosynthesis-derived carbon from the plant, while supporting the plant diet with mineral nutrients, mostly phosphorus and nitrogen. Present in both natural and human-controlled agricultural niches, mycorrhizas offer several ecosystem services. Mycorrhizal fungi are in fact well-known drivers

A. Genre (✉) · S. Capitanio · P. Bonfante
Dipartimento di Scienze della Vita e Biologia dei Sistemi, Università di Torino, Torino, Italy
e-mail: andrea.genre@unito.it

Y.-P. Hsueh, M. Blackwell (eds.), *Fungal Associations*, The Mycota 9,
https://doi.org/10.1007/978-3-031-41648-4_9

of carbon sequestration and soil particle aggregation. They also have a major impact on the composition of microbial and plant communities, as originally demonstrated in a seminal study by Van der Heijden et al. (1998) where the belowground diversity of arbuscular mycorrhizal fungi emerged as a major factor in the maintenance of plant biodiversity. Consensus is growing on the role of biodiversity in ecosystem functioning, and—in this respect—mycorrhizal symbioses are considered to give a major contribution to the mechanisms that support the whole biosphere and its functioning (Eisenhauer et al. 2022). The link between mycorrhizas and planet biodiversity is quite well defined: 90% of extant vascular and non-vascular plants can form mycorrhizas and, among them, angiosperms account for the largest and most diverse group of mycorrhizal species (85–90%). They include annual herbs and perennial trees as well as the majority of staple crops such as rice, wheat, maize, potato, sweet potato, tomato, and cassava. The number of fungal mycorrhizal species is more difficult to ascertain and is currently estimated to account for around 10% of known fungal species (Hyde 2022), belonging to the three major phyla of Ascomycota, Basidiomycota, and Mucoromycota.

Box 9.1 Main Types of Mycorrhizal Symbioses

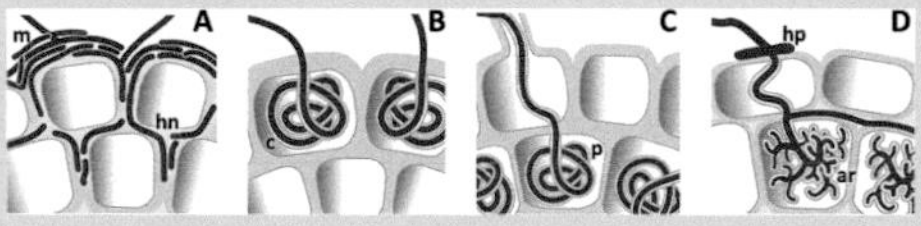

The figure illustrates four major types of mycorrhizal interactions. In ectomycorrhizas (A), the epidermal cells of lateral roots are targeted by extraradical hyphae that proliferate on the surface of the whole root tip, engulfing it in a pseudoparenchymatous tissue known as the mantle (m). Inner mantle hyphae further penetrate between epidermal cells, proliferating between epidermal and outer cortical cells to generate the so-called Hartig net (hn), a dense mesh of hyphae hosted in an expanded apoplast.

In ericoid mycorrhizas (B), extraradical hyphae directly penetrate epidermal cell walls without the development of apparent adhesion structures. Convoluted hyphal coils (c) are then produced inside each epidermal cell, enclosed within the symbiotic interface compartment and the perifungal membrane (8).

In orchids (C), mycorrhizal symbiosis involves both germinating seeds (protocorms) and seedling roots, but the best described colonization process is in protocorms. Here, epidermal hair cells are the main penetration site for external hyphae, which then reach the cortical parenchyma to form large intracellular hyphal coils, or pelotons (p), within a membrane-delimited symbiotic interface.

Lastly, in arbuscular mycorrhizas (D), presymbiotic hyphae contact the epidermal cells of young lateral roots forming an adhesion structure called the hyphopodium (hp). The hyphopodium, often flattened and branched, adheres strongly to the epidermal cell walls. A penetrating hypha then develops in this area, crosses the epidermis and outer cortical cell layers with both loose intracellular coils and branched hyphae, and finally reaches inner cortical cells, where tree-like arbuscules are formed (ar). Also in this case, all intracellular hyphae and arbuscules are accommodated in the symbiotic interface. This interface, outlined by the periarbuscular membrane, is believed to be the main site of nutrient exchange, thanks to its extensive surface.

These diverse mycorrhizal types involve different groups of fungi: Basidiomycota and Ascomycota are the preferential partners in ectomycorrhizas and ericoid mycorrhizas, while Basidiomycota taxa associate with orchids and

(continued)

Box 9.1 (continued)

Glomeromycotina develop arbuscular mycorrhizas (Genre et al. 2020).

The resulting plant-fungal diversity, supported by over 400 million years of co-evolution, has resulted in four main mycorrhizal types, defined according to their major morphological traits (Box 9.1): ectomycorrhizas (ECM), ericoid mycorrhizas (ERM), orchid mycorrhizas (ORM), and arbuscular mycorrhizas (AM). Their specific cell-to-cell interactions were initially described by ultrastructural observations between 1970s and 1990s, and these pioneering studies provided the foundation for subsequent investigations on signaling and gene regulation in these symbioses (Bonfante 2018). This corpus of studies revealed that several cellular and molecular traits are shared among the four types of mycorrhizas, even if they originated independently at different times during plant evolution (Genre et al. 2020) and have an uneven phylogenetic distribution (Tedersoo et al. 2020): AMs are the dominant type (involving 72% of plant species), followed by ORM (10%), ECM (2%) and ERM (1.4%). Nevertheless, this distribution between plant species does not mirror the real ecological success of each mycorrhizal type. In fact, ECMs dominate the vast boreal forests, despite the low number of gymnosperm and angiosperm species involved. By contrast, AMs—with their extremely broad host ranges—are present across most land ecosystems, from arid sub-arctic environments to sub-tropical rainforests. A reason for the ecological success of AMs can be found in their evolutionary history. AMs are in fact acknowledged as the most ancient form of symbiosis, with the earliest fossil evidence of such plant–fungus associations dating back to 400 million years ago. In fact, the study of the uniquely preserved ecosystem of the Rhynie chert revealed that extinct plants such as *Aglaophyton majus* hosted intracellular structures that are very similar to modern arbuscules (Remy et al. 1994).

AMs are found in the roots of wild plants and of most crops, raising therefore the current interest of scientists, producers and politicians. Understanding the mechanisms that regulate plant-fungal interplay has in fact emerged as a social responsibility for solving the dilemma of saving space for wild nature preservation (Wilson 2016) or growing plants to feed a growing human population.

In the framework of this complex ecological and ethical scenario, we propose a synopsis of the recent literature on the interactions between AM fungi and their hosts, with a particular focus on the intimate contact that develops between plant cells and fungal hyphae, in terms of molecular signaling, nutrient exchange, and cell organization. Novel technologies, such as genome sequencing, high-throughput transcriptomics and proteomics as well as live sub-cellular imaging, are providing the tools to decipher the genetic, molecular, and cellular bases of such plant–fungal interactions.

9.2 An Insight in the Biology of Arbuscular Mycorrhizal Fungi

Arbuscular Mycorrhizal (AM) fungi are the most relevant representatives of Glomeromycotina, a subphylum of the Mucoromycota, which also embraces *Geosiphon pyriformis*, a rare fungus hosting cyanobacteria (Schüßler 2012). Even if their ranking as Glomeromycota or Glomeromycotina is still controversial (Tedersoo et al. 2018; Spatafora et al. 2016), the assignment of AM fungi to Mucoromycota satisfies morphological, genomic, and functional parameters. As in other Mucoromycota, they are aseptate and multinucleated fungi which often host endobacteria in their cytoplasm (Bonfante and Venice 2020), but also possess unique features: they are described as asexual, obligate symbionts, which depend on the reduced carbon obtained by the host plants, while they provide plants with minerals. Glomeromycotina which contain different orders interact with a huge number of plants:

liverworts, ferns, gymnosperms (from *Ginkgo biloba* to *Cupressus*), and angiosperms (from trees such as poplar to herbs and shrubs). It is also relevant to remind that AM fungi are probably not alone to colonize plants producing intracellular branched hyphae: Mucoromycotina "fine root endophytes," previously known as *Glomus tenue*, are a group of soil fungi that form endosymbioses with different land plants (Sinanaj et al. 2021), even if significant questions on their distribution and functions are still open.

Genomics has improved some of our views on AM fungal biology: AMF genomes are among the largest in the fungal kingdom, ranging from 125 Mbp for the A1 strain of *Rhizophagus irregularis* (Tisserant et al. 2013) to the unexpected sizes of 570 Mbp of *Gigaspora rosea* (Morin et al. 2019) and 770 Mbp of *G. margarita* (Venice et al. 2020). By contrast, the gene number is comparable among the AM sequenced species (around 25,000–26,000 genes), even if in the largest genomes, the size was inflated by a huge amount of transposable elements, which occupy around 70% of their genomes. However, *Paraglomus occultum* which belongs to the early diverging order of Paraglomales has a very reduced genome, i.e. 39.6 Mb, fewer genes, and lacks the genes involved in the production of fatty acids and sugars (see below), suggesting that the MRCA of Glomeromycotina was already an obligate plant biotroph (Malar et al. 2022). As for some plant pathogens, AM genomes encode for a large number of effectors, that might guarantee the communication between AM fungi and their hosts, and for a limited presence of plant cell wall degrading enzymes. This trait may be involved in eluding host defense and maintaining the viability of host cells. AM fungi are fully dependent on plant-derived lipids, since their genomes lack a cytosolic Fatty Acid Synthase complex that is common to fungi (FASI), but possess a mitochondrial FASII for the synthesis of lipoic acid (reviewed in Bonfante and Venice 2020). This feature, together with the apparent lack of the pathway for thiamine biosynthesis and the limited uptake of sugars from soil (Tisserant et al. 2013), contributes to the signature of AMF as obligate biotrophs. However, *R. irregularis* could be grown and reproduced axenically on a medium added with fatty acids, among which myristate (Kameoka et al. 2019; Sugiura et al. 2020). Some AM fungal genomes have revealed the presence of pathways leading to the synthesis of secondary metabolites, as polyketides (Venice et al. 2020b). The multinucleate status of AM fungi, on the one hand, complicates transformation procedures and, on the other, raises many debates on their level of genetic variability. Genome sequences and an in-depth analysis of MAT-loci, which are related to fungal sexuality events, revealed that while most of the isolates are homokaryotic, i.e., containing nuclei of one genetically similar nucleotype, a few are dikaryotic, i.e. containing nuclei of two genetically divergent nucleotypes (Ropars et al. 2016; Kokkoris et al. 2021). Furthermore, the availability of multiple sequenced strains from the same species revealed an impressive amount of intra-specific variability (the *R. irregularis* strains only share around 8000 conserved genes), such that the concept of "pangenomes" has been suggested to describe AM genomes (Mathieu et al. 2018).

In conclusion, current views indicate a long evolutionary story of AM fungi mirroring their capacity to adapt to their plant hosts and changing environments, and to interact with endobacteria. However, due to the multinucleate status of AM fungi, their genes cannot be easily tested for their functionality, with the rare exception of the fungal genes involved in Pi uptake (Xie et al. 2022). As a consequence, our current physiological knowledge is mostly based on the host plant side, while many of the striking biological features of AM fungi (i.e., obligate biotrophy, multinucleated condition, unknown sexual status) make them remain a rather intractable experimental material (Lanfranco and Bonfante 2023).

9.3 The Development of Arbuscular Mycorrhiza

The establishment of AM symbiosis (Fig. 9.1) has been elucidated in detail and is largely conserved,

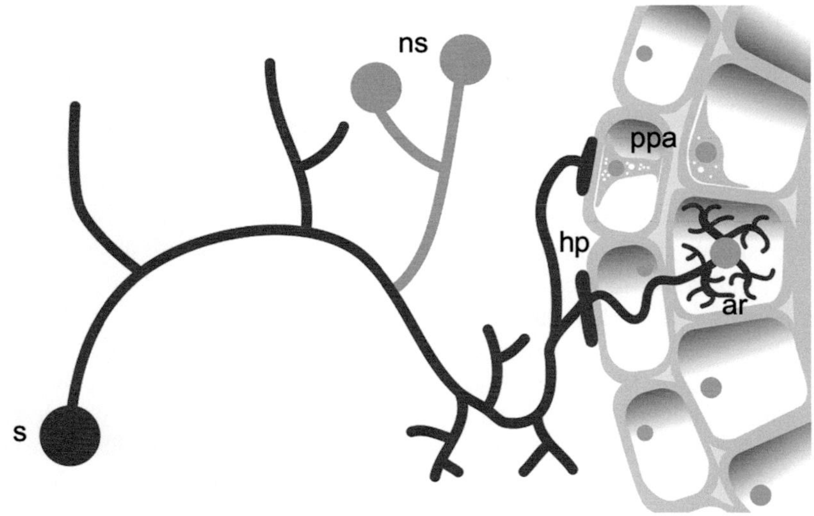

Fig. 9.1 Arbuscular mycorrhiza development starts with the germination of resting spores (s) in the rhizosphere. This generates a loose explorative mycelium, whose hyphae eventually reach the host root surface, where they develop adhering hyphopodia (hp). Hyphopodium perception triggers the assembly of a prepenetration apparatus (ppa) in the underlying epidermal and cortical cells, which builds the symbiotic interface where intracellular hyphae and arbuscules (ar) are hosted. Symbiosis functionality allows the extraradical mycelium to generate new spores (ns)

with minor variations depending on the plant and fungal partners involved (Choi et al. 2018). Fungal spore germination is activated in response to optimal moisture and temperature conditions and generates a loose mycelium that grows for a few days at the expense of spore-stored lipids (Genre et al. 2020). The resulting hyphae explore the surrounding soil in search of a host root. Following reciprocal plant-fungus recognition through an exchange of diffusible molecules, fungal hyphae contacting the root epidermis develop swollen and often branched adhesion structures called hyphopodia. In response to both chemical and physical stimuli produced by the hyphopodium, the contacted epidermal cell develops an intracellular accommodation structure, called the prepenetration apparatus (PPA) (Genre et al. 2005), a columnar cytoplasmic aggregation that concentrates the whole exocytic pathway and anticipates the route of the penetrating hypha across the epidermal cell. PPA-associated exocytosis drives the biogenesis of the symbiotic interface, a novel cell compartment surrounded by an extension of the host plasma membrane, that contains all intracellular hyphae (Bonfante 2001). Once the epidermal layer is crossed, hyphae can branch and grow both inter- and intracellularly to reach the inner tissues. The PPA mechanism is replicated in each colonized cell and reaches the highest complexity and extension in the deepest cortical cells where hyphae branch repeatedly to generate the core structure of AM symbiosis: arbuscules (Luginbuehl and Oldroyd 2017). Arbuscules are highly branched intracellular hyphae that can occupy up to 90% of the host cell volume. Arbuscules are ephemeral structures that collapse and degenerate within a few days (Gutjahr and Parniske 2013; Kobae and Hata 2010), with the host cell regaining its previous organization.

While root colonization proceeds, the extraradical mycelium also develops extensively in the surrounding soil, fed by plant-derived sugars and lipids. Besides scavenging water and mineral nutrients to be delivered to the host plant, the extraradical mycelium is also responsible for the production of a new generation of spores.

9.4 Signaling as a Prerequisite for Symbiosis

AM fungi perceive the vicinity of a host via root-exuded molecules that induce spore germination and hyphal branching (Buée et al. 2000; Nagahashi and Douds 2004). The most studied plant symbiotic signals are carotenoid-derived phytohormones called strigolactones (SL), which have a primary role in plant development (Al-Babili and Bouwmeester 2015; Ruyter-Spira et al. 2013). AM fungi sense SLs in root exudates (Fig. 9.2) at concentrations as low as 10 nM. Fungal responses to GR24, a synthetic molecule commonly used to study SL actions, include stimulation of mitochondrial activity, a rapid increase in ATP and NADH, and nuclear proliferation (Akiyama et al. 2005; Besserer et al. 2006, 2008; Akiyama et al. 2010; Salvioli et al. 2016). GR24 exposure also causes a sharp increase in Ca^{2+} concentration in the fungal cytoplasm (Moscatiello et al. 2014). Although fungal SL receptors remain unknown, such observations suggest that SLs are perceived via a Ca^{2+}-mediated signaling pathway and trigger a cellular and molecular prelude to root colonization (Besserer et al. 2006, 2008; Bonfante and Genre 2015).

AM fungi also release signal molecules (Fig. 9.2) that trigger plant symbiotic responses (Bonfante and Requena 2011), including transcriptional regulation, nucleus-associated Ca^{2+} signals, starch accumulation in roots, and lateral root formation (Lanfranco et al. 2016). Repeated oscillations in nuclear Ca^{2+} concentration (spiking) have been observed in the root epidermal cells contacted by AM fungal hyphopodia, but also when the same cells were treated with exudates from germinated AM fungal spores (Chabaud et al. 2011). Similarly, the expression of the early symbiotic gene ENOD11 in *M. truncatula* is upregulated upon both fungal contact (Chabaud et al. 2002) and the perception of fungal exudates (Kosuta et al. 2003).

Different N-acetylglucosamine oligosaccharides have been characterized in AM fungal exudates as bio-active molecules responsible for such plant responses. They include tetra- and penta-chito-oligosaccharides (CO4 and CO5)

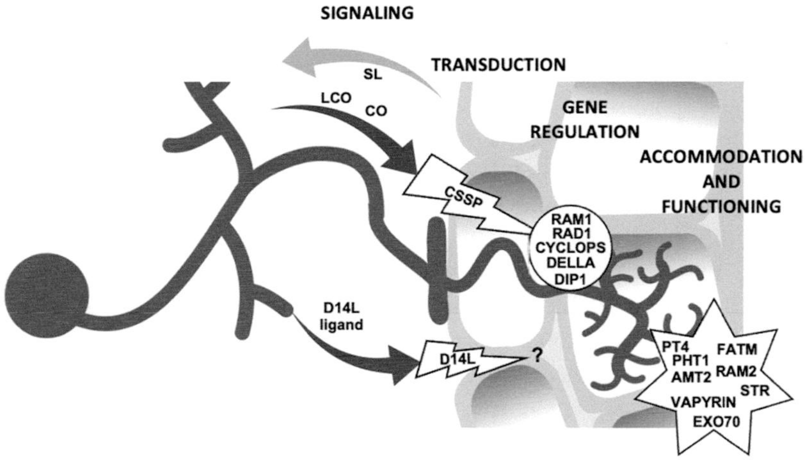

Fig. 9.2 Root colonization by AM fungi involves a sequence of molecular mechanisms. Firstly, an exchange of signals ensures reciprocal recognition and triggers symbiotic responses: root secreted strigolactones (SL) activate fungal metabolism and branching; Myc factors (LCO and CO) trigger the common symbiotic signaling pathway (CSSP), in parallel with CSSP-independent signal transduction mediated by D14L. Downstream of the CSSP, several transcription factors (RAM1, RAD1, CYCLOPS, DELLA, DIP1) control the expression of a number of genes that are required for fungal accommodation and arbuscule functioning, such as transporters (PT4, PHT1, AMT2, STR), lipid biosynthetic enzymes (FATM, RAM2), and structural proteins related to the symbiotic interface assembly (VAPYRIN, EXO70)

(Genre et al. 2013) as well as lipo-chito-oligosaccharides (LCOs), which are very similar to nodulation (Nod) factors released by nitrogen-fixing rhizobia (Maillet et al. 2011). When applied as purified molecules, such Myc factors mimic the perception of fungal exudates in the host roots, including nuclear Ca^{2+} spiking and the regulation of symbiosis-related genes. Interestingly, the release of CO4 and CO5 in *R. irregularis* exudate is boosted upon GR24 treatment (Genre et al. 2013), suggesting the existence of a positive loop between plant and fungal signal perception and production of these oligosaccharides (Lanfranco et al. 2016; Volpe et al. 2023).

Recent investigations are shedding light on the role of alternative receptor complexes, with central roles played by CERK1 and MYR1 LysM receptor-like kinases (Miyata et al. 2014; He et al. 2019) in the plant perception of different fungal-derived molecules, such as short- and long-chain COs (Zhang et al. 2021), leading to the activation of symbiotic or defense-related plant responses (Chiu and Paszkowsky 2021).

The study of plant signaling mechanisms involved in the perception of AM fungal signals has been developed in legumes such as *Medicago truncatula*, largely following the research on rhizobial Nod factor signaling. Such comparative investigations have revealed the existence of a so-called common symbiotic signaling pathway (CSSP), which includes several genes that are essential for both symbioses (Oldroyd 2013; Gobbato 2015). Evidence that the same genes are also involved in diverse symbiotic, pathogenic, and parasitic plant interactions is accumulating (Genre and Russo 2016).

The CSSP (Fig. 9.3) starts on the plant cell membrane, with a malectin-like domain (MLD) leucine-rich repeat (LRR) receptor-like kinase (known as SYMRK, in *Lotus japonicus*). SYMRK is believed to be able to form alternative complexes with either Nod factor or Myc factor receptors. In its cytoplasmic domain, SYMRK interacts with a MAP kinase kinase (Chen et al. 2012), and HMGR1, a 3-hydroxy-3-methylglutaryl-CoA reductase involved in mevalonate synthesis. Indeed, mevalonate has been demonstrated to trigger downstream symbiotic responses such as nuclear Ca^{2+} spiking and ENOD11 expression (Venkateshwaran et al. 2015).

The remaining known CSSP proteins are localized to either the nuclear envelope or the nucleoplasm: three nucleoporins—NUP85, NUP133, and NENA (Kanamori et al. 2006; Saito et al. 2007; Groth et al. 2010)—are likely involved in nuclear targeting of downstream CSSP actors, such as the nuclear envelope localized ion channel CASTOR/POLLUX (Charpentier et al. 2008), the Ca^{2+} channel CNGC15 (Charpentier et al. 2016), and the SERCA-type ATPase MCA8 (Capoen et al. 2011). The nuclear envelope lumen is a major Ca^{2+} store, and indeed Ca^{2+}-mediated signaling is a central hub in the CSSP, with the induction of characteristic oscillations (spiking) in nuclear Ca^{2+} concentration (Genre and Russo 2016). Ca^{2+}-spiking is interpreted by the nucleoplasm-localized Ca^{2+}-and-calmodulin-dependent protein kinase CCaMK (Mitra et al. 2004). Its activation regulates gene expression through its interacting partner CYCLOPS (Miwa et al. 2006) and the transcription factors NSP1, NSP2, NIN, and RAM1 (Oldroyd 2013; Genre and Russo 2016).

However, this largely characterized pathway is probably not exclusive (Bonfante and Requena 2011) and emerging evidence suggests the presence of additional, CSSP-independent signaling. On this line, Gutjahr et al. (2015) identified loss of responsiveness to AM fungi in a rice mutant, which was also mirrored by the absence of physical contact and of characteristic transcriptional responses to AM fungal diffusible signals. The responsible gene, DWARF 14 LIKE (D14L), encodes an alpha/beta-fold hydrolase involved in the perception of karrikins and unrelated to the recognition of known Myc factors, demonstrating the existence of parallel fungal recognition strategies in AM host plants.

Despite the remaining unclear aspects, over 20 years of research have conclusively demonstrated that a common genetic basis underpins AM colonization in 72% of land plants, through the perception of fungal signals and the

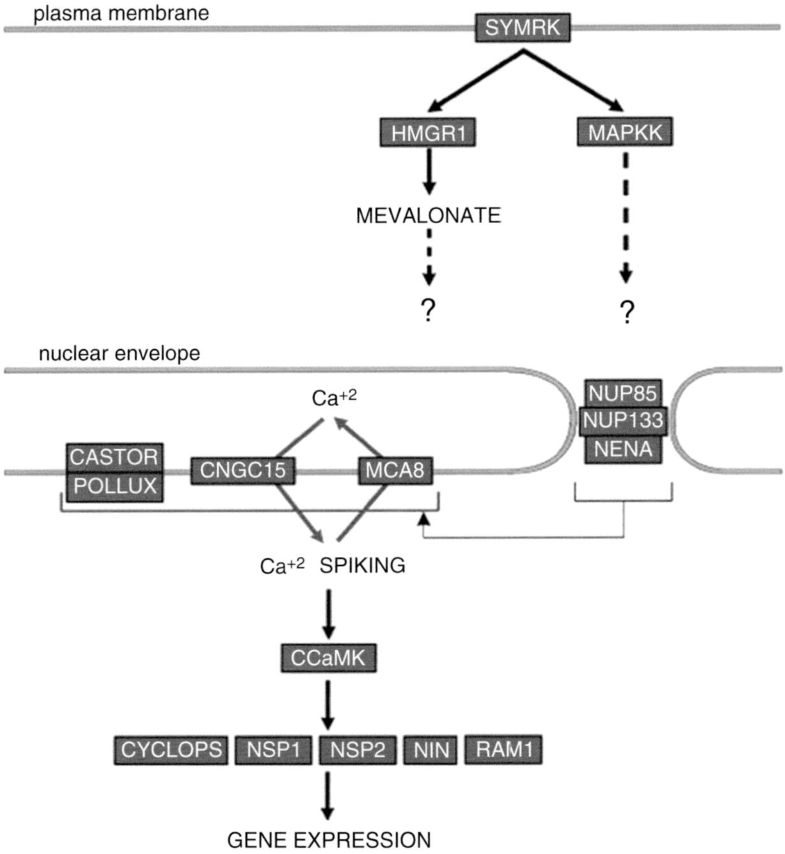

Fig. 9.3 Scheme of the "common symbiotic signaling pathway" (CSSP), mediating the perception of AM fungal signals. The plasma membrane bound leucine-rich repeat (LRR) receptor-like kinase SYMRK is a central element on receptor complexes for Nod factors or Myc factors. In its cytoplasmic domain, SYMRK interacts with a MAP kinase kinase, and HMGR1, an enzyme involved in mevalonate biosynthesis. Downstream this membrane-associated module act three nucleoporins—NUP85, NUP133, and NENA—likely involved in nuclear targeting of the ion channel CASTOR/POLLUX, the Ca^{2+} channel CNGC15, and the SERCA-type ATPase MCA8. In turn, these generate Ca^{2+}-mediated signals (spiking) in the nucleoplasm. Ca^{2+}-spiking is interpreted by the nuclear Ca^{2+}-and-calmodulin-dependent protein kinase CCaMK, its interacting partner CYCLOPS and the transcription factors NSP1, NSP2, NIN, and RAM1, leading to gene regulation

activation of conserved accommodations strategies.

9.5 The Symbiotic Interface

The perifungal or periarbuscular membrane surrounding intracellular hyphae and arbuscules, respectively, (Harrison 2012) outlines the so-called symbiotic interface, a novel cell compartment containing a thin layer of unstructured cell wall materials (Bonfante 2001; Gutjahr and Parniske 2013; Balestrini and Bonfante 2014) where the fungus is hosted and where most of the signal and nutrient exchanges are believed to occur (Luginbuehl and Oldroyd 2017; Choi et al. 2018).

A number of plant genes are required for arbuscule accommodation and symbiotic interface development and functioning. Besides cell reorganization-related proteins, such as Vesicle-Associated Membrane Proteins (Ivanov et al. 2012), EXO70I (Zhang et al. 2015), proteases (Takeda et al. 2009; Rech et al. 2013), Vapyrin

(Pumplin et al. 2010), arbuscule-associated genes include lipid biosynthetic enzymes such as FATM and RAM2 (Bravo et al. 2017) and a specific set of membrane-associated proteins that reside on the periarbuscular membrane throughout arbuscule functioning and are directly involved in nutrient exchange. These include a proton ATPase (Krajinski et al. 2014; Wang et al. 2014), ATP-binding cassette (ABC) transporters, Stunted Arbuscule (STR) and STR2 (Zhang et al. 2010), phosphate transporters PT4 (Javot et al. 2007) and PHT1 (Yang et al. 2012), ammonium transporters AMT2 (Guether et al. 2009). Interestingly, trafficking of some of these proteins, such as phosphate transporters, to the periarbuscular membrane requires their genes to be expressed during arbuscule branching (Pumplin et al. 2012), leading to the hypothesis that PAM construction is achieved by synchronizing the massive reorientation of exocytosis with the transcription of periarbuscular membrane-resident proteins (Gutjahr and Parniske 2013). In line with this, a few plant transcription factors required for AM symbiosis and arbuscule development have been identified, such as CYCLOPS (Horváth et al. 2011), the gibberellin repressor protein DELLAs (Takeda et al. 2015), Reduced Arbuscular Mycorrhiza1 (RAM1; Gobbato et al. 2012), Required for Arbuscule Development1 (RAD1; Xue et al. 2015), and DELLA-Interacting Protein1 (DIP1; Yu et al. 2014). Additional evidence suggests a model where DELLA proteins regulate arbuscule development through modulation of RAM1 and RAD1 that in turn regulate genes required to support arbuscule branching (Park et al. 2015).

Arbuscule formation and functioning is also expected to be under fungal control; however, our understanding of such processes is limited. Fungal effectors are thought to promote compatibility or to suppress plant defense responses by interfering with metabolism or signaling pathways, in analogy with other plant–microbe interactions (Lo Presti et al. 2015). Indeed, Secreted Protein 7 (Kloppholz et al. 2011) is secreted into the host cell and localizes to the plant nucleus, where it counteracts the plant immune response by interacting with the pathogenesis-related transcription factor Ethylene Response Factor 19. In this frame, AM fungal genomes code for a large number of putative secreted proteins (Sędzielewska-Toro and Delaux 2016), and a few of them have been given a possible role in the accommodation of fungal structures (Tsuzuki et al. 2016; Fiorilli et al. 2016). Even if the mechanisms of action for these putative secreted proteins have not been elucidated yet, the functional characterization of the periarbuscular interface is currently the focus of AM research, and important advances are expected in the coming years.

9.6 Evolutionary and Developmental Perspectives on Fungal Accommodation by Plant Tissues

It was recently shown (Russo et al. 2019a, b; Carotenuto et al. 2019a, b) that prepenetration responses culminating in interface construction include cell cycle reactivation in cortical cells, with the onset of both anticlinal cell divisions and recursive endoreduplication events in advance of arbuscule accommodation (Russo and Genre 2021).

On the one hand, the combined use of gene expression and confocal microscopy has highlighted the appearance of ectopic cell divisions in the inner cortex a few hours after hyphopodium formation (Fig. 9.4), when colonizing hyphae are normally confined to the root epidermis (Russo et al. 2019a, b). Sparse couples of "split cells" (half the length of the surrounding parenchymal cells) were consistently observed in the inner root cortex of diverse AM host plants (Russo et al. 2019a). The occurrence of cell division in the fully differentiated tissue was confirmed by the direct observation of dividing cells as early as 48 hours post hyphopodium formation in *Daucus carota* expressing a GFP fusion with TPLATE (Russo et al. 2019a), an adaptin-related protein that accumulates on the cell plate membrane and plasmalemma at the cortical division zone (Van Damme et al. 2006).

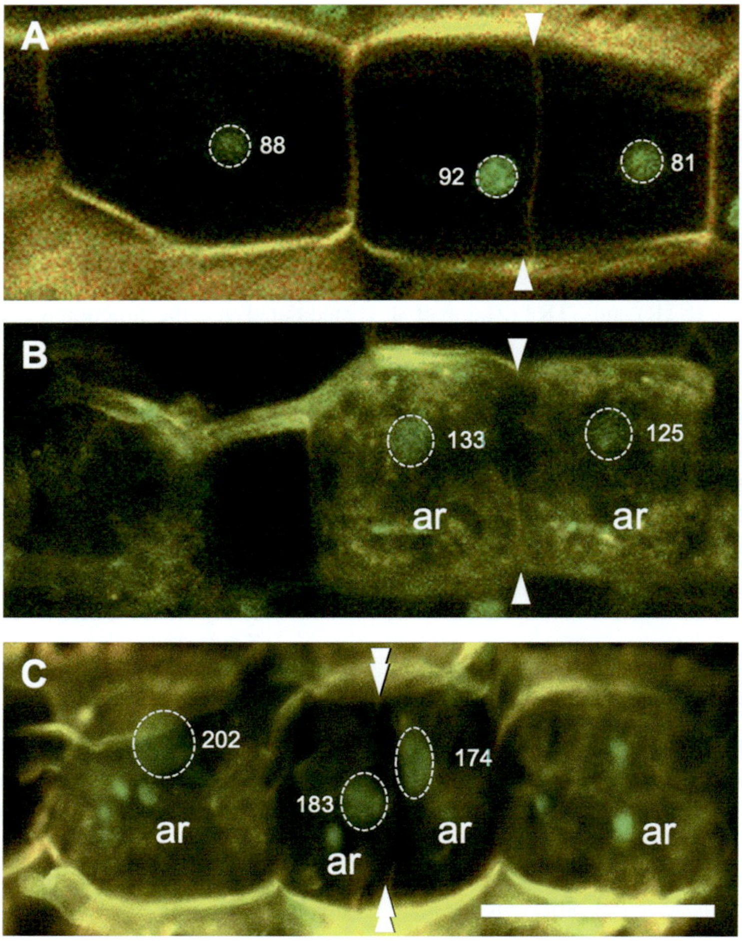

Fig. 9.4 Confocal images of DAPI (green, nuclei) and SR2200 (gold, cell walls) stained sections of *M. truncatula* (cv "Jemalong" A17) roots colonized by the AM fungus *G. margarita* (Strain BEG34). The images are representative of the induction of cell cycle-related processes for arbuscule accommodation. (**a**) Cell division (arrowheads) generates "split cells" in the inner cortex during early root colonization. (**b**) Arbuscule (ar) accommodation in split cells is preceded by an increase in cell ploidy, represented by the numbers indicating nuclear volume in μm^3 (estimated by image analysis according to Carotenuto et al. 2019b). (**c**) Recursive events of both cell division and endoreduplication may lead to higher ploidy and couples of shorter "split cells" (double arrowheads). All four cells in this image host an arbuscule. Bar = 50 μm

On the other hand, flow cytometry, confocal imaging, and gene expression studies revealed the diffuse occurrence of endoreduplication (DNA duplication in the absence of cell division) in the root cortex (Carotenuto et al. 2019b), allowing the precise localization of inner cortical cells with different levels of increased ploidy in the colonized areas of the root (Carotenuto et al. 2019a).

Altogether, these observations outlined a diffuse context of cell cycle reactivation in association with AM fungal accommodation (Russo and Genre 2021). Firstly, cell divisions in the inner cortex have been observed when intraradical hyphae were limited to epidermal and outer cortical layers but not in later stages; by contrast, recursive endoreduplication cycles appear to be active for a longer period of time, with arbusculated and neighboring cells reaching levels of 128 and even 256C ploidy, corresponding to up to 6 cycles of endoreduplication. In more detail, endoreduplication appeared to surge at the front of fungal expansion and reach the highest peaks in the central area of infection units, suggesting the existence of a proportion between ploidy and the abundance (or age) of intraradical fungal structures. Importantly, Carotenuto et al. (2019b) also observed that the couples of "split" cortical cells derived from cell division often displayed different ploidy levels, with higher ploidy in cells that were closer to the fungus or hosting an older arbuscule. This confirms that cell division takes place before endoreduplication, or at least that endoreduplication can proceed after cell division.

Secondly, the observation that both cell division and ploidy increase at a distance from arbuscules or colonizing hyphae suggests the existence of a yet unidentified signaling process reactivate the cell cycle before fungal arrival.

In addition, the concentration of both ectopic cell divisions and endocycle events to the inner cortex envisages a remarkable correlation with the accommodation of arbuscules, which normally develop in the same cell layer. Cell proliferation, with its limited occurrence, appears to have a secondary role, if any, in the generation of additional space for arbuscule accommodation. By contrast, the sparse cell divisions observed in AM colonized areas might relate to the developmental fate of cortical cells. In the roots of most plants, in fact, cortical cell differentiation is determined by an endocycle that doubles their DNA content from 2C to 4C (Cebolla et al. 1999; Edgar et al. 2014) with a consequent size increase (Robinson et al. 2018). In line with that, *in situ* studies of cell ploidy in uncolonized roots of *M. truncatula* (Carotenuto et al. 2019a) revealed that most cortical cells had 4C nuclei, while a few of them displayed 8C and 16C ploidy levels. Even if experimentally challenging, it would be very interesting to investigate if there is a relationship between initial cell ploidy and the occurrence of ectopic cell division in early AM interaction.

Besides tissue differentiation, endoreduplication is also common in plant interactions with diverse microbes: replicating DNA produces multiple copies of each gene, intensifying cell responsiveness to microbial signals. Examples are numerous from pathogens and parasites (de Almeida Engler and Gheysen 2013; Chandran and Wildermuth 2016; Wildermuth et al. 2017) to symbionts (Suzaki et al. 2014; Lace and Ott 2018). Furthermore, endoreduplication-related cell enlargement is typically associated with the accommodation of several microbes, and specifically with arbuscules in AM (Balestrini and Bonfante 2014; Heck et al. 2016).

The requirement of a specific rearrangement in the host cell organization for arbuscule accommodation is apparent from a simple observation of the structural and functional complexity of the periarbuscular interface (Luginbuehl and Oldroyd 2017; Choi et al. 2018; Roth et al. 2019; Ivanov et al. 2019), compared to the tunnel-like interface hosting linear hyphae in outer root tissues. In fact, while epidermal and outer cortical PPAs are structured as roughly linear cytoplasmic bridges across the vacuole, the PPAs that generate periarbuscular interfaces are much more complex and extensive, appearing as large accumulations of cytoplasm that extend from the hyphal penetration site and occupy most of the host cell central volume (Genre et al. 2008). Such a massive, centripetally oriented exocytic event has striking ultrastructural and molecular similarities with the assembly of the cell plate on the cell equatorial plane at the end of mitosis, and indeed an evolutionary correlation between symbiotic interface biogenesis and cell plate deposition has been envisaged in both AM (Russo et al. 2019b) and N-fixing nodulation (Downie 2014). In support of this hypothesis, *in vivo* imaging of GFP-TPLATE fusions revealed a strong accumulation of TPLATE at sites of PPA assembly and at sites of cell-to-cell hyphal passage, where the perifungal membrane fuses with the plasmalemma in striking analogy with cell plate fusion with the cell membrane at the end of mitosis (Russo et al. 2019b).

If the recruitment of cell division processes to assemble the extensive symbiotic interface surrounding each arbuscule now appears more convincing, developmental restraints could contribute to explain why sparse cell division and diffuse endoreduplication are limited to the cortex. Dong et al. (2021) have recently highlighted that an SHR-SCR module (known to regulate cortex/endodermis initial cell division in the root meristem) maintains its activity and is required for cell cycle reactivation in legume inner cortex during nodule organogenesis (Suzaki et al. 2014). Even if analogous studies in rice (which does not form root nodules, but hosts AM fungi) did not confirm SHR-SCR expression in cortical cells, it is reasonable to speculate that analogous mechanisms involving meristematic transcription factors maintain a disposition to reactivate the cell cycle in the inner cortical cells. This peculiarity is likely related to the evolution of root branching (Xiao et al. 2019), but appears to have later been co-opted in several plant interactions, from N-fixing symbioses (Dong et al. 2021) to nematode parasitism (de Almeida Engler and Gheysen 2013), where both cell division and endoreduplication are required for microbe accommodation. While such processes involve the formation of new organs (i.e., lateral roots, N-fixing nodules or nematode-hosting cysts), their occurrence in AM, where organogenesis is absent, appears puzzling; even more so, if we consider that AM symbiosis appeared in land plants before the evolution of true roots (Strullu-Derrien et al. 2014, 2018).

By discussing the developmental and evolutionary context where cell cycle processes interweave with AM symbiosis, a scenario emerges (Fig. 9.1) in which the perception of AM fungal colonization in outer root tissues triggers a so-far unknown intraradical signaling process activating cell cycle-related processes ahead of the penetrating intraradical hyphae. Inner cortical cells, due to their peculiar developmental status deploy two downstream responses: a few of them (possibly depending on their ploidy) complete mitosis, splitting into two smaller cells, as cell elongation is very limited in a mature tissue (Russo et al. 2019a, b); the remaining majority of inner cortical cells enter the endocycle, duplicating their DNA content up to several times, continuously stimulated by the approaching fungal symbiont—in fact, endoreduplication also extends to those cells that had divided earlier (Carotenuto et al. 2019a, b). Such a model implies that host cells largely anticipate and direct fungal colonization, in line with previous suggestions that the plant holds substantial control over symbiosis development (Gutjahr and Parniske 2013).

9.7 Conclusions

We currently have no information on how the earliest land plants acquired the ability to host a symbiotic fungus inside their cells. One can speculate that initial surface interactions provided an advantageous exchange of nutrients, pressing toward more intimate contacts, such as the penetration of fungal hyphae between the plant cells and eventually inside their lumen. In this context, creating *de novo* a fully functional symbiotic interface—as in modern plants—appears unrealistic. By contrast, stimulating cell divisions in differentiated organs could have been a more amenable strategy to generate both crack openings in the surface tissues (an entry route that is conserved in many extant plant–microbe interactions; Ibáñez et al. 2017) and irregular intercellular spaces in the inner ones, producing a protected niche for the fungus. The subsequent re-routing of cell plate formation toward the creation of a more efficient symbiotic interface appears achievable, especially in the light of the current findings, and the observation of split cells in some of the earliest fossils of AM hosts indicates that this is indeed an ancient response associated with fungal accommodation (Strullu-Derrien et al. 2018).

Based on the information originated by genome sequencing, it will be interesting to focus future investigation on AM fungi (such as Paraglomerales) and plants (such as bryophytes) belonging to more ancient taxa, in order to characterize their root colonization process: are hyphopodia crucial for root penetration or could crack-entry play a role in such interactions as it does in other plant–fungus and plant–bacterium interactions? Is the PPA model a common hallmark of fungal accommodation? Is arbuscule morphology conserved, alongside the expression of known markers of their functionality? In conclusion, a combination of cellular and molecular biology applied to plant-fungal couples different from the model organisms that have focused research so far may offer new ideas on how AM fungi became the favorite guests for 72% of land plant species.

References

Akiyama K, Matsuzaki K, Hayashi H (2005) Plant sesquiterpenes induce hyphal branching in arbuscular mycorrhizal fungi. Nature 435:824–827

Akiyama K, Ogasawara S, Hayashi H (2010) Structural requirement of strigolactones for hyphal branching in AM fungi. Plant Cell Physiol 51:1104–1117

Al-Babili S, Bouwmeester HJ (2015) Strigolactones, a novel carotenoid-derived plant hormone. Annu Rev Plant Biol 66:161–118

Balestrini R, Bonfante P (2014) Cell wall remodeling in mycorrhizal symbiosis: a way towards biotrophism. Front Plant Sci 5:237

Bar-On YM, Phillips R, Milo R (2018) The biomass distribution on Earth. Proc Natl Acad Sci U S A 115: 6506–6511

Besserer A, Puech-Pages V, Kiefer P et al (2006) Strigolactones stimulate arbuscular mycorrhizal fungi by activating mitochondria. PLoS Biol 4:1239–1247

Besserer A, Becard G, Roux C et al (2008) GR24, a synthetic analogue of strigolactones, stimulates mitosis and growth of the arbuscular mycorrhizal fungus Gigaspora rosea by boosting its energetic metabolism. Plant Physiol 148:402–413

Bonfante P (2001) At the interface between mycorrhizal fungi and plants: the structural organization of cell wall, plasma membrane and cytoskeleton. In: Hock B (ed) Fungal associations. The mycota, vol 9. Springer, Berlin, pp 45–61

Bonfante P (2018) The future has roots in the past: the ideas and scientists that shaped mycorrhizal research. New Phytol 220:982–995

Bonfante P, Genre A (2015) Arbuscular mycorrhizal dialogues: do you speak 'plantish' or 'fungish'? Trends Plant Sci 20:150–154

Bonfante P, Requena N (2011) Dating in the dark: how roots respond to fungal signals to establish arbuscular mycorrhizal symbiosis. Curr Opin Plant Biol 14:451–457

Bonfante P, Venice F (2020) Mucoromycota: going to the roots of plant interacting fungi. Fungal Biol Rev 34: 100–113

Bravo A, Brands M, Wewer V et al (2017) Arbuscular mycorrhiza-specific enzymes FatM and RAM2 fine-tune lipid biosynthesis to promote development of arbuscular mycorrhiza. New Phytol 214:1631–1645

Buée M, Rossignol M, Jauneau A et al (2000) The pre-symbiotic growth of arbuscular mycorrhizal fungi is induced by a branching factor partially purified from plant root exudates. Mol Plant-Microbe Interact 13: 693–698

Capoen W, Sun J, Wysham D, Otegui MS et al (2011) Nuclear membranes control symbiotic calcium signalling of legumes. Proc Natl Acad Sci U S A 108:14348–14353

Carotenuto G, Volpe V, Russo G et al (2019a) Local endoreduplication as a feature of intracellular fungal accommodation in arbuscular mycorrhizas. New Phytol 223:430–446

Carotenuto G, Sciascia I, Oddi L et al (2019b) Size matters: three methods for estimating nuclear size in mycorrhizal roots of *Medicago truncatula* by image analysis. BMC Plant Biol 19:180

Cebolla A, Vinardell JM, Kiss E et al (1999) The mitotic inhibitor ccs52 is required for endoreduplication and ploidy-dependent cell enlargement in plants. EMBO J 18:4476–4484

Chabaud M, Venard C, Defaux-Petras A et al (2002) Targeted inoculation of *Medicago truncatula* in vitro root cultures reveals MtENOD11 expression during early stages of infection by arbuscular mycorrhizal fungi. New Phytol 156:265–273

Chabaud M, Genre A, Sieberer BJ et al (2011) Arbuscular mycorrhizal hyphopodia and germinated spore exudates trigger Ca^{2+} spiking in the legume and non legume root epidermis. New Phytol 189:347–355

Chandran D, Wildermuth MC (2016) Modulation of host endocycle during plant-biotroph interactions. Enzyme 40:65–103

Charpentier M, Bredemeier R, Wanner G, Takeda N, Schleiff E, Parniske M (2008) Lotus japonicus CASTOR and POLLUX are ion channels essential for perinuclear calcium spiking in legume root endosymbiosis. Plant Cell 20:3467–3479

Charpentier M, Sun J, Vaz Martins T, Radhakrishnan GV, Findlay K, Soumpourou E, Thouin J, Véry AA, Sanders D, Morris RJ, Oldroyd GE (2016) Nuclear-localized cyclic nucleotide-gated channels mediate symbiotic calcium oscillations. Science 352:1102–1105

Chen T, Zhu H, Ke D et al (2012) A MAP kinase kinase interacts with SymRK and regulates nodule organogenesis in *Lotus japonicus*. Plant Cell 24:823–838

Chiu CH, Paszkowsky U (2021) How membrane receptors tread the fine balance between symbiosis and immunity signaling. Proc Natl Acad Sci U S A 118:e2106567118

Choi J, Summers W, Paszkowski U (2018) Mechanisms underlying establishment of arbuscular mycorrhizal symbioses. Annu Rev Phytopathol 56:135–160

de Almeida Engler J, Gheysen G (2013) Nematode-induced endoreduplication in plant host cells: why and how? Mol Plant-Microbe Interact 26:17–24

Dong W, Zhu Y, Chang H et al (2021) An SHR-SCR module specifies legume cortical cell fate to enable nodulation. Nature 589:586–590

Downie JA (2014) Legume nodulation. Curr Biol 24:184–190

Edgar B, Zielke N, Gutierrez C (2014) Endocycles: a recurrent evolutionary innovation for post-mitotic cell growth. Nat Rev Mol Cell Biol 15:197–210

Eisenhauer N, Bonfante P, Buscot F et al (2022) Biotic interactions as mediators of context-dependent biodiversity-ecosystem functioning relationships. Res Ideas Outcomes 8:e85873

Fiorilli V, Belmondo S, Khouja HR et al (2016) RiPEIP1, a gene from the arbuscular mycorrhizal fungus Rhizophagus irregularis, is preferentially expressed in planta and may be involved in root colonization. Mycorrhiza 26:609–621. https://doi.org/10.1007/s00572-016-0697-0

Genre A, Russo G (2016) Does a common pathway transduce symbiotic signals in plant-microbe interactions. Front Plant Sci 7:96

Genre A, Chabaud M, Timmers T et al (2005) Arbuscular mycorrhizal fungi elicit a novel intracellular apparatus in *Medicago truncatula* root epidermal cells before infection. Plant Cell 17:3489–3499

Genre A, Chabaud M, Faccio A et al (2008) Prepenetration apparatus assembly precedes and predicts the colonization patterns of arbuscular mycorrhizal fungi within the root cortex of both *Medicago truncatula* and *Daucus carota*. Plant Cell 20:1407–1420

Genre A, Chabaud M, Balzergue C et al (2013) Short-chain chitin oligomers from arbuscular mycorrhizal fungi trigger nuclear Ca^{2+} spiking in *Medicago truncatula* roots and their production is enhanced by strigolactone. New Phytol 198:190–202

Genre A, Lanfranco L, Perotto S et al (2020) Unique and common traits in mycorrhizal symbioses. Nat Rev Microbiol 18:649–660

Gobbato E (2015) Recent developments in arbuscular mycorrhizal signaling. Curr Opin Plant Biol 26:1–7

Gobbato E, Marsh JF, Vernié T, Wang E, Maillet F, Kim J, Miller JB, Sun J, Bano SA, Ratet P, Mysore KS, Dénarié J, Schultze M, Oldroyd GE (2012) A GRAS-type transcription factor with a specific function in mycorrhizal signaling. Curr Biol 22(23):2236–2241

Groth M, Takeda N, Perry J et al (2010) NENA, a *Lotus japonicus* homologue of Sec13, is required for rhizodermal infection by arbuscular mycorrhizal fungi but dispensable for cortical endosymbiotic development. Plant Cell 22:2509–2526

Guether M, Neuhauser B, Balestrini R et al (2009) A mycorrhizal-specific ammonium transporter from lotus japonicus acquires nitrogen released by arbuscular mycorrhizal fungi. Plant Physiol 150:73–83

Gutjahr C, Parniske M (2013) Cell and developmental biology of arbuscular mycorrhiza symbiosis. Annu Rev Cell Develop Biol 29:593–617

Gutjahr C, Gobbato E, Choi J et al (2015) Rice perception of symbiotic arbuscular mycorrhizal fungi requires the karrikin receptor complex. Science 350:1521–1524

Harrison MJ (2012) Cellular programs for arbuscular mycorrhizal symbiosis. Curr Opin Plant Biol 15:691–698

He J, Zhang C, Dai H et al (2019) A LysM receptor heteromer mediates perception of arbuscular mycorrhizal symbiotic signal in rice. Mol Plant 12:1561–1576

Heck C, Kuhn H, Heidt S et al (2016) Symbiotic fungi control plant root cortex development through the novel GRAS transcription factor MIG1. Curr Biol 26: 2770–2778

Horváth B, Yeun LH, Domonkos A et al (2011) *Medicago truncatula* IPD3 is a member of the common symbiotic signaling pathway required for rhizobial and mycorrhizal symbioses. Mol Plant-Microbe Interact 24:1345–1358

Hyde KD (2022) The numbers of fungi. Fungal Divers 114:1

Ibáñez F, Wall L, Fabra A (2017) Starting points in plant-bacteria nitrogen- fixing symbioses: intercellular invasion of the roots. J Exp Bot 68:1905–1918

Ivanov S, Fedorova EE, Limpens E et al (2012) Rhizobium-legume symbiosis shares an exocytotic pathway required for arbuscule formation. Proc Natl Acad Sci U S A 109:8316–8321

Ivanov S, Austin J, Berg RH et al (2019) Extensive membrane systems at the host-arbuscular mycorrhizal fungus interface. Nat Plants 5:194–203

Javot H, Penmetsa RV, Terzaghi N et al (2007) A Medicago truncatula phosphate transporter indispensable for the arbuscular mycorrhizal symbiosis. Proc Natl Acad Sci U S A 104:1720–1725

Kameoka H, Tsutsui I, Saito K (2019) Stimulation of asymbiotic sporulation in arbuscular mycorrhizal fungi by fatty acids. Nat Microbiol 4:1654–1660

Kanamori N, Madsen LH, Radutoiu S et al (2006) A nucleoporin is required for induction of Ca^{2+} spiking in legume nodule development and essential for rhizobial and fungal symbiosis. Proc Natl Acad Sci U S A 103:359–364

Kloppholz S, Kuhn H, Requena N (2011) A secreted fungal effector of *Glomus intraradices* promotes symbiotic biotrophy. Curr Biol 21:1204–1209

Kobae Y, Hata S (2010) Dynamics of periarbuscular membranes visualized with a fluorescent phosphate transporter in arbuscular mycorrhizal roots of rice. Plant Cell Physiol 51:341–353

Kokkoris V, Chagnon PL, Yildirir G et al (2021) Host identity influences nuclear dynamics in arbuscular mycorrhizal fungi. Curr Biol 31:1531–1538

Kosuta S, Chabaud M, Lougnon G et al (2003) A diffusible factor from arbuscular mycorrhizal fungi induces symbiosis-specific *MtENOD11* expression in roots of *Medicago truncatula*. Plant Physiol 131:952–962

Krajinski F, Courty PE, Sieh D et al (2014) The H+-ATPase HA1 of Medicago truncatula is essential for phosphate transport and plant growth during arbuscular mycorrhizal symbiosis. Plant Cell 26:1808–1817

Lace B, Ott T (2018) Commonalities and differences in controlling multipartite intracellular infections of legume roots by symbiotic microbes. Plant Cell Physiol 59:666–677

Lanfranco L, Bonfante P (2023) Lessons from arbuscular mycorrhizal fungal genomes. Curr Opin Microbiol 75:102357

Lanfranco L, Bonfante P, Genre A (2016) The mutualistic interaction between plants and arbuscular mycorrhizal fungi. Microbiol Spectr 4:1–20

Lo Presti L, Lanver D, Schweizer G, Shigeyuki T, Liang L, Tollot M, Zuccaro A, Kahmann R (2015) Fungal effectors and plant susceptibility. Annu Rev Phytopathol 66:513–545

Luginbuehl LH, Oldroyd GE (2017) Understanding the arbuscule at the heart of endomycorrhizal symbioses in plants. Curr Biol 27:R952–R963

Maillet F, Poinsot V, André O, Puech-Pagès V, Haouy A, Gueunier M, Dénarié J (2011) Fungal lipochitooligosaccharide symbiotic signals in arbuscular mycorrhiza. Nature 469:58–63

Malar C, Wang M, Stajich Y et al (2022) Early branching arbuscular mycorrhizal fungus Paraglomus occultum carries a small and repeat-poor genome compared to relatives in the Glomeromycotina. Microb Genom 8: 000810

Mathieu S, Cusant L, Roux C et al (2018) Arbuscular mycorrhizal fungi: intraspecific diversity and pangenomes. New Phytol 220:1129–1134

Mitra RM, Gleason CA, Edwards A et al (2004) A Ca^{2+}/calmodulin-dependent protein kinase required for symbiotic nodule development: gene identification by transcript-based cloning. Proc Natl Acad Sci U S A 101:4701–4705

Miwa H, Sun J, Oldroyd GE, Downie JA (2006) Analysis of calcium spiking using a cameleon calcium sensor reveals that nodulation gene expression is regulated by calcium spike number and the developmental status of the cell. Plant J 48:883–894

Miyata K, Kozaki T, Kouzai Y et al (2014) The bifunctional plant receptor, Os CERK1, regulates both chitin-triggered immunity and arbuscular mycorrhizal symbiosis in rice. Plant Cell Physiol 55:1864–1872

Morin ES, Miyauchi H, San Clemente ECH et al (2019) Comparative genomics of *Rhizophagus irregularis, R. cerebriforme, R. diaphanus* and *Gigaspora rosea* highlights specific genetic features in Glomeromycotina. New Phytol 222:1584–1598

Moscatiello R, Sello S, Novero M et al (2014) The intracellular delivery of TAT-aequorin reveals calcium-mediated sensing of environmental and symbiotic signals by the arbuscular mycorrhizal fungus *Gigaspora margarita*. New Phytol 203:1012–1020

Nagahashi G, Douds DD (2004) Isolated root caps, border cells, and mucilage from host roots stimulate hyphal branching of the arbuscular mycorrhizal fungus, Gigaspora gigantea. Mycol Res 108:1079–1088

Oldroyd GED (2013) Speak, friend, and enter: signalling systems that promote beneficial symbiotic associations in plants. Nat Rev Microbiol 11:252–263

Park H-J, Floss DS, Levesque-Tremblay V et al (2015) Hyphal branching during arbuscule development requires *Reduced Arbuscular Mycorrhizal*. Plant Physiol 69:2774–2788

Pumplin N, Mondo SJ, Topp S, Starker CG, Gantt JS, Harrison MJ (2010) *Medicago truncatula* Vapyrin is a novel protein required for arbuscular mycorrhizal symbiosis. Plant J 61:482–494

Pumplin N, Zhang X, Noar RD et al (2012) Polar localization of a symbiosis-specific phosphate transporter is mediated by a transient reorientation of secretion. Proc Natl Acad Sci U S A 109:665–672

Rech SS, Heidt S, Requena N (2013) A tandem Kunitz protease inhibitor (KPI106)-serine carboxypeptidase (SCP1) controls mycorrhiza establishment and arbuscule development in Medicago truncatula. Plant J 75:711–725

Remy W, Taylor TN, Hass H et al (1994) Four hundred-million-year-old vesicular arbuscular mycorrhizae. Proc Natl Acad Sci U S A 91(25):11841–11843

Robinson DO, Coate JE, Singh A et al (2018) Ploidy and size at multiple scales in the arabidopsis sepal. Plant Cell 30:2308–2329

Ropars J, Toro KS, Noel J et al (2016) Evidence for the sexual origin of heterokaryosis in arbuscular mycorrhizal fungi. Nat Microbiol 1:16033

Roth R, Hillmer S, Funaya C et al (2019) Arbuscular cell invasion coincides with extracellular vesicles and membrane tubules. Nat Plants 5:204–211

Russo G, Genre A (2021) Divide and be conquered - cell cycle reactivation in arbuscular mycorrhizal symbiosis. Front Plant Sci 12:753265

Russo G, Carotenuto G, Fiorilli V et al (2019a) TPLATE recruitment reveals endocytic dynamics at sites of symbiotic interface assembly in arbuscular mycorrhizal interactions. Front Plant Sci 10:1628

Russo G, Carotenuto G, Fiorilli V (2019b) Ectopic activation of cortical cell division during the accommodation of arbuscular mycorrhizal fungi. New Phytol 221: 1036–1048

Ruyter-Spira C, Al-Babili S, van der Krol S et al (2013) The biology of strigolactones. Trends Plant Sci 18:72–83

Saito K, Yoshikawa M, Yano K et al (2007) NUCLEOPORIN85 is required for calcium spiking, fungal and bacterial symbioses, and seed production in *Lotus japonicus*. Plant Cell 19:610–624

Salvioli A, Ghignone S, Novero M et al (2016) Symbiosis with an endobacterium increases the fitness of a mycorrhizal fungus, raising its bioenergetic potential. ISME J 10:130–144

Schüßler A (2012) The *Geosiphon–Nostoc* endosymbiosis and its role as a model for arbuscular mycorrhiza research. In: Hock B (ed) Fungal associations. The mycota, vol 9. Springer, Berlin, pp 77–91

Sędzielewska-Toro KA, Delaux P-M (2016) Mycorrhizal symbioses: today and tomorrow. New Phytol 209:917–920

Sinanaj B, Hoysted GA, Pressel S et al (2021) Critical research challenges facing Mucoromycotina 'fine root endophytes'. New Phytol 232:1528–1534

Spatafora JW, Chang Y, Benny GL et al (2016) A phylum-level phylogenetic classification of zygomycete fungi based on genome-scale data. Mycologia 108:1028–1046

Strullu-Derrien C, Kenrick P, Pressel S et al (2014) Fungal associations in Horneophyton ligneri from the Rhynie Chert (c. 407 million year old) closely resemble those in extant lower land plants: novel insights into ancestral plant-fungus symbioses. New Phytol 203:964–979

Strullu-Derrien C, Selosse MA, Kenrick P et al (2018) The origin and evolution of mycorrhizal symbioses: from palaeomycology to phylogenomics. New Phytol 220: 1012–1030

Sugiura Y, Akiyama R, Tanaka S et al (2020) Myristate can be used as a carbon and energy source for the asymbiotic growth of arbuscular mycorrhizal fungi. Proc Natl Acad Sci U S A 117:25779–25788

Suzaki T, Ito M, Yoro E et al (2014) Endoreduplication-mediated initiation of symbiotic organ development in *Lotus japonicus*. Development 141:2441–2445

Takeda S, Asamizu E, Tabata S et al (2009) Apoplastic plant subtilase support arbuscular mycorrhiza development in Lotus japonicus. Plant J 58:766–777

Takeda N, Handa Y, Tsuzuki S et al (2015) Gibberellins interfere with symbiosis signaling and gene expression and alter colonization by arbuscular mycorrhizal fungi in *Lotus japonicus*. Plant Physiol 167:545–557

Tedersoo L, Sánchez-Ramírez S, Kõljalg U et al (2018) High-level classification of the *Fungi* and a tool for evolutionary ecological analyses. Fungal Divers 90: 135–159

Tedersoo L, Bahram M, Zob M (2020) How mycorrhizal associations drive plan population and community. Science 367:eaba1223

Tisserant E, Malbreil M, Kuo A et al (2013) Genome of an arbuscular mycorrhizal fungus provides insight into the oldest plant symbiosis. Proc Natl Acad Sci U S A 110: 20117–20122

Tsuzuki S, Handa Y, Takeda N, Kawaguchi M (2016) Strigolactone-induced putative secreted protein 1 is required for the establishment of symbiosis by the arbuscular mycorrhizal fungus Rhizophagus irregularis. Mol Plant-Microbe Interact 29:277–286

Van Damme D, Coutuer S, De Rycke R et al (2006) Somatic cytokinesis and pollen maturation in Arabidopsis depend on TPLATE, which has domains similar to coat proteins. Plant Cell 18:3502–3518

van der Heijden M, Klironomos J, Ursic M et al (1998) Mycorrhizal fungal diversity determines plant biodiversity, ecosystem variability and productivity. Nature 396:69–72

Venice F, Ghignone A, Salvioli A et al (2020) At the nexus of three kingdoms: the genome of the mycorrhizal fungus *Gigaspora margarita* provides insights into plant, endobacterial and fungal interactions. Environ Microbiol 22:122–141

Venice F, Desirò A, Silva G et al (2020b) The mosaic architecture of NRPS-PKS in the arbuscular mycorrhizal fungus *Gigaspora margarita* shows a domain with bacterial signature. Front Microbiol 11:581313

Venkateshwaran M, Jayaraman D, Chabaud M et al (2015) A role for the mevalonate pathway in early plant symbiotic signaling. Proc Natl Acad Sci U S A 112:9781–9786

Volpe V, Chialva M, Mazzarella T, Crosino A, Capitanio S, Costamagna L, Koehler W, Genre A (2023) Long-lasting impact of chitooligosaccharide application on strigolactone biosynthesis and fungal accommodation promotes arbuscular mycorrhiza in *Medicago truncatula*. New Phytol 237:2316–2331

Wang E, Yu N, Bano SA et al (2014) A H^+-ATPase that energizes nutrient uptake during mycorrhizal symbioses in rice and Medicago truncatula. Plant Cell 26:1818–1183

Wildermuth MC, Steinwand MA, McRae AG et al (2017) Adapted biotroph manipulation of plant cell ploidy. Annu Rev Phytopathol 55:537–564

Wilson EO (2016) We must keep every scrap of nature in and around our cities. Nature holds the key to our aesthetic, intellectual, cognitive and even spiritual satisfaction. Landscape 18:26–35

Xiao TT, van Velzen R, Kulikova O, Franken C, Bisseling T (2019) Lateral root formation involving cell division in both pericycle, cortex and endodermis is a common and ancestral trait in seed plants. Development 146: dev182592

Xie X, Lai W, Che X, Wang S et al (2022) A SPX domain-containing phosphate transporter from *Rhizophagus irregularis* handles phosphate homeostasis at symbiotic interface of arbuscular mycorrhizas. New Phytol 234:650–671

Xue L, Cui H, Buer B, Vijayakumar V, Delaux PM, Junkermann S, Bucher M (2015) Network of GRAS transcription factors involved in the control of arbuscule development in *Lotus japonicus*. Plant Physiol 167:854–871

Yang SY, Grønlund M, Jakobsen I et al (2012) Non-redundant regulation of rice arbuscular mycorrhizal symbiosis by two members of the PHOSPHATE TRANSPORTER1 gene family. Plant Cell 24:4236–4251

Yu N, Luo D, Zhang X et al (2014) A DELLA protein complex controls the arbuscular mycorrhizal symbiosis in plants. Cell Res 24:130–133

Zhang Q, Blaylock LA, Harrison MJ (2010) Two Medicago truncatula half-ABC transporters are essential for arbuscule development in arbuscular mycorrhizal symbiosis. Plant Cell 22:1483–1497

Zhang X, Pumplin N, Ivanov S, Harrison MJ (2015) EXO70I is required for development of a sub-domain of the periarbuscular membrane during arbuscular mycorrhizal symbiosis. Curr Biol 25:2189–2195

Zhang C, He J, Dai H et al (2021) Discriminating symbiosis and immunity signals by receptor competition in rice. Proc Natl Acad Sci U S A 118:e2023738118

Masters of Manipulation: How Our Molecular Understanding of Model Symbiotic Fungi and Their Hosts Is Changing the Face of "Mutualism"

10

Jonathan M. Plett, Annegret Kohler, and Francis Martin

Abstract

Ectomycorrhizal (ECM) fungi are soil-borne organisms found predominantly in forest environments where they colonize the roots of trees and shrubs to gain nutrients that support growth. Classed as mutualistic fungi, this group of organisms also supports their host plants through a variety of mechanisms. However, to gain a foothold in these competitive environments, ECM fungi have evolved the ability to manipulate both their host plants and other organisms to carve a significant niche. In this chapter, using well-established and developing models, we will consider the role of ECM fungal signals involved in the structuring of their environment during the colonization of soil and plant tissues. We will also give an overview of the different signaling network hubs that ECM fungi must overcome or control to establish a nutrient exchange between both partners. We will conclude with the outlook in this area of research and future directions that are necessary to further our understanding of how fungal mutualism evolved and how to manage these symbioses under variable climate scenarios.

Keywords

Mycorrhiza · Effector · Symbiosis · Climate · Microbiome · Pathogen

10.1 Mutually Dependent: Plant-Fungal Mutualism Supports Ecosystem Stability

One of the fundamental factors that plants must overcome to adapt/survive is the ability to acquire sufficient nutrients from the soil to complement carbon fixed through photosynthesis. In temperate forests, approximately 90% of available nutrients are locked into organic polymers (Wild 1988), while in high latitude ecosystems this percentage can rise higher than 95% (Chalot and Brun 1998). As plants favor the use of inorganic nutrients, a pool that is typically insufficient to support the plant life in these environments, another mechanism must exist to supply the nutrient needs in these ecosystems. In general, it is theorized that most plants rely on symbioses with mycorrhizal fungi, bacteria, and upon the activity of rhizospheric arthropods and saprotrophic fungi that recycle nutrients (Fig. 10.1; Ma et al. 2021; Yan et al. 2022). A large effort has focused on the role of mycorrhizal fungi in these environments; although this group of fungi is typically poor in

J. M. Plett (✉)
Hawkesbury Institute for the Environment, Western Sydney University, Richmond, NSW, Australia
e-mail: J.Plett@westernsydney.edu.au

A. Kohler · F. Martin
UMR INRA/UHP 1136, Interactions Arbres/Micro-organismes, Centre INRA de Nancy, Champenoux, France

Y.-P. Hsueh, M. Blackwell (eds.), *Fungal Associations*, The Mycota 9,
https://doi.org/10.1007/978-3-031-41648-4_10

the enzymes necessary to scavenge humus-bound nutrients, they do encode genes for secreted proteins and enzymatic pathways involved in habitat engineering and in nutrient acquisition necessary to meet the needs of the plants throughout their lifecycles (Kohler et al. 2015; Miyauchi et al. 2020; Wang et al. 2020). The complexity of unraveling the interplay between the two partners has challenged researchers for over a century. In the last decade, the research community has made significant advances in understanding how mycorrhizal fungi structure their environment to facilitate symbiosis—but are current findings commensurate with original hypotheses of how mutualisms were thought to function? Or are we seeing a blurring of fungal lifestyle boundaries in between pathogenic/saprotrophic/mycorrhizal fungi?

One class of mycorrhizal fungi, typically found in forest environments, are the ectomycorrhizal (ECM) fungi. ECM fungi are not a phylogenetically distinct group, but an assemblage of very different fungal species that have independently developed a mutualistic symbiotic lifestyle (Van der Heijden et al. 2015; Lutzoni et al. 2018; Strullu-Derrien et al. 2018). The hyper-diverse lineages from which these fungi have evolved, including white- and brown-rot fungi, soil/litter decomposers, and root endophytes, have led to an incredible richness of genotypic variation that can be plumbed to understand the molecular foundation upon which mutualism can evolve (Martin et al. 2016). Unlike other mycorrhizal fungal types (e.g., arbuscular mycorrhiza [AM] or orchid mycorrhiza), ECM fungi are not obligate biotrophs and can survive independently on soil substrates without a host, although they will be severely limited in their speed of growth and reproduction as they are not efficient soil organic matter decomposers and would thus be limited by the low concentration of soluble carbohydrates available in forest soils. Given the high density of ECM fungal spores, propagules, and hyphae within the soils of forest ecosystems, most of these fungi are not long without receptive hosts and most lateral roots of trees and shrubs in the upper soil profiles are colonized within days following emergence (Taylor et al. 2000; Ruess et al. 2003; Adams et al. 2006). Physically, symbiosis establishment between ECM fungal hyphae and a favorable host involves growth of hyphae around a nascent short root to form a sheath (mantle) followed by an inward growth in-between root cells into the plant apoplastic space (forming the Hartig net; Horan et al. 1988; Selosse et al. 2006). The Hartig net can range in depth from just around the root epidermal cells (in hardwood species) to several cell layers deep within the root parenchyma (in conifer hosts). Mechanistically, the formation of these hybrid tissues involves hyphal division while concurrently loosing septation and apical coherence while physical connections between plant cells are lost and plant cell shapes change to more ovoid, regular structures (Köttke and Oberwinkler 1986).

The underlying mechanisms that physically shape the mycorrhizal root and control the flow of nutrients inherent to the mutualism are based on an ongoing signaling dialogue between the two partners involving metabolites, nucleic acids, and proteins. A number of these signaling pathways and developmental gene networks are common among ECM fungi, suggesting that, despite the disparate fungal lineages from which they have evolved, there may be a common mutualistic “toolbox.” This symbiosis toolbox was originally hypothesized to be unique to mutualistic organisms and not bear resemblance to the methods used by pathogenic microbes to colonize a host. Similarly, methods of nutrient acquisition from the soil were hypothesized to be distinct from saprotrophic ancestors, thereby increasing reliance of the fungus on the host for carbon (Lindahl and Tunlid 2015). Although our understanding of the physiological mechanisms and signaling pathways at play during mutualistic development and functioning need further refinement, the past decade has brought exciting discoveries in this area that challenge long-held notions of what defines a “mutualist.”

Since the first published genome of an ECM fungus (Martin et al. 2008), the number of ECM genome sequencing programs completed, or in progress, is growing rapidly (Kohler et al. 2015;

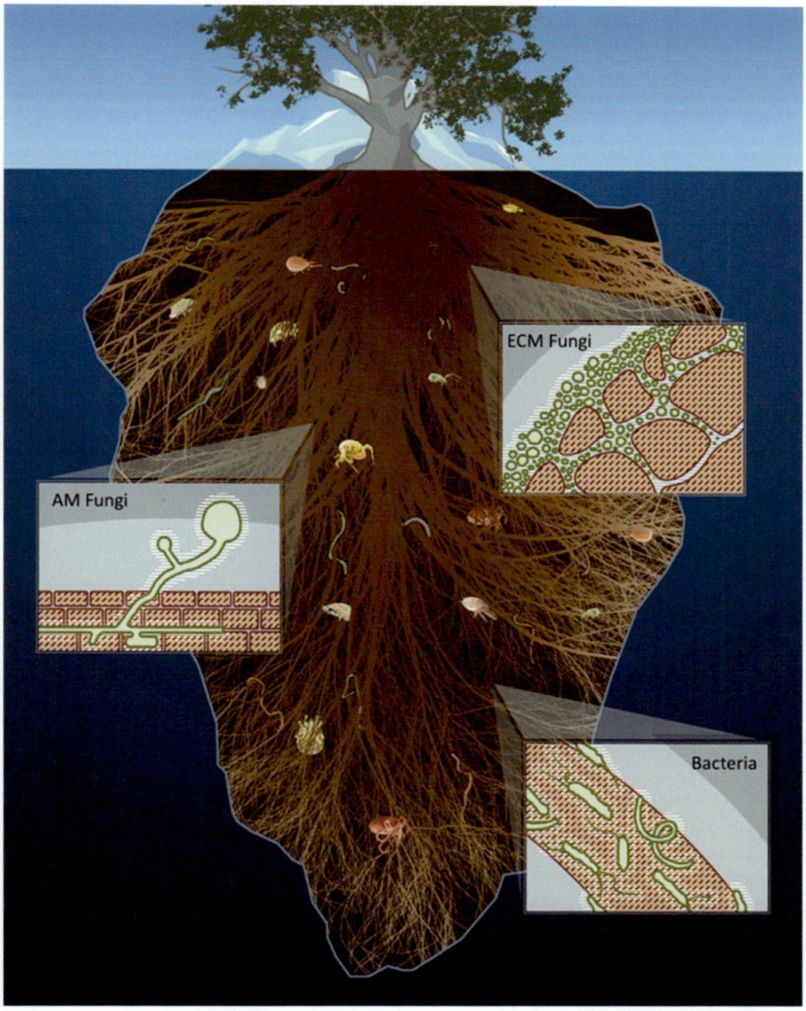

Fig. 10.1 The rhizospheric zoo. Like the quintessential iceberg metaphor, often what we see above ground does not prepare us for the diversity of life below ground, interacting with plant roots. Organisms ranging from mycorrhizal fungi (e.g., AM and ECM fungi), bacteria, mites, and nematodes all affect the root system and the greater plant. Graphic Credit: David Laugesen

Miyauchi et al. 2020; Looney et al. 2022; Tang et al. 2021; Lebreton et al. 2022; Wu et al. 2022) as are application of the first molecular tools necessary to manipulate these fungi to better study the role of individual genes or gene products (e.g., RNA silencing; Kang et al. 2020; Chowdhury et al. 2022). Sequencing projects of ECM plant hosts such as *Populus* (Tuskan et al. 2006), *Picea* (Nystedt et al. 2013), *Eucalyptus* (Myburg et al. 2014; Wang et al. 2020), *Pinus* (Zimin et al. 2014), *Quercus* (Plomion et al. 2018), and *Fagus* (Mishra et al. 2018, 2022) allow for transcriptomic and proteomic analysis of signaling events in both partners during the establishment of symbiosis. Using these techniques it appears that ontogenic and metabolic programs that lead to the development of symbiosis are driven by both the differential expression of transcription factors (Charvet-Candela et al. 2002; Montanini et al. 2011; Garcia et al. 2015; Law et al. 2022) and transduction pathways conserved with other fungi (i.e., transcription factors also found in saprotrophic fungi). In addition, the expression of novel symbiosis-specific gene networks is critical to these early stages (Paparokidou et al. 2021; Ruytinx et al. 2021; Rivera Pérez et al. 2022; Tang et al. 2021). Table 10.1 summarizes recent ECM transcriptome studies.

Despite the complex factors that regulate the establishment and functioning of these symbioses, the interaction between ECM fungi

Table 10.1 Contemporary published works investigating the transcriptional control of the colonization of plant tissues by ectomycorrhizal fungi

Ectomycorrhizal fungus	Host tree	Experimental conditions	References	Year
Cenococcum geophilum	*Pinus sylvestris*	Mature mycorrhizal root tips	Peter et al. (2016)	*2016*
Suillus sp.	*Pinus* sp.	Mature mycorrhizal root tips	Liao et al. (2016)	*2016*
Hebeloma cylindrosporum	*Pinus pinaster*	Early ectomycorrhizal development	Doré et al. (2017)	2017
Tuber melanosporum, Tuber magnatum	*Quercus robur, Corylus avellana*	Symbiosis-induced genes, comparative transcriptomics	Murat et al. (2018)	2018
Suillus luteus	*Pinus massoniana*	Tolerance to aluminium stress	Liu et al. (2020)	2020
10 ectomycorrhizal fungi	diverse host plants	Phylostratigraphy approach to identify conserved and species-specific symbiosis-induced genes	Miyauch et al. (2020)	2020
Tricholoma vaccinum	*Picea abies*		Abdulsalam et al. (2021)	2021
Paxillus involutus	*Pinus sylvestris*	Phosphate availability	Paparokidou et al. (2021)	2021
Pisolithus tinctorius	*Quercus suber*	Colonization process; transcriptome and metabolome	Sebastiana et al. (2021)	2021
Laccaria bicolor	*Populus tremula x alba*	Time course of ectomycorrhizal development	Ruytinx et al. (2021)	2021
Lactarius species	*Pinus*	Symbiosis-induced genes, in particular proteases, comparative transcriptomics	Tang et al. (2021)	2021
Pisolithus microcarpus	*Eucalyptus grandis*	Intra-species variations	Plett et al. (2021)	2021
Fungal community including *Cortinarius* and other ECM	*Picea abies*	Metatranscriptome of ectomycorrhizal roots	Schneider et al. (2021)	2021
Diverse ectomycorrhizal fungi	*Fagus sylvatica*	Nitrogen-Uptake, metatranscriptome	Rivera Pérez et al. (2022)	2022
Pisolithus microcarpus	*Eucalyptus grandis*	Role of abscisic acid during colonization	Hill et al. (2022)	2022
Pisolithus microcarpus	*Eucalyptus grandis*	Role of small RNAs	Wong-Bajracharya et al. (2022)	2022
Tricholoma matsutake	*Pinus densiflora*	Symbiosis-induced genes, in particular MiSSPs	Sakamoto et al. (2022)	2022

and their hosts continues to be crucial to the health and sustainability of forest ecosystems globally. For example, recent work by Anthony and colleagues (Anthony et al. 2022) was key in demonstrating that across Europe, the composition of ECM fungal communities in root systems was linked to forest growth. Further, there is now renewed attention in the use of ECM fungi during restoration forestry (Policelli et al. 2020), although the implementation of this worldwide is hampered by several knowledge gaps in how genotypic variation in mycorrhizal individuals and communities will affect the overall outcome of intentional inoculation strategies (Stuart and Plett 2020). The aim of this chapter is to highlight the most recent work into the understanding of signaling within the rhizosphere during the establishment of ECM symbiosis and to illustrate the way comparative and functional genomics informs our understanding about the mechanisms used by mutualistic organisms during the establishment of a relationship between an ECM fungus, its hosts, and the organisms that surround them.

10.2 Finding a Host: Plant Communication to the Teaming Masses

The soil near plant roots is inhabited by a high diversity of bacteria, fungi, and invertebrate species (Fig. 10.1; Buée et al. 2009). Within this niche, called the rhizosphere, mycorrhizal fungi must perceive the presence of host plants while filtering the signals from other organisms, both beneficial and detrimental, around them. In this section, we will address how ECM fungi and their hosts signal in the "microbial zoo" of the rhizosphere.

Prospecting hyphae originating from spore germination, from nascent fungal colonies, or from colonized roots sense the presence of a root by perceiving concentration gradients of different nutrients and metabolites in soil (Fries et al. 1989). Characterizing the diversity of molecules in the complex environment of rhizospheric soil in the stage just prior to fungal/root contact, termed pre-symbiosis for the purposes of this chapter, is very difficult as soils are complex matrices and determining the timing of when fungal hyphae come into physical contact with the plant is nearly impossible. In recent years, however, adaptations of different techniques from the medical field have led to new understanding of compounds and proteins secreted from plant roots and fungi during this stage. One of the most interesting techniques is the use of microdialysis. Used originally to determine the activity of metabolic pathways in animal tissues *in situ* (Ungerstedt 1991; Westerink 1995), microdialysis utilizes fine capillary probes that, through the process of diffusion, are able to take up nutrients, metabolites, and proteins from complex tissues or matrices. Microdialysis has now seen novel applications in forestry settings to assess nutrient bioavailability and dissolved organic compounds in real time with minimal disturbance (Randewig et al. 2019), as well as to emulate the function of root or hyphal tissues to determine the impact of different chemicals on the formation of soil microbiomes (Buckley et al. 2020, 2022). It is unlikely that the large majority of signals produced by roots are meant to specifically signal ECM fungi (as they could be also used by saprotrophic or pathogenic fungi to recognize the presence of a plant), but rather that ECM fungi have "taken advantage" of these concentration gradients to find host plants. Compounds from two groups found within the root exudate, the flavonoids and the strigolactones, have received special attention in this regard due to their effect on the physiology of mycorrhizal fungi.

Over 4000 different flavones have been isolated from root exudates, each differently functionalized. It is thought that a number of these flavonoids have a role as chemical attractants or repellants to mycorrhizal fungi. Early work demonstrated that a range of different flavonoids induce germination in the spores of AM and ECM fungi (Fries et al. 1989; Kikuchi et al. 2007). Further, flavonoids from *Eucalyptus* root exudates stimulate the growth and branching of the ECM fungus *Pisolithus microcarpus*, morphological changes which likely facilitate the possibility that a hyphal tip will encounter a root (Köttke and Oberwinkler 1986; Horan et al. 1988; Lagrange et al. 2001; Martin et al. 2007). More recently, the use of microdialysis in this latter interaction has further supported an increase in phenylpropanoids secreted during the first few hours of pre-symbiosis and physical interaction between *P. microcarpus* and *E. grandis* (Plett et al. 2021). In this case, the authors found that there were significant increases in flavylium, phenylpropanoic acid, coumarin, and tannin within the first 48 h of interaction. In addition to phenylpropanoids, there were dozens of other metabolites produced at higher concentration in the rhizosphere, but the majority were novel and therefore identification of the compound or the class of compound was impossible. This indicates that a richness of plant signals released into the rhizosphere remain to be characterized, a fact that currently limits our understanding of these early stages of the interaction. Interestingly, as opposed to the AM symbiosis where colonized roots can decrease their exudation of certain flavonoid compounds (e.g., genistein; Larose et al. 2002), flavonoids are found enriched in rhizospheric soil following the establishment of functional mutualistic symbiosis (Wong-Bajracharya et al. 2020). This latter finding was unanticipated as, energetically, we would expect the loss of a chemoattractant in the exudates of roots already colonized by a mutualistic fungus. This strategy would mutually benefit both organisms: it is metabolically expensive for the plant to produce these secondary metabolites when already benefiting from a mutualistic exchange while it would be counterproductive for free living mycorrhizal mycelium to search out a colonized host root system. Further work needs to be undertaken in this area to determine if possible, changes in the class of flavonoids produced by bare roots as opposed to colonized roots differentially attract ECM fungi.

It is likely that flavonoids do not act alone, but rather with other root exudate compounds. Strigolactones, a class of sesquiterpene lactones produced by a wide range of mono- and dicotyledonous plants found in the root exudates (Machin et al. 2020). In AM fungi, recognition of these compounds is required for host specificity (Akiyama et al. 2005), and they play important roles in the control of AM hyphal growth and branching in the proximity to the root (Bouwmeester et al. 2003; Akiyama et al. 2005; Besserer et al. 2008; Bonfante and Genre 2010). Recent studies suggest that strigolactones originated as AM symbiosis-inducing rhizosphere signaling molecules and were later recruited as plant hormones (Kodama et al. 2022). Strigolactone concentrations decrease rapidly in the root exudates of plants that are not receptive to colonization, although strigolactone production seems to be linked more closely to nutrient status of the root (Akiyama and Hayashi 2008). The role of these compounds in ECM fungi, however, remains elusive. While strigolactones play a role in fungal growth in AM fungi, this has not been found for two previously tested ECM fungi (i.e., *L. bicolor* and *P. involutus*; Steinkellner et al. 2007). While mutants have yet to be created in the strigolactone biosynthesis pathway to directly test the role of strigolactones in ECM colonization of host roots, recent work suggests that elements of the larger biochemical pathway (i.e., that of carotenoids) are facilitative to the mycorrhization process (Hill et al. 2022). In this latter paper, across a time course of colonization, one of the main plant pathways that was induced across the integration of fungal hyphae into plant roots was the carotenoid biosynthesis pathway. In the early stages of colonization, gene expression and metabolic analysis suggests that a receptive host prioritizes flux away from the strigolactone pathway in favor of production of abscisic acid (ABA). Addition of ABA, meanwhile, facilitates the integration of the fungus within the root as denoted by a Hartig net with a greater surface area. The biochemical regulation of this pathway changes toward the end of colonization when genes encoding the enzymes for strigolactone biosynthesis, CCD7 and CCD8, are slightly induced (Hill et al. 2022). This finding further reinforces the notion that the root secretome changes between the early stages of ECM colonization and later, functional stages. These results would also suggest that the signaling programs in plants that facilitate one type of mycorrhizal lineage may not be utilized by another.

While we can identify secreted signals from host plants, our understanding of the mechanisms by which root exudates are sensed by the ECM fungus has not appreciably advanced in the last decade. Early work demonstrated that upon receipt of different signals from a plant host, *L. bicolor* genes involved in transduction pathways were among the most abundant of the mRNAs in ectomycorrhizal tissues (Voiblet et al. 2001). The regulated genes correspond to the different sub-units of G-proteins (e.g., a fungal G-protein α subunit), response regulators (e.g., Ras protein kinases), and calcium binding proteins (e.g., calmodulin). A Ras cDNA has also been described in *L. bicolor* and may be involved in the control of cell growth and proliferation of hyphae (Sundaram et al. 2001). Similarly, the *T. melanosporum* genome encodes the signaling genes documented in other filamentous fungi that are involved in pathways controlling stress response, filamentous growth, virulence, and mating. During the interaction with the host plant most signaling transcripts are strongly expressed, although very few are specific to mycorrhizal root tips (Martin et al. 2010). With ever increasing fungal genomes (Miyauchi et al. 2020) and the ability to transform ECM fungi with gene reporter constructs (Kemppainen and Pardo 2021), this area deserves special attention as understanding the breadth of perception systems in ECM fungi will enable us to narrow our focus on what plant-based compounds are critical to the early stages of this interaction.

10.3 Deconstruction of a Root: Formation of the Symbiotic Interface

Upon physical contact between ECM hyphae and a host root, the fungus must attach to the plant tissue and form an interface with the host that will protect hyphae from plant-based defenses but also allow for the continued exchange of signaling molecules as well as nutrients. To some degree, ECM fungi are at a "disadvantage" compared to other lifestyles of fungi at this stage of symbiosis because convergent evolution within this lifestyle has led to the loss or reduction of several plant cell wall active enzyme families (Kohler et al. 2015; Miyauchi et al. 2020). For instance, the genome of *L. bicolor* only encodes one endoglucanase GH5 with a fungal cellulose-binding module while there are no cellulases from families GH6 and GH7. Instead, ECM fungi are reliant on a range of proteins and compounds that attach to, and then gently restructure (i.e., not degrade), the plant cell wall to form a more cohesive structure known as the symbiotic interface. Over the last decade, we have seen a range of proteins and enzymes involved in this step characterized for the first time.

As the formation of mycorrhizal root tips never occurs in a static environment as roots are constantly growing, ECM fungal hyphae must excrete a mucilage of different compounds to attach firmly to the root. While a range of adhesins or oligosaccharides are involved in this step, new proteins are also being discovered within the fungal tree of life that can modify the chemical properties of plant cell wall surfaces that improve adhesion. Within the Pezizomycotina, comparative genomics identified a conserved sequence motif (C-CXXXC-C-C-C-C-C) within surface-active protein that, when tested, changed the hydrophobicity of a surface to promote hyphal attachment regardless of fungal lifestyle (Zhao et al. 2021). In the endophytic fungus *Serendipita indica* (also known as *Piriformospora indica*; Sebacinales), secretion of the protein Fungal Glycan-Binding 1 alters cell wall composition to facilitate host colonization (Wawra et al. 2016). Within mutualistic fungi, a range of secreted hydrophobins also may support attachment to host surfaces during colonization due to their amphipathic structure; they accumulate at the surface of hyphae with their hydrophobic domains directed outward, aiding aggregation of hyphae or adhesion to hydrophobic host surfaces (Casarrubia et al. 2018). These proteins also may enable the fungus to avoid plant defenses by coating hyphal surfaces to mask the antigenicity of fungal β-1,3-D glucans and chitin, known elicitors of the plant defense system (Sharp et al. 1984a, b; Roby et al. 1987; Templeton et al. 1994). Due to their incredible stability, hydrophobin layers have also been proposed to protect the fungal hyphae against plant proteases and reactive oxygen species (Wösten and Wessels 1997).

While our current models concerning the activity of hydrophobins have been largely informed by their role in pathogenic fungi (e.g., Moonjely et al. 2018; Yu et al. 2020), recent publications on insect mutualists (Lovett et al. 2022), ericoid mycorrhizal fungi (Casarrubia et al. 2018), saprotrophs (Cai et al. 2020), as well as ECM fungi are beginning to broaden our understanding of this class. To date, the hydrophobins encoded by ECM fungi have been studied in a restricted set of lineages. In *Pisolithus microcarpus* (originally identified as *P. tinctorius* prior to phylogenetic reconstruction of the genus; Martin et al. 2002), early work demonstrated that the hydrophobins *hydPt-1* and *hydPt-2* were induced during the early stages of fungal aggregation on roots (Tagu et al. 1996; Voiblet et al. 2001; Duplessis et al. 2005). Similarly, increased accumulation of hydrophobin transcripts was observed in *Paxillus involutus/ Betula pendula* ectomycorrhiza (Le Quéré et al. 2006). The genome of *L. bicolor* S238N-H82 encodes 12 hydrophobin genes (Martin et al. 2008), some of which exhibit interesting differences when compared to classical hydrophobin genes (Plett et al. 2012a). A first defining characteristic of hydrophobins encoded by *L. bicolor* is the level of hydrophobicity within the core of the consensus sequence: hydrophobicity between cysteine doublets is much greater when compared to other class I (e.g., SC3 from *Schizophyllum commune*) or even to class II hydrophobins (e.g., HFBI from *Trichoderma reesei*) (Kim et al. 2005; Jensen et al. 2010). A second characteristic observed in a subset of the hydrophobins encoded by *L. bicolor* (*LbH3*, *LbH11*, and *LbH14*) is an altered frequency of cysteine residues (Jensen et al. 2010; Plett et al. 2012a), suggesting that the mature protein has an altered function as seen in other fungi that encode non-canonical hydrophobins (Kershaw and Talbot 1998). There are also interesting differences in the panel of hydrophobins expressed during root colonization whereby the variety of hydrophobins expressed depend upon the plant host being colonized. In the *L. bicolor* model, if the fungus is able to colonize a plant species at a high level (used as a proxy for degree of host receptivity), far fewer hydrophobins are expressed as opposed to a plant species with much lower levels of colonization (i.e., poorly receptive host; Plett et al. 2012a). As hydrophobins are regulated by changes in the external environment (e.g., during host colonization; Wessels 1996), are produced to hide antigenic hyphal compounds from the plant immune system (Aimanianda et al. 2009), or act as a base for enzymes that loosen connections between plant tissues (Corvis et al. 2005, 2006, 2007; Qin et al. 2007; Zhao et al. 2007; Wang et al. 2010), it is possible that the large variety of hydrophobins encoded by *L. bicolor* are a necessary adaptation to aid in the colonization of a wide range of host species. Counter to this hypothesis, however, was the finding that the majority of *L. bicolor* hydrophobins are under purifying selection suggesting that maintenance of a high number of hydrophobins in ECM fungi is not evolutionarily advantageous (Plett et al. 2012a). While it has been found that hydrophobin expression correlates to the pathogenesis of animal pathogens, in this same link it has not been found as a causal phenotype in other plant-colonizing fungi (Parta et al. 1994; Thau et al. 1994; Brasier et al. 1995; Bowden et al. 1996). Therefore, further work is needed to understand the role of these compounds during host colonization and in possibly being a key determinant in host specificity.

Other proteins may also support adhesion of fungal hyphae to the root. One class that may play a role are the Symbiosis-Regulated Acidic Mannoproteins (SRAPs; Martin et al. 1999, 2008). While a large number of different SRAP classes have been identified in fungi (Hilbert et al. 1991; Burgess et al. 1995), one class in particular, the 32 kDa class of SRAPs has received attention in their role as potential adhesins based on timing of expression, an encoded cell adhesion RGD motif, and its location on the outer hyphal cell wall. In the model of *P. microcarpus* (previously mis-identified as *P. tinctorius*), expression of *PmSRAP32-1* is especially high at 3 days post contact between fungal hyphae and *Eucalyptus globus* roots, a time period when hyphae are aggregating around plant roots. The *PmSRAP32* class is composed of at least 6 isoforms (Kohler et al. 2015) that range in pI from 4.5 to 5.5 and are secreted to the plant/hyphal interface during colonization of the plant root (De Carvalho 1994). RGD motifs such as those found in SRAPs are thought to be involved in adhesion as RGD containing proteins from the fungus *Lentinus edodes* induce flocculation in yeast (Kondoh et al. 1995) and aggregation of hyphae in *Schizophyllum commune* (Yasuda and Shishido 1997). However, the role of SRAP32 proteins in cell adhesion has yet to be proven and more recent work would suggest that RGD motifs could have alternative functions. In plant pathogens, the rust fungus *Uromyces vignae* uses RGD containing proteins (RGDS and

RGDSP) to control and repress the defense reaction by the plant host by disrupting the connection between the plasma membrane and the plant cell wall, a connection necessary for the proper maintenance of plant cell wall-associated defenses (Hostetter 2000). Similarly, in mammalian systems, these proteins have been found to bind extracellular proteins to facilitate cell entry (Chernyavska et al. 2019; Mészáros et al. 2021). Another class of secreted fungal proteins that could affect fungal aggregation on the root surface are the Mycorrhiza-induced Small Secreted Proteins (MiSSPs). The ECM fungus *L. bicolor* encodes a small secreted protein of 8 kDa (MiSSP8) that undergoes post-translational modification and is involved in the formation of the fungal mantle during host colonization (Pellegrin et al. 2019). The mechanism by which it supports this mantle formation is currently unknown, but it may be associated with a repetitive motif found in the protein that is common across fungi of multiple lifestyles. Therefore, further work needs to be done to elucidate the full function of these proteins during symbiosis, but these studies show a high degree of similarity among mutualistic fungi and saprotrophic and pathogenic microbes in adhesion to their substrates/hosts.

Following adhesion, ingrowth of ECM fungal hyphae into the root apoplastic space to form the Hartig net requires alterations to plant cell wall chemistry and the connections that hold cells together in a cohesive mass. Early work in this area demonstrated that, while plant cell wall active enzymes were regulated during host colonization in two lineages of ECM fungi (i.e., *L. bicolor* and *T. melanosporum*), only two classes were in common: the GH5-CBM1 glucanase and a GH28 pectinase (Martin et al. 2008, 2010). This specific expression suggested at the time that these two proteins may play a key role during the establishment of symbiosis (Martin et al. 2010). The GH5-CBM1 protein of *L. bicolor* has since been characterized (Zhang et al. 2018a) as well as a candidate GH5 protein from *Tricholoma matsutake* (Onuma et al. 2019). The latter protein was characterized as being active during non-symbiotic growth in the fungus and to hydrolyze β-glucan, lichenan, and CMC-Na. The *L. bicolor* GH5, meanwhile, was secreted during the stages of fungal ingrowth into the root and found to localize to the periphery of the hyphae, and RNAi knockdown mutants of *L. bicolor* for this gene led to a reduction in the development of mycorrhizal root tips. While this activity was key to *L. bicolor,* its broader applicability to other ECM lineages is questionable as impairment of this activity in *Tricholoma matsutake* did not affect colonization of the host: in fact, lack of the β-1,4-endoglucanase activity in *T. matsutake* led to improved host growth rate (Murata et al. 2021). These differences highlight a key point that needs refining in our understanding of ECM evolution: while there are many instances of convergent evolution in the function of ECM mutualisms, there will be distinct differences between different genera due to the number of times that this lifestyle has independently evolved in the fungal tree of life.

The activity of LbGH5-CBM1 was found to operate against a variety of different substrates including cellulose, mannans, and galactomannans. This activity was commensurate with a β-1,4 endoglucanase as the hydrolysis of these substrates led to release of cellobiose and cellotriose. Conversely, this enzyme was inactive against crystalline cellulose, xyloglucan, xylan, arabinoxylan, arabinan, and pectin (Zhang et al. 2018a). As these latter components in the cell wall, especially pectin, are necessary for cell integrity and cell–cell connections, other mechanisms must be encoded by ECM fungi to enable full apoplastic invasion. *L. bicolor* encodes two possible mechanisms to alter pectin: a GH28 and a series of pectin methyltransferases (PMEs) that may contribute to loosening cell–cell connections. Similar to LbGH5-CBM1, the *L. bicolor* enzyme LbGH28 production was necessary for elaboration of a full Hartig net, and it was localized both to the hyphal cells and to the surface of plant cells during symbiosis (Zhang et al. 2022). As predicted based on sequence analysis, this protein was active against pectins as well as polygalacturonic acid but was unable to use rhamnogalacturonan, galactan, or pectic galactan as substrates. In a concurrent publication, it was shown that the gene encoding

LbPME1 exhibited higher expression coinciding with PME activity during formation of the mycorrhizal root tip (Chowdhury et al. 2022). Repression of this fungal protein led to a reduction in Hartig net formation while over-expression of *LbPME1* led to a better developed Hartig net as well as altered esterification of pectin in plant cell walls. Therefore, it was concluded that PMEs form a necessary component of ECM fungal cell wall active enzymes during colonization. Beyond these two examples, ECM fungi likely use a range of other peptidases and proteases to manipulate host cell walls. *Lactarius deliciosus*, for example, encodes nearly 13× the average copy number of fungalysin proteases, many of which are regulated during the interaction between *L. deliciosus* and *Pinus taeda* (Tang et al. 2021; Lebreton et al. 2022), proteases that could be involved during the integration step. ECM fungi may also utilize non-enzymatic routes to gain a foothold in the apoplastic space (e.g., through the release of auxin which may alter proton levels; Gay et al. 1994). Future work in this area comparing the mechanisms in different ECM evolutionary lineages will provide valuable information concerning the diversity of enzymes utilized during this stage of colonization.

10.4 Manipulating Host Hub Signaling Pathways: ECM Colonization Alters Root Function Directly and Indirectly

While the growth and development of ECM fungi are physically disruptive to the host plant, ECM fungi also wield a level of control over host physiological processes. During colonization, the plant cell displays a range of different signaling and biosynthetic pathways that are differentially regulated. These include altered hormone profiles and sensitivity, attenuated immune process, altered secondary metabolism, and induction of nutrient pathways and movement. While identifying which pathways are changed during colonization has been studied for many decades, the mechanistic understanding of *how* these alterations are orchestrated has lagged. With the production of genomic resources for a range of mycorrhizal fungi and their hosts, however, we are now seeing significant advances in this area. In this section, we will focus on our advances in understanding how host hormonal and immunity pathways are altered.

One of the earliest studied plant hormones and its role in mycorrhizal establishment was auxin (Rupp and Mudge 1985; Gay et al. 1994; Gea et al. 1994; Kaska et al. 1999). In a post-genomic era, host roots in contact with *L. bicolor* were found to exhibit altered expression of genes involved in the polar transport of auxin (*PtaPIN* and *PtaAUX* genes), auxin conjugation (*PtaGH3*), and auxin signaling (*PtaIAA*; Felten et al. 2009; Vayssières et al. 2015). Similarly, the interaction between *Tricholoma vaccinum* and spruce was shown to alter auxin signaling and movement and that this occurred only in a compatible host/fungal interaction and not in non-host plant roots (Krause et al. 2015). Mirroring this, changes to root architecture induced by the presence of the ECM fungus *Tuber borchii* required an intact auxin signaling network (Splivallo et al. 2009). We now have a better understanding of how other hormones are affected during this process. The hormones ethylene (ET) and jasmonic acid (JA) have been found to be inhibitory to the colonization process of *L. bicolor* on its host *Populus* through signal transduction proteins including *ETHYLENE RESPONSE FACTOR 1* (Plett et al. 2014). These responses, meanwhile, are opposed by salicylic acid (SA). Further work in this model system has shown that most plant hormones inhibit mycorrhizal colonization of the host including gibberellic acid (GA; Basso et al. 2020). GA treatment was not found to alter the ability of the fungus to penetrate into the root but did decrease the overall number of lateral roots colonized in the *L. bicolor:Populus* model. The authors hypothesized, based on transcriptomic data, that GA may inhibit *L. bicolor* hyphae from attaching to the root surface due to downregulation of glycoprotein synthesis. Basso et al. (2020) also showed that the colonization process altered not only biosynthesis for a range of plant hormones, but also sensitivity to the hormones produced. Within the first 2 weeks of colonization, root sensitivity to JA was found to decrease while sensitivity to GA, SA, and ET was increased based on the expression of marker genes (Basso et al. 2020). The means by which hormone sensitivity is altered still require further investigation. Early clues suggest that one such mechanism may be endogenous to the plant: work with *P. microcarpus* colonizing the roots of *E. grandis* found that the host protein methylation enzyme EgPRMT10 plays a key role in the regulation of hormone signaling related to ET, JA, and auxin (Plett et al. 2019). Therefore, one mechanism by which hormone sensitivity is altered in host plants during ECM colonization may be through post-secondary modification of proteins.

Evidence also suggests that hormonal sensitivity may be altered directly due to fungal activity. The most direct manner is through production of certain biochemical mimics/agonists. For example, one lineage of ECM fungi is able to produce abundant indolic compounds that, while they are not auxin or auxin-like, affect plant-based auxin signaling. This molecule—hypaphorine—is a tryptophan betaine excreted by *Pisolithus* hyphae and is accumulated abundantly in hyphal cells in contact with host roots. Hypaphorine is able to stimulate the expression of auxin-regulated genes in *Eucalyptus*, and it was hypothesized that hypaphorine might act as an auxin-like compound (Nehls et al. 1998). Work following this initial publication demonstrated that hypaphorine is likely not an auxin mimic but, more surprisingly, counteracts auxin action on tap root elongation or root hair elongation (Ditengou and Lapeyrie 2000; Ditengou et al. 2000). It is probable therefore that hypaphorine acts in the plant auxin pathway by competing with auxins for auxin-binding proteins and receptors (Kawano et al. 2001). In the early publications, hypaphorine was thought to be exclusive to *Pisolithus*. Since that time, however, it has been found to be produced by a wide range of fungi that are not classified as mutualists as well as in marine organisms and in human milk (Vit et al. 2015) indicating that it is a metabolite that has been co-opted through the evolutionary process to have a wide range of roles and not be a specific signal of mutualism.

ECM fungi may also alter hormonal pathways indirectly through the production of Mycorrhiza-Induced Small Secreted Proteins (MiSSPs). Upon the release of the genome of *L. bicolor*, it was found that over a dozen small secreted proteins (SSPs) with no known homology were up-regulated during the late stages of colonization (Martin et al. 2008). The size of these proteins, their predicted secretion signal, and their expression in symbiotic tissues were a surprise as they were reminiscent of "effector" proteins utilized by pathogenic organisms to control host cell function. It was questioned at the time whether or not the MiSSPs up-regulated during mutualistic exchanges between *L. bicolor* and its host may act in a similar manner to pathogenic effector proteins or if they would have unique functions to mutualism. The first of these MiSSPs to be characterized, LbMiSSP7, was induced by root exudates (both host and non-host) before physical contact between the two organisms and continues at a high level through to a mature mycorrhizal root tip (Plett and Martin 2011). LbMiSSP7 was determined to be one of the essential signals for controlling the in-growth of fungal hyphae into the apoplastic space of the root as RNAi mutant lines of *L. bicolor* with lowered production of LbMiSSP7 were not able to establish a functional mycorrhizal root tip (Plett et al. 2011). The site of action of LbMiSSP7 is in the nucleus of the plant cell where it interacts with multiple JA co-receptors leading to the down-regulation of certain JA signaling pathways (Plett et al. 2014; Fig. 10.2). This understanding was further refined by Daguerre et al. (2020), who found that PtJAZ6 interacts with two poplar transcription factor families (PtMYC2 and PtJAM1). In the absence of JA, these two transcription factors repress transcription of JA-regulated genes. When LbMiSSP7 localizes to the host nucleus, however, it strengthens the binding of PtJAZ6 to PtMYC2.1 and inhibits the dimerization of PtJAZ6 with other JAZ proteins. DNA Affinity Purification sequencing (DAP-seq) analysis has identified that genes directly activated by MYC2.1 and MYC2.2 contain domains that

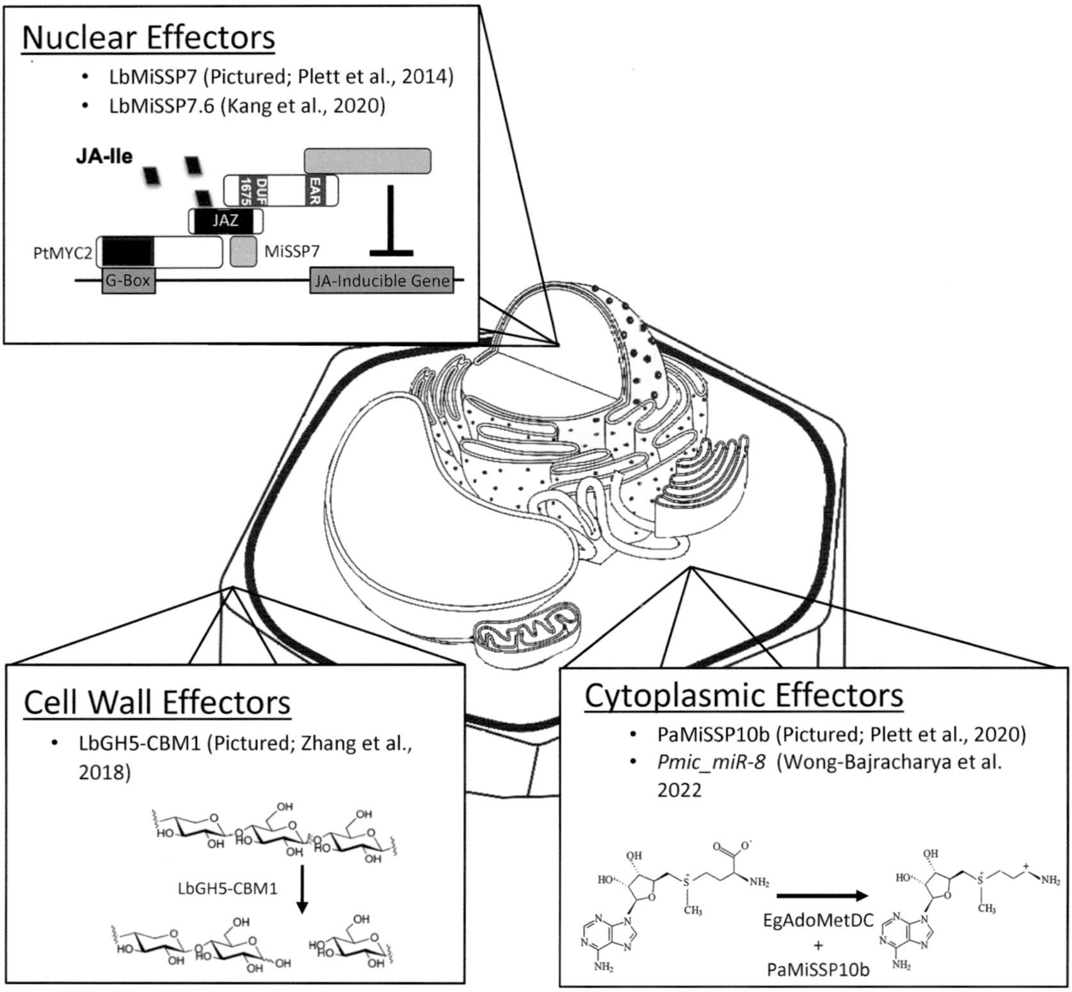

Fig. 10.2 Graphical representation of a host cell and a selection of pathways impacted by ECM MiSSPs. To date, effector-like MiSSPs have been found in all ECM lineages although only a handful have been functionally characterized. Of those that have been studied, processes in the host nucleus, cytoplasm, and cell wall have all been described

would suggest roles as additional transcription factors, receptor-like kinases, and many defense-related proteins, including terpene synthases (TPS; Marques-Galvez et al. 2022). Monoterpenes produced by some of these TPSs repress ECM formation, suggesting that in addition to JA inhibition LbMiSSP7 improves ECM symbiosis through repression of plant terpene biosynthesis. Another characterized MiSSP from *L. bicolor,* LbMiSSP7.6, was also found to enter into host cells and localize to the nucleus (Kang et al. 2020). This protein that interacts with transcription factors belongs to the Myb/SANT-like DNA-binding domain family and, as such, could alter the signaling of a number of hormonal pathways. Similarly, SECRETED PROTEIN7 (SP7) from the AM fungus *Glomus intraradices* also appears to enter the plant cell and localize to the plant nucleus where it interacts with Ethylene Response Factor19 (ERF19) to repress plant defense signaling (Kloppholz et al. 2011). The occurrence small secreted proteins in mycorrhizal fungi that target plant hormone signaling pathways to promote colonization demonstrates that mutualists use another mechanism found in other lifestyles of fungi to target and manipulate conserved "hub" pathways during the colonization of plant tissues (Plett and Martin 2018; Plett and Plett 2022).

Beyond hormone-controlled pathways associated with plant defense, the researchers have asked if ECM fungi are able to control other aspects of plant immunity. Of the MiSSPs characterized to date that affect non-hormonal pathways, PaMiSSP10b of *Pisolithus albus* was found to interact with the polyamine pathway of *E. grandis* (Plett et al. 2020; Fig. 10.2). Polyamines are a general class of plant metabolites that function in plant defense through reinforcement of the cell wall, induction of the hypersensitive response through, or in addition to, the production of hydrogen peroxide (Jiménez-Bremont et al. 2014). Due to their role in defense against pathogens, disease-causing organisms have evolved small, secreted effector proteins that target polyamine biosynthesis in the plant (Hewezi et al. 2010; Stes et al. 2011). Simple polyamines are produced through the activity of S-adenosyl methionine decarboxylase (AdoMetDC) proteins, which utilize adenosyl methionine from the Yang cycle. PaMiSSP10b was found to interact with EgAdoMetDC, but when compared to pathogenic effectors that impact the polyamine pathway, PaMiSSP10b was found to increase the activity of the enzyme (Plett et al. 2020). This activity led to the synthesis of more polyamines and a reduction in reactive oxygen species in the host root and that, as opposed to pathogenic systems, but similar to AM model systems (Sannazzaro et al. 2007), was positive for ECM colonization (Plett et al. 2020). Two other classes of fungal signaling compounds beyond the aforementioned MiSSPs have also been found that support host colonization: Proteolytic Cleavage Products (PCPs) and micro-RNA. PCPs may play a similar role during colonization as MiSSPs: these small peptides of <100 amino acids are formed due to the activity of cleaved larger precursor proteins. Villalobos Solis et al. (2020) suggest that a number of *L. bicolor* PCPs have putative functions in repression of reactive oxygen species, although this was not functionally demonstrated. miRNAs, meanwhile, are short strands of RNA that can transfer between plant-associated fungi and their hosts through vesicles (Peres da Silva et al. 2015; Wang and Dean 2020) where they directly target messenger RNA (mRNA) in the host cell (Wang et al. 2017). The formation of a hybrid double-stranded RNA complex between the miRNA and the full-length host mRNA is then identified and degraded stopping translation of the mRNA (Weiberg et al. 2015). Initially characterized only in pathogenic microbes, *P. microcarpus* was the first ECM fungus found to encode miRNA. Of the 11 identified miRNAs, between five and six were induced by the colonization process (Wong-Bajracharya et al. 2022; Fig. 10.2). Of these, *Pmic_miR-8* was characterized and found to enter into host cells where it targeted the transcription of several NB-ARC genes. While this family of plant genes have many function, one of their primary roles is associated with the perception of invading organisms. Once a foreign tissue is detected, this gene family is then responsible for the signaling that leads to tissue and cellular resistance (van der Biezen and Jones 1998). Wong-Bajracharya et al. (2022) found that *Pmic_miR-8* was necessary for the formation of mycorrhizal root tips and the inhibition of this signal led to increased NB-ARC transcription and enabled the root to grow away from the fungus. Therefore, we are beginning to see a range of previously identified (i.e., MiSSP) and novel (e.g., PCP, miRNA) fungal signals used to alter the signaling or sensitivity of plant immune pathways, a view that is currently still limited by the few numbers of systems in which functional characterization has been undertaken.

10.5 Mycorrhizal Fungi Influence the Developmental Programs of Soil Bacteria

Like their effect on plants, mycorrhizal fungi excrete different metabolites, hormones, and proteins into the soil that act as rhizospheric signals (Toljander et al. 2007; Gorka et al. 2019;

Wong-Bajracharya et al. 2020). It would appear, based on current research, that these signals may play a role in structuring rhizospheric microbial populations on and around the hyphae of ECM fungi. Research concerning bacterial populations in this zone has been gaining popularity since Garbaye (1994) described these "mycorrhizal helper bacteria" such that we now have a more in-depth understanding of fungal–bacterial interactions, interactions that are more widespread than originally expected (Frey-Klett et al. 2007). Interestingly, the bacterial communities that associate with a certain ECM fungi or ECM colonized root tips tend to be similar, suggesting that the ECM fungus may play a role in "cultivating" or attracting specific genera (Nurmiaho-Lassila et al. 1997; Frey-Klett et al. 2005; Stuart et al. 2022). This is similar to results from AM fungi (Emmett et al. 2021) and saprotrophic fungi (Liu et al. 2018). Using microscopic techniques, purification, and isolation as well as large-scale sequencing, these bacterial populations are being described, with the dominant genera being the *Pseudomonas*, *Pedomicrobium, Rhodococcus, Streptomycetes, Burkholderia*, *Variibacter,* and *Bacillus* (Abdel-Fattah and Mohamedin 2000; Ochsenreiter et al. 2003; Schrey et al. 2005; Bomberg and Timonen 2007; Schrey and Tarkka 2008; Li et al. 2018). Bacteria from the listed genera promote the formation of a mutualistic relationship between the fungus and plant roots by complementing the functions of mycorrhizal root tips including nutrient acquisition and biological control of host plants (Artursson et al. 2006; Frey-Klett et al. 2007; Tarkka and Frey-Klett 2008). For example, *L. bicolor* interacts with *Pseudomonas fluorescens*, certain strains of which improve the growth and survival of *L. bicolor* hyphae (Brulé et al. 2001; Deveau et al. 2007) and feed the fungus with thiamine, an essential cofactor of several central enzymes of carbon metabolism (Tarkka et al. 2009). In the research by Stuart et al. (2022), *Pisolithus albus* hyphae modified the soil bacterial community such that phosphate solubilization through glucose dehydrogenases was favored, but not bacterial phosphatases. This solubilization profile complemented the fungal "host" which produced enzymes largely for the organic phosphorus mineralization. In *Tuber indicum,* bacterial communities shifted around fungal hyphae to perform different roles in amino acid metabolism and nutrient transport while functions associated with carbohydrate metabolism were decreased, all changes expected to promote ECM host colonization (Li et al. 2018).

As the genera of bacteria associated with the ECM fungi appear to be fungal lineage-specific, how do the fungi cultivate these specific bacterial populations? One possible mechanism of control could be through the production of proteins that disrupt or modify quorum sensing between certain bacterial species or that act as antibiotics. Quorum sensing is a mechanism by which bacteria can sense their population density and thereby their growth rate as a population. First identified in *Vibrio fischeri*, an aquatic bacterium, quorum sensing is often mediated by small molecules called autoinducers (AIs) that act to transduce messages between bacteria (Fuqua et al. 1994). While there are a number of different molecules that act as AIs (e.g., diketopiperazine, furanosyl borate diester, or γ-butyrolactone), the most common class of molecules that act as AIs are *N*-acyl homoserine lactones (AHLs). AHLs are typically synthesized at a basal level by each bacterium which, upon reaching a certain critical level in the medium surrounding the bacteria, induce the expression of different suites of genes which restrict growth of the colony or serve a number of other purposes (Walker et al. 2003). Quorum sensing molecules, however, can be made by more organisms than just bacteria. For example, ECM fungi produce compounds that interfere with bacterial quorum sensing systems (Uroz and Heinonsalo 2008). These compounds exhibited AHL lactonase activity, enzymes that convert AHLs to the inactive *N*-acyl homoserine (AHS) derivative. The discovery of a mechanism by which ECM fungi can disrupt AHL signaling through inactivation of AHLs to AHSs, or the inhibition of their perception, could well be a strategy developed by the fungi to interfere with the growth of populations of bacteria with deleterious functions. Through this mechanism the fungus could control the surrounding bacterial community and may be a pathway that would partially explain the reason why only certain bacteria are present in or on mycorrhizal root tips (Nurmiaho-Lassila et al. 1997; Frey-Klett et al. 2005). Another mechanism could be through the production of antimicrobial compounds. In work using six ECM fungal species that colonize *Pinus,* secretions were found to consistently inhibit Gram-positive bacteria while they did not affect Gram-negative bacterial genera (Shirakawa et al. 2019). This was found to correlate with microbiome studies of colonized roots demonstrating that bacterial communities associated with *Pinus* ECM root tips were predominated by Gram-negatives. Beyond these two mechanisms there are likely others at play. Within AM symbioses, for example, fructose secreted from fungal hyphae activates phosphate solubilizing bacteria (Zhang et al. 2018b), while in pathogenic systems and saprotrophic models, small, secreted proteins directly alter microbial assembly in the hyphosphere (Contreras et al. 2018; Kombrink et al. 2019; Snelders et al. 2021, 2022). Therefore, it will be interesting in the coming years if similar pathways to manipulate ECM-associated microbial communities will be uncovered.

10.6 Master Manipulators: Using Our Understanding of ECM Fungal Communication Strategies to, in Turn, Manipulate Them for Sustainable Outcomes

It has been a decade since we last updated this research discussion (Plett et al. 2012b), and our views on the genetic and molecular underpinnings that form a "mutualistic" fungus continue to evolve. Based on the work covered above, we are finding how very similar ECM fungi are to pathogens in their methods of communicating with, and orchestrating colonization of, a host plant. While ECM fungi continue to receive far less research attention in comparison with pathogenic systems, the work we cover above demonstrates how the last decade has seen a sustained increased effort to establish genomic and molecular resources so that we can better understand this complex biological system. Consistently, it is becoming clear that nearly all mechanisms that evolved in pathogenic systems to communicate with hosts are also found in mutualistic systems, especially in the case of conserved "hubs" within protein networks of the host plant and rhizospheric community upon which fungal proteins/metabolites/nucleic acids may act to achieve host colonization, while avoiding/manipulating the plant immune system. As we have covered in this chapter, we would summarize past research to suggest that ECM fungi likely first target the plant structure by using a variety of methods to manipulate the physical characteristics of the host tissue to adhere and gain access to the tissues. The mechanisms used here mirror both those found in pathogenic fungi and those used by saprotrophic fungi to colonize the surface of a substrate. ECM fungal proteins then attack the second hub which we would suggest is that of defense-related hormone signaling (e.g., ethylene, salicylic acid) to create localized weakening of the plant immune system which would then allow for the beginning of hyphal penetration into the root apoplastic space. Concurrently with these actions, research groups around the world have found that ECM fungi also appear to cultivate "helper" bacteria in the rhizosphere that aid the fungus by providing extra nutrients (e.g., thiamine), by fending off pathogenic fungi and bacteria, etc. In this step ECM fungi play upon bacterial sensory interaction hubs such as quorum sensing or directly through providing nutrition or by killing using antimicrobial compounds. These processes are similar to those of pathogenic fungi which also manipulate the microbiome surrounding them to hinder plant immunity and to clear a niche in which the pathogen can establish; saprotrophic fungi also produce host distinct hyphospheric microbiomes. Therefore, increasingly, we can see that the molecular definition of fungal lifestyles continues to blur along a saprotrophism-symbiosis-pathogenesis continuum.

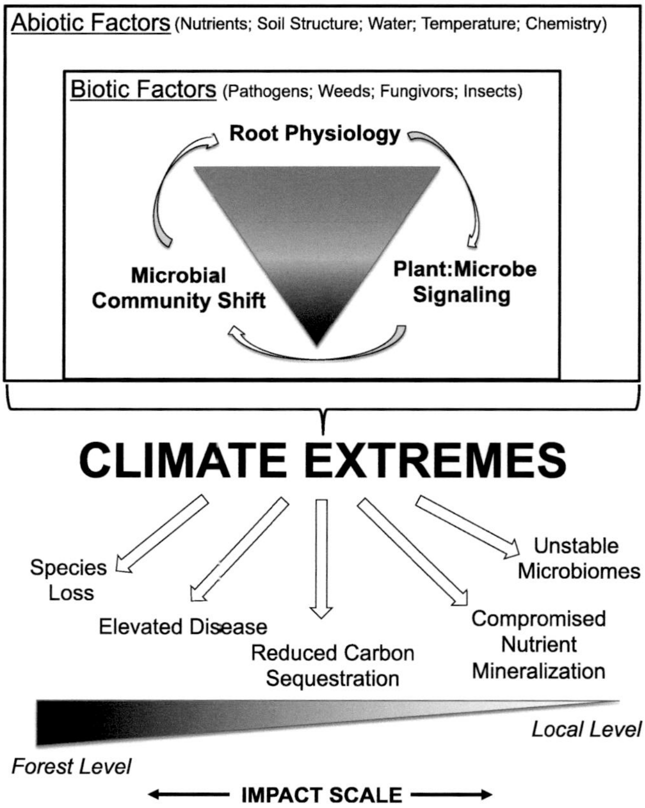

Fig. 10.3 Climate change impacts symbiotic triangle. The establishment of mutualistic mycorrhizal symbiosis is a factor of receptive root physiological status, active plant:microbe signaling and the interaction between microbes. These processes, in turn, are impacted by both biotic and abiotic factors. This delicate balance is threatened by climate extremes with likely outcomes including disrupted nutrient flow, increased disease, and eventual plant and microbial species loss

The precision and tuning of this process in keeping the plant host healthy and alive is likely the means by which ECM fungi can still be classified as "beneficial." However, as the mutualistic interaction requires a finely tuned, and likely precarious, dialogue between two partners in order to function properly (Fig. 10.3), interruption at any one of the signaling hubs described above is likely to greatly change the nature of the interaction between the organisms—if not completely abolish the mutualism. Interruption of these signals could come about through a variety of mechanisms including alterations due to abiotic conditions from global climate change and increasing anthropomorphic infringement on natural environments (Fig. 10.3). Similarly, differences in biotic environments as increasingly globalized movement of humans, plants, and animals have led to changes in many natural environments that could impact these relationships. We would hypothesize that disruption of these ECM–plant interactions could cause untold damage to the sustainability of forest habitats and the ecosystems they support (Fig. 10.3), continued and sustained effort should be dedicated to identifying common and divergent molecular "toolboxes" related to the ECM lifestyle. Within these toolboxes we should delve deeper into how to foster healthy communication between ECM fungi and their hosts *and* the soil/hyphospheric microbial community. We must also continue to broaden our complement of model study systems and look at the impact of new climate parameters upon small-scale ecosystems and, hopefully, on population dynamics within whole ecosystems. We have access to an increasingly large array of genomic and functional resources that are providing us with candidate biomarkers that could be used under natural settings to evaluate ECM functioning *in situ*. By studying these systems, and their adaptive plasticity under different parameters, we will know better how ECM fungi will fare under new climatic conditions and know how to better manage the microbiome of forests.

Acknowledgments Research in Jonathan Plett's lab is sponsored by the Australian Research Council, Forestry & Wood Products Australia and by the National Institute for Forest Products Innovation. Research in Francis Martin's and Annegret Kohler's labs is sponsored by the US Department of Energy, Office of Science, Biological and Environmental Research as part of the Plant–Microbe Interfaces Scientific Focus Area (https://pmi.ornl.gov) and the Laboratory of Excellence ARBRE (grant no. ANR-11-LABX-0002_ARBRE).

References

Abdel-Fattah GM, Mohamedin AH (2000) Interactions between a vesicular–arbuscular mycorrhizal fungus (*Glomus intraradices*) and *Streptomyces coelicolor* and their effects on sorghum plants grown in soil amended with chitin of brawn scales. Biol Fertil Soils 32:401–409

Abdulsalam O, Wagner K, Wirth S, Kunert M, David A, Kallenbach M, Boland W, Kothe E, Krause K (2021) Phytohormones and volatile organic compounds, like geosmin, in the ectomycorrhiza of *Tricholoma vaccinum* and Norway spruce (*Picea abies*). Mycorrhiza 31(2):173–188

Adams F, Reddell P, Webb MJ, Shipton WA (2006) Arbuscular mycorrhizas and ectomycorrhias on *Eucalyptus grandis* (*Myrtaceae*) trees and seedlings in native forests of tropical north-eastern Australia. Aust J Bot 54:271–281

Aimanianda V, Bayry J, Bozza S, Kniemeyer O, Perruccio K, Elluru SR, Clavaud C, Paris S, Brakhage AA, Kaveri SV, Romani L, Latge J-P (2009) Surface hydrophobin prevents immune recognition of airborne fungal spores. Nature 460:1117–1121

Akiyama K, Hayashi H (2008) Plastid-derived strigolactones show the way to roots for symbionts and parasites. New Phytol 178:695–698

Akiyama K, Matsuzaki K, Hayashi H (2005) Plant sesquiterpenes induce hyphal branching in arbuscular mycorrhizal fungi. Nature 435:824–827

Anthony MA, Crowther TW, Van Der Linde S et al (2022) Forest tree growth is linked to mycorrhizal fungal composition and function across Europe. ISME J 16: 1327–1336

Artursson V, Finlay RD, Jansson JK (2006) Interactions between arbuscular mycorrhizal fungi and bacteria and their potential for stimulating plant growth. Environ Microbiol 8:1–10

Basso V, Kohler A, Miyauchi S et al (2020) An ectomycorrhizal fungus alters sensitivity to jasmonate, salicylate, gibberellin, and ethylene in host roots. Plant Cell Environ 43:1047–1068

Besserer A, Becard G, Jauneau A, Roux C, Sejalon-Delmas N (2008) GR24, a synthetic analog of strigolactones, stimulates the mitosis and growth of the arbuscular mycorrhizal fungus *Gigaspora rosea* by boosting its energy metabolism. Plant Physiol 148: 402–413

Bomberg M, Timonen S (2007) Distribution of Cren- and Euryarchaeota in Scots pine mycorrhizospheres and boreal forest humus. Microb Ecol 54:406–416

Bonfante P, Genre A (2010) Mechanisms underlying beneficial plant-fungus interactions in mycorrhizal symbiosis. Nat Commun 1:48

Bouwmeester HJ, Matusova R, Zhongkui S, Beale MH (2003) Secondary metabolite signalling in host-parasitic plant interactions. Curr Opin Plant Biol 6: 358–364

Bowden CG, Smalley E, Guries RP, Hubbes M, Temple B, Horgen PA (1996) Lack of association between cerato-ulmin production and virulence in *Ophiostoma novo-ulmi*. MPMI 9:556–564

Brasier CM, Kirk SA, Tegli S (1995) Naturally occurring noncerato-ulmin producing mutants of *Ophiostoma novo-ulmi* are pathogenic but lack aerial mycelium. Mycol Res 99:436–440

Brulé C, Frey-Klett P, Pierrat JC, Courrier S, Gérard F, Lemoine MC, Rousselet JL, Sommer G, Garbaye J (2001) Survival in the soil of the ectomycorrhizal fungus *Laccaria bicolor* and the effects of a mycorrhiza helper *Pseudomonas fluorescens*. Soil Biol Biochem 33:1683–1694

Buckley S, Brackin R, Jämtgård S et al (2020) Microdialysis in soil environments: Current practice and future perspectives. Soil Biol Biochem 143: 107743

Buckley S, Brackin R, Näsholm T et al (2022) The influence of sucrose on soil nitrogen availability–a root exudate simulation using microdialysis. Geoderma 409:115645

Buée M, De Boer W, Martin F, van Overbeek L, Jurkevitch E (2009) The rhizosphere zoo: an overview of plant-associated communities of microorganisms, including phages, bacteria, archaea and fungi, and of some of their structuring factors. Plant Soil 321:189–212

Burgess T, Laurent P, Dell B, Malajczuk N, Martin F (1995) Effect of fungal-isolate aggressivity on the biosynthesis of symbiosis-related polypeptides in differentiating eucalypt ectomycorrhiza. Planta 195: 408–417

Cai F, Gao R, Zhao Z et al (2020) Evolutionary compromises in fungal fitness: hydrophobins can hinder the adverse dispersal of conidiospores and challenge their survival. ISME J 14:2610–2624

Casarrubia S, Daghino S, Kohler A et al (2018) The hydrophobin-like OmSSP1 may be an effector in the ericoid mycorrhizal symbiosis Front. Plant Sci 9:546

Chalot M, Brun A (1998) Physiology of organic nitrogen acquisition by ectomycorrhizal fungi and ectomycorrhizas. FEMS Microbiol Ref 22:21–44

Charvet-Candela V, Hitchin S, Ernst D et al (2002) Characterization of an Aux/IAA cDNA upregulated in *Pinus pinaster* roots in response to colonization by the ectomycorrhizal fungus *Hebeloma cylindrosporum*. New Phytol 154:769–777

Chernyavska M, Schmid M, Freitag PC et al (2019) Unravelling receptor and RGD motif dependence of retargeted adenoviral vectors using advanced tumor model systems. Sci Rep 9:1–13

Chowdhury J, Kemppainen M, Delhomme N et al (2022) *Laccaria bicolor* pectin methylesterases are involved in ectomycorrhiza development with *Populus tremula*× *Populus tremuloides*. New Phytol. https://doi.org/10.1111/nph.18358

Contreras G, Braun MS, Schäfer H et al (2018) Recombinant Afusin C, an anionic fungal CSαβ defensin from *Aspergillus fumigatus*, exhibits antimicrobial activity against gram-positive bacteria. PLoS One 13: pe0205509

Corvis Y, Walcarius A, Rink R, Mrabet NT, Rogalska E (2005) Preparing catalytic surfaces for sensing applications by immobilizing enzymes via hydrophobin layers. Anal Chem 77:1622–1630

Corvis Y, Brezesinski G, Rink R, Walcarius A, Van der Heyden A, Mutelet F, Rogalska E (2006) Analytical investigation of the interactions between SC3 hydrophobin and lipid layers: elaborating of nanostructured matrixes for immobilizing redox systems. Anal Chem 78:4850–4864

Corvis Y, Trzcinska K, Rink R, Bilkova P, Gorecka E, Bilewicz R, Rogalska E (2007) Electron donor-acceptor fullerene derivative retained on electrodes using SC3 hydrophobin. J Phys Chem C 111:1176–1179

Daguerre Y, Basso V, Wittulski S, Schellenberger R, Meyer L, Bailly J, Kohler A, Plett JM, Martin F, Veneault-Fourrey C (2020) The mutualism effector MiSSP7 of *Laccaria bicolor* alters the interactions between the poplar JAZ6 protein and its associated proteins. Sci Rep 10:20362

De Carvalho D (1994) Contribution à l'étude des protéines régulées par la symbiose ectomycorhizienne. Ph.D. thesis. Ecole Nationale de Génie Rurale et Forestier, Nancy

Deveau A, Palin B, Delaruelle C, Peter M, Kohler A, Pierrat JC, Sarniguet A, Garbaye J, Martin F, Frey-Klett P (2007) The mycorrhiza helper *Pseudomonas Fluorescens* BBc6R8 has a specific priming effect on the growth, morphology and gene expression of the ectomycorrhizal fungus *Laccaria bicolor* S238N. New Phytol 175:743–755

Ditengou FA, Lapeyrie F (2000) Hypaphorine from the ectomycorrhizal fungus *Pisolithus tinctorius* counteracts activities of indole-3-acetic acid and ethylene but not synthetic auxins in eucalypt seedlings. Mol Plant-Microbe Interact 13:151–158

Ditengou FA, Béguiristain T, Lapeyrie F (2000) Root hair elongation is inhibited by hypaphorine, the indole alkaloid from the ectomycorrhizal fungus *Pisolithus tinctorius*, and restored by IAA. Planta 211:722–728

Doré J, Kohler A, Dubost A, Hundley H, Singan V, Peng Y, Kuo A, Grigoriev IV, Martin F, Marmeisse R, Gay G (2017) The ectomycorrhizal basidiomycete *Hebeloma cylindrosporum* undergoes early waves of transcriptional reprogramming prior to symbiotic structures differentiation. Environ Microbiol 19(3):1338–1354

Duplessis S, Courty PE, Tagu D, Martin F (2005) Transcript patterns associated with ectomycorrhiza development in *Eucalyptus globulus* and *Pisolithus microcarpus*. New Phytol 165:599–611

Emmett BD, Lévesque-Tremblay V, Harrison MJ (2021) Conserved and reproducible bacterial communities associate with extraradical hyphae of arbuscular mycorrhizal fungi. ISME J 15:2276–2288

Felten J, Kohler A, Morin E, Bhalerao RP, Palme K, Martin F, Ditengou FA, Legue V (2009) The ectomycorrhizal fungus *Laccaria bicolor* stimulates lateral root formation in poplar and *Arabidopsis* through auxin transport and signaling. Plant Physiol 151:1991–2005

Frey-Klett P, Chavatte M, Clausse M-L, Courrier S, Le Roux C, Raaijmakers J, Martinotti MG, Pierrat J-C, Garbaye J (2005) Ectomycorrhizal symbiosis affects functional diversity of rhizosphere fluorescent pseudomonads. New Phytol 165:317–328

Frey-Klett P, Garbaye J, Tarkka MT (2007) The mycorrhiza helper bacteria revisited. New Phytol 176:22–36

Fries N, Serck-Hanssen K, Dimberg LH, Theander O (1989) In vitro enhancement of spore germination and early hyphal growth of a vesicular-arbuscular mycorrhizal fungus by host root exudates and plant flavonoids. Symbiosis 7:243–255

Fuqua WC, Winans SC, Greenberg EP (1994) Quorum sensing in bacteria: the LuxR-LuxI family of cell density-responsive transcriptional regulators. J Bacteriol 176(2):269–275

Garbaye J (1994) Helper bacteria: a new dimension to the mycorrhizal symbiosis. New Phytol 128:197–210

Garcia K, Delaux PM, Cope KR et al (2015) Molecular signals required for the establishment and maintenance of ectomycorrhizal symbioses. New Phytol 208:79–87

Gay G, Normand L, Marmeisse R, Sotta B, Debaud JC (1994) Auxin overproducer mutants of *Hebeloma cylindrosporum* Romagnési have increased mycorrhizal activity. New Phytol 128:645

Gea L, Normand L, Vian B, Gay G (1994) Structural aspects of ectomycorrhiza of *Pinus pinaster* (Ait.) Sol. formed by an IAA-overproducer mutant of *Hebeloma cylindrosporum* Romagnési. New Phytol 128:659

Gorka S, Dietrich M, Mayerhofer W et al (2019) Rapid transfer of plant photosynthates to soil bacteria via ectomycorrhizal hyphae and its interaction with nitrogen availability. Front Microbiol 10:168

Hewezi T, Howe PJ, Maier TR et al (2010) Arabidopsis spermidine synthase is targeted by an effector protein of the cyst nematode *Heterodera schachtii*. Plant Physiol 152:968–984

Hilbert JL, Costa G, Martin F (1991) Ectomycorrhizal synthesis and polypeptide changes during the early stage of eucalypt mycorrhiza development. Plant Physiol 97:977–984

Hill RA, Wong-Bajracharya J, Anwar S et al (2022) Abscisic acid supports colonization of *Eucalyptus grandis* roots by the mutualistic ectomycorrhizal fungus *Pisolithus microcarpus*. New Phytol 233:966–982

Horan DP, Chilvers GA, Lapeyrie F (1988) Time sequence of the infection process in eucalypt ectomycorrhizas. New Phytol 109:451–458

Hostetter MK (2000) RGD-mediated adhesion in fungal pathogens of humans, plants and insects. Curr Opin Microbiol 3(4):344–348

Jensen BG, Andersen MR, Pedersen MH, Frisvad JC, Sondergaard I (2010) Hydrophobins from *Aspergillus* species cannot be clearly divided into two classes. BMC Res Notes 3:344

Jiménez-Bremont JF, Marina M, Guerrero-Gonzalez MDLL et al (2014) Physiological and molecular implications of plant polyamine metabolism during biotic interactions. Front Plant Sci 5:95

Kang H, Chen X, Kemppainen M, Pardo AG, Veneault-Fourrey C, Kohler A, Martin FM (2020) The small secreted effector protein MiSSP7.6 of *Laccaria bicolor* is required for the establishment of ectomycorrhizal symbiosis. Environ Microbiol 22:1435–1446

Kaska DD, Myllylä R, Cooper JB (1999) Auxin transport inhibitors act through ethylene to regulate dichotomous branching of lateral root meristems in pine. New Phytol 142:49

Kawano T, Kawano N, Hosoya H, Lapeyrie F (2001) Fungal auxin antagonist hypaphorine competitively inhibits indole-3-acetic acid-dependent superoxide generation by horseradish peroxidase. Biochem Biophys Res Commun 288:546–551

Kemppainen M, Pardo A (2021) Nucleus-directed fluorescent reporter system for promoter studies in the ectomycorrhizal fungus *Laccaria bicolor*. J Microbiol Meth 190:p106341

Kershaw MJ, Talbot NJ (1998) Hydrophobins and repellents: proteins with fundamental roles in fungal morphogenesis. Fungal Genet Biol 23:18–33

Kikuchi K, Matsushita N, Suzuki K, Hogetsu T (2007) Flavonoids induce germination of basidiospores of the ectomycorrhizal fungus *Suillus bovinus*. Mycorrhiza 17:563–570

Kim S, Ahn I-P, Rho H-S, Lee Y-H (2005) *MPH1*, a *Magnaporthe grisea* hydrophobin gene, is required for fungal development and plant colonization. Mol Microbiol 57:1224–1237

Kloppholz S, Kuhn H, Requena N (2011) A secreted fungal effector of *Glomus intraradices* promotes symbiotic biotrophy. Curr Biol 21:1204–1209

Kodama K, Rich MK, Yoda A et al (2022) An ancestral function of strigolactones as symbiotic rhizosphere signals. Nat Commun 13:3974

Kohler A, Kuo A, Nagy LG, Morin E, Barry KW, Buscot F, Canbäck B, Choi C, Cichocki N, Clum A, Colpaert J (2015) Convergent losses of decay mechanisms and rapid turnover of symbiosis genes in mycorrhizal mutualists. Nat Genet 47:410–415

Kombrink A, Tayyrov A, Essig A, Stöckli M, Micheller S, Hintze J, Van Heuvel Y, Dürig N, Lin CW, Kallio PT, Aebi M (2019) Induction of antibacterial proteins and peptides in the coprophilous mushroom *Coprinopsis cinerea* in response to bacteria. ISME J 13(3):588–602

Kondoh O, Muto A, Kajiwara S, Takagi J, Saito Y, Shishido K (1995) Fruiting body-specific cDNA, mfbAc, from the mushroom *Lentinus edodes* encodes a high-molecular-weight cell-adhesion protein containing an Arg-Gly-Asp motif. Gene 154:31–37

Köttke I, Oberwinkler F (1986) Root-Fungus interactions observed on initial stages of mantle formation and Hartig net establishment in mycorrhizas of *Amanita muscaria* on *Picea abies* in pure culture. Can J Bot 64:2348–2354

Krause K, Henke C, Asiimwe T, Ulbricht A, Klemmer S, Schachtschabel D, Boland W, Kothe E (2015) Biosynthesis and secretion of indole-3-acetic acid and its morphological effects on *Tricholoma vaccinum*-spruce ectomycorrhiza. Appl Environ Microbiol 81(20): 7003–7011

Lagrange H, Jay-Allemand C, Lapeyrie F (2001) Rutin, the phenolglycoside from eucalyptus root exudates, stimulates *Pisolithus* hyphal growth at picomolar concentrations. New Phytol 149:349–355

Larose G, Chênevert R, Moutoglis P, Gagné S, Piché Y, Vierheilig H (2002) Flavonoid levels in roots of *Medicago sativa* are modulated by the developmental stage of the symbiosis and the root colonizing arbuscular mycorrhizal fungus. J Plant Physiol 159(12):1329–1339

Law SR, Serrano AR, Daguerre Y, Sundh J, Schneider AN, Stangl ZR, Castro D, Grabherr M, Näsholm T, Street NR, Hurry V (2022) Metatranscriptomics captures dynamic shifts in mycorrhizal coordination in boreal forests. Proc Natl Acad Sci U S A 119(26): e2118852119

Le Quéré A, Astrup Eriksen K, Rajashekar B, Schützendübel A, Canbäck B, Johansson T, Tunlid A (2006) Screening for rapidly evolving genes in the ectomycorrhizal fungus *Paxillus involutus* using cDNA microarrays. Mol Ecol 15(2):535–550

Lebreton A, Tang N, Kuo A, LaButti K, Andreopoulos W, Drula E, Miyauchi S, Barry K, Clum A, Lipzen A, Mousain D (2022) Comparative genomics reveals a dynamic genome evolution in the ectomycorrhizal milk-cap (*Lactarius*) mushrooms. New Phytol 235: 306–319

Li Q, Yan L, Ye L, Zhou J, Zhang B, Peng W, Zhang X, Li X (2018) Chinese black truffle (*Tuber indicum*) alters the ectomycorrhizosphere and endoectomycosphere

microbiome and metabolic profiles of the host tree *Quercus aliena*. Front Microbiol 9:2202
Liao HL, Chen Y, Vilgalys R (2016) Metatranscriptomic study of common and host-specific patterns of gene expression between pines and their symbiotic ectomycorrhizal fungi in the genus *Suillus*. PLoS Genet 12(10):e1006348
Lindahl BD, Tunlid A (2015) Ectomycorrhizal fungi–potential organic matter decomposers, yet not saprotrophs. New Phytol 205(4):1443–1447
Liu Y, Sun Q, Li J, Lian B (2018) Bacterial diversity among the fruit bodies of ectomycorrhizal and saprophytic fungi and their corresponding hyphosphere soils. Sci Rep 8(1):1–10
Liu H, Chen H, Ding G, Li K, Ren Q (2020) Identification of candidate genes conferring tolerance to aluminum stress in *Pinus massoniana* inoculated with ectomycorrhizal fungus. BMC Plant Biol 20(1):1–3
Looney B, Miyauchi S, Morin E, Drula E, Courty PE, Kohler A, Kuo A, LaButti K, Pangilinan J, Lipzen A, Riley R (2022) Evolutionary transition to the ectomycorrhizal habit in the genomes of a hyperdiverse lineage of mushroom-forming fungi. New Phytol 233(5):2294–2309
Lovett B, Kasson MT, Gandier JA (2022) Ecology drives the observed spectrum of hydrophobin protein diversity across Kingdom Fungi. bioRxiv. https://doi.org/10.1101/2022.08.19.504535
Lutzoni F, Nowak MD, Alfaro ME, Reeb V, Miadlikowska J, Krug M, Arnold AE, Lewis LA, Swofford DL, Hibbett D, Hilu K (2018) Contemporaneous radiations of fungi and plants linked to symbiosis. Nat Commun 9(1):1–11
Ma X, Zhu B, Nie Y, Liu Y, Kuzyakov Y (2021) Root and mycorrhizal strategies for nutrient acquisition in forests under nitrogen deposition: a meta-analysis. Soil Biol Biochem 163:108418
Machin DC, Hamon-Josse M, Bennett T (2020) Fellowship of the rings: a saga of strigolactones and other small signals. New Phytol 225(2):621–636
Marques-Galvez JE, Basso V, Kohler A, Barry K, Keymanesh K, Johnson J, Singan V, Grigoriev I, Vilgalys R, Martin F, Veneault-Fourrey C (2022) The establishment of *Populus* x *Laccaria bicolor* ectomycorrhiza requires the inactivation of MYC2 coordinated defense response with a key role for root terpene synthases. bioRxiv. https://doi.org/10.1101/2022.09.06.505662
Martin F, Laurent P, de Carvalho D, Voiblet C, Balestrini R, Bonfante P, Tagu D (1999) Cell wall proteins of the ectomycorrhizal basidiomycete Pisolithus tinctorius: identification, function and expression in symbiosis. Fungal Genet Biol 27:161–174
Martin F, Díez J, Dell B, Delaruelle C (2002) Phylogeography of the ectomycorrhizal *Pisolithus* species as inferred from nuclear ribosomal DNA ITS sequences. New Phytol 153(2):345–357
Martin F, Perotto S, Bonfante P (2007) Mycorrhizal fungi: a fungal community at the interface between soil and roots. In: Pinton R, Varanini Z, Nannipieri P (eds) The rhizosphere. Biochemistry and organic substances at the soil-plant interface. Marcel Dekker, New York, pp 263–296
Martin F, Aerts A, Ahrén D, Brun A, Danchin EGJ et al (2008) The genome of *Laccaria bicolor* provides insights into mycorrhizal symbiosis. Nature 452:88–92
Martin F, Kohler A, Murat C, Balestrini R, Coutinho PM et al (2010) Périgord black truffle genome uncovers evolutionary origins and mechanisms of symbiosis. Nature 464:1033–1038
Martin F, Kohler A, Murat C, Veneault-Fourrey C, Hibbett DS (2016) Unearthing the roots of ectomycorrhizal symbioses. Nat Rev Microbiol 14(12):760–773
Mészáros B, Sámano-Sánchez H, Alvarado-Valverde J, Čalyševa J, Martínez-Pérez E, Alves R, Shields DC, Kumar M, Rippmann F, Chemes LB, Gibson TJ (2021) Short linear motif candidates in the cell entry system used by SARS-CoV-2 and their potential therapeutic implications. Science. Signals 14(665):peabd0334
Mishra B, Gupta DK, Pfenninger M, Hickler T, Langer E, Nam B, Paule J, Sharma R, Ulaszewski B, Warmbier J, Burczyk J (2018) A reference genome of the European beech (*Fagus sylvatica* L.). Gigascience 7(6):giy063
Mishra B, Ulaszewski B, Meger J, Aury JM, Bodénès C, Lesur-Kupin I, Pfenninger M, Da Silva C, Gupta DK, Guichoux E, Heer K (2022) A chromosome-level genome assembly of the European beech (*Fagus sylvatica*) reveals anomalies for organelle DNA integration, repeat content and distribution of SNPs. Front Genet 12:691058
Miyauchi S, Kiss E, Kuo A, Drula E, Kohler A, Sánchez-García M, Morin E, Andreopoulos B, Barry KW, Bonito G, Buée M (2020) Large-scale genome sequencing of mycorrhizal fungi provides insights into the early evolution of symbiotic traits. Nat Commun 11(1):1–17
Montanini B, Levati E, Bolchi A, Kohler A, Morin E, Tisserant E, Martin F, Ottonello S (2011) Genome-wide search and functional identification of transcription factors in the mycorrhizal fungus *Tuber melanosporum*. New Phytol 189:736–750
Moonjely S, Keyhani NO, Bidochka MJ (2018) Hydrophobins contribute to root colonization and stress responses in the rhizosphere-competent insect pathogenic fungus *Beauveria bassiana*. Microbiology 164(4):517–528
Murat C, Payen T, Noel B, Kuo A, Morin E, Chen J, Kohler A, Krizsán K, Balestrini R, Da Silva C, Montanini B et al (2018) Pezizomycetes genomes reveal the molecular basis of ectomycorrhizal truffle lifestyle. Nat Ecol Evol 2(12):1956–1965
Murata H, Nakano S, Yamanaka T, Shimokawa T, Abe T, Ichida H, Hayashi Y, Tahara K (2021) Argon-ion beam induced mutants of the ectomycorrhizal agaricomycete *Tricholoma matsutake* defective in β-1,

4-endoglucanase activity promote the seedling growth of *Pinus densiflora* in vitro. Botany 99(3):139–149

Myburg AA, Grattapaglia D, Tuskan GA, Hellsten U, Hayes RD, Grimwood J, Jenkins J, Lindquist E, Tice H, Bauer D, Goodstein DM (2014) The genome of *Eucalyptus grandis*. Nature 510(7505):356–362

Nehls U, Béguiristain T, Ditengou F, Lapeyrie F, Martin F (1998) The expression of a symbiosis-regulated gene in eucalypt roots is regulated by auxins and hypaphorine, the tryptophan betaine of the ectomycorrhizal basidiomycete Pisolithus tinctorius. Planta 207:296–302

Nurmiaho-Lassila EL, Timonen S, Haahtela K, Sen R (1997) Bacterial colonization patterns of intact *Pinus sylvestris* mycorrhizospheres in dry pine forest soil: an electron microscopy study. Can J Microbiol 43:1017–1035

Nystedt B, Street NR, Wetterbom A, Zuccolo A, Lin YC, Scofield DG, Vezzi F, Delhomme N, Giacomello S, Alexeyenko A, Vicedomini R (2013) The Norway spruce genome sequence and conifer genome evolution. Nature 497(7451):579–584

Ochsenreiter T, Selezi D, Quaiser A, Bonch-Osmolovskaya L, Schleper C (2003) Diversity and abundance of Crenarchaeota in terrestrial habitats studied by 16S RNA surveys and real time PCR. Environ Microbiol 5:787–797

Onuma H, Hara K, Sugita K, Kano A, Fukuta Y, Shirasaka N (2019) Purification and characterization of a glycoside hydrolase family 5 endoglucanase from *Tricholoma matsutake* grown on barley based solid-state medium. J Biosci Bioeng 128(6):669–676

Paparokidou C, Leake JR, Beerling DJ, Rolfe SA (2021) Phosphate availability and ectomycorrhizal symbiosis with Pinus sylvestris have independent effects on the *Paxillus involutus* transcriptome. Mycorrhiza 31(1): 69–83

Parta M, Chang Y, Rulong S, Pinto-DaSilva P, Kwon-Chung KJ (1994) HYP1, a hydrophobin gene from Aspergillus fuigatus, complements the rodletless phenotype in *Aspergillus nidulans*. Infect Immun 62: 4389–4395

Pellegrin C, Daguerre Y, Ruytinx J, Guinet F, Kemppainen M, Frey NFD, Puech-Pagès V, Hecker A, Pardo AG, Martin FM, Veneault-Fourrey C (2019) *Laccaria bicolor* MiSSP8 is a small-secreted protein decisive for the establishment of the ectomycorrhizal symbiosis. Environ Microbiol 21(10):3765–3779

Peres da Silva R, Puccia R, Rodrigues ML, Oliveira DL, Joffe LS, César GV, Nimrichter L, Goldenberg S, Alves LR (2015) Extracellular vesicle-mediated export of fungal RNA. Sci Rep 5(1):1–12

Peter M, Kohler A, Ohm RA, Kuo A, Krützmann J, Morin E, Arend M, Barry KW, Binder M, Choi C, Clum A (2016) Ectomycorrhizal ecology is imprinted in the genome of the dominant symbiotic fungus *Cenococcum geophilum*. Nat Commun 7(1):1–5

Plett JM, Martin F (2011) Blurred boundaries: lifestyle lessons from fungal ectomycorrhizal genomes. Trends Genet 27:14–22

Plett JM, Martin FM (2018) Know your enemy, embrace your friend: using omics to understand how plants respond differently to pathogenic and mutualistic microorganisms. Plant J 93(4):729–746

Plett JM, Plett KL (2022) Leveraging genomics to understand the broader role of fungal small secreted proteins in niche colonization and nutrition. ISME Comm 2:49

Plett JM, Kemppainen M, Kale SD, Kohler A, Legue V, Brun A, Tyler B, Pardo A, Martin F (2011) A secreted effector protein of *Laccaria bicolor* is required for symbiosis development. Curr Biol 21:1197–1203

Plett JM, Gibon J, Kohler A, Duffy K, Hoegger PJ, Velagapudi R, Han J, Kües U, Grigoriev IV, Martin F (2012a) Phylogenetic, genomic organization and expression analysis of hydrophobin genes in the ectomycorrhizal basidiomycete *Laccaria bicolor*. Fungal Genet Biol 49(3):199–209

Plett JM, Kohler A, Martin F (2012b) De-constructing a mutualist: how the molecular blueprints of model symbiotic fungi are changing our understanding of mutualism. In: Hock B (ed) Fungal Associations. The Mycota, vol 9. Springer, Berlin, pp 93–117

Plett JM, Khachane A, Ouassou M, Sundberg B, Kohler A, Martin F (2014) Ethylene and jasmonic acid act as negative modulators during mutualistic symbiosis between *Laccaria bicolor* and *Populus* roots. New Phytol 202(1):270–286

Plett KL, Raposo AE, Anderson IC, Piller SC, Plett JM (2019) Protein arginine methyltransferase expression affects ectomycorrhizal symbiosis and the regulation of hormone signalling pathways. MPMI 32(10): 1291–1302

Plett JM, Plett KL, Wong-Bajracharya J, de Freitas PM, Dutra-Costa M, Kohler A, Martin F, Anderson IC (2020) Mycorrhizal effector PaMiSSP10b alters polyamine biosynthesis in *Eucalyptus* root cells and promotes root colonization. New Phytol 228(2): 712–727

Plett KL, Buckley S, Plett JM, Anderson IC, Lundberg-Felten J, Jämtgård S (2021, 1817) Novel microdialysis technique reveals a dramatic shift in metabolite secretion during the early stages of the interaction between the ectomycorrhizal fungus *Pisolithus microcarpus* and its host *Eucalyptus grandis*. Microorganisms 9(9)

Plomion C, Aury JM, Amselem J, Leroy T, Murat F, Duplessis S, Faye S, Francillonne N, Labadie K, Le Provost G, Lesur I et al (2018) Oak genome reveals facets of long lifespan. Nat Plants 4(7):440–452

Policelli N, Horton TR, Hudon AT, Patterson TR, Bhatnagar JM (2020) Back to roots: the role of ectomycorrhizal fungi in boreal and temperate forest restoration. Front For Glob Change 3:97

Qin M, Wang LK, Feng XZ, Yang YL, Wang R, Wang C, Yu L, Shao B. and Qiao MQ (2007) Bioactive surface modification of mica and poly (dimethylsiloxane) with

hydrophobins for protein immobilization. *Langmuir*, 23(8):4465–4471
Randewig D, Marshall JD, Näsholm T, Jämtgård S (2019) Combining microdialysis with metabolomics to characterize the *in situ* composition of dissolved organic compounds in boreal forest soil. Soil Biol Biochem 136:107530
Rivera Pérez CA, Janz D, Schneider D, Daniel R, Polle A (2022) Transcriptional landscape of ectomycorrhizal fungi and their host provides insight into N uptake from forest soil. Msystems 7(1):e00957–e00921
Roby D, Gadelle A, Toppan A (1987) Chitin oligosaccharides as elicitors of chitinase activity in melon plants. Biochem Biophys Res Commun 143: 885–892
Ruess RW, Hendrick RL, Burton AJ, Pregitzer KS, Sveinbjornsson B, Allen ME, Maurer GE (2003) Coupling fine root dynamics with ecosystem carbon cycling in black spruce forests of interior Alaska. Ecol Monogr 73:643–662
Rupp LA, Mudge KW (1985) Ethephon and auxin induce mycorrhiza-like changes in the morphology of root organ cultures of Mugo pine. Physiol Plant 64:316
Ruytinx J, Miyauchi S, Hartmann-Wittulsky S, de Freitas PM, Guinet F, Churin JL, Put C, Le Tacon F, Veneault-Fourrey C, Martin F, Kohler A (2021) A transcriptomic atlas of the ectomycorrhizal fungus *Laccaria bicolor*. Microorganisms 9(12):2612
Sakamoto Y, Sato S, Takizawa M, Narimatsu M (2022) Identification of upregulated genes in *Tricholoma matsutake* mycorrhiza. FEMS Microbiol Letters 369(1): fnac085
Sannazzaro AI, Echiverria M, Alberto EO, Ruiz OA, Menendez AB (2007) Modulation of polyamine balance in *Lotus glaber* by salinity and arbuscular mycorrhiza. Plant Physiol Biochem 45:39–46
Schneider AN, Sundh J, Sundström G, Richau K, Delhomme N, Grabherr M, Hurry V, Street NR (2021) Comparative fungal community analyses using metatranscriptomics and internal transcribed spacer amplicon sequencing from Norway spruce. Msystems 6(1):e00884–e00820
Schrey SD, Tarkka MT (2008) Friends and foes: streptomycetes as modulators of plant disease and symbiosis. Antonie Leeuwenhoek 94:11–19
Schrey SD, Schellhammer M, Ecke M, Hampp R, Tarkka MT (2005) Mycorrhiza helper bacterium Streptomyces AcH 505 induces differential gene expression in the ectomycorrhizal fungus Amanita muscaria. New Phytol 168:205–216
Sebastiana M, Gargallo-Garriga A, Sardans J, Pérez-Trujillo M, Monteiro F, Figueiredo A, Maia M, Nascimento R, Silva MS, Ferreira AN, Cordeiro C (2021) Metabolomics and transcriptomics to decipher molecular mechanisms underlying ectomycorrhizal root colonization of an oak tree. Sci Rep 11(1):1–6
Selosse MA, Richard F, He X, Simard SW (2006) Mycorrhizal networks: des liaisons dangereuses? Trends Ecol Evol 21:621–628
Sharp JA, McNeil M, Albersheim P (1984a) The primary structure of one elicitor active and seven elicitor-inactive hexa (ß-D-glucopyranolsyl)-D-glucitols isolated from the mycelia walls of Phytophthora megasperma f.sp. glycinea. J Biol Chem 25:11321–11326
Sharp JA, Valent B, Albersheim P (1984b) Purification and partial characterization of a ß-glucan fragment that elicits phytoalexin accumulation in soybean. J Biol Chem 25:11312–11320
Shirakawa M, Uehara I, Tanaka M (2019) Mycorrhizosphere bacterial communities and their sensitivity to antibacterial activity of ectomycorrhizal fungi. Microb Environ pME18146
Snelders NC, Petti GC, van den Berg GC, Seidl MF, Thomma BP (2021) An ancient antimicrobial protein co-opted by a fungal plant pathogen for in planta mycobiome manipulation. Proc Natl Acad Sci 118:49
Snelders NC, Boshoven JC, Song Y, Schmitz N, Fiorin GL, Rovenich H, van den Berg GC, Torres DE, Petti GC, Prockl Z, Faino L et al (2022) A highly polymorphic effector protein promotes fungal virulence through suppression of plant-associated Actinobacteria. New Phytol. https://doi.org/10.1111/nph.18576
Splivallo R, Fischer U, Gobel C, Feussner I, Karlovsky P (2009) Truffles regulate plant root morphogenesis via the production of auxin and ethylene. Plant Phys 150: 2018–2029
Steinkellner S, Lendzemo V, Langer I, Schweiger P, Khaosaad T, Toussaint JP, Vierheilig H (2007) Flavonoids and strigolactones in root exudates as signals in symbiotic and pathogenic plant-fungus interactions. Molecules 12(7):1290–1306
Stes E, Biondi S, Holsters M, Vereecke D (2011) Bacterial and plant signal integration via D3-type cyclins enhances symptom development in the *Arabidopsis-Rhodococcus fascians* interactions. Plant Physiol 156: 712–725
Strullu-Derrien C, Selosse MA, Kenrick P, Martin FM (2018) The origin and evolution of mycorrhizal symbioses: from palaeomycology to phylogenomics. New Phytol 220(4):1012–1030
Stuart EK, Plett KL (2020) Digging deeper: in search of the mechanisms of carbon and nitrogen exchange in ectomycorrhizal symbioses. Front Plant Sci 10:1658
Stuart EK, Castañeda-Gómez L, Macdonald CA, Wong-Bajracharya J, Anderson IC, Carrillo Y, Plett JM, Plett KL (2022) Species-level identity of *Pisolithus* influences soil phosphorus availability for host plants and is moderated by nitrogen status, but not CO_2. Soil Biol Biochem 165:108520
Sundaram S, KimSJ SH, McQuattie CJ, Hiremath ST, Podila GK (2001) Isolation and characterization of a symbiosis-regulated *ras* from the ectomycorrhizal fungus *Laccaria bicolor*. Mol Plant-Microbe Interact 14: 618–628
Tagu D, Nasse B, Martin F (1996) Cloning and characterization of hydrophobin-encoding cDNAs from the

ectomycorrhizal basidiomycete *Pisolithus tinctorius*. Gene 168:93–97

Tang N, Lebreton A, Xu W, Dai Y, Yu F, Martin FM (2021) Transcriptome profiling reveals differential gene expression of secreted proteases and highly specific gene repertoires involved *in Lactarius–Pinus* symbioses. Front Plant Sci 2021:1775

Tarkka MT, Frey-Klett P (2008) Mycorrhiza helper bacteria. In: Varma A (ed) Mycorrhiza, 3rd edn. Springer, Berlin, pp 113–132

Tarkka MT, Sarniguet A, Frey-Klett P (2009) Inter-kingdom encounters: recent advances in molecular bacterium–fungus interactions. Curr Genet 55:233–243

Taylor AFS, Martin F, Read DJ (2000) Fungal diversity in ectomycorrhizal communities of Norway spruce {*Picea abies* (L.) Karst.} and beech (*Fagus sylvatica* L.) along north-south transects in Europe. Ecol Stud 142:343–365

Templeton MD, Rikkerink EHA, Beever RE (1994) Small, cysteine-rich proteins and recognition in fungal-plant interactions. Mol Plant-Microbe Interact 7:320–325

Thau N, Monod M, Cretani B, Rolland C, Tronchin G et al (1994) Rodletless mutants of *Aspergillus fumigatus*. Infect Immun 62:4380–4388

Toljander JF, Lindahl BD, Paul LR, Elfstrand M, Finlay RD (2007) Influence of arbuscular mycorrhizal mycelial exudates on soil bacterial growth and community structure. FEMS Microbiol Ecol 61(2):295–304

Tuskan GA, Difazio S, Jansson S, Bohlmann J, Grigoriev I et al (2006) The genome of black cottonwood, *Populus trichocarpa*. Science 313:1596–1604

Ungerstedt U (1991) Microdialysis—principles and applications for studies in animals and man. J Int Med 230(4):365–373

Uroz S, Heinonsalo J (2008) Inactivation of N-acyl homoserine lactone quorum sensing molecules by forest soil fungi. FEMS Microbiol Ecol 65:271–278

van der Biezen EA, Jones JD (1998) The NB-ARC domain: a novel signalling motif shared by plant resistance gene products and regulators of cell death in animals. Curr Biol 8:R226–R228

van der Heijden MG, Martin FM, Selosse MA, Sanders IR (2015) Mycorrhizal ecology and evolution: the past, the present, and the future. New Phytol 205(4): 1406–1423

Vayssières A, Pěnčík A, Felten J, Kohler A, Ljung K, Martin F, Legué V (2015) Development of the poplar-*Laccaria bicolor* ectomycorrhiza modifies root auxin metabolism, signaling, and response. Plant Physiol 169(1):890–902

Villalobos Solis MI, Poudel S, Bonnot C, Shrestha HK, Hettich RL, Veneault-Fourrey C, Martin F, Abraham PE (2020) A viable new strategy for the discovery of peptide proteolytic cleavage products in plant-microbe interactions. MPMI 33(10):1177–1188

Vit A, Misson L, Blankenfeldt W, Seebeck FP (2015) Ergothioneine biosynthetic methyltransferase EgtD reveals the structural basis of aromatic amino acid betaine biosynthesis. Chem Biochem 16:119–125

Voiblet C, Duplessis S, Encelot N, Martin F (2001) Identification of symbiosis-regulated genes in *Eucalyptus globulus-Pisolithus tinctorius* ectomycorrhiza by differential hybridization of arrayed cDNAs. Plant J 25: 181–192

Walker TS, Bais HP, Grotewold E, Vivanco JM (2003) Root exudation and rhizosphere biology. Plant Physiol 132:44–51

Wang M, Dean RA (2020) Movement of small RNAs in and between plants and fungi. Mol Plant Pathol 21(4): 589–601

Wang X, Wang H, Huang Y, Zhao Z, Qin X, Wang Y, Miao Z, Chen Q, Qiao M (2010) Noncovalently functionalized multi-wall carbon nanotubes in aqueous solution using the hydrophobin HFBI and their electroanalytical application. Biosens Bioelectron 26: 1104–1108

Wang M, Weiberg A, Dellota E Jr, Yamane D, Jin H (2017) *Botrytis* small RNA Bc-siR37 suppresses plant defense genes by cross-kingdom RNAi. RNA Biol 14(4):421–428

Wang W, Das A, Kainer D, Schalamun M, Morales-Suarez A, Schwessinger B, Lanfear R (2020) The draft nuclear genome assembly of *Eucalyptus pauciflora*: a pipeline for comparing *de novo* assemblies. Gigasci 9(1):pgiz160

Wang T, Persson P, Tunlid A (2020) A widespread mechanism in ectomycorrhizal fungi to access nitrogen from mineral-associated proteins. Environ Microbiol 23(10):5837–5849

Wawra S, Fesel P, Widmer H, Timm M, Seibel J, Leson L, Kesseler L, Nostadt R, Hilbert M, Langen G, Zuccaro A (2016) The fungal-specific β-glucan-binding lectin FGB1 alters cell-wall composition and suppresses glucan-triggered immunity in plants. Nat Commun 7(1):1–11

Weiberg A, Bellinger M, Jin H (2015) Conversations between kingdoms: small RNAs. Curr Opin Biotechnol 32:207–215

Wessels JGH (1996) Hydrophobins: proteins that change the nature of the fungal surface. Adv Microb Physiol 38:1–45

Westerink BH (1995) Brain microdialysis and its application for the study of animal behaviour. Behav Brain Res 70(2):103–124

Wild A (1988) Plant nutrients in soil. In: Wild A (ed) Russel's soil conditions and plant growth, 11th edn. Longman Scientific and Technical, Harlow, pp 695–743

Wong-Bajracharya J, Castaneda-Gomez L, Plett KL, Anderson IC, Carrillo Y, Plett JM (2020) Untangling the effect of roots and mutualistic ectomycorrhizal fungi on soil metabolite profiles under ambient and elevated carbon dioxide. Soil Biol Biochem 151: 108021

Wong-Bajracharya J, Singan VR, Monti R, Plett KL, Ng V, Grigoriev IV, Martin FM, Anderson IC, Plett

JM (2022) The ectomycorrhizal fungus *Pisolithus microcarpus* encodes a microRNA involved in cross-kingdom gene silencing during symbiosis. Proc Natl Acad Sci U S A 119(3):e2103527119

Wösten HAB, Wessels JGH (1997) Hydrophorins, from molecular structure to multiple functions in fungal development. Mycoscience 38:363–374

Wu G, Miyauchi S, Morin E, Kuo A, Drula E, Varga T, Kohler A, Feng B, Cao Y, Lipzen A, Daum C (2022) Evolutionary innovations through gain and loss of genes in the ectomycorrhizal Boletales. New Phytol 233(3):1383–1400

Yan H, Freschet GT, Wang H, Hogan JA, Li S, Valverde-Barrantes OJ, Fu X, Wang R, Dai X, Jiang L, Meng S (2022) Mycorrhizal symbiosis pathway and edaphic fertility frame root economics space among tree species. New Phytol 234(5):1639–1653

Yasuda T, Shishido K (1997) Aggregation of yeast cells induced by the Arg-Gly-Asp motif-containing fragment of high-molecular mass cell-adhesion protein MFBA, derived from the basidiomycetous mushroom Lentinus edodes. FEMS Microbiol Lett 154:195–200

Yu C, Dou K, Wang S, Wu Q, Ni M, Zhang T, Lu Z, Tang J, Chen J (2020) Elicitor hydrophobin Hyd1 interacts with Ubiquilin1-like to induce maize systemic resistance. J Integr Plant Biol 62(4):509–526

Zhang F, Anasontzis GE, Labourel A, Champion C, Haon M, Kemppainen M, Commun C, Deveau A, Pardo A, Veneault-Fourrey C, Kohler A (2018a) The ectomycorrhizal basidiomycete *Laccaria bicolor* releases a secreted β-1,4 endoglucanase that plays a key role in symbiosis development. New Phytol 220(4):1309–1321

Zhang L, Feng G, Declerck S (2018b) Signal beyond nutrient, fructose, exuded by an arbuscular mycorrhizal fungus triggers phytate mineralization by a phosphate solubilizing bacterium. ISME J 12(10):2339–2351

Zhang F, Labourel A, Haon M, Kemppainen M, Da Silva ME, Brouilly N, Veneault-Fourrey C, Kohler A, Rosso MN, Pardo A, Henrissat B (2022) The ectomycorrhizal basidiomycete *Laccaria bicolor* releases a GH28 polygalacturonase that plays a key role in symbiosis establishment. New Phytol 233(6):2534–2547

Zhao Z-X, Qiao M-Q, Yin F, Shao B, Wu B-Y, Wang Y-Y, Wang X-S, Qin X, Li S, Yu L, Chen Q (2007) Amperometric glucose biosensor based on self-assembly hydrophobin with high efficiency of enzyme utilization. Biosens Bioelectron 22:3021–3027

Zhao Z, Cai F, Gao R, Ding M, Jiang S, Chen P et al (2021) At least three families of hyphosphere small secreted cysteine-rich proteins can optimize surface properties to a moderately hydrophilic state suitable for fungal attachment. Environ Microbiol 23(10): 5750–5768

Zimin A, Stevens KA, Crepeau MW, Holtz-Morris A, Koriabine M, Marçais G, Puiu D, Roberts M, Wegrzyn JL, de Jong PJ, Neale DB (2014) Sequencing and assembly of the 22-Gb loblolly pine genome. Genetics 196(3):875–890

Nematode-Trapping Fungi and *Caenorhabditis elegans* as a Model System for Predator–Prey Interactions

11

Guillermo Vidal-Diez de Ulzurrun, Sheng-Chian Juan, Tzu-Hsiang Lin, and Yen-Ping Hsueh

Abstract

All habitats on earth host various sympatric organisms that interact in different ways. Predator–prey dynamics represent one such mode of interaction, which are shaped by coevolution of the interacting organisms, resulting in complex signal exchanges and eliciting a diversity of hunting and escape strategies. A predatory lifestyle has evolved independently in major phyla of the fungal kingdom. These carnivorous fungi use diverse strategies to hunt fast-moving nematodes, and they are of particular interest due to their potential as biocontrol agents against parasitic nematodes. The fascinating biology of carnivorous fungi has interested scientists for decades. Over the past few years, significant progress has been made in understanding the molecular mechanisms underlying predation by nematode-trapping fungi. Given the extensive research tools and resources that are available for the model nematode *Caenorhabditis elegans*, the microscopic fungal–nematode predator–prey system represents an exceptional model for exploring coevolution and predator–prey dynamics. Herein, we summarize the current knowledge of the predator–prey interactions between nematode-trapping fungi and *C. elegans*, focusing on the predation process and the molecular mechanisms underlying it. We envision that, given recent advances in experimental techniques developed for nematode-trapping fungi, scientists will continue to unravel how these carnivorous fungi have coevolved with their nematode prey.

Keywords

Nematode-trapping fungi · *Arthrobotrys oligospora* · *Caenorhabditis elegans* · Predator–prey interactions · Co-evolution

The Meiosporic State of *Arthrobotrys*

For many years only the conidial state of *Arthrobotrys* species was known until Donald H. Pfister neglected a Petri dish on the back of a lab bench for several months. Pfister, a mycologist who studies discomycetes (ascomycetes with apothecial ascomata), was culturing a species of *Orbilia* from ascospores derived from tiny cup-like apothecia collected on dung. A serendipitous discovery came when he eventually examined the culture plate and found he had grown an *Arthrobotrys* species, the conidial state of a fungus well known for its nematode-trapping ability.

(continued)

G. Vidal-Diez de Ulzurrun · S.-C. Juan · T.-H. Lin · Y.-P. Hsueh (✉)
Institute of Molecular Biology, Academia Sinica, Taipei, Taiwan
e-mail: pinghsueh@gate.sinica.edu.tw

Y.-P. Hsueh, M. Blackwell (eds.), *Fungal Associations*, The Mycota 9,
https://doi.org/10.1007/978-3-031-41648-4_11

Pfister developed the discovery, linking *Orbilia* and *Arthrobotrys*, which he compared to the mythical twins Castor and Pollux. He presented his findings in his presidential address to the Mycological Society of America (Pfister 1997).

Molecular methods have allowed for the in-depth study of these ascosporic fungi and their diverse conidial states. We now know that *Orbilia* is distinct from other discomycetes, and species of *Orbilia* and related genera are placed in a separate class, the Orbiliomycetes. Several lineages found within Orbiliomycetes correspond to morphologically, ecologically, and physiologically diverse conidial states. In addition to nematode-trapping states such as *Arthrobotrys*, *Dactylellina*, and *Drechslerella*, some non-nematode-trapping fungi are also members of the Orbiliomycetes such as species of the aeroaquatic fungus genus *Helicoon*. Conidia of this genus and others disperse in water, buoyed by air bubbles trapped in the developing conidia.

11.1 Introduction

Various organisms coinhabit many kinds of environments, creating biodiverse ecosystems in which they all interact in many ways. Some biological interactions result in benefits for the interacting species, whereas others result in positive outcomes for only one of the organisms. Indeed, some interactions may even be detrimental, such as in predation–prey interactions when one species consumes another. Although predators are often assumed to be fast-moving and aggressive, some predators rely on very subtle techniques to lure and capture their prey. For instance, some make traps, produce attractive odors to lure their prey, or use toxins and venoms to paralyze and kill, all representing strategies that are the result of years of coevolution with their prey. Complex communication systems involving various kinds of signals between predator and prey play critical roles in these interactions. Thus, predation poses a challenging biological question, and several aspects of this core behavior remain to be answered.

Fungi and nematodes are found in most known ecosystems and display all types of interactions, ranging from mutualism to predation (Yang et al. 2020). Their interactions are complex and change in response to different biotic and abiotic factors. Both groups of organisms may be harmful to humans, animals, and plants, so understanding and controlling their dynamics is crucial. Among the interactions between nematodes and fungi, we find those whereby sessile fungi hunt fast-moving nematodes particularly interesting. The reason for this resides in the potential of nematophagous fungi to act as biocontrol agents against parasitic nematodes, such as those causing weight loss in livestock and disease in plants (Lopez-Llorca et al. 2007). These fungi represent an environmentally friendly alternative to classical pesticides, but unfortunately they fail to establish in most agricultural soils (Jaffee 2003). Thus, to harness the nematocidal potential of fungi in complex environments, we need to learn more about their predatory behavior and the molecular mechanisms underlying it.

Nematophagous fungi have evolved complex predatory strategies, such as producing nematocidal toxins, developing conidia that attach to their prey, and forming traps that lure and capture nematodes (Li et al. 2015). The distinctive morphologies of the fungal traps and the complexity of the trapping processes have made nematode-trapping fungi (NTF) fascinating for many scientists. Trap morphology varies among different fungal species, which either use adhesives or mechanical force to capture nematodes (Nordbring-Hertz et al. 2011). The trapping process involves various signals between predator and prey in several steps. NTF, such as the ascomycete *Arthrobotrys oligospora* (Orbiliales), use signals to attract nematode to their mycelia. Upon sensing the prey, they begin making adhesive traps. Traps capture the nematodes, which are then paralyzed, before being pierced by the fungal hyphae that infect and consume them from the inside. Although several genes and pathways have been described that play roles in prey sensing and trap formation, the molecular mechanisms underlying fungal predation remain largely unknown. The plasticity of the predatory behavior observed in wild isolates of the NTF *A. oligospora* only corroborates the complexity and diversity of factors involved in this trait (Yang et al. 2020). Additionally, the responses of nematode prey to the predatory behavior of NTF are still poorly understood, hindering a complete understanding of this predator–prey system.

Nevertheless, the scale at which nematodes and fungi interact means their dynamics can be reproduced under laboratory conditions, with full predator–prey interactions occurring in a Petri dish. Both organisms can reproduce easily and rapidly in laboratories. Thus, several generations can be raised within a limited timeframe while tracking their genetic background, which facilitates coevolutionary analyses. The genomes of several nematophagous fungi have been sequenced recently, further informing genomic and genetic studies (de Ulzurrun et al. 2020; Lee et al. 2021). Additionally, the model nematode *Caenorhabditis elegans* is a common prey item of various NTF, itself bringing a wide range of tools and resources to studies of these predator–prey dynamics at the molecular level.

In this chapter, we summarize the current knowledge of the interactions between nematodes and fungi. We outline the ecology of both organisms, including their habitats, diversity, and the types of interactions they form in natural ecosystems. This chapter pays special attention to nematophagous fungi and particularly to NTF. Therefore, we thoroughly describe the trapping process, as well as the known molecular mechanisms underlying each of its stages. In parallel, we review the strategies adopted by *C. elegans* to avoid predation and fungal infection. We hope that this chapter sheds some light on the widely unappreciated predator–prey interactions between fungi and nematodes and that it will encourage scientists to explore this amazing inter-kingdom system.

11.2 Carnivorism in the Fungal Kingdom

Fungi are found in most known ecosystems, where they coexist with a wide variety of organisms, sometimes in synergistic relationships with their neighbors. Some fungi, however, actively prey on other organisms to obtain additional nutrients, and these are carnivorous fungi. Carnivorous fungi include amoebophagous fungi, such as *Acaulopage dichotoma* (phylum Zoopagomycota) which traps and feeds on free-living amoebas (Corsaro et al. 2018), fungi that prey on insects, such as *Laccaria bicolor* (phylum Basidiomycota) that paralyzes and eats springtails (Klironomos and Hart 2001), and nematophagous fungi in several different phyla that prey on nematodes (Li et al. 2015).

The most studied group within the carnivorous fungi is the nematophagous fungi due to the diverse strategies that they employ to capture their prey, the wide range of habitats in which these interactions take place, and their potential in agricultural control measures. More than 700 species of nematophagous fungi have been described in various fungal lineages, namely the Ascomycota, Basidiomycota, Zoopagomycota, and Chytridiomycota (Li et al. 2015) (Fig. 11.1). Although these fungi share common prey, their strategies for luring and capturing nematodes differ greatly. For instance, endoparasitic fungi in the Zoopagomycota use conidia that attach to the nematode cuticle and pierce it to infect their prey from the inside (Lopez-Llorca et al. 2007). Other fungi, such as the ascomycete *Metapochonia rubescens* (Clavicipitaceae), attack nematode eggs, breaking the eggshell using a special structure called an appressorium (Lopez-Llorca et al. 2007). Another method employed by these fungi to capture nematodes is the production of nematocidal toxins, such as those produced by hyphae of the edible mushroom *Pleurotus ostreatus* (Barron 2003). This fungus usually extracts nutrients from wood, but under starvation conditions it produces toxins that can paralyze and kill nematodes (Lee et al. 2021). Upon contact between *P. ostreatus* and prey,

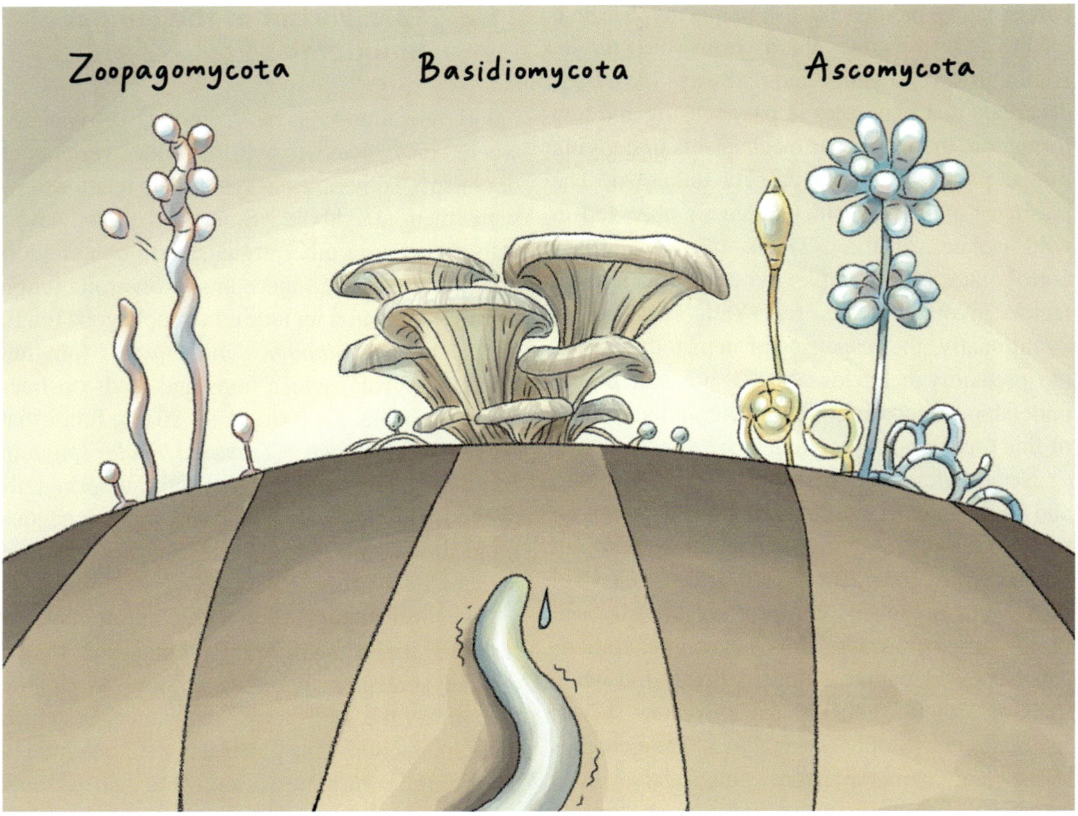

Fig. 11.1 Nematophagous fungi rely on different strategies to kill and consume their nematode prey. Endoparasitic fungi in phylum Zoopagomycota use spores and conidia that attach to the nematode cuticle to pierce it and infect their prey. Basidiomycetes such as *Pleurotus ostreatus* release nematocidal toxins that paralyze and kill nematodes. Nematode-trapping fungi in phylum Ascomycota lure nematodes to sophisticated trapping devices to capture and consume them

fungal-derived toxins enter the nematode through its sensory cilia leading to hypercontraction and calcium influx of pharyngeal and body wall muscles, ultimately causing necrosis of the neuromuscular system (Lee et al. 2020). Another group of nematophagous fungi, i.e., the NTF, develop complex and distinctive physiological structures aimed at trapping nematodes (Nordbring-Hertz et al. 2011).

Arthrobotrys oligospora was first described in 1888 (Zopf 1888). Under nutrient-deprived conditions and in the presence of nematodes, this fungus produces adhesive, three-dimensional networks known as "traps" to catch nematodes (Fig. 11.2). This finding inspired scientists to identify and study NTF worldwide. To date, more than 200 species of NTF have been described, and they are broadly distributed from tropical to polar regions (Nordbring-Hertz et al. 2011). For example, NTF within the *Orbiliales* evolved from a common ancestor that first preyed on nematodes to fulfill its nitrogen needs in nitrogen-poor environments (Barron 2003; Ahrén and Tunlid 2003; Li et al. 2005). Traps come in a variety of shapes and they can be classified into two main kinds: mechanical traps [constricting rings such as those produced by *Drechslerella stenobrocha* (Orbiliales) (Liu et al. 2012)] and adhesive traps (adhesive nets, hyphae, knobs and columns, and non-constraining rings) (Nordbring-Hertz et al. 2011).

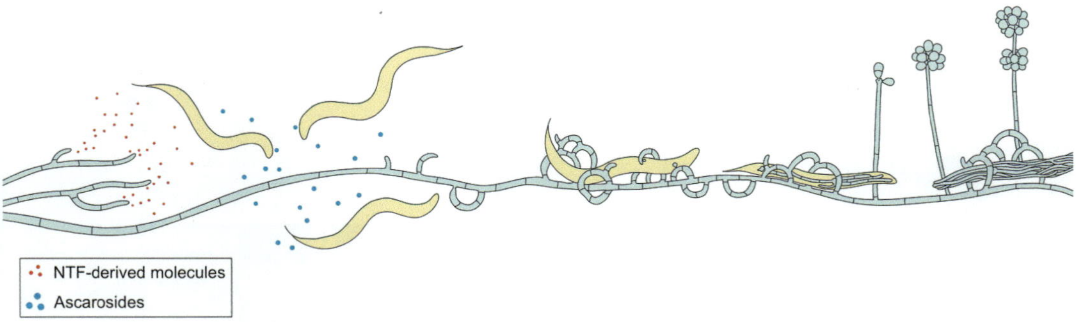

Fig. 11.2 Summary of the trapping process of adhesive net-forming nematode-trapping fungi. First (from left), nematodes are attracted to the fungal mycelium (attraction). The fungus senses signals emanating from the nematodes, which triggers trap formation (sensing). The nematodes are then captured and paralyzed by the adhesive traps (adhesion) and, subsequently, their cuticles are penetrated by the fungus (penetration). Once inside, the fungus starts an infection process that culminates with digestion of the nematode (digestion). In the end, the fungus starts to form conidiophores and conidia (conidiation)

11.3 The Trapping Process of Nematode-Trapping Fungi

In most natural ecosystems, nematodes and NTF cohabit without harming each other. However, a combination of specific biotic and abiotic factors may alter those dynamics and convert them into a predator–prey system (Nordbring-Hertz et al. 2011). Some abiotic factors that trigger the carnivorous behavior of NTF are starvation, carbon dioxide levels (Bartnicki-Garcia et al. 1964), and the availability of carbon and nitrogen sources (Scholler and Rubner 1994). In addition to environmental factors, biotic stimuli are required to induce trap formation, with direct contact between living nematodes and fungal hyphae being the most effective inducer (Åhman et al. 2002; Su et al. 2017; Xie et al. 2010). Early studies showed that nematode extracts and culture filtrates such as nemin can induce trap morphogenesis in NTF (Lawton 1957; Pramer and Stoll 1959). Additionally, factors non-specific to nematodes, such as dipeptides like phenylalanine-valine (Nordbring-Hertz 1973; Friman et al. 1985) or urea and diketopiperazines secreted by bacteria (Li et al. 2011, 2016; Wang et al. 2014), also resulted in trap formation. A group of small molecules widely conserved among nematodes and that function as sex pheromones in *C. elegans*, i.e., ascarosides, are potent inducers of trap formation by NTF (Hsueh et al. 2013). Several chemically synthesized ascarosides have been shown to induce trap morphogenesis in *A. oligospora* (Hsueh et al. 2013), suggesting that these molecules may be the key signal emanating from nematodes that is recognized by NTF. In an effort to uncover the pathways by which nematode and environmental signals are sensed, recent studies have revealed that conserved signal transduction pathways—including the G-protein signaling pathway, MAPK cascade, cAMP-PKA pathway, and AMPK pathway—are at least partly involved (Yang et al. 2020; Kuo et al. 2020; Chen et al. 2021, 2022; Wang et al. 2022) (Fig. 11.3). G-protein signaling is critical for sensing of environments in the fungal kingdom (Alvaro and Thorner 2016). In *A. oligospora*, deletion of GPB1, a homolog of G-protein β subunit STE4 involved in the pheromone signaling pathway in the ascomycete *Saccharomyces cerevisiae*, abolished trap formation (Yang et al. 2020). G-protein β subunit *STE4* is upstream of the Fus3/Kss1 cascade in *S. cerevisiae*. Studies have shown that components in the Fus3/Kss1 cascade of *A. oligospora* include MAPK Fus3, its upstream kinase MAPKK Ste7, and the downstream transcription factor Ste12 are also required for nematode-sensing and trap formation (Chen et al. 2021) (Fig. 11.3). Deletion of any of these

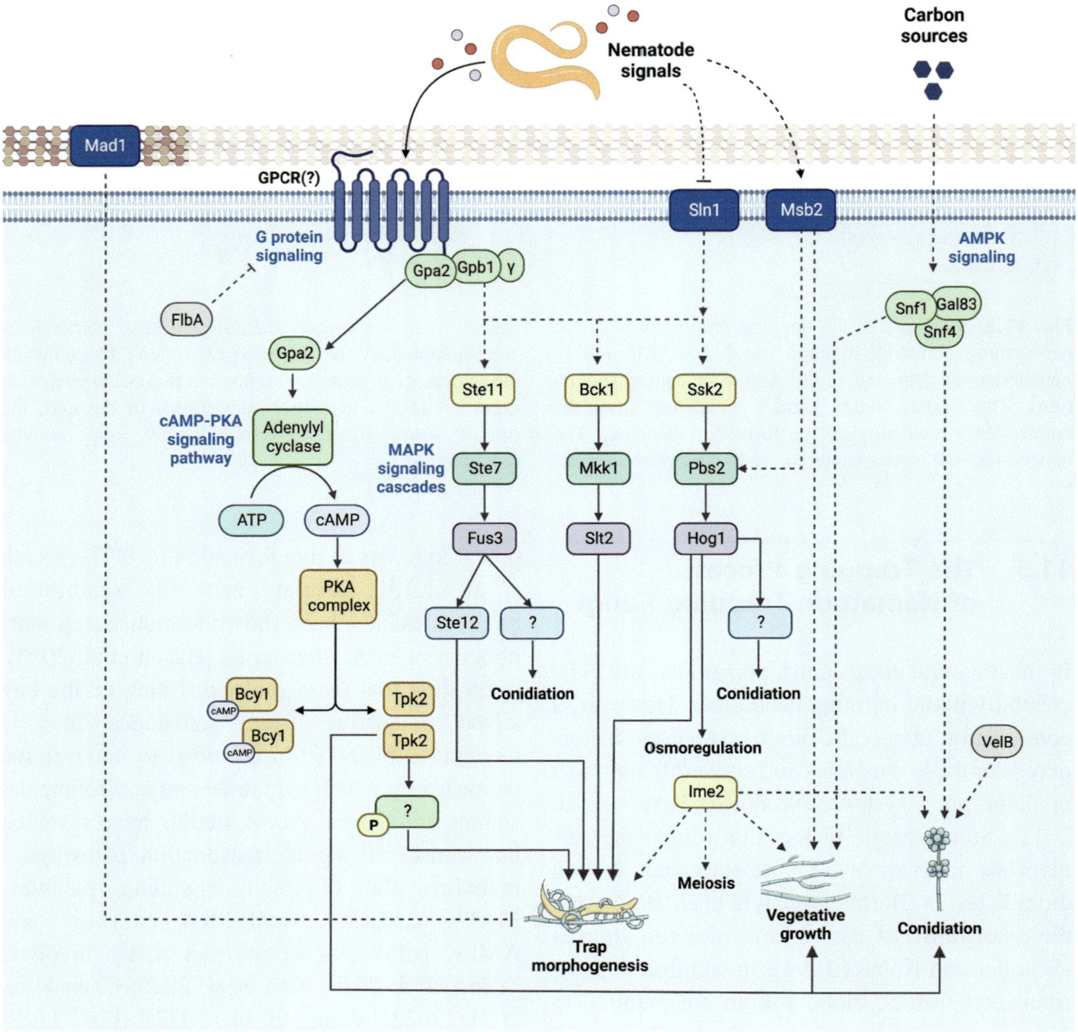

Fig. 11.3 Overall signaling pathways involved in trap morphogenesis, vegetative growth, and conidiation processes in *Arthrobotrys oligospora*

components resulted in defects in colony growth and trap development. Additionally, mutants lacking *STE7* and *FUS3* displayed conidiation defects, supporting that there might be another downstream target of Fus3 responsible for conidiation. Another MAPK pathway involved in nematode sensing is the high osmolarity glycerol (HOG) pathway, which is involved in the yeast response to altered osmolarity in the environment (Rodríguez-Peña et al. 2010). The functions of *HOG1* orthologs differ among fungal species. For example, MAPK *HOG1* controls the osmoadaptive response in *S. cerevisiae* (Reed et al. 1987), whereas its ortholog in *Neurospora crassa* is involved in nutrient sensing (Huberman et al. 2017). Deletion of *HOG1* and *MSB2* (a sensor responsible for osmolarity (Tatebayashi et al. 2007)) orthologs in *A. oligospora* resulted in diminished trap formation ability (Kuo et al. 2020). Moreover, a *hog1* deletion mutant also exhibited defective conidiation, indicating that there might be downstream targets for *HOG1* that regulate signaling transduction for conidiation (Fig. 11.3). Aside from the MAPK pathways, a non-classical pathway may be involved in prey sensing. This pathway has been identified in several fungi and it is known to regulate meiosis, vegetative growth, sporulation, and environmental adaptation (Irniger 2011). In *A. oligospora*, *Ao*Ime2 was identified and characterized as playing roles in multiple developmental stages as a meiosis regulator. Mutant lines in which the *Ao*Ime2 gene has been deleted exhibit defects in meiosis and display a lower number of nuclei in hyphae and conidia, as well as fewer traps with lower trapping ability (Xie et al. 2020) (Fig. 11.3). In the cAMP-PKA pathway, deletion of the PKA catalytic subunit gene *TPK2* resulted in complete abolition of trap formation as well as conidiation defects (Chen et al. 2022). Addition of cAMP restored the trap-forming ability of a *gpa2* mutant but not a *gpb1* mutant, indicating that *GPA2* regulates the cAMP-PKA pathway whereas *GPB1* is involved in the MAPK pathway. Taken together, these results indicate that the cAMP-PKA pathway plays a crucial role in sensing nematodes and trap morphogenesis among NTF and that G-protein coupled receptors (GPCR) operating upstream of the MAPK and cAMP-PKA signaling pathways might act as the receptors for sensing prey (Chen et al. 2022) (Fig. 11.3).

In addition to sensing prey, detection of other energy sources is also crucial for NTF to determine if it is necessary to initiate trap morphogenesis. AMP-activated protein kinase (AMPK) is conserved among eukaryotes and plays a role as an energy sensor, especially for sensing carbon sources (Mihaylova and Shaw 2011; Hedbacker and Carlson 2008). In *S. cerevisiae*, the AMPK complex comprises three subunits: catalytic α-subunit SNF1; regulatory β-subunits GAL83, SIP1, and SIP2; and γ-subunit SNF4. A recent study identified their homologs in *A. oligospora*, i.e., the α-subunit *Ao*SNF1, β-subunit *Ao*GAL83, and γ-subunit *Ao*SNF4 and demonstrated that this pathway plays a role in carbon source utilization, sporulation, lipid synthetic regulation, and pathogenicity (Wang et al. 2022) (Fig. 11.3).

Contact between a predatory fungus and nematode prey is crucial to triggering predation and trap morphogenesis. Furthermore, traps only work upon direct contact with living and motile nematodes, since contact with dead or immobile nematodes does not trigger trap development (Tunlid et al. 1992). Consequently, sessile fungi need to induce nematode contact by attracting them. Nematodes are attracted to the mycelia of *Arthrobotrys* and *Monacrosporium* species (Balan et al. 1974; Field and Webster 1977; Saxena et al. 1987; Warcup 1957), and mycelia hosting traps have been shown to lure even more prospective prey (Jansson 1982). One of the most attractive compounds used by *A. oligospora* to attract nematodes is 3-methyl-2-butenoate (MMB), which mimics the sex and food cue of nematodes (Hsueh et al. 2017). Aside from MMB, a methylated salicylic acid 6-MSA deployed at hyphal tips by another NTF *A. flagrans* attracts *C. elegans* (Yu et al. 2021). Interestingly, 6-MSA is an upstream product of the biosynthetic pathway for arthrosporol, an extract from *A. oligospora* that negatively regulates trap formation. Ascarosides inhibit polyketide synthases involved in the biosynthesis of arthrosporol (ArtA to ArtD), thereby reducing its production and triggering trap formation. Moreover, expression of artA-cluster genes increases after nematodes have been trapped, thus avoiding unnecessary trap formation (Yu et al. 2021).

When an NTF switches to predatory behavior, trap formation is initiated at various sites along its vegetative hyphae (Nordbring-Hertz et al. 1989). The adhesive nets formed by *A. oligospora* are built via several steps involving hyphal fusion events. The initial cells of fungal traps (trap initials) are distinguishable at an early stage due to their possession of numerous dense bodies (Veenhuis et al. 1985a). These cytosolic organelles are absent in other fungal cells and bud from special regions of the endoplasmic reticulum. The role of these dense bodies remains unknown, but they are thought to be involved in prey penetration since their number significantly declines upon a trap piercing the nematode cuticle (Veenhuis et al. 1989). Traps initially grow perpendicular to the parental hypha (Fig. 11.4a) and then, as they elongate, they arch either backward or forward until they fuse with another vegetative hyphal cell (Nordbring-Hertz 2004) (Fig. 11.4b). As the trap initial cells elongate and curve, they subdivide into three cells to give rise to a loop (Fig. 11.4b), which is formed in about 1–2 h and has a diameter of about 25μm. Each cell in a loop can act as a new trap initial (Fig. 11.4c), subsequently resulting in a new loop that anastomoses to either other vegetative hyphae or other existing loops (Fig. 11.4d). Since these loops are randomly formed in all possible directions, the final traps of *A. oligospora* resemble intricate three-dimensional networks (Fig. 11.4e). A recent study revealed a potential gene involved in the cell recognition and fusion steps of trap formation. This gene, *sofT*, is an ortholog of SO (responsible for cell fusion in *N. crassa* (Fleißner et al. 2005)) and it is required for anastomosis in the NTF *A. flagrans* (Youssar et al. 2019). Adhesive networks developed by *A. flagrans* mutants lacking *sofT* fail to recognize and fuse to basal or neighboring cells, resulting in incomplete loops. The striatin-interacting phosphatase and kinase (STRIPAK) complex, a conserved signaling hub in eukaryotes (Hwang and Pallas 2014; Kück et al. 2019) that regulates multicellular development, cell fusion, septation, and pathogenicity in filamentous fungi (Kück et al. 2016), is also involved in trap formation. In *A. flagrans*, one of its components named SipC is involved in growth, trap morphogenesis, and conidiation (Wernet et al. 2022). Sipc deficiency results in the formation of unusual but still functional traps. One phenotype displayed by *sipc* deletion mutants results in trap initials growing back into vegetative hyphae due to delayed actin formation, which accumulates at the septa separating the trap cells. That study also demonstrated that long microtubules lie on the inner side of the trap, with disruption of these microtubules resulting in unusual trap morphogenesis. Additionally, a recent work studied regulators of G-protein signaling (RGSs) that negatively regulate G-protein signaling pathways in *A. oligospora* (Ma et al. 2021). Seven putative RGSs were identified and their deletions were found to reduce trap formation. Among them, the *AoflbA* mutant line not only exhibited the strongest phenotype but also lacked conidiation ability.

Ultimately, the adhesive nets contain numerous dense bodies that increase in number as the trap matures (Nordbring-Hertz 2004; Veenhuis et al. 1985b). Apart from dense bodies, trap cells differ from the rest of the mycelium in that they are surrounded by an extracellular adhesive layer, formed by polymer fibrils composed of proteins and carbohydrates (Tunlid et al. 1992; Veenhuis et al. 1985b; Tunlid and Jansson 1991). When a trap contacts a nematode, the structure of this adhesive layer changes; the polymer fibrils that are initially loosely packed and randomly distributed become densely packed and reorganize to align perpendicularly to the nematode surface (Veenhuis et al. 1985a; Tunlid and Jansson 1991). These structural changes result in a bridging structure connecting the nematode's cuticle to the cell wall of the fungal trap. Since this process occurs at multiple sites along the space between the two organisms, the trap cell wall becomes effectively anchored to the nematode cuticle, thereby immobilizing the prey. The adhesion between fungal adhesive traps and the nematode cuticle was long assumed to be mediated by lectins and, more specifically, by an N-acetylgalactosamine-specific lectin (AOL) present in the traps of *A. oligospora* (Tunlid et al. 1992; Nordbring-Hertz and Mattiasson 1979). However, further analyses have shown that the predatory ability of AOL-deficient mutants was similar to that observed for the wild type (Balogh et al. 2003). More recent studies on the polymers of the adhesive layer have indicated that adhesion is mediated by proteins or protein-containing substances (Wang et al. 2015). The entomopathogenic fungus *Metarhizium anisopliae* (sometimes as *M. robertsii*) (Clavicipitaceae) produces proteins called MAD1 and MAD2 adhesins to enable it to stick to the surface of insects. A homolog of MAD1 in *A. oligospora*, *Ao*Mad1, has been identified as a factor contributing to its pathogenicity (Liang et al. 2015). *Ao*Mad1 is located at the cell wall,

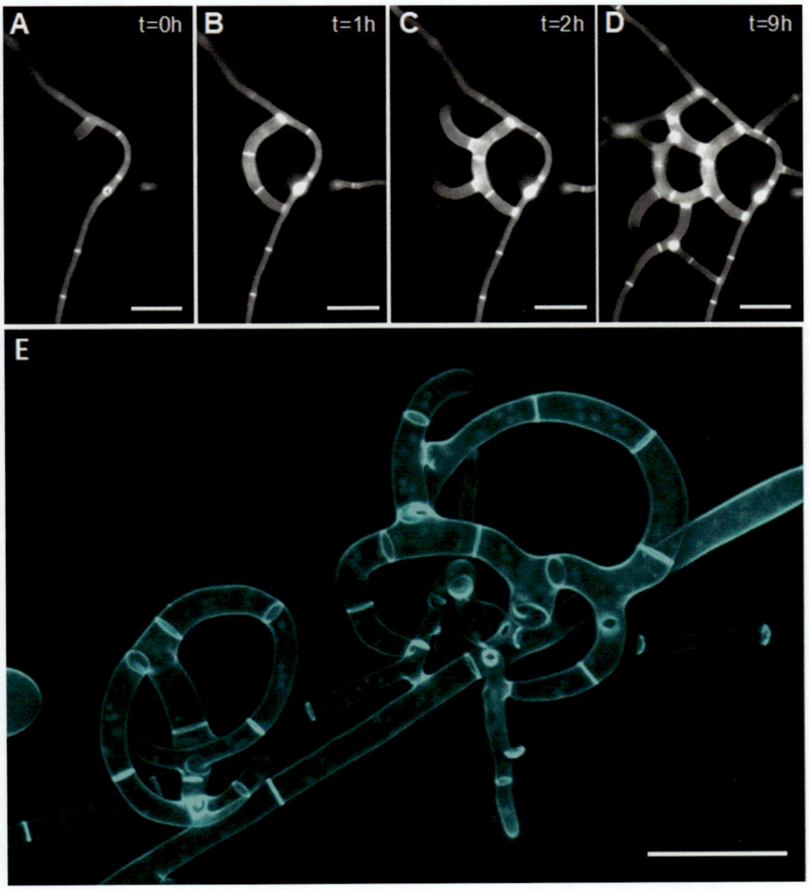

Fig. 11.4 Trap morphogenesis by *Arthrobotrys oligospora*. A trap grows perpendicular to a vegetative hypha (**a**), forming an initial three-cell loop by fusing back into the parent hypha (**b**). Cells of the first loop act as trap initials (**c**), giving rise to new loops (**d**). The final trap configuration comprises several loops that have grown in different directions, resulting in three-dimensional adhesive nets (**e**). Fungal cell walls in fig. (**a**) to (**d**) were stained by SCRI Renaissance 2200 (SR2200) and were photographed by MD ImagXpress Micro L. Fungal cell wall in fig. (**e**) was stained by Calcofluor White (CFW) and was photographed by ZEISS LSM 980 inverted confocal microscope. Scale bars, 25μm

and mutants in which the respective gene has been deleted lack the extracellular layer; as such they exhibit a smooth trap surface. However, the traps of those mutants were still functional, and, in fact, the number of traps increased in *Aomad1* deletion mutants. These results suggest that key additional components contributing to the adhesiveness of trap cells of *A. oligospora* remain to be defined.

The immobilized nematode is still protected by its cuticle so, in order to feed on their prey, NTF first have to break through this barrier. The nematode cuticle consists of three layers of proteins and nematode-specific collagens, and its structure varies at different life-stages and among species (Cox et al. 1981; Page and Johnstone 2007). NTF deploy both enzymatic degradation and mechanical pressure to penetrate this barrier (Nordbring-Hertz et al. 2011; Yang et al. 2007). Enzymes for degrading the cuticle include collagenases (Lopez-Llorca et al. 2007), hydrolytic enzymes such as those identified in *Arthrobotrys* species (Schenck et al. 1975; Tosi et al. 2002), and serine proteases (Yang et al. 2007). Indeed, serine protease inhibitors suppress the ability of fungi to kill nematodes (Wang et al. 2014), and the nematocidal activity of these proteases varies among nematodes, indicating they exhibit specific host affinity (Tunlid et al. 1994a; Wang et al. 2006). Interestingly, the genome of *A. oligospora* contains lots of protease genes (Yang et al. 2011) similar to those involved in host infection possessed by other pathogenic fungi (Monod et al. 2002). NTF also secrete collagenase and subtilisin-like extracellular serine proteases (PII) to degrade the cuticle (Tosi et al. 2002; Tunlid et al. 1994a; Minglian et al. 2004). The protease inhibitor phenylmethylsulfonyl fluoride (PMSF) was found to significantly reduce the ability of NTF to immobilize nematodes (Tunlid and Jansson 1991; Tunlid et al. 1994b). However, PII gene expression was not observed during the cuticle penetration stage, but during colonization and digestion, implying that PII may be used to decompose the cuticle from the inside. Serine proteases are highly expressed in trophic hyphae growing inside infected nematodes, indicating that these enzymes might play a role in digesting nematode prey (Åhman et al. 2002). Aside from serine proteases, a fungal lectin, AOL is also accumulated in trophic hyphae (Rosen et al. 1997). However, further functional studies on these highly expressed enzymes will be required to identify critical players for nematode digestions. In addition, several genes are known to be involved in serine protease production in *A. oligospora*, including protein kinases such as non-classical MAPK Ime2 and AMPK Snf1 (Wang et al. 2022; Xie et al. 2020), the RGSs *Ao*Rgs and *Ao*FlbA (Ma et al. 2021), and velvet protein *Ao*VelB (Zhang et al. 2019).

Penetration of the cuticle is mediated by the formation of a penetration tube, a structure that emerges from the trap cells which decompose the nematode cuticle and pierce into the worm body. The newly-forming cell wall can pass through the original trap cell wall due to partial hydrolysis, resulting in a small penetration tube (Veenhuis et al. 1985b). After penetration of the nematodes, an enlarged hyphal structure can be observed. This enlarged hyphal structure was named the infection bulb, which was first observed and described in *A. oligospora* in 1955 (Shepherd 1955). To investigate the detailed invasion process further, we constructed a cytosolic-GFP labeled *A. oligospora* strain to conduct live-imaging. We observed that the penetration tube became anchored into the worm body within 10 minutes after contacting *C. elegans* (Fig. 11.5a). Then, the invasion hypha started to swell and developed into large spherical infection bulb within 20 minutes (Fig. 11.5b) As the infection bulb matured, trophic hyphae developed and grew rapidly inside the nematodes. (Fig. 11.5d and e).

NTF conidium formation may also be affected by interactions with their prey. For instance, under nutrient-limited conditions, *A. oligospora* forms very few nodes of conidia on conidiophores on the plate surface and tend to produce the majority of conidiophores on aerial hyphae. In contrast, after consuming nematodes, it produces multiple nodal conidia on the plate surface rather than in the air. Though several works exploring nematode-sensing signaling pathways have shown that mutants lacking Fus3, Ste7, Slt2, *Ao*FlbA, and *Ao*Snf1 displayed impaired conidiation (Chen et al. 2021; Wang et al. 2022; Ma et al. 2021; Zhen et al. 2018), very little is known about how NTF regulate the conidiation process in response to the presence of nematodes. A recent work showed that the velvet proteins VelB and VosA regulate the conidation process in NTF (Park and Yu 2012). In *A. oligospora*, deletion of *AoVelB* and *AoVosA* resulted in no difference in growth rate or conidiation, but the *AoVelB* deletion mutants completely lacked conidiation ability even though they could still form normal conidiophores (Zhang et al. 2019). Significantly, AoVelB is also involved in trap formation and serine protease production. Ime2 that is involved in trap morphogenesis and serine protease production also affects conidium formation (Xie et al. 2020). *Aoime2* deletion mutants produce fewer and abnormal conidia with a reduced number of nuclei, and sporulation-related genes are down-regulated, revealing a conserved role for Ime2 in conidium regulation. These studies suggest that there might be an intriguing link between conidium and trap formation.

The plasticity of the predatory activities observed among wild isolates of *A. oligospora* hints at a strong genetic factor driving this diversity (Yang et al. 2020). Possibilities include significant differences in trap morphology and efficiency, as well as in the response to trap-triggering stimuli among strains. The complexity of nematode and NTF interactions has been further observed in natural ecosystems. In nature, traps are rarely observed, unlike the propensity for trapping behavior exhibited by NTF in the laboratory, which ultimately hinders their use as

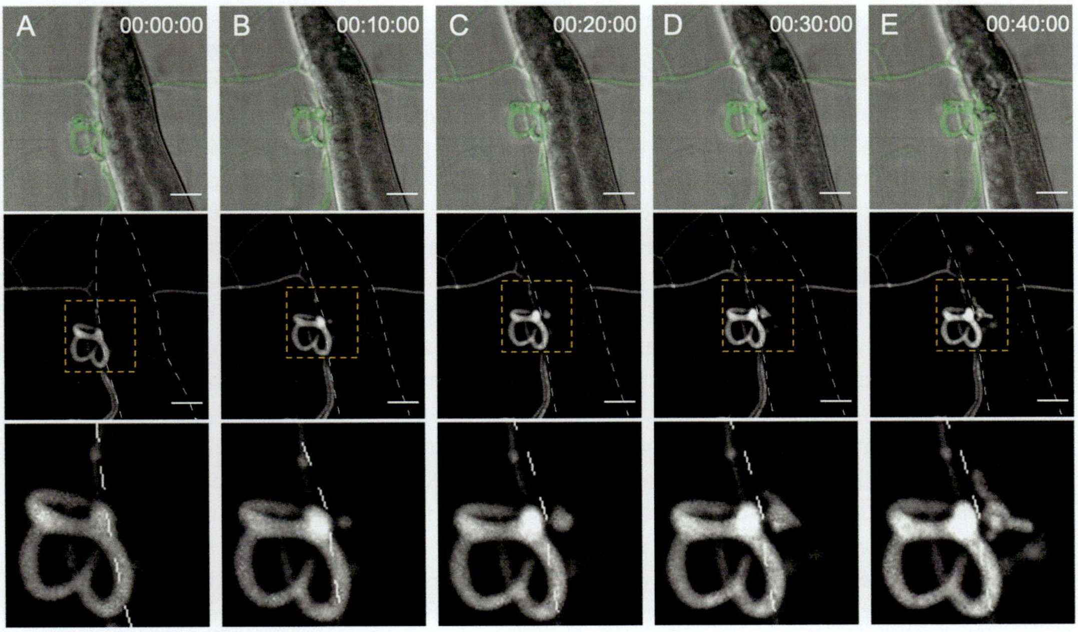

Fig. 11.5 Invasion process of cytosolic-GFP labeled *Arthrobotrys oligospora* growing into *Caenorhabditis elegans*. Penetration tube grew into the nematode (**a**) and the infection bulb developed well within 20 min (**b** and **c**), which eventually formed trophic hyphae to colonize and consume nematode (**d** and **e**). Scale bars, 30μm. The invasion process was recorded by Andor Revolution WD with Lambda 20X water lens. Images were analyzed with MetaMorph

biocontrol agents against nematodes. More general biological questions such as when and why NTF evolved to develop traps remain to be addressed. In order to answer some of these questions, whole genomes of several nematophagous fungi have been sequenced, including the network-making NTF *A. oligospora* strain TWF154 and ATCC24927 (Yang et al. 2011, 2020) and *A. flagrans* strain CIM1 (Zhang et al. 2020a), the knob-making NTF *Dactylellina haptotyla* strain CBS 200.50 (as *Monacrosporium haptotylum*) (Meerupati et al. 2013), the constricting ring-producing NTF *D. stenobrocha* strain CGMCC 3.6768 (Liu et al. 2014), the endoparasitic nematophagous fungus *Esteya vermicola* strain CBS115803 (Ophiostomataceae) (Wang et al. 2018), and the toxin-producing *P. ostreatus* strain PC9 (Lee et al. 2021). We believe these genomic resources, combined with additional molecular mechanistic studies will enable scientists to unravel further the mysteries behind the interactions between NTF and nematodes.

11.4 *Caenorhabditis elegans* as a Model to Study Fungal–Nematode Interactions

Phylum Nematoda is the most abundant on the planet, occurring all over the world and cohabiting with a diversity of organisms. It has been estimated that over a million nematode species inhabit the Earth, and yet only 30,000 nematodes have been described to date (Kiontke and Fitch 2013; Lorenzen 1994). A large proportion of nematodes are parasitic, causing severe diseases in humans, animals, and plants (Bethony et al. 2006; Jasmer et al. 2003) with those affecting plants resulting in significant agricultural losses (Jones et al. 2013; Trudgill and Blok 2001). For example, root-knot nematodes and another >4000 plant-parasitic nematodes cost at least US$80 billion per year in crop losses (Nicol et al. 2011).

Soil nematodes can be found worldwide, with the highest nematode density being found in the tundra and the lowest in deserts, ranging from 2000 to 80 worms per 100 g of soil, respectively (van den Hoogen et al. 2019). To survive in these environments, they have diversified considerably in terms of morphological features such as length (from millimeters to several meters), anatomy (cuticular decorations, vulva, and mouthparts), lifecycle (hermaphroditism, larval stages, and parasitism), and feeding habits (bacterivory, fungivory, herbivory, omnivory, and predatory) (Kiontke and Fitch 2013; Ridall and Ingels 2021; Yeates et al. 1993). All of these nematodes interact with their respective neighbors. For example, the predatory nematode *Pristionchus* preys on other nematodes or even on their own species, and the roundworm *Ascaris* infects 1.2 billion people through a fecal-oral cycle. *Oscheius* is one of the most common soil nematodes, feeding on bacteria (Dold and Holland 2011; Felix 2006; Sommer 2006) and playing a central role in soil ecosystems, so that it is used as a biological indicator of soil environments (Ferris 2010; Neher 2001).

The interactions between fungi and nematodes are diverse and cover the full range, spanning predator to prey (Yang et al. 2020; Baynes et al. 2012; Giannakis and Sanders 1990). Fungivorous nematodes play diverse roles in ecosystems (Baynes et al. 2012). The nematodes *Aphelenchoides* spp. and *Aphelenchus avenae* feed on pathogenic fungi, thus aiding host plants fight fungal infections, and they have, therefore, proven of interest as biocontrol agents (Lagerlof et al. 2011; Okada 2006). Other nematodes feed on plant-beneficial fungi such as of the genera *Gliocladium* and *Trichoderma* (Hypocreales), thereby limiting the beneficial effects that these fungi have on plants (Hasna et al. 2007). On the other hand, nematodes can also be the host or prey of fungi. The nematode-pathogenic fungi *Aspergillus spp. Fusarium* spp., and *Penicillium* spp. can infect various species of nematodes (Zhang et al. 2020b). Additionally, other fungi (such as NTF) are carnivorous, actively preying on nematodes using complex strategies. These nematophagous fungi share their ecological niche with the well-developed model nematode *C. elegans*, providing an unrivalled opportunity to study predator–prey coevolution in detail using the power of the comprehensive toolbox associated with this model organism.

Some characteristics, alluded to earlier, that make a model organism ideal for laboratory work are a short lifecycle, ease of cultivation, small size, and large populations. *C. elegans* possesses all of these characteristics, which is why it has become such a well-studied model organism. *C. elegans* can be fed on *Escherichia coli* both in agar plates and liquid cultures, making it easy to culture. This nematode is primarily self-fertilized, hermaphroditic, and generates brood sizes of ~300 within a 3-day lifecycle (Corsi et al. 2015), enabling researchers to conduct genomics analyses easily. In addition, *C. elegans* has a fully differentiated body of 1 mm in length comprising only a thousand or so cells, of which 302 in adult hermaphrodites are sensory-, inter-, and motor neurons (White et al. 1986). Furthermore, a unique characteristic that differentiates these nematodes from other such organisms is that they have transparent bodies throughout their lifecycle, thus permitting scientists to observe their insides through a microscope (Corsi et al. 2015) (Fig. 11.6).

For all of these reasons, Sydney Brenner and his collaborators began developing *C. elegans* as a novel model organism in the 1960s. They conducted the first EMS mutagenesis screening and characterized its genetics in 1973 (Brenner 1974). Decades of extensive study since then have led to major scientific findings, including the genetic regulation of development and programmed cell death (Sulston and Horvitz 1977; Ellis and Horvitz 1986; Yuan and Horvitz 1992), the mechanism of RNA interference (Fire et al. 1998), and construction of a fluorescent protein as a reporter (Chalfie et al. 1994). Moreover, *C. elegans* became the first multicellular organism to have its whole genome sequenced, representing the strongest genetic resource yet and opening up a new era in genetic research (C. elegans Sequencing Consortium 1998).

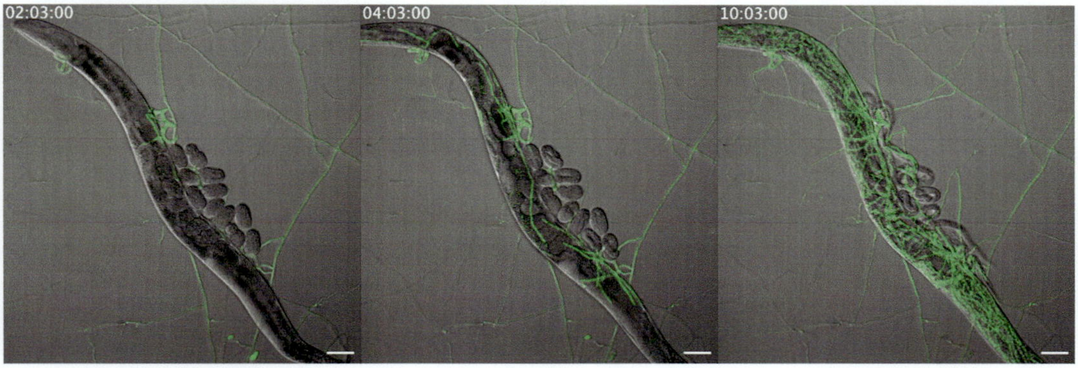

Fig. 11.6 Live-imaging of a cytosolic-GFP labeled *Arthrobotrys oligospora* growing inside *Caenorhabditis elegans*. The colonization process was recorded by Andor Revolution WD and was analyzed with MetaMorph. Scale bars, 50μm

Today, WormBase (https://wormbase.org/) provides an annotated *C. elegans* genome with 49,176 genes, including 19,981 protein-coding genes. This extensive scientific resource collates knowledge on *C. elegans*, including gene descriptions, expression data, single-cell neuronal expression maps, ontology, protein features, and related publications.

Its early establishment as a model species resulted in several methods and tools being developed to facilitate the study of *C. elegans* in laboratories. For instance, microinjection techniques allow researchers to conduct genetic manipulation on these worms by injecting DNA (including PCR products or plasmids) directly into their gonads, resulting in extrachromosomal arrays that remain heritable (Stinchcomb et al. 1985). The diverse applications of CRISPR-Cas gene-editing systems are also widely being deployed (Dickinson and Goldstein 2016). *C. elegans* also provides considerable resources for neurobiology research, with the whole-animal and sex-specific connectomes, as well as a neurotransmitter map, quantitatively illustrating information flow (Cook et al. 2019; Gendrel et al. 2016; Pereira et al. 2015; Serrano-Saiz et al. 2013). The mating behavior of male *C. elegans* is mostly encode by male-specific neurons. Males have 385 neurons while hermaphrodites only have 302 neurons and 81 out of the 91 male specific neurons are involved in the mating circuit (Lints and Hall 2009). The sexual dimorphic nociceptive behavior arose from sex-specific connection in sensory to interneuron that triggers dimorphic interneuron responses (Pechuk et al. 2022). Recently, whole-brain GCaMP activity recording *in vivo* was achieved by means of a polychromatic neuronal identification tool (Yemini et al. 2021).

Other resources include extensive databases, such as the *Caenorhabditis* natural diversity resource (CaeNDR). This database comprises >1500 wild-isolate strains collected worldwide and their respective genomes, enabling study of natural variations and the performance of genome-wide association mapping with web-based tools (https://caendr.org) (Cook et al. 2017). From an evolutionary and ecological perspective, CaeNDR alone has wild-isolate strains covering large genetic and geographic distributions. Population genetics and genome-wide association studies on these wild isolates have revealed that natural variations influence several *C. elegans* behaviors, such as foraging, pheromone responses, starvation resistance, and metabolism (Cook et al. 2017; Fox et al. 2022; Webster et al. 2022; Greene et al. 2016; Lee et al. 2019). In nature, wild *C. elegans* are often found in the dauer stage that occurs in response to stress, starvation, and high population density (Barriere and Felix 2007), indicating that the growth conditions of *C. elegans* are far from optimal in their natural habitats. These habitats include both urban and rural areas, and several substrates such as rotten fruits and flowers, soil, arthropods, and mollusks, supporting that *C. elegans* forms different types of biological interactions with various species in nature (Frezal and Felix 2015). Overall then, the efforts expended in studying *C. elegans* have resulted in a diversity of tools and rich resources, rendering it an excellent model for studying its interactions with other organisms, including fungi.

11.5 The Behavior of *C. elegans* and Its Responses to Biological Interactions

C. elegans is often the target of infection by pathogenic viruses, bacteria, and fungi, as well as being the prey of NTF. Together with the various tools and resources specific to *C. elegans*, this susceptibility makes it an excellent model for studying pathogen–host and predator–prey interactions. Some of these pathogens are specific to *C. elegans*, e.g., *Orsay* virus only infects this species and no other in the genus *Caenorhabditis* (Felix et al. 2011). *Myzocytiopsis humicola*, one of the oomycetes with chitin contained in cellulin granules, is a natural pathogen that has been isolated mainly from wild *C. elegans* strains. Zoospores of this oomycete develop adhesive buds that attach to the nematode cuticle, which then penetrate the nematode body through narrow germ tubes. Each zoospore produces a thallus that becomes compartmentalized at maturity to appear as a string of pearl-like structures to fill the entire organism (Osman et al. 2018). Infection triggers a pathogen-specific transcriptomic response in *C. elegans*, characterized by upregulation of the chitinase-like (*chil*) gene family upon exposure to this pathogen (Osman et al. 2018). Further research has revealed that oomycete recognition by *C. elegans* is mediated by chemosensory neurons that modulate upregulation of *chil* genes. This allows the nematode to limit zoospore attachment and increase its survival via the defense mechanism of inducing *chil* genes in the epidermis (Fasseas et al. 2021).

Two major fungal pathogens that infect *C. elegans*, as well as other nematodes, in nature are *Drechmeria* and *Harposporium*. The ascomycete *Drechmeria coniospora* (Ophiocordycipitaceae) differs from oomycetes in that its conidia target natural openings in the nematode, such as the mouth and vulva rather than penetrating the cuticle (Felix and Duveau 2012; Schulenburg and Felix 2017). Infection by *Drechmeria* kills the nematodes within 4 days and, once inside its host, the fungus releases thousands of spores through the cuticle to the outside environment (Jansson 1994). Transcriptomic characterization of *C. elegans* during *Drechmeria* infection has revealed a fundamental defense response that involves antimicrobial peptides, including two major families of neuropeptide-like proteins (NLPs) and structurally-related caenacins (CNCs) (Pujol et al. 2008).

Pathogen–host interactions have been studied extensively using *C. elegans* and their natural pathogens. Comprehensive analyses of transcriptomic data have suggested that genes encoding antimicrobial peptides (AMPs) are upregulated in response to both bacterial and fungal infections *in vitro*, demonstrating antimicrobial ability at least under those conditions (Engelmann et al. 2011; Couillault et al. 2004). In addition, several conserved pathways regulating the innate immune response in *C. elegans* have been identified, including the p38 and PMK-1 MAPK pathways, as well as the DAF-2/DAF-16 and TGF-β pathways, which are activated following both bacterial and fungal infections, though some genes only play a role in pathogen-specific responses (Pujol et al. 2008; Engelmann et al. 2011; Couillault et al. 2004; Kawli and Tan 2008; Zugasti and Ewbank 2009). In order to avoid pathogen infections, *C. elegans* has acquired avoidance strategies against bacterial pathogens such as *Pseudomonas aeruginosa* strain PA14. This avoidance behavior is inherited across up to four generations by establishing transgenerational epigenetic memory via a PIWI-interacting RNA (piRNA) mechanism that regulates neuronal gene expression (Moore et al. 2019). This learned behavior involves foreign small RNA from the PA14 strain pairing with the nematode's *maco-1* gene to elicit avoidance behavior (Kaletsky et al. 2020). This and other stress-induced learning behaviors might release *C. elegans* from the stress of heat-shock, starvation, and infections to increase survival and reproductive fitness (Moore et al. 2019; Wan et al. 2021; Zhang et al. 2021; Rechavi et al. 2014).

The predator–prey interactions of NTF and nematodes rely on bilateral communication. The prey *C. elegans* produces ascarosides pheromones, small molecules defined as glycosides of the dideoxysugar ascarylose that differ mainly in the length of the ascarylose carbon chain (Ludewig and Schroeder 2013). They regulate dauer formation, mating, development, and social interactions (Butcher et al. 2007; Srinivasan et al. 2008; Macosko et al. 2009; Kim et al. 2009). Interestingly, NTF such as *A. oligospora* can sense several types of ascarosides such as ascr#1, ascr#3, and ascr#7 triggering the most significant trap-producing response (Hsueh et al. 2017). The predator *A. oligospora* produces several volatile organic compounds that are attractive to worms, including MMB (Hsueh et al. 2017). This fatal attractant is a thousand-fold stronger than the well-studied *C. elegans* attractant isoamyl alcohol, eliciting both sex- and stage-specific nematode attraction and disrupting mating behavior. This finding indicates that MMB might mimic a *Caenorhabditis* sex hormone to lure the nematodes (Hsueh et al. 2017). Behavioral responses to MMB are regulated by the *C. elegans* AWC olfactory neuron. Forward genetic screening has identified two mutants that exhibit strongly reduced chemotaxis toward MMB, both of which possessed mutations in the same transcription factor, odr-7, implying that chemotactic MMB attraction might involve multiple redundant receptors (Hsueh et al. 2017).

Predator–prey interactions are universal across the tree of life, playing a critical role in stabilizing ecosystems and maintaining diversity and population sizes. To survive, prey species have evolved strategies to lower predation, but predatory species have in turn coevolved more efficient predatory behaviors. The resulting feedback between predator and prey species has resulted in an evolutionary arms race (Burak et al. 2018). Classical studies of predator–prey systems were focused on population dynamics, with various mathematical models showing that population size oscillates between species (Wilks and Lotka 1941; Volterra 1926). For instance, the lynx–hare cycle showed a counterclockwise pattern, whereby the prey population peaks prior to that of the predator (Elton and Nicholson 1942), rotifer–algal populations exhibit an antiphase pattern (Becks et al. 2010; Yoshida et al. 2003), and the phage-bacteria model uncovered cryptic cycles (Bohannan and Lenski 1999). Recent studies have indicated that evolution of one species, as well as the coevolution of two species, can shape the pattern of predator–prey cycles (Cortez and Weitz 2014). Over the past few decades, scientists have studied predator–prey interactions at the population level, with these interactions leading to an arms race and often prompting behavioral changes. Nevertheless, the molecular details of such evolutionary races remain elusive. Recently, a study on lion–zebra and cheetah–impala interactions demonstrated that the predators have greater muscle power, resulting in better athletic capabilities, whereas the prey have better maneuverability, giving them the ability to evade their major predator and match their locomotion. That study has provided physiological evidence that predator–prey coevolution leads to a sustainable success rate (both in terms of capture and evasion), and thus survival of both species (Wilson et al. 2018).

The predator–prey system of the NTF *A. oligospora* and the nematode *C. elegans* facilitates study of their molecular interactions from the perspectives of both predator and prey. New tools developed for *A. oligospora*—such as genetic knockout, RNA sequencing, and fluorescence imaging—permit several aspects of this fungus to be studied in detail not previously attainable. Together with the discovery of an efficient model strain of NTF (Yang et al. 2020) and the long-established resources available for *C. elegans*, these tools will serve as the perfect basis for studying the enigmatic molecular mechanism driving the nematode-trapping process.

Moreover, how NTF sense their prey and respond to these signals, as well as how the prey responds to predators, including chemotaxis and defense responses, remain to be explored. We believe that this unique predator–prey model might shed light on the molecular details of cross-kingdom communication and how predator–prey interactions have coevolved.

Acknowledgments The authors thank Meredith Blackwell for providing the overview of this chapter.

References

Åhman J et al (2002) Improving the pathogenicity of a nematode-trapping fungus by genetic engineering of a subtilisin with nematotoxic activity. Appl Environ Microbiol 68:3408–3415

Ahrén D, Tunlid A (2003) Evolution of parasitism in nematode-trapping fungi. J Nematol 35:194–197

Alvaro CG, Thorner J (2016) Heterotrimeric G protein-coupled receptor signaling in yeast mating pheromone response. J Biol Chem 291:7788–7795

Balan J, Križková L, Nemec P, Vollek V (1974) Production of nematode-attracting and nematicidal substances by predacious fungi. Folia Microbiol 19:512–519

Balogh J, Tunlid A, Rosén S (2003) Deletion of a lectin gene does not affect the phenotype of the nematode-trapping fungus *Arthrobotrys oligospora*. Fungal Genet Biol 39:128–135

Barriere A, Felix MA (2007) Temporal dynamics and linkage disequilibrium in natural Caenorhabditis elegans populations. Genetics 176:999–1011

Barron GL (2003) Predatory fungi, wood decay, and the carbon cycle. Biodiversity 4:3–9

Bartnicki-Garcia S, Eren J, Pramer D (1964) Carbon dioxide-dependent morphogenesis in *Arthrobotrys conoides*. Nature 204:804–804

Baynes MA et al (2012) A mutualistic interaction between a fungivorous nematode and a fungus within the endophytic community of Bromus tectorum. Fungal Ecol 5: 610–623

Becks L, Ellner SP, Jones LE, Hairston NG Jr (2010) Reduction of adaptive genetic diversity radically alters eco-evolutionary community dynamics. Ecol Lett 13: 989–997

Bethony J et al (2006) Soil-transmitted helminth infections: ascariasis, trichuriasis, and hookworm. Lancet 367:1521–1532

Bohannan BJM, Lenski RE (1999) Effect of prey heterogeneity on the response of a model food chain to resource enrichment. Am Nat 153:73–82

Brenner S (1974) The genetics of *Caenorhabditis elegans*. Genetics 77:71–94

Burak MK, Monk JD, Schmitz OJ (2018) Eco-evolutionary dynamics: the predator-prey adaptive play and the ecological theater. Yale J Biol Med 91:481–489

Butcher RA, Fujita M, Schroeder FC, Clardy J (2007) Small-molecule pheromones that control dauer development in Caenorhabditis elegans. Nat Chem Biol 3: 420–422

C. elegans Sequencing Consortium (1998) Genome sequence of the nematode C. elegans: a platform for investigating biology. *Science* 282:2012–2018

Chalfie M, Tu Y, Euskirchen G, Ward WW, Prasher DC (1994) Green fluorescent protein as a marker for gene-expression. Science 263:802–805

Chen S-A, Lin H-C, Schroeder FC, Hsueh Y-P (2021) Prey sensing and response in a nematode-trapping fungus is governed by the MAPK pheromone response pathway. Genetics 217:iyaa008

Chen S-A, Lin H-C, Hsueh Y-P (2022) The cAMP-PKA pathway regulates prey sensing and trap morphogenesis in the nematode-trapping fungus Arthrobotrys oligospora. G3 Genes|Genomes|Genetics. https://doi.org/10.1093/g3journal/jkac217

Cook DE, Zdraljevic S, Roberts JP, Andersen EC (2017) CeNDR, the Caenorhabditis elegans natural diversity resource. Nucleic Acids Res 45:D650–D657

Cook SJ et al (2019) Whole-animal connectomes of both Caenorhabditis elegans sexes. Nature 571:63

Corsaro D et al (2018) New insights from molecular phylogenetics of amoebophagous fungi (Zoopagomycota, Zoopagales). Parasitol Res 117: 157–167

Corsi AK, Wightman B, Chalfie M (2015) A transparent window into biology: a primer on Caenorhabditis elegans. Genetics 200:387–407

Cortez MH, Weitz JS (2014) Coevolution can reverse predator-prey cycles. Proc Natl Acad Sci U S A 111: 7486–7491

Couillault C et al (2004) TLR-independent control of innate immunity in Caenorhabditis elegans by the TIR domain adaptor protein TIR-1, an ortholog of human SARM. Nat Immunol 5:488–494

Cox GN, Kusch M, Edgar RS (1981) Cuticle of *Caenorhabditis elegans*: its isolation and partial characterization. J Cell Biol 90:7–17

de Ulzurrun GV-D, Lee Y-Y, Stajich JE, Schwarz EM, Hsueh Y-P (2020) Genomic analyses of two Italian oyster mushroom *Pleurotus pulmonarius* strains. *bioRxiv*. https://doi.org/10.1101/2020.09.03.281089

Dickinson DJ, Goldstein B (2016) CRISPR-based methods for Caenorhabditis elegans genome engineering. Genetics 202:885–901

Dold C, Holland CV (2011) Ascaris and ascariasis. Microbes Infect 13:632–637

Ellis HM, Horvitz HR (1986) Genetic-control of programmed cell-death in the nematode C-elegans. Cell 44:817–829

Elton C, Nicholson M (1942) Fluctuations in numbers of the muskrat (Ondatra zibethica) in Canada. J Anim Ecol 11:96–126

Engelmann I et al (2011) A comprehensive analysis of gene expression changes provoked by bacterial and fungal infection in *C. elegans*. PLoS One 6:e19055

Fasseas MK et al (2021) Chemosensory neurons modulate the response to oomycete recognition in Caenorhabditis elegans. Cell Rep 34:108604

Felix MA (2006) Oscheius tipulae. WormBook, pp 1–8. 10.1895/wormbook.1.119.1

Felix MA, Duveau F (2012) Population dynamics and habitat sharing of natural populations of Caenorhabditis elegans and C. briggsae. BMC Biol 10:59

Felix MA et al (2011) Natural and experimental infection of Caenorhabditis nematodes by novel viruses related to nodaviruses. PLoS Biol 9:e1000586

Ferris H (2010) Contribution of nematodes to the structure and function of the soil food web. J Nematol 42:63–67

Field JI, Webster J (1977) Traps of predacious fungi attract nematodes. Trans Br Mycol Soc 68:467–469

Fire A et al (1998) Potent and specific genetic interference by double-stranded RNA in Caenorhabditis elegans. Nature 391:806–811

Fleißner A et al (2005) The so locus is required for vegetative cell fusion and postfertilization events in Neurospora crassa. Eukaryot Cell 4:920–930

Fox BW et al (2022) C. elegans as a model for inter-individual variation in metabolism. Nature 607:571

Frezal L, Felix MA (2015) *C. elegans* outside the Petri dish. Elife 4

Friman E, Olsson S, Nordbring-Hertz B (1985) Heavy trap formation by *Arthrobotrys oligospora* in liquid culture. FEMS Microbiol Ecol 31:17–21

Gendrel M, Atlas EG, Hobert O (2016) A cellular and regulatory map of the GABAergic nervous system of C. elegans. Elife 5

Giannakis N, Sanders FE (1990) Interactions between mycophagous nematodes, mycorrhizal and other soil fungi. Agr Ecosyst Environ 29:163–167

Greene JS, Dobosiewicz M, Butcher RA, McGrath PT, Bergmann CI (2016) Regulatory changes in two chemoreceptor genes contribute to a Caenorhabditis elegans QTL for foraging behavior. eLife 5

Hasna MK, Insunza V, Lagerlöf J, Rämert B (2007) Food attraction and population growth of fungivorous nematodes with different fungi. Ann Appl Biol 151: 175–182

Hedbacker K, Carlson M (2008) SNF1/AMPK pathways in yeast. Front Biosci 13:2408

Hsueh YP, Mahanti P, Schroeder FC, Sternberg PW (2013) Nematode-trapping fungi eavesdrop on nematode pheromones. Curr Biol 23:83–86

Hsueh Y-P et al (2017) Nematophagous fungus Arthrobotrys oligospora mimics olfactory cues of sex and food to lure its nematode prey. Elife 6

Huberman LB, Coradetti ST, Glass NL (2017) Network of nutrient-sensing pathways and a conserved kinase cascade integrate osmolarity and carbon sensing in Neurospora crassa. Proc Natl Acad Sci U S A 114: E8665–E8674

Hwang J, Pallas DC (2014) STRIPAK complexes: structure, biological function, and involvement in human diseases. Int J Biochem Cell Biol 47:118–148

Irniger S (2011) The Ime2 protein kinase family in fungi: more duties than just meiosis. Mol Microbiol 80:1–13

Jaffee BA (2003) Correlations between most probable number and activity of nematode-trapping fungi. Phytopathology 93:1599–1605

Jansson HB (1982) Predacity by nematophagous fungi and its relation to the attraction of nematodes. Microb Ecol 8:233–240

Jansson HB (1994) Adhesion of conidia of drechmeria-coniospora to Caenorhabditis-elegans wild-type and mutants. J Nematol 26:430–435

Jasmer DP, Goverse A, Smant G (2003) Parasitic nematode interactions with mammals and plants. Annu Rev Phytopathol 41:245–270

Jones JT et al (2013) Top 10 plant-parasitic nematodes in molecular plant pathology. Mol Plant Pathol 14:946–961

Kaletsky R et al (2020) C. elegans interprets bacterial non-coding RNAs to learn pathogenic avoidance. Nature 586:445–451

Kawli T, Tan MW (2008) Neuroendocrine signals modulate the innate immunity of Caenorhabditis elegans through insulin signaling. Nat Immunol 9:1415–1424

Kim K et al (2009) Two chemoreceptors mediate developmental effects of dauer pheromone in C-elegans. Science 326:994–998

Kiontke K, Fitch DH (2013) Nematodes. Curr Biol 23: R862–R864

Klironomos JN, Hart MM (2001) Food-web dynamics. Animal nitrogen swap for plant carbon. Nature 410: 651–652

Kück U, Beier AM, Teichert I (2016) The composition and function of the striatin-interacting phosphatases and kinases (STRIPAK) complex in fungi. Fungal Genet Biol 90:31–38

Kück U, Radchenko D, Teichert I (2019) STRIPAK, a highly conserved signaling complex, controls multiple eukaryotic cellular and developmental processes and is linked with human diseases. Biol Chem 400:1005–1022

Kuo C-Y, Chen S-A, Hsueh Y-P (2020) The high osmolarity glycerol (HOG) pathway functions in osmosensing, trap morphogenesis and conidiation of the nematode-trapping fungus Arthrobotrys oligospora. J Fungi 6:191

Lagerlof J, Insunza V, Lundegardh B, Ramert B (2011) Interaction between a fungal plant disease, fungivorous nematodes and compost suppressiveness. Acta Agric Scand Sect B Plant Soil Sci 61:372–377

J. R. Lawton (nee Ress) JR (1957) The formation of constricting rings in nematode-catching hyphomycetes grown in pure culture. J Exp Bot 8:50–54

Lee D et al (2019) Selection and gene flow shape niche-associated variation in pheromone response. Nat Ecol Evol 3:1455–1463

Lee C-H et al (2020) Sensory cilia as the Achilles heel of nematodes when attacked by carnivorous mushrooms. Proc Natl Acad Sci 117:6014–6022

Lee YY, Vidal-Diez de Ulzurrun G, Schwarz EM, Stajich JE, Hsueh YP (2021) Genome sequence of the oyster mushroom Pleurotus ostreatus strain PC9. G3 (Bethesda) 11

Li Y et al (2005) Phylogenetics and evolution of nematode-trapping fungi (Orbiliales) estimated from nuclear and protein coding genes. Mycologia 97: 1034–1046

Li L et al (2011) Induction of trap formation in nematode-trapping fungi by a bacterium. FEMS Microbiol Lett 322:157–165

Li J et al (2015) Molecular mechanisms of nematode-nematophagous microbe interactions: basis for biological control of plant-parasitic nematodes. Annu Rev Phytopathol 53:67–95

Li L et al (2016) Nematode-trapping fungi and fungus-associated bacteria interactions: the role of bacterial diketopiperazines and biofilms on *Arthrobotrys oligospora* surface in hyphal morphogenesis. Environ Microbiol 18:3827–3839

Liang L et al (2015) A proposed adhesin AoMad1 helps nematode-trapping fungus Arthrobotrys oligospora recognizing host signals for life-style switching. Fungal Genet Biol 81:172–181

Lints R, Hall DH (2009) Male neuronal support cells, overview. WormAtlas. doi:10.3908/wormatlas.2.9

Liu K, Tian J, Xiang M, Liu X (2012) How carnivorous fungi use three-celled constricting rings to trap nematodes. Protein Cell 3:325–328

Liu K et al (2014) Drechslerella stenobrocha genome illustrates the mechanism of constricting rings and the origin of nematode predation in fungi. BMC Genomics 15:114

Lopez-Llorca LV, Maciá-Vicente JG, Jansson HB (2007) Mode of action and interactions of nematophagous fungi. In: Ciancio A, Mukerji KG (eds) Integrated management and biocontrol of vegetable and grain crops nematodes, vol 2. Springer Netherlands, Dordrecht, pp 51–76

Lorenzen S (1994) The phylogenetic systematics of freeliving nematodes. Ray Society, London

Ludewig AH, Schroeder FC (2013) Ascaroside signaling in *C. elegans*. WormBook, pp 1–22. 10.1895/wormbook.1.155.1

Ma N et al (2021) Functional analysis of seven regulators of G protein signaling (RGSs) in the nematode-trapping fungus Arthrobotrys oligospora. Virulence 12:1825–1840

Macosko EZ et al (2009) A hub-and-spoke circuit drives pheromone attraction and social behaviour in C-elegans. Nature 458:1171–U1110

Meerupati T et al (2013) Genomic mechanisms accounting for the adaptation to parasitism in nematode-trapping fungi. PLoS Genet 9:e1003909

Mihaylova MM, Shaw RJ (2011) The AMPK signalling pathway coordinates cell growth, autophagy and metabolism. Nat Cell Biol 13:1016–1023

Minglian Z, Minghe M, Keqin Z (2004) Characterization of a neutral serine protease and its full-length cDNA from the nematode-trapping fungus Arthrobotrys oligospora. Mycologia 96:16–22

Monod M et al (2002) Secreted proteases from pathogenic fungi. Int J Med Microbiol 292:405–419

Moore RS, Kaletsky R, Murphy CT (2019) Piwi/PRG-1 argonaute and TGF-beta mediate transgenerational learned pathogenic avoidance. Cell 177:1827–1841 e1812

Neher DA (2001) Role of nematodes in soil health and their use as indicators. J Nematol 33:161–168

Nicol JM et al. (2011) Current nematode threats to world agriculture. In: Genomics and molecular genetics of plant-nematode interactions, pp 21–43. doi: https://doi.org/10.1007/978-94-007-0434-3_2

Nordbring-Hertz B (1973) Peptide-induced morphogenesis in the nematode-trapping fungus *Arthrobotrys oligospora*. Physiol Plant 29:223–233

Nordbring-Hertz B (2004) Morphogenesis in the nematode-trapping fungus *Arthrobotrys oligospora*--an extensive plasticity of infection structures. Mycologist 18:125–133

Nordbring-Hertz B, Mattiasson B (1979) Action of a nematode-trapping fungus shows lectin-mediated host–microorganism interaction. Nature 281:477–479

Nordbring-Hertz B, Friman E, Veenhuis M (1989) Hyphal fusion during initial stages of trap formation in *Arthrobotrys oligospora*. A Van Leeuw J Microb 55: 237–244

Nordbring-Hertz B, Jansson H-B, Tunlid A (2011) Nematophagous fungi. In: eLS. Wiley, Chichester, pp 1–13. https://doi.org/10.1002/9780470015902.a0000374.pub3

Okada H (2006) Ecology of fungivorous nematodes and their use for suppression of plant diseases. Bulletin of the National Agricultural Research Center for Tohoku Region (Japan)

Osman GA et al (2018) Natural infection of C. elegans by an oomycete reveals a new pathogen-specific immune response. Curr Biol 28:640–648 e645

Page AP, Johnstone IL (2007) The cuticle. WormBook, pp 1–15. https://doi.org/10.1895/wormbook.1.138.1

Park H-S, Yu J-H (2012) Genetic control of asexual sporulation in filamentous fungi. Curr Opin Microbiol 15: 669–677

Pechuk V et al (2022) Reprogramming the topology of the nociceptive circuit in C elegans reshapes sexual behavior. Curr Biol. https://doi.org/10.1016/j.cub.2022.08.038

Pereira L et al (2015) A cellular and regulatory map of the cholinergic nervous system of C. elegans. Elife 4

Pfister DH (1997) Castor, Pollux and life histories of fungi. *Mycologia* 89:1–23

Pramer D, Stoll NR (1959) Nemin: a morphogenic substance causing trap formation by predaceous fungi. Science 129:966–967

Pujol N et al (2008) Anti-fungal innate immunity in C. elegans is enhanced by evolutionary diversification of antimicrobial peptides. PLoS Pathog, e1000105 4

Rechavi O et al (2014) Starvation-induced transgenerational inheritance of small RNAs in C. elegans. Cell 158:277–287

Reed RH, Chudek JA, Foster R, Gadd GM (1987) Osmotic significance of glycerol accumulation in exponentially growing yeasts. Appl Environ Microbiol 53:2119–2123

Ridall A, Ingels J (2021) Suitability of free-living marine nematodes as bioindicators: status and future considerations. Front Mar Sci 8

Rodríguez-Peña JM, García R, Nombela C, Arroyo J (2010) The high-osmolarity glycerol (HOG) and cell wall integrity (CWI) signalling pathways interplay: a yeast dialogue between MAPK routes. Yeast 27:495–502

Rosen S, Sjollema K, Veenhuis M, Tunlid A (1997) A cytoplasmic lectin produced by the fungus *Arthrobotrys oligospora* functions as a storage protein during saprophytic and parasitic growth. Microbiol-Uk 143:2593–2604

Saxena G, Dayal R, Mukerji KG (1987) Interaction of nematodes with nematophagus fungi: induction of trap formation, attraction and detection of attractants. FEMS Microbiol Lett 45:319–327

Schenck NC, Graham SO, Green NE (1975) Temperature and light effect on contamination and spore germination of vesicular-arbuscular mycorrhizal fungi. Mycologia 67:1189–1192

Scholler M, Rubner A (1994) Predacious activity of the nematode-destroying fungus *Arthrobotrys oligospora* in dependence of the medium composition. Microbiol Res 149:145–149

Schulenburg H, Felix MA (2017) The natural biotic environment of Caenorhabditis elegans. Genetics 206:55–86

Serrano-Saiz E et al (2013) Modular control of glutamatergic neuronal identity in C. elegans by distinct homeodomain proteins. Cell 155:659–673

Shepherd AM (1955) Formation of the infection bulb in Arthrobotrys oligospora Fresenius. Nature 175:475–475

Sommer RJ (2006) Pristionchus pacificus. WormBook, pp 1–8. https://doi.org/10.1895/wormbook.1.102.1

Srinivasan J et al (2008) A blend of small molecules regulates both mating and development in Caenorhabditis elegans. Nature 454:1115–U1146

Stinchcomb DT, Shaw JE, Carr SH, Hirsh D (1985) Extrachromosomal DNA transformation of Caenorhabditis elegans. Mol Cell Biol 5:3484–3496

Su H et al (2017) Trapping devices of nematode-trapping fungi: formation, evolution, and genomic perspectives. Biol Rev Camb Philos Soc 92:357–368

Sulston JE, Horvitz HR (1977) Post-embryonic cell lineages of the nematode, Caenorhabditis elegans. Dev Biol 56:110–156

Tatebayashi K et al (2007) Transmembrane mucins Hkr1 and Msb2 are putative osmosensors in the SHO1 branch of yeast HOG pathway. EMBO J 26:3521–3533

Tosi S, Annovazzi L, Tosi I, Iadarola P, Caretta G (2002) Collagenase production in an antarctic strain of *Arthrobotrys tortor Jarowaja*. Mycopathologia 153: 157–162

Trudgill DL, Blok VC (2001) Apomictic, polyphagous root-knot nematodes: exceptionally successful and damaging biotrophic root pathogens. Annu Rev Phytopathol 39:53–77

Tunlid A, Jansson S (1991) Proteases and their involvement in the infection and immobilization of nematodes by the nematophagous fungus Arthrobotrys oligospora. Appl Environ Microbiol 57:2868–2872

Tunlid A, Jansson HB, Nordbring-Hertz B (1992) Fungal attachment to nematodes. Mycol Res 96:401–412

Tunlid A, Rosén S, Ek B, Rask L (1994a) Purification and characterization of an extracellular serine protease from the nematode-trapping fungus *Arhtrobotrys oligospora*. Microbiol-Uk 140:1687–1695

Tunlid A, Rosén S, Ek B, Rask L (1994b) Purification and characterization of an extracellular serine protease from the nematode-trapping fungus Arthrobotrys oligospora. Microbiology 140:1687–1695

van den Hoogen J et al (2019) Soil nematode abundance and functional group composition at a global scale. Nature 572:194–198

Veenhuis M, Nordbring-Hertz B, Harder W (1985a) Development and fate of electron-dense microbodies in trap cells of the nematophagous fungus *Arthrobotrys oligospora*. A Van Leeuw J Microb 51:399–407

Veenhuis M, Nordbring-Hertz B, Harder W (1985b) An electron-microscopical analysis of capture and initial stages of penetration of nematodes by *Arthrobotrys oligospora*. A Van Leeuw J Microb 51:385–398

Veenhuis M, Van Wijk C, Wyss U, Nordbring-Hertz B, Harder W (1989) Significance of electron dense microbodies in trap cells of the nematophagous fungus *Arthrobotrys oligospora*. A Van Leeuw J Microb 56: 251–261

Volterra V (1926) Variazioni e fluttuazioni del numero d'individui in specie animali conviventi (Società anonima tipografica "Leonardo da Vinci")

Wan QL et al (2021) N-6-methyldeoxyadenine and histone methylation mediate transgenerational survival advantages induced by hormetic heat stress. Sci Adv 7

Wang RB, Yang JK, Lin C, Zhang Y, Zhang KQ (2006) Purification and characterization of an extracellular serine protease from the nematode-trapping fungus *Dactylella shizishanna*. Lett Appl Microbiol 42:589–594

Wang X et al (2014) Bacteria can mobilize nematode-trapping fungi to kill nematodes. Nat Commun 5:5776

Wang R, Wang J, Yang X (2015) The extracellular bioactive substances of *Arthrobotrys oligospora* during the nematode-trapping process. Biol Control 86:60–65

Wang R et al (2018) Comparative genomic analyses reveal the features for adaptation to nematodes in fungi. DNA Res 25:245–256

Wang W, Zhao Y, Bai N, Zhang K-Q, Yang J (2022) AMPK is involved in regulating the utilization of carbon sources, conidiation, pathogenicity, and stress response of the nematode-trapping fungus arthrobotrys oligospora. Microbiol Spectr. e02225-02222

Warcup JH (1957) Studies on the occurrence and activity of fungi in a wheat-field soil. Trans Br Mycol Soc 40: 237–259

Webster AK et al (2022) Using population selection and sequencing to characterize natural variation of starvation resistance in Caenorhabditis elegans. Elife 11

Wernet V, Wäckerle J, Fischer R (2022) The STRIPAK component SipC is involved in morphology and cell-fate determination in the nematode-trapping fungus Duddingtonia flagrans. Genetics 220:iyab153

White JG, Southgate E, Thomson JN, Brenner S (1986) The structure of the nervous-system of the nematode caenorhabditis-elegans. Philos T R Soc B 314:1–340

Wilks S, Lotka AJ (1941) Théorie analytique des associations biologiques. Bull Am Math Soc 47:532–534

Wilson AM et al (2018) Biomechanics of predator-prey arms race in lion, zebra, cheetah and impala. Nature 554:183–188

Xie H et al (2010) Trap induction and trapping in eight nematode-trapping fungi (Orbiliaceae) as affected by juvenile stage of Caenorhabditis elegans. Mycopathologia 169:467–473

Xie M et al (2020) Protein kinase Ime2 is required for mycelial growth, conidiation, osmoregulation, and pathogenicity in nematode-trapping fungus Arthrobotrys oligospora. Front Microbiol 10:3065

Yang J, Tian B, Liang L, Zhang KQ (2007) Extracellular enzymes and the pathogenesis of nematophagous fungi. Appl Microbiol Biotechnol 75:21–31

Yang J et al (2011) Genomic and proteomic analyses of the fungus Arthrobotrys oligospora provide insights into nematode-trap formation. PLoS Pathog 7:e1002179

Yang C-T et al (2020) Natural diversity in the predatory behavior facilitates the establishment of a robust model strain for nematode-trapping fungi. Proc Natl Acad Sci U S A 117:6762–6770

Yeates GW, Bongers T, De Goede RG, Freckman DW, Georgieva SS (1993) Feeding habits in soil nematode families and genera-an outline for soil ecologists. J Nematol 25:315–331

Yemini E et al (2021) NeuroPAL: a multicolor atlas for whole-brain neuronal identification in C. elegans. Cell 184:272

Yoshida T, Jones LE, Ellner SP, Fussmann GF, Hairston NG Jr (2003) Rapid evolution drives ecological dynamics in a predator-prey system. Nature 424:303–306

Youssar L et al (2019) Intercellular communication is required for trap formation in the nematode-trapping fungus Duddingtonia flagrans. PLoS Genet 15: e1008029

Yu X et al (2021) Fatal attraction of Caenorhabditis elegans to predatory fungi through 6-methyl-salicylic acid. Nat Commun 12:1–10

Yuan JY, Horvitz HR (1992) The Caenorhabditis-elegans cell-death gene ced-4 encodes a novel protein and is expressed during the period of extensive programmed cell-death. Development 116:309–320

Zhang G et al (2019) The velvet proteins VosA and VelB play different roles in conidiation, trap formation, and pathogenicity in the nematode-trapping fungus Arthrobotrys oligospora. Front Microbiol 10:1917

Zhang W et al (2020a) Comparative genome and transcriptome analysis of the nematode-trapping fungus Duddingtonia flagrans reveals high pathogenicity during nematode infection. Biol Control 143:104159

Zhang Y et al (2020b) Fungi-nematode interactions: diversity, ecology, and biocontrol prospects in agriculture. J Fungi 6

Zhang Q et al (2021) The memory of neuronal mitochondrial stress is inherited transgenerationally via elevated mitochondrial DNA levels. Nat Cell Biol 23:870

Zhen Z et al (2018) MAP kinase Slt2 orthologs play similar roles in conidiation, trap formation, and pathogenicity in two nematode-trapping fungi. Fungal Genet Biol 116:42–50

Zopf WF (1888) Zur Kenntniss der Infections-Krankheiten niederer Thiere und Pflanzen, vol 52. E. Blochmann

Zugasti O, Ewbank JJ (2009) Neuroimmune regulation of antimicrobial peptide expression by a noncanonical TGF-beta signaling pathway in Caenorhabditis elegans epidermis. Nat Immunol 10:249–256

12 When a Mind Is Not Its Own: Mechanisms of Host Behavior Control by Parasitic Fungi

Carolyn Elya

Abstract

Many animal parasites can alter the behavior of animals in ways that benefit their own fitness. Insects are a frequent target for behavior manipulation by various parasites, especially fungi. Despite many decades of observing altered behavior in insects parasitized by fungi, our understanding of the mechanisms underpinning this behavior manipulation remains largely incomplete. This chapter explores our current understanding of the mechanistic basis of behavior manipulations by these so-called zombie fungi, key considerations in continuing to make progress in our understanding, and where I think the next advances will come.

Keywords

Behavior manipulation · Entomopathogens · Zombie fungi · Cadaver transmission · Active host transmission · Summit disease · *Entomophthora* · *Ophiocordyceps* · *Massospora*

C. Elya (✉)
Department of Organismic and Evolutionary Biology, Harvard University, Cambridge, MA, USA
e-mail: cnelya@g.harvard.edu

12.1 Introduction

Insects are recognized as the largest and most varied group of eukaryotic organisms on the planet, with 900,000 recognized species (Grimaldi et al. 2005) and an expected 5.5 million total species (Stork 2018). The fungal kingdom is also incredibly diverse and species-rich. The number of fungal species has been proposed to be anywhere between 2 and 12 million (Hawksworth and Lücking 2017; Wu et al. 2019) (though this remains a controversial topic of active discussion, see Bhunjun et al. (2022)). It is therefore, perhaps, unsurprising that the majority of fungal phyla contain entomopathogens, species that rely on insects for nutrition (Araújo and Hughes 2016), and that the ability to use insects as food has arisen multiple times over the earth's evolutionary past (Hughes 2011). Indeed, approximately 65% of all insect orders have been observed to be infected by fungi (Araújo and Hughes 2016). Entomopathogenic fungi use a variety of strategies in utilizing insects for nutrition. Some entomopathogenic fungi are generalists: they are able to infect and draw resources from a variety of host species [e.g., *Metarhizium robertsii*, *Beauveria bassiana* (St Leger and Wang 2020; Ortiz-Urquiza 2021)], leveraging the large available insect biomass in their environment. Others are specialists, targeting a narrow range of host species. Many specialist entomopathogenic fungi have evolved to manipulate host behavior. By doing so, the

Y.-P. Hsueh, M. Blackwell (eds.), *Fungal Associations*, The Mycota 9,
https://doi.org/10.1007/978-3-031-41648-4_12

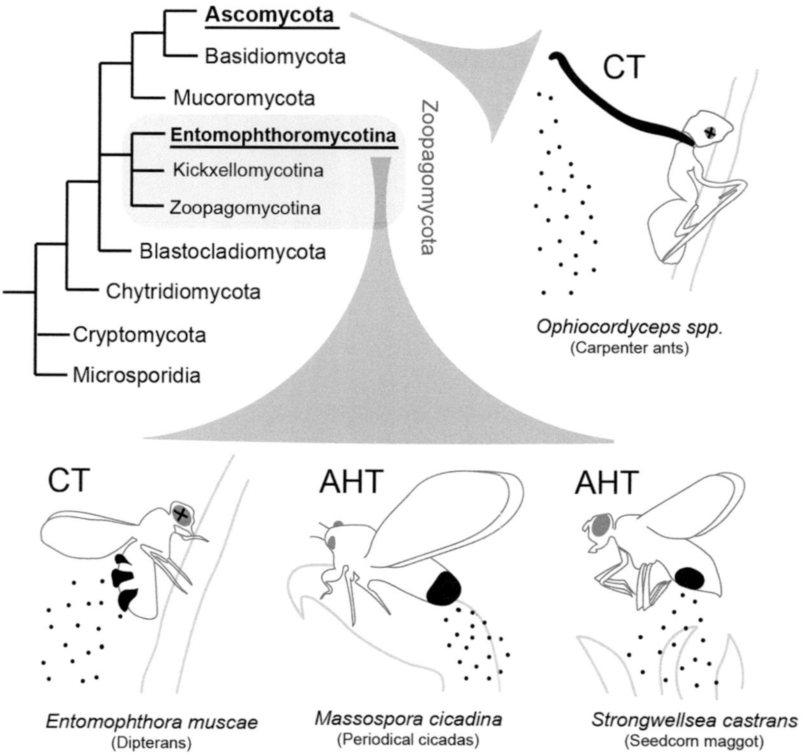

Fig. 12.1 Four examples of zombie fungi, their mode of dispersal and their relative phylogenetic placement within the fungal kingdom. Clockwise from upper right: *Ophiocordyceps* spp., aka zombie ant fungus, infect carpenter ants and spread via cadaver transmission (CT); *Strongwellsea castrans* infects *Delia platura* and spreads via active host transmission (AHT); *Massospora cicadina* infects 13- and 17-year periodical cicadas and spreads via AHT; *Entomophthora muscae* infects various dipterans (true flies) and spreads via CT; fungal phylogeny after Spatafora et al. (2016). Underlined clades (Phylum Ascomycota and Subphylum Entomophthoromycotina) are home to most zombie fungi. Funnels adjacent to these clades indicate taxonomic placement for the examples shown in this figure. *Ophiocordyceps* is an ascomycete fungus (Order Hypocreales); all other examples are in phylum Zoopagomycota (Order Entomophthorales). Tree branch lengths are not proportional to phylogenetic distance

fungus increases the likelihood that it will encounter a suitable new host that lies within its target host range. These behavior-manipulating fungi, also referred to as "zombie fungi," have been observed predominantly in two fungal phyla, Ascomycota and Zoopagomycota (Hughes et al. 2016). Like entomopathogenicity itself, the ability to hijack insect behavior has evolved more than once, at least twice within the subphylum Entomophthoromycotina and several times among Ascomycota (Hughes et al. 2016) (Fig. 12.1).

Broadly speaking, zombie fungi utilize one of the two strategies to access new hosts: cadaver transmission (CT) or active host transmission (AHT). In cadaver-transmitting systems, the fungus coerces the host to die in an elevated location and then sporulates from the dead host from this high vantage point. In active host transmitting systems, the fungus alters the abdominal structure of a living host (either completely replacing it with fungal tissue or creating an opening through which fungal tissue can emerge) and uses the movement of the host to its advantage for dispersing spores.

Perhaps the most widely known example of a cadaver-transmitting fungus is the *Ophiocordyceps*-infected zombie ant (Fig. 12.1).

Though often referred to as a single phenomenon, there are actually many different species of *Ophiocordyceps* that each target a different host ant species and elicit their own variation on a behavioral theme (Evans et al. 2011; de Bekker et al. 2014; Loreto et al. 2018; Sakolrak et al. 2018). In these systems, *Ophiocordyceps*-infected ant hosts show abnormal behaviors late in infection. Whereas healthy ants are usually found in the nest or on foraging trails, zombie ants aberrantly leave their nests and wander off the foraging trail to climb vegetation. After their climb, zombie ants affix themselves to the plant substrate, typically by biting down and subsequent "lock-jaw," then die (Hughes et al. 2011). Curiously, death in this manner occurs with specific circadian timing, with zombie ants exhibiting the final climbing and biting behavior specifically around solar noon (Hughes et al. 2011). Over several days, *Ophiocordyceps* erupts from the dead ant's head, producing a fruiting body that will eject infectious fungal spores into the environment (Hughes et al. 2011). Since ants tend to die elevated above foraging trails, this ideally positions *Ophiocordyceps* to infect additional ants. The majority of described behavior-manipulating entomopathogens rely on similar behavior manipulations and cadaver transmission to complete their life cycle.

There are far fewer examples of active host transmitting zombie fungi. Among these is *Massospora cicadina* which infects periodical cicadas, insects that only emerge from dormancy in the soil every 13 or 17 years (Fig. 12.1). Cicada nymphs acquire *Massospora* infection via resting spores deposited in the soil (Lovett et al. 2020). Once the infected adult emerges, *Massospora* grows in its posterior end, ultimately sloughing off the posterior segments of the abdomen and replacing these tissues with a fungal plug (Lovett et al. 2020). *Massospora* then disseminates spores to other cicadas during flight and copulation attempts by the infected individuals, effectively using the host as a mobile transmission vector (Lovett et al. 2020).

Zombie fungi and their pitied insect victims have been documented in the scientific literature since the mid-1800s (Cohn 1855; Fawcett 1886; Thaxter 1888). Ever since this time they have fascinated scholars and non-scholars alike, recently making several appearances in the popular media (Nuwer By Rachel Nuwer on May 27 2021; Yong 2017; Langley 2018; Lu 2019; Zimmer 2019; Iati 2021). Yet despite longstanding fascination in zombie fungi, we are only just now beginning to uncover the mechanisms by which they elicit behaviors in their unwitting hosts.

Call It *Ophiocordyceps*

Once there was a genus of 400 species of insect pathogens, *Cordyceps*, classified in the order Clavicipitales. Use of DNA sequencing and phylogenetic analysis to study the insect associated species of *Cordyceps* revealed that the ecology was more ancient than previously realized and that these insect pathogenic species did not comprise a monophyletic genus. The species previously placed in a single genus now are classified in three families (Clavicipitaceae, Cordycipitaceae, Ophiocordycipitaceae) of Hypocreales reflecting their independent lineages. The phylogenetic classification of these species now includes more than 20 genera and combines ascospore- and conidium-producing species within common genera based on morphological and DNA criteria. Some species now placed in the genus *Ophiocordyceps* are insect pathogens highlighted in this chapter and in current dystopian fiction. Others are known throughout the world in traditional medicine. Less well known are yeast-like members of the *Ophiocordyceps* lineage, including yeast-like symbionts in mutualistic associations with bacteria and insects of the order Hemiptera. The evolutionary pathway from virulent pathogen to obligate symbiont remains to be explained.

12.2 Common Themes in Behavioral Manipulation by Zombie Fungi

There are five major recurring themes in the behaviors elicited by zombie fungi: hyperactivity, summit disease, perimortem postural changes, circadian timing, and sexual attraction (Table 12.1). Here, I will summarize these behaviors and discuss the mechanisms by which they are thought to occur. Where possible, I will discuss occurrences of similar behaviors that are elicited by non-fungal parasites and available mechanistic evidence for these systems.

Table 12.1 Recurring themes in behaviors elicited by zombie fungi

Behavior	Description	Possible host targets	Hypothesized fungal action	Examples	References
Hyperactivity	Increase in locomotion	Nutrient depletion, locomotion circuits	Secretion of fungal effectors (metabolites, SSPs)	*Ophiocordyceps* spp. *Massospora cicadina*	Will et al. (2020), Trinh et al. (2021), Cooley et al. (2018), Boyce et al. (2019)
Summit disease	Climbing to an elevated location right before death	Phototactic and/or gravitactic pathways	Secretion of fungal effectors (metabolites, SSPs)	*Ophiocordyceps spp. Entomophthora muscae Entomophaga grylli Pandora formicae Eryniopsis lampyridarum*	Hughes et al. (2011), Krasnoff et al. (1995), Pickford and Riegert (1964), Csata et al. (2021), Steinkraus et al. (2017)
Postural changes	Specific positioning of body parts to maintain cadaver's elevation and/or accommodate fungal dispersal	Invasion and/or restriction of musculature	Mechanical force; modulation of motor neuron activity	*Ophiocordyceps* spp. *Entomophthora muscae Entomophaga grylli Pandora formicae Eryniopsis lampyridarum*	Hughes et al. (2011), Will et al. (2020), Trinh et al. (2021), Krasnoff et al. (1995), Pickford and Riegert (1964), Csata et al. (2021), Steinkraus et al. (2017)
Circadian timing	Behaviors only occur within specific time window relative to environmental cues	Host circadian network, pathways underlying behaviors that occur with circadian timing	Secretion of fungal effectors, perhaps according to fungal molecular clock	*Ophiocordyceps* spp. *Entomophthora* spp. *Entomophaga grylli Entomophaga maimaiga Erynia neoaphidis*	Hughes et al. (2011), Milner et al. (1984), Krasnoff et al. (1995), Elya et al. (2018), Pickford and Riegert (1964), Nielsen and Hajek (2006), Dustan (1924)
Sexual attraction	Increased frequency of attempted mating with infected individuals	Olfactory and gustatory systems, hormonal axes	Production of volatile chemical attractants, stimulant metabolites, hormonal shifts	*Massospora cicadina Entomophthora muscae*	Boyce et al. (2019), Naundrup et al. (2021)

12.2.1 Hyperactivity

Hyperactivity, alternatively referred to as enhanced locomotor activity or ELA, refers to a sustained increase in locomotion and is a behavior observed within hosts infected by both CT and AHT fungi. Carpenter ants infected with *Ophiocordyceps* exhibit hyperactivity prior to their final climb (Trinh et al. 2021). *Massospora*-infected cicadas exhibit hyperactivity after their posterior abdomen has been replaced with a spore-releasing fungal plug (Boyce et al. 2019). In both instances, hyperactivity likely plays a role in positioning the host for spore dispersal. For the ant, hyperactivity may play a role in driving the ant away from its nest and to an elevated location from which it is ideal to eject spores. For the cicada, hyperactivity is hypothesized to increase the number of conspecifics encountered, and thus increase the number of newly infected hosts (Cooley et al. 2018; Lovett et al. 2020).

The molecular drivers of zombie fungus-induced hyperactivity are unclear, but a variety of mechanisms have been proposed. First, hyperactivity could result from more general changes in host physiology. Consistent with their mechanism of action (i.e., consuming host resources for fuel), zombie fungi have been observed to elicit starvation-like states in their host (Elya et al. 2018; Will et al. 2020). As many insects have been observed to increase locomotion during times of starvation (Browne and Evans 1960; Leonard 1970; C. P. Wheater 1991; Yang et al. 2015), hyperactivity could potentially arise from general nutrient depletion.

Alternatively, hyperactivity could be elicited via compounds produced and secreted by the fungus. Genomic, transcriptomic, and metabolomic studies in zombie ants have identified several candidates. In *Ophiocordyceps spp.*-infected ants, two independent studies in distinct experimental systems examined gene expression at the point of manipulation (i.e., biting). The authors observed upregulation of a putative enterotoxin as well as several putative small secreted proteins (SSPs) (de Bekker et al. 2015; Will et al. 2020), the latter of which have been repeatedly implicated in mediating interactions between fungi and their host plants and animals (Kim et al. 2016). Genomic analysis has found that orthologues of the enterotoxin are conserved only among ant-manipulating members of genus *Ophiocordyceps* (de Bekker et al. 2017a). Additionally, many of the SSPs appear to be species-specific, suggesting that they contribute to the unique pathology of these fungi (de Bekker et al. 2017a). These studies also observed a dramatic upregulation of genes in a cluster predicted to synthesize an aflatrem-like alkaloid at the point of manipulation (de Bekker et al. 2015; Will et al. 2020). Aflatrem has known neurobiological activity, often observed to cause tremors and seizures in livestock (Valdes et al. 1985). In cicadas, a recent study investigating the metabolome of *Massospora*-infected individuals found the amphetamine relative cathinone (Boyce et al. 2019). Amphetamines have similar structures to the hormones adrenaline and epinephrine and have been well established to exhibit stimulatory effects on a variety of animals (Randrup and Munkvad 1967). Overall, many exciting compounds have been identified as correlates of hyperactivity. It will be up to future experiments to demonstrate their causality.

Hyperactivity is also elicited in insect hosts by several non-fungal parasites. While the specificity of our mechanistic insight varies among these systems, what we have learned so far suggests that several diverse mechanisms, rather than a common pathway, underlie this behavior alteration. In crickets infected with hairworms (Phylum Nematomorpha) and sandhoppers infected with mermithids (Phylum Nematoda), hyperactivity occurs prior to a suicidal plunge into water (Herbison et al. 2019). Proteomics studies in these systems have led to the hypothesis that hyperactivity occurs as a result of dysregulation of energy utilization and/or neurophysiology (e.g., synaptic vesicle packaging and release, maintenance of neuronal compartments) (Herbison et al. 2019). In several species of lepidopteran larvae infected with nucleopolyhedroviruses (NPVs), pre-death hyperactivity has been elegantly demonstrated to depend on a virally-encoded protein tyrosine phosphatase (*ptp*) (Kamita et al. 2005; van Houte et al. 2012). However, a recent study demonstrated that *ptp* is dispensable for hyperactivity in at least one case (Kokusho and Katsuma 2021). Thus, even among NPVs it appears that there may be diverse mechanisms underlying hyperactivity.

In contrast to the increased activity seen in zombie crickets, sandhoppers, and caterpillars, jewel wasps (*Ampulex compressa*) evoke hypoactivity (reduced locomotion) in their cockroach hosts. After an initial non-specific sting that serves to temporarily paralyze its cockroach host, the wasp targets a second injection into the cockroach brain (Haspel et al. 2003). Following this second injection the cockroach engages in approximately 25 min of grooming behaviors and fails to initiate walking behavior (Haspel et al. 2003). Work in this system has found that the venom cocktail for the second injection is a complex mixture of small molecules, peptides and proteins (Arvidson et al. 2019), that specifically targets the insect central complex (a conserved brain region important for coordinating and planning motor output) (Haspel et al. 2003). These behavioral effects are mediated by D1-like dopamine receptors in the cockroach (Nordio et al. 2022), inducing host hyperactivity, and, more broadly, alter host activity levels, apparently, a frequent strategy employed by behavior-manipulating parasites.

12.2.2 Summit Disease

Summit disease, also referred to as tree-top disease or Wipfelkrankheit (Hofmann 1891), refers to the behavior in which an insect climbs to an elevated location immediately prior to its death. The insect then perishes at this high perch, typically held in place by its mouthparts (Hughes et al. 2011; Csata et al. 2021), limbs (Carruthers et al. 1997), and/or fungal holdfasts (Chung et al. 2017). Summit disease is elicited by several fungal parasites including *Ophiocordyceps* (Hughes et al. 2011) and *Entomophthora muscae* (Krasnoff et al. 1995) and is one of the most consistent behavioral motifs observed in cadaver-transmitting fungal parasites (Table 12.1). While the mechanistic basis of this behavior remains to be determined, there is a short list of host pathways hypothesized to serve as fungal targets to elicit this behavior. The first of these is phototaxis, orientation, and movement with respect to light. In field studies, *Ophiocordyceps*-infected ants have been observed to die more frequently in sunny versus shaded regions of observation plots (Andriolli et al. 2019) and to perish with their heads oriented toward opening in the forest canopy (Chung et al. 2017). These observations suggest that light plays an important role in the end-of-life behaviors exhibited by zombie ants. As light appears to influence where ants summit and die, the host phototaxis pathway has been hypothesized to be involved in summiting behavior. Gravitaxis, the orientation and movement in response to earth's gravitational pull, is another pathway proposed to play a role in summit behavior. The hallmark of summit behavior is death following elevation, and achieving elevation requires navigating against the earth's gravitational pull. Thus, it seems reasonable that fungal parasites might drive summiting behavior via altering their host's gravitactic preferences or circuitry.

How fungal parasites might alter phototaxis and gravity sensing in summiting or other host pathways is still unclear, although, again, several hypotheses have arisen from multi-omic approaches in *Ophiocordyceps*-ant systems. As previously discussed, a putative fungal enterotoxin, several SSPs, and a gene cluster responsible for synthesizing a related compound to the tremorgenic aflatrem have been found to be highly expressed during *Ophiocordyceps*-elicited biting behavior in ants (de Bekker et al. 2015; Will et al. 2020). Immediately preceding biting, ants had exhibited both hyperactivity and summiting behaviors. It seems likely that these behaviors likely overlap temporally, thus the compounds implicated in manipulation could be important for driving one or both of hyperactivity and summiting.

There are also several non-fungal parasites that elicit summit disease which can provide insight into the mechanistic underpinnings of this behavior. The trematode *Dicrocoelium dendriticum* infects ants and coerces them to climb to the top of surrounding flora that, like *Ophiocordyceps*-infected ants, they attach via biting (van Paridon et al. 2017). Recent structural analysis of *D. dendriticum*-infected ants revealed that at least one of the infected host's resident worms consistently invades the host's subesophageal ganglion (SOG) (Martín-Vega et al. 2018). Among other functions, the SOG contains descending neurons, neurons that send information from the brain to the ventral nerve cord (the insect analog of the vertebrate spinal cord), that are important for coordinating motor output, such as walking (Hsu and Bhandawat 2016). The location of the worm by itself is unlikely to drive a behavior as complex as summiting but does place the worm in an ideal position for altering the activity of host neurons either through chemical or mechanical action. Notably, *Ophiocordyceps* does not invade ant brain tissue (Fredericksen et al. 2017); however, like *D. dendriticum*, some species of zombie fungi (e.g., *Entomophthora muscae, Entomophaga grylli, Pandora formicae, Strongwellsea castrans*) have been observed to occupy the host brain (Humber 1976; Brobyn and Wilding 1977, 1983; Funk et al. 1993; Elya et al. 2018; Csata et al. 2021). That brain occupancy is not a consistent feature of behavior-manipulating parasites likely reflects that there are multiple routes (e.g., direct and indirect) to modifying host behavior.

As with parasite-induced hyperactivity, our most thorough understanding of a mechanism underlying parasite-elicited summit disease comes from baculoviruses. Several baculoviruses drive summiting in their larval hosts immediately before death, which is thought to facilitate the enhanced dispersal of viral particles (Vasconcelos et al. 1996). Elegant work in *Lymantria dispar* NPV determined that the gene ecdysteroid uridine 4′-diphosphate glucosyltransferase (*egt*) was necessary and sufficient to elicit summit disease in spongy moths (Hoover et al. 2011). The enzyme encoded by *egt* catalyzes a reaction that leads to the inactivation of the larval hormone 20-hydroxyecdysone (20E) blocking pupation as a result (O'Reilly and Miller 1989). Thus, *egt* is hypothesized to drive summiting behavior in spongy moths partly by blocking the normal occurrence of pupation: infected caterpillars continue feeding on plants rather than returning to the ground to undergo metamorphosis, leading them to be elevated at death. However, *egt* does not appear to be required to drive summiting behavior in other NPV-larval systems (Ros et al. 2015). In addition, various studies have shown that light cues and perception thereof are required for summiting, and that summiting hosts have aberrant expression of circadian genes (Han et al. 2017; Bhattarai et al. 2018). All told, it seems likely that there are several overlapping pathways involved in summit disease that vary even among related viruses.

12.2.3 Perimortem Postural Changes

Many zombie fungus-infected insects exhibit perimortem postural changes ranging from the positioning of mouth parts to the splaying of legs and wings. In cadaver-transmitting systems, these postural changes typically occur after summiting but before death. In *Ophiocordyceps*-infected ants, for example, manipulated individuals maintain elevation by biting onto an available substrate immediately prior to death (Hughes et al. 2016). These animals exhibit lockjaw, which recent work suggests may be achieved by penetration of the muscles by fungal tissue as well as the formation of a fungal cage-like structure surrounding mandibular muscle (Fredericksen et al. 2017; Mangold et al. 2019). In *Entomophthora*-infected flies, a manipulated individual extends its proboscis and is then "glued" where standing via secretions that are thought to be produced by the fungi (Brobyn and Wilding 1983). Proboscis extension has been proposed to arise from mechanical force (i.e., the body cavity has become so full that the proboscis cannot remain contracted due to steric hindrance) (Brobyn and Wilding 1983). More recent observations have reported that the proboscis both extends and retracts in the dying fly, suggesting that proboscis movement could be driven either by impingement upon underlying muscles or by directly or indirectly perturbing the action of motor neurons that directly innervate musculature and drive changes in contractility (Elya et al. 2018).

After summiting and adherence in place, zombie flies and beetles exhibit a further postural change, positioning their wings up and away from the dorsal abdomen (Krasnoff et al. 1995; Steinkraus et al. 2017; Elya et al. 2018). The repositioning of the wings serves to create a direct "line of fire" for the fungal spores that will be ejected from the dorsal abdomen. In the case of flies, wing-raising is rapid, typically fully executed in 15 min (Krasnoff et al. 1995). Wing-raising in zombie flies seems that it could possibly arise by several different mechanisms. The fungus could impinge on musculature directly as it grows in the body cavity or could indirectly alter posture by rapid uptake of remaining liquid (dehydration). Alternatively, the fungus could alter the activity of motor neurons, either directly or indirectly. In the case of *Eryniopsis lampyridarum*-infected beetles with wings that take many hours to raise, a mechanical explanation is the more likely.

Similar postural changes have been observed in insects parasitized by non-fungal organisms. Ants infected by *D. dendriticum* also bite down to maintain an elevated position. These ants do not die in place, but instead remain motionless *in situ* for several days (Martín-Vega et al. 2018). This behavior is thought to benefit the trematode by making infected ants more likely to be consumed by their grazing herbivores, which can then spread the parasite via droppings back to the primary snail host (Goater and Colwell 2007). A mechanistic understanding of biting behavior in *D. dendriticum*-infected ants is lacking, although the presence of flukes in the SOG may play a role (Martín-Vega et al. 2018).

12.2.4 Circadian Timing

An extremely common theme among zombie fungi is that they drive changes in host behavior (and ultimately cause death) with stereotyped timing. For example, flies infected with *E. muscae* only exhibit summit disease and subsequent postural changes at sunset (Krasnoff et al. 1995; Elya et al. 2018) and, in the field, as mentioned earlier, ants infected with *Ophiocordyceps* summit and die around solar noon (Hughes et al. 2011). How this timing is achieved is particularly curious as there are likely two molecular clocks at play in each case—one belonging to each the host and the fungus. Molecular clocks are cellular mechanisms that can maintain circadian rhythms in the absence of environmental cues, or Zeitgebers, such as light or temperature. Clocks synchronize, or entrain, to environmental signals to align with the earth's daily cycles and are usually driven by core transcription/translation feedback loops. Clocks have been demonstrated in organisms ranging from bacteria to animals, though their core molecular components vary along mostly phylogenetic lines (Dunlap and Loros 2017). Details of how timing from the clock confers various behaviors and phenotypes are continuously emerging.

In the case of *E. muscae*-infected flies, the favored explanation of how this timing is achieved is that it is not dictated by the host clock. This is derived from the two observations. First, flies infected with *E. muscae* in constant darkness (referred to as "free-running conditions") do not die with rhythmic timing, even though their fly hosts are known to maintain circadian periodicity in the absence of environmental cues (Krasnoff et al. 1995). This suggests that the host clock alone is insufficient to drive the timing of death. Second, flies housed in darkness 72 h after fungal exposure do show rhythmic timing of death, which suggests that there is some machinery present capable of driving timed death in the absence of immediate cues, but that this machinery can only entrain after the onset of infection (Krasnoff et al. 1995). The most likely identity of this mystery machinery is a fungal clock. Though it is very likely that *E. muscae* has a clock of its own, this has not yet been experimentally demonstrated.

On the other hand, the prevailing hypothesis for timing in *Ophiocordyceps*-infected ants is that this is driven by the host clock, which in turn is manipulated by the fungus (de Bekker and Das 2022). Transcriptomic studies have shown that the expression of various clock genes are perturbed at the moment of manipulation (de Bekker et al. 2015; Will et al. 2020) and that foraging rhythms observed in healthy ants are essentially absent in infected individuals (Trinh et al. 2021). The observation that *Ophiocordyceps* continues to produce transcripts in a circadian manner under free-running conditions strongly supports the notion that this fungus has its own molecular timekeeping mechanism (de Bekker et al. 2017b). Details as to how fungal clock activity is linked to host behavior are lacking, but it seems reasonable to hypothesize that host behavior might be altered by the secretion of compounds that alter host physiology, ranging from specifically in neuronal populations to more generally to trigger internal state changes, leading to behavioral outputs.

12.2.5 Sexual Attraction

Some zombie fungi have been observed to alter host sexual behavior, a change that is hypothesized to increase the likelihood of transmission. In the case of *Massospora,* the fungus appears to alter sexual behaviors in infected animals: infected male cicadas produce female-like wing flicks (Cooley et al. 2018) or tolerate physical contact by male conspecifics (Cooley 1999), behaviors that do not occur in healthy males. An increase in sexual behaviors could be an attempt by the host to increase reproductive output before death, but, curiously, these behaviors are only observed in *Massospora*-infected males that actively transmit to another round of hosts and not in infected males that produce resting spores (Cooley et al. 2018). This observation suggests that hypersexuality is driven by the fungus rather than the host. Increased sexual behavior could be related to fungal production of cathinone (Boyce et al. 2019), as amphetamine intake has been associated with increased sexual behaviors in other animals (Frohmader et al. 2010). Alternatively, the fungus could modulate hormonal levels in the host leading to alterations in sexual behavior. At present, the mechanistic basis of this manipulation is unknown.

Entomophthora muscae, on the other hand, enhances transmission by appealing to healthy hosts (Moller 1993; Watson and Petersen 1993; Zurek et al. 2002; Naundrup et al. 2021). Male house flies show a remarkable attraction to late-stage female cadavers, an effect that appears to be mediated by the combined effects of fungal-produced compounds (notably sesquiterpenes), visual cues related to the characteristic death pose as well as endogenous female pheromones (Naundrup et al. 2022). Here, there is no host fitness benefit from attracting mates, as the host is dead, strongly arguing that, again, this is a fungus-driven manipulation rather than a host response to infection. Overall, this work is consistent with many previous observations of fungi employing chemical mimicry to attract hosts, to serve as either transmission vectors or prey (Ngugi and Scherm 2006; Matsuura et al. 2009; Hsueh et al. 2017).

Insect sexual behaviors have also been observed to change in the presence of non-fungal parasites, though we have little insight into their mechanistic underpinnings. One likely manipulated behavior occurs in corn earworm female moths infected with the Hz-2V virus with a primary site of replication in reproductive tissues (Burand et al. 2005). These females produce up to seven times the amount of pheromones made by uninfected females and exhibit hypersexual behaviors, continuing to initiate contact with males and resume calling to attract male conspecifics after mating, behaviors which are typically not observed in healthy individuals (Burand et al. 2005). In addition, healthy females that mated with infected males were more likely to resume calling after copulation than if they mated with uninfected males (Burand and Tan 2006), suggesting that there is some alteration to male reproductive chemistry to dampen normal female refractory period post-mating. As copulation requires close physical contact with a partner, these changes in female pheromone production and sexual behavior are likely to drive increased transmission of the virus to male partners and their subsequent consorts. However, it is possible that these effects could be driven by the host in response to infection as an attempt to increase reproductive output before sterilization takes hold. That the host may be driving these changes might be supported by the observation that infection by another virus, Gonad-specific virus (GSV), also leads to increased pheromonal production (Raina et al. 2000), although, unlike Hz-2 V females, GSV-infected females refuse to mate (Raina et al. 2000). There are other instances of infection by parasites being correlated with increased sexual behavior (milkweed beetle *Labidomera clivicollis* and mite *Chrysomelboai labidomerae*, mite *Unionicola ypsilophora* and midge *Paratrichocladius rufiventris,* cricket *Gryllus texensis* and iridovirus IIV-6/CrIV (McLACHLAN 1999; Abbot and Dill 2001; Adamo et al. 2014)), but, again, it is not yet clear which organisms are driving these changes. The argument has been made that any increase in sexual behavior only benefits the parasite, since these parasites induce sterility in their hosts. However, I do not think we can formally exclude the possibility that these sterile hosts fail to "recognize" their impotency and are driving their own futile attempts to reproduce.

12.3 General Considerations for Understanding Manipulated Behaviors

The five themes I introduced early in this chapter (see **Common themes in behavioral manipulation by zombie fungi**) encapsulate many of the behaviors elicited by zombie fungi, but they are by no means an exhaustive list. In addition, there are likely many more parasitic puppeteers to be described in our world, and more manipulated behaviors out there to be discovered. However, as we continue to broaden our understanding in this realm, it is critical to be guided by parsimony. Just because a behavior occurs in an infected animal and has some apparent level of sophistication, we should not assume that this behavior is manipulated by the pathogen. At a minimum, a truly manipulated behavior is one that confers an adaptive advantage to the parasite and is specifically elicited as a result of infection by that parasite, not just a general sickness response of the host. In general, it is wise to remain skeptical that a behavior reflects manipulation unless there is overwhelming evidence to suggest this is the case. As there are already great pieces on this topic (Poulin 1995; Bhattarai et al. 2021), it will not be fully discussed here.

If all signs point to a behavior being truly the result of parasitic manipulation, it is important to check our innate tendencies to ascribe agency or intent of the hijacking organism, and to keep in mind that all these organisms developed their present methods of host behavior manipulation following broad evolutionary principles. That is, while it may be our impulse to look at a behavior hijacking system and be initially struck by its seeming cleverness or insidiousness, we should resist the urge to embrace complex hypotheses explaining manipulated behaviors in favor of considering simpler ones first. While chemical signaling may underlie a behavior in some cases, mechanical forces and physical limitations of the fungus and animal can be just as crucial in driving behavior. Along these lines, *a priori* reasoning predicts that manipulated behaviors are most likely to occur by ectopic activation of existing host circuitry, not by cobbling together novel circuits. Importantly, all the behaviors that a manipulated host exhibits are things the host can already do—for example, a fly can climb, an ant can bite, a cicada can mate. The fungus is not causing new behaviors to occur but altering the timing of these behaviors to align with promoting the fitness of the fungus rather than the host.

In addition to allowing ourselves to be guided by parsimony, it is also important to avoid assuming that just because two behaviors look alike that they are driven by the same or similar underlying mechanisms (de Bekker et al. 2021). That is, parallel behaviors do not necessarily reflect parallel pathways. This does not mean to suggest that each parasite uses a completely different mechanism to drive similar behavior, but it means that we cannot take for granted that just because we have figured out a mechanism in one system that it will be generally applicable to others. Behavior manipulations that look the same but are driven by different parasites probably reflect one of two possibilities: either these behaviors are so advantageous to the pathogen that they are heavily selected for, or these behaviors are relatively easy to elicit (either there are many ways to drive the same behavior, so there are many local minima that a parasite might stumble across and/or the behavior is governed by readily manipulable circuit elements). Understanding the mechanistic underpinnings of similar behaviors in different systems can help us differentiate between these possibilities. As much as possible, research efforts should be geared toward parallel progress in multiple systems to allow for this comparative analysis.

Finally, as we pursue a deeper understanding of the mechanistic basis of parasite manipulated behaviors, it's important to keep in mind that these behaviors are likely to be the product of alterations in many parallel pathways or many systems targeted simultaneously (Adamo 2013). That is, while it is tempting to hypothesize simple feed-forward mechanisms, evolution is not a linear process, and the robustness of zombified behaviors is very likely a reflection of redundancy in their mechanisms of action. We may find examples where a single gene appears to have a large effect on behavior [e.g., *egt* in baculovirus (Hoover et al. 2011)], but it is very likely that what remains to be discovered are smaller pieces of the whole manipulation story. Ultimately, though this may be a more challenging course to understanding these compelling phenomena, perhaps it will be more satisfying when we finally crack the case.

12.4 Onward and Upward: Prospects in Understanding Behaviors Driven by Zombie Fungi

Despite our keen interest and immense efforts, we still have a long way to go toward fully understanding how zombie fungi alter the behavior of their hosts. Although demonstrating the mechanistic basis of manipulated behavior has proved challenging for many reasons, the future of the field is brighter than ever before.

Historically, one of the main impediments to our mechanistic understanding of the behaviors elicited by zombie fungi has been the lack of experimentally tractable systems. However, notable progress has been made on at least two fronts: both *Ophiocordyceps* and *Entomophthora muscae* infections can now take place entirely in a laboratory setting (Elya et al. 2018; Will et al. 2020). A recently isolated strain of *E. muscae* infects the fruit fly *Drosophila melanogaster* (Elya et al. 2018), an insect that has been extensively studied and developed as a model organism to study neurobiology and behavior. In the *E. muscae*-fruit fly system, researchers can leverage the *D. melanogaster* genetic toolkit address a virtually endless number of hypotheses, including whether gravitactic or phototactic pathways are involved in summiting behavior and whether the fly's clock is involved in determining the time of its death. In addition, the universal genomic editing tool CRISPR-Cas9 holds great promise to be applied to lab-grown organisms, such as *E. muscae* and *Ophiocordyceps*, to test the causal role of fungal genes in myriad host–pathogen interactions, including manipulated behaviors. Taking advantage of these laboratory systems will be key in demonstrating causal relationships between host and fungal genotypes and behavioral phenotype and allow us to improve our understanding in ways that simply were not possible before.

Even for systems that are not yet laboratory ready, the ongoing improvement of existing and development of new technologies has opened research avenues previously inaccessible. For example, long read and chromatin conformation capture sequencing is enabling the assembly of bloated, repeat-rich entomophthoralean genomes (Hajek et al. 2022), and improved metabolomics detection technology recently allowed the discovery of mind-altering compounds in *Massospora*-infected cicadas (Boyce et al. 2019). In addition, improved imaging methods have offered insight into the structure of fungal and host tissues (Fredericksen et al. 2017) and offer the future discovery of morphological correlates of parasitized insects through additional structural analyses. The recent discoveries in the field have in turn generated a level of excitement that seems likely to translate to expansion of the zombie fungus field. With additional brains and hands working together on these systems, it seems inevitable that we will encounter additional culturing breakthroughs to bring new systems into the laboratory. And so, our efforts will continue to feed forward.

Another challenge within the realm of behavior manipulation has been the difficulty of observing and analyzing zombified behaviors efficiently. Many of these behaviors are not only subtle but also occur clustered in daily bursts, making them challenging to study by traditional analog ethological methods. The widespread use of computational methods to track and analyze animal behavior offers a huge opportunity to help move our understanding forward. For lab-tractable zombie systems, automated ethology can help us with establishing objective criteria for staging organisms, which can allow us to perform comparative analyses of metabolites and gene expression and allow us to run behavioral screens under different abiotic conditions (in non-model hosts) or in various host genotypes (in model hosts). Even for systems that are not yet lab-tractable, we can marry ethological platforms with field work to expand our understanding of typical host behavior (e.g., better understand circadian patterns) and test specific mechanistic hypotheses (e.g., does the application of cathinone or psilocybin to healthy cicadas recapitulate either of the locomotion or mating phenotypes observed in *Massospora*-infected individuals?) We have already begun to see the promise of these approaches in the *Ophiocordyceps*-ant system (Trinh et al. 2021).

Ultimately, the prospect of understanding how zombie fungi elicit behavioral changes offers much more than just satiating scientific curiosity. Taking advantage of "zombie" insects as model systems (i.e., studying hijacked behaviors) will serve to complement traditional neuroscience approaches (i.e., studying intrinsic behaviors) and will serve to enrich our understanding of how behavior is generated and executed in animals.

References

Abbot P, Dill LM (2001) Sexually transmitted parasites and sexual selection in the milkweed leaf beetle, Labidomera clivicollis. Oikos 92:91–100. https://doi.org/10.1034/j.1600-0706.2001.920111.x

Adamo SA (2013) Parasites: evolution's neurobiologists. J Exp Biol 216:3–10. https://doi.org/10.1242/jeb.073601

Adamo SA, Kovalko I, Easy RH, Stoltz D (2014) A viral aphrodisiac in the cricket Gryllus texensis. J Exp Biol 217:1970–1976. https://doi.org/10.1242/jeb.103408

Andriolli FS, Ishikawa NK, Vargas-Isla R et al (2019) Do zombie ant fungi turn their hosts into light seekers? Behav Ecol 30:609–616. https://doi.org/10.1093/beheco/ary198

Araújo JPM, Hughes DP (2016) Diversity of entomopathogenic fungi: which groups conquered the insect body? Adv Genet 94:1–39. https://doi.org/10.1016/bs.adgen.2016.01.001

Arvidson R, Kaiser M, Lee SS et al (2019) Parasitoid jewel wasp mounts multipronged neurochemical attack to hijack a host brain. Mol Cell Proteomics 18:99–114. https://doi.org/10.1074/mcp.RA118.000908

Bhattarai UR, Li F, Katuwal Bhattarai M et al (2018) Phototransduction and circadian entrainment are the key pathways in the signaling mechanism for the baculovirus induced tree-top disease in the lepidopteran larvae. Sci Rep 8:17528. https://doi.org/10.1038/s41598-018-35885-4

Bhattarai UR, Doherty J-F, Dowle E, Gemmell NJ (2021) The adaptiveness of host behavioural manipulation assessed using Tinbergen's four questions. Trends Parasitol. https://doi.org/10.1016/j.pt.2021.01.006

Bhunjun C, Senanayake I, Hyde KD, et al (eds) (2022) Numbers of fungal species. Fungal Divers 114

Boyce GR, Gluck-Thaler E, Slot JC et al (2019) Psychoactive plant- and mushroom-associated alkaloids from two behavior modifying cicada pathogens. Fungal Ecol 41:147–164. https://doi.org/10.1016/j.funeco.2019.06.002

Brobyn PJ, Wilding N (1977) Invasive and developmental processes of Entomophthora species infecting aphids. Trans Br Mycol Soc 69:349–366. https://doi.org/10.1016/S0007-1536(77)80072-7

Brobyn PJ, Wilding N (1983) Invasive and developmental processes of Entomophthora muscae infecting houseflies (Musca domestica). Trans Br Mycol Soc 80:1–8. https://doi.org/10.1016/S0007-1536(83)80157-0

Browne LB, Evans DR (1960) Locomotor activity of the blowfly as a function of feeding and starvation. J Insect Physiol 4:27–37. https://doi.org/10.1016/0022-1910(60)90065-2

Burand JP, Tan W (2006) Mate preference and mating behavior of male Helicoverpa zea (Lepidoptera: Noctuidae) infected with the sexually transmitted insect virus Hz-2V. Ann Entomol Soc Am 99:969–973. https://doi.org/10.1603/0013-8746(2006)99[969:MPAMBO]2.0.CO;2

Burand JP, Tan W, Kim W et al (2005) Infection with the insect virus Hz-2v alters mating behavior and pheromone production in female Helicoverpa zea moths. J Insect Sci 5:6. https://doi.org/10.1093/jis/5.1.6

Carruthers RI, Ramos ME, Larkin TS et al (1997) The Entomophaga Grylli (Fresenius) Batko species complex: its biology, ecology, and use for biological control of pest grasshoppers1. Mem Entomol Soc Can 129:329–353. https://doi.org/10.4039/entm129171329-1

Chung T-Y, Sun P-F, Kuo J-I et al (2017) Zombie ant heads are oriented relative to solar cues. Fungal Ecol 25:22–28. https://doi.org/10.1016/j.funeco.2016.10.003

Cohn F (1855) Empusa muscae und die Krankheit der Stubenfliegen. Ein Beitrag zur Lehre von den durch parasitische Pilze charakterisierten Epidemien. Nova acta Academiae Caesareae Leopoldino-Carolinae Germanicae Naturae Curiosorum 25:299–360

Cooley JR (1999) Sexual behavior in North American cicadas of the genera Magicicada and Okanagana. PhD, University of Michigan

Cooley JR, Marshall DC, Hill KBR (2018) A specialized fungal parasite (Massospora cicadina) hijacks the sexual signals of periodical cicadas (Hemiptera: Cicadidae: Magicicada). Sci Rep 8:1432. https://doi.org/10.1038/s41598-018-19813-0

Csata E, Billen J, Barbu-Tudoran L, Markó B (2021) Inside Pandora's box: development of the lethal myrmecopathogenic fungus Pandora formicae within its ant host. Fungal Ecol 50:101022. https://doi.org/10.1016/j.funeco.2020.101022

de Bekker C, Das B (2022) Hijacking time: How Ophiocordyceps fungi could be using ant host clocks to manipulate behavior. Parasite Immunol e12909. https://doi.org/10.1111/pim.12909

de Bekker C, Quevillon LE, Smith PB et al (2014) Species-specific ant brain manipulation by a specialized fungal parasite. BMC Evol Biol 14:166. https://doi.org/10.1186/s12862-014-0166-3

de Bekker C, Ohm RA, Loreto RG et al (2015) Gene expression during zombie ant biting behavior reflects the complexity underlying fungal parasitic behavioral manipulation. BMC Genomics 16:620. https://doi.org/10.1186/s12864-015-1812-x

de Bekker C, Ohm RA, Evans HC et al (2017a) Ant-infecting Ophiocordyceps genomes reveal a high diversity of potential behavioral manipulation genes and a possible major role for enterotoxins. Sci Rep 7:12508. https://doi.org/10.1038/s41598-017-12863-w

de Bekker C, Will I, Hughes DP et al (2017b) Daily rhythms and enrichment patterns in the transcriptome of the behavior-manipulating parasite Ophiocordyceps kimflemingiae. PLoS One 12:e0187170. https://doi.org/10.1371/journal.pone.0187170

de Bekker C, Beckerson WC, Elya C (2021) Mechanisms behind the madness: how do zombie-making fungal entomopathogens affect host behavior to increase transmission? MBio 12:e01872–e01821. https://doi.org/10.1128/mBio.01872-21

Dunlap JC, Loros JJ (2017) Making time: conservation of biological clocks from fungi to animals. Microbiol Spectr 5. https://doi.org/10.1128/microbiolspec.FUNK-0039-2016

Dustan AG (1924) Studies on a new species of empusa parasitic on the green apple bug (Lygus communis var. novascotiensis Knight) in the Annapolis Valley. Acad Entomol Soc, pp 14–36

Elya C, Lok TC, Spencer QE et al (2018) Robust manipulation of the behavior of Drosophila melanogaster by a fungal pathogen in the laboratory. elife 7. https://doi.org/10.7554/eLife.34414

Evans HC, Elliot SL, Hughes DP (2011) Hidden diversity behind the zombie-ant fungus Ophiocordyceps unilateralis: four new species described from carpenter ants in Minas Gerais, Brazil. PLoS One 6:e17024. https://doi.org/10.1371/journal.pone.0017024

Fawcett W (1886) XXXII.—an entomogenous fungus. Ann Mag Nat Hist 18:316–318. https://doi.org/10.1080/00222938609459974

Fredericksen MA, Zhang Y, Hazen ML et al (2017) Three-dimensional visualization and a deep-learning model reveal complex fungal parasite networks in behaviorally manipulated ants. Proc Natl Acad Sci U S A. https://doi.org/10.1073/pnas.1711673114

Frohmader KS, Wiskerke J, Wise RA et al (2010) Methamphetamine acts on subpopulations of neurons regulating sexual behavior in male rats. Neuroscience 166:771–784. https://doi.org/10.1016/j.neuroscience.2009.12.070

Funk CJ, Ramoska WA, Bechtel DB (1993) Histopathology of Entomophaga grylli Pathotype 2 Infections in Melanoplus differentialis. J Invertebr Pathol 61:196–202. https://doi.org/10.1006/jipa.1993.1035

Goater CP, Colwell DD (2007) Epidemiological characteristics of an invading parasite: Dicrocoelium dendriticum in sympatric wapiti and beef cattle in southern Alberta, Canada. J Parasitol 93:491–494. https://doi.org/10.1645/GE-1060R.1

Grimaldi D, Engel MS, Engel MS, Senior Curator and Professor Michael S Engel (2005) Evolution of the Insects. Cambridge University Press

Hajek A, Gryganskyi A, Stajich J (2022) Entomophaga maimaiga ARSEF 7190 v1.0. https://mycocosm.jgi.doe.gov/Entmai1/Entmai1.info.html. Accessed 12 Aug 2022

Han Y, van Houte S, van Oers MM, Ros VID (2017) Timely trigger of caterpillar zombie behaviour: temporal requirements for light in baculovirus-induced tree-top disease. Parasitology 1–6. https://doi.org/10.1017/S0031182017001822

Haspel G, Rosenberg LA, Libersat F (2003) Direct injection of venom by a predatory wasp into cockroach brain. J Neurobiol 56:287–292. https://doi.org/10.1002/neu.10238

Hawksworth DL, Lücking R (2017) Fungal diversity revisited: 2.2 to 3.8 million species. Microbiol Spectr 5. https://doi.org/10.1128/microbiolspec.FUNK-0052-2016

Herbison R, Evans S, Doherty J-F et al (2019) A molecular war: convergent and ontogenetic evidence for adaptive host manipulation in related parasites infecting divergent hosts. Proc Biol Sci 286: 20191827. https://doi.org/10.1098/rspb.2019.1827

Hofmann O (1891) Insektentotende Pilze mit besonderer Berucksichtigung der Nonne. Zeitschrift fur Naturwissenschaften 64:384–385

Hoover K, Grove M, Gardner M et al (2011) A gene for an extended phenotype. Science 333:1401. https://doi.org/10.1126/science.1209199

Hsu CT, Bhandawat V (2016) Organization of descending neurons in Drosophila melanogaster. Sci Rep 6:20259. https://doi.org/10.1038/srep20259

Hsueh Y-P, Gronquist MR, Schwarz EM et al (2017) Nematophagous fungus Arthrobotrys oligospora mimics olfactory cues of sex and food to lure its nematode prey. elife 6. https://doi.org/10.7554/eLife.20023

Hughes B (2011) Conquer or die!: Wellington's Veterans and the liberation of the new world. Bloomsbury Publishing

Hughes DP, Andersen SB, Hywel-Jones NL et al (2011) Behavioral mechanisms and morphological symptoms of zombie ants dying from fungal infection. BMC Ecol 11:13. https://doi.org/10.1186/1472-6785-11-13

Hughes DP, Araújo J, Loreto R et al (2016) From so simple a beginning: the evolution of behavioral manipulation by fungi. In: Advances in genetics. Academic

Humber RA (1976) The systematics of the genus Strongwellsea (Zygomycetes: Entomophthorales). Mycologia 68:1042–1060

Iati M (2021) A fungus could turn some cicadas into sex-crazed "salt shakers of death." *The Washington Post*

Kamita SG, Nagasaka K, Chua JW et al (2005) A baculovirus-encoded protein tyrosine phosphatase gene induces enhanced locomotory activity in a lepidopteran host. Proc Natl Acad Sci U S A 102:2584–2589. https://doi.org/10.1073/pnas.0409457102

Kim K-T, Jeon J, Choi J et al (2016) Kingdom-wide analysis of fungal small secreted proteins (SSPs) reveals their potential role in host association. Front Plant Sci 7:186. https://doi.org/10.3389/fpls.2016.00186

Kokusho R, Katsuma S (2021) Bombyx mori nucleopolyhedrovirus ptp and egt genes are dispensable for triggering enhanced locomotory activity and climbing behavior in Bombyx mandarina larvae. J Invertebr Pathol 107604. https://doi.org/10.1016/j.jip.2021.107604

Krasnoff SB, Watson DW, Gibson DM, Kwan EC (1995) Behavioral effects of the entomopathogenic fungus, Entomophthora muscae on its host Musca domestica: Postural changes in dying hosts and gated pattern of mortality. J Insect Physiol 41:895–903. https://doi.org/10.1016/0022-1910(95)00026-Q

Langley L (2018) These zombie flies are being mind-controlled by a fungus. Natl Geogr

Leonard DE (1970) Effects of starvation on behaviour, number of larval instars, and developmental rate of Porthetria dispar. J Insect Physiol 16:25–31. https://doi.org/10.1016/0022-1910(70)90109-5

Loreto RG, Araújo JPM, Kepler RM et al (2018) Evidence for convergent evolution of host parasitic manipulation in response to environmental conditions: convergent evolution of host manipulation. Evolution 174:424. https://doi.org/10.1111/evo.13489

Lovett B, Macias A, Stajich JE et al (2020) Behavioral betrayal: how select fungal parasites enlist living insects to do their bidding. PLoS Pathog 16: e1008598. https://doi.org/10.1371/journal.ppat.1008598

Lu J (2019) How a parasitic fungus turns ants into "zombies." National Geographic

Mangold CA, Ishler MJ, Loreto RG et al (2019) Zombie ant death grip due to hypercontracted mandibular muscles. J Exp Biol 222. https://doi.org/10.1242/jeb.200683

Martín-Vega D, Garbout A, Ahmed F et al (2018) 3D virtual histology at the host/parasite interface: visualisation of the master manipulator, Dicrocoelium dendriticum, in the brain of its ant host. Sci Rep 8: 8587. https://doi.org/10.1038/s41598-018-26977-2

Matsuura K, Yashiro T, Shimizu K et al (2009) Cuckoo fungus mimics termite eggs by producing the cellulose-digesting enzyme beta-glucosidase. Curr Biol 19:30–36. https://doi.org/10.1016/j.cub.2008.11.030

McLACHLAN A (1999) Parasites promote mating success: the case of a midge and a mite. Anim Behav 57: 1199–1205. https://doi.org/10.1006/anbe.1999.1087

Milner RJ, Holdom DG, Glare TR (1984) Diurnal patterns of mortality in aphids infected by entomophthoran fungi. Entomol Exp Appl 36:37–42. https://doi.org/10.1111/j.1570-7458.1984.tb03404.x

Moller AP (1993) A fungus infecting domestic flies manipulates sexual behaviour of its host. Behav Ecol Sociobiol 33. https://doi.org/10.1007/bf00170255

Naundrup A, Bohman B, Kwadha C et al (2021) A pathogenic fungus uses volatiles to entice male flies into fatal matings with infected female cadavers. bioRxiv:2021.10.21.465334

Naundrup A, Bohman B, Kwadha CA et al (2022) Pathogenic fungus uses volatiles to entice male flies into fatal matings with infected female cadavers. ISME J 16: 2388–2397. https://doi.org/10.1038/s41396-022-01284-x

Ngugi HK, Scherm H (2006) Mimicry in plant-parasitic fungi. FEMS Microbiol Lett 257:171–176. https://doi.org/10.1111/j.1574-6968.2006.00168.x

Nielsen C, Hajek AE (2006) Diurnal pattern of death and sporulation in Entomophaga maimaiga-infected

Lymantria dispar. Entomol Exp Appl 118:237–243. https://doi.org/10.1111/j.1570-7458.2006.00382.x
Nordio S, Kaiser M, Adams ME, Libersat F (2022) Parasitoid wasp venom manipulates host innate behavior via subtype-specific dopamine receptor activation. J Exp Biol 225. https://doi.org/10.1242/jeb.243674
Nuwer R (By Rachel Nuwer on May 27 2021) How this zombie fungus turns cicadas into horror-movie sex bots. Scientific American
O'Reilly DR, Miller LK (1989) A baculovirus blocks insect molting by producing ecdysteroid UDP-glucosyl transferase. Science 245:1110–1112. https://doi.org/10.1126/science.2505387
Ortiz-Urquiza A (2021) The split personality of Beauveria bassiana: understanding the molecular basis of fungal parasitism and mutualism. mSystems 6:e0076621. https://doi.org/10.1128/mSystems.00766-21
Pickford R, Riegert PW (1964) The fungous disease caused by Entomophthora grylli Fres., and its effects on grasshopper populations in Saskatchewan in 1963. Can Entomol 96:1158–1166
Poulin R (1995) "Adaptive" changes in the behaviour of parasitized animals: a critical review. Int J Parasitol 25: 1371–1383. https://doi.org/10.1016/0020-7519(95)00100-x
Raina AK, Adams JR, Lupiani B et al (2000) Further characterization of the gonad-specific virus of corn earworm, Helicoverpa zea. J Invertebr Pathol 76:6–12. https://doi.org/10.1006/jipa.2000.4942
Randrup A, Munkvad I (1967) Stereotyped activities produced by amphetamine in several animal species and man. Psychopharmacologia 11:300–310. https://doi.org/10.1007/BF00404607
Ros VID, van Houte S, Hemerik L, van Oers MM (2015) Baculovirus-induced tree-top disease: how extended is the role of egt as a gene for the extended phenotype? Mol Ecol 24:249–258. https://doi.org/10.1111/mec.13019
Sakolrak B, Blatrix R, Sangwanit U, Kobmoo N (2018) Experimental infection of the ant Polyrhachis furcata with Ophiocordyceps reveals specificity of behavioural manipulation. Fungal Ecol 33:122–124. https://doi.org/10.1016/j.funeco.2018.03.001
Spatafora JW, Chang Y, Benny GL et al (2016) A phylum-level phylogenetic classification of zygomycete fungi based on genome-scale data. Mycologia 108:1028–1046. https://doi.org/10.3852/16-042
St Leger RJ, Wang JB (2020) Metarhizium: jack of all trades, master of many. Open Biol 10:200307. https://doi.org/10.1098/rsob.200307
Steinkraus DC, Hajek AE, Liebherr JK (2017) Zombie soldier beetles: epizootics in the goldenrod soldier beetle, Chauliognathus pensylvanicus (Coleoptera: Cantharidae) caused by Eryniopsis lampyridarum (Entomophthoromycotina: Entomophthoraceae). J Invertebr Pathol 148:51–59. https://doi.org/10.1016/j.jip.2017.05.002
Stork NE (2018) How many species of insects and other terrestrial arthropods are there on earth? Annu Rev Entomol 63:31–45. https://doi.org/10.1146/annurev-ento-020117-043348
Thaxter R (1888) The entomophthoraeae of the United States
Trinh T, Ouellete R, de Bekker C (2021) Getting lost: the fungal hijacking of ant foraging behaviour in space and time. Anim Behav 181:165–184
Valdes JJ, Cameron JE, Cole RJ (1985) Aflatrem: a tremorgenic mycotoxin with acute neurotoxic effects. Environ Health Perspect 62:459–463. https://doi.org/10.1289/ehp.8562459
van Houte S, Ros VID, Mastenbroek TG et al (2012) Protein tyrosine phosphatase-induced hyperactivity is a conserved strategy of a subset of baculoviruses to manipulate lepidopteran host behavior. PLoS One 7: e46933. https://doi.org/10.1371/journal.pone.0046933
van Paridon BJ, Gilleard JS, Colwell DD, Goater CP (2017) Life cycle, host utilization, and ecological fitting for invasive lancet liver fluke, Dicrocoelium dendriticum, emerging in Southern Alberta, Canada. J Parasitol 103:207–212. https://doi.org/10.1645/16-140
Vasconcelos SD, Cory JS, Wilson KR et al (1996) Modified behavior in Baculovirus-infected Lepidopteran larvae and its impact on the spatial distribution of inoculum. Biol Control 7:299–306. https://doi.org/10.1006/bcon.1996.0098
Watson DW, Petersen JJ (1993) Sexual activity of male musca domestica (Diptera: Muscidae) infected with Entomophthora muscae (Entomophthoraceae: Entomophthorales). Biol Control 3:22–26. https://doi.org/10.1006/bcon.1993.1004
Wheater CP (1991) Effect of starvation on locomotor activity in some predacious Coleoptera (Carabidae, Staphylinidae). Coleopt Bull 45:371–378
Will I, Das B, Trinh T, et al (2020) Genetic underpinnings of host manipulation by Ophiocordyceps as revealed by comparative transcriptomics. G3: Genes, Genomes, Genetics 2020.01.03.893917
Wu B, Hussain M, Zhang W et al (2019) Current insights into fungal species diversity and perspective on naming the environmental DNA sequences of fungi. Mycology 10:127–140. https://doi.org/10.1080/21501203.2019.1614106
Yang Z, Yu Y, Zhang V et al (2015) Octopamine mediates starvation-induced hyperactivity in adult Drosophila. Proc Natl Acad Sci U S A. https://doi.org/10.1073/pnas.1417838112
Yong E (2017) How the zombie fungus takes over ants' bodies to control their minds. The Atlantic
Zimmer C (2019) After this fungus turns ants into zombies, their bodies explode. *The New York Times*
Zurek L, Wes Watson D, Krasnoff SB, Schal C (2002) Effect of the entomopathogenic fungus, Entomophthora muscae (Zygomycetes: Entomophthoraceae), on sex pheromone and other cuticular hydrocarbons of the house fly, Musca domestica. J Invertebr Pathol 80:171–176

Genetics and Infection Biology of the Entomopathogenic Fungi 13

Song Hong, Junmei Shang, Yanlei Sun, and Chengshu Wang

Abstract

Insects and fungi, both megadiverse groups of organisms, have a long history of associations with each other. Among the fungi, Hypocreales (Ascomycota) and Entomophthoromycotina (Zoopagomycota), contain the most common pathogens of insects. Ascomycete entomopathogenic fungi (EPF) are mostly well-studied and have been found to be either heterothallically or homothallically sexual with a divergent number of genes at the mating-type loci of different species. So far more than 100 strains of >60 EPF species have been genome sequenced which demonstrated substantial variations in genome size and gene content among each other. However, convergent evolution of fungal entomopathogenicity has been coined, especially in terms of the similar expansion of protease and chitinase gene families in EPF for targeting the protein- and chitin-rich insect cuticles. Functional genetic studies have unveiled an array of EPF genes involved in mediating host recognition and interactions with insect hosts, especially mechanisms of invading or evading host immunity to facilitate fungal colonization of insect body cavities. The discovery of small molecules and their essential roles in mediating EPF associations with insect hosts also have been advanced. In addition to advancing the mechanisms of fungus–insect interactions, these works highly benefit the development of cost-effective mycoinsecticides and the protection of ecologically and economically important beneficial insects.

Keywords

Entomopathogenic fungi · Genome variation · Divergent sexuality · Convergent evolution · Insect-host specificity · Cuticle penetration · Hemocoel colonization · Chemical biology

13.1 Introduction

The largest number of eukaryotes on earth are insects, while fungi are second in estimated species; exact numbers, however, remain a matter of debate (Blackwell 2011; Mora et al. 2011;

S. Hong · J. Shang · Y. Sun
Key Laboratory of Insect Developmental and Evolutionary Biology, CAS Center for Excellence in Molecular Plant Sciences, Shanghai Institute of Plant Physiology and Ecology, Chinese Academy of Sciences, Shanghai, China

C. Wang (✉)
Key Laboratory of Insect Developmental and Evolutionary Biology, CAS Center for Excellence in Molecular Plant Sciences, Shanghai Institute of Plant Physiology and Ecology, Chinese Academy of Sciences, Shanghai, China

School of Life Science and Technology, ShanghaiTech University, Shanghai, China
e-mail: wangcs@sippe.ac.cn

Y.-P. Hsueh, M. Blackwell (eds.), *Fungal Associations*, The Mycota 9,
https://doi.org/10.1007/978-3-031-41648-4_13

Hawksworth and Lücking 2017; Hyde 2022). Fungi and insects have close associations with each other including fungal infection of insects, and numerous fungus–insect interactions including insects using fungi as food and vectoring fungal propagules for dispersal or transmission. Of these, the fungal species that infect and kill insects play a pivotal role in manipulating insect population density in different environments (Savory et al. 2015; Lovett and St Leger 2017; St. Leger 2021). Species of entomopathogenic fungi (EPF) such as the ascomycetes *Beauveria bassiana*, *Metarhizium anisopliae*, and *M. acridum* have long been under development for use as biocontrol agents for eco-friendly management of insect pests (de Faria and Wraight 2007; Wang and Wang 2017). Species of the fungi in Entomophthorales (Entomophthoromycotina) like *Entomophaga maimaiga* have also been successfully used in biological control of the gypsy moth (*Lymantria dispar*) in North America after introduction of the pathogen from Europe (Andreadis and Weseloh 1990).

By using the model organism *Drosophila melanogaster* as a fungal infection model, the basic knowledge of host antifungal immune response has been advanced, especially for the discovery and elucidation of the innate antifungal Toll immune pathway (Ferrandon et al. 2007; Buchon et al. 2014). The investigation of fungal associations with insect hosts thus has both applied and basic research importance. This chapter will summarize the advances of EPF genetics and infection biology largely over the last decade or so.

13.2 Fungal Species that Infect Insects

There are more than 1500 species of fungi reported to infect insects (Vega et al. 2009; Vega and Kaya 2012; Shang et al. 2015; Araújo and Hughes 2016). These EPF are largely distributed in phylum Ascomycota and subphylum Entomophthoromycotina (Zoopagomycota), followed by species in order of descending numbers of EPF of Microsporidia (Cryptomycota), Blastocladiomycota, Basidiomycota, and Mucoromycota as well as the fungus-like oomycete insect pathogens (Humber 2008). Except for the *per os* infection of microsporidian parasites, other EPF infect insects largely through cuticle penetration (Wang and Wang 2017; Sheng et al. 2022).

Species of ascomycetes, especially the cordycipitoid fungi in Hypocreales (Sung et al. 2007) such as *Beauveria* (Cordycipitaceae), *Metarhizium* (Clavicipitaceae), and *Ophiocordyceps* (Ophiocordycipitaceae) are common EPF, and new species continue to be reported frequently (Humber 2017; Imoulan et al. 2017; Mongkolsamrit et al. 2020; Wang et al. 2020a). Phylogenetic analysis of these hypocrealean pathogens suggests that some EPF generally have closer relationships to the plant pathogenic fungi than to mammalian pathogens, implying that some EPF diverged from plant pathogens (Spatafora et al. 2007; Shang et al. 2016; Wang and Wang 2017). In particular, the hypocrealean EPF such as *Metarhizium* and *Beauveria* species have recently been explored as plant endophytes or rhizosphere fungi for mediating nutrient exchanges with plants to promote plant growth and defenses against pests (Behie et al. 2012; Moonjely et al. 2016; Sheng et al. 2022; Luo et al. 2023) and as biofertilizers (Litwin et al. 2020).

In contrast, there are relatively fewer species of fungi other than Hypocreales and Entomophthoromycotina known to kill insects. Exceptions include certain microsporidian species that infect silkworms (*Nosema bombycis*), honeybees (*Nocardia cereanae* and *N. apis*), and shrimp (e.g., *Nos. granulosis*) that have caused huge economic and ecological losses. Another microsporidian species *Antonospora locustae* (previously called *Nos. locustae*) has been used successfully in the biological control of locust pests (Pan et al. 2018). Although many species of insects are hosts for microsporidia, few are killed. Except for a few *Septobasidium* and *Uredinella* species, it is rare to find basidiomycete fungi that can parasitize insects, although seldom kill them. The species *Sporodiniella umbellata*

(Mucorales) is one of the few known species of Mucoromycota known to be a specialized pathogen of insects (Humber 2008).

Obligate zoosporic parasites in *Coelomomyces* and *Catenaria* (Blastocladiomycota) largely infect mosquito larvae or *Daphnia* hosts in freshwater (Gleason et al. 2010). It has been interestingly found that some *Coelomomyces* species such as *C. psorophorae* require two phylogenetically distinct hosts to complete their full life cycle: the diploid (or sporophyte) phase is in mosquitos, while the haploid (or gametophyte) phase is in a copepod or ostracod host (obligate alternate host) (Whisler et al. 1975; Gleason et al. 2010). Reported oomycete insect pathogens, such as the species *Leptolegnia chapmanii*, *Lagenidium giganteum*, and *Pythium guiyangense*, also infect mosquito larvae in water (Shen et al. 2020). Of these, *P. guiyangense* has been tried for mass production and application for mosquito control (Shen et al. 2019). *Pythium guiyangense* kills mosquitos whereas many *Pythium* species are devastating plant pathogens (Shen et al. 2019), another example of a close association between EPF and plant pathogens.

13.3 Fungi Versus Oomycetes

Oomycetes and fungi inhabit similar aquatic and terrestrial habitats. Both groups have members with filamentous growth and sexual and asexual reproduction by spores. Both groups are heterotrophic with lifestyles that include saprobes, parasites, and pathogens of plants and animals. They have osmotrophic nutrition with excreted extracellular enzymes to break down complex molecules in the surrounding environment from which they absorb smaller molecules.

Oomycetes were long regarded as "lower fungi," but with technological advances differences between the two groups became clear. Primary differences include diploid nuclei in the oomycete somatic body and often haploid or alternating haploid and diploid nuclei in the body of the fungi described in this chapter. Oomycete cell walls are largely composed of cellulose (1,3) and (1,6)-β-glucans, and hydroxyproline. The cell walls of fungi are quite different from those of oomycetes, composed of β (1,3)-D-glucan and chitin. The cell wall differences also provide markers to selectively target each group when developing pesticides and therapeutics (Cortés et al. 2019; Usman et al. 2023).

The differences between oomycetes and fungi reflect the different evolutionary histories of the two groups. Although the heterotrophic oomycetes lack plastids, they evolved within a lineage that includes photosynthetic brown and golden-brown algae and diatoms (Thakur et al. 2019). Fungi and animals are opisthokonts, more closely related to each other than to other organisms such as oomycetes and green plants. Increasing evidence from genomics strongly supports that plant pathogenicity of oomycetes such as *Phytophthora* spp. is the result of horizontal gene transfer from plant pathogenic fungi (Richards et al. 2006; Thakur et al. 2019; RoyChowdhury et al. 2022).

As in the case of certain plant pathogens, one of the typical features of EPF is their host specificity. In addition to the intracellular and highly insect-host-specific microsporidian parasites, the zoosporic, and entomophthoralean EPF and oomycetes are also highly host-specific, targeting certain species or a narrow range of insect hosts (Roy et al. 2006; Arnesen et al. 2018). The diverse ascomycete EPF include specialists and generalists that kill a wide range of different orders of insect hosts. In particular, the zombie-ant fungi, *Ophiocordyceps* species, are highly host-specific even to the level of one fungus for one ant species (Kobmoo et al. 2012; Luangsa-Ard et al. 2018). The divergent *Metarhizium* species, however, evolved as either specialists such as *M. acridum* and *M. album* or generalists such as *M. robertsii*, *M. anisopliae*, and *M. brunneum* (Hu et al. 2014; Mongkolsamrit et al. 2020). *Metarhizium* host recognition is typically detected by observing the formation of appressoria by the spores germinated on host cuticles (Shang et al. 2021). For example, *M. acridum* conidia can only form appressoria on locust hosts but not on cockroaches. In

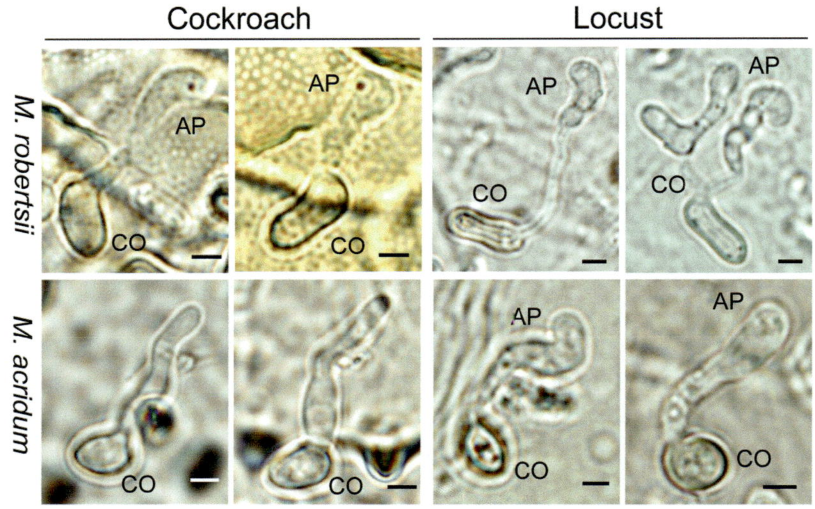

Fig. 13.1 Microscopic imaging of *Metarhizium* spore differentiation on different insects. The conidia (CO) of the generalist *M. robertsii* and locust-specific species *M. acridum* were inoculated on the hind wings of cockroach and locust for 18 h. It was evident that *M. robertsii* could form the infection structure appressoria (AP) on both insects while the specialist could only form appressoria on the locust host, not the non-host cockroach. Bar, 5 μm

contrast, the generalist species *M. robertsii* can produce appressoria on both insects (Fig. 13.1). Variation in host preference and virulence has been used in evaluating biocontrol agents of different strains of the same EPF species, e.g., the same strain of *B. bassiana* may have varied abilities to kill different species of insects (Wraight et al. 2010; Mei et al. 2020).

A typical feature of some EPF species that has brought widespread interest to these fungi is their ability to manipulate host behavior. Some entomophthoralean EPF cause "submit disease" to drive the dying insects to climb upward to the top of plants or alter host copulation behaviors to benefit the transmission of fungal spores (Roy et al. 2006; Hansen and De Fine Licht 2019; Lovett et al. 2020). The infection of some *Ophiocordyceps* species also causes climbing behaviors of infected ants and insects before the formation of incapacitated zombies (de Bekker 2019). Phylogenetic analysis of zombie-ant fungi suggests that these fungi might have diverged from non-manipulating and beetle-infecting ancestors (Araujo and Hughes 2019). Except for the implications of zombie-ant fungi in controlling host muscles and evident invasion of the host nervous system by *Entomophthora muscae* (Fredericksen et al. 2017; Elya et al. 2018), mechanisms in controlling the behaviors of dying insects remain objects of research. Entomophthoralean *Massospora* species infect insects (largely cicadas) that remain active while sporulating and spread their infective spores in flight (Boyce et al. 2019). Continuing research on infections by the diverse EPF should answer questions about the similar behaviors of the distinct hosts.

13.4 Divergent Reproduction Controls

Similar to other fungi, many EPF species undergo asexual, sexual, and/or parasexual reproduction at different developmental stages. Except for *C. militaris* and *Ophiocordyceps* fungi, the sexual state of other cordycipitoid fungi is largely cryptic and rare in nature (Hu et al. 2013; Zheng et al. 2013; Mei et al. 2020). However, genomic analysis revealed that the haploids of *Cordyceps* sensu lato largely contain the idiomorphic mating type (MAT) *MAT1-1* or *MAT1-2* locus, indicating that these EPF species may undergo heterothallic sexual cycles (Fig. 13.2a). Intriguingly, the highly duplicated genome of *O. sinensis* contains both the *MAT1-1* and *MAT1-2* loci (Fig. 13.2b). This fungus is homothallic (Hu et al. 2013; Zheng et al. 2013; Xia et al. 2017a). It is noteworthy that both the *MAT1-1* and *MAT1-2* idiomorphs have also been detected in the single spore isolate of *M. majus* (Hu et al. 2014). Since the sexual state of this fungus remains unclear, it is still hard to conclude that *M. major* is a homothallic fungus. For the two strains of *M. acridum* with full genomes sequenced, CQMa102 is a clear *MAT1-2*-type haploid (Gao et al. 2011); however, the repetitive strain ARSEF 324 was suggested to be diploid/heterozygous with the alternative insertion of transposons at the *MAT1-2* loci (Nielsen et al. 2021).

In addition to the reproductive style difference between the closely-related EPF, the structures of their *MAT* loci are also highly divergent from each other (Fig. 13.2a). For example, the *MAT1-1* loci of *Metarhizium* species contain three genes, i.e., *MAT1-1-1* (encoding an α-domain protein), *MAT1-1-2* (encoding a protein that contains the conserved proline, proline, and phenylalanine residues) and *MAT1-1-3* [(encoding a high mobility group (HMG) domain protein], while the *MAT1-1* loci of *Cordyceps* and *Beauveria* species only have *MAT1-1-1* and *MAT1-1-2*. The *MAT1-2* loci of the haploid EPF strains also differ from each other. The *MAT1-2* idiomorph of *C. militaris* contains only *MAT1-2-1* (encoding an HMG domain protein) whereas an additional *MAT1-2-3* (encoding a function unknown protein) gene is present at the *MAT1-2* locus of *M. acridum* while an extra of *MAT1-2-8* (functionally unknown) occurs in *B. bassiana* (Valero-Jimenez et al. 2016). Intriguingly, the sequenced *MAT1-2* haploid of *C. cicadae* contains a pseudo *MAT1-1-1* gene that encodes a protein short of the critical α-domain (Lu et al. 2017). Gene transcription analysis revealed that this pseudogene is silent during fungal development, and the deletion of both this gene and *MAT1-2-1* did not prevent the formation of asexual-like fruiting bodies.

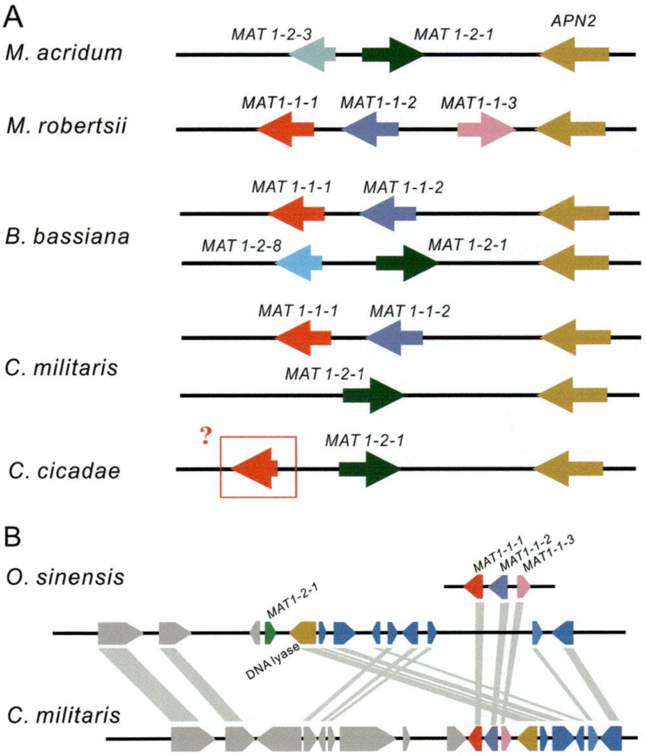

Fig. 13.2 Schematic structuring of the *MAT* loci encoded in different species of hypocrealean ascomycete entomopathogenic fungi. (**a**) Divergent *MAT* structures of the selected heterothallic species. (**b**) Syntenic relationship of the *MAT* loci between the homothallic *O. sinensis* and heterothallic *C. militaris*. The *MAT1-1* idiomorph comprises the α-domain gene (*MAT1-1-1*), the proline-proline-phenylalanine-rich domain gene (*MAT1-1-2*), and with or without an HMG-box domain gene (*MAT1-1-3*). The *MAT1-2* idiomorph includes an HMG domain gene (*MAT1-2-1*) with or without a small open reading frame (e.g., *MAT1-2-8* in *B. bassiana*) of unclear function

Despite the essential role of *MAT* loci in controlling fungal sexuality, the induction of fruiting body formation revealed that both the *MAT1-1* and *MAT1-2* isolates of *C. militaris* could fruit without the production of sexual perithecia and ascospores, which suggests that fungal fruiting is independent of mating (Zheng et al. 2011). Similar asexual fruiting has been observed in ascomycetes such as *Morchella sextelata* (Liu et al. 2022). However, the pairing test of the *MAT1-1* and *MAT1-2* deletion mutants failed to obtain fruiting bodies of *C. militaris* (Lu et al. 2016). Otherwise, the cross of the Δ*MAT1-1-1* with MAT1-2 isolates abolished the production of sexual perithecia and ascospores while the mating of Δ*MAT1-1-2* and *MAT1-2* led to fruiting body formation with barren perithecia (Lu et al. 2016). It is still enigmatic that the *MAT1-1/MAT1-2* ratio of the *C. militaris* conidia formed by newly collected wild and sexual samples failed to reach a 50:50 rate but is highly biased toward the *MAT1-1* type (Zheng et al. 2011). Unequal prevalence of mating types also occurs after the sub-culturing of *C. militaris* strains, which was found in association with fungal culture degeneration (Chen et al. 2017).

Even infrequently occurring, the sexual state of *B. bassiana* (*Cordyceps bassiana*) has been observed and was inducible under laboratory conditions (Li et al. 2001). Population genomic studies revealed that *B. bassiana* largely maintains clonal reproduction in nature, and the *MAT1-1/MAT1-2* ratio of the isolates collected from forest environments of southern China was close to 50% (Mei et al. 2020). In contrast, the mating types were highly biased toward the *MAT1-1* type (65%) for the *B. bassiana* population sampled in Denmark (Meyling et al. 2009), and extremely skewed (*MAT1-1/MAT1-2* = 146/4) for the populations sampled in Cuba (Ramos et al. 2017). Variation of *MAT* ratios was also found among sub-populations of *B. bassiana* (Cai et al. 2022). Likewise, the population survey of *Metarhizium* species also suggests a dominance of clonal reproduction of these fungi. Interestingly, however, the distribution pattern of *MAT1-1/MAT1-2* ratios is associated with the isolate haplogroups (Rehner 2020). For example, the haplogroup 1 of *M. robertsii* isolates ($n = 216$) all belong to the *MAT1-1* genotype while the ratios of the haplogroup 2 (*MAT1-1/MAT1-2* = 70/16) and 3 (*MAT1-1/MAT1-2* = 6/58) isolates are biased to be either *MAT1-1* or *MAT1-2* idiomorph. The isolates of *M. brunneum* collected in the USA are skewed to be *MAT1-2* (30/37) genotype (Rehner 2020). The cause and effect of disparate mating-type distributions in fungal populations are still puzzling. As for entomophthoralean EPF, except for the known sexual form of zygospores, the *MAT* loci and reproductive control of these fungi are still unclear due to the complexity of genome architecture in these fungi.

13.5 Genomic Advances of the Entomopathogenic Fungi

By taking advantage of the technical advances in genome sequencing, genomic studies of EPF species have thrived for more than 10 years (Wang et al. 2016; Wang and Wang 2017). To date, there are more than 110 strains belonging to 65 species of EPF with genome sequences reported (Table 13.1). Of these most belong to the hypocrealean ascomycete pathogens followed by the entomophthoralean and microsporidian species. A few species of oomycete insect pathogens have also been sequenced with different genome sizes, while none of the blastocladian insect parasites has been sequenced. Even though fungal entomopathogenicity has diverged multiple times (Wang and Wang 2017), it is still surprising to see the considerable differences in genome sizes and contents among the inter- and intra-phylum of these fungi (Table 13.1). In particular, the genome sizes of the entomophthoralean species such as *Massospora platypediae* (>2 gigabytes, Gb), *Massospora cicadina* (ca. 1.5 Gb), and *Entomophthora maimaiga* (ca. 1.2 Gb) reach the gigabyte levels whereas those of the strict intracellular microsporidia EPE are usually less than 10 megabytes (Mb). Huge genome size variations are also present among the intra-phylum species. For example, small genomes are also present for entomophthoralean fungi such as

Table 13.1 Information of the entomopathogenic fungal or oomycete species and strains with sequenced genomes

Species	Strain	Genome size (Mb)	Gene number	NCBI Accessions	References
Ascomycetes					
Akanthomyces lecanii (Lecanicillium lecanii)	RCEF 1005	35.6	11,030	AZHF00000000	Shang et al. (2016)
Akanthomyces lecanii (Lecanicillium lecanii)	UM487	32.6	8126	LUKN00000000	NP
Aschersonia aleyrodi (Moelleriella libera)	RCEF 2490	30.9	8461	AZGY00000000	Shang et al. (2016)
Aschersonia badia (Hypocrella siamensis)	MTCC 10142	28.8	9292	JMQE00000000	Agrawal et al. (2016)
Ascosphaera apis	ARSEF 7405	20.4	6442	AZGZ00000000	Shang et al. (2016)
Beauveria bassiana	ARSEF 2860	33.7	10,366	ADAH00000000	Xiao et al. (2012)
Beauveria bassiana	D1-5	34.7	11,861	ANFO00000000	NP
Beauveria bassiana	ARSEF 5078	34.6	10,831	JTCZ00000000	Valero-Jimenez et al. (2016)
Beauveria bassiana	ARSEF 1520	35.1	10,599	JTCW00000000	
Beauveria bassiana	ARSEF 4305	34.7	10,513	JTCY00000000	
Beauveria bassiana	ARSEF 8028	34.7	10,210	JRHA00000000	
Beauveria bassiana	ARSEF 2579	36.7	10,753	JTCX00000000	
Beauveria bassiana	ARSEF 3097	32.2	NA	JABAOI000000000	NP
Beauveria bassiana	JEF-007	36.4	10,695	MRVG00000000	Lee et al. (2018)
Beauveria bassiana	ERL836	35.45	10,399	PPTI00000000	NP
Beauveria bassiana	JEF-350	35.66	11,225	JAAZTN000000000	Kim et al. (2022)
Beauveria bassiana	BV-ECA0	32.8	9593	JACVNG000000000	Castro-Vásquez et al. (2022)
Beauveria bassiana	BV-ECA1	33	9636	JACVNF000000000	
Beauveria bassiana	BV-ECA13	32.8	9566	JACVNE000000000	
Beauveria bassiana	BV-ECA26	32.16	9465	JACVND000000000	
Beauveria bassiana	BV-ECA27	32.97	9648	JACVNC000000000	
Beauveria bassiana	BV-ECA31	33	9546	JACVNB000000000	
Beauveria bassiana	BV-ECA43	34.5	10,210	JACVNA000000000	
Beauveria bassiana	BV-ECA44	34.8	10,453	JACVMZ000000000	
Beauveria brongniartii	RCEF 3172	32.4	9595	AZHA00000000	Shang et al. (2016)
Beauveria felina	SYSU-MS7908	29.4	NA	JAEMHR000000000	NP
Beauveria rudraprayagi	MTCC 8017	36.6	NA	JMNB00000000	NP
Beauveria pseudobassiana	KACC 47484	34.51	NA	LUDX00000000	NP
Beauveria pseudobassiana	RGM 2184	34.5	8469	JAKJXD000000000	Altimira et al. (2022)
Cordyceps cicadae	CC02	33.3	9701	MWMN00000000	Lu et al. (2017)
Cordyceps cicadae	BA-001	33.3	9627	AEIW00000000	NP
Cordyceps militaris	Cm01	32.2	9684	AEVU00000000	Zheng et al. (2011)
Cordyceps militaris	ATCC 34164	33.6	9371	PRJNA323705	Kramer and Nodwell (2017)
Cordyceps militaris	HN01	32.3	10,095	MQTM00000000	Chen et al. (2019)
Cordyceps guangdongensis	GD15	29.05	9150	NRQP00000000	Zhang et al. (2018)
Keithomyces neogunnii (Metacordyceps neogunnii)	Cg7.2a	32.6	8641	JAIOWQ000000000	Long et al. (2022)
Hirsutella thompsonii	MTCC 3556	34.6	9756	APKB00000000	Agrawal et al. (2015)
Hirsutella thompsonii	MTCC 6686	34.7	10,259	APKU00000000	

(continued)

Table 13.1 (continued)

Species	Strain	Genome size (Mb)	Gene number	NCBI Accessions	References
Isaria fumosorosea (Cordyceps fumosorosea)	ARSEF 2679	33.5	10,060	AZHB00000000	Shang et al. (2016)
Isaria javanica (Cordyceps javanica)	IJ1G	34.93	11,441	SPUK00000000	Lin et al. (2019)
Isaria javanica (Cordyceps javanica)	IJ2G	34.97	11,143	SPUL00000000	
Metarhizium acridum	CQMa 102	39.4	9849	ADNI00000000	Gao et al. (2011)
Metarhizium acridum	ARSEF 324H1	44.7	12,536	JAGRQK000000000	Nielsen et al. (2021)
Metarhizium acridum	ARSEF 324H2	39.2	10,197	JAGRQL000000000	
Metarhizium album	ARSEF 1941	30.5	8472	AZHE00000000	Hu et al. (2014)
Metarhizium anisopliae	BRIP 53293	38.6	11,415	APNB00000000	Pattemore et al. (2014)
Metarhizium anisopliae	BRIP 53284	37.9	11,456	APNC00000000	NP
Metarhizium anisopliae	ARSEF 549	38.5	10,891	AZNF00000000	Hu et al. (2014)
Metarhizium anisopliae	CQMa421	45.3	NA	PRJNA644049	NP
Metarhizium anisopliae	JEF-290	42.9	13,646	JABUBV000000000	Lee et al. (2019)
Metarhizium anisopliae	E6	38.4	10,817	JNNZ00000000	Staats et al. (2014)
Metarhizium brunneum	ARSEF 3297	37.1	10,689	AZNG00000000	Hu et al. (2014)
Metarhizium brunneum	ARSEF 4556	37.4	11,420	PRJNA608152	Saud et al. (2021)
Metarhizium cylindrosporum	RCEF 6700	32.1	8596	PRJNA817331	NP
Metarhizium guniujiangensis	RCEF 0880	30.1	8652	PRJNA817331	NP
Metarhizium frigidum	ARSEF 4124	50.03	13,547	PRJNA272623	NP
Metarhizium guizhouense	ARSEF 977	43.2	11,787	AZNH00000000	Hu et al. (2014)
Metarhizium humberi	ESALQ1638	38.3	10,633	JACEFI000000000	Iwanicki et al. (2022)
Metarhizium majus	ARSEF 297	42.1	11,535	AZNE00000000	Hu et al. (2014)
Metarhizium rileyi	RCEF 4871	32	8764	AZHC00000000	Shang et al. (2016)
Metarhizium robertsii	ARSEF 23	41.7	11,689	ADNJ02000000	Gao et al. (2011)
Metarhizium robertsii	ARSEF 2575	39.7	12,384	JELW00000000	NP
Ophiocordyceps sinensis	CO 18	120	6972	ANOV00000000	Hu et al. (2013)
Ophiocordyceps sinensis	1229	114	6972	LKHE00000000	Li et al. (2016)
Ophiocordyceps sinensis	ZJB12195	101.1	9961	LWBQ00000000	NP
Ophiocordyceps sinensis	IOZ07	110.8	8916	JAAVMX000000000	Shu et al. (2020)
Ophiocordyceps sinensis	CC1406–203	119.2	8621	NGJJ00000000	Liu et al. (2020b)
Ophiocordyceps sinensis	Nyingchi	124.8	7939	PRJNA382001	Xia et al. (2017a)
Ophiocordyceps camponoti-floridani	EC05	30.5	7455	JAACLJ000000000	Will et al. (2020)
Ophiocordyceps camponoti-leonardi	BCC 80369	37.91	7046	PDHP00000000	Kobmoo et al. (2018)
Ophiocordyceps camponoti-saundersi	BCC 79314	49.27	6946	PDHQ00000000	
Ophiocordyceps polyrhachis-furcata	BCC 54312	43	6799	LKCN00000000	
Ophiocordyceps camponoti-rufipedis	Map-16	21.9	7621	NJES00000000	de Bekker et al. (2015), de Bekker et al. (2017)
Ophiocordyceps unilateralis	1348a	22.2	7995	NJEU00000000	

(continued)

Table 13.1 (continued)

Species	Strain	Genome size (Mb)	Gene number	NCBI Accessions	References
Ophiocordyceps unilateralis	Map-64	23.3	8174	NJET00000000	
Ophiocordyceps unilateralis	SC16a	23.9	8629	LAZP00000000	
Ophiocordyceps subramanianii	1346	32.3	11,275	NA	
Ophiocordyceps unilateralis	CCMB 661	30.4	NA	JACJUF000000000	NP
Ophiocordyceps polyrhachis-furcat	BCC 54312	43.3	10,146	LKCN00000000	Wichadakul et al. (2015)
Ophiocordyceps xuefengensis	HCMA 001	23.97	NA	JACYFI000000000	NP
Sporothrix insectorum	RCEF 264	34.7	9496	AZHD00000000	Shang et al. (2016)
Tolypocladium inflatum	NRRL 8044	30.4	9998	AOHE00000000	Bushley et al. (2013)
Torrubiella hemipterigena	BCC 1449	28.5	11,471	CDHN00000000	Horn et al. (2015)
Microsporidia					
Nosema ceranae	BRL01	7.86	2614	ACOL00000000	Cornman et al. (2009)
Nosema ceranae	PA08 1199	5.69	3209	JPQZ00000000	Pelin et al. (2015)
Nosema ceranae	BRL	8.82	2288	MUP00000000	Huang et al. (2021)
Nosema bombycis	CQ1	15.69	4458	ACJZ00000000	Pan et al. (2013)
Nosema antheraeae	YY	6.6	3413	NA	
Nosema apis	BRL 01	8.57	2771	ANPH00000000	Chen et al. (2013)
Antonospora locustae (Nosema locustae)	CLX	3.17	1857	MPSM00000000	Chen et al. (2020b)
Nosema granulosis	Ou3-Ou53	8.86	3639	SBJO00000000	Cormier et al. (2021)
Edhazardia aedis	USNM 41457	51.35	4190	AFBI00000000	Desjardins et al. (2015)
Vavraia culicis	Floridensis	6.12	2773	AEUG00000000	
Tubulinosema ratisbonensis	Franzen	7.56	3016	RCSS00000000	NP
Entomophthorales					
Basidiobolus meristosporus	CBS 931.73	88.8	16,111	MCFE00000000	Mondo et al. (2017), Arnesen et al. (2018)
Basidiobolus heterosporus	B8920	44.05	NA	JNET00000000	NP
Conidiobolus coronatus	NRRL 28638	39.9	10,635	JXYT00000000	Chang et al. (2015)
Conidiobolus incongruus	B7586	93.9	NA	JNEM00000000	NP
Neoconidiobolus thromboides (Conidiobolus thromboides)	FSU 785	24.4	8867	PRJNA196084	NP
Capillidium heterosporum (Conidiobolus heterosporus)	CGMCC3.15857	33.4	10,857	WBMQ00000000	Wang et al. (2020b)
Entomophthora muscae	Berkeley	1228	21,712	PRJNA479887	Wang et al. (2018)
Entomophthora muscae	Air source	24.29	NA	FCOS00000000	NP
Entomophthora maimaiga	ARSEF 7190	1177.4	14,701	MycoCosm, JGI	NP
Massospora cicadina	MCPNR19	1488.88	5453	JAKSZP000000000	Boyce et al. (2019)
Massospora platypediae	NA	2054.3	NA	QKRY000000000	
Massospora cicadina	MICH 231384	766.6	NA	QMCF00000000	NP
Zoophthora radicans	ATCC 208865	655.2	14,479	MycoCosm, JGI	NP

(continued)

Table 13.1 (continued)

Species	Strain	Genome size (Mb)	Gene number	NCBI Accessions	References
Mucoromycetes					
Sporodiniella umbellata	MES 1446	26.24	10,781	MycoCosm, JGI	NP
Oomycetes					
Pythium guiyangense	Su	110	30,943	QXDM00000000	Shen et al. (2019)
Lagenidium giganteum	ARSEF 373	126.3	NA	NSDO00000000	NP
Lagenidium giganteum	MTLA-03	70.79	NA	PSQM00000000	NP

NA not available, *NP* not published

Neoconidiobolus thromboides (previously *Conidiobolus thromboides*, 24.4 Mb) and *Capillidium heterosporum* (previously *Conidiobolus heterosporus*, 33.4 Mb). Except for the highly repetitive (>40%) and expanded genomes of *Ophiocordyceps sinensis* (>100 Mb) (Hu et al. 2013; Xia et al. 2017a; Shu et al. 2020), the ascomycete EPF have typical genome sizes of 30–50 Mb similar to most filamentous plant and saprobic fungi. Due to the losses of essential metabolic pathways for the synthesis of amino acids, nucleotides, and even ATP (Nakjang et al. 2013), the compact genomes of obligate entomopathogenic microsporidia are largely smaller than those of other EPF. However, size variations also exist between the microsporidian parasites of mosquitos from a few megabytes (e.g., *Vavraia culicis*, 6.12 Mb) to >50 Mb (e.g., *Edhazardia aedis*, 51.3 Mb).

Until the employment of the third generation of sequencing techniques, it was technically difficult and costly to obtain the full genome sequences of EPF, as with other eukaryotes. Among these genome-sequenced EPF, only two ascomycete strains, i.e., *Metarhizium brunneum* ARSEF 4556 (Saud et al. 2021) and *Cordyceps militaris* ATCC 34164 (Kramer and Nodwell 2017) have been sequenced and assembled with seven chromosomes each from telomere to telomere. Surely, it does not mean that the closely-related cordycipitoid EPF would have the same number of chromosomes. At least, similar to the presence or absence of conditionally dispensable (CD) chromosomes among the strains of *Fusarium oxysporum* (Ma et al. 2010), CD-like chromosome loss was also observed in *M. anisopliae* (Wang et al. 2003). It is a bit surprising that the small-sized microsporidian genomes have higher numbers of chromosomes than these two ascomycete EPF. For example, the locust parasite *Antonospora locustae* has 17 chromosomes (Chen et al. 2020b), and the silkworm microsporidia *Nosema bombycis* has 18 chromosomes (Pan et al. 2013). It is also noteworthy that, in contrast to the approximately 50% GC content of filamentous fungal genomes, the genomes of some microsporidian parasites are highly AT rich, e.g., a ratio of 74% AT for the genome of the honeybee parasite *N. ceranae* (Cornman et al. 2009), 78% AT for the mosquito parasite *E. aedis* (Desjardins et al. 2015), and 69% AT for the genome of *N. bombycis* (Pan et al. 2013). Even the genome of *A. locustae* having an overall ratio of 42% GC content, the ultra-low GC region (ca. 20% of 25 kb) is present in most chromosomes (Chen et al. 2020b). The selection and function of this kind of evolutionary output remain elusive. Similar to the other early diverged fungi, the genome of the entomophthoralean fungus *Basidiobolus meristosporus* is rich with methylated adenines (i.e., N6-methyldeoxyadenine, 6 mA), a modification associated with the magnitude of gene expressions (Mondo et al. 2017).

Comparative genomic analysis of ascomycete EPF revealed that fungal entomopathogenicity is characteristically associated with their genomic expansion of protease and chitinase protein families to degrade the protein- and chitin-rich insect cuticles (Shang et al. 2016). Serine proteases are also rich in the entomophthoralean pathogens (Arnesen et al. 2018), and in the oomycete pathogen *Pythium guiyangense* (Shen et al. 2019). In contrast, plant pathogens typically evolved the expanded families of carbohydrate-active enzyme genes for targeting plant cell walls. However, an array of putative effector genes is similarly present in the genomes of EPF and plant pathogens (Shang et al. 2016; Wang 2020).

Regarding the evolution of EPF host specificity, comparative and phylogenomic analysis of *Metarhizium* species with different host ranges revealed a speciation trajectory of *Metarhizium* with species diverging from specialists to generalists via intermediate transitional species with narrow host ranges (Hu et al. 2014). During this process, extensive protein family expansions occurred largely through gene duplications, which includes obtaining secondary metabolic gene clusters in the generalist *Metarhizium* species such as the biosynthetic gene cluster (BGC) of non-selective insecticidal destruxins (Wang et al. 2012; Hu et al. 2014). Duplication of polyketide synthase (PKS) BGC in *Metarhizium* species can benefit the fungi to adapt to diverse environments for stress resistance (Chen et al. 2015; Zeng et al. 2018). In particular, the generalists evolved higher numbers of G-protein coupled receptors (GPCRs) for the recognition of diverse hosts than the specialist species did (Wang et al. 2016; Shang et al. 2021; Shang et al. 2022).

The genome-wide analysis of the index of repeat-induced point (RIP) mutations for each *Metarhizium* species suggested that the RIP genome-defense mechanism against gene duplication would occur in specialist but not generalist *Metarhizium* species (Hu et al. 2014; Nielsen et al. 2021). Since the RIP process occurs only after fertilization and immediately prior to meiosis (van Wyk et al. 2021), the findings suggest that the specialist *Metarhizium* would maintain the ability to reproduce sexually (St Leger and Wang 2020). In support to some extent, *C. militaris* that commonly perform sexual reproduction is short of highly similar paralogous genes and maintains a high-value RIP index (Zheng et al. 2011). In contrast, the largely asexual relative *B. bassiana* lacks the RIP mechanism of genome defense and encodes many duplicated paralogues (Xiao et al. 2012; Valero-Jimenez et al. 2016). It is interesting to note that, unexpectedly, the *Aedes aegypti*-specific microsporidian *Edhazardia aedis* has a much larger genome and many more genes (51.3 Mb; 4190 genes) than those of *Vavraia culicis* (6.1 Mb; 2773 genes), a microsporidian generalist of mosquitos. Nevertheless, *E. aedis* apparently diverged earlier than did the generalist *V. culicis* (Desjardins et al. 2015).

As shown in Table 13.1, genomes of multiple strains have been sequenced for quite a few EPF species. Besides the alternative *MAT* idiomorph of sequenced strains, the data reveal the varied degree of genome size and gene content differences among strains. In particular, the genome size of *Massospora cicadina* strain MCPNR19 (ca. 1.5 Gb) almost doubles that of the strain MICH 231384 of *M. cicadina* (ca. 766.6 Mb). A similar pattern is observed between the two sequenced strains of the oomycete *Lagenidium giganteum* (Table 13.1). For the two sequenced strains of *M. acridum*, strain CQMa102 has a clear haploid genomic architecture (Gao et al. 2011). In contrast, the sequence data of *M. acridum* ARSEF 324 strain were assembled into two hybrid genomes H1 and H2, which have different genome sizes and gene contents from each other (Nielsen et al. 2021). Overall, comparative genomic studies of inter- and intra-species of EPF are still required. Taken together with functional genetic investigations, the diverse biology of these fungi can then be mapped to their genomic blueprints.

13.6 Molecular Interactions Among Fungi and Insects

By benefiting from the advent of EPF genomics as well as tractable transformation techniques, functional genetic studies have been greatly increased in recent years to unveil the molecular associations between fungi and insects (Wang and Wang 2017; Li and Xia 2022; Tang et al. 2022). By using the species of *B. bassiana*, *M. robertsii*, *M. anisopliae*, and *M. acridum* as models, main efforts have focused on the mechanisms in invading or evading host immunity involved in fungal penetration of insect cuticles and colonization of insect body cavities.

Shortly after spores land on and adhere to insect cuticles, the spores of EPF germinate to initiate host recognition, followed by penetration of cuticles (Shang et al. 2023). A finely-tuned cascade has been unveiled in *M. robertsii* to distinguish cuticle and insect-hemocoel micro-niches during fungal infection of insects (Zhang et al. 2021). In observations of this process, GPCRs have been demonstrated to have differential expression in different *Metarhizium* species when induced on different insects (Gao et al. 2011). Loss-of-function investigation of different GPCRs in *M. robertsii* revealed that alternative members contribute at varied levels to the fungal recognition of different insect species, which was reflected by the ratio of appressorium formation. Interestingly, one of the ten-GPCR gene deletion mutants lost its ability to form appressoria on all hosts examined due to the impairment of the nucleus translocation of the mitogen-activated protein kinase (MAPK) Mero-Fus3 (Shang et al. 2021). This behavior is consistent with the finding that Mero-Fus3 has been verified to be only one of four MAPKs required for mediating appressorium formation in *M. robertsii* (Chen et al. 2016). The histone lysine methyltransferase of *M. robertsii* is also involved in the epigenetic regulation of appressorium formation (Lai et al. 2020). Coupled with host recognition, EPF will secrete different proteases and chitinases into the micro-niche of mucilage for host cuticle degradation (Ortiz-Urquiza and Keyhani 2013; Zhang et al. 2021). The formation of mucilage was unveiled in *Metarhizium* species through the production of the exopolysaccharide galactosaminogalactan (GAG) by a five-gene cluster. On top of the function of adhesin Mad1 (Wang and St Leger 2007a), GAG production can additively facilitate spore adherence to cuticles and the formation of biofilms on hydrophobic surfaces (Mei et al. 2021).

Aside from the differentiation of infection structures and secretion of cuticle-degradation enzymes (Gao et al. 2020), the generation of proper turgor pressure within infection cells is also a critical prerequisite for the EPF penetration of insect cuticles, which is similar to plant pathogens such as *Magnaporthe oryzae* (Wilson and Talbot 2009). However, in contrast to the requirement of glycogen in *M. oryzae* (Badaruddin et al. 2013), the degradation of triacylglycerol (TAG) instead of glycogen is the main source for the solute buildup in *M. robertsii* (Huang et al. 2019; Li et al. 2022). Thus, the genes involved in TAG biogenesis and degradation (into glycerol) are critically needed in *M. robertsii* for cuticle penetration and therefore pathogenicity (Gao et al. 2013, 2016; Chen et al. 2018a).

Similar to animal cells, TAG is stored in liquid droplets (LDs) in fungi. The key LD-surface protein, perilipin-like protein Mpl1, was found in *Metarhizium* to contribute to maintaining the homeostasis of LDs (Wang and St Leger 2007b). Insect cuticle surfaces are nutrient poor, which will induce autophagic responses at the early stage of EPF infection (Duan et al. 2013; Ying and Feng 2019). LD isolations followed by lipidomics and proteomics analysis of *M. robertsii* uncovered the potential mechanisms of a short period of TAG increase in fungal cells (Chen et al. 2018a, b). In particular, upon nitrogen starvation, the core enzyme MrATG8 and the stress-related heat shock proteins (HSPs) were activated, whereas the turnover of lipases and the ATG proteins such as ATG1, ATG7, ATG11, and ATG20 involved in the internalization of LDs into vacuoles were downregulated when referred to fungal growth in a nutrient-rich medium (Fig. 13.3). In addition, the species of phospholipids were differentially regulated by different proteins (Chen et al. 2018a). Consistent with this notion, the machinery of microlipophagy (i.e., LD entry into vacuoles without the function of autophagosomes) instead of macroautophagy, will be followed after TAG accumulation to degrade LDs for building up the appressorial turgor pressure (Li et al. 2022).

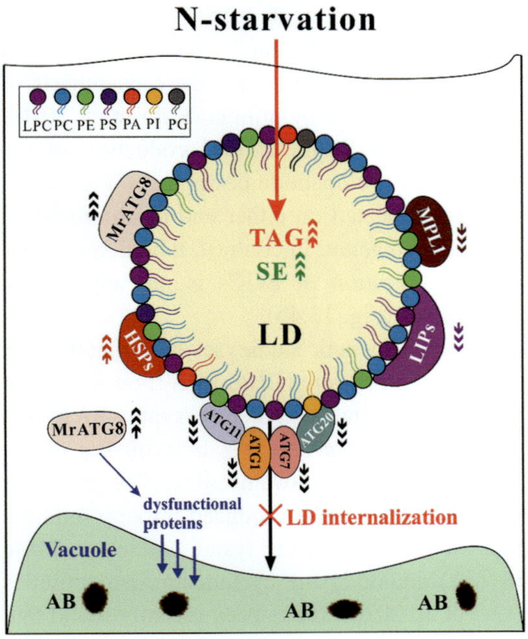

Fig. 13.3 Lipid metabolism implicated at the early stage of *M. robertsii* infections. *TAG* triacylglycerol, *SE* steryl esters, *AB* autophagic body, *LC* lysophosphatidylcholine, *PC* phosphatidylcholine, *PE* phosphatidylethanolamine, *PS* phosphatidylserine, *PA* phosphatidic acid, *PI* phosphatidylinositol, *PG* phosphatidylglycerol, *HSPs* heat shock proteins, *LIPs* lipases, *LD* lipid droplets. Up arrows, increase; down arrows, decrease

It is noteworthy that the microbiotas assembled on insect cuticles have great potential in mediating colonization resistance against fungal parasite infections (Bruner-Montero et al. 2021; Pessotti et al. 2021; Hong et al. 2022; Janke et al. 2022). Fungal pathogens have evolved additional abilities or strategies by producing antimicrobial peptides and/or antibiotic compounds to outcompete insect surface microorganisms to facilitate fungal penetration of insect cuticles (Sun et al. 2022a, b; Hong et al. 2023).

Once reaching insect hemocoels, EPF cells encounter the extensive host cellular and humoral immune defenses (Wang and Wang 2017; Qu and Wang 2018). Different studies have revealed the diverse strategies employed by EPF to counteract insect antifungal immune responses. For example, cell wall remodeling to reduce glucan content and mask cell wall components with a collagen-like protein Mcl1, occur to evade cellular immunity once the cells of *M. anisopliae* enter the caterpillar body cavity (Wang and St Leger 2006). Otherwise, similar to the function of the LysM effectors of plant pathogens (Kombrink and Thomma 2013), the LysM-like proteins encoded by *B. bassiana* can bind the cell wall chitin or chitin oligomers to evade insect immunity (Cen et al. 2017). The study has also shown that *B. bassiana* can employ the microRNA-like RNA to target and suppress the Toll receptor ligand Spätzle 4 of mosquitos to invade host immunity (Cui et al. 2019). The M35 family of metalloprotease of *M. robertsii* requires degrading host prophenoloxidases to suppress insect cuticular and hemocyte melanization activities (Huang et al. 2020). Consistent with the finding of the requirement of Niemann-Pick disease type C2 (Npc2) proteins for antibacterial response in *Drosophila melanogaster* (Shi et al. 2012), an insect-like *Npc2* horizontally obtained in *M. robertsii* was required for fungal virulence to sequester host cholesterol for fungal sterol biosynthesis during its quick replication in insect hemocoels (Zhao et al. 2014).

As indicated above, similar to plant pathogens, EPF genomes also encode an array of effector-like proteins (Shang et al. 2016). However, due to the lack of plant apoplast structures and the short/lack of the intracellular nucleotide-binding and oligomerization domain NOD-like receptors (NLRs) in animal cells (Jones et al. 2016; Wang 2020), the effector-like interactive mechanisms have been questioned in fungus–animal interactions (Lowe and Howlett 2012). Increasing evidence, however, has shown that the pathogenic fungi of insects, nematodes, and mammals may similarly employ the secreted effectors as virulence factors to suppress host immunity (Stuart et al. 2013; Lopes Fischer et al. 2020; Wang 2020; Fischer and Requena 2022). For example, the secreted effector-like proteins of the human pathogens *Candida albicans* and *Cryptococcus neoformans* can manipulate host immune responses via toll-like receptors (Wang et al. 2019b; Dang et al. 2022). A cytolytic peptide candidalysin secreted by *C. albicans* can damage epithelial cell membranes by pore formation (Moyes et al. 2016). It is anticipated that the

EPF effectors may similarly effect in invading and/or evading insect-host immunity.

13.7 Chemical Genetics and Biology of the Entomopathogenic Fungi

Apart from the functional insights of EPF genes/enzymes involved in mediating the interactions with insect hosts summarized above, different studies have shown that the small molecules produced by EPF secondary metabolic BGCs also contribute to mediating fungal infection of insect hosts. Prior to chemical genetic studies, a plethora of bioactive metabolites have been identified from different EPF species, including polyketides, non-ribosomal peptides, alkaloids, pyridones, and terpenes (Zhang et al. 2020a). The early purposes of compound identifications were largely the searches for novel chemicals and drug screenings (Newman and Cragg 2020). Except for antibiotic activity assays, the biological functions of fungal metabolites were considered dispensable and largely ignored. The lack of gene/genomic sequence information and tractable genetic transformation systems also limited the chemical, genetic, and biological function studies of EPF in the early days.

Different numbers of BGCs have been predicted from each new EPF genome acquired, which vary from ca. 30 to >60 BGCs in *Cordyceps* and *Metarhizium* species (Shang et al. 2016; Chen et al. 2020a; Zhang et al. 2020b). Big gaps still exist in terms of the connections between the identified BGCs and compounds that have been structurally identified and between the predicted BGCs and unknown compounds that can be biosynthesized. For example, the seven chromosomes of *M. brunneum* encode 66 BGCs, while only seven conserved BGCs have been genetically elucidated, completely or in part in different *Metarhizium* species (Fig. 13.4a). Likewise, only three out of 35 BGCs have been characterized in *C. militaris* (Fig. 13.4b). The distribution patterns follow to some extent the telomere/sub-telomere localizations of BGCs that have been reported in other fungi (Keller et al. 2005; Palmer and Keller 2010). In particular, the homologous PKS-terpene *Sub* cluster for the biosynthesis of the meroterpene subglutinols (Kato et al. 2016), and the *DtxS* BGC for the production of the insecticidal cyclopeptide destruxins (Wang et al. 2012) are located on either end of chromosome 7 of *M. brunneum*. In contrast, the telomere/sub-telomere location of BGCs is not apparent in *C. militaris* (Fig. 13.4b).

Chemical genetic studies revealed that the *Swn* BGC contains a core NRPS-PKS (non-ribosomal peptide synthetase-polyketide synthase) hybrid enzyme gene *SwnK* for the biosynthesis of the alkaloid mycotoxin swainsonine. The precursor pipecolic acid can be produced through multiple pathways, including the lysine aminotransferase (SwnA) and/or lysine cyclodeaminase activities (Luo et al. 2020). It has been recently found that an *Upm* supercluster of *Metarhizium* species contains three BGCs including the intertwined pseurotin/ovalicin BGCs that are mesosyntencially conserved to the pseurotin/fumagillin BGCs of *Aspergillus fumigatus*. The former, however, has the additional ustilaginoidin BGC, and the supercluster is jointly controlled by a master regulator to mediate the biosynthesis of four classes of compounds, i.e., ustilaginoidin D, indigotides, pseurotin A, and hydroxyl-ovalicin (Sun et al. 2022b). The *SerS* (also called *MaNPS1*) BGC was evident for the biosynthesis of the cyclic heptapeptide serinocyclins (Krasnoff et al. 2007; Moon et al. 2008). However, the detailed biosynthetic mechanism of serinocyclins remains unclear to date. In particular, the enzymes involved in production of four non-proteinogenic amino acids: 1-aminocyclopropane-1-carboxylic acid, hydroxyproline, γ-hydroxylysine, and β-alanine, are still unclear. In contrast, the *DtxS* BGC has been well characterized to produce the hexapeptide destruxins with the functions of four genes, and the core NRPS DtxS1 contains six modules for the linear assembly of individual substrate into cyclopeptides (Wang et al. 2012). Differently, the hexapeptide beauvericin is biosynthesized by an NRPS BbBEAS that contains only two modules to mediate the

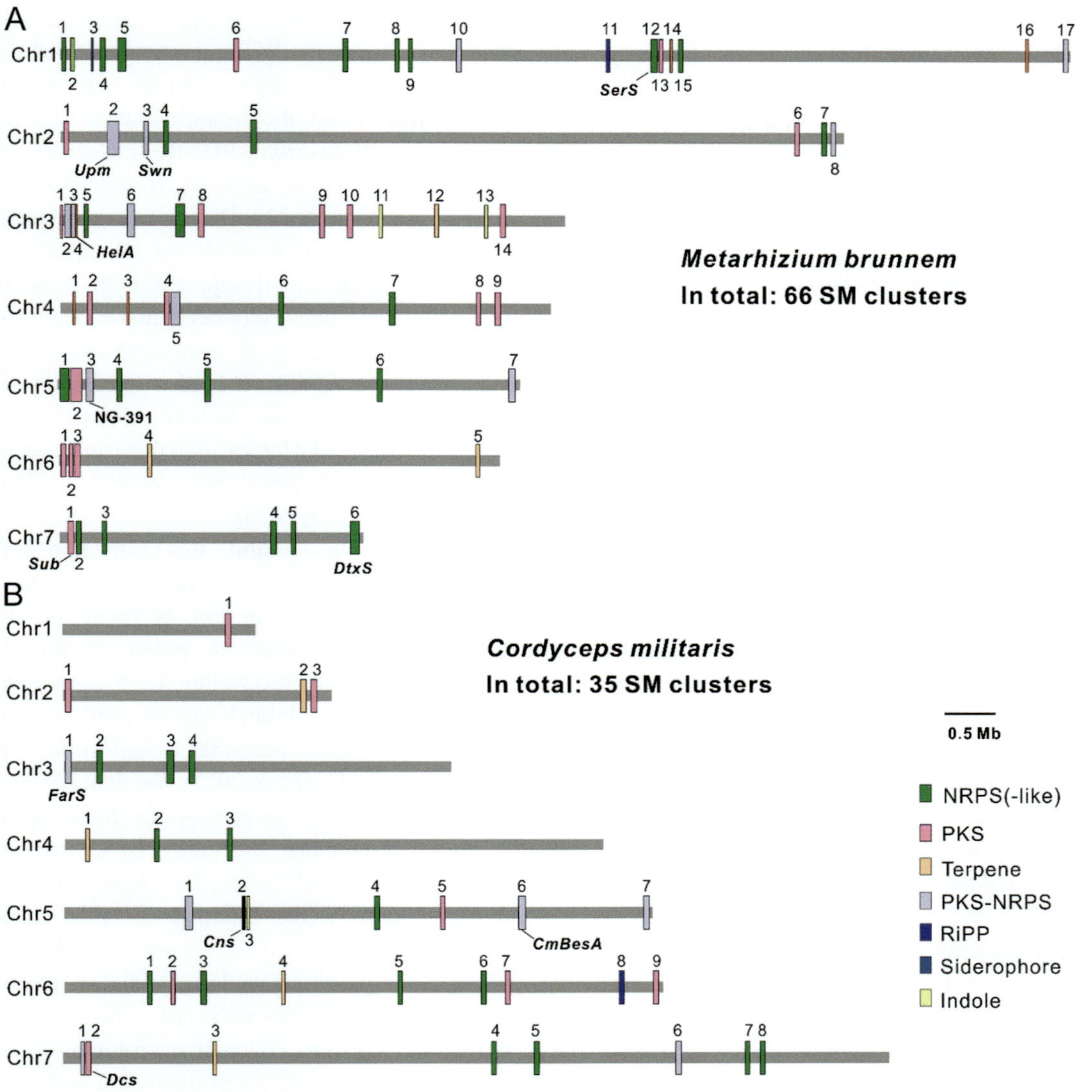

Fig. 13.4 Mapping of the secondary metabolic (SM) gene clusters on the chromosomes of *M. brunneum* (**a**) and *C. militaris* (**b**) strains with their genomes finished. The sequential digits show the number of biosynthetic gene clusters encoded on individual chromosomes of each species. The clusters highlighted in bold with named genes have been characterized for biosynthetic mechanism and/or chemical biology, which have been indicated in this context

interactive biosynthesis of beauvericin from a putative N-methyl-dipeptidol intermediate (Xu et al. 2008).

Although it cannot be predicted by different programs such as antiSMASH (Blin et al. 2021), the *Cns* cluster containing four genes for the coupled biosynthesis of the adenosine anticancer compounds cordycepin and pentostatin has been characterized in *C. militaris*, interestingly, which follows a protector-protégé strategy that the production of the deaminase inhibitor pentostatin can protect the stability of cordycepin from deamination into 3-deoxyinosine (Xia et al. 2017b). The cyclodepsipeptide beauveriolides can by produced by the related species *B. bassiana*, *B. brongniartii*, and *C. militaris* that all contain the conserved *BesA-BesB* NRPS-PKS BGC. However, uncommon chemical diversity has been found among these species; the rather divergent *B. brongniartii* and *C. militaris* species produce the same analogs such as beauveriolides I, II, and III, whereas *B. bassiana* produces the main derivatives of beauveriolides Ba, Ka, and Ca (Yin et al. 2020). The conserved *Dcs* PKS cluster is also present in *B. bassiana*, *C. militaris*, and *C. fumosorosea*, which is responsible for the production of nonanolides such as decarestrictine C1. It was interesting to find a thioesterase functioning as a polyketide cyclase to catalyze the formation of medium-ring lactones (Gao et al. 2021). Biosynthetic elucidation of the bibenzoquinone oosporein revealed the function of the *OpS* gene cluster for the sequential biosynthesis of the intermediates orsellinic acid, benzenetriol, benzenetetrol, and finally the dimerized pigment oosporein. The pathway is tightly controlled by a pathway-specific transcription factor OpS3 (Feng et al. 2015). It is noteworthy that the genes located outside some BGCs of entomopathogens can also function as tailoring enzymes in catalyzing the biosynthesis of derivatives. For example, the tandemly-linked glycosyltransferase-methyltransferase gene pair *BbGT1/MT1* outside the *TenS* BGC catalyze the methylglucosylation of 15-hydroxytenellin, the product of the *TenS* cluster in *B. bassiana*, into glycosides (Chen et al. 2021). The *BbGT1/MT1* orthologues *MrGT1/MT1* function outside the *Upm* supercluster in *M. robertsii* for the production of indigotides, i.e., the methylglucosylated monomeric naphtho-γ-pyrones (Sun et al. 2022b).

By using the null mutants obtained in chemical genetic studies, insect bioassays with the wild type (WT) and mutant strains of *B. bassiana* or *Metarhizium* species facilitated the identification of which metabolites are required for EPF infection of insect hosts. For example, virulence contributions have been found for the cyclopeptides beauvericin (Xu et al. 2008), destruxins (Wang et al. 2012), cyclosporine (Yang et al. 2018), and beauveriolides (Yin et al. 2020); the bibenzoquinone oosporein (Feng et al. 2015; Mc Namara et al. 2019); and 2-pyridones tenellin analogs (Chen et al. 2021). On top of the evident insecticidal activities, survival assays using the WT and mutants of different fungal species against alternative insect species also found the non-obvious virulence contributions of destruxins (Donzelli et al. 2012) and tenellin (Eley et al. 2007) for *M. robertsii* and *B. bassiana*, respectively. Dtx A was implicated in targeting insect immunophilins (Wang et al. 2019a), and Dtxs could suppress both cellular and humoral immune responses to assist the propagation of fungal cells in insects (Wang et al. 2012; Han et al. 2022). It is also interesting to find that the *DtxS* cluster is only present in the generalist species of *Metarhizium* that can infect a wide range of insect hosts (Wang et al. 2012; Hu et al. 2014; Xu et al. 2015).

Oosporein can inhibit the activation of prophenoloxidases and transcription of antimicrobial peptide genes in insects (Feng et al. 2015). Consistent with its potent antibiotic activities, oosporein production in *B. bassiana* has the ability to inhibit bacterial proliferation within insects before and after fungal killing of the insect hosts (Fan et al. 2017; Wei et al. 2017). Interestingly, it has recently been found that *M. robertsii* contains the *HelA* BGC that is conserved to that of *Aspergillus fumigatus* for the production of the triterpene helvolic acid (HA) (Lv et al. 2017). HA is a potent antibiotic and abundantly accumulated in conidial spores to facilitate fungal infection of insects by "outwitting" the host cuticular microbiotas (Sun et al. 2022a). Likewise, the production of multiple classes of compounds with antibiotic activities by the *Upm* BGC can assist *M. robertsii* to counteract different antagonistic bacteria to facilitate fungal adaptations to diverse niches (Sun et al. 2022b). In short, it is not surprising that fungal small molecules play essential roles in mediating EPF associations with insect hosts, an area that deserves further investigation.

13.8 Summary and Perspectives

In summary, the genetics and molecular biology of fungal associations with insects have been well explored and advanced in the past decade or so. The obtaining of the high-quality genomes of representative EPF species has facilitated the understanding of the evolution of fungal entomopathogenicity and the functional genetic investigations of fungus–insect interactions. However, compared to the well-advanced effector mechanisms of fungus–plant interactions (Jones et al. 2016), the virulence requirement of effectors in fungal suppression of insect immunity is still poorly understood (Wang 2020). Future efforts should be directed toward this area. Following the de novo genomic studies of different EPF species, population genomic, and pan-genomic investigations by re-sequencing of diverse intra- and interspecific genomes are still needed to unveil the genetic diversity as well as the hidden drivers of selection leading to the varied degree of host specificity or host preference among different strains or species of entomopathogens. Consistent with the protective functions of human skin and plant phyllosphere microbiotas (Liu et al. 2020a; Harris-Tryon and Grice 2022), recent studies have shown that insect ectomicrobiomes can similarly mediate the colonization resistance against fungal parasites (Hong et al. 2022; Janke et al. 2022). Future investigation of fungus–insect interactions will have to be expanded beyond bilateral regimes to better understand the mechanisms of multilateral interactions among fungi, bacteria, and insects. The knowledge obtained may benefit the development of cost-effective mycoinsecticides as well as the protection of ecologically and economically important beneficial insects like bees and silkworms.

Acknowledgment The authors are highly appreciative of Drs. Meredith Blackwell and Yen-Peng Hsueh for their help in preparing this chapter. This work was supported by the National Key R&D Program of China (2022YFD1400700 and 2022YFD1400500).

References

Agrawal Y, Khatri I, Subramanian S, Shenoy BD (2015) Genome sequence, comparative analysis, and evolutionary insights into chitinases of entomopathogenic fungus *Hirsutella thompsonii*. Genome Biol Evol 7: 916–930

Agrawal Y, Narwani T, Subramanian S (2016) Genome sequence and comparative analysis of clavicipitaceous insect-pathogenic fungus *Aschersonia badia* with *Metarhizium* spp. BMC Genomics 17:367

Altimira F, Arias-Aravena M, Jian L et al (2022) Genomic and experimental analysis of the insecticidal factors secreted by the entomopathogenic fungus *Beauveria pseudobassiana* RGM 2184. J Fungi (Basel) 8:253

Andreadis TG, Weseloh RM (1990) Discovery of *Entomophaga maimaiga* in North American gypsy moth, *Lymantria dispar*. Proc Natl Acad Sci U S A 87:2461–2465

Araújo JP, Hughes DP (2016) Diversity of entomopathogenic fungi: which groups conquered the insect body? Adv Genet 94:1–39

Araujo JPM, Hughes DP (2019) Zombie-ant fungi emerged from non-manipulating, beetle-infecting ancestors. Curr Biol 29:3735–3738

Arnesen JA, Malagocka J, Gryganskyi A, Grigoriev IV, Voigt K, Stajich JE, De Fine Licht HH (2018) Early diverging insect-pathogenic fungi of the order Entomophthorales possess diverse and unique subtilisin-like serine proteases. G3 (Bethesda) 8: 3311–3319

Badaruddin M, Holcombe LJ, Wilson RA, Wang ZY, Kershaw MJ, Talbot NJ (2013) Glycogen metabolic genes are involved in trehalose-6-phosphate synthase-mediated regulation of pathogenicity by the rice blast fungus *Magnaporthe oryzae*. PLoS Pathog 9: e1003604

Behie SW, Zelisko PM, Bidochka MJ (2012) Endophytic insect-parasitic fungi translocate nitrogen directly from insects to plants. Science 336:1576–1577

Blackwell M (2011) The Fungi: 1, 2, 3 ... 5.1 million species? Am J Bot 98:426–438

Blin K, Shaw S, Kloosterman AM, Charlop-Powers Z, van Wezel GP, Medema MH, Weber T (2021) antiSMASH 6.0: improving cluster detection and comparison capabilities. Nucleic Acids Res 49:W29–w35

Boyce GR, Gluck-Thaler E, Slot JC et al (2019) Psychoactive plant- and mushroom-associated alkaloids from two behavior modifying cicada pathogens. Fungal Ecol 41:147–164

Bruner-Montero G, Wood M, Horn HA, Gemperline E, Li L, Currie CR (2021) Symbiont-mediated protection of *Acromyrmex* leaf-cutter ants from the entomopathogenic fungus *Metarhizium anisopliae*. mBio 12:e0188521

Buchon N, Silverman N, Cherry S (2014) Immunity in *Drosophila melanogaster*--from microbial recognition to whole-organism physiology. Nat Rev Immunol 14: 796–810

Bushley KE, Raja R, Jaiswal P et al (2013) The genome of *Tolypocladium inflatum*: evolution, organization, and expression of the cyclosporin biosynthetic gene cluster. PLoS Genet 9:e1003496

Cai Y, Nie Y, Gao Y, Huang B (2022) Natural populations of the entomopathogenic fungus *Beauveria bassiana* in Chinese forest ecosystems are diverse and reveal equal

frequencies of mating types within phylogenetic species. Fungal Ecol 56:101139
Castro-Vásquez R, Solano-González S, Molina-Bravo R, Montero-Astúa M (2022) Draft genome sequences of eight isolates of *Beauveria bassiana* of neotropical origin. Microbiol Resour Announc 11:e00267–e00222
Cen K, Li B, Lu YZ, Zhang SW, Wang CS (2017) Divergent LysM effectors contribute to the virulence of *Beauveria bassiana* by evasion of insect immune defenses. PLoS Pathog 13:e1006604
Chang Y, Wang SS, Sekimoto S et al (2015) Phylogenomic analyses indicate that early fungi evolved digesting cell walls of algal ancestors of land plants. Genome Biol Evol 7:1590–1601
Chen Y, Pettis JS, Zhao Y et al (2013) Genome sequencing and comparative genomics of honey bee microsporidia, *Nosema apis* reveal novel insights into host-parasite interactions. BMC Genomics 14:451
Chen YX, Feng P, Shang YF, Xu YJ, Wang CS (2015) Biosynthesis of non-melanin pigment by a divergent polyketide synthase in *Metarhizium robertsii*. Fungal Genet Biol 81:142–149
Chen X, Xu C, Qian Y et al (2016) MAPK cascade-mediated regulation of pathogenicity, conidiation and tolerance to abiotic stresses in the entomopathogenic fungus *Metarhizium robertsii*. Environ Microbiol 18: 1048–1062
Chen A, Wang Y, Shao Y, Huang B (2017) A novel technique for rejuvenation of degenerated caterpillar medicinal mushroom, *Cordyceps militaris* (Ascomycetes), a valued traditional Chinese medicine. Int J Med Mushrooms 19:87–91
Chen YX, Li B, Cen K, Lu YZ, Zhang SW, Wang CS (2018a) Diverse effect of phosphatidylcholine biosynthetic genes on phospholipid homeostasis, cell autophagy and fungal developments in *Metarhizium robertsii*. Environ Microbiol 20:293–304
Chen YX, Cen K, Lu Y, Zhang S, Shang YF, Wang CS (2018b) Nitrogen-starvation triggers cellular accumulation of triacylglycerol in *Metarhizium robertsii*. Fungal Biol 122:410–419
Chen Y, Wu Y, Liu L et al (2019) Study of the whole genome, methylome and transcriptome of *Cordyceps militaris*. Sci Rep 9:898
Chen B, Sun YL, Luo FF, Wang CS (2020a) Bioactive metabolites and potential mycotoxins produced by *Cordyceps* fungi: a review of safety. Toxins 12:410
Chen L, Gao X, Li R et al (2020b) Complete genome of a unicellular parasite (*Antonospora locustae*) and transcriptional interactions with its host locust. Microb Genom 6:mgen000421
Chen B, Sun YL, Li SQ, Yin Y, Wang CS (2021) Inductive production of the iron-chelating 2-pyridones benefits the producing fungus to compete for diverse niches. mBio 12:e0327921
Cormier A, Chebbi MA, Giraud I et al (2021) Comparative genomics of strictly vertically transmitted, feminizing microsporidia endosymbionts of amphipod crustaceans. Genome Biol Evol 13:evaa245
Cornman RS, Chen YP, Schatz MC et al (2009) Genomic analyses of the microsporidian *Nosema ceranae*, an emergent pathogen of honey bees. PLoS Pathog 5: e1000466
Cortés JCG, Curto M, Carvalho VSD, Pérez P, Ribas JC (2019) The fungal cell wall as a target for the development of new antifungal therapies. Biotechnol Adv 37: 107352
Cui C, Wang Y, Liu J, Zhao J, Sun P, Wang S (2019) A fungal pathogen deploys a small silencing RNA that attenuates mosquito immunity and facilitates infection. Nat Commun 10:4298
Dang EV, Lei S, Radkov A, Volk RF, Zaro BW, Madhani HD (2022) Secreted fungal virulence effector triggers allergic inflammation via TLR4. Nature 608:161–167
de Bekker C (2019) *Ophiocordyceps*-ant interactions as an integrative model to understand the molecular basis of parasitic behavioral manipulation. Curr Opin Insect Sci 33:19–24
de Bekker C, Ohm RA, Loreto RG et al (2015) Gene expression during zombie ant biting behavior reflects the complexity underlying fungal parasitic behavioral manipulation. BMC Genomics 16:620
de Bekker C, Ohm RA, Evans HC, Brachmann A, Hughes DP (2017) Ant-infecting *Ophiocordyceps* genomes reveal a high diversity of potential behavioral manipulation genes and a possible major role for enterotoxins. Sci Rep 7:12508
de Faria MR, Wraight SP (2007) Mycoinsecticides and mycoacaricides: a comprehensive list with worldwide coverage and international classification of formulation types. Biol Control 43:237–256
Desjardins CA, Sanscrainte ND, Goldberg JM et al (2015) Contrasting host-pathogen interactions and genome evolution in two generalist and specialist microsporidian pathogens of mosquitoes. Nat Commun 6:7121
Donzelli BGG, Krasnoff SB, Moon YS, Churchill AC, Gibson DM (2012) Genetic basis of destruxin production in the entomopathogen *Metarhizium robertsii*. Curr Genet 58:105–116
Duan ZB, Chen YX, Huang W, Shang YF, Chen PL, Wang CS (2013) Linkage of autophagy to fungal development, lipid storage and virulence in *Metarhizium robertsii*. Autophagy 9:538–549
Eley KL, Halo LM, Song Z et al (2007) Biosynthesis of the 2-pyridone tenellin in the insect pathogenic fungus *Beauveria bassiana*. Chembiochem 8:289–297
Elya C, Lok TC, Spencer QE, McCausland H, Martinez CC, Eisen M (2018) Robust manipulation of the behavior of *Drosophila melanogaster* by a fungal pathogen in the laboratory. elife 7:e34414
Fan Y, Liu X, Keyhani NO, Tang G, Pei Y, Zhang W, Tong S (2017) Regulatory cascade and biological activity of *Beauveria bassiana* oosporein that limits bacterial growth after host death. Proc Natl Acad Sci U S A 114:E1578–E1586
Feng P, Shang Y, Cen K, Wang C (2015) Fungal biosynthesis of the bibenzoquinone oosporein to evade insect

immunity. Proc Natl Acad Sci U S A 112:11365–11370

Ferrandon D, Imler JL, Hetru C, Hoffmann JA (2007) The *Drosophila* systemic immune response: sensing and signalling during bacterial and fungal infections. Nat Rev Immunol 7:862–874

Fischer R, Requena N (2022) Small-secreted proteins as virulence factors in nematode-trapping fungi. Trends Microbiol 30:615–617

Fredericksen MA, Zhang Y, Hazen ML, Loreto RG, Mangold CA, Chen DZ, Hughes DP (2017) Three-dimensional visualization and a deep-learning model reveal complex fungal parasite networks in behaviorally manipulated ants. Proc Natl Acad Sci U S A 114:12590–12595

Gao Q, Jin K, Ying SH et al (2011) Genome sequencing and comparative transcriptomics of the model entomopathogenic fungi *Metarhizium anisopliae* and *M. acridum*. PLoS Genet 7:e1001264

Gao Q, Shang YF, Huang W, Wang CS (2013) Glycerol-3-phosphate acyltransferase contributes to triacylglycerol biosynthesis, lipid droplet formation, and host invasion in *Metarhizium robertsii*. Appl Environ Microbiol 79:7646–7653

Gao Q, Lu Y, Yao H, Xu YJ, Huang W, Wang C (2016) Phospholipid homeostasis maintains cell polarity, development and virulence in *Metarhizium robertsii*. Environ Microbiol 18:3976–3990

Gao BJ, Mou YN, Tong SM, Ying SH, Feng MG (2020) Subtilisin-like Pr1 proteases marking evolution of pathogenicity in a wide-spectrum insect-pathogenic fungus. Virulence 11:365–380

Gao DW, Jamieson CS, Wang G, Yan Y, Zhou J, Houk KN, Tang Y (2021) A polyketide cyclase that forms medium-ring lactones. J Am Chem Soc 143:80–84

Gleason FH, Marano AV, Johnson P, Martin WW (2010) Blastocladian parasites of invertebrates. Fungal Biol Rev 24:56–67

Han P, Gong Q, Fan J, Abbas M, Chen D, Zhang J (2022) Destruxin A inhibits scavenger receptor B mediated melanization in *Aphis citricola*. Pest Manag Sci 78: 1915–1924

Hansen AN, De Fine Licht HH (2019) Why are there so few examples of entomopathogenic fungi that manipulate host sexual behaviors? Fungal Ecol 38:21–27

Harris-Tryon TA, Grice EA (2022) Microbiota and maintenance of skin barrier function. Science 376:940–945

Hawksworth DL, Lücking R (2017) Fungal diversity revisited: 2.2 to 3.8 million species. Microbiol Spectr 5:FUNK-0052-2016

Hong S, Sun Y, Sun D, Wang CS (2022) Microbiome assembly on *Drosophila* body surfaces benefits the flies to combat fungal infections. iScience 25:104408

Hong S, Sun YL, Chen HM, Wang CS (2023) Suppression of the insect cuticular microbiomes by a fungal defensin to facilitate parasite infection. ISME J 17:1–11

Horn F, Habel A, Scharf DH et al (2015) Draft genome sequence and gene annotation of the entomopathogenic fungus *Verticillium hemipterigenum*. Genome Announc 3:e0143914

Hu X, Zhang YJ, Xiao GH et al (2013) Genome survey uncovers the secrets of sex and lifestyle in caterpillar fungus. Chin Sci Bull 58:2846–2854

Hu X, Xiao G, Zheng P et al (2014) Trajectory and genomic determinants of fungal-pathogen speciation and host adaptation. Proc Natl Acad Sci U S A 111: 16796–16801

Huang W, Hong S, Tang G, Lu Y, Wang C (2019) Unveiling the function and regulation control of the DUF3129 family proteins in fungal infection of hosts. Philos Trans R Soc Lond Ser B Biol Sci 374:20180321

Huang A, Lu M, Ling E, Li P, Wang CS (2020) A M35 family metalloprotease is required for fungal virulence against insects by inactivating host prophenoloxidases and beyond. Virulence 11:222–237

Huang Q, Wu ZH, Li WF et al (2021) Genome and evolutionary analysis of *Nosema ceranae*: A microsporidian parasite of honey bees. Front Microbiol 12:645353

Humber RA (2008) Evolution of entomopathogenicity in fungi. J Invertebr Pathol 98:262–266

Humber RA (2017) Back to the future: cordycipitoid fungi in a postgenomic world. Mycology 8:267–275

Hyde KD (2022) The numbers of fungi. Fungal Divers 114:1

Imoulan A, Hussain M, Kirk PM, El Meziane A, Yao Y-J (2017) Entomopathogenic fungus *Beauveria*: Host specificity, ecology and significance of morpho-molecular characterization in accurate taxonomic classification. J Asia-Pacif Entomol 20:1204–1212

Iwanicki NSA, Botelho A, Klingen I, Junior ID, Rossmann S, Lysoe E (2022) Genomic signatures and insights into host niche adaptation of the entomopathogenic fungus *Metarhizium humberi*. G3 12:jkab416

Janke RS, Kaftan F, Niehs SP et al (2022) Bacterial ectosymbionts in cuticular organs chemically protect a beetle during molting stages. ISME J 16:2691–2701

Jones JD, Vance RE, Dangl JL (2016) Intracellular innate immune surveillance devices in plants and animals. Science 354:aaf6395

Kato H, Tsunematsu Y, Yamamoto T, Namiki T, Kishimoto S, Noguchi H, Watanabe K (2016) New natural products isolated from *Metarhizium robertsii* ARSEF 23 by chemical screening and identification of the gene cluster through engineered biosynthesis in *Aspergillus nidulans* A1145. J Antibiot 69:561–566

Keller NP, Turner G, Bennett JW (2005) Fungal secondary metabolism - from biochemistry to genomics. Nat Rev Microbiol 3:937–947

Kim JC, Park SE, Lee SJ, Kim JS (2022) Whole-genome sequence of *Beauveria bassiana* JEF-350, a strain with high insecticidal activity against melon thrips (*Thrips palmi*). Microbiol Resour Announc 11:e0047022

Kobmoo N, Mongkolsamrit S, Tasanathai K, Thanakitpipattana D, Luangsa-Ard JJ (2012) Molecular phylogenies reveal host-specific divergence of

Ophiocordyceps unilateralis sensu lato following its host ants. Mol Ecol 21:3022–3031
Kobmoo N, Wichadakul D, Arnamnart N, Rodriguez De La Vega RC, Luangsa-Ard JJ, Giraud T (2018) A genome scan of diversifying selection in Ophiocordyceps zombie-ant fungi suggests a role for enterotoxins in co-evolution and host specificity. Mol Ecol 27:3582–3598
Kombrink A, Thomma BP (2013) LysM effectors: secreted proteins supporting fungal life. PLoS Pathog 9:e1003769
Kramer GJ, Nodwell JR (2017) Chromosome level assembly and secondary metabolite potential of the parasitic fungus *Cordyceps militaris*. BMC Genomics 18:912
Krasnoff SB, Keresztes I, Gillilan RE, Szebenyi DM, Donzelli BG, Churchill AC, Gibson DM (2007) Serinocyclins A and B, cyclic heptapeptides from *Metarhizium anisopliae*. J Nat Prod 70:1919–1924
Lai Y, Cao X, Chen J, Wang L, Wei G, Wang S (2020) Coordinated regulation of infection-related morphogenesis by the KMT2-Cre1-Hyd4 regulatory pathway to facilitate fungal infection. Sci Adv 6:eaaz1659
Lee SJ, Lee MR, Kim S et al (2018) Genomic analysis of the insect-killing fungus *Beauveria bassiana* JEF-007 as a biopesticide. Sci Rep 8:12388
Lee MR, Li D, Lee SJ et al (2019) Use of *Metarhizum aniopliae* s.l. to control soil-dwelling longhorned tick, *Haemaphysalis longicornis*. J Invertebr Pathol 166: 107230
Li J, Xia Y (2022) Host-pathogen interactions between *Metarhizium* spp. and locusts. J Fungi 8:602
Li ZZ, Li CR, Huang B, Fan MZ (2001) Discovery and demonstration of the teleomorph of *Beauveria bassiana* (Bals.) Vuill., an important entomogenous fungus. Chin Sci Bull 46:751–753
Li Y, Hsiang T, Yang RH et al (2016) Comparison of different sequencing and assembly strategies for a repeat-rich fungal genome, *Ophiocordyceps sinensis*. J Microbiol Methods 128:1–6
Li B, Song S, Wei X, Tang G, Wang CS (2022) Activation of microlipophagy during early infection of insect hosts by *Metarhizium robertsii*. Autophagy 18:608–623
Lin R, Zhang X, Xin B et al (2019) Genome sequence of *Isaria javanica* and comparative genome analysis insights into family S53 peptidase evolution in fungal entomopathogens. Appl Microbiol Biotechnol 103: 7111–7128
Litwin A, Nowak M, Rozalska S (2020) Entomopathogenic fungi: unconventional applications. Rev Environ Sci Biotechnol 19:23–42
Liu H, Brettell LE, Singh B (2020a) Linking the phyllosphere microbiome to plant health. Trends Plant Sci 25:841–844
Liu J, Guo L, Li Z et al (2020b) Genomic analyses reveal evolutionary and geologic context for the plateau fungus *Ophiocordyceps sinensis*. Chin Med 15:107
Liu Q, Qu S, He G, Wei J, Dong C (2022) Mating-type genes play an important role in fruiting body development in *Morchella sextelata*. J Fungi (Basel) 8:564
Long L, Liu Z, Deng C, Li C, Wu L, Hou B, Lin Q (2022) Genomic Sequence and Transcriptome Analysis of the Medicinal Fungus Keithomyces neogunnii. Genome Biol Evol 14:evac033
Lopes Fischer N, Naseer N, Shin S, Brodsky IE (2020) Effector-triggered immunity and pathogen sensing in metazoans. Nat Microbiol 5:14–26
Lovett B, St Leger RJ (2017) The insect pathogens. Microbiol Spectr 5:FUNK-0001-2016
Lovett B, Macias A, Stajich JE, Cooley J, Eilenberg J, de Fine Licht HH, Kasson MT (2020) Behavioral betrayal: How select fungal parasites enlist living insects to do their bidding. PLoS Pathog 16:e1008598
Lowe RG, Howlett BJ (2012) Indifferent, affectionate, or deceitful: lifestyles and secretomes of fungi. PLoS Pathog 8:e1002515
Lu Y, Xia Y, Luo F, Dong C, Wang C (2016) Functional convergence and divergence of mating-type genes fulfilling in *Cordyceps militaris*. Fungal Genet Biol 88: 35–43
Lu Y, Luo F, Cen K et al (2017) Omics data reveal the unusual asexual-fruiting nature and secondary metabolic potentials of the medicinal fungus *Cordyceps cicadae*. BMC Genomics 18:668
Luangsa-Ard J, Tasanathai K, Thanakitpipattana D, Khonsanit A, Stadler M (2018) Novel and interesting *Ophiocordyceps* spp. (Ophiocordycipitaceae, Hypocreales) with superficial perithecia from Thailand. Stud Mycol 89:125–142
Luo FF, Hong S, Chen B et al (2020) Unveiling of swainsonine biosynthesis via a multi-branched pathway in fungi. ACS Chem Biol 15:2476–2484
Luo F, Tang G, Hong S, Gong T, Xin X-F, Wang CS (2023) Promotion of *Arabidopsis* immune responses by a rhizosphere fungus via supply of pipecolic acid to plants and selective augment of phytoalexins. Sci China Life Sci 66:1119–1133
Lv JM, Hu D, Gao H et al (2017) Biosynthesis of helvolic acid and identification of an unusual C-4-demethylation process distinct from sterol biosynthesis. Nat Commun 8:1644
Ma LJ, van der Does HC, Borkovich KA et al (2010) Comparative genomics reveals mobile pathogenicity chromosomes in *Fusarium*. Nature 464:367–373
Mc Namara L, Dolan SK, Walsh JMD, Stephens JC, Glare TR, Kavanagh K, Griffin CT (2019) Oosporein, an abundant metabolite in *Beauveria caledonica*, with a feedback induction mechanism and a role in insect virulence. Fungal Biol 123:601–610
Mei LJ, Chen M, Shang Y et al (2020) Population genomics and evolution of a fungal pathogen after releasing exotic strains to control insect pests for 20 years. ISME J 14:1422–1434
Mei L, Wang X, Yin Y, Tang G, Wang CS (2021) Conservative production of galactosaminogalactan in *Metarhizium* is responsible for appressorium mucilage

production and topical infection of insect hosts. PLoS Pathog 17:e1009656

Meyling NV, Lubeck M, Buckley EP, Eilenberg J, Rehner SA (2009) Community composition, host range and genetic structure of the fungal entomopathogen *Beauveria* in adjoining agricultural and seminatural habitats. Mol Ecol 18:1282–1293

Mondo SJ, Dannebaum RO, Kuo RC et al (2017) Widespread adenine N6-methylation of active genes in fungi. Nat Genet 49:964–968

Mongkolsamrit S, Khonsanit A, Thanakitpipattana D et al (2020) Revisiting *Metarhizium* and the description of new species from Thailand. Stud Mycol 95:171–251

Moon YS, Donzelli BG, Krasnoff SB et al (2008) *Agrobacterium*-mediated disruption of a nonribosomal peptide synthetase gene in the invertebrate pathogen *Metarhizium anisopliae* reveals a peptide spore factor. Appl Environ Microbiol 74:4366–4380

Moonjely S, Barelli L, Bidochka MJ (2016) Insect pathogenic fungi as endophytes. Adv Genet 94:107–135

Mora C, Tittensor DP, Adl S, Simpson AG, Worm B (2011) How many species are there on Earth and in the ocean? PLoS Biol 9:e1001127

Moyes DL, Wilson D, Richardson JP et al (2016) Candidalysin is a fungal peptide toxin critical for mucosal infection. Nature 532:64–68

Nakjang S, Williams TA, Heinz E et al (2013) Reduction and expansion in microsporidian genome evolution: new insights from comparative genomics. Genome Biol Evol 5:2285–2303

Newman DJ, Cragg GM (2020) Natural products as sources of new drugs over the nearly four decades from 01/1981 to 09/2019. J Nat Prod 83:770–803

Nielsen KN, Salgado JFM, Natsopoulou ME, Kristensen T, Stajich JE, De Fine Licht HH (2021) Diploidy within a haploid genus of entomopathogenic fungi. Genome Biol Evol 13:evab158

Ortiz-Urquiza A, Keyhani NO (2013) Action on the surface: entomopathogenic fungi versus the insect cuticle. Insects 4:357–374

Palmer JM, Keller NP (2010) Secondary metabolism in fungi: does chromosomal location matter? Curr Opin Microbiol 13:431–436

Pan G, Xu J, Li T et al (2013) Comparative genomics of parasitic silkworm microsporidia reveal an association between genome expansion and host adaptation. BMC Genomics 14:186

Pan G, Bao J, Ma Z et al (2018) Invertebrate host responses to microsporidia infections. Dev Comp Immunol 83:104–113

Pattemore JA, Hane JK, Williams AH, Wilson BA, Stodart BJ, Ash GJ (2014) The genome sequence of the biocontrol fungus *Metarhizium anisopliae* and comparative genomics of *Metarhizium* species. BMC Genomics 15:660

Pelin A, Selman M, Aris-Brosou S, Farinelli L, Corradi N (2015) Genome analyses suggest the presence of polyploidy and recent human-driven expansions in eight global populations of the honeybee pathogen *Nosema ceranae*. Environ Microbiol 17:4443–4458

Pessotti RC, Hansen BL, Reaso JN, Ceja-Navarro JA, El-Hifnawi L, Brodie EL, Traxler MF (2021) Multiple lineages of *Streptomyces* produce antimicrobials within passalid beetle galleries across eastern North America. elife 10:e65091

Qu S, Wang S (2018) Interaction of entomopathogenic fungi with the host immune system. Dev Comp Immunol 83:96–103

Ramos Y, Portal O, Lysoe E, Meyling NV, Klingen I (2017) Diversity and abundance of *Beauveria bassiana* in soils, stink bugs and plant tissues of common bean from organic and conventional fields. J Invertebr Pathol 150:114–120

Rehner SA (2020) Genetic structure of *Metarhizium* species in western USA: finite populations composed of divergent clonal lineages with limited evidence for recent recombination. J Invertebr Pathol 177:107491

Richards TA, Dacks JB, Jenkinson JM, Thornton CR, Talbot NJ (2006) Evolution of filamentous plant pathogens: gene exchange across eukaryotic kingdoms. Curr Biol 16:1857–1864

Roy HE, Steinkraus DC, Eilenberg J, Hajek AE, Pell JK (2006) Bizarre interactions and endgames: entomopathogenic fungi and their arthropod hosts. Annu Rev Entomol 51:331–357

RoyChowdhury M, Sternhagen J, Xin Y, Lou B, Li X, Li C (2022) Evolution of pathogenicity in obligate fungal pathogens and allied genera. PeerJ 10:e13794

Saud Z, Kortsinoglou AM, Kouvelis VN, Butt TM (2021) Telomere length de novo assembly of all 7 chromosomes and mitogenome sequencing of the model entomopathogenic fungus, *Metarhizium brunneum*, by means of a novel assembly pipeline. BMC Genomics 22:87

Savory F, Leonard G, Richards TA (2015) The role of horizontal gene transfer in the evolution of the oomycetes. PLoS Pathog 11:e1004805

Shang Y, Feng P, Wang C (2015) Fungi that infect insects: altering host behavior and beyond. PLoS Pathog 11: e1005037

Shang YF, Xiao GH, Zheng P, Cen K, Zhan S, Wang CS (2016) Divergent and convergent evolution of fungal pathogenicity. Genome Biol Evol 8:1374–1387

Shang JM, Shang YF, Tang GR, Wang CS (2021) Identification of a key G-protein coupled receptor in mediating appressorium formation and fungal virulence against insects. Sci China Life Sci 64:466–477

Shang J, Tang G, Lu M, Wang C (2022) Host and environmental sensing by entomopathogenic fungi to infect hosts. Curr Clin Microbiol Rep 9:69–74

Shang JM, Tang GR, Yang J, Lu M, Wang CZ, Wang CS (2023) Sensing of a spore surface protein by a *Drosophila* chemosensory protein induces behavioral defense against fungal parasitic infections. Curr Biol 33:276–286

Shen D, Tang Z, Wang C et al (2019) Infection mechanisms and putative effector repertoire of the

mosquito pathogenic oomycete *Pythium guiyangense* uncovered by genomic analysis. PLoS Genet 15: e1008116
Shen D, Nyawira KT, Xia A (2020) New discoveries and applications of mosquito fungal pathogens. Curr Opin Insect Sci 40:111–116
Sheng H, McNamara PJ, St Leger RJ (2022) *Metarhizium*: an opportunistic middleman for multitrophic lifestyles. Curr Opin Microbiol 69:102176
Shi XZ, Zhong X, Yu XQ (2012) *Drosophila melanogaster* NPC2 proteins bind bacterial cell wall components and may function in immune signal pathways. Insect Biochem Mol Biol 42:545–556
Shu R, Zhang J, Meng Q et al (2020) A new high-quality draft genome assembly of the Chinese Cordyceps *Ophiocordyceps sinensis*. Genome Biol Evol 12: 1074–1079
Spatafora JW, Sung GH, Sung JM, Hywel-Jones NL, White JF Jr (2007) Phylogenetic evidence for an animal pathogen origin of ergot and the grass endophytes. Mol Ecol 16:1701–1711
St Leger RJ, Wang JB (2020) *Metarhizium*: jack of all trades, master of many. Open Biol 10:200307
St. Leger RJ (2021) Insects and their pathogens in a changing climate. J Invertebr Pathol 184:107644
Staats CC, Junges A, Guedes RL et al (2014) Comparative genome analysis of entomopathogenic fungi reveals a complex set of secreted proteins. BMC Genomics 15: 822
Stuart LM, Paquette N, Boyer L (2013) Effector-triggered versus pattern-triggered immunity: how animals sense pathogens. Nat Rev Immunol 13:199–206
Sun YL, Hong S, Chen HM, Yin Y, Wang CS (2022a) Production of helvolic acid in *Metarhizium* contributes to fungal infection of insects by bacteriostatic inhibition of the host cuticular microbiomes. Microbiol Spectr 10:e0262022
Sun YL, Chen B, Li XL, Yin Y, Wang CS (2022b) Orchestrated biosynthesis of the secondary metabolite cocktails enables the producing fungus to combat diverse bacteria. mBio 13:e0180022
Sung GH, Hywel-Jones NL, Sung JM, Luangsa-Ard JJ, Shrestha B, Spatafora JW (2007) Phylogenetic classification of *Cordyceps* and the clavicipitaceous fungi. Stud Mycol 57:5–59
Tang D, Tang X, Fang W (2022) New downstream signaling branches of the mitogen-activated protein kinase cascades identified in the insect pathogenic and plant symbiotic fungus *Metarhizium robertsii*. Front Fungal Biol 3:911366
Thakur R, Shiratori T, Ishida KI (2019) Taxon-rich multigene phylogenetic analyses resolve the phylogenetic relationship among deep-branching Stramenopiles. Protist 170:125682
Usman S, Ge X, Xu Y et al (2023) Loss of phosphomannose isomerase impairs growth, perturbs cell wall integrity, and reduces virulence of *Fusarium oxysporum* f. sp. *cubense* on banana plants. J Fungi (Basel) 9:478
Valero-Jimenez CA, Faino L, Spring In't Veld D, Smit S, Zwaan BJ, van Kan JA (2016) Comparative genomics of *Beauveria bassiana*: uncovering signatures of virulence against mosquitoes. BMC Genomics 17:986
van Wyk S, Wingfield BD, De Vos L, van der Merwe NA, Steenkamp ET (2021) Genome-wide analyses of repeat-induced point mutations in the Ascomycota. Front Microbiol 11:622368
Vega FE, Kaya HK (2012) Insect pathology, 2nd edn. Academic, San Diego, CA
Vega FE, Goettel MS, Blackwell M et al (2009) Fungal entomopathogens: new insights on their ecology. Fungal Ecol 2:149–159
Wang CS (2020) Grand challenges in the research of fungal interactions with animals. Front Fungal Biol 1: 602032
Wang C, St Leger RJ (2006) A collagenous protective coat enables *Metarhizium anisopliae* to evade insect immune responses. Proc Natl Acad Sci U S A 103: 6647–6652
Wang C, St Leger RJ (2007a) The MAD1 adhesin of *Metarhizium anisopliae* links adhesion with blastospore production and virulence to insects, and the MAD2 adhesin enables attachment to plants. Eukaryot Cell 6:808–816
Wang CS, St Leger RJ (2007b) The *Metarhizium anisopliae* perilipin homolog MPL1 regulates lipid metabolism, appressorial turgor pressure, and virulence. J Biol Chem 282:21110–21115
Wang CS, Wang SB (2017) Insect pathogenic fungi: genomics, molecular interactions, and genetic improvements. Annu Rev Entomol 62:73–90
Wang C, Skrobek A, Butt TM (2003) Concurrence of losing a chromosome and the ability to produce destruxins in a mutant of *Metarhizium anisopliae*. FEMS Microbiol Lett 226:373–378
Wang B, Kang Q, Lu Y, Bai LQ, Wang CS (2012) Unveiling the biosynthetic puzzle of destruxins in *Metarhizium* species. Proc Natl Acad Sci U S A 109: 1287–1292
Wang JB, St Leger RJ, Wang C (2016) Advances in genomics of entomopathogenic fungi. Adv Genet 94: 67–105
Wang Y, Stata M, Wang W, Stajich JE, White MM, Moncalvo JM (2018) Comparative genomics reveals the core gene toolbox for the fungus-insect symbiosis. mBio 9:e0063618
Wang J, Weng Q, Hu Q (2019) Effects of Destruxin A on silkworm's immunophilins. Toxins (Basel) 11:349
Wang W, Deng Z, Wu H et al (2019b) A small secreted protein triggers a TLR2/4-dependent inflammatory response during invasive *Candida albicans* infection. Nat Commun 10:1015
Wang Y-B, Wang Y, Fan Q et al (2020a) Multigene phylogeny of the family Cordycipitaceae (Hypocreales): new taxa and the new systematic position of the Chinese cordycipitoid fungus *Paecilomyces hepiali*. Fungal Divers 103:1–46

Wang Y, Nie Y, Yu D, Xie X, Qin L, Yang Y, Huang B (2020b) Genome-wide study of saprotrophy-related genes in the basal fungus *Conidiobolus heterosporus*. Appl Microbiol Biotechnol 104:6261–6272

Wei G, Lai Y, Wang G, Chen H, Li F, Wang S (2017) Insect pathogenic fungus interacts with the gut microbiota to accelerate mosquito mortality. Proc Natl Acad Sci U S A 114:5994–5999

Whisler HC, Zebold SL, Shemanchuk JA (1975) Life history of *Coelomomyces psorophorae*. Proc Natl Acad Sci U S A 72:693–696

Wichadakul D, Kobmoo N, Ingsriswang S, Tangphatsornruang S, Chantasingh D, Luangsa-Ard JJ, Eurwilaichitr L (2015) Insights from the genome of *Ophiocordyceps polyrhachis-furcata* to pathogenicity and host specificity in insect fungi. BMC Genomics 16:881

Will I, Das B, Trinh T, Brachmann A, Ohm RA, de Bekker C (2020) Genetic Underpinnings of Host Manipulation by *Ophiocordyceps* as Revealed by Comparative Transcriptomics. G3 (Bethesda) 10:2275–2296

Wilson RA, Talbot NJ (2009) Under pressure: investigating the biology of plant infection by *Magnaporthe oryzae*. Nat Rev Microbiol 7:185–195

Wraight SP, Ramos ME, Avery PB, Jaronski ST, Vandenberg JD (2010) Comparative virulence of *Beauveria bassiana* isolates against lepidopteran pests of vegetable crops. J Invertebr Pathol 103:186–199

Xia EH, Yang DR, Jiang JJ et al (2017a) The caterpillar fungus, *Ophiocordyceps sinensis*, genome provides insights into highland adaptation of fungal pathogenicity. Sci Rep 7:1806

Xia YL, Luo FF, Shang YF, Chen PL, Lu YZ, Wang CS (2017b) Fungal cordycepin biosynthesis is coupled with the production of the safeguard molecule pentostatin. Cell Chem Biol 24:1479–1489

Xiao G, Ying SH, Zheng P et al (2012) Genomic perspectives on the evolution of fungal entomopathogenicity in *Beauveria bassiana*. Sci Rep 2:483

Xu Y, Orozco R, Wijeratne EM, Gunatilaka AA, Stock SP, Molnar I (2008) Biosynthesis of the cyclooligomer depsipeptide beauvericin, a virulence factor of the entomopathogenic fungus *Beauveria bassiana*. Chem Biol 15:898–907

Xu Y-J, Luo F, Gao Q, Shang Y, Wang C (2015) Metabolomics reveals insect metabolic responses associated with fungal infection. Anal Bioanal Chem 407:4815–4821

Yang X, Feng P, Yin Y, Bushley K, Spatafora JW, Wang C (2018) Cyclosporine biosynthesis in *Tolypocladium inflatum* benefits fungal adaptation to the environment. MBio 9:e01211–e01218

Yin Y, Chen B, Song SX, Li B, Yang XQ, Wang CS (2020) Production of diverse beauveriolide analogs in closely related fungi: a rare case of fungal chemodiversity. mSphere 5:e0066720

Ying SH, Feng MG (2019) Insight into vital role of autophagy in sustaining biological control potential of fungal pathogens against pest insects and nematodes. Virulence 10:429–437

Zeng G, Zhang P, Zhang Q et al (2018) Duplication of a *Pks* gene cluster and subsequent functional diversification facilitate environmental adaptation in *Metarhizium* species. PLoS Genet 14:e1007472

Zhang C, Deng W, Yan W, Li T (2018) Whole genome sequence of an edible and potential medicinal fungus, *Cordyceps guangdongensis*. G3 (Bethesda) 8:1863–1870

Zhang L, Fasoyin OE, Molnar I, Xu Y (2020a) Secondary metabolites from hypocrealean entomopathogenic fungi: novel bioactive compounds. Nat Prod Rep 37:1181–1206

Zhang L, Yue Q, Wang C, Xu Y, Molnar I (2020b) Secondary metabolites from hypocrealean entomopathogenic fungi: genomics as a tool to elucidate the encoded parvome. Nat Prod Rep 37:1164–1180

Zhang X, Meng Y, Huang Y, Zhang D, Fang W (2021) A novel cascade allows *Metarhizium robertsii* to distinguish cuticle and hemocoel microenvironments during infection of insects. PLoS Biol 19:e3001360

Zhao H, Xu C, Lu HL, Chen X, St Leger RJ, Fang W (2014) Host-to-pathogen gene transfer facilitated infection of insects by a pathogenic fungus. PLoS Pathog 10:e1004009

Zheng P, Xia Y, Xiao G et al (2011) Genome sequence of the insect pathogenic fungus *Cordyceps militaris*, a valued traditional Chinese medicine. Genome Biol 12:R116

Zheng P, Xia Y, Zhang S, Wang C (2013) Genetics of *Cordyceps* and related fungi. Appl Microbiol Biotechnol 97:2797–2804

Xylaria Sclerotia Formed Within Termite Nests: A Review of Their Biology and Human Uses 14

Huei-Mei Hsieh and Yu-Ming Ju

Abstract

Certain *Xylaria* species form massive sclerotia of several centimeters in diameter within the underground nests of macrotermitine termites after the termites abandon the nest. Vernacularly the sclerotia are called Wulingshen in China or Nilamanga and Puttumanga in southern India. *Xylaria nigripes* once was commonly regarded as the sole producer of the massive sclerotia, which are in fact produced by at least six *Xylaria* species, including *X. nigripes* and close relatives. *Xylaria* sclerotia have been used in folk medicine for treating a wide range of ailments, whereas dried mycelia from cultures are used mainly for soothing anxiety, depression, insomnia, and some neurological disorders. Here, we review the biological functions, pharmacological properties, and beneficial implications for human health from cultured materials prepared from mycelia or stromata produced in cultures of sclerotium-forming *Xylaria* species.

Keywords

Ascomycota · Macrotermitinae · Wuling capsule · Wulingshen · Xylariaceae · Medicinal properties

H.-M. Hsieh · Y.-M. Ju (✉)
Institute of Plant and Microbial Biology, Academia Sinica, Taipei, Taiwan
e-mail: yumingju@gate.sinica.edu.tw

14.1 Introduction

Massive sclerotia of several centimeters in diameter formed within abandoned subterranean termite nests are known in traditional medicine in China as Wulingshen and Puttumanga in India. The sclerotia are produced by fungi of the genus *Xylaria* Hill ex Schrank (Xylariaceae, Sordariomycetes, Ascomycota). The nests are built by macrotermitine termites (subfamily Macrotermitinae), the fungus-growing termites, which have a broad geographical distribution in tropical and subtropical regions of Asia and Africa (Aanen et al. 2009; Mueller et al. 2005). The fungi cultivated by macrotermitine termites are species of the genus *Termitomyces* Heim (Lyophyllaceae, Agaricomycetes, Basidiomycota), which form a mutually symbiotic relationship with the termites (Batra and Batra 1979; Sangvichien and Taylor-Hawksworth 2001). *Xylaria* species are not cultivated by termites but enter termite nests as stowaways where they are competitors of *Termitomyces* (Visser et al. 2009). Nevertheless, the exact interactions among *Xylaria* species, termites, and *Termitomyces* mushrooms remain to be clarified (Rogers et al. 2005).

Xylaria sclerotia (Fig. 14.1a, b) formed within subterranean termite nests are massive compared to those produced by other fungal species. Sclerotia are resistant structures composed of tightly compacted mycelium. Sclerotia of the *Xylaria* species discussed here vary in shape from spherical to ovoid or even irregular and are

Y.-P. Hsueh, M. Blackwell (eds.), *Fungal Associations*, The Mycota 9,
https://doi.org/10.1007/978-3-031-41648-4_14

Fig. 14.1 *Xylaria nigripes*, *X. neonigripes*, and *X. furcata*. A–E. *X. nigripes*. A. Sclerotia. B. Vertical section of sclerotium showing a white interior. C. Colony producing stromata on a 9-cm Petri dish containing oatmeal agar; the culture was obtained from a sclerotium shown in A. D. Part of a stroma produced from the colony in C showing a grayish powdery conidial mass on the surface. E. Stromata produced in nature. F. *X. neonigripes*, which can be separated from *X. nigripes* by the narrowed sterile stromatal apex. Stromata are produced in nature. G. *X. furcata*. Stromata are produced in nature. Bars: A, B, E–G = 1 cm; D = 2 mm

several centimeters in diameter. A compacted mass of tasteless white hyphae comprises a core covered by a peridium that is black or dark brown, and smooth on the surface while fresh but wrinkled after drying (Ju et al. 2022). The top of the sclerotium has a narrow-broken point, indicating its former connection point with the attached base of the parent stroma (Ju et al. 2022). Fungal stromata are similar to sclerotia because they are composed of tightly organized mycelium, but they differ in function because they are the site of ascospore and conidium production in *Xylaria*.

The British mycologist Miles J. Berkeley was probably the first to describe massive sclerotia from termite nests as *Sclerotium stipitatum* Berk. & Currey, based on a specimen collected from southern Lara State, India, formerly known as Travancore. It is called Puttumanga in Tamil language, which means termite-hill mango (Berkeley 1862). However, the producer of the sclerotia was unclear for the next 44 years until Petch (1906) studied massive sclerotia from termite nests in Sri Lanka. He placed them under a bell jar on a damp filter paper, where ascospore-producing stromata of a *Xylaria* species were produced. This was the first time massive sclerotia from termite nests were connected to a *Xylaria* species. It should be noted that Petch (1913) referred to the *Xylaria* species as *X. nigripes* (Klotzsch) M. C. Cooke but his concept of *X. nigripes* was broad, with *X. escharoidea* (Berk.) Sacc. and six other names listed as synonyms. *Xylaria escharoidea* is a distinct species and is more commonly encountered than *X. nigripes* (Rogers et al. 2005). Many specimens labeled as *X. nigripes* should be recognized as *X. escharoidea* because their ascospores possess a distinctive pore-like germination site.

Massive *Xylaria* sclerotia have been consumed as herbal medicine in China. Medicinal uses of these sclerotia from termite nests were first reported in "Revised local gazetteer of the district of Guan-xian" from Sichuan, China (Zhuang and Zheng 1886), where the name Wulingshen was introduced for the first time. The medicinal effect of Wulingshen in the gazetteer is brief, only mentioning it for nourishing kidneys. The gazetteer mentioned that Wulingshen is also called Lei-zhen-zi, which literally means fruits of thunder, apparently because sclerotia hanging from the roof of nesting chambers twisted as thunder struck, an extremely difficult if not impossible observation. The name Lei-zhen-zi seemingly implies that the formation of Wulingshen coincides with the thunderstorm season in China; 84% of the thunderstorms occur during May and September (Zhang et al. 2017). Field studies are needed to confirm whether there is a connection between the formation of Wulingshen and thunderstorms. Interestingly, a connection between thunderbolts and fungi was also noted by Lowy (1974) who recorded that *Amanita muscaria* (L.) Lam. was believed to fruit on the ground in places where thunderbolts struck in the Roman, Hindu, and Mayan cultures. Chinese literature was rather vague as to which species produced Wulingshen until Wu (1984) pointed out that Wulingshen was produced by *X. nigripes* and an unnamed *Xylaria* species.

Wulingshen is said in folklore to be effective in eliminating dampness in body condition, easing infantile convulsions, stopping palpitations, promoting lactation, nourishing hearts and kidneys, and curing insomnia, nose bleeds, and hemorrhaging after childbirth (Ying et al. 1987). More recently, traditional uses of wild-collected sclerotia in China have progressed to the production of commercial products. Today, dried mycelial powder, Wuling powder, prepared from *Xylaria* cultures, is encapsulated and referred to as Wuling capsule by Pharmacopoeia Commission of China (2020). Wuling capsule was approved by the China Food and Drug Administration (CFDA) with the authorized document number Z19990048 and has become a leading Chinese patent medicine for dealing with insomnia. Production of Wuling powder is subjected to standardized cultivation and preparation, and its containing mannitol and adenosine have been proposed for assaying the product. The recommended dosage of Wuling capsule for insomnia is three capsules (0.33 g each), followed by three doses per day taken orally for 20 continuous days (Feng et al. 2016).

Information on *Xylaria* sclerotia and Wulingshen is easily available in internet searches, but most of the information concerns the efficacy of Wulingshen for treating illnesses or the high price of Wulingshen (currently several hundred US dollars per Kg in dry weight). Here, we will discuss compounds or functional extracts produced from cultivated mycelium (used interchangeably with Wuling capsule) and stromata, and their potential applications to human health.

14.2 Known Habitat and Localities of Massive *Xylaria* Sclerotia

Petch (1913) mentioned that massive *Xylaria* sclerotia had been collected from India, Sri Lanka, Java in Indonesia, and Africa. The close association between nests of macrotermitine termites and massive *Xylaria* sclerotia strongly suggests the geographic distribution of the sclerotium-producing *Xylaria* species coincides with that of macrotermitine termites in tropical and subtropical regions of Asia and Africa.

When the macrotermitine termites properly manage the fungus gardens within termite nests, the dominant fungi are the *Termitomyces* mushrooms, the cultivated crop of termites. Once the termite nests are abandoned by the termites, *Xylaria* species soon gain the chance to start exploiting the nutrients of the fungus gardens by producing hyphae profusely. Massive sclerotia are produced by sclerotium-forming *Xylaria* species under proper conditions in the fungus gardens, which are gradually shrunk to disappearance after the nutrient is completely depleted. At this stage, massive sclerotia are often found hanging from the roof of the underground chambers as if they were formed in the air, a phenomenon that Zhuang and Zheng (1886) had noted in the original description of Wulingshen. Although massive *Xylaria* sclerotia are produced in abandoned termite nests, they are not always found when the abandoned nests are excavated; the proper conditions under which massive sclerotia are formed remain to be discovered. Petch (1913), based on his field experiences, believed that massive sclerotia were only formed in dry situations or when the nests were abandoned in the dry season. To our knowledge, massive sclerotia are only collected from nature, not formed in artificial conditions.

14.3 *Xylaria* Species Producing Massive Sclerotia in Termite Nests: Systematics and Relationship with Termites

Xylaria nigripes was once considered to be the sole producer of massive sclerotia found in termite nests. However, several studies indicate that this may not be the case. The studies by Wu (1984) and Chen et al. (1988) revealed more than one *Xylaria* species capable of producing massive sclerotia, and Deepna Latha et al. (2015) linked an Indian *Xylaria* sclerotium to *X. acuminatilongissima* Y.-M. Ju & H.-M. Hsieh by using ITS (Nuc rDNA internal transcribed spacers = ITS1-5.8S-ITS2) sequence. Wulingshen, a valuable herbal medicine in Sichuan Province, China, can be procured from traditional Chinese medicine markets there. Ju et al. (2022) obtained ITS sequences from 54 Wulingshen samples and concluded that they belonged to six different *Xylaria* species, two of which, *X. nigripes* and *X. neonigripes* Y.-M. Ju, H.-M. Hsieh & X.-S. He are shown in Fig. 14.1e, f. The result was further supported by sequences of three protein-coding loci: β-tubulin (*β-TUB*) and α-actin (*α-ACT*), and RNA polymerase II second largest subunit (*RPB2*). The six *Xylaria* species connected with Wulingshen from Sichuan, China, are *X. nigripes*, *X. rogersionigripes* Y.-M. Ju, H.-M. Hsieh & X.-S. He, *X. neonigripes*, *X. subescharoidea*, and two unnamed species (*X.* sp. A, *X.* sp. B) that are known only in their sclerotial forms. About one-half (51.9%) of the massive sclerotia were from *X. rogersionigripes*, followed by 20.4% from *X. nigripes*, 13.0% from *X. neonigripes*, 11.1% from *X.* sp. A, 1.9% from *X. subescharoidea*, and 1.9% from *X.* sp. B.

There are 43 known *Xylaria* species associated with termite nests in the world (Hsieh et al. 2020; Ju and Hsieh 2007; Ju et al. 2022; Rogers et al.

2005; Wangsawat et al. 2021a), which can be divided into two general groups by their stromatal morphology: the *X. nigripes* group (Fig. 14.1e, f) and the *X. furcata* Fr. group (Fig. 14.1g). Species in the *X. nigripes* group have cylindrical, unbranched stromata that are more robust, ca. 4–8 mm diameter at the perithecium-bearing part. Species in the *X. furcata* group have antler-like, dichotomously branched stromata that are delicate, ca. 1–2 mm diameter at the perithecium-bearing part. It should be emphasized that not every *Xylaria* species associated with termite nests can produce massive sclerotia, but all known sclerotium-producing species belong to the *X. nigripes* group.

The production of massive sclerotia is unique among *Xylaria* species. Are the sclerotium-producing *Xylaria* species phylogenetically related? Ju et al. (2022) included the sclerotium-producing *Xylaria* species in phylogenetic analyses based on sequences of *β-TUB*, *α-ACT*, and *RPB2* in the context of more than a hundred *Xylaria* species from all known substrate types, with which *Xylaria* species are associated. The phylogenetic trees had topologies that were similar to those shown in Hsieh et al. (2010) and U'Ren et al. (2016), with the species associated with termite nests grouped into a single clade. Interestingly, within this clade, the six Wulingshen-producing species mentioned above, together with *X. acuminatilongissima*, *X. escharoidea*, and *X. cirrata* Pat., formed a subclade, suggesting that the sclerotium-forming habit arose only once within *Xylaria* (Fig. 14.2). *Xylaria acuminatilongissima* was linked to a massive sclerotium from Kerala, India (Deepna Latha et al. 2015), while *X. escharoidea* and *X. cirrata* have not been linked to massive sclerotia. It is possible that the latter two species may have lost the ability to form massive sclerotia.

In the studies reviewed below, the Xylarias studied were often assumed to be *X. nigripes*.

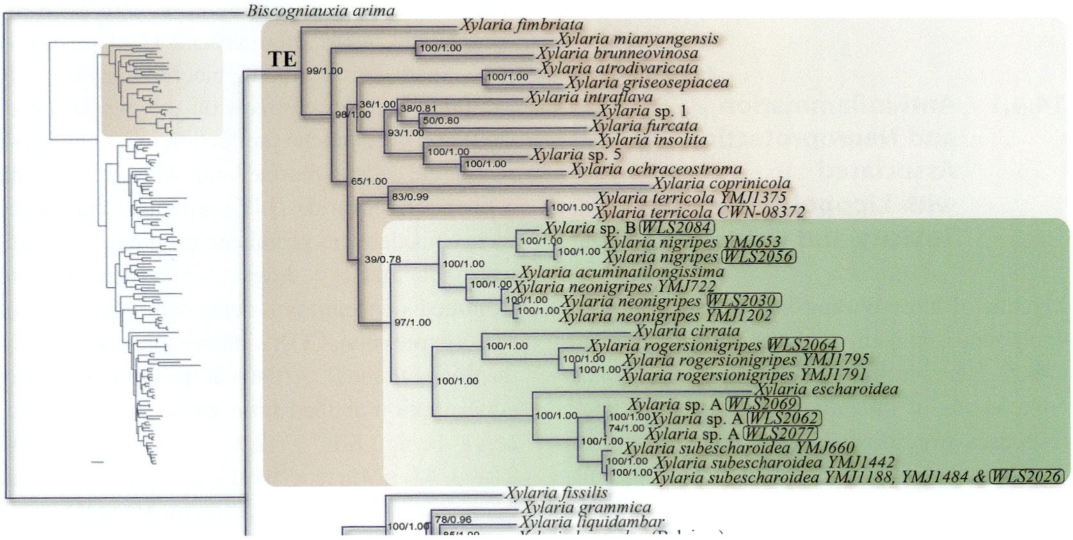

Fig. 14.2 Phylogenetic tree, generated by Maximum-Likelihood analysis based on sequences of *β-TUB*, *α-ACT*, and *RPB2*, is modified from Fig. 2 in Ju et al. (2022), where the complete tree was fully presented in Supplementary Fig. 1 of Ju et al. (2022) but is shown only schematically on the left, with the area within the rectangle enlarged to show the detail of the clade denoted by TE. Within the TE clade, the subclade in green shade encompassed the Wulingshen-producing species, i.e., *X. neonigripes*, *X. nigripes*, *X. rogersionigripes*, *X. subescharoidea*, *X.* sp. A, and *X.* sp. B. Wulingshen samples prefixed with "WLS" are framed. While *X. acuminatilongissima* was linked to a *Xylaria* sclerotium in India (Deepna Latha et al. 2015), *X. escharoidea* and *X. cirrata* are the only species that have not been linked to sclerotia. Numbers at internodes represent bootstrap values and are immediately followed by the posterior probability values greater than 50% with Bayesian inference analysis

ITS sequences from some of them have been deposited in GenBank, and the identities of these species were reassessed by comparing their sequences with those of the sclerotium-forming *Xylaria* species studied by Ju et al. (2022). In these cases, the results from the comparisons are presented here; otherwise, original identifications are retained.

14.4 Possible Functions of Compounds Isolated from Cultured Sclerotium-Forming *Xylaria* Species

Information about compound-producing *Xylaria* species, including their morphological state for compound production, associated GenBank number, culture or specimen deposition, and applied experimental cell lines, and functions of various compounds are listed in Table 14.1. A brief introduction to applied cell lines is also denoted in Table 14.1.

14.4.1 Anti-inflammation and Neuroprotection Associated with Lipopolysaccharide Toxicity and Oxidative Stress

Mycelia of most Wulingshen cultures synthesize compounds known to have anti-inflammatory effects. Here we briefly describe several signaling components of inflammation and the microglial responses towards these signaling components. These include inducible nitric oxide synthase (iNOS), cyclooxygenase (COX2), tumor necrosis factor TNF-α, interleukin IL-1β, nitric oxide (NO), and prostaglandin E2 (PGE2), which are widely known for signaling in inflammation (Bachiller et al. 2018; Kawahara et al. 2015; Rojo et al. 2008). The activation of inflammatory molecules is known to have a deleterious role in neurodegenerative diseases by expanding the inflammatory response and increasing neuronal damage (Bachiller et al. 2018). Microglia, a population of macrophage cells in the central nervous system (CNS), are broadly distributed throughout the brain and the spinal cord and are crucial for a healthy CNS (Ginhoux et al. 2013). Microglial response is pathology dependent, affects immunity and metabolism, and plays an important role in the progression of neurodegenerative diseases, such as Parkinson's disease, Alzheimer's disease, and frontotemporal dementia (Bachiller et al. 2018). Main pathways including JAK/STATs and NFkB govern microglial pro-inflammatory response, and an uncontrolled inflammatory response by microglial cells can lead to a detrimental outcome (Bachiller et al. 2018). An important concern for selecting bioactive compounds for anti-inflammatory effects concerns their cytotoxic effects on cells treated with these compounds. Only the bioactive compounds not causing significant toxicity on cells are reviewed herein.

Eleven compounds were isolated from ethyl acetate extracts of the liquid fermentation of the *X. nigripes* culture YMJ 653 (GenBank accession number EU179868) (Chang et al. 2017). Of the compounds tested, nigriterpene C, a sesquiterpene, and formannoxin alcohol significantly exhibited inhibitory effects on the expression of iNOS, COX2, and the production of NO by BV-2 microglial cells of murine brain when treated with a lipopolysaccharide (LPS) stimulant and hence were considered as promising candidates for anti-neuroinflammatory drugs. Nigriterpene C also exhibited concentration-dependent inhibition of iNOS expression, COX2 expression, and NO production with the IC_{50} (the half maximal inhibitory concentration applied for a given biological process) being 8.1 ± 2.3, 16.6 ± 5.5, and 21.7 ± 4.9 μM, respectively, and therefore could be more potent than formannoxin alcohol.

Ergosterol peroxide and ergostarien-3β-ol isolated from a hexane fraction of dried stromata of cultivated *X. nigripes* exhibited anti-inflammatory activities with ergosterol peroxide being stronger in the inhibition of several pro-inflammatory mediators in RAW 264.7 macrophages treated with LPS (Liaw et al. 2017). The pro-inflammatory mediators include NO, IL-1β, TNF-α, IL-6, and PGE_2. Ergosterol peroxide (25 μmol/L) also greatly suppressed the

Table 14.1 Information of compound-producing *Xylaria* species

Compound names or extracts	Species name	Suggested or confirmed name	Morphological state	Strain number	Deposition Location	GenBank no.	Function	Cell line/ microorganism	References
(3β,5α,6β)-Stigmastane-3,5,6-triol	*X. nigripes*	N/A	Mycelia	N/A	ZJPC[a]	N/A	Cytotoxicity against osteosarcoma cells	U2OS[b]	Xiong et al. (2016)
(3β,5α,6β, 22E)-6-Methoxyergosta-7,22-diene-3,5-diol	*X. nigripes*	N/A	Mycelia	N/A	ZJPC[a]	N/A	Neuroprotection by an increase in cell viability	SH-SY5Y[c]	Xiong et al. (2016)
(3β,5α,6β, 22E)-6-Methoxyergosta-7,22-diene-3,5-diol	*X. nigripes*	N/A	Mycelia	N/A	ZJPC[a]	N/A	Cytotoxicity against lung cancer cells	A549[d]	Xiong et al. (2016)
(3β,5α,6β, 22E)-6-Methoxyergosta-7,22-diene-3,5-diol	*X. nigripes*	N/A	Mycelia	N/A	ZJPC[a]	N/A	Cytotoxicity against osteosarcoma cells	U2OS[b]	Xiong et al. (2016)
(3β,5α,8α,22E)-Ergosta-6,22-diene-3,5,8-triol	*X. nigripes*	N/A	Mycelia	N/A	ZJPC[a]	N/A	Anti-neuroinflammation	BV-2[e]	Xiong et al. (2016)
4,8-Dihydroxy-3,4-dihydronaphthalen-1(2H)-one	*X. escharoidea*	*X. escharoidea*	Mycelia	TF18	N/A	MN509048	Antimicrobial activity	*S. aureus, B. subtilis, P. aeruginosa, and C. albicans*	Nagam et al. (2021)
(6S)- Agroclavine N-oxide	*X. nigripes*	N/A	Dried stromata	CUI 20090818	KIB[f]	N/A	Cholesterol ester transfer protein inhibitor	N/A	Hu and Li (2017)
8,9-Didehydro-10-hydroxy-6,8-dimethylergolin	*X. nigripes*	N/A	Dried stromata	CUI 20090818	KIB[f]	N/A	Cholesterol ester transfer protein inhibitor	N/A	Hu and Li (2017)
Agroclavine	*X. nigripes*	N/A	Dried stromata	CUI 20090818	KIB[f]	N/A	Inhibitory activity on acetylcholinesterase	N/A	Hu and Li (2017)
Agroclavine	*X. nigripes*	N/A	Dried stromata	CUI 20090818	KIB[f]	N/A	Cholesterol ester transfer protein inhibitor	N/A	Hu and Li (2017)
Ergostarien-3β-ol	*X. nigripes*	*X. rogersionigripes*	Cultivated stromata	KJ-XN-07-1	KJBC[g]	JQ927570	Anti-inflammation	RAW 264.7[h]	Liaw et al. (2017)
Ergosterol D	*X. nigripes*	N/A	Mycelia	N/A	ZJPC[a]	N/A	Cytotoxicity against osteosarcoma cells	U2OS[b]	Xiong et al. (2016)

(continued)

Table 14.1 (continued)

Compound names or extracts	Species name	Suggested or confirmed name	Morphological state	Strain number	Deposition Location	GenBank no.	Function	Cell line/ microorganism	References
Ergosterol peroxide	*X. nigripes*	*X. rogersionigripes*	Cultivated stromata	KJ-XN-07-1	KJBC[g]	JQ927570	Anti-inflammation	RAW 264.7[h]	Liaw et al. (2017)
Fimbriether B	*X. fimbriata*	*X. fimbriata*	Mycelia	YMJ491	IPMB[i]	GU324753	Anti-inflammation	RAW 264.7[h]	Chen et al. (2019)
Fimbriether E	*X. fimbriata*	*X. fimbriata*	Mycelia	YMJ491	IPMB[i]	GU324753	Anti-inflammation	RAW 264.7[h]	Chen et al. (2019)
Fimbriether G	*X. fimbriata*	*X. fimbriata*	Mycelia	YMJ491	IPMB[i]	GU324753	Anti-inflammation	RAW 264.7[h]	Chen et al. (2019)
Formannoxin alcohol	*X. nigripes*	*X. nigripes*	Mycelia	YMJ653	IPMB[i]	EU179868	Anti-neuroinflammation	BV-2[e]	Chang et al. (2017)
Nigriterpene C	*X. nigripes*	*X. nigripes*	Mycelia	YMJ653	IPMB[i]	EU179868	Anti-neuroinflammation	BV-2[e]	Chang et al. (2017)
Pollenopyrrosides A	*X. nigripes*	N/A	Mycelia	N/A	ZJPC[a]	N/A	Inhibition of oxidative stress	A7r5[j]	Li et al. (2015)
Pollenopyrrosides B	*X. nigripes*	N/A	Mycelia	N/A	ZJPC[a]	N/A	Inhibition of oxidative stress	A7r5[j]	Li et al. (2015)
(R)-5-Methylmellein	*X. nigripes*	N/A	Mycelia	N/A	ZJPC[a]	N/A	Inhibitor against monoamine oxidase A	N/A	Huang et al. (2019)
Water soluble, nondigestible polysaccharides CXNP	*X. nigripes*	*X. rogersionigripes*	Cultivated stromata	N/A	KJBC[g]	KJ627787	Anti-inflammation	RAW 264.7[h]	Chen et al. (2018)
Water soluble, nondigestible polysaccharides TXNP	*X. nigripes*	*X. rogersionigripes*	Cultivated stromata	N/A	KJBC[g]	KJ627786	Anti-inflammation	RAW 264.7[h]	Chen et al. (2018)
Water soluble, nondigestible polysaccharides F1 of XN	*X. nigripes*	*X. rogersionigripes*	Cultivated stromata	N/A	KJBC[g]	KJ627786	Anti-inflammation	RAW 264.7[h]	Jen et al. (2021)

Water soluble, nondigestible polysaccharides F2 of XN	*X. nigripes*	*X. rogersionigripes*	Cultivated stromata	N/A	KJBC[g]	KJ627786	Anti-inflammation	RAW 264.7[h]	Jen et al. (2021)
Xylanigripone A	*X. nigripes*	N/A	Dried stromata	CUI 20090818	KIB[f]	N/A	Cholesterol ester transfer protein inhibitor	N/A	Hu and Li (2017)
Xylanigripone C	*X. nigripes*	N/A	Dried stromata	CUI 20090818	KIB[f]	N/A	Inhibitory activity on acetylcholinesterase	N/A	Hu and Li (2017)
Xylapyrrosides A	*X. nigripes*	N/A	Mycelia	N/A	ZJPC[a]	N/A	Inhibition of oxidative stress	A7r5[j]	Li et al. (2015)
Xylapyrrosides B	*X. nigripes*	N/A	Mycelia	N/A	ZJPC[a]	N/A	Inhibition of oxidative stress	A7r5[j]	Li et al. (2015)
Xylarinaps A	*X. nigripes*	N/A	Mycelia	CGBWSHF 00611	SCUN[k]	N/A	Antioxidation and neuroprotection by an increase in cell viability	PC12[l]	Li et al. (2021)
Xylarinaps B	*X. nigripes*	N/A	Mycelia	CGBWSHF 00611	SCUN[k]	N/A	Antioxidation and neuroprotection by an increase in cell viability	PC12[l]	Li et al. (2021)
Xylarinaps D	*X. nigripes*	N/A	Mycelia	CGBWSHF 00611	SCUN[k]	N/A	Antioxidation and neuroprotection by an increase in cell viability	PC12[l]	Li et al. (2021)
Xylarinaps E	*X. nigripes*	N/A	Mycelia	CGBWSHF 00611	SCUN[k]	N/A	Antioxidation and neuroprotection by an increase in cell viability	PC12[l]	Li et al. (2021)
Xylarinig A	*X. nigripes*	N/A	Mycelia	CGBWSHF 00611	SCUN[k]	N/A	Neuroprotection by an increase in cell viability	PC12[l]	Li et al. (2022)
Xylarinig B	*X. nigripes*	N/A	Mycelia	CGBWSHF 00611	SCUN[k]	N/A	Neuroprotection by an increase in cell viability	PC12[l]	Li et al. (2022)

(continued)

Table 14.1 (continued)

Compound names or extracts	Species name	Suggested or confirmed name	Morphological state	Strain number	Deposition Location	GenBank no.	Function	Cell line/ microorganism	References
Xylarinig C	*X. nigripes*	N/A	Mycelia	CGBWSHF 00611	SCUN[k]	N/A	Neuroprotection by an increase in cell viability	PC12[l]	Li et al. (2022)

[a]ZJPC: Zhejiang Jolly Pharmaceutical Co., China
[b]U2OS: Derived from sarcoma of human tibia and one of its applications is in the study of bone tumorigenicity (Lauvrak et al. 2013)
[c]SH-SY5Y: A subline of the SK-N-SH cell line which was from neuroblastoma of bone marrow and widely used in the research of Parkinson's disease (Xicoy et al. 2017)
[d]A549: Isolated from human lung adenocarcinoma and its applications include drug metabolism and tumorigenicity (Croce et al. 1999; Foster et al. 1998)
[e]BV-2: Derived from murine microglia and immortalized by transduction of neonatal primary microglia with v-raf/v-myc carrying J2 retrovirus and used for the study of neuroinflammation and neurodegenerative disorders (Timmerman et al. 2018)
[f]KIB: Kunming Institute of Botany, CAS, China
[g]KJBC: Kang Jian Biotech Corp., Ltd., Taiwan
[h]RAW 264.7: Monocyte/macrophage-like, established from a tumor of BALB/c mice induced with Abelson murine leukemia virus, and the cells increase nitric oxide production and phagocytosis upon lipopolysaccharide stimulation (Taciak et al. 2018)
[i]IPMB: Institute of Plant and Microbial Botany, AS, Taiwan
[j]A7r5: Derived from aortic smooth muscle cells of rat and can be used for the study of vascular oxidative stress (Liao et al. 2018; Zhou et al. 2020)
[k]SCUN: South-Central University for Nationalities, China
[l]PC12: Cells originated from a pheochromocytoma of rat adrenal medulla and are used for studies on neurotoxicity, neuroprotection, neurosecretion, neuroinflammation, and synaptogenesis (Wiatrak et al. 2020)

expressions of iNOS, COX2, and NF-κB, and hence the mechanism of the inhibition was involved in the regulation of the NF-κB pathway. BLAST searches using ITS indicated that the so-called *X. nigripes* culture (GenBank JQ927570) used in this study actually showed a 99.62% similarity to *X. rogersionigripes* (YMJ1148).

Two water soluble, nondigestible polysaccharides, TXNP and CXNP, extracted from stromata formed in two *X. nigripes* cultures with GenBank accession numbers KJ627786 and KJ627787, were obtained by following steps including water extraction, ethanol precipitation, and treatments of α-amylase and α-glucosidase; these two samples were then compared for anti-inflammatory activities (Chen et al. 2018). TXNP and CXNP were dose-dependent in suppressing NO, IL-1β, TNF-α, and PGE2 productions as well as iNOS, COX-2, and NF-κB expressions in LPS-induced RAW264.7 macrophages. TXNP predominantly composed of glucose (97.8%) had better suppression of these inflammatory mediators than CXNP composed of mannose (41.4%), glucose (17.1%), and xylose (41.5%). Both of the ITS sequences of the two cultures KJ627786 and KJ627787 matched that of *X. rogersionigripes* (YMJ1148) with which they shared 100% and 99.81% similarities, respectively. Water soluble, nondigestible polysaccharides from the strain with the GenBank accession no. KJ627786 were separated and purified into two bioactive polysaccharide components F1 and F2, with the former being 885.2 kDa and composed of predominantly glucan (95.5%) and the latter being 24.5 kDa and a composition of mannose (13.8%), glucose (71.9%), and galactose (14.3%) (Jen et al. 2021). F1 exhibited better anti-inflammatory activities, suppressing the production of NO, TNF-α, and IL-6 more strongly, than F2. The combination of F1 and F2 demonstrated synergistic effects in the expression of anti-inflammatory activities that were comparable to that of the polysaccharide before being fractioned.

Nineteen compounds, including nine pyrrole-containing alkaloids and six steroids, were isolated from an ethanol extract of the fermented mycelia of *X. nigripes* (Xiong et al. 2016). One of the six steroids, i.e., (3β,5α,8α,22E)-ergosta-6,22-diene-3,5,8-triol exhibited an inhibitory effect on NO production in the LPS-induced BV-2 microglial cells, and its value of IC_{50} was 27.6 μM, comparable to N^Gmonomethyl-L-arginine (IC_{50} = 14.4 μm), an inhibitor of NO synthase. Another steroid, (3β,5α,6β, 22E)-6-methoxyergosta-7,22-diene-3,5-diol, at the concentration of 10 μM, exhibited a significant neuroprotective effect on human SH-SY5Y neuroblastoma cells, showing a 21% increase in cell viability by attenuating β-amyloid$_{25-35}$-induced cell damage.

Four spirocyclic pyrrole alkaloids, including pollenopyrrosides A and B and xylapyrrosides A and B, were isolated from the ethanol extracts of cultured mycelia of *X. nigripes* (Li et al. 2015). In addition, total syntheses for three of the isolated compounds, i.e., pollenopyrroside B, xylapyrrosides A and B, were achieved, and three new analogs, xylapyrrosides A1, A2, and B1 were also produced during the syntheses. The isolated and synthesized pyrrole alkaloids mentioned above as well as an antioxidant flavonoid (t)-catechin hydrate inhibited cytotoxicity on A7r5 rat vascular smooth muscle cells, treated with tert-Butyl hydroperoxide (tBHP) for induction of oxidative stress. Cell viability dropped to 59.6% after tBHP was employed but increased to a range of 85.3%–103.8% after the addition of pyrrole alkaloids tested at 100 μM. The antioxidant effects of the pyrrole alkaloids are concentration-dependent with pollenopyrroside B and xylapyrroside A2 being stronger in the suppression of the tBHP-induced apoptosis. In another study by Verano and Tan (2017), both synthesized pollenopyrroside A and xylapyrroside A exhibited an inhibition effect with IC_{50} values being 17 μM and 11 μM, respectively, on the production of reactive oxygen species (ROS) induced by the treatment of high glucose on rat mesangial cells. Their analogs, the newly synthesized 2-OH pollenopyrroside A and 2-OH xylapyrroside, are highly valuable as antioxidants with IC_{50} values of 0.52 μM and 0.27 μM, respectively. Both studies bring new insights for profound exploration of the unique spiro-alkaloids in drug discovery.

Four out of the five pairs of naphthalenone derivative enantiomers, i.e., xylarinaps A, B, D, and E, reported in Li et al. (2021) and four new resorcinol derivatives, i.e., (−)-xylarinig A, (+)-xylarinig A, xylarinigs B and C, reported in Li et al. (2022) were isolated from ethyl acetate extracts of the solid fermentation of *X. nigripes*, exhibiting neuroprotective effects from the damage of PC12 cells after induction with oxygen and glucose deprivation (OGD). The OGD-induced apoptosis of PC12 cells was significantly decreased after the treatment of the xylarinaps A, B, D, and E, and (−)-xylarinig A, (+)-xylarinig A, xylarinigs B and C. Superoxide dismutase (SOD) and glutathione peroxidase (GSH-Px) alleviate oxidative stress through detoxification of free radicals and hence are important components of the major antioxidant systems (Battin and Brumaghim 2009; Fukai and Ushio-Fukai 2011). The application of xylarinaps A, B, D, and E also resulted in a decrease of the malondialdehyde (MDA) level and increases of the SOD and GSH-Px levels, which indicates their potential usage in reducing oxidative stress (Li et al. 2021). The concentrations of enhanced cell viability and inhibited apoptosis were effective at 0.1 and 1 μmol/L in studies of Li et al. (2021, 2022).

Although sclerotium-forming ability has not been reported in *X. fimbriata*, a species associated with termite nests, anti-inflammatory function was also detected from its fermented broth, where three isoprenyl phenolic ethers, fimbriethers B, E, and G, were isolated from its ethyl acetate extracts of the isolate YMJ491 (GenBank GU324753) (Chen et al. 2019). Fimbriethers B, E, and G at a concentration of 100 μM inhibited NO production in cells of LPS-induced murine macrophage RAW 264.7 with the E_{max} values of inhibitory activity being 31.3 ± 1.3%, 38.9 ± 0.1%, and 49.7 ± 0.5%, respectively. Also, in the study, the methyl benzoate moiety and stereochemistry of C-3′ are considered as the possible sites for significant biological activity.

14.4.2 Cytotoxic Effects Against Cancer Cell Lines

Three of the six steroids from Xiong et al. (2016), i.e., (3β, 22E)-ergosta-7,9(11),22-trien-3-ol (ergosterol D), (3β,5α,6β, 22E)-6-methoxyergosta-7,22-diene-3,5-diol, and (3β,5α,6β)-stigmastane-3,5,6-triol, exhibited remarkable cytotoxic effects against human U2OS osteosarcoma cells, with IC_{50} values being 0.93 μM, 6.0 μM, and 13.2 μM, respectively. (3β,5α,6β, 22E)-6-methoxyergosta-7,22-diene-3,5-diol also exhibited cytotoxicity against A549 human lung carcinoma cells (IC_{50} = 11.9 μm). Staurosporine was the positive control with IC_{50} values being 0.01 μM and 0.003 μM in the cell line test of U2OS and A549, respectively.

14.4.3 Antimicrobial Properties

4,8-Dihydroxy-3,4-dihydronaphthalen-1(2H)-one isolated from methanol extract from *X. escharoidea* exhibited antimicrobial activity against pathogenic organisms including *Staphylococcus aureus*, *Bacillus subtilis*, *Pseudomonas aeruginosa*, and *Candida albicans* on par with standards of ampicillin and amphotericin B (Nagam et al. 2021). Moreover, molecular docking studies revealed that 4,8-dihydroxy-3,4-dihydronaphthalen-1(2H)-one was a prominent antibacterial agent. The activity was demonstrated from a marked interaction with key residues of binding pockets on DNA binding domain ($AgrA_C$) of accessory gene regulator protein A (AgrA) of *S. aureus*. The compound displayed three binding interactions, namely the 8-OH bond with Glu188, the 4-OH bond with Glu144, and the >C = O bond with Lys146. *Xylaria escharoidea* is not a known sclerotium-forming species but is mentioned here because it is in the same species group as other sclerotium-forming *Xylaria* species and occurs in termite mounds (Ju et al. 2022).

14.4.4 Enzyme Inhibitors

Ergot alkaloids are produced by various fungi in the phylum Ascomycota, but they have been reported only in *Dicyma* Boulanger species (Florea et al. 2017) and *X. nigripes* (Hu and Li 2017) within Xylariales. Among six ergot alkaloids isolated from dried stromata of *X. nigripes* by chloroform/methanol extraction, both xylanigripone C and agroclavine exhibited inhibition of acetylcholinesterase (AChE) activity with the former being stronger at a concentration of 50 μM and an 84% inhibition rate of the positive control tacrine (Hu and Li 2017). The function of AChE involves the termination of impulse transmission by rapid hydrolysis of the neurotransmitter acetylcholine in numerous cholinergic pathways in the central and peripheral nervous systems; a few reversible inhibitors of AChE are approved drugs in pharmacotherapy of Alzheimer's disease (Čolović et al. 2013). The understanding of the reversibility of xylanigripone C inhibition on AChE will be helpful for understanding the medicinal utilization of the compound. Hu and Li (2017) also showed that four ergot alkaloids, including xylanigripone A, agroclavine, 8,9-didehydro-10-hydroxy-6,8-dimethylergolin, and (6S)-agroclavine N-oxide, at the concentration of 160 μM compared to a water control, resulted in a decrease in the activities of cholesterol ester transfer protein (CETP) at rates of 49%, 73%, 36.5%, and 58.5%, respectively. CETP is a plasma protein for facilitating the transport of cholesteryl esters and triglycerides between lipoproteins. The development of CETP inhibitors aimed at reducing the risk of coronary heart disease has had a long and difficult course, and only Anacetrapib has gone into the planned completion in clinical trials among the four considered inhibitors (Tall and Rader 2018).

Monoamine oxidase inhibitors (MAOIs) are capable of blocking the monoamine oxidase enzymes, which break down different types of neurotransmitters including norepinephrine, serotonin, dopamine, and tyramine from the brain, thus retaining the availability of these neurotransmitters; patients with neurological disorders, such as Parkinson's disease and multiple system atrophy, can benefit from treatment of MAOIs (Sub Laban and Saadabadi 2022). Furthermore, MAOIs form a separate class from other antidepressants and usually are not the first choice in treating mental health disorders as a result of several dietary restrictions, side effects, and safety concerns. Two types of monoamine oxidase, MAO-A and MAO-B are present, with the former mostly distributed in the placenta, gut, and liver, while the latter in the brain, liver, and platelets. A phenolic compound, (R)-5-methylmellein, extracted from mycelia of cultured *X. nigripes*, exhibited activity as a selective inhibitor against MAOI-A with IC_{50} of 4.6 μM, which, however, was less potent than the known MAO-A inhibiter clorgyline with IC_{50} of 0.025 μM (Huang et al. 2019). Also, a variety of (R)-5-methylmellein analogs were synthesized, and an (R)-5-methylmellein analog (13aR) was the most potent and selective MAO-A inhibitor with IC_{50} of 0.06 μm among the 36 analogs tested.

14.5 Protection Effects of Wuling Capsule or Fermented Mycelia of Sclerotium-Producing *Xylaria* Species

14.5.1 Regulation of Anxiety and Depression

The prevalence of depression in patients with epilepsy varied in several studies, but significant numbers of depressed patients were revealed (Fiest et al. 2013; Jackson and Turkington 2005). Furthermore, a small subset of epileptic patients suffered from lifetime depression (Fiest et al. 2013). In a study of 104 patients with comorbidities of epilepsy and depression, Wuling capsule reduced symptoms of depression (Peng et al. 2015). Compared to the placebo group, the group taking Wuling capsule had improvements in comorbid depression, including a total effective rate for depression, scales designed to assess quality of life, and concerns about future seizures.

For both placebo and test groups, differences in seizure frequency and severity were not observed, and most adverse effects were mild and comparable. In a study of 64 cases of Type 2 diabetes mellitus, Wuling capsule ameliorated depression and improved the state of inflammation by decreasing concentrations of cytokines (IL-6 and TNF-α) that contribute to inflammation and reduced oxidative stress by increasing the concentration of SOD and decreasing the concentration of MDA (Wang et al. 2020).

Major depressive disorder (MDD), is a mental disorder that causes a persistent sadness or depressed mood, anhedonia or decreased interest in engaging activities, feeling of emptiness or hopelessness, low motivation, a difficulty in concentration, appetite changes, psychomotor retardation or irritability, disturbed sleep, apathetic, negative, or even suicidal thoughts; MDD was ranked as the third cause of the burden of disease worldwide in 2008 by WHO, which projected that MDD would rank first by 2030 (Bains and Abdijadid 2022). A systematic review of clinical data of MDD with a total of 880 cases from year 2006 to 2015 and a subsequent meta-analysis of the data based on randomized controlled trials revealed that the cotreatment of Wuling capsule and an antidepressant exhibited better efficacy than antidepressant monotherapy in symptomatic improvement at last-observation endpoint as well as study-defined response and remission, while there was no significant difference between Wuling capsule and some antidepressants tested as monotherapies (Zheng et al. 2016). Furthermore, the numbers of adverse drug reactions, including nausea, constipation, tachycardia, and dry mouth were less reported in patients with Wuling capsule treatment when compared to those with antidepressant monotherapy.

Post-stroke depression (PSD), an important complication of stroke leading to increased disability and mortality, occurs in a significant number of patients and is associated with the severity of impairment in activities of daily living, social functioning, and cognitive function (Robinson and Jorge 2016). A systematic review and meta-analysis of clinical data of PSD with a total of 1378 cases from 2006 to 2013 based on randomized controlled trials revealed that treatment of Wuling capsule singly when compared to no treatment, or a cotreatment of Wuling capsule and an antidepressant when compared to several antidepressant therapies alone, were more effective in reducing PSD estimated from diagnostic scores and response rates (Peng et al. 2014).

The small-scale clinical studies of the Wuling capsule reviewed above could help evaluate the feasibility, time required, possible adverse events, and effective size of the subject pool for a full-scale project.

14.5.2 Possible Mechanism for the Alleviation of Depression

Brain-derived neurotrophic factor (BDNF) plays a critical role in cognition, and its relation to PSD is easily recognized because BNDF serum levels are lower in PSD patients than in post-stroke patients without PSD; certain antidepressants could enhance BDNF expression in the brain, resulting in a relaxation of depression symptoms (Zhang and Liao 2020). The therapeutic effect can be eliminated in animals with the BDNF gene deletion (Zhang and Liao 2020). Wuling capsule at a dose of 100 mg kg^{-1}b.wt. exhibited an alleviation effect on the level of depression as well as the impairment of hippocampal-dependent learning and memory in PSD model rats, the same as the treatment of the antidepressant escitalopram at a dose of 0.2 mg kg^{-1}b.wt. for 21 days (Li et al. 2011). Interestingly, the recovery function of Wuling capsule on depression behaviors and cognitive deficits was not related to the expression of BDNF, as determined via immunohistochemical staining and mRNA amplification, which remained low after Wuling capsule treatment. Similarly, in a study of hippocampal neurogenesis of rats under chronic unpredictable mild stress, Wuling capsule exhibited a mitigation effect on depressed behaviors, and the antidepressant effect was related to enhancing the expression of the protein connexin-43 rather than through BDNF mediation (Li et al. 2010).

Translocator protein (TSPO), a five-transmembrane structure located at outer and inner mitochondrial membrane contact sites, is critical for brain steroidogenesis and modulates behaviors related to anxiety and depression (Barron et al. 2021). In a study of learned helplessness (LH) model of mice with depression, TSPO and certain mitophagy-related proteins, including voltage-dependent anion channel 1 (VDAC1), PTEN-induced kinase PINK1 and Beclin 1, significantly decreased in LH model mice (Li et al. 2016b). Also, kinesin KIFC2, relevant to mitochondrial transport was downregulated in the LH mice. Mechanistic steps of a PINK1/Parkin-directed pathway linking mitochondrial damage, ubiquitylation, and autophagy have demonstrated the PINK1/Parkin-mediated mitophagy is dependent on VDAC1 (Geisler et al. 2010). In a study of LH model mice with depression induced by inescapable e-shock, Wuling powder treatment with a concentration ranging from 0.5 to 2.0 g/kg significantly ameliorated the helpless behavior, and shortened the time of latency to eat and the immobility time during the forced swimming test (Li et al. 2016a). This study also demonstrated that Wuling powder exerted neuroprotective effects by altering the expression level of the proteins related to the TPSO mediated-mitophagy signaling pathway in mesencephalon of mice. Wuling powder significantly upregulated the expression level of TPSO, VDAC1, PINK1, Beclin 1, and KIFC2, which were significantly decreased in LH mice when compared to the controls. On the other hand, in LH mice the increased expression level of Parkin, possibly due to the recruiting effect of Parkin to the damaged mitochondria, was downregulated via Wuling powder treatment. Wuling powder also significantly elevated the integrated optical density of Synaptogyrin-3-postitive nerve fibers in ventral striatum of LH mice exhibiting a decreased optical density and hence expressing anhedonia to initiate feeding in novelty-suppressed feeding test and depression.

NO is synthesized from L-arginine by nitric oxide synthase throughout the CNS including spinal cord, retina, and brain (Wiesinger 2001). The role of NO as a neurotransmitter and neuromodulator in the nervous system includes modulation of synaptic plasticity, perception of pain, learning, aggression, and depression; control of the synthesis of NO is the key to regulating its activity (Esplugues 2002). Evidence revealed that the inhibition of neuronal nitric oxide synthase in rat hippocampus induced antidepressant-like effects, an implication of neurobiological function of endogenous hippocampal NO in times of stress and depression (Joca and Guimarães 2006). NO activates guanylate cyclase, and the activation enhances the formation of intracellular messenger cyclic guanosine monophosphate (cGMP) and the effects of cGMP were regulated via mediators, including cGMP-degrading phosphodiesterases which play a role in degradation of cGMP as well as determination of the shape of cGMP signal (Russwurm and Koesling 2004). L-arginine-nitric oxide-cGMP signaling pathway has been reported in the involvement of the pathogenesis of depression (Calabresi et al. 1999; Dhir and Kulkarni 2007). In a study of chronic unpredictable mild stress in rats, a powder prepared from Wuling mycelia exhibited antidepressant-like activity via a suppression of behavioral changes, including sucrose preference, food consumption, and crossing number (locomotor activity and willingness during an open-field test); the activity was possibly involved in the L-arginine-nitric oxide-cGMP signaling pathway (Tan et al. 2016). The antidepressant-like effect of Wuling powder was significantly reversed with the pretreatment of L-arginine and sildenafil (phosphodiesterase 5 inhibitor) but was significantly enhanced with pretreatment of 7-nitroindazole (a specific neuronal nitric oxide synthase inhibitor) and methylene blue (direct inhibitor of both nitric oxide synthase and soluble guanylate cyclase). Hence, the antidepressant-like effect may be dependent on the inhibition of NO synthesis and mediated through the reduction in cGMP.

14.5.3 Regulation of Inflammatory Mediators, Reduced Apoptosis, and Neuroprotection

Both the water extract and the ethanol extract prepared from a culture obtained from the specimen KJ-XN-07-01 of *X. nigripes* exhibited immunomodulatory properties, including a dose-dependent inhibitory effect on NO, PGE2, and cytokine (IL-1β, IL-6, TNF-α, and IFN-γ) production in LPS-stimulated macrophages; water extract was more potent than ethanol extract (Ko et al. 2011). Compared with the ethanol extract, the water extract exhibited a greater inhibition on iNOS and COX2 expressions and a stronger suppression on the LPS-induced IκB-α phosphorylation and NF-κB activation. Furthermore, the ethanol extract was higher in total flavonoid and phenol content but lower in β-glucan content than the water extract. These results revealed that the mechanism of anti-inflammatory action could be mediated by inhibiting iNOS and COX2 expression via NF-κB signaling pathway, and the potent anti-inflammatory activities of water extract could be associated with the β-glucan content. Divate and Chung (2017) reported that both hot water and 70% ethanol extracts of *X. nigripes* mycelia exerted protection against inflammatory stimuli from LPS in RAW264.7 cells by reducing the concentrations of NO, TNF-α, IL-6, and COX2. Both extracts exhibited the anti-inflammation potential to increase phagocytic activity and concentration of anti-inflammatory cytokine IL-10 of RAW 264.7 cells against diseased-related stimuli from *E. coli*. The study also revealed that *X. nigripes* mycelia possessed immunomodulatory activities by enhancing splenocyte proliferation against stimulation of mitogens, concanavalin A or LPS in mice spleens as well as increasing cytotoxicity of splenic natural killer cells against YAC-1 lymphoma cells.

Both the ethanol extract and the hot water extract from cultured mycelia of *X. nigripes* exerted neuroprotection against H_2O_2-induced cell damage of PC12 cells by inhibiting the release of lactate dehydrogenase to reduce cell injury, decreasing DNA damage, and rebuilding mitochondrial membrane potential (Divate et al. 2017c). Also, abnormal apoptosis resulting from the H_2O_2-induction was arrested by both the applied extracts, which regulated Bcl-2 family, the regulators of apoptosis by upregulation of Bcl-2, the pro-survival member and downregulation of Bax and caspase 3, the pro-apoptotic members. Compared to the hot water extract, the ethanol extract exhibited better protective effects on mitochondrial dysfunctions in cells with oxidative damage, and higher antioxidant activity by scavenging DPPH (2,2-diphenyl-1-picrylhydrazyl) radicals, inhibiting lipid oxidation, and reducing power. Both extracts contain substantial amounts of flavonoids, phenols, and polysaccharides with phenolic contents being significantly higher in the ethanol extract. In a study of rats with ischemic stroke induced by middle cerebral artery occlusion (MCAO), neuroprotection effects of the water extract prepared from *X. nigripes* were reported (Sun et al. 2017). The neuroprotective effects included significantly reduced neuronal injury, relieved neurobehavioral deficits, and alleviated MCAO-induced reduction of the expression levels of (i) heme oxygenase-1 (HO-1), (ii) nuclear factor-erythroid 2-related factor 2 (Nrf2), and (iii) neurons and astrocyte marker genes in the ischemic brain. HO-1 and Nrf2 are two major regulators during the processes of anti-oxidation, anti-inflammation, and anti-apoptosis. The markers are glial fibrillary acidic protein, nerve growth factor receptor, BDNF, neurotrophic tyrosine receptor kinase type 1 and type 2. Moreover, the increased levels of ROS, MDA, and cleaved caspase 3 via MCAO induction were reduced by the administration of the water extract. As the finding revealed, the neuroprotective effect of *X. nigripes* following ischemic insult in the brain is mediated by the anti-oxidative pathway Nrf2/HO-1. The neuropathological features of Alzheimer's disease (AD) include the accumulation of misfolded β-amyloid (Aβ) peptides in extracellular plaques and phosphorylated tau (P-tau) protein as paired filaments (PHFs) in intraneuronal neurofibrillary tangles; Aβ42, P-tau, and total tubulin associated unit (T-tau) are core cerebrospinal fluid biomarkers in AD pathology (Janelidze et al. 2020). In clinical treatment of AD, Wuling capsule combined with Oxiracetam significantly improved cognitive function, activities of daily living and quality of life, and reduced inflammatory reaction as measured by high sensitivity

C-reactive protein, IL-1β, and TNF-α when compared to Oxiracetam monotherapy (Fu et al. 2021). Moreover, the reduced Aβ1-42, and increased T-tau and P-tau181 levels in AD patients, when compared to healthy subjects, were ameliorated via the treatment of Wuling capsule combined with Oxiracetam and the ameliorated effects are significantly better than Oxiracetam monotherapy.

In a study of rats with seizure development and impairment in learning ability induced by pentylenetetrazole injection 12 times within 48 d, treatment with Wuling mycelia exhibited neuroprotection effects, including (i) significant shortening of the duration and alleviation of behavioral changes upon stimulated seizure, and (ii) the postponement of latency to the onset of myoclonic seizures and generalized seizure after specific times of treatment (Chen et al. 2012a). Wuling treatment was also beneficial via the effects on reduction of frequency of working error and reference memory error and partial reversal of impairment of learning ability. The pentylenetetrazole induction in rats also resulted in a decrease in histamine contents in brain including the regions of hippocampus, thalamus, and hypothalamus, which were elevated with the administration of Wuling mycelia. Histamine has been shown to modulate inflammatory processes and regulate microglia and astrocytes in the brain, and a reduction of its synthesis has been associated with the disorders of Tourette's syndrome and obsessive-compulsive disorder (Carthy and Ellender 2021).

Aqueous extract prepared from *X. nigripes* exhibited protective effects against hepatotoxicity induced via carbon tetrachloride (CCl_4) in mice (Song et al. 2011). Intragastrication of the aqueous extract significantly inhibited the elevation of serum glutamate oxalate transaminase, serum glutamate pyruvate transaminase, and the levels of liver thiobarbituric acid reactive substances. It also resulted in an increase in the liver SOD, catalase, and GSH-Px concentrations as well as serum total antioxidant activity in the CCl_4-induced hepatotoxicity mice. Furthermore, liver conditions, as revealed histologically, were obviously ameliorated with the treatment of the extract, which was substantially high in the concentrations of epicatechin, catechin, and coumaric acid among analyzed polyphenolic compounds. The effective protection of the extract could be through the effects of antioxidants on reducing oxidative stress.

X. nigripes purchased from a culture center (BCRC 34219) was grown on various solid substrates, including wheat bran, soy meal, and a mixture with ratios of wheat bran to soy meal of 4:1, 1:1, and 1:4; the fermented materials were then extracted with hot water or 70% ethanol for the comparison of bioactivities (Divate et al. 2017b). In general, ethanol extraction was more effective than hot water extraction for the acquisition of active compounds. The ethanol extract from the wheat bran fermentation was the most potent in exhibiting antioxidant activity, especially DPPH radical scavenging and reducing power as well as anti-inflammation activities, measured by the inhibition of COX2 activity when compared to the hot water extract and the extracts from other growth substrates. The ethanol extract from the substrate containing equal amounts of wheat bran and soy meal was significantly more effective in the prevention of the H_2O_2-induced damage in neuronal cells of PC12 when compared to the water extract and the extracts from other growth substrates.

Solid-state fermentation was applied to achieve synergistic effects of the bioactivities of two components, i.e., *X. nigripes* mycelium (YMJ653, EU179868) and growth substrates prepared from whole grains, including brown rice, red beans, mung beans, or soybeans (Divate et al. 2017a). Compared to unfermented grains, the grains with the mycelial fermentation exhibited greater antioxidative, anti-inflammatory, and neuroprotective potential. The red bean fermentation was more potent than the other fermentation substrates. The IC_{50} for the ethanol extract of the fermented and unfermented red beans on DPPH scavenging activity was 2.0 ± 0.1 and 2.7 ± 0.1 mg/mL, respectively, and on lipid peroxidation inhibition was 0.97 ± 0.1 and 3.44 ± 0.1 mg/mL, respectively. The ethanol extract of the fermented and unfermented red beans exhibited $36.5 \pm 0.5\%$ and

24.9 ± 1.5% of inhibition on COX2, respectively; and 27.2 ± 1.1% and 25.4 ± 2.2% of neuronal protection in PC12 cells, respectively. Furthermore, lipid peroxidation inhibition and reducing power both exhibited a significant correlation with the total phenolic contents of the tested extracts.

The ethyl acetate extracts prepared from *X. chaiyaphumensis* Wangsawat et al., *X. siamemsis* Wangsawat et al., *X. subintraflava* Wangsawat et al., *X. thienhirunae* Wangsawat et al., and *X. vinacea* Wangsawat et al. exhibited antioxidant properties, and *X. vinacea* was the most potent with IC_{50} of 0.194 mg/mL and 0.020 mg/mL, as measured via DPPH and ABTS [2,2′-azino-bis (3-ethylbenzothiazoline-6-sulfonic acid)] assays, respectively (Wangsawat et al. 2021b). Although these five *Xylaria* species are associated with termite nests, their ability to form sclerotia has not been reported.

14.5.4 Sedative Effects and Regulation of Insomnia Disorder

The amino acid, γ-aminobutyric acid (GABA), synthesized in the cytoplasm of the presynaptic neuron from glutamate by the enzyme glutamate decarboxylase, is the primary inhibitory neurotransmitter within the adult CNS, and primarily functions to reduce neuronal excitability by inhibiting nerve transmission (Allen et al. 2022; Jewett and Sharma 2022). Many illnesses, such as generalized anxiety, schizophrenia, autism spectrum disorder, seizures, and epilepsy have been associated with low concentrations of GABA and numerous drugs that modulate GABA signaling have been applied to clinical situations (Allen et al. 2022). The balance between inhibitory neuronal transmission via GABA and excitatory neuronal transmission via glutamate is of great importance for proper cell membrane stability and neurologic function (Allen et al. 2022). GABA was detected in Wuling capsule (He and Liu 2010). The product GABA-rich Zylaria™ with the inclusion of mycelial extracts of *X. nigripes* benefits health through its natural and gentle tranquillizing effect on human bodies and has been marketed for supporting sleep quality and relaxation (Deshmukh et al. 2021).

Wuling powder in combination with nembutal sodium, typically used as a sedative, results in shortening sleep onset and prolonging sleep duration when compared to a single treatment of nembutal sodium in mice (Ma et al. 1999). The administration of Wuling powder adjunctive resulted in increased concentrations of glutamate and GABA in the brain and elevations in the activity of glutamate decarboxylase, responsible for GABA synthesis and the binding affinity of GABA receptors in the cerebral cortex (Ma et al. 1999). Hence, it is suggestive that Wuling powder facilitated the entry of glutamate and GABA into brain, to activate GABA receptors, and thus exert its sedative and sleep-promoting properties.

A systematic review of clinical data for 1850 individuals with insomnia disorder from 2008 to 2021 and subsequent meta-analysis of the data based on randomized controlled trials revealed that sleep quality, assessed as Pittsburgh Sleep Quality (PSQI), can be safely and effectively ameliorated by the treatment of Wuling capsule monotherapy (Zhou et al. 2022). Additionally, the cotreatment of Wuling capsule with a conventional medication from benzodiazepine drugs, sedating antidepressants, or cognitive behavior therapy was more effective in PSQI than the conventional treatment alone. In the study, the severity of insomnia disorder was measured via Sleep Dysfunction Rating Scale and the scale was significantly reduced by Wuling capsule when compared to the control, benzodiazepines. Adverse events also were significantly reduced in the trial of Wuling capsule as monotherapy when compared to benzodiazepines and in the trial of Wuling capsule as adjuvant therapy with sedating antidepressant when compared to sedating antidepressant monotherapy. The treatment duration, 8 weeks was better than 4 weeks for the ameliorating effects of Wuling capsule on insomnia disorder. The use of Wuling capsule as a treatment of insomnia disorder required careful recommendation, due to the high heterogeneity, possibly from differences in age and course of disease among 18 studies, as well as the high risk of bias in the included trials with some being insufficient in the information of generating random number table and stratified blocked randomization as described in the study.

14.5.5 Regulation of Memory Impairment

Rapid eye movement (REM) sleep which occurs in mammals and birds, is a unique phase of sleep. REM sleep behavior disorder (RBD) is characterized by dream enactment and the loss of muscle atonia during REM sleep with dreams often involving violent and aggressive action. Several synucleinopathies, including Parkinson's disease, Lewy Body Dementia, and multiple system atrophy, are frequently associated with the RBD type classified as "secondary/symptomatic" (Matar et al. 2021). CREB (cAMP-response element binding protein), a transcription factor, promotes the recruitment of the co-activator protein CBP, a CREB-binding protein required for both short- and long-term memory formation (Chen et al. 2010). The function of CREB is largely regulated by the phosphorylation at Ser133 by protein kinase A (PKA), and the activation of PKA/CREB signaling pathway in the hippocampus is critically important in spatial memory formation (Mizuno et al. 2002).

The water extract prepared from *X. nigripes* mitigated an impairment of spatial learning and memory which was induced in Sprague Dawley rats deprived of REM sleep (Zhao et al. 2014). Also, a reduced phosphorylation of CREB was detected in the hippocampi of rats deprived of REM sleep, but the intragastric administration of the water extract reversed the reduced phosphorylation. Phosphorylation of CREB was suggested as one of the mechanisms for mitigation of the impairment.

14.5.6 Regulation of Insulin Sensitivity

Insulin receptor substrate (IRS) molecules are significant mediators in insulin signaling, and defects in muscle IRS-1 expression and function have been reported in insulin-resistant states such as Type 2 diabetes mellitus (T2DM) and obesity (Sesti et al. 2001). Glucose transporter type 4 (GLUT4) is responsible for the transportation of glucose across the plasma membrane, and one of the potential causes for resistance to insulin-stimulated glucose transport may be related to defective intracellular signaling of GLUT4 translocation in skeletal muscle from stored intracellular vesicles to active components of the plasma membrane, a possible inherent impairment in the muscle cells as T2DM is a heritable disease; research shows that T2DM conditions can be improved or reversed by the increase of GLUT4 levels (Vargas et al. 2022).

The aqueous extract prepared from stromatal powder of *X. nigripes* significantly exhibited a hypoglycemic effect on normal rats and rats with insulin resistance (SIIR) resulting from an administration of steroid (Chen et al. 2015). The decreases in plasma glucose concentration were 26% (from 177.6 ± 12.5 to 133.3 ± 29.7 mg/dL) and 21% (from 128.6 ± 12.5 to 100.9 ± 10.7 mg/dL), respectively, in SIIR and normal rats. The hypoglycemic effect on both groups of rats was also accompanied by increases in concentrations of IRS-1 and GLUT4. Furthermore, the hypoglycemic effect of the extract on SIIR rats was abolished when a three-day's injection of serotonin antagonist α-p-chlorophenylalanine was applied and followed by the administration of the extract. Increased insulin sensitivity and the consequence of a serotonin-related hypoglycemic effect may be related to changes in serotonin regulation after administration of the extract (Chen et al. 2015).

14.6 Amino Acids, Minerals, Nucleosides, and Others from Sclerotia, Mycelia From Cultures, and Wuling Capsule

Amino acids, including aspartic acid, glutamic acid, serine, histidine, glycine, arginine, proline, alanine, threonine, methionine, valine, phenylalanine, isoleucine, and leucine, had been identified from sclerotia and mycelia of a sclerotium-producing *Xylaria* species as well as in Wuling capsule (Chen et al. 1990; He and Liu 2010). Additionally, cystine and lysine were detected in sclerotia and mycelia (Chen et al. 1990). The detection of tyrosine was reported from mycelia (Chen et al. 1990). Mannitol and glucose were reported from sclerotia and mycelia; additionally, fructose, lactose, sucrose, and raffinose were identified from mycelia (Chen et al. 1990). The minerals, Fe, Mn, Sn, Cr, Ni, Mo, V, Cu, Zn, Se, Mg, B, Ca, K, Na, and P, were detected both in sclerotia and in mycelia (Chen et al. 1990). Nucleosides of Wuling capsule include adenosine, uridine, and guanosine (Lu et al. 2011). Other identified components of Wuling capsule are 5-methylmellein, 5-hydroxymellein, 5-carboxylmellein, and genistein (Chen et al. 2012b; Lu et al. 2011).

14.7 Epilogue

With more massive *Xylaria* sclerotia being analyzed, it is likely that there will be more sclerotium-producing *Xylaria* species discovered. This work needs to be coupled with the activities of collecting fresh stromata from the wild, from which morphological characteristics can be described and cultures can be obtained for further study. There have been many research reports on Wulingshen or Wulingshen-derived materials, mainly on medicinal ingredients, pharmacology, and clinical applications. In these research reports, most of the research materials used were identified as *X. nigripes*. Some of the DNA sequences deposited in GenBank of the National Center for Biotechnology Information (NCBI) were labeled as derived from *X. nigripes*. We noticed that most of these sequences were in fact from *X. rogersionigripes* rather than *X. nigripes* (Ju et al. 2022). Therefore, attention should be paid to which species of sclerotium-producing *Xylaria* species are used in research in the published studies on Wulingshen. Wulingshen sold at traditional Chinese medicine markets is frequently mixed with massive sclerotia produced by more than one *Xylaria* species. Are the medicinal ingredients of Wulingshen produced by different *Xylaria* species similar? Chemical components of different sclerotium-producing *Xylaria* species should be assessed for the appropriateness of consuming Wulingshen in mixtures. Wuling capsule, which has been sold with popularity in Chinese drugstores, is based on the dried mycelia obtained from a *Xylaria* species. In some published articles related to Wuling capsule, *X. nigripes* was assumed to be the research species, but this requires reconfirmation (Ju et al. 2022).

References

Aanen DK, de Fine Licht HH, Debets AJ, Kerstes NA, Hoekstra RF, Boomsma JJ (2009) High symbiont relatedness stabilizes mutualistic cooperation in fungus-growing termites. Science 326:1103–1106

Allen MJ, Sabir S, Sharma S. (2022) GABA receptor. In: StatPearls [Internet]. StatPearls Publishing, Treasure Island, FL. Available via https://www.ncbi.nlm.nih.gov/books/NBK526124/. Accessed 15 Aug 2022

Bachiller S, Jiménez-Ferrer I, Paulus A, Yang YY, Swanberg M, Deierborg T, Boza-Serrano A (2018) Microglia in neurological diseases: a road map to brain-disease dependent-inflammatory response. Front Cell Neurosci 12

Bains N, Abdijadid S. (2022) Major depressive disorder [Internet]. Treasure Island, FL: StatPearls Publishing. Accessed 15 Aug 2022

Barron AM, Higuchi M, Hattori S, Kito S, Suhara T, Ji B (2021) Regulation of anxiety and depression by mitochondrial translocator protein-mediated steroidogenesis: the role of neurons. Mol Neurobiol 58:550–563

Batra LR, Batra SWT (1979) Termite-fungus mutualism. In: Batra LR (ed) Insect-fungus symbiosis: nutrition, mutualism and commensalism. Allanheld, Osmun & Co., Montclair, NJ, pp 117–163

Battin EE, Brumaghim JL (2009) Antioxidant activity of sulfur and selenium: a review of reactive oxygen species scavenging, glutathione peroxidase, and metal-binding antioxidant mechanisms. Cell Biochem Biophys 55:1–23

Berkeley MJ (1862) On two tuberiform vegetable productions from Travancore. Trans Linn Soc Lond 23:91–92

Calabresi P, Gubellini P, Centonze D, Sancesario G, Morello M, Giorgi M, Pisani A, Bernardi G (1999) A critical role of the nitric oxide/cGMP pathway in corticostriatal long-term depression. J Neurosci 19: 2489–2499

Carthy E, Ellender T (2021) Histamine, neuroinflammation and neurodevelopment: a review. Front Neurosci-Switz 15(680214):1–26

Chang JC, Hsiao G, Lin RK, Kuo YH, Ju YM, Lee TH (2017) Bioactive constituents from the termite nest-

derived medicinal fungus *Xylaria nigripes*. J Nat Prod 80:38–44
Chen W-R, Li Z-T, Shi M-Z, Ni J-Y (1988) Study on biological characteristics of fungal Chinese medicine Wulingshen (in Chinese). Chin J Mod App Pharm 5: 10–13
Chen W-R, Fang H-F, Ma Z-Z, Ding R-R (1990) Chemical constituents of *Xylaria nigripes* (in Chinese). Chin Pharm J 25:647–649
Chen GQ, Zou XY, Watanabe H, van Deursen JM, Shen J (2010) CREB binding protein is required for both short-term and long-term memory formation. J Neurosci 30:13066–13077
Chen GF, Ren GL, Zhang LS, Hu XY (2012a) Effects of Wuling mycelia on pentylenetetrazole-induced epilepsy in rats (in Chinese). Zhejiang Da Xue Xue Bao Yi Xue Ban 41:647–652
Chen Y, Lu J, Zhu M, Luo L (2012b) Determination of multiple constituents in Wuling capsules by HPLC simultaneously (in Chinese). China J Chin Mat Med 37:218–221
Chen YI, Tzeng CY, Cheng YW, Hsu TH, Ho WJ, Liang ZC, Hsieh CW, Tzen JTC, Chang SL (2015) The involvement of serotonin in the hypoglycemic effects produced by administration of the aqueous extract of *Xylaria nigripes* with steroid-induced insulin-resistant rats. Phytother Res 29:770–776
Chen CF, Su CH, Lai MN, Ng LT (2018) Differences in water soluble non-digestible polysaccharides and anti-inflammatory activities of fruiting bodies from two cultivated *Xylaria nigripes* strains. Int J Biol Macromol 116:728–734
Chen MC, Wang GJ, Kuo YH, Chiang YR, Cho TY, Ju YM, Lee TH (2019) Isoprenyl phenolic ethers from the termite nest-derived medicinal fungus *Xylaria fimbriata*. J Food Drug Anal 27:111–117
Čolović MB, Krstić DZ, Lazarević-Pašti TD, Bondžić AM, Vasić VM (2013) Acetylcholinesterase inhibitors: pharmacology and toxicology. Curr Neuropharmacol 11:315–335
Croce MV, Colussi AG, Price MR, Segal-Eiras A (1999) Identification and characterization of different subpopulations in a human lung adenocarcinoma cell line (A549). Pathol Oncol Res 5:197–204
Deepna Latha KP, Veluthoor S, Manimohan P (2015) On the taxonomic identity of a fungal morph used in traditional medicine in Kerala State, India. Phytotaxa 201:287–295
Deshmukh SK, Sridhar KR, Gupta MK (2021) Application of selected species of the genus *Xylaria* in traditional medicine. In: Deshmukh SK, Sridhar KR (eds) Advances in macrofungi: pharmaceuticals and cosmeceuticals. CRC Press, Boca Raton (FL), pp 122–136
Dhir A, Kulkarni SK (2007) Involvement of L-arginine-nitric oxide-cyclic guanosine monophosphate pathway in the antidepressant-like effect of venlafaxine in mice. Prog Neuro-Psychoph 31:921–925
Divate RD, Chung YC (2017) In vitro and in vivo assessment of anti-inflammatory and immunomodulatory activities of *Xylaria nigripes* mycelium. J Funct Foods 35:81–89
Divate RD, Wang CC, Chou ST, Chang CT, Wang PM, Chung YC (2017a) Production of *Xylaria nigripes*-fermented grains by solid-state fermentation and an assessment of their resulting bioactivity. Lwt-Food Sci Technol 81:18–25
Divate RD, Wang CC, Chou ST, Chang CT, Wang PM, Chung YC (2017b) Using wheat bran and soybean meal as solid state fermentation substances for the production of *Xylaria nigripes* with bioactivities. J Taiwan Inst Chem E 70:127–133
Divate RD, Wang PM, Wang CC, Chou ST, Chang CT, Chung YC (2017c) Protective effect of medicinal fungus *Xylaria nigripes* mycelia extracts against hydrogen peroxide-induced apoptosis in PC12 cells. Int J Immunopath Ph 30:105–112
Esplugues JV (2002) NO as a signalling molecule in the nervous system. Br J Pharmacol 135:1079–1095
Feng DD, Tang T, Lin XP, Yang ZY, Yang S, Xia ZA, Wang Y, Zheng P, Wang Y, Zhang CH (2016) Nine traditional Chinese herbal formulas for the treatment of depression: an ethnopharmacology, phytochemistry, and pharmacology review. Neuropsychiatr Dis Treat 12:2387–2402
Fiest KM, Dykeman J, Patten SB, Wiebe S, Kaplan GG, Maxwell CJ, Bulloch AGM, Jette N (2013) Depression in epilepsy: a systematic review and meta-analysis. Neurology 80:590–599
Florea S, Panaccione DG, Schardl CL (2017) Ergot alkaloids of the family Clavicipitaceae. Phytopathology 107:504–518
Foster KA, Oster CG, Mayer MM, Avery ML, Audus KL (1998) Characterization of the A549 cell line as a type II pulmonary epithelial cell model for drug metabolism. Exp Cell Res 243:359–366
Fu Y, Tong J, Di L, Zhang J, Liu X (2021) Clinical observation of Wuling capsules combined with Oxiracetam in the treatment of Alzheimer's disease. China Pharm 30:69–71
Fukai T, Ushio-Fukai M (2011) Superoxide dismutases: role in redox signaling, vascular function, and diseases. Antioxid Redox Sign 15:1583–1606
Geisler S, Holmström KM, Skujat D, Fiesel FC, Rothfuss OC, Kahle PJ, Springer W (2010) PINK1/Parkin-mediated mitophagy is dependent on VDAC1 and p62/SQSTM1. Nat Cell Biol 12:119–131
Ginhoux F, Lim S, Hoeffel G, Low D, Huber T (2013) Origin and differentiation of microglia. Front Cell Neurosci 7(45):1–14
He X, Liu P (2010) Determination of 14 kinds of amino acids in Wuling capsule by HPLC-FLD (in Chinese). Chin Tradit Pat Med 32:1358–1361
Hsieh H-M, Lin C-R, Fang M-J, Rogers JD, Fournier J, Lechat C, Ju Y-M (2010) Phylogenetic status of *Xylaria* subgen. *Pseudoxylaria* among taxa of the subfamily Xylarioideae (Xylariaceae) and phylogeny of

the taxa involved in the subfamily. Mol Phylogenet Evol 54:957–969

Hsieh H-M, Chou J-C, Ju Y-M (2020) *Xylaria insolita* and *X. subescharoidea*: two newly described species collected from a termite nesting site in Hua-lien, Taiwan. Bot Stud 61(11):1–9

Hu DB, Li M (2017) Three new ergot alkaloids from the fruiting bodies of *Xylaria nigripes* (Kl.) Sacc. Chem Biodivers 14:e1600173:1–7

Huang C, Xiong J, Guan HD, Wang CH, Lei XS, Hu JF (2019) Discovery, synthesis, biological evaluation and molecular docking study of (R)-5-methylmellein and its analogs as selective monoamine oxidase A inhibitors. Bioorgan Med Chem 27:2027–2040

Jackson MJ, Turkington D (2005) Depression and anxiety in epilepsy. J Neurol Neurosur Ps 76:i45–i47

Janelidze S, Mattsson N, Palmqvist S, Smith R, Beach TG, Serrano GE, Chai XY, Proctor NK, Eichenlaub U, Zetterberg H et al (2020) Plasma P-tau181 in Alzheimer's disease: relationship to other biomarkers, differential diagnosis, neuropathology and longitudinal progression to Alzheimer's dementia. Nat Med 26: 379–386

Jen CI, Su CH, Lu MK, Lai MN, Ng LT (2021) Synergistic anti-inflammatory effects of different polysaccharide components from *Xylaria nigripes*. J Food Biochem 45(e13694):1–11

Jewett BE, Sharma S. (2022) Physiology, GABA. In: StatPearls [Internet]. StatPearls Publishing, Treasure Island, FL. Available via https://www.ncbi.nlm.nih.gov/pubmed/30020683. Accessed 15 Aug 2022

Joca SRL, Guimarães FS (2006) Inhibition of neuronal nitric oxide synthase in the rat hippocampus induces antidepressant-like effects. Psychopharmacology 185: 298–305

Ju Y-M, Hsieh H-M (2007) *Xylaria* species associated with nests of *Odontotermes formosanus* in Taiwan. Mycologia 99:936–957

Ju Y-M, Hsieh H-M, He X-S (2022) Wulingshen, the massive *Xylaria* sclerotia used as traditional Chinese medicine, is produced by multiple species. Mycologia 114:175–189

Kawahara K, Hohjoh H, Inazumi T, Tsuchiya S, Sugimoto Y (2015) Prostaglandin E2-induced inflammation: Relevance of prostaglandin E receptors. Biochim Biophys Acta 1851:414–421

Ko HJ, Song AR, Lai MN, Ng LT (2011) Immunomodulatory properties of *Xylaria nigripes* in peritoneal macrophage cells of Balb/c mice. J Ethnopharmacol 138: 762–768

Lauvrak SU, Munthe E, Kresse SH, Stratford EW, Namlos HM, Meza-Zepeda LA, Myklebost O (2013) Functional characterisation of osteosarcoma cell lines and identification of mRNAs and miRNAs associated with aggressive cancer phenotypes. Br J Cancer 109:2228–2236

Li D-Q, Li X-J, Duan J-F, Cai W (2010) Wuling capsule promotes hippocampal neurogenesis by improving expression of connexin 43 in rats exposed to chronic unpredictable mild stress (in Chinese). Zhong Xi Yi Jie He Xue Bao 8:662–669

Li DQ, Li XJ, Duan JF, Cai W (2011) Effects of Wuling capsule on hippocampal dependent cognitive changes in post-stroke depression rats. Int J Pharmacol 7:50–57

Li M, Xiong J, Huang Y, Wang LJ, Tang Y, Yang GX, Liu XH, Wei BG, Fan H, Zhao Y et al (2015) Xylapyrrosides A and B, two rare sugar-morpholine spiroketal pyrrole-derived alkaloids from *Xylaria nigripes*: isolation, complete structure elucidation, and total syntheses. Tetrahedron 71:5285–5295

Li DM, Zheng J, Wang MY, Feng L, Liu YY, Yang N, Zuo PP (2016a) Wuling powder prevents the depression-like behavior in learned helplessness mice model through improving the TSPO mediated mitophagy. J Ethnopharmacol 186:181–188

Li DM, Zheng J, Wang MY, Feng L, Ren ZL, Liu YY, Yang N, Zuo PP (2016b) Changes of TSPO-mediated mitophagy signaling pathway in learned helplessness mice. Psychiatry Res 245:141–147

Li J, Li LQ, Long HP, Liu J, Jiang YP, Xue Y, Wang WX, Tan GS, Gong ZC, Liu JK (2021) Xylarinaps A-E, five pairs of naphthalenone derivatives with neuroprotective activities from *Xylaria nigripes*. Phytochemistry 186(112729):1–9

Li LQ, Li J, Long HP, Liu JK, Wang X (2022) Four new resorcinol derivatives with neuroprotective activities from *Xylaria nigripes*. Nat Prod Res 36:1522–1528

Liao W, Fan HB, Wu JP (2018) Egg white-derived antihypertensive peptide IRW (Ile-Arg-Trp) inhibits angiotensin II-stimulated migration of vascular smooth muscle cell via angiotensin type I receptor. J Agric Food Chem 66:5133–5138

Liaw CC, Wu SJ, Chen CF, Lai MN, Ng LT (2017) Anti-inflammatory activity and bioactive constituents of cultivated fruiting bodies of *Xylaria nigripes* (ascomycetes), a Chinese medicinal fungus. Int J Med Mushrooms 19:915–924

Lowy B (1974) *Amanita muscaria* and the thunderbolt legend in Guatemala and Mexico. Mycologia 66:188–191

Lu J-X, Zhu M, Chen Y, Zhang P, Fang L (2011) HPLC specific chromatogram of chemical constituents of Wuling capsules (in Chinese). Yao Wu Fen Xi Za Zhi 4:764–767

Ma Z-Z, Zhang P-P, Chen W-R (1999) Studies on the sedative and sleeping effects of Wuling mycelia and its pharmacological mechanism. Chinese Pharm J 34: 374–347

Matar E, McCarter SJ, St Louis EK, Lewis SJG (2021) Current concepts and controversies in the management of REM sleep behavior disorder. Neurotherapeutics 18:107–123

Mizuno M, Yamada K, Maekawa N, Saito K, Seishima M, Nabeshima T (2002) CREB phosphorylation as a molecular marker of memory processing in the hippocampus for spatial learning. Behav Brain Res 133:135–141

Mueller UG, Gerardo NM, Aanen DK, Six DL, Schultz TR (2005) The evolution of agriculture in insects. Annu Rev Ecol Evol S 36:563–595

Nagam V, Aluru R, Shoaib M, Dong GR, Li Z, Pallaval VB, Ni JF (2021) Diversity of fungal isolates from fungus-growing termite *Macrotermes barneyi* and characterization of bioactive compound from *Xylaria escharoidea*. Insect Sci 28:392–402

Peng L, Zhang X, Kang DY, Liu XT, Hong Q (2014) Effectiveness and safety of Wuling capsule for post stroke depression: a systematic review. Complement Ther Med 22:549–566

Peng WF, Wang X, Hong Z, Zhu GX, Li BM, Li Z, Ding MP, Geng Z, Jin Z, Miao L et al (2015) The anti-depression effect of *Xylaria nigripes* in patients with epilepsy: a multicenter randomized double-blind study. Seizure-Eur J Epilep 29:26–33

Petch T (1906) The fungi of certain termite nests. Ann Roy Bot Gard (Peradeniya) 3:185–270

Petch T (1913) Termite fungi: a résumé. Ann Roy Bot Gard (Peradeniya) 5:303–341

Pharmacopoeia Commission of China (2020) Pharmacopoeia of the People's Republic of China, the 2020 edition, vol. 1 (in Chinese). China Medical Science Press, Beijing, 1902 p + index

Robinson RG, Jorge RE (2016) Post-stroke depression: a review. Am J Psychiat 173:221–231

Rogers JD, Ju Y-M, Lehmann J (2005) Some *Xylaria* species on termite nests. Mycologia 97:914–923

Rojo LE, Fernandez JA, Maccioni AA, Jimenez JM, Maccioni RB (2008) Neuroinflammation: implications for the pathogenesis and molecular diagnosis of Alzheimer's disease. Arch Med Res 39:1–16

Russwurm M, Koesling D (2004) NO activation of guanylyl cyclase. EMBO J 23:4443–4450

Sangvichien E, Taylor-Hawksworth PA (2001) *Termitomyces* mushrooms: a tropical delicacy. Mycologist 15:31–33

Sesti G, Federici M, Hribal ML, Lauro D, Sbraccia P, Lauro R (2001) Defects of the insulin receptor substrate (IRS) system in human metabolic disorders. FASEB J 15:2099–2111

Song AR, Ko HJ, Lai MN, Ng LT (2011) Protective effects of Wu-Ling-Shen (*Xylaria nigripes*) on carbon tetrachloride-induced hepatotoxicity in mice. Immunopharm Immunot 33:454–460

Sub Laban T, Saadabadi A. (2022) Monoamine oxidase inhibitors (MAOI) [Internet]. Treasure Island, FL: StatPearls Publishing. Accessed 15 Aug 2022

Sun RZ, Zhang YH, He SS, Li R, Xue F, Chen YC, Yan WJ, Wang HN, Peng ZW (2017) *Xylaria nigripes* protects mice against cerebral ischemic injury by activating Nrf2/HO-1 pathway. Int J Clin Exp Med 10:11636–11645

Taciak B, Bialasek M, Braniewska A, Sas Z, Sawicka P, Kiraga L, Rygiel T, Krol M (2018) Evaluation of phenotypic and functional stability of RAW 264.7 cell line through serial passages. PLoS One 13: e0198943:1–13

Tall AR, Rader DJ (2018) The trials and tribulations of CETP Inhibitors. Circ Res 122:106–112

Tan YF, Liao ZL, Qiu YJ, Zhu JP, Yu EY (2016) Possible involvement of L-arginine-nitric oxide (NO)-cyclic guanosine monophosphate (cGMP) signaling pathway in the antidepressant-like effect of Wuling mycelia powder in rat. Biomed Pharmacother 78:60–65

Timmerman R, Burm SM, Bajramovic JJ (2018) An overview of in vitro methods to study microglia. Front Cell Neurosci 12(242):1–12

U'Ren JM, Miadlikowska J, Zimmerman NB, Lutzoni F, Stajich JE, Arnold AE (2016) Contributions of North American endophytes to the phylogeny, ecology, and taxonomy of Xylariaceae (Sordariomycetes, Ascomycota). Mol Phylogenet Evol 98:210–232

Vargas E, Podder V, Carrillo Sepulveda MA (2022) Physiology, glucose transporter type 4 [Internet]. StatPearls Publishing, Treasure Island, FL. Accessed 15 Aug 2022

Verano AL, Tan DS (2017) Family-level stereoselective synthesis and biological evaluation of pyrrolomorpholine spiroketal natural product antioxidants. Chem Sci 8:3687–3693

Visser AA, Ros VI, De Beer ZW, Debets AJ, Hartog E, Kuyper TW, Laessoe T, Slippers B, Aanen DK (2009) Levels of specificity of *Xylaria* species associated with fungus-growing termites: a phylogenetic approach. Mol Ecol 18:553–567

Wang HP, Chen HZ, Gao Y, Wang SJ, Wang X, Tang XM, Fang W, Shi XY, Yao J, Chen Q (2020) The effect of wuling capsule on depression in Type 2 diabetic patients. Biosci Rep 40:BSR20191260:1–9

Wangsawat N, Ju Y-M, Phosri C, Whalley AJS, Suwannasai N (2021a) Twelve new taxa of *Xylaria* associated with termite nests and soil from northeast Thailand. Biology-Basel 10(575):1–33

Wangsawat N, Nahar L, Sarker SD, Phosri C, Evans AR, Whalley AJS, Choowongkomon K, Suwannasai N (2021b) Antioxidant activity and cytotoxicity against cancer cell lines of the extracts from novel *Xylaria* species associated with termite nests and LC-MS analysis. Antioxidants-Basel 10(1557):1–16

Wiatrak B, Kubis-Kubiak A, Piwowar A, Barg E (2020) PC12 cell line: cell types, coating of culture vessels, differentiation and other culture conditions. Cells-Basel 9(958):1–15

Wiesinger H (2001) Arginine metabolism and the synthesis of nitric oxide in the nervous system. Prog Neurobiol 64:365–391

Wu J-L (1984) Fungal Chinese medicine—Wulingshen (in Chinese). J Chengdu Univ Tradit Chin Med 1984: 33–35

Xicoy H, Wieringa B, Martens GJM (2017) The SH-SY5Y cell line in Parkinson's disease research: a systematic review. Mol Neurodegener 12(10):1–11

Xiong J, Huang Y, Wu XY, Liu XH, Fan H, Wang W, Zhao Y, Yang GX, Zhang HY, Hu JF (2016) Chemical constituents from the fermented mycelia of the medicinal fungus *Xylaria nigripes*. Helv Chim Acta 99:83–89

Ying J-Z, Mao X-L, Ma Q-M, Zong Y-C, Wen H-A (1987) Icons of medicinal fungi from China. Science Press, Beijing, p 575

Zhang E, Liao P (2020) Brain-derived neurotrophic factor and post-stroke depression. J Neurosci Res 98:537–548

Zhang QH, Ni X, Zhang FQ (2017) Decreasing trend in severe weather occurrence over China during the past 50 years. Sci Rep-Uk 7:42310:1–8

Zhao ZQ, Li YP, Chen HY, Huang LQ, Zhao F, Yu Q, Xiang ZH, Zhao ZX (2014) *Xylaria nigripes* mitigates spatial memory impairment induced by rapid eye movement sleep deprivation. Int J Clin Exp Med 7: 356–362

Zheng W, Zhang Y-F, Zhong H-Q, Mai S-M, Yang X-H, Xiang Y-T (2016) Wuling capsule for major depression disorder: a meta-analysis of randomized controlled trials. E Asian Arch Psychiat 26:87–97

Zhou B, Qiu Y, Wu N, Chen AD, Zhou H, Chen Q, Kang YM, Li YH, Zhu GQ (2020) FNDC5 attenuates oxidative stress and NLRP3 inflammasome activation in vascular smooth muscle cells via activating the AMPK-SIRT1 signal pathway. Oxidative Med Cell Longev 2020:6384803:1–15

Zhou HF, Zhao Y, Peng WH, Han WB, Wang DY, Wang ZC, Ren XX, Pan GZ, Lin Q, Wang X (2022) Efficacy and safety of Wuling capsule for insomnia disorder: a systematic review and meta-analysis of randomized controlled trials. Sleep Med 93:1–14

Zhuang S-H, Zheng T-S (1886) Revised local gazetteer of the district of Guan-xian (in Chinese), vol 12. Guan County Magistrate, Guanxian, p 162

GPSR Compliance

The European Union's (EU) General Product Safety Regulation (GPSR) is a set of rules that requires consumer products to be safe and our obligations to ensure this.

If you have any concerns about our products, you can contact us on ProductSafety@springernature.com

In case Publisher is established outside the EU, the EU authorized representative is:

Springer Nature Customer Service Center GmbH
Europaplatz 3
69115 Heidelberg, Germany

Batch number: 08104546

Printed by Printforce, the Netherlands